Super Maths

The Revision Book

Abhas Saini

S. CHAND
PUBLISHING

S CHAND AND COMPANY LIMITED

(ISO 9001 Certified Company)

S. CHAND PUBLISHING

S Chand And Company Limited

(ISO 9001 Certified Company)

Head Office: D-92, Sector – 02, Noida 201301, U.P. (India), Ph. 91-120-4682700

Registered Office: A-27, 2nd Floor, Mohan Co-operative Industrial Estate, New Delhi – 110 044

Phone: 011-49731800

www.**schandpublishing.com**; e-mail: **info@schandpublishing.com**

Marketing Offices:

Chennai : Ph: 044-2363 2120, chennai@schandpublishing.com
Guwahati : Ph: 0361-4066 369, guwahati@schandpublishing.com
Hyderabad : Ph: 040-4018 6018, hyderabad@schandpublishing.com
Jalandhar : Ph: 0181-4645 630, jalandhar@schandpublishing.com
Kolkata : Ph: 033-2335 7458 kolkata@schandpublishing.com
Lucknow : Ph: 0522-400 3633, lucknow@schandpublishing.com
Mumbai : Ph: 022-2500 0297, mumbai@schandpublishing.com
Patna : Ph: 0612-226 0011, patna@schandpublishing.com

S. CHAND'S Seal of Trust

In our endeavour to protect you against counterfeit/fake books, we have pasted a hologram over the cover of this book. The hologram displays full visible effect, emboss effect, relief effect, mirror lens effect, pearl effect, motion effect, animated text, kinetic effect, concealed effect, micro structure, multicolour small text 'S.CHAND', nanotext ' 50 micron' 'ORIGINAL' 'S,CHAND', mirror strip '6.5 mm', mirror lens 3 mm with text 'SC', microtext 'OK', scratch strip '7 mm', color sparkling effect under scratch and QR code size 10 mm x 10 mm, etc.

A fake hologram does not display ALL these effects.

First Edition 2025

ISBN : 978-93-6958-179-5 **Product Code:** H12SMR61MATH40ENAA25O

PRINTED IN INDIA

By Vikas Publishing House Private Limited, Plot 20/4, Site-IV, Industrial Area Sahibabad, Ghaziabad – 201 010 and Published by S Chand And Company Limited, A-27, 2nd Floor, Mohan Co-operative Industrial Estate, New Delhi – 110 044

PREFACE

This book is specifically crafted to meet the rigorous demands of competitive exams such as SSC, Railway, Banking, and other related tests in mathematics. It offers a clear and concise presentation of essential mathematical theory alongside a carefully curated set of problems, ensuring that every key theorem and direct result is covered without unnecessary repetition.

Key features of this book include:

- Theory + Problems – A balanced blend of succinct theoretical explanations and corresponding problems that allow you to immediately apply and reinforce concepts.
- All Theorems/Direct Results Covered – Every significant theorem and result needed for SSC, Railway, Banking, and similar competitive exams is compiled in one place, ensuring comprehensive coverage of the syllabus.
- Short & Concise – The material is presented in a clear, compact manner, enabling you to quickly grasp complex topics.
- No Repetition – The content is streamlined to avoid redundancy, making your study sessions more focused and effective.
- Practice 1000 Questions, 10 Times – Instead of overwhelming you with an excessive number of problems, the book emphasizes deep learning through the repeated practice of 1000 carefully selected questions.
- Best for Revision – This book is your ideal companion for quick reviews and thorough revision before the exam.

May this book serve as an invaluable resource on your journey to mastering mathematics and achieving success in SSC, Railway, Banking, and other competitive examinations.

Happy studying!

With Best Wishes
ABHAS SAINI

CONTENTS

1. Percentage 1–22
2. Profit, Loss & Discount 23–46
3. Ratio & Proportion 47–59
4. Mixture & Alligation 60–67
5. Partnership 68–75
6. Simple Interest & Compound Interest 76–94
7. Time Work & Pipe Cistern 95–111
8. Speed, Time & Distance 112–123
9. Race 124–128
10. Boats and Streams 129–133
11. Average 134–143
12. Ages 144–149
13. Number System 150–172
14. LCM & HCF 173–179
15. Algebra 180–205
16. Surds and Indices 206–214
17. Simplification 215–220
18. Trigonometry 221–244

19. Height and Distance 245–252

20. Geometry 253–304

21. Mensuration 2D & 3D 305–349

22. Coordinate Geometry 350–357

23. Permutation & Combination 358–373

24. Probability 374–403

25. Statistics 404–411

26. Data Interpretation 412–421

1 Percentage

Percentage: - In Mathematics, a percentage is a number or ratio that can be expressed as a fraction of 100.

प्रतिशत:- गणित में प्रतिशत एक संख्या या अनुपात है जिसे 100 के भिन्न के रूप में व्यक्त किया जा सकता है।

Percentage to Fraction Table
(भिन्न तालिका का प्रतिशत)

$100\% = \frac{1}{1}$ $95\% = \frac{19}{20}$ $90\% = \frac{9}{10}$

$85\% = \frac{17}{20}$ $80\% = \frac{4}{5}$ $75\% = \frac{3}{4}$

$70\% = \frac{7}{10}$ $65\% = \frac{13}{20}$ $60\% = \frac{3}{5}$

$55\% = \frac{11}{20}$ $50\% = \frac{1}{2}$ $45\% = \frac{9}{20}$

$40\% = \frac{2}{5}$ $35\% = \frac{7}{20}$ $30\% = \frac{3}{10}$

$25\% = \frac{1}{4}$ $20\% = \frac{1}{5}$ $15\% = \frac{3}{20}$

$10\% = \frac{1}{10}$ $5\% = \frac{1}{20}$

Fraction to Percentage Table
(भिन्न से प्रतिशत तालिका)

$\frac{1}{3} = 33.33\% = 33\frac{1}{3}\%$ $\frac{2}{3} = 66.66\% = 66\frac{2}{3}\%$

$\frac{1}{6} = 16.66\% = 16\frac{2}{3}\%$ $\frac{5}{6} = 83.33\% = 83\frac{1}{3}\%$

$\frac{1}{7} = 14.28\%$ $\frac{2}{7} = 28.57\%$

$\frac{3}{7} = 42.85\%$ $\frac{4}{7} = 57.14\%$

$\frac{1}{8} = 12.5\%$ $\frac{3}{8} = 37.5\%$ $\frac{5}{8} = 62.5\%$

$\frac{7}{8} = 87.5\%$

$\frac{1}{9} = 11.11\%$ $\frac{2}{9} = 22.22\%$ $\frac{4}{9} = 44.44\%$

$\frac{5}{9} = 55.55\%$ $\frac{7}{9} = 77.77\%$ $\frac{8}{9} = 88.88\%$

$\frac{1}{11} = 9.09\%$ $\frac{2}{11} = 18.18\%$ $\frac{3}{11} = 27.27\%$

$\frac{4}{11} = 36.36\%$ $\frac{5}{11} = 45.45\%$ $\frac{6}{11} = 54.54\%$

$\frac{7}{11} = 63.63\%$ $\frac{8}{11} = 72.72\%$ $\frac{9}{11} = 81.81\%$

$\frac{10}{11} = 90.90\%$ $\frac{1}{12} = 8.33\%$

$\frac{1}{14} = 7.14\%$ $\frac{1}{15} = 6.66\%$ $\frac{1}{16} = 6.25\%$

Successive Percentage Change
(क्रमिक प्रतिशत परिवर्तन)

1. If $x\%$ and $y\%$ change occurs successively then
   यदि $x\%$ और $y\%$ परिवर्तन क्रमिक रूप से होता है
   Overall Change% $= x + y + \frac{xy}{100}$ %
   समग्र परिवर्तन% $= x + y + \frac{xy}{100}$ %
2. If $x\%$, $y\%$ and $z\%$ change occurs successively then
   यदि $x\%$, $y\%$ और $z\%$ परिवर्तन क्रमिक रूप से होता है
   Overall Change%
   $= x + y + z + \frac{xy + yz + zx}{100} + \frac{xyz}{10000}$ %
   समग्र परिवर्तन%
   $= x + y + z + \frac{xy + yz + zx}{100} + \frac{xyz}{10000}$ %

Some More Important Facts (कुछ और महत्वपूर्ण तथ्य)

1. Price × Consumption = Expenditure
   कीमत × खपत = व्यय
2. Income = Expenditure + Savings
   आय = व्यय + बचत
3. Earning = Times × Wages
   कमाई = समय × मजदूरी
4. Net Income = Gross Income - Income Tax
   शुद्ध आय = सकल आय - आयकर
5. $n(A \cup B) = n(A) + n(B) - n(A \cap B)$
6. $n(A \cup B \cup C) = n(A) + n(B) + n(C) - n(A \cap B) - n(B \cap C) - n(C \cap A) + n(A \cap B \cap C)$

Type - 1 (Basic)

1. 0.625 is what percentage of $1\frac{1}{4}$?
0.625 1% का कितना प्रतिशत है?

Solution :

$0.625 = x\%$ of $\frac{5}{4}$

$\frac{625}{1000} = \frac{x}{100} \times \frac{5}{4}$

$x = \frac{625 \times 4}{50} = 50$

2. A number which when increased by 16% becomes 1914. The number is
एक संख्या जिसे 16% बढ़ाने पर 1914 हो जाती है। वह संख्या है

Solution :

Let the number be x

$16\% = \frac{4}{25}$

$x \times \frac{29}{25} = 1914$

$x = \frac{1914 \times 25}{29}$

$x = 1650$

3. A is 80% more than B and C is $48\frac{4}{7}\%$ less than the sum of A and B. By what percent is C less than A?
A, B से 80% अधिक है और C, A और B के योग से $48\frac{4}{7}\%$ कम है। C, A से कितने प्रतिशत कम है?

Solution :

$A : B = 9 : 5 = 45 : 25$ $C : (A + B) = 36 :$

$70 = 36 : 70$

$C : A = 36 : 45$

$\frac{9}{45} = 20\%$ less

4. If decreasing 180 by $x\%$ gives the same result as increasing 60 by $x\%$, then $x\%$ of 410 will be more than $(x + 20)\%$ of 210 (correct to two decimal places) by?
यदि 180 को $x\%$ से घटाने पर 60 को $x\%$ बढ़ाने के समान परिणाम मिलता है, तो 410 का $x\%$ 210 के $(x + 20)\%$ (दो दशमलव स्थानों तक सही) से कितना अधिक होगा?

Solution :

$180(100 - x) = 60(100 + x)$

$300 - 3x = 100 + x$

$x = 50$

50% of 410 more than 70% of 210

$205 : 147$

$\frac{58}{147} = 39.46\%$

5. For a number greater than one, the difference between itself and its reciprocal is 20% of the sum of itself and its reciprocal. By how much percentage (Nearest to an integer) is the square of the number less than its cube?
एक से बड़ी संख्या के लिए, उसके और उसके व्युत्क्रम के बीच का अंतर उसके और उसके व्युत्क्रम के योग का 20% है। किसी संख्या का वर्ग उसके घन से कितने प्रतिशत (पूर्णांक के निकटतम) कम है?

Solution :

$\left(x - \frac{1}{x}\right) = 20\%$ of $\left(x + \frac{1}{x}\right)$

$5x - \frac{5}{x} = x + \frac{1}{x}$

$4x = \frac{6}{x}$

$x^2 = \frac{3}{2}$

$x = \frac{\sqrt{3}}{2}$

$\frac{3}{2} : \frac{3\sqrt{3}}{2\sqrt{2}}$

$\sqrt{2} : \sqrt{3}$

$(\sqrt{3} - \sqrt{2}) \times \frac{100}{\sqrt{3}} = 18\%$ approx

6. Two Numbers are less than a third Number by 37% and 30% respectively. By what percentage is the second number more than the first?
दो संख्याएँ तीसरी संख्या से क्रमशः 37% और 30% कम हैं। दूसरी संख्या पहली से कितने प्रतिशत अधिक है?

Solution :

$A : B : C = 63 : 70 : 100$

$\frac{7}{63} = \frac{100}{9}\%$

7. The income of A is 45% more than the income of B and the income of C is 60% less than the sum of the incomes of A and B. The income of D is 20% more than that of C. If the difference between the incomes of B and D is 13200, then the income of C is?
A की आय B की आय से 45% अधिक है और C की आय A और B की आय के योग से 60% कम है। D की आय C की आय से 20% अधिक है। यदि इनके बीच का अंतर है B और D की आय 13200 है, तो C की आय क्या है?

Solution :

$A : B = 29 : 20$

$C : (A + B) = 2 : 5$

$D : C = 6 : 5$

$A : B : C : D = 725 : 500 : 490 : 588$

$D - B = 88$ units $= 13200$

1 unit $= 150$

$C = 490$ units $= 73500$ Ans.

8. A number is mistakenly multiplied by $\frac{7}{5}$ instead of being multiplied by $\frac{3}{2}$. What is the percentage change in the result due to this mistake?
एक संख्या को गलती से $\frac{3}{2}$ से गुणा करने के बजाय $\frac{7}{5}$ से गुणा कर दिया जाता है। इस गलती के कारण परिणाम में कितने प्रतिशत परिवर्तन हुआ?

Solution :

$\frac{3}{2} : \frac{7}{5}$

$15 : 14$

Decreased by $\frac{100}{15}\% = 6\frac{2}{3}\%$

Type - 2 (Inc/Dec Based Ques)

9. A person's salary was increased by 50% and subsequently decreased by 50%. How much percentage does he lose or gain?
एक व्यक्ति का वेतन 50% बढ़ाया गया और बाद में 50% कम कर दिया गया। उसे कितने प्रतिशत की हानि या लाभ हुआ?

Solution :

First increase by $x\%$ and then decrease by $x\%$

Net change $= -\frac{x^2}{100}$

$-\frac{50^2}{100} =$ decreased by 25%

10. The number first increased by 20% and then increased by 25%. The number, so obtained, is now decreased by 30%. What is the net increase or decrease percent in the original number (nearest to an integer)?
संख्या में पहले 20% की वृद्धि हुई और फिर 25% की वृद्धि हुई। इस प्रकार प्राप्त संख्या अब 30% कम हो गई है। मूल संख्या (एक पूर्णांक के निकटतम) में शुद्ध वृद्धि या कमी प्रतिशत क्या है?

Solution :

First increase by 20% and then increase by 25%

$a + b + \frac{ab}{100}$

$20 + 25 + 5 = 50\%$

Then decrease by 30%

50 – 30 – 15

5% Overall increase

Or

$\frac{6}{5} \times \frac{5}{4} \times \frac{7}{10} = \frac{21}{20}$

$\frac{1}{20}$ = 5% overall increase

11. If a number is first increased by 15%, then reduced by 15%, it results in 782. If the same number is first reduced by 25%, then increased by 25% and again reduced by 20%, then what will be the resulting number?
यदि किसी संख्या को पहले 15% बढ़ाया जाता है, फिर 15% घटाया जाता है, तो परिणाम 782 होता है। यदि उसी संख्या को पहले 25% घटाया जाता है, फिर 25% बढ़ाया जाता है और फिर 20% कम किया जाता है, तो परिणाम क्या होगा संख्या?

Solution :

Let the number be x

$x \times \frac{23}{20} \times \frac{17}{20} = 782$

$x = 800$

$800 \times \frac{3}{4} \times \frac{5}{4} \times \frac{4}{5} = 600$

12. The price of a commodity increases 15% every year. If the difference between its prices at the end of third and fourth year is 193.20, then what is the 90% of its price at the end of second year?
एक वस्तु की कीमत हर साल 15% बढ़ जाती है। यदि तीसरे और चौथे वर्ष के अंत में इसकी कीमतों के बीच का अंतर 193.20 है, तो दूसरे वर्ष के अंत में इसकी कीमत का 90% क्या है?

Solution :

According to the question,

Difference between the 3rd and 4th year

= 15%

15% = 193.20

100% = 1288

2nd year Price $\times \frac{23}{20}$ = 3rd year Price

2nd year Price = $1288 \times \frac{20}{23} = 1120$

90% of 1120 = 1008

Type -3 Price, Consumption & Expenditure (कीमत, उपभोग एवं व्यय)

13. The price of sugar has increased by 18%. A person wants to increase the expenditure by 12% only. By what percent, correct to one decimal place, should he decrease his consumption?
चीनी की कीमत में 18% का इजाफा हुआ है. एक व्यक्ति व्यय में केवल 12% की वृद्धि करना चाहता है। दशमलव स्थान तक सही, उसे कितने प्रतिशत तक अपनी खपत कम करनी चाहिए?

Solution :

Price × Consumption = Expenditure

Change in Price = 50:59

Change in Expenditure = 25:28

Change in Consumption

$\frac{59}{28} : \frac{50}{25}$

59 : 56

Change in Consumption = $\frac{300}{59}$ = 5.1% approx

14. Due to an increase of 20% in the price of rice, the consumer decreases his consumption by 5%. What is the percentage change in total expenditure?
चावल की कीमत में 20 प्रतिशत की वृद्धि के कारण उपभोक्ता की खपत 5 प्रतिशत कम हो जाती है। कुल व्यय में प्रतिशत परिवर्तन क्या है?

Solution :

$P \times C = E$

Change in Price = 5:6

Change in Consumption = 20:19

Change in Expenditure = 100 : 114

14% Net change

15. If the price of rice is raised by 28%, then by how much per cent should a person reduce his consumption of rice, so that his expenditure remains the same?

यदि चावल की कीमत 28% बढ़ जाती है, तो एक व्यक्ति को चावल की खपत में कितने प्रतिशत की कमी करनी चाहिए, ताकि उसका खर्च वही रहे?

Solution :

Change in Price = 50:57

Expenditure remains the same

Change in Consumption = 57:50

$$\frac{700}{57} = 12.28\%$$

16. If price of rice is increased by 25 percent, then by what percentage consumption of rice should be decreased so expenditure incurred on rice gets decreased by 40 percent?

यदि चावल की कीमत 25 प्रतिशत बढ़ जाती है, तो चावल की खपत में कितने प्रतिशत की कमी की जानी चाहिए ताकि चावल पर होने वाला व्यय 40 प्रतिशत कम हो जाए?

Solution :

Price Change = 4:5

Expenditure Change = 5:3

Change in Consumption

$$\frac{5}{4} : \frac{3}{5}$$

25 : 12

$$\frac{13 \times 100}{25} = 52\% \text{ decrease}$$

17. When the price of sugar increased by 28%, a family reduced the consumption per month such that the expenditure on sugar was only 12% more than the earlier one. If the family consumed 18.4kg sugar per month earlier, what is its new consumption of sugar per month?

जब चीनी की कीमत 28% बढ़ गई, तो एक परिवार ने प्रति माह खपत इस तरह कम कर दी कि चीनी पर खर्च पहले की तुलना में केवल 12% अधिक हो गया। यदि परिवार पहले प्रति माह 18.4 किलोग्राम चीनी की खपत करता था, तो उसकी प्रति माह चीनी की नई खपत क्या है?

Solution :

Price Change = 25:32

Expenditure Change = 25:28

Change in Consumption

$$\frac{25}{25} : \frac{28}{32}$$

$\Rightarrow 8 : 7$

8 = 18.4kg

7 = 16.1kg

18. The reduction of 15% in the price of salt enables a person to buy 2kg more for Rs. 272. The reduced price of salt per kg (in Rs) is?

नमक की कीमत में 15% की कमी से एक व्यक्ति 2 रुपये में 2 किलो अधिक खरीद सकता है। 272. प्रति किलोग्राम नमक की घटी हुई कीमत (रुपये में) है?

Solution :

Change in Price = 20:17

Change in Consumption = 17:20

$$3 = 2\text{kg}\ ,\ 20 = \frac{40}{3}\text{kg}$$

$$\text{Reduced Price} = \frac{272}{\frac{40}{3}} = 20.4 \text{ Rs/kg}$$

19. Every month, a man consumes 20kg of rice and 8kg of wheat. The price of rice is 20% of the price of wheat, resulting in a total expenditure of ₹300 per month for rice and wheat combined. If the price of wheat is increased by 20%, what percentage reduction in rice consumption is needed to maintain the same expenditure of ₹300? (Given that price of rice and consumption of wheat is constant.)

हर महीने एक आदमी 20 किलो चावल और 8 किलो गेहूं खाता है। चावल की कीमत गेहूं की कीमत का 20% है, जिसके परिणामस्वरूप चावल और गेहूं को मिलाकर प्रति माह ₹ 300 का कुल खर्च होता है। यदि गेहूं की कीमत 20% बढ़ जाती है, तो ₹ 300 के समान व्यय को बनाए रखने के लिए चावल की खपत में कितने प्रतिशत की कमी की आवश्यकता है? (यह देखते हुए कि चावल की कीमत और गेहूं की खपत स्थिर है।)

Solution :

Price of wheat = $5x$

New Price of wheat = $6x$

Price of Rice = x $20x+8\times5x = 20y+8\times6x$

$20y = 60x - 48x = 12x$

$\Rightarrow y = \dfrac{3x}{5}$

$x : \dfrac{3x}{5} \Rightarrow 5 : 3$

$\dfrac{2}{5}$ = 40% reduction in Price of Rice to get same expenditure

20. In a manufacturing unit, it was noted that the price of raw material has increased by 25% and the labour cost has gone up from 30% of the cost of raw material to 38% of the cost of the raw material. What percentage of the consumption of raw material should be reduced to keep the cost the same as that before the increase?

एक विनिर्माण इकाई में, यह देखा गया कि कच्चे माल की कीमत में 25% की वृद्धि हुई है और श्रम लागत कच्चे माल की लागत के 30% से बढ़कर कच्चे माल की लागत का 38% हो गई है। लागत को वृद्धि से पहले के समान रखने के लिए कच्चे माल की खपत का कितना प्रतिशत कम किया जाए?

Solution :

Price = 20:25

Labour cost = 6 : 9.5

Consumption of Raw Material = Cost of Raw Material + Labour Cost

Consumption = 26 : 34.5

Change in Consumption to keep the cost same

34.5 : 26

$\dfrac{8.5 \times 100}{34.5} = 24.63\%$

Type - 4 (Income, Exp Savings)

21. Radha saves 25% of her income. If her expenditure increases by 20% and her income increases by 29%, then her savings increased by ?

राधा अपनी आय का 25% बचाती है। यदि उसका व्यय 20% बढ़ जाता है और उसकी आय 29% बढ़ जाती है, तो उसकी बचत बढ़ जाती है?

Solution :

Exp Inc Saving

20 x

29

3 : 1

$20 \times 3 + x = 29 \times 4$

$x = 116 - 60 = 56\%$

22. Radha saves $x\%$ of her income. If her income increases by 28% and the expenditure increases by 20%, then her savings increase by 40%. What is the value of x?

राधा अपनी आय का $x\%$ बचाती है। यदि उसकी आय 28% बढ़ जाती है और व्यय 20% बढ़ जाता है, तो उसकी बचत 40% बढ़ जाती है। एक्स का मूल्य क्या है ?

Solution :

Exp Inc Saving

20 40

28

12 : 8

3 : 2

$\dfrac{2}{5}$ = 40% Savings

23. The monthly expenses of a person are $66\frac{2}{3}$ % more than her monthly savings. If her monthly income increases by 44% and her monthly expenses increase by 60%, then there is an increase of Rs. 1,040 in her monthly savings. What is the initial expenditure (in Rs.)?

एक व्यक्ति का मासिक खर्च उसकी मासिक बचत से $66\frac{2}{3}$ % अधिक है। यदि उसकी मासिक आय 44% बढ़ जाती है और उसका मासिक खर्च 60% बढ़ जाता है, तो रुपये की वृद्धि होती है। उसकी मासिक बचत 1,040 रुपये है। प्रारंभिक व्यय (रुपये में) क्या है?

Solution :

Exp + Savings = Income

5 + 3 = 8

New Income = 8 × 1.44 = 11.52

New Exp = $5 \times 1.6 = 8$

New Savings = 3.52

Inc in Savings = $3.52 - 3 = 0.52$

0.52 units = 1040

1 unit = 2000

5 units = 10, 000

Initial Expenditure = ₹10,000

24. The income of A is 80% of B's income and the expenditure of A is 60% of B's expenditure. If the income of A is equal to 90% of B's expenditure, then by what percentage are the savings of A more than B's savings?

A की आय B की आय का 80% है और A का व्यय B के व्यय का 60% है। यदि A की आय B के व्यय के 90% के बराबर है, तो A की बचत B की बचत से कितने प्रतिशत अधिक है?

Solution :

Income $A : B = 4 : 5$, Exp $A : B = 3 : 5$

Inc A : Exp $B = 9 : 10$, Inc 180 : 225

Exp 120 : 200 , Sav 60 : 25 , $\frac{35}{25} = 140\%$

25. The monthly salary of a person was Rs. 50,000. He used to spend on family expenses (E), Taxes (T), charity (C), and the rest were his savings, E was 60% of the income, T was 20% of E, and C was 15% of T. When his salary got raised by 40%, he maintained the percentage level of E, but T becomes 30% of E, and C becomes 20% of T. The difference between two savings (in Rs) is?

एक व्यक्ति का मासिक वेतन रु. 50,000. वह पारिवारिक खर्च (ई), कर (टी), दान (सी) पर खर्च करता था, और बाकी उसकी बचत थी, ई आय का 60% था, टी ई का 20% था, और सी टी का 15% था। जब उसका वेतन 40% बढ़ गया, तो उसने E का प्रतिशत स्तर बनाए रखा, लेकिन T, E का 30% हो गया और C, T का 20% हो गया। दो बचतों के बीच का अंतर (रुपये में) क्या है?

Solution :

Let the initial salary be $1000x$.

$E = 60\%$ of $1000x = 600x$

$T = 20\%$ of $600x = 120x$

$C = 15\%$ of $120x = 18x$

Total Expense = $600x + 120x + 18x = 738x$

Initial savings = $1000x - 738x = 262x$

After salary got raised by 40%,

New salary = $1400x$

New $E = 60\%$ of $1400x = 840x$

New $T = 30\%$ of $840x = 252x$

New $C = 20\%$ of $252x = 50.4x$

New Total Expense = $840x + 252x + 50.4x = 1142.4x$

New Savings = $1400x - 1142.4x = 257.6x$

Difference in savings = $262x - 257.6x = 4.4x$

$1000x$ = Rs. 50, 000

$x = 50$

Difference in savings = $4.4x = 4.4 \times 50$ = Rs. 220

$\therefore$ The difference between the two savings is Rs. 220.

26. The sum of the salaries of A and B together is Rs. 43,000. A spends 95% of his salary and B spends 80% of his salary. If now their savings are the same, what is B's salary (in Rs.)?

A और B के वेतन का योग मिलाकर रु. 43000. A अपने वेतन का 95% खर्च करता है और B अपने वेतन का 80% खर्च करता है। यदि अब उनकी बचत समान है, तो B का वेतन (रुपये में) क्या है?

Solution :

Savings are Equal

5% of $A = 20\%$ of B

$A : B = 4 : 1$

5 = 43,000

1 = 8,600

27. Surekha spends 24% of her monthly income on groceries, 16% of it on rent, and 55% of the remaining on children's education and others. If she saves Rs 5,940 in a month, then how much (in Rs) does she spend on rent?
सुरेखा अपनी मासिक आय का 24% किराने के सामान पर, 16% किराए पर और शेष 55% बच्चों की शिक्षा और अन्य पर खर्च करती है। यदि वह एक महीने में 5,940 रुपये बचाती है, तो वह किराए पर कितना (रुपये में) खर्च करती है?

Solution :

$$x \times \frac{3}{5} \times \frac{9}{20} = 5940$$

$$x = 22000$$

$$22000 \times 16\% = 3520$$

28. Total savings of X and Y is 40% of their total income. Their average expenditure is Rs 21,000. What is the total salary of X and Y?
X और Y की कुल बचत उनकी कुल आय का 40% है। इनका औसत खर्च 21,000 रुपये है. X और Y का कुल वेतन कितना है?

Solution :

$$60\% = 2 \times 21000 = 42000$$

$$100\% = 70000$$

29. The monthly income of a person was Rs 80,000 and his monthly expenditure was Rs 45,000. Next year, his income increased by 16% and his expenditure increased by 8%. Find his New savings.
एक व्यक्ति की मासिक आय 80,000 रुपये थी और उसका मासिक व्यय 45,000 रुपये था। अगले वर्ष उनकी आय में 16% की वृद्धि हुई और उनके व्यय में 8% की वृद्धि हुई। उसकी नई बचत का पता लगाएं।

Solution :

$$\text{New Income} = 80k \times \frac{29}{25} = 92800$$

$$\text{New Exp} = 45k \times \frac{27}{25} = 48600$$

$$\text{New Savings} = 92800 - 48600 = 44200$$

30. P's income is 150 percent of Q's income and P's expenditure is 60 percent of Q's expenditure. If Q saves 20 percent of his income, then how much percentage does P save of his own income?
P की आय Q की आय का 150 प्रतिशत है और P का व्यय Q के व्यय का 60 प्रतिशत है। यदि Q अपनी आय का 20 प्रतिशत बचाता है, तो P अपनी आय का कितना प्रतिशत बचाता है?

Solution :

	Income	Exp	Savings
Q	100	80	20
P	150	48	102

$$\frac{102}{150} = 68\%$$

Type - 5 (Tax/Commission Based Questions) (टैक्स/कमीशन आधारित प्रश्न))

31. Aman's annual income has increased by Rs. 20 Lakhs but tax on income that he has to pay has reduced from 20% to 16%. He now pays the same amount of tax as earlier. What is his new total income (in Rs lakhs)?
अमन की वार्षिक आय में रु। 20 लाख की वृद्धि हुई है। लेकिन आय पर उसे जो टैक्स देना होता है, वह 20% से घटकर 16% हो गया है। अब वह पहले जितना ही टैक्स चुकाते हैं। उसकी नई कुल आय (लाख रुपये में) क्या है?

Solution :

$$x \times 20\% = (x + 20 \text{ lakh}) \times 16\%$$

$$4\% \text{ of } x = 20 \text{ lakh} \times 16\%$$

$$x = 80 \text{ lakh}$$

$$\text{New Income} = 80 + 20 = 100 \text{ lakh}$$

32. When income tax is 8 paisa in a rupee, a person's net income is Rs 3,09,580. What will be his net income (in Rs) if the income tax is raised to 12 paisa?
जब आयकर एक रुपये में 8 पैसे होता है, तो एक व्यक्ति की शुद्ध आय 3,09,580 रुपये होती है। यदि आयकर 12 पैसे तक बढ़ा दिया जाए तो उसकी शुद्ध आय (रुपये में) क्या होगी?

Solution :

$$\text{Total Income} = 309580 \times \frac{100}{92}$$

Net Income after paying 12 paise tax on 1 rupee

$$309580 \times \frac{100}{92} \times \frac{82}{100} = 275930$$

33. Suhani pays tax at the rate of 30% on her entire income of ₹ 90,000 and Ritika pays tax at the rate of 40% on her entire income of ₹ y. If the total rate of tax on their combined income becomes 37%, then what is the value of y?

सुहानी अपनी ₹ 90,000 की पूरी आय पर 30% की दर से कर का भुगतान करती है और रितिका अपनी ₹ y की पूरी आय पर 40% की दर से कर का भुगतान करती है। यदि उनकी संयुक्त आय पर कर की कुल दर 37% हो जाती है, तो y का मूल्य क्या है?

Solution :

30   40

    37

3   :   7

$3 = 90k$

$1 = 30k$

$7 = 210k$

34. If the income tax is increased by 19% then Net income is decreased by 6%. Find the rate of income Tax?

यदि आयकर 19% बढ़ जाता है तो शुद्ध आय 6% कम हो जाती है। आयकर की दर ज्ञात करें?

Solution :

Let the total income tax rate be x

So, according to the question

19% of $x = 6\%$ of $(100 - x)$

$19x = 600 - 6x$

$25x = 600$

$x = \frac{600}{25} = 24$

$\therefore$ Rate $= 24\%$

35. An agent gets 5% commission on all sales up to Rs. 10,000 and 4% commission on all sales exceeding this amount. He gives Rs. 31,100 from his sales to his company after deducting his commission. Find his total sales.

एक एजेंट को 10000 रुपये तक की सभी बिक्री पर 5% कमीशन और इस राशि से अधिक की सभी बिक्री पर 4% कमीशन मिलता है। वह रुपये देता है। अपना कमीशन काटने के बाद अपनी कंपनी को बिक्री से 31100 रुपये। उसकी कुल बिक्री ज्ञात कीजिए।

Solution :

1% of 10k = 100

$96\% = 31100 + 100 = 31200$

$1\% = 325$

$100\% = 32500$

36. A salesman is allowed 5.5% commission on the sales made by him up to Rs. 10,000 and additional bonus of 0.5% on sale over Rs. 10,000. If his total earnings are Rs. 1990. Find the total sales made by him?

एक सेल्समैन को उसके द्वारा की गई बिक्री पर 5.5% कमीशन की अनुमति है। 10000 रुपये और 10000 रुपये से अधिक की बिक्री पर 0.5% का अतिरिक्त बोनस। यदि उसकी कुल कमाई रुपये है। 1990 उसके द्वारा की गई कुल बिक्री ज्ञात कीजिए?

Solution :

0.5% of 10k = 50

$6\% = 1990 + 50 = 2040$

$1\% = 340$

$100\% = 34000$

37. A company allows 11% commission on total sales. If the salesman is appointed on a fixed salary of Rs. 8200 and 5% commission on the sales more than Rs. 20,000, then the salesman gets Rs. 1200 more. Find the total sales made by the salesman.

एक कंपनी कुल बिक्री पर 11% कमीशन की अनुमति देती है। यदि सेल्समैन को रुपये के निश्चित वेतन पर नियुक्त किया गया है। 8200 रुपये से अधिक की बिक्री पर 5% कमीशन। 20000, तो सेल्समैन को 1200 रुपये और विक्रेता द्वारा की गई कुल बिक्री ज्ञात कीजिए।

Solution :

11% of $x - 8200 - 5\%(x - 20k) = 1200$

6% of $x = 1200 + 8200 - 1000$

6% of $x = 8400$

$x = 140000$

Type - 6 (Election Based Questions)

38. In an election between 2 parties A and B, A gets 37% of total votes cast and thus lost by 338 votes. The total number of casted votes is (assume none of the votes was declared invalid)

दो पार्टियों A और B के बीच एक चुनाव में, A को कुल वोटों का 37% वोट मिले और इस तरह वह 338 वोटों से हार गया। डाले गए वोटों की कुल संख्या है (मान लें कि कोई भी वोट अवैध घोषित नहीं किया गया)

Solution :

winner = 63%

looser = 37%

26% = 338

100% = 1300

39. In an election, a candidate secures 42% of the votes polled but is defeated by his only opponent by a majority of 3080 votes, because of 1400 invalid votes. The percentage of invalid votes is........

एक चुनाव में, एक उम्मीदवार को डाले गए वोटों में से 42% वोट मिलते हैं, लेकिन 1400 अमान्य वोटों के कारण वह अपने एकमात्र प्रतिद्वंद्वी से 3080 वोटों के बहुमत से हार जाता है। अवैध मतों का प्रतिशत .... है

Solution :

Total Votes = 100%

Looser = 42%

Winner = 42% + 3080

Equate Valid Votes:

$100\% - 1400 = 42\% + 42\% + 3080$

$100\% - 84\% = 3080 + 1400$

$16\% = 4480$

$100\% = 28000$

Percentage of Invalid Votes:

$$\frac{1400}{28000} = 5\%$$

40. In an election between two candidates, 65% of the voters cast their votes, out of which 3% of the votes were decided to be invalid. A candidate got 81965 votes which are 65% of the total valid votes. What is the total number of votes enrolled in that election?

दो उम्मीदवारों के बीच एक चुनाव में, 65% मतदाताओं ने अपने वोट डाले, जिनमें से 3% वोट अवैध घोषित किए गए। एक उम्मीदवार को 81965 वोट मिले जो कुल वैध वोटों का 65% है। उस चुनाव में दर्ज वोटों की कुल संख्या कितनी है?

Solution :

65% of Valid Votes = 81965

Valid Votes = 126100

$$\text{Total Votes} \times \frac{97}{100} \times \frac{65}{100} = 126100$$

Total Votes = 200000

41. In an election between two candidates, 5% of the registered voters did not cast their vote. 10% of the votes were found to be either invalid or of NOTA. The winning candidate received 60% votes in his favour and won the election by 17271 votes. Find the number of registered voters.

दो उम्मीदवारों के बीच चुनाव में, पंजीकृत मतदाताओं में से 5% ने अपना वोट नहीं डाला। 10% वोट या तो अवैध पाए गए या नोटा के। विजयी उम्मीदवार को अपने पक्ष में 60% वोट मिले और वह 17271 वोटों से चुनाव जीत गया। पंजीकृत मतदाताओं की संख्या ज्ञात कीजिए।

Solution :

Total 100%

Valid Votes = 85%

winner = 60%

looser = 60% − 17271

$85\% = 60\% + 60\% - 17271$

$45\% = 17271$

$100\% = 38380$

42. In an election, 2% persons enrolled in the voter list did not participate and 500 votes were invalid. Two candidates A and B fought the election, and A defeated B by 200 votes. If 43% of the persons enrolled in the voter list casted their votes in favour of A, then what is the number of the total casted votes?

एक चुनाव में, मतदाता सूची में नामांकित 2% व्यक्तियों ने भाग नहीं लिया और 500 वोट अवैध थे। दो उम्मीदवारों A और B ने चुनाव लड़ा और A ने B को 200 वोटों से हरा दिया। यदि मतदाता सूची में नामांकित 43% व्यक्तियों ने A के पक्ष में अपना वोट डाला, तो कुल डाले गए वोटों की संख्या क्या है?

Solution :

Total Votes = 100%

Casted Votes = 98%

Valid Votes = 98% - 500

A = 43%

Solution :

B = 43% - 200

98% - 500 = 43% + 43% - 200

98% - 86% = 500 - 200

12% = 300

100% = 2500 Total Votes

43. In an election between two candidates, one candidate got 60% of the valid votes. If the total number of votes polled were 8850, and 10% of the votes were invalid, then find the number of valid votes that the other candidate got.
दो उम्मीदवारों के बीच एक चुनाव में, एक उम्मीदवार को 60% वैध वोट मिले। यदि डाले गए वोटों की कुल संख्या 8850 थी, और 10% वोट अवैध थे, तो दूसरे उम्मीदवार को मिले वैध वोटों की संख्या ज्ञात कीजिए।

Solution :

Total 100%

Invalid Votes = 10%

Valid Votes = 100 - 10 = 90%

Candidate B got 40% of Valid votes = 40% of

90 = 36%

100% = 8850

36% = 3186 Votes

Or $8850 \times \frac{9}{10} \times \frac{2}{5} = 3186$

44. Three candidates were participating in an election. The person at third place got 20% of the total votes while the difference between the votes of the winner and the first runner up was 20% of the total votes. If the difference between the votes of the first runner up and the second runner up was 37,000, how many votes did the winner receive?
एक चुनाव में तीन उम्मीदवार भाग ले रहे थे। तीसरे स्थान पर रहे व्यक्ति को कुल वोटों का 20% वोट मिले जबकि विजेता और प्रथम उपविजेता के वोटों के बीच का अंतर कुल वोटों का 20% था। यदि प्रथम रनर अप और दूसरे रनर अप के वोटों के बीच का अंतर 37,000 था, तो विजेता को कितने वोट मिले?

Solution :

3rd = 20%

Difference between 1st & 2nd = 20%

1st = 50%

2nd = 30%

5 : 3 : 2

1 = 37000

5 = 185000

45. Raju, Ravi and Ashok contested an election. 5% votes polled were invalid. Raju got 30% of the total votes. Ravi got 32% of the total votes. The winner got 5136 more votes than the person who received the least number of votes. Find the total number of votes polled.
राजू, रवि और अशोक ने एक चुनाव लड़ा। डाले गए 5% वोट अवैध थे. राजू को कुल वोटों का 30% वोट मिले। रवि को कुल वोटों का 32% वोट मिले। विजेता को सबसे कम वोट पाने वाले व्यक्ति से 5136 अधिक वोट मिले। डाले गए वोटों की कुल संख्या ज्ञात कीजिए।

Solution :

Total Votes = 100%

Valid Votes = 95%

Raju = 30%

Ravi = 32%

Ashok = 30% + 5136

30% × 5136 + 32% + 30% = 95%

3% = 5136

1% = 1712

100% = 171200 Total Votes

46. In an election, a voter has choices to vote for one candidate out of three candidates. There is also an option 'None Of The Above' (NOTA) to reject all candidates. The three candidates get votes that are 6 times NOTA. The winner gets votes that are twice the second runnerup's votes. The first runner-up gets 900 votes more than NOTA and defeats the second runner-up by 150 votes. What is the total number of votes polled?

एक चुनाव में, एक मतदाता के पास तीन उम्मीदवारों में से एक उम्मीदवार को वोट देने का विकल्प होता है। सभी उम्मीदवारों को अस्वीकार करने के लिए 'उपरोक्त में से कोई नहीं' (नोटा) विकल्प भी है। तीनों उम्मीदवारों को नोटा से 6 गुना ज्यादा वोट मिले। विजेता को दूसरे रनरअप के वोटों से दोगुने वोट मिलते हैं। प्रथम रनर-अप को नोटा से 900 वोट अधिक मिलते हैं और वह दूसरे रनर-अप को 150 वोटों से हरा देता है। कुल पड़े वोटों की संख्या कितनी है?

Solution :

Total Candidate votes : NOTA = 6:1

winner (x) = 2 × second runnerup (z)

$x = 2z$

first runner up (y) = 900 + NOTA

$y - z = 150$

$x + y + z = 6$ NOTA

$2z + 150 + z + z = 6(y - 900)$

$4z + 150 = 6(150 + z - 900)$

$4z + 150 = 900 + 6z - 5400$

$2z = 5400 - 900 + 150$

$2z = 4650$

$z = 2325$

Total votes = $x + y + z +$ NOTA

$2z + 150 + z + z + (150 + z - 900)$

$5z - 600$

$5 \times 2325 - 600 = 11025$

Total votes = 11025

47. In an election between A and B, every fifth vote polled was marked as invalid by the machine. In the remaining votes, A wins the election with a margin of 2500 votes or 10% of the total votes polled, over B. If 90% of the invalid votes would have been in favour of B, then which of the following would have been the result of the election?

A और B के बीच चुनाव में, डाले गए प्रत्येक पांचवें वोट को मशीन द्वारा अवैध घोषित कर दिया गया। शेष वोटों में, A, B से 2500 वोटों या कुल डाले गए वोटों के 10% के अंतर से चुनाव जीतता है। यदि 90% अवैध वोट B के पक्ष में होते, तो निम्नलिखित में से कौन सा वोट होता? चुनाव का नतीजा?

Solution :

Invalid votes = $\frac{1}{5}$ = 20% votes

valid votes = 80%

difference between A & B = 10%

A : B = 45% : 35%

if 90% of invalid votes, voted to B

then A:B = 45% : 35% + 18%

45% : 53%

B would have won by 8%

48. In an election between two candidates, the defeated candidate secured 42% of the valid votes polled and lost the election by 2545800 votes. If 365500 votes were declared invalid and 30% people did not cast their vote, what was the approximate number of people (in millions) in the electorate who did NOT cast their votes?

दो उम्मीदवारों के बीच एक चुनाव में, पराजित उम्मीदवार को कुल वैध मतों का 42% प्राप्त हुआ और वह 2545800 मतों से चुनाव हार गया। यदि 365500 वोट अवैध घोषित कर दिए गए और 30% लोगों ने अपना वोट नहीं डाला, तो मतदाताओं में उन लोगों (लाखों में) की अनुमानित संख्या क्या थी जिन्होंने अपना वोट नहीं डाला?

Solution :

Looser = 42%

winner = 42% + 2545800

42% + 42% + 2545800 = 100%

16% = 2545800

100% = 15911250

Total Casted Votes = 15911250 + 365500 = 16276750

70% (Casted Votes) = 16276750

30% (Not Casted) = 6975750

Approx 7 million

49. In an election between two candidates, y% of the voters did not vote 10% of the votes cast were declared invalid, while all the valid votes were cast in favour of either of the two candidates. The candidate who got 59.375% of the valid votes cast was declared elected by 2484 votes. If the number of people eligible to vote in that election was 16,000, what is the value of y?

दो उम्मीदवारों के बीच चुनाव में, y% मतदाताओं ने वोट नहीं दिया। डाले गए 10% वोट अवैध घोषित कर दिए गए, जबकि सभी वैध वोट दोनों उम्मीदवारों में से किसी एक के पक्ष में पड़े थे। जिस उम्मीदवार को कुल वैध मतों में से 59.375% वोट मिले, उसे 2484 वोटों से निर्वाचित घोषित किया गया। यदि उस चुनाव में मतदान करने के योग्य लोगों की संख्या 16,000 थी, तो y का मान क्या है?

Solution :

Winner = 59.375%

Looser = 100% - 59.375% = 40.625%

59.375% - 40.625% = 18.75%

18.75% = 2484

100% = 13248

$16000 \times \frac{9}{10} \times \left(100 - \frac{x}{100}\right) = 13248$

$(100 - x) \ /100 = 92/100$

100 - x = 92

x = 100 - 92 = 8%

50. In an election between P, R and S, 1/10 of the total votes polled are invalid. Votes polled in favour of R and S are half of the total votes polled, and they get equal votes. Due to a fault in the machine, half of the votes of S and half of the invalid votes are counted additionally in favour of R. What is the overall percentage of votes secured by R due to the fault?

P, R और S के बीच चुनाव में, डाले गए कुल वोटों में से 1/10 वोट अवैध हैं। आर और एस के पक्ष में पड़े वोट कुल पड़े वोटों के आधे हैं और उन्हें बराबर वोट मिले हैं। मशीन में खराबी के कारण, S के आधे वोट और आधे अवैध वोट R के पक्ष में अतिरिक्त रूप से गिने जाते हैं। गलती के कारण R द्वारा प्राप्त वोटों का कुल प्रतिशत क्या है?

Solution :

Total Votes 100

Invalid Votes 10

Valid Votes 90

$R = S = \frac{100}{4} = 25$

P = 40

$R = 25 + \frac{25}{2} + \frac{10}{2}$

$R = 25 + 12.5 + 5$

$R = 42.5\%$

51. In an election, 12% of the voters in the voter list did not cast their votes, whereas 84 voters cast their ballot papers blank. There were only two candidates: Ramagya and Shravan. The winner, Ramagya, was supported by 54% of all the voters in the list. He got 1456 more votes than his rival, Shravan. Find the total number of voters in the list.

एक चुनाव में, मतदाता सूची में 12% मतदाताओं ने अपना वोट नहीं डाला, जबकि 84 मतदाताओं ने अपने मतपत्र खाली डाले। केवल दो उम्मीदवार थे: रामाज्ञा और श्रवण। विजेता रामाज्ञा को सूची के सभी मतदाताओं में से 54% का समर्थन प्राप्त था। उन्हें अपने प्रतिद्वंदी श्रवण से 1456 वोट ज्यादा मिले. सूची में मतदाताओं की कुल संख्या ज्ञात कीजिए।

Solution :

Not casted = 12%

84 votes casted blank ballot papers

Valid votes = $88\% - 84$

Ramagya = 54%

Shravan = $54\% - 1456$

$54\% + 54\% - 1456 = 88\% - 84$

$108\% - 88\% = 1456 - 84$

$20\% = 1372$

$100\% = 6860$ Votes

52. In an election between two candidates, the defeated candidate secured 42% of the valid votes polled and lost the election by 7,68,400 votes. If 82,560 votes were declared invalid and 20% people did NOT cast their vote, then the invalid votes were what percentage (rounded off to 1 decimal place) of the votes which people did NOT cast?

दो उम्मीदवारों के बीच हुए चुनाव में, पराजित उम्मीदवार को कुल वैध वोटों में से 42% वोट मिले और वह 7,68,400 वोटों से चुनाव हार गया। यदि 82,560 वोट अवैध घोषित कर दिए गए और 20% लोगों ने अपना वोट नहीं डाला, तो अवैध वोट उन वोटों का कितना प्रतिशत (दशमलव स्थान तक पूर्णांकित) थे जो लोगों ने नहीं डाले?

Solution :

Total 100%

Valid Votes 80% − 82560

looser 42% of valid votes

winner 58% of valid votes

16% of valid votes = 768400

valid votes = 4802500

80% = 4802500 + 82560

80% = 4885060 ⇒ 20% = 1221265

$$\frac{82560}{1221265} = 6.76\% \text{ (6.8\% apprx)}$$

∴ The invalid votes were 6.8% of the votes which people did NOT cast.

Type - 7(Area/Volume Based Questions)

53. If each side of a rectangle is decreased by 10%, then the area will decrease by ?
यदि एक आयत की प्रत्येक भुजा 10% कम हो जाती है। तो क्षेत्रफल कम हो जायेगा?

Solution :

$L \sim 10 : 9$

$B \sim 10 : 9$

$A \sim 100 : 81$

19% decrease in Area

54. If each edge of a cube is increased by 10%, then the percentage increase in its surface area is ?
यदि किसी घन के प्रत्येक किनारे को 10% बढ़ा दिया जाए, तो उसके सतह क्षेत्र में प्रतिशत वृद्धि होगी?

Solution :

Side ~ 10 : 11
Area ~ 100 : 121
21% increase in Area

55. If the area of a square is decreased by 19%, then the diagonal of the square is decreased by ?
यदि किसी वर्ग का क्षेत्रफल 19% कम हो जाता है, तो वर्ग का विकर्ण कितना कम हो जाता है?

Solution :

Area ~ 100 : 81

Diagonal ~ $\sqrt{100} : \sqrt{81} \sim 10 : 9$

10% decrease

56. If the length of a rectangle is increased by 12% and the breadth is decreased by 8%, the net effect on the area is ?
यदि एक आयत की लंबाई 12% बढ़ा दी जाए और चौड़ाई 8% कम कर दी जाए, तो क्षेत्रफल पर शुद्ध प्रभाव क्या होगा?

Solution :

$L \sim 25 : 28$

$B \sim 25 : 23$

$A \sim 625 : 644$

$\frac{19}{625} \sim 3.04\%$ increase

57. The length and breadth of a cuboid are increased by 10% and 20%, respectively, and its height is decreased by 20%. The percentage increase in the volume of the cuboid is ?
एक घनाभ की लंबाई और चौड़ाई क्रमश: 10% और 20% बढ़ जाती है, और इसकी ऊंचाई 20% कम हो जाती है। घनाभ के आयतन में प्रतिशत वृद्धि है?

Solution :

$L \sim 10 : 11$

$B \sim 5 : 6$

$H \sim 5 : 4$

$V \sim 250 : 264$

$\frac{14}{250} \sim 5.6\%$ increase

Type - 8 (Examination Based Questions) (परीक्षा आधारित प्रश्न)

58. In an examination, Shreya scored 84 out of 90 in Mathematics, 45 out of 50 in science, 23 out of 25 in Computer Science and 68 out of 80 in English. In which subject did Shreya score the highest percentage?

एक परीक्षा में, श्रेया ने गणित में 90 में से 84 अंक, विज्ञान में 50 में से 45 अंक, कंप्यूटर विज्ञान में 25 में से 23 अंक और अंग्रेजी में 80 में से 68 अंक प्राप्त किये। श्रेया ने किस विषय में सर्वाधिक प्रतिशत अंक प्राप्त किये?

Solution :

Maths $\sim 93.33\%$

Science $\sim 90\%$

CS $\sim 92\%$

Eng $\sim 85\%$

In Maths she scored the highest percentage

59. In an examination of 180 marks of subject, 75% marks are allocated for theory and the rest for project work. Veena scored 60% in theory. What percentage of marks must she score in the project to achieve an overall 62% marks in the subject?

180 अंकों के विषय की परीक्षा में 75% अंक थ्योरी के लिए और बाकी प्रोजेक्ट वर्क के लिए आवंटित किए जाते हैं। वीणा ने थ्योरी में 60% अंक प्राप्त किये। विषय में कुल 62% अंक प्राप्त करने के लिए उसे प्रोजेक्ट में कितने प्रतिशत अंक प्राप्त करने होंगे?

Solution :

60   x

     62

3   :   1

$60 \times 3 + x = 62 \times 4$

$x = 68\%$

60. 35% of the students in a college are girls and the rest are boys. The total number of students in the college is 2800. 75% of the boys and 85% of the girls passed the final examination. The percentage of total students who passed the final examination is:

एक कॉलेज में 35% छात्र लड़कियाँ हैं और शेष लड़के हैं। कॉलेज में छात्रों की कुल संख्या 2800 है। 75% लड़कों और 85% लड़कियों ने अंतिम परीक्षा उत्तीर्ण की। अंतिम परीक्षा में उत्तीर्ण होने वाले कुल छात्रों का प्रतिशत है:

Solution :

75   85

     x

65   :   35

13   :   7

$75 \times 13 + 85 \times 7 = 20x$

$x = 78.5\%$

61. The ratio of the number of boys to that of girls in a village is 5 : 3. If 40% of the boys and 60% of the girls appeared in an examination, then the ratio of the number of boys and girls who appeared in the examination to the number who did not appear in the same examination, is:

एक गाँव में लड़कों की संख्या का लड़कियों की संख्या से अनुपात 5: 3 है। यदि 40% लड़के और 60% लड़कियाँ एक परीक्षा में उपस्थित हुए, तो परीक्षा में बैठने वाले लड़कों और लड़कियों की संख्या का अनुपात उस संख्या की परीक्षा जो उसी परीक्षा में उपस्थित नहीं हुई:

Solution :

40   60

     x

5   :   3

47.5 : 52.5

95 : 105

19 : 21

62. In an examination a candidate had to sit for three papers A, B and C. The candidate secured 75% marks in Paper A, 80% marks in Paper B, and 60% marks in Paper C. If the weightage assigned to Papers A, B and C were 40%, 50% and 10%, respectively, then find the weighted percentage of marks obtained by the candidate, when all three papers were taken together.

एक परीक्षा में एक उम्मीदवार को तीन पेपर A, B और C में बैठना था। उम्मीदवार ने पेपर A में 75% अंक, पेपर B में 80% अंक और पेपर C में 60% अंक प्राप्त किए। यदि पेपर A को वेटेज दिया गया है, बी और सी क्रमश: 40%, 50% और 10% थे, तो उम्मीदवार द्वारा प्राप्त अंकों का भारित प्रतिशत ज्ञात करें, जब तीनों पेपर एक साथ लिए गए थे।

Solution :

$A = 40 \times \frac{3}{4} = 30$

$B = 50 \times \frac{4}{5} = 40$

$C = 10 \times \frac{3}{5} = 6$

Weighted Percentage $= 76\%$

63. In an examination, A obtained 10% more marks than B, B obtained 20% more marks than C. If A obtained 272 more marks than C, then the marks obtained by B is ?

एक परीक्षा में, A ने B से 10% अधिक अंक प्राप्त किए, B ने C से 20% अधिक अंक प्राप्त किए। यदि A ने C से 272 अधिक अंक प्राप्त किए, तो B द्वारा प्राप्त अंक हैं?

Solution :

$A : B = 11 : 10$

$B : C = 6 : 5$

$A : B : C = 66 : 60 : 50$

$16 = 272$

$1 = 17$

$B = 60 = 1020$ marks

64. An examination was held on five subjects where the maximum marks for each subject were 100. X scored 82, 97, 88 and 91 marks in four subjects. If his total marks were 90% then, find his marks in the fifth subject.

पाँच विषयों पर एक परीक्षा आयोजित की गई थी जहाँ प्रत्येक विषय के लिए अधिकतम अंक 100 थे। X ने चार विषयों में 82, 97, 88 और 91 अंक प्राप्त किए। यदि उसके कुल अंक 90% थे, तो पांचवें विषय में उसके अंक ज्ञात कीजिए।

Solution :

$82 + 97 + 88 + 91 + x = 90 \times 5 = 450$

$x = 92$

65. In an examination, 48% of candidates passed in science and 56% failed in mathematics. If 32% failed in both subjects, then what percent passed in both subjects?

एक परीक्षा में, 48% उम्मीदवार विज्ञान में उत्तीर्ण हुए और 56% गणित में असफल रहे। यदि 32% दोनों विषयों में अनुत्तीर्ण हुए, तो दोनों विषयों में कितने प्रतिशत उत्तीर्ण हुए?

Solution :

Passed In Science $n(A) = 48\%$

Passed In Maths $n(B) = 100 - 56 = 44\%$

Passed $n(A \cup B) = 100 - 32 = 68\%$

$n(A \cup B) = n(A) + n(B) - n(A \cap B)$

$68 = 48 + 44 - x$

$x = 92 - 68 = 24\%$

24% of students passed in both subjects

66. In an examination, 45% of all the students who appeared are boys and the rest are girls. If 60% of the boys and 70% of the girls passed, then what is the percentage of students who failed?

एक परीक्षा में उपस्थित सभी विद्यार्थियों में से 45% लड़के हैं और शेष लड़कियाँ हैं। यदि 60% लड़के और 70% लड़कियाँ उत्तीर्ण हुईं, तो अनुत्तीर्ण छात्रों का प्रतिशत क्या है?

Solution :

60   70

x

45   :   55

9   :   11

$60 \times 9 + 70 \times 11 = 20x$

$x = 65.5\%$

$100 - 65.5\% = 34.5\%$ failed

67. Geeta scored 30% and failed by 50 marks, while Sandeep who scored 45% marks, got 25 marks more than the minimum marks required to pass the examination. How many marks did Vimal get if he scored 64% marks?

गीता ने 30% अंक प्राप्त किए और 50 अंकों से अनुत्तीर्ण हो गई, जबकि संदीप, जिसने 45% अंक प्राप्त किए, को परीक्षा उत्तीर्ण करने के लिए आवश्यक न्यूनतम अंकों से 25 अंक अधिक मिले। यदि विमल ने 64% अंक प्राप्त किये तो उसे कितने अंक प्राप्त हुए?

Solution :

$30\% + 50 = 45\% - 25$

$15\% = 75$

$1\% = 5$

$64\% = 320$ marks

68. A certain number of students from school X appeared in an examination and 30% students failed. 150% more students than those from school X, appeared in the same examination from school Y. If 80% of the total number of students who appeared from X and Y passed, then what is the percentage of students who failed from Y?

स्कूल X से एक निश्चित संख्या में छात्र एक परीक्षा में उपस्थित हुए और 30% छात्र असफल रहे। स्कूल X के छात्रों की तुलना में 150% अधिक छात्र, स्कूल Y से समान परीक्षा में उपस्थित हुए। यदि X और Y से परीक्षा देने वाले छात्रों की कुल संख्या में से 80% उत्तीर्ण हुए, तो Y से अनुत्तीर्ण होने वाले छात्रों का प्रतिशत क्या है?

Solution :

Let Students in X = $100x$

Failed students in X = $30x$

Passes students in X = $70x$

Students in Y = $100x \times \frac{5}{2} = 250x$

Total number of students (X+Y) = $100x + 250x = 350x$

Number of passes students in both schools = $80\% \times 350x = 280x$

Passed students in Y = $280x - 70x = 210x$

Failed students in Y = $250x - 210x = 40x$

Percentage of failed students in school Y = $\frac{40x}{250x} \times 100\% = 16\%$

$\therefore$ The percentage of students who failed from school Y is 16%.

69. Some students (only boys and girls) from different schools appeared for an Olympiad exam. 20% of the boys and 15% of the girls failed the exam. The number of boys who passed the exam was 70 more than that of the girls who passed the exam. A total of 90 students failed. Find the number of students that appeared for the exam.

विभिन्न स्कूलों के कुछ छात्र (केवल लड़के और लड़कियाँ) ओलंपियाड परीक्षा के लिए उपस्थित हुए। 20% लड़के और 15% लड़कियाँ परीक्षा में असफल रहे। परीक्षा उत्तीर्ण करने वाले लड़कों की संख्या परीक्षा उत्तीर्ण करने वाली लड़कियों की संख्या से 70 अधिक थी। कुल 90 छात्र फेल हो गये. परीक्षा में बैठने वाले छात्रों की संख्या ज्ञात कीजिए।

Solution :

Let boys = x and girls = y

$20\% \text{ of } x + 15\% \text{ of } y = 90 \quad (1)$

$80\% \text{ of } x - 85\% \text{ of } y = 70 \quad (2)$

Solving both equations:

$y = 200$

$x = 300$

Total students = 500

Type – 9 (Population Based Questions)

70. The population of a town has increased by 5% at the end of the first year and decreased by 4% at the end of the second year. If the population at the end of the second year was 55,12,248, then the population at the beginning of the first year was?

एक कस्बे की जनसंख्या पहले वर्ष के अंत में 5% बढ़ गई और दूसरे वर्ष के अंत में 4% घट गई। यदि दूसरे वर्ष के अंत में जनसंख्या 55,12,248 थी, तो पहले वर्ष की शुरुआत में जनसंख्या थी?

Solution :

$$5512248 \times \frac{25}{24} \times \frac{20}{21} = 5468500$$

71. The present population of a village is 15,280. If the number of males increases by 25% and the number of females increases by 15%, then the population will become 18,428. The difference between present population of males and females in the village is:

एक गाँव की वर्तमान जनसंख्या 15,280 है। यदि पुरुषों की संख्या 25% बढ़ जाती है और महिलाओं की संख्या 15% बढ़ जाती है, तो जनसंख्या 18,428 हो जाएगी। गाँव में पुरुषों और महिलाओं की वर्तमान जनसंख्या के बीच का अंतर है?

Solution :

Increase by 3148

25% of 15280 = 3820

3820 − 3148 = 672

10% of females = 672

100% of females = 6720

Males = 8560

Difference = 8560 − 6720 = 1840

72. The population of a town is 1,25,000. If the annual birth rate is 12.7% and the annual death rate is 2.7%, how much will the population increase after 3 years?

एक कस्बे की जनसंख्या 1,25,000 है। यदि वार्षिक जन्म दर 12.7% है और वार्षिक मृत्यु दर 2.7% है, तो 3 वर्ष बाद जनसंख्या में कितनी वृद्धि होगी?

Solution :

Net Change in a year = 12.7 − 2.7 = 10%

every year

In 3 years Net change is = 33.1%

125000 × 33.1% = 41375

73. The population of town B is 300% more than that of town A. For the next two years, the population of A increases by $x\%$ per year and that of B decreases by the same percentage per year. After 2 years, if the population of A and B become equal, then the value of x is:

शहर B की जनसंख्या शहर A की तुलना में 300% अधिक है। अगले दो वर्षों के लिए, A की जनसंख्या प्रति वर्ष $x\%$ बढ़ जाती है और B की जनसंख्या प्रति वर्ष समान प्रतिशत कम हो जाती है। 2 वर्षों के बाद, यदि A और B की जनसंख्या बराबर हो जाती है, तो x का मान है

Solution :

$1 \times (100 + x)^2 = 4(100 - x)^2$

$\frac{100 + x}{100 - x} = 2$

$100 - x = 200 - 2x$

$x = 100$

Type – 10 (Miscellaneous)(मिश्रित)

74. Sudha decided to donate 12% of her monthly income to an orphanage. On the day of donation, she changed her decision and donated a sum of Rs. 4800 which was equal to 80% of what she had decided earlier. What is 27% of her monthly income?

सुधा ने अपनी मासिक आय का 12% एक अनाथालय को दान करने का फैसला किया। दान के दिन, उसने अपना निर्णय बदल दिया और रुपये की राशि दान की। 4800 जो कि उसके द्वारा पहले तय किये गये मूल्य के 80% के बराबर था। उसकी मासिक आय का 27% क्या है?

Solution :

80% = 4800

1% = 60

100% = 6000

12% = 6000

1% = 500

27% = 13500

75. A fruit seller sells 45% of the oranges that he has along with one more orange to a customer. He then sells 20% of the remaining oranges and 2 more oranges to a second customer. He then sells 90% of the now remaining oranges to a third customer and is still left with 5 oranges. How many oranges did the fruit seller have initially?

एक फल विक्रेता अपने पास मौजूद संतरे का 45% और एक और संतरा ग्राहक को बेचता है। फिर वह शेष संतरे का 20% और 2 और संतरे दूसरे ग्राहक को बेचता है। फिर वह बचे हुए संतरों का 90% तीसरे ग्राहक को बेच देता है और उसके पास अभी भी 5 संतरे बचे हैं। फल विक्रेता के पास प्रारंभ में कितने संतरे थे?

Solution :

Reverse Process

10% = 5 oranges

100% = 50 Oranges

80% = 52 100% = 65

55% = 66

100% = 120 oranges Total

76. A man ordered 10 physics books and some chemistry books. The price of a chemistry book is twice the price of a Physics book. While preparing the bill, the clerk interchanged the number of physics and chemistry books by mistake. Which decreased the bill by 12.5%. The ratio of the number of physics books to the number of chemistry books in the original order is?

एक आदमी ने 10 भौतिकी की किताबें और कुछ रसायन विज्ञान की किताबें ऑर्डर कीं। रसायन विज्ञान की किताब की कीमत भौतिकी की किताब की कीमत से दोगुनी है। बिल बनाते समय क्लर्क ने गलती से फिजिक्स और केमिस्ट्री की किताबों की संख्या आपस में बदल दी। जिससे बिल में 12.5% की कमी आई। मूल क्रम में भौतिकी की पुस्तकों की संख्या और रसायन विज्ञान की पुस्तकों की संख्या का अनुपात क्या है?

Solution :

Let price of Physics Book = ₹ y & Number of Chemistry book = x

$(10y + x \cdot 2y) \cdot \frac{7}{8} = 10 \cdot 2y + x \cdot y$

$70y + 14xy = 160y + 8xy$

$6xy = 90y$

$x = 15$

No. of Phy Book : Chem Book = 10 : 15 = 2 : 3

77. Varun gave 60% of the money he had to his wife. He also gave 20% of the remaining amount to each of his three children. Three-fifth of the amount now left was spent on miscellaneous items, and the remaining amount of Rs 9,600 was deposited in the bank. How much money did Varun have initially?

वरुण ने अपने पास मौजूद धन का 60% अपनी पत्नी को दे दिया। उन्होंने शेष राशि का 20% अपने तीन बच्चों में से प्रत्येक को दिया। अब बची हुई राशि का तीन-पाँचवाँ भाग विविध मदों पर खर्च किया गया, और शेष 9,600 रुपये की राशि बैंक में जमा कर दी गई। वरुण के पास शुरू में कितने पैसे थे?

Solution :

Let Total Money be $100x$

wife = $60x$

Children = 20% of $40x = 8x$

3 children = $24x$

$\frac{2}{5}$th of remaining

$16x = 9600$

$100x = 150000$

78. In a constituency, 55% of the total number of voters are males and the rest are females. If 40% of the males are illiterate and 40% of the females are literate, then by what percentage is the number of illiterate females more than that of the illiterate males (correct to the one decimal place)?

एक निर्वाचन क्षेत्र में कुल मतदाताओं में से 55% पुरुष हैं और शेष महिलाएँ हैं। यदि 40% पुरुष निरक्षर हैं और 40% महिलाएँ साक्षर हैं, तो निरक्षर महिलाओं की संख्या, निरक्षर पुरुषों की तुलना में कितने प्रतिशत अधिक है (दशमलव के एक स्थान तक सही)?

Solution :

Illiterate Males = $55 \times 40\% = 22$

Illiterate Females = $45 \times 60\% = 27$

27 : 22

The percentage is $\frac{500}{22} = 22.72\%$

79. In a workshop with 80 students and 10 resource persons, sweets were distributed at the end of the workshop. The amount spent for distributing sweets to each student was 30% of the total number of students and the amount spent towards each resource person was 40% of the total number of students. The total amount spent for sweets distribution in the workshop was.

80 छात्रों और 10 संसाधन व्यक्तियों के साथ एक कार्यशाला में, कार्यशाला के अंत में मिठाइयाँ वितरित की गईं। प्रत्येक छात्र को मिठाई बांटने पर खर्च की गई राशि छात्रों की कुल संख्या का 30% थी और प्रत्येक संसाधन व्यक्ति पर खर्च की गई राशि छात्रों की कुल संख्या का 40% थी। कार्यशाला में मिठाई वितरण पर कुल कितनी राशि खर्च हुई?

Solution :

Amt spend on Each student = $80 \times 30\% = 24$

Amt spend on Each resource = $80 \times 40\% = 32$

Total amount = $24 \times 80 + 32 \times 10 = 2240$

80. A crate of fruits contains one spoiled fruit for every 25 fruits. 60% of the spoiled fruits were sold. If the seller had sold 48 spoiled fruits, then the number of fruits in the crate were.
फलों के एक टोकरे में प्रत्येक 25 फलों के लिए एक खराब फल होता है। 60% खराब फल बिक गए। यदि विक्रेता ने 48 खराब फल बेचे थे, तो टोकरे में फलों की संख्या कितनी थी?

Solution :

60% = 48

100% = 80

1 spoiled fruit = 25 Fruits

80 spoiled fruits = 2000 Fruits

81. A, B, and C donate 8%, 7%, and 9% of their salaries respectively to a charitable trust. The salaries of A and B are the same, and the difference between their donations is ₹259. The total donation of A and B is ₹1,185 more than that of C. The total donation of A and C is what percentage of the total salaries of A, B, and C? (Correct to one decimal place)
A, B और C अपने वेतन का क्रमश: 8%, 7% और 9% एक धर्मार्थ ट्रस्ट को दान करते हैं। A और B का वेतन समान है और उनके दान के बीच का अंतर ₹259 है। A और B का कुल दान C की तुलना में ₹1,185 अधिक है। A और C का कुल दान A, B और C के कुल वेतन का कितना प्रतिशत है? (दशमलव के एक स्थान तक सही)

Solution :

Let salaries of A, B be 100%

8% − 7% = 259

1% = 259

8% = 2072

15% = 3885

100% = 25900

Donation of C = 3885 − 1185 = 2700

9% = 2700

100% = 30, 000

Solution :

Total donation of A and C = 2072 + 2700 = 4772

Total salaries of A, B, and C = 25900+25900+30000 = 81800

$\therefore$ Required percentage = $\frac{4772}{81800} \times 100$ = 5.8%

82. A sports-goods shop has tennis balls of 3 colors: red, green, and white. The number of white balls is 60% more than the number of red balls, and the number of green balls is 12.5% less than the number of white balls. If the total number of balls is 120, then how many green balls are there?
एक खेल-सामान की दुकान में 3 रंगों की टेनिस गेंदें हैं - लाल, हरा और सफेद। सफेद गेंदों की संख्या लाल गेंदों की संख्या से 60% अधिक है और हरी गेंदों की संख्या जबकि गेंदों की संख्या से 12.5% कम है। यदि गेंदों की कुल संख्या 120 है, तो कितनी हरी गेंदें हैं?

Solution :

According to the question,

White Balls : Red Balls = 160 : 100 = 8 : 5

White Balls : Green Balls = 100 : 87.5 = 8 : 7

Ratio of Red, Green, and White balls = 5 : 7 : 8

Green balls = $120 \times \frac{7}{20} = 42$

$\therefore$ There are 42 green balls.

83. A man started off a business with a certain capital amount. In the first year, he earned 60% profit and donated 50% of the total capital (initial amount + profit). He followed the same procedure with the remaining capital after the second and the third year. If at the end of the three years, he is left with Rs. 15,360, what was the initial amount (in Rs.) with which the man started his business?
एक व्यक्ति ने एक निश्चित पूंजी राशि के साथ एक व्यवसाय शुरू किया। पहले वर्ष में, उन्होंने 60% लाभ कमाया और कुल पूंजी का 50% (प्रारंभिक राशि लाभ) दान कर दिया। उन्होंने

दूसरे और तीसरे वर्ष के बाद शेष पूंजी के साथ भी यही प्रक्रिया अपनाई। यदि तीन वर्ष के अंत में उसके पास रु. 15,360, प्रारंभिक राशि (रुपये में) क्या थी जिसके साथ आदमी ने अपना व्यवसाय शुरू किया?

Solution :

Reverse Process

$\frac{8}{5} \times \frac{1}{2} = \frac{4}{5}$

$15360 \times \frac{5}{4} \times \frac{5}{4} \times \frac{5}{4} = 30,\ 000$

84. Out of the total toffees, 1/10 are wasted. 35% of the rest are given to S and 1/8 of the total are given to N. If the toffees with N are 10 more than the wasted toffees, then how many toffees were given to S?

कुल टॉफ़ी में से 1/10 बर्बाद हो जाती हैं। शेष का 35% S को दिया जाता है और कुल का 1/8 भाग N को दिया जाता है। यदि N के पास वाली टॉफियाँ बर्बाद हुई टॉफियों से 10 अधिक हैं, तो S को कितनी टॉफियाँ दी गईं?

Solution :

Total Toffee = 80

Wasted = $\frac{1}{10}$ of 80 = 8

$N = \frac{1}{8}$ of 80 = 10

$N = 10 +$ Wasted, then

$2 = 10$

$1 = 5$

80 = 400 Total Toffee

$S = 400 \times \frac{9}{10} \times \frac{7}{20} = 126$

$S = 126$ Toffees

85. What percentage of the numbers from 101 to 1000 have 9 in the unit's digit?

101 से 1000 तक कितने प्रतिशत संख्याओं का इकाई अंक 9 है?

Solution :

Total Numbers = 1000 − 101 + 1 = 900

Numbers from 101 to 200 where 9 is at unit digit are = 109, 119, 129, 139, 149, 159, 169, 179, 189, 199

So on for further (200.....1000)

Unit digit with 9 = 9 × 10 = 90 numbers

Percentage = $\frac{90}{900} = 10\%$

86. If the word PHOTOGRAPH is spelt with 'F' in place of 'PH', then what would be the percentage of reduction in the number of letters?

यदि PHOTOGRAPH शब्द को 'PH' के स्थान पर 'F' से लिखा जाए तो अक्षरों की संख्या में कितने प्रतिशत की कमी होगी?

Solution :

Total word in PHOTOGRAPH is = 10

If PH is replaced by F, then the total word in

FOTOGRAF is = 8

Required percentage = $\frac{(10-8)}{10} \times 100 =$ 20%

87. Two-thirds of the number of employees of a company are males and the rest are females. If 3/8 of the male employees and 2/5 of the female employees are temporary employees and the total number of permanent employees is 740, then 7/15 of the total number of employees exceeds the number of temporary female employees by:

किसी कंपनी के कर्मचारियों की संख्या में दो-तिहाई पुरुष हैं और शेष महिलाएँ हैं। यदि 3/8 पुरुष कर्मचारी और 2/5 महिला कर्मचारी अस्थायी कर्मचारी हैं और स्थायी कर्मचारियों की कुल संख्या 740 है, तो कुल कर्मचारियों की संख्या का 7/15 अस्थायी महिला कर्मचारियों की संख्या से कितना अधिक है?

Solution :

Let the total employees be $60x$

Number of males = $40x$

Number of females = $20x$

Number of permanent males = $40x \times \frac{5}{8} = 25x$

Number of permanent females = $20x \times \frac{3}{5} =$ $12x$

Number of temporary females = $8x$

According to the question

$25x + 12x = 740$

$37x = 740$

$x = 20$

Total employees = $60x = 1200$

$\frac{7}{15}$ of the total number = $1200 \times \frac{7}{15} = 560$

Number of temporary females = $8x = 8 \times 20 = 160$

$\therefore$ Required difference = 560 – 160 = 400

88. At the beginning of day 1 of a month, Rajesh has 500 eggs. He sells 20% of the eggs by the end of the day and added b% of the eggs at the beginning of the next day and sold 20% of the eggs at the end of the next day. This pattern continued up to the end of the third day of the month when he is left with 1024 eggs. The value of b is equal to:
महीने के पहले दिन की शुरुआत में, राजेश के पास 500 अंडे हैं। वह दिन के अंत तक 20% अंडे बेचता है और अगले दिन की शुरुआत में b% अंडे जोड़ता है और अगले दिन के अंत में 20% अंडे बेचता है। यह क्रम महीने के तीसरे दिन के अंत तक जारी रहा जब उसके पास 1024 अंडे बचे। b का मान इसके बराबर है:

Solution :

$$500 \times \left(\frac{4}{5}\right)^3 \times \left(1 + \frac{b}{100}\right)^2 = 1024$$

$$\left(1 + \frac{b}{100}\right)^2 = 4$$

$$1 + \frac{b}{100} = 2$$

$$b = 100$$

89. A stock portfolio consists of four stocks. Stock A represents 20% of the portfolio and has a return of 6%. Stock B represents 30% of the portfolio and has a return of 8%. Stock C represents 20% of the portfolio and has a return of 4%. Stock D represents the remaining 30% of the portfolio and has a negative-return of 5%. What is the average return of the portfolio?
एक स्टॉक पोर्टफोलियो में चार स्टॉक हैं। स्टॉक A पोर्टफोलियो का 20% दर्शाता है और इसका रिटर्न 6% है। स्टॉक B पोर्टफोलियो का 30% दर्शाता है और इसका रिटर्न 8% है। स्टॉक C पोर्टफोलियो का 20% दर्शाता है और इसका रिटर्न 4% है। स्टॉक D पोर्टफोलियो के शेष 30% को दर्शाता है और इसका नेगेटिव-रिटर्न 5% है। पोर्टफोलियो का औसत रिटर्न क्या है?

Solution :

Average Return = $20 \times 6\% + 30 \times 8\% + 20 \times 4\% + 30 \times (-5\%)$

= 1.2 + 2.4 + 0.8 - 1.5

= 2.9% Return

2 Profit, Loss & Discount

Profit / Loss & Discount
(लाभ, हानि एवं छूट)

1. Cost price (CP) - The price at which goods are or have been bought by a merchant or retailer.
   लागत मूल्य (सीपी) - वह मूल्य जिस पर माल किसी व्यापारी या खुदरा विक्रेता द्वारा खरीदा जाता है या खरीदा गया है।

2. Selling price (SP) - The price at which something is sold.
   विक्रय मूल्य (एसपी) - वह मूल्य जिस पर कुछ बेचा जाता है।

3. Marked price/List price/Label price (MP) - This is the price at which product is intended to be sold.
   अंकित मूल्य/सूची मूल्य/लेबल मूल्य (एमपी) - यह वह मूल्य है जिस पर उत्पाद बेचा जाना है।

4. Profit - When selling price is more than the cost price.
   लाभ - जब विक्रय मूल्य लागत मूल्य से अधिक हो।

Profit = SP - CP

लाभ = विक्रय मूल्य - क्रय मूल्य

Profit % = $\frac{Profit}{CP} \times 100$

लाभ % = $\frac{\text{लाभ}}{\text{क्रय मूल्य}} \times 100$

Loss - When cost price is more than the selling price.

Loss = CP - SP

Loss % = $\frac{\text{Loss}}{\text{CP}} \times 100$

हानि - जब लागत मूल्य विक्रय मूल्य से अधिक हो।

हानि = क्रय मूल्य - विक्रय मूल्य

हानि % = $\frac{\text{हानि}}{\text{क्रय मूल्य}} \times 100$

Discount - It is the amount that seller offers to the customer to increase his sales.

Discount = MP - SP

Discount % = $\frac{\text{Discount}}{\text{MP}} \times 100$

छूट - यह वह राशि है जो विक्रेता अपनी बिक्री बढ़ाने के लिए ग्राहक को प्रदान करता है।

छूट = अंकित मूल्य - विक्रय मूल्य

छूट % = $\frac{\text{छूट}}{\text{अंकित मूल्य}} \times 100$

Relation Between CP, MP, Profit/Loss, and Discount

सीपी, एमपी, लाभ / हानि के बीच संबंध

$$\frac{CP}{MP} = \frac{(100\ D)}{(100 \pm P/L)}$$

$$\frac{\text{सीपी}}{\text{एमपी}} = \frac{(100 - \text{डी})}{(100 \pm \text{पी / एल})}$$

- When two products are sold at the same selling price, one at a loss of p% and the other at a profit of p%, then net result is always loss, which is equal to:

  Net result $= \left(\frac{p^2}{100}\right)\%$

  जब दो उत्पाद समान बिक्री मूल्य पर बेचे जाते हैं, एक p% की हानि और दूसरा p% की लाभ पर, तो शुद्ध परिणाम हमेशा हानि है, जो बराबर होता है:

  शुद्ध परिणाम $= \left(\frac{p^2}{100}\right)\%$

- When two products are sold at the same selling price, one at a loss/profit of x% and the other at a loss/profit of y%, then net result is equal to: Net result $= \left\{\frac{100+(x+y)+2xy}{200+x+y}\right\}\%$

  जब दो उत्पाद समान बिक्री मूल्य पर बेचे जाते हैं, एक x% की हानि/लाभ और दूसरा y% की हानि/लाभ पर, तो शुद्ध परिणाम बराबर होता है:

  शुद्ध परिणाम = $\left\{\frac{100+(x+y)+2xy}{200+x+y}\right\}\%$

Type - 1 (Basic)

1. 40% of the cost price of an article is equal to 30% of its selling price. The approximate percentage of profit or loss on the cost price is:
किसी वस्तु के लागत मूल्य का 40% उसके विक्रय मूल्य के 30% के बराबर है। लागत मूल्य पर लाभ या हानि का अनुमानित प्रतिशत है:

Solution :

According to the question, 40% of CP = 30% of SP

$$\frac{CP}{SP} = \frac{30}{40} = \frac{3}{4} + 1$$

$$\therefore \text{ Profit percentage } = \frac{1}{3} \times 100 = 33.33\%$$

2. A shopkeeper marks his articles at 30% above the cost price and allows the purchaser a discount of 20% for cash buying. What profit percent does he make?
एक दुकानदार अपनी वस्तुओं पर लागत मूल्य से 30% अधिक मूल्य अंकित करता है और खरीदार को नकद खरीद पर 20% की छूट देता है। वह कितने प्रतिशत लाभ कमाता है?

Solution :

$$\text{CP/MP} = \frac{100 - d}{100 + p}$$

$$\frac{10}{13} = \frac{100 - 20}{100 + p}$$

$$\frac{10}{13} = \frac{80}{100 + p}$$

$$100 + p = 104$$

$$p\% = 4\%$$

3. A shopkeeper bought 40 pieces of an article at a rate of 50 per item. He sold 35 pieces with 20% profit. The remaining 5 pieces were found to be damaged, and he sold them with 10% loss. Find his overall profit percentage.
एक दुकानदार ने 50 प्रति वस्तु की दर से एक वस्तु के 40 टुकड़े खरीदे। उसने 20% लाभ पर 35 नगें बेचीं। शेष 5 टुकड़े क्षतिग्रस्त पाए गए और उसने उन्हें 10% हानि के साथ बेच दिया। उसका कुल लाभ प्रतिशत ज्ञात कीजिये।

Solution :

Total cost of 40 pieces $= 40 \times 50 =$ ₹2000

$$\text{Total selling price} = 35 \times 50 \times \frac{120}{100} + 5 \times 50 \times \frac{90}{100}$$

$$= 2100 + 225$$

$= $ ₹2325

$$\therefore \text{ profit\%} = \frac{325}{2000} \times 100 = 16.25\%$$

4. A shopkeeper bought 60 pencils at a rate of 4 for ₹5 and another 60 pencils at a rate of 2 for ₹3. He mixed all the pencils and sold them at a rate of 3 for ₹4. Find his gain or loss percentage.
एक दुकानदार ने ₹5 में 4 की दर से 60 पेंसिलें खरीदीं और ₹3 में 2 की दर से अन्य 60 पेंसिलें खरीदीं। उसने सभी पेंसिलों को मिलाया और उन्हें ₹4 में 3 की दर से बेच दिया। उसका लाभ या हानि प्रतिशत ज्ञात कीजिये।

Solution :

4 Pencil at ₹ 5

60 Pencil at ₹ 75

2 Pencil at ₹ 3

60 Pencil at ₹ 90

Total CP = 75 + 90 = 165

Sold 3 Pencil at ₹ 4

SP of 120 Pencil = ₹ 160

$$165 : 160 = 33 : 32$$

$$\frac{1}{33} = 3\frac{1}{3}\% \text{ Loss}$$

5. A shopkeeper bought a table for ₹4600 and a chair for ₹1800. He sells the table with 10% gain and the chair with 6% gain. Find the overall gain percentage.
एक दुकानदार ने ₹4600 में एक मेज और ₹1800 में एक कुर्सी खरीदी। वह मेज को 10% लाभ पर और कुर्सी को 6% लाभ पर बेचता है। कुल लाभ प्रतिशत ज्ञात कीजिये।

Solution :

10   6

x

4600 : 1800

23 : 9

$10 \times 23 + 6 \times 9 = (23 + 9)x$

$230 + 54 = 32x$

$x = \frac{284}{32} = \frac{71}{8}\%$

6. A person sold an article at a loss of 18%. Had he sold it for ₹960 more, he would have gained 12%. If the article is sold for ₹3840, then how much is the profit percentage?

एक व्यक्ति ने एक वस्तु को 18% की हानि पर बेचा। यदि उसने इसे ₹960 अधिक में बेचा होता, तो उसे 12% का लाभ होता। यदि वस्तु ₹3840 में बेची जाती है, तो लाभ प्रतिशत कितना है?

Solution :

CP = 100%

Old SP = 82%

New SP = 112%

$82\% + 960 = 112\%$

$30\% = 960$

$100\% = 3200$

CP = 3200, SP = 3840

Profit % $= \frac{640 \times 100}{3200}$

Profit % $= 20\%$

7. In a medical transaction, 17 times the cost price is equal to 8 times the sum of the cost price and the selling price. What is the gain or loss percentage?

एक चिकित्सा लेनदेन में, लागत मूल्य का 17 गुना लागत मूल्य और विक्रय मूल्य के योग के 8 गुना के बराबर है। लाभ या हानि प्रतिशत क्या है?

Solution :

$17CP = 8(CP + SP)$

$17CP = 8CP + 8SP$

$9CP = 8SP$

$CP/SP = \frac{8}{9}$ Profit = 1

Profit percentage $= \frac{1}{8} \times 100 = 12.5\%$

8. An article is sold at a certain price. If it is sold at $33\frac{1}{3}\%$ of this price, there is a loss of $33\frac{1}{3}\%$. What is the percentage profit if the article is sold at 80% of its original selling price?

एक वस्तु एक निश्चित कीमत पर बेची जाती है। यदि इसे इस कीमत के $33\frac{1}{3}\%$ पर बेचा जाता है, तो $33\frac{1}{3}\%$ की हानि होती है। यदि वस्तु को उसके मूल विक्रय मूल्य के 80% पर बेचा जाता है तो प्रतिशत लाभ क्या है?

Solution :

$33\frac{1}{3}\%$ of SP $= 66\frac{2}{3}\%$ of CP

$\frac{SP}{3} = \frac{2CP}{3}$

$\frac{CP}{SP} = \frac{1}{2}$

$1 \rightarrow 1.6$

60% Profit

9. Oranges are bought for ₹119.60 and sold at the rate of ₹41.40 per dozen. The profit percentage is:

संतरे ₹119.60 में खरीदे जाते हैं और ₹41.40 प्रति दर्जन की दर से बेचे जाते हैं। लाभ प्रतिशत है:

Solution :

CP of 52 oranges = 119.60

CP of 1 orange $= \frac{119.60}{52} =$ ₹2.3

SP of 12 oranges = ₹41.40

SP of 1 orange $= \frac{41.40}{12} =$ ₹3.45

Profit percentage $= \frac{3.45 - 2.3}{2.3} \times 100$

$= \frac{1150}{23} = 50\%$

10. The marked price of an article is ₹400. After allowing a discount of 20% on the marked price, a shopkeeper makes a loss of ₹48. His loss percentage is:

एक वस्तु का अंकित मूल्य ₹400। अंकित मूल्य पर 20% की छूट देने के बाद, एक दुकानदार को ₹48 का नुकसान होता है। उसका हानि प्रतिशत है:

Solution :

According to the question,

$$SP = 80 \times \frac{400}{100} = ₹320$$

$$CP = 320 + 48 = ₹368$$

$$\text{Loss percentage} = \frac{48 \times 100}{368} = 13\frac{1}{23}\%$$

11. The selling price of an article is ₹2300 and the profit percentage is 25%. If the article is sold again for ₹1656, then what is the loss/gain percentage?
एक वस्तु का विक्रय मूल्य ₹2300 और लाभ प्रतिशत 25% है। यदि वस्तु पुनः ₹1656 में बेची जाती है, तो हानि/लाभ प्रतिशत क्या है?

Solution :

$$CP = 2300 \times \frac{4}{5} = 1840$$

$$SP = 1656$$

$$\text{Loss} = 184$$

$$\text{Loss percentage} = \frac{184}{1840} \times 100 = 10\%$$

12. A shopkeeper bought a computer for ₹42000 and a printer for ₹3000. If he sold both the items together for ₹51750, then what is his profit percentage?
एक दुकानदार ने ₹42000 में एक कंप्यूटर और ₹3000 में एक प्रिंटर खरीदा। यदि उसने दोनों वस्तुओं को एक साथ ₹51750 में बेचा, तो उसका लाभ प्रतिशत क्या है?

Solution :

$$\text{Total CP} = 42000 + 3000 = ₹45000$$

$$\text{SP} = ₹51750$$

$$\text{Profit} = ₹6750$$

$$\text{Profit percentage} = \frac{6750}{45000} \times 100 = 15\%$$

13. The cost price of an article is 25% less than its selling price. What is the profit or loss percentage?
एक वस्तु का लागत मूल्य उसके विक्रय मूल्य से 25% कम है। लाभ या हानि प्रतिशत क्या है?

Solution :

$$\text{CP} : \text{SP} = 3 : 4$$

$$\text{Profit percentage} = \frac{1}{3} \times 100 = 33\frac{1}{3}\%$$

14. Radha purchased a computer table for ₹10000 and a center table for ₹5000. She sold the computer table with 8% profit. With what profit percentage should she sell the center table so as to gain 10% on the whole transaction?
राधा ने ₹10000 में एक कंप्यूटर टेबल और ₹5000 में एक सेंटर टेबल खरीदी। उसने 8% लाभ के साथ कंप्यूटर टेबल बेची। उसे सेंटर टेबल को कितने प्रतिशत लाभ पर बेचना चाहिए ताकि पूरे लेनदेन पर 10% का लाभ हो?

Solution :

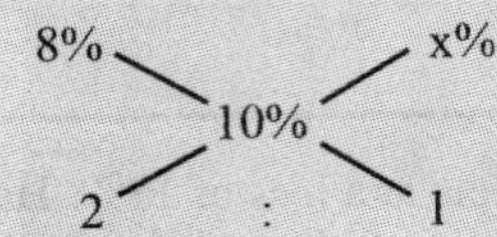

$$8 \times 2 + x = 3 \times 10$$

$$16 + x = 30$$

$$x = 14\%$$

15. On selling an article for ₹246.80, the gain is 20% more than the amount of loss incurred on selling it for ₹216. If the article is sold for ₹220.75, then what is the gain/loss percentage (correct to the nearest integer)?
एक वस्तु को ₹246.80 में बेचने पर लाभ, ₹216 में बेचने पर हुई हानि की राशि से 20% अधिक है। यदि वस्तु ₹220.75 में बेची जाती है, तो लाभ/हानि प्रतिशत (निकटतम पूर्णांक तक सही) क्या है?

Solution :

$$\text{Profit} = \text{Loss} \times \frac{6}{5} \quad (20\% = \tfrac{1}{5})$$

According to the question:

$$(246.80 - CP) \times 5 = (CP - 216) \times 6$$

$$11CP = 5 \times 246.80 + 6 \times 216$$

$$11CP = 2530$$

$$CP = 230$$

$$SP = ₹220.75, \; CP = ₹230$$

$$\text{Loss percentage} = \frac{9.25}{230} \times 100$$

$$= 4.02\%$$

$$= 4\% \text{ (approximately)}$$

16. If the selling price is tripled and the cost price doubled, the profit would become 65%. What is the percent profit (in %)?

यदि विक्रय मूल्य तीन गुना और लागत मूल्य दोगुना कर दिया जाए तो लाभ 65% हो जाएगा। प्रतिशत लाभ (% में) क्या है?

Solution :

$2CP \times \frac{33}{20} = 3SP$

$33CP = 30SP$

$CP/SP = \frac{10}{11}$

10% profit

17. On a certain item, profit is 120%. If the cost price increases by 10%, then what will be the new profit margin (in %) if the selling price remains the same?

एक निश्चित वस्तु पर लाभ 120% है। यदि लागत मूल्य 10% बढ़ जाता है तो विक्रय मूल्य समान रहने पर नया लाभ मार्जिन (% में) क्या होगा?

Solution :

100 → 220 (old CP:SP)

110 → 220 (new CP:SP)

100% profit

18. Some fruits are bought at 15 for ₹140 and an equal number of fruits at 10 for ₹120. If all the fruits are sold at ₹132 per dozen, then what is the profit percent in the entire transaction?

कुछ फल ₹140 में 15 और उतने ही फल ₹120 में 10 पर खरीदे जाते हैं। यदि सभी फल ₹132 प्रति दर्जन पर बेचे जाते हैं, तो पूरे सौदे में लाभ प्रतिशत क्या है?

Solution :

15 Fruits bought at ₹ 140

then, 30 Fruits Cost = ₹ 280

10 Fruits bought at ₹ 120

then, 30 Fruits Cost = ₹ 360

60 Fruits Cost = 280 + 360 = 640

Acc to Ques,

12 Fruits Sold at 132

then, 60 Sold at 660

640 : 600

Required Profit % $= \frac{20}{640} = \frac{25}{8}\%$

P% $= 3\frac{1}{3}\%$

19. The marked price of an article is ₹480. A person buys it at two successive discounts of 15% and 10%. He sells it at a profit of 25%. What is his profit (in ₹)?

एक वस्तु का अंकित मूल्य ₹480 है। एक व्यक्ति इसे 15% और 10% की दो क्रमिक छूट पर खरीदता है। वह इसे 25% के लाभ पर बेचता है। उसका लाभ (₹ में) क्या है?

Solution :

Marked price of article = ₹480

Cost price after successive discounts = $480 \times \frac{85}{100} \times \frac{90}{100} =$ ₹367.2

Selling price of article at 25% profit = $367.2 \times \frac{125}{100} =$ ₹459

Required profit = 459 − 367.2 = ₹91.80

20. A shopkeeper bought 40 dozen eggs at ₹1.40 each. 80 eggs were broken during transportation. He sold the remaining eggs at ₹1.60 each. What is his profit or loss percentage?

एक दुकानदार ने ₹1.40 प्रति अंडे की दर से 40 दर्जन अंडे खरीदे। परिवहन में 80 अंडे टूट गये। उसने शेष अंडे ₹1.60 प्रति अंडे पर बेच दिए। उसका लाभ या हानि प्रतिशत क्या है?

Solution :

According to the question:

Cost price of 40 dozen eggs = $40 \times 12 \times 1.40 =$ ₹672

∵ 80 eggs were broken

Remaining eggs = $40 \times 12 - 80 = 480 - 80 = 400$

Selling price of 400 eggs at ₹1.60 each = $400 \times 1.6 =$ ₹640

Loss = 672 − 640 = ₹32

Loss percentage $= \frac{32}{672} \times 100$

$= \frac{100}{21} = 4\frac{16}{21}\%$

21. An article was sold at a profit of 14%. Had it been sold for ₹121 less, a loss of 8% would have been incurred. If the same article would have been sold for ₹536.25, then the profit/loss percent would have been?
एक वस्तु 14% के लाभ पर बेची गई। यदि इसे ₹121 कम में बेचा जाता तो 8% की हानि होती। यदि वही वस्तु ₹536.25 में बेची जाती तो लाभ/हानि प्रतिशत कितना होता?

Solution :

Let the cost price of the article = 100%

$(14+8)\% = 121$

$22\% = 121$

$2\% = 11$

$CP = 100\% = 550$

Loss = $550 - 536.25 = 13.75$

Loss percentage = $\frac{13.75}{550} \times 100 = 2.5\%$

22. A sells an item at 20% profit to B. B sells the same at 10% profit to C and receives ₹132000. Had C purchased the same item from A, he would have spent 5% less than what he spent with B. What profit would A have made then?
A एक वस्तु को 20% लाभ पर B को बेचता है, B उसे 10% लाभ पर C को बेचता है और ₹132000 प्राप्त करता है। यदि C ने वही वस्तु A से खरीदी होती, तो उसने B के साथ खर्च की तुलना में 5% कम खर्च किया होता। A ने क्या लाभ कमाया होगा?

Solution :

Let CP of the item for A = ₹100

$A(100) + (20\%) = B(120) + (10\%) = C(132)$

132 units → ₹132000

1 unit → ₹1000

If C purchased the same from A:

He spent = $132 \times \frac{95}{100} = 125.4$ units

Then profit of A = $125.4 - 100 = 25.4$ units

$= 25.4 \times 1000 =$ ₹25400

23. 30 dozens of walnuts were bought for ₹14400, and 32 packets of walnuts (with 20 walnuts in each packet) were bought for ₹57600. If the walnuts are mixed together and sold at ₹432 (with 5 walnuts per packet), what is the profit percentage?
30 दर्जन अखरोट ₹14400 में खरीदे गए और 32 पैकेट अखरोट (प्रत्येक पैकेट में 20 अखरोट) ₹57600 में खरीदे गए। यदि अखरोटों को एक साथ मिलाकर ₹432 (एक पैकेट में 5 अखरोट) पर बेचा जाए तो लाभ प्रतिशत कितना हुआ?

Solution :

30 dozen = $30 \times 12 = 360$ walnuts

32 packets (20 walnuts per packet)

$= 32 \times 20 = 640$ walnuts

Total walnuts = $360 + 640 = 1000$

Cost price of 1000 walnuts

= ₹14400 + 57600 = ₹72000

Selling price of 5 walnuts = ₹432

Selling price of 1000 walnuts

$= \frac{432 \times 1000}{5} =$ ₹86400

Profit = ₹86400 − ₹72000 = ₹14400

Profit percentage = $\frac{14400}{72000} \times 100 = 20\%$

24. A shopkeeper allows 10% discount on the marked price of an article and still gains 17%. If he gives 15% discount on the marked price, then his profit percent is:
एक दुकानदार किसी वस्तु के अंकित मूल्य पर 10% की छूट देता है और फिर भी उसे 17% का लाभ होता है। यदि वह अंकित मूल्य पर 15% की छूट देता है, तो उसका लाभ प्रतिशत क्या है?

Solution :

Trick: Let profit percentage = x $\frac{MP}{CP} = \frac{100+P\%}{100-D\%}$

$\frac{117}{90} = \frac{100+x}{85}$

$$\frac{13}{2} = \frac{100+x}{17}$$

$$200 + 2x = 221$$

$$x = 10.5\%$$

25. By selling an article for ₹1170, Elisa suffers as much loss as she would have gained by selling it at a profit of 22%. If she sells it for ₹1450, then her profit/loss percent (correct to one decimal place) is:

एक वस्तु को ₹1170 में बेचने पर एलिसा को उतनी ही हानि होती है जितनी उसे 22% के लाभ पर बेचने पर प्राप्त होती। यदि वह इसे ₹1450 में बेचती है, तो उसका लाभ/हानि प्रतिशत (एक दशमलव स्थान तक सही) क्या है?

Solution :

Cost price of the article $= \frac{1170 \times 100}{78} =$ ₹1500

Loss percentage $= \frac{CP - SP}{CP} \times 100$

$\frac{1500 - 1450}{1500} \times 100$

$= \frac{50}{1500} \times 100 = 3.3\%$

26. Radha marks her goods 25% above the cost price. She sells 35% of goods at the marked price, 40% at 15% discount, and the remaining at 20% discount. What is her overall percentage gain?

राधा अपने सामान पर लागत मूल्य से 25% अधिक अंकित करती है। वह 35% सामान अंकित मूल्य पर, 40% 15% छूट पर और शेष 20% छूट पर बेचती है। उसका कुल प्रतिशत लाभ क्या है?

Solution :

$35 : 40 : 25$

$$35 \times \frac{5}{4} + 40 \times \frac{5}{4} \times \frac{17}{20} + 25 \times \frac{5}{4} \times \frac{4}{5}$$

$$= \frac{174}{4} + \frac{85}{2} + 25 = 111.25$$

Overall percentage gain = 11.2%

27. The profit made by selling an article for ₹8800 is equal to the amount of loss incurred on selling the same article for ₹7200. What will be the profit percent if it was sold for ₹9600?

एक वस्तु को ₹8800 में बेचने पर हुआ लाभ उसी वस्तु को ₹7200 में बेचने पर हुई हानि की राशि के बराबर है। यदि इसे ₹9600 में बेचा गया तो लाभ प्रतिशत क्या होगा?

Solution :

Let the Cost Price (CP) of the article be ₹x.

Profit = Selling Price − Cost Price = $8800 - x$

Loss = Cost Price − Selling Price = $x - 7200$

$8800 - x = x - 7200$

$x = \frac{8800 + 7200}{2} =$ ₹8000 Profit = Selling Price − Cost Price

Profit = 9600 − 8000 = ₹1600

Profit percentage $= \frac{1600}{8000} \times 100 = 20\%$

28. A dealer buys an article at a discount of 20% on its list price and marks it 25% above the list price. If he allows a 20% discount on the new list price, then what is his profit percent?

एक डीलर एक वस्तु को उसके सूची मूल्य पर 20% की छूट पर खरीदता है और उस पर सूची मूल्य से 25% अधिक मूल्य अंकित करता है। यदि वह नई सूची मूल्य पर 20% की छूट देता है, तो उसका लाभ प्रतिशत क्या है?

Solution :

Let MP = ₹100

CP $= 100 \times \frac{4}{5} =$ ₹80

Markup 25%: $100 \times \frac{5}{4} =$ ₹125

SP $= 125 \times \frac{4}{5} =$ ₹100

Ratio: $80 : 100 = 4 : 5$

Profit percentage $= \frac{1}{4} \times 100 = 25\%$

29. An article is listed at ₹5000 and two successive discounts of 12% and 12% are given on it. How much will the seller gain or lose if he gives a simple discount of 24%?

एक वस्तु ₹5000 पर सूचीबद्ध है और उस पर 12% और 12% की दो क्रमिक छूट दी गई हैं। यदि विक्रेता 24% की साधारण छूट देता है तो उसे कितना लाभ या हानि होगी?

Solution :

SP of the article with equivalent discount:

$$5000 \times \frac{88}{100} \times \frac{88}{100}$$

$= 88 \times 44$

$= ₹3872$

SP of the article with only 24% discount:

$5000 \times \frac{76}{100}$

$= ₹3800$

Loss $= 3872 - 3800 = ₹72$

30. A seller buys an item of marked price of ₹5000 at two successive discounts of 20% and 5%. He spends ₹200 on its repair and sells it for ₹5000. What is his profit or loss percentage?

एक विक्रेता ₹5000 के अंकित मूल्य की एक वस्तु 20% और 5% की दो क्रमिक छूट पर खरीदता है। वह इसकी मरम्मत पर ₹200 खर्च करता है और इसे ₹5000 में बेचता है। उसका लाभ/हानि प्रतिशत क्या है?

Solution :

Two successive discounts on marked price:

$5000 \times \frac{100-20}{100} \times \frac{100-5}{100}$

$5000 \times \frac{80}{100} \times \frac{95}{100} = ₹3800$

Total cost price after repair $= 3800 + 200 =$ ₹4000

Profit = SP − CP

Profit $= 5000 - 4000 = ₹1000$

Profit percentage $= \frac{1000}{4000} \times 100 = 25\%$

31. An article is marked at 100% above its cost price. After allowing two successive discounts of 5% and 20% respectively on the marked price, it was sold at $x\%$ profit. What is the value of x?

एक वस्तु पर उसके लागत मूल्य से 100% अधिक मूल्य अंकित है। अंकित मूल्य पर क्रमशः 5% और 20% की दो क्रमिक छूट देने के बाद, इसे $x\%$ लाभ पर बेचा गया। x का मूल्य क्या है?

Solution :

Let the cost price of the article = ₹100

Marked price = ₹200

Selling price $= 200 \times \frac{95}{100} \times \frac{80}{100} = ₹152$

Profit percentage $(x) = 152 - 100 = 52\%$

32. The selling price of a mobile phone is ₹59620 and it was sold at 8.4% profit. The cost price (in ₹) of the mobile phone is:

एक मोबाइल फोन का विक्रय मूल्य ₹59620 है और इसे 8.4% लाभ पर बेचा गया था। मोबाइल फोन का लागत मूल्य (₹ में) है:

Solution :

Cost price $= \frac{59620 \times 100}{108.4} = ₹55000$

33. An article is sold at a profit of $13\frac{1}{4}\%$. Had it been sold for ₹76.70 more, then the profit would have been $16\frac{1}{5}\%$. 50% of the cost price of the article (in ₹) is:

एक वस्तु को $13\frac{1}{4}\%$ के लाभ पर बेचा जाता है। यदि इसे ₹76.70 अधिक पर बेचा जाता, तो लाभ $16\frac{1}{5}\%$ होता। वस्तु के लागत मूल्य का 50% (₹ में) है:

Solution :

Difference in profit percentages:

$\frac{81}{5}\% - \frac{53}{4}\% = \frac{59}{20}\%$

$\frac{59}{20}\% = ₹76.70$

$1\% = ₹26$

$100\% = ₹2600$

50% of CP = ₹1300

34. The marked price of an item is ₹5800. If Ravi earns a profit of 25% after allowing a discount of 20%, then the cost price of the item is:

एक वस्तु का अंकित मूल्य ₹5800 है। यदि रवि 20% की छूट देने के बाद 25% का लाभ कमाता है, तो वस्तु का लागत मूल्य है:

Solution :

Cost price $= 5800 \times \frac{4}{5} \times \frac{4}{5} = ₹3712$

35. An almirah was sold at a profit of 15%. If its cost had been 5% less and it had been sold for ₹1470 less, then the profit would have been 10%. What is the cost price of the almirah?

एक अलमारी 15% के लाभ पर बेची गई। यदि इसकी लागत 5% कम होती और इसे ₹1470 रुपए कम में बेचा जाता, तो लाभ 10% होता। अलमारी का लागत मूल्य क्या है?

Solution :

Let, $(CP_1) = x$

According to the question

Selling price $(SP_1) = \frac{115x}{100} = \frac{23x}{20}$

New cost price $(CP_2) = x \times \frac{95}{100} = \frac{19x}{20}$

New selling price $(SP_2) = \frac{19x}{20} \times \frac{110}{100} = \frac{209x}{200}$

According to the question:

$\frac{23x}{20} - \frac{209x}{200} = ₹1470$

$\frac{230x - 209x}{200} = ₹1470$

$21x = 1470 \times 200$

$x = ₹14000$

36. A shopkeeper marked every item 25% above the cost price and allowed 10% discount. Shruti, being a regular customer, got an additional 5% discount on the bill and paid ₹2394 for the item purchased. What is the cost price of the item (in ₹)?

एक दुकानदार ने प्रत्येक वस्तु पर लागत मूल्य से 25% अधिक अंकित किया और 10% की छूट दी। नियमित ग्राहक होने के कारण श्रुति को बिल पर 5% अतिरिक्त छूट मिली और उसने खरीदी गई वस्तु के लिए ₹2394 का भुगतान किया। वस्तु का लागत मूल्य (₹ में) क्या है?

Solution :

Cost price $= 2394 \times \frac{20}{19} \times \frac{10}{9} \times \frac{4}{5} = ₹2240$

37. Surbhi sold an article for ₹176 after giving 12% discount on its marked price. Had she not given any discount, she would have earned a profit of 25%. What is the cost price (in ₹) of the article?

सुरभि ने एक वस्तु को उसके अंकित मूल्य पर 12% छूट देने के बाद ₹176 में बेचा। यदि उसने कोई छूट न दी होती, तो उसने 25% का लाभ कमाया होता। वस्तु का मूल्य (₹ में) क्या है?

Solution :

Let the marked price of the article = ₹100

According to the question:

$\frac{\text{MP}}{\text{SP}} = \frac{100 \times 2}{88 \times 2} = \frac{200}{176}$

Cost price: $CP \times \frac{125}{100} = 200$

$CP = ₹160$

38. An article is sold at 20% profit instead of 5% loss, and the man gains ₹120 more. What is the cost price?

एक वस्तु को 5% हानि के बजाय 20% लाभ पर बेचा जाता है, और आदमी को ₹120 अधिक का लाभ होता है। लागत मूल्य क्या है?

Solution :

Ratio: 95 : 100 : 120

Gap of 25% = ₹120

5% = ₹24

100% = ₹480

39. Anu sold an article for ₹480 at some profit. Had she sold it for ₹400, then there would have been a loss equal to one-third of the initial profit. What was the cost price of the article?

अनु ने कुछ लाभ पर एक वस्तु ₹480 में बेची। यदि उसने इसे ₹400 में बेचा होता, तो प्रारंभिक लाभ के एक तिहाई के बराबर हानि होती। वस्तु का लागत मूल्य क्या था?

Solution :

As per the question:

Suppose profit on selling the article for ₹480 = ₹x

Loss on selling it for ₹400 = ₹$\frac{x}{3}$

According to the question:

$480 - x = 400 + \frac{x}{3}$

$1440 - 3x = 1200 + x$

$4x = 240$

$x = ₹60$

Cost price $= 480 - x = 480 - 60 = ₹420$

40. A manufacturer sells cooking gas stoves to shopkeepers at 10% profit, and in turn, they sell the stoves to customers at 15% profit. If a customer buys a stove for ₹7590, what is its manufacturing cost?

एक निर्माता दुकानदारों को 10% लाभ पर रसोई गैस स्टोव बेचता है, और बदले में वे 15% लाभ कमाने के लिए ग्राहक को रसोई गैस स्टोव बेचते हैं। यदि किसी ग्राहक को ₹7590 में रसोई गैस स्टोव मिलता है, तो इसकी विनिर्माण लागत क्या है?

Solution :

Let the manufacturing cost of a gas stove = 100 units

Cost price for the shopkeeper = 110 units

Cost price for the customer = $110 \times \frac{115}{100} = \frac{11 \times 23}{2}$

$\because \frac{11 \times 23}{2} = 7590$

1 unit = ₹60

100 units = ₹6000

41. One-third of goods are sold at a 15% profit, 25% of the goods are sold at a 20% profit, and the rest at a 20% loss. If the total profit of ₹138.50 is earned on the whole transaction, then the value (in ₹) of the goods is:

एक तिहाई माल 15% लाभ पर बेचा जाता है, 25% माल 20% लाभ पर बेचा जाता है और शेष 20% हानि पर बेचा जाता है। यदि पूरे लेनदेन पर ₹138.50 का कुल लाभ अर्जित किया जाता है, तो माल का मूल्य (₹ में) है:

Solution :

Let total goods = 12

$4 \times 15\% + 3 \times 20\% - 5 \times 20\%$

$= 60\% + 60\% - 100\% = 20\%$

20% = ₹138.50

100% = ₹692.50

Value of goods = ₹692.50 × 12 = ₹8310

42. Sushma bought 6 tables and 12 chairs for ₹12000. She sold the tables at a profit of 15% and the chairs at a loss of 10%. If her total gain was ₹300, then the total cost of the tables was:

सुषमा ने ₹12000 में 6 मेज और 12 कुर्सियाँ खरीदीं। उसने मेजों को 15% के लाभ पर और कुर्सियों को 10% की हानि पर बेच दिया। यदि उसका कुल लाभ ₹300 था, तो तालिकाओं की कुल लागत थी:

Solution :

Let the cost price of a table ₹x and the cost price of a chair ₹y.

According to the question:

$6x + 12y = ₹12000$ (i)

$\frac{90x}{100} - \frac{120y}{100} = ₹300$

$9x - 12y = ₹3000$ (ii)

Adding equations (i) and (ii):

$15x = ₹15000 \Rightarrow x = ₹1000$

Total cost price of the tables = $6x$ = ₹6000

43. Vijay bought a fridge and a washing machine together for ₹38200. He sold the fridge at a profit of 15% and the washing machine at a loss of 24%, and both were sold at the same price. The cost price of the washing machine is:

विजय ने ₹38200 में एक फ्रिज और एक वॉशिंग मशीन एक साथ खरीदी। उसने फ्रिज को 15% के लाभ पर और वॉशिंग मशीन को 24% की हानि पर बेचा और दोनों को समान कीमत पर बेचा। वॉशिंग मशीन का लागत मूल्य है:

Solution :

Let the cost price of fridge x and washing machine be x and $38200 - x$ respectively.

According to the question:

Then, $x \times = (38200 - x) \times \frac{76}{100}$

$115x = 38200 \times 76 - 76x$

$191x = ₹38200 \times 76$

$x = 200 \times 76$

$x = ₹15200$

Cost price of the washing machine = ₹38200− ₹15200 = ₹23000

44. An article is sold at a profit of 30%. If both the cost price and selling price of the article are decreased by ₹100, the profit now would be 45%. The original cost price of the article is:

एक वस्तु 30% के लाभ पर बेची जाती है। यदि वस्तु का लागत मूल्य और विक्रय मूल्य दोनों ₹100 कम कर दिया जाए, तो अब लाभ 45% होगा। वस्तु का मूल लागत मूल्य है:

Solution :

Let the cost price of the article = ₹x.

Selling price = $\frac{13}{10}x$

Profit % = $\frac{SP - CP}{CP} \times 100$

$45 = \frac{\left(\frac{13x}{10} - 100\right) - (x - 100)}{x - 100}$

$45(x - 100) = \left(\frac{13x}{10} - 100 - x + 100\right) \times$

100

$45(x - 100) = 30x$

$15x = ₹4500$

$x = ₹300$

45. A car dealer purchased an old car for ₹108500 and spent some amount on its maintenance. He sold it for ₹156250, thereby earning a profit of 25%. How much money did he spend on the maintenance of the car?

एक कार डीलर ने ₹108500 में एक पुरानी कार खरीदी और उसके रखरखाव पर कुछ राशि खर्च की। उसने इसे ₹156250 में बेच दिया, जिससे 25% का लाभ हुआ। उसने कार के रखरखाव पर कितना पैसा खर्च किया?

Solution :

Suppose he spent ₹x on the maintenance of the car.

Cost price of the car = (₹108500 + x)

Selling price = (₹108500 + x) × $\frac{125}{100}$

$₹156250 = (₹108500 + x) \times \frac{5}{4}$

$₹125000 - ₹108500 = x$

$x = ₹16500$

46. The profit earned by selling an article for ₹832 is equal to the loss incurred when the article is sold for ₹448. What should be the selling price (in ₹) to make a profit of 10%?

एक वस्तु को ₹832 में बेचने से अर्जित लाभ उस वस्तु को ₹448 में बेचने पर हुई हानि के बराबर है। 10% का लाभ कमाने के लिए विक्रय मूल्य (₹ में) क्या होना चाहिए?

Solution :

$\frac{832 - 448}{2} = ₹192$

Cost price (CP) = ₹448 + 192 = ₹640

Selling price (SP) = 640 × $\frac{11}{10}$ = ₹704

47. By selling an article for ₹131.25, a trader gains as much percent as the number representing the cost price of the article. In order to earn 40% profit, at what price (in ₹) should he sell the article?

एक वस्तु को ₹131.25 में बेचने पर, एक व्यापारी को वस्तु की लागत मूल्य का प्रतिनिधित्व करने वाली संख्या के बराबर प्रतिशत का लाभ होता है। 40% लाभ अर्जित करने के लिए, उसे वस्तु को किस कीमत पर (₹ में) बेचना चाहिए?

Solution :

$x \times \frac{100 + x}{100} = ₹131.25$

$100x + x^2 = 13125$

$(x + 175)(x - 75) = 0$

$x = ₹75$

Selling price for 40% profit = 75 × $\frac{140}{100}$ = ₹105

48. Anil bought two articles A and B at a total cost of ₹10000. He sold article A at 15% profit and article B at 10% loss. In the whole deal, he made no profit or no loss. Find the selling price of article A.

अनिल ने ₹10000 की कुल लागत पर दो वस्तुएँ A और B खरीदीं। उसने वस्तु A को 15% लाभ पर और वस्तु B को 10% हानि पर बेच दिया। पूरे सौदे में उसे न तो कोई लाभ हुआ और न ही कोई हानि। वस्तु A का विक्रय मूल्य ज्ञात कीजिए।

Solution :

15% of A = 10% of B

$A : B = 2 : 3$

5 = ₹10000

2 = ₹4000

$4000 \times \frac{23}{20} = 4600$

49. Amit sold an article for ₹7000 and incurred a loss. Had he sold it for ₹8750, his gain would have been three-fourths of the amount of loss he incurred. At

what price should he sell the article to get 10% profit?
अमित ने एक वस्तु ₹7000 में बेची और हानि उठाई। यदि उसने इसे ₹8750 में बेचा होता, तो उसका लाभ उसकी हानि की राशि का तीन-चौथाई होता। 10% लाभ प्राप्त करने के लिए उसे वस्तु को किस कीमत पर बेचना चाहिए?

Solution :

Loss = $(CP - ₹7000)$

According to the question:

$\text{Profit} = \text{Loss} \times \frac{3}{4}$

$4 \times (₹8750 - CP) = 3 \times (CP - ₹7000)$

$35000 - 4CP = 3CP - 21000$

$7CP = ₹56000$

$\Rightarrow CP = ₹8000$

Selling price for 10% profit = $₹8000 \times \frac{110}{100} =$ ₹8800

50. The cost price of 33 books is the same as the selling price of 'x' books. If the profit is 10%, then the value of 'x' is:
33 पुस्तकों का लागत मूल्य 'x' पुस्तकों के विक्रय मूल्य के समान है। यदि लाभ 10% है, तो 'x' का मान है:

Solution :

Profit on books = 10%

Selling price of x books = $\frac{x \times 110}{100}$

According to the question:

$\frac{x \times 110}{100} = 33$

Solving: $x = \frac{33 \times 100}{110} = 30$

Trick:

$33CP = xSP,$

$\frac{CP}{SP} = \frac{x}{33}$

$110\% = 33$

$100\% = 30$

$\therefore x = 30$

51. Cost of 4 pens, 6 notebooks, and 9 files is ₹305. Cost of 3 pens, 4 notebooks, and 2 files is ₹145. What is the cost (in ₹) of 5 pens, 8 notebooks, and 16 files?
4 पेन, 6 नोटबुक और 9 फाइलों की लागत ₹305 है। 3 पेन, 4 नोटबुक और 2 फाइलों की लागत ₹145 है। 5 पेन, 8 नोटबुक और 16 फाइलों की लागत (₹ में) क्या है?

Solution :

4 pens + 6 notebooks + 9 files = ₹305 (i)

pens + 4 notebooks + 2 files = ₹145 (ii)

On multiplying by 2 in equation (i)

8 pens + 12 notebooks + 18 files = ₹610 (iii)

From the equation (iii) to equation (ii)

5 pens + 8 notebooks + 16 files = ₹610 − ₹145 = ₹465

52. A trader buys jowar at ₹30 per kg. 20% of the grain gets wasted. He plans to sell the remaining grain so that he makes 40% overall profit. At what price (in ₹ per kg) should he sell the grain?
एक व्यापारी ₹30 प्रति किलोग्राम पर ज्वार खरीदता है। 20% अनाज बर्बाद हो जाता है। वह शेष अनाज बेचने की योजना बना रहा है ताकि उसे कुल मिलाकर 40% लाभ हो। उसे अनाज किस दाम पर (₹ प्रति किलोग्राम में) बेचना चाहिए?

Solution :

Let total grain = 100 kg Cost price = 30×100=3000

After 20% grain gets wasted

remaining grain = 80kg

Suppose he should sell the remaining grains at x per kg.

Total Selling price = 80 x

As per question,

$\therefore 80x = 3000 \times \frac{140}{100}$

$x = 30 \times \frac{140}{80} = ₹52.5$

53. An oil refinery buys oil at ₹3600 per barrel. There is 10% wastage. If the refinery wants to earn 5% profit, then at what price should it sell including 8% tax on selling price (in ₹ per barrel)?

एक तेल रिफाइनरी ₹3600 प्रति बैरल पर तेल खरीदती है। 10% बर्बादी होती है। यदि रिफाइनरी 5% लाभ कमाना चाहती है, तो उसे बिक्री मूल्य पर 8% कर सहित किस कीमत पर बेचना चाहिए? (₹ प्रति बैरल में)

Solution :

Price of one barrel oil = ₹3600

10% of the oil is wasted, then the price of one barrel of oil = $\frac{3600 \times 100}{90}$

= ₹4000

According to the question,

New cost price of oil = $4000 \times \frac{105}{100} \times \frac{108}{100}$

= ₹4536 Per barrel

54. Sulekha bought 36 kg of sugar for ₹1040. She sold it at a profit equal to the selling price of 10 kg of it. What is the selling price (in ₹) for 5 kg of sugar?
सुलेखा ने ₹1040 में 36 किलोग्राम चीनी खरीदी। उसने इसे 10 किलोग्राम के विक्रय मूल्य के बराबर लाभ पर बेच दिया। 5 किलोग्राम चीनी का विक्रय मूल्य (₹ में) क्या है?

Solution :

CP of 36kg Sugar = ₹ 1040

According to the question,

$SP \times 10 = SP \times 36 - CP \times 36$

$CP \times 36 = 26 \times SP$

$\frac{CP}{SP} = \frac{26}{36} = \frac{13}{18}$

CP = 13, SP = 1040

SP = 18 ⇒ 1440

Now, SP of 5 kg of sugar = $\frac{1440 \times 5}{36}$

SP (5kg sugar) = 200

∴ The selling price of 5 kg sugar is ₹200

55. The selling price of one article after allowing a discount of 15% on its cost price is the same as the selling price of another article after allowing a discount of 25% on its cost price. If the sum of the cost prices of both articles is ₹640, find the selling price of each article.
एक वस्तु के लागत मूल्य पर 15% की छूट देने के बाद उसका विक्रय मूल्य, उसके लागत मूल्य पर 25% की छूट देने के बाद किसी अन्य वस्तु के विक्रय मूल्य के समान है। यदि दोनों वस्तुओं के क्रय मूल्यों का योग ₹640 है, तो प्रत्येक वस्तु का विक्रय मूल्य ज्ञात कीजिए।

Solution :

According to the question:

$CP_1 \times \frac{85}{100} = CP_2 \times \frac{75}{100}$ $CP_1 \times \frac{17}{20} = CP_2 \times \frac{15}{20}$

$\frac{CP_1}{CP_2} = \frac{15}{17}$

Total cost price = ₹640:

32 units = ₹640, 1 unit = ₹20.

Cost price of the first article = 15 × ₹20 = ₹300.

Selling price = ₹300 × $\frac{17}{20}$ = ₹255.

56. A shopkeeper purchased 10 chairs at the rate of ₹1450 per chair and spent ₹600 as cartage. For how much should he sell each chair to earn 10% profit?
एक दुकानदार ने ₹1450 प्रति कुर्सी की दर से 10 कुर्सियाँ खरीदीं और ₹600 गाड़ी पर खर्च किए। 10% लाभ कमाने के लिए उसे प्रत्येक कुर्सी को कितने में बेचना चाहिए?

Solution :

Total expenditure on chairs = 1450 × 10 + 600 = ₹15100

Total selling price of chair after 10% gain = $\left[\frac{1500}{100} \times (110 + 10)\right] = 16610$

Selling price of each chair = 16610 = ₹1661

57. Oil equal to 20% of the weight of groundnut is extracted in a mill. The matter left after extraction is sold as cattle feed at the rate of ₹12.5/kg. The groundnuts are bought at ₹20/kg. The processing cost is ₹5/kg. At what price (per kg) should the oil be sold to earn 20% profit on total costs (Total cost = Cost of groundnuts and Processing costs)?
एक मिल में मूंगफली के वजन के 20% के बराबर तेल निकाला जाता है। निष्कर्षण के बाद बचा हुआ पदार्थ ₹12.5/किग्रा की दर से पशु आहार के रूप में बेचा जाता है। मूंगफली ₹20/किग्रा

के हिसाब से खरीदी जाती है। प्रसंस्करण लागत ₹5/किग्रा है। कुल लागत पर 20% लाभ कमाने के लिए तेल को किस कीमत (प्रति किलोग्राम) पर बेचा जाना चाहिए?

Solution :

Suppose total weight of groundnut = 100 kg.

Total oil extracted from groundnut = $100 \times \frac{20}{100} = 20$ kg.

Substances after extraction = $100 - 20 = 80$ kg.

Total cost of groundnut = $100 \times (20 + 5) =$ ₹2500.

Total price after 20% profit = ₹2500 × $\frac{120}{100} =$ ₹3000.

Total value of the substance = 80 × ₹12.5 = ₹1000.

Required price of oil = $\frac{₹3000 - ₹1000}{20} =$ ₹100 kg

58. A vendor sells a coconut at ₹24 and suffers 24% loss. If he wants to make 14% profit, then at what price (in ₹) should he sell it?

एक विक्रेता एक नारियल को ₹24 में बेचता है और उसे 24% हानि होती है। यदि वह 14% लाभ कमाना चाहता है, तो उसे किस कीमत पर बेचना चाहिए?

Solution :

Let the cost price of the coconut = ₹x.

According to the question:

$$x \times \frac{100 - 24}{100} = ₹24$$

$$x = ₹24 \times \frac{100}{76}$$

Selling price after 14% profit = $24 \times ₹\frac{100}{76} \times \frac{114}{100}$

Trick:

76% = ₹24, 19% = ₹6, 114% = ₹36

59. Sudha bought 80 articles at the same price. She sold some of them at 8% profit and the remaining at 12% loss, resulting in an overall profit of 6%. The number of items sold at 8% profit is:

सुधा ने समान कीमत पर 80 वस्तुएँ खरीदीं। उसने उनमें से कुछ को 8% लाभ पर और शेष को 12% हानि पर बेच दिया, जिसके परिणामस्वरूप कुल मिलाकर 6% का लाभ हुआ। 8% लाभ पर बेची गई वस्तुओं की संख्या ज्ञात कीजिए:

Solution :

By allegation rate:

+8% –12%

+6%

18% : 2%

Ratio = 18 : 2 = 9 : 1

Number of articles sold at 8% profit = $80 \times \frac{9}{10} = 72$.

Trick:

Let x articles be sold at 8% profit:

$$x \times 8\% - (80 - x) \times 12\% = 80 \times 6\%$$

$$2x - 240 + 3x = 120$$

$$5x = 360 \quad \Rightarrow x = 72$$

60. The difference between the selling prices of some articles if sold for ₹12 per article instead of ₹49 per article is ₹150. If the cost price of these articles is ₹250, find the selling price of 21 articles if 20% profit is earned.

यदि कुछ वस्तुओं को ₹49 प्रति वस्तु के बजाय ₹12 प्रति वस्तु पर बेचा जाए तो उनके विक्रय मूल्य के बीच का अंतर ₹150 है। यदि इन वस्तुओं का क्रय मूल्य ₹250 है, तो 20% लाभ अर्जित करने पर 21 वस्तुओं का विक्रय मूल्य ज्ञात कीजिए।

Solution :

Let the total number of articles = x.

$12x - 9x =$ ₹150

$3x =$ ₹150

$\Rightarrow x =$ ₹50

Cost price of an article = ₹$\frac{250}{50} =$ ₹5.

Selling price of an article = ₹5 × $\frac{120}{100} =$ ₹6.

Selling price of 21 articles = ₹6 × 21 = ₹126.

61. A man bought 2 articles for ₹3000 each. He sold one article at 5% profit and the other at 10% loss. What is the total profit or loss percentage?
एक आदमी ने ₹3000 में 2 वस्तुएँ खरीदीं। उसने एक वस्तु को 5% लाभ पर और दूसरी को 10% हानि पर बेचा। कुल लाभ या हानि प्रतिशत क्या है?

Solution :

$$\frac{10-5}{2} = \frac{5}{2} = 2.5\% \text{ Profit}$$

62. Three articles are bought at ₹180 each. One of them is sold at a loss of 10%. If the other two articles are sold so as to gain 25% on the whole transaction, what is the gain percentage on the two articles?
तीन वस्तुएँ ₹180 प्रत्येक पर खरीदी जाती हैं। उनमें से एक को 10% की हानि पर बेचा जाता है। यदि अन्य दो वस्तुएँ इस प्रकार बेची जाती हैं कि पूरे लेन-देन पर 25% का लाभ प्राप्त हो, तो दोनों वस्तुओं पर लाभ प्रतिशत क्या है?

Solution :

$$\frac{-10+x+x}{3} = 25$$

$$-10+2x = 75$$

$$2x = 85$$

$$x = 42.5\%$$

63. A trader bought two articles for ₹490. He sold one at a loss of 20% and the other at a profit of 16%. If the selling price of both articles is the same, what is the cost price (in ₹) of the article sold at 20% loss?
एक व्यापारी ने ₹490 में दो वस्तुएँ खरीदीं। उसने एक को 20% की हानि पर और दूसरे को 16% के लाभ पर बेचा। यदि दोनों वस्तुओं का विक्रय मूल्य समान है, तो 20% हानि पर बेची गई वस्तु का लागत मूल्य (₹ में) क्या है?

Solution :

Let the cost price ratio be $A : B$.

Selling prices are equal:

$$A \times \frac{4}{5} = B \times \frac{29}{25}$$

$$A : B = 29 : 20$$

49 units = ₹490

1 unit = ₹10

Cost price of $A = 29 \times$ ₹10 = ₹290

64. Aditya sells two wrist watches from his personal collection for ₹12600 each. On the first watch, he gains 26%, and on the second, he loses 10%. Find the overall gain or loss percentage.
आदित्य अपने निजी संग्रह से दो कलाई घड़ियाँ प्रत्येक ₹12600 में बेचते हैं। पहली घड़ी में उसे 26% का लाभ होता है और दूसरी घड़ी में 10% की हानि होती है। कुल लाभ या हानि प्रतिशत ज्ञात कीजिए।

Solution :

Net profit/loss:

$$\text{Net } P/L\% = \frac{100(26-10) - 2 \times 26 \times 10}{200 + 26 - 10}$$

$$\text{Simplifying: } \frac{1600-520}{216} = \frac{1080}{216} = 5\%$$

65. A man sells two articles for ₹620 each, one at a profit of 24% and the other at a loss of 24%. What is his loss/profit percentage in the whole transaction?
एक आदमी दो वस्तुएँ प्रत्येक ₹620 में बेचता है, एक 24% के लाभ पर और दूसरी 24% की हानि पर। पूरे लेन-देन में उसका हानि/लाभ प्रतिशत क्या है?

Solution :

Direct Result (loss %):

$$\text{Loss } \% = \frac{x^2}{100}$$

$$= \frac{24^2}{100} = 5.76\% \text{ loss}$$

66. Abhi bought two articles for ₹624. He sold one at a loss of 14% and the other at a profit of 14%. If the selling price of both the articles is equal, then what is the difference between their cost price (in ₹)?
अभि ने ₹624 में दो वस्तुएँ खरीदीं। उसने एक को 14% की हानि पर और दूसरे को 14% के लाभ पर बेचा। यदि दोनों वस्तुओं का विक्रय मूल्य बराबर है, तो उनके लागत मूल्य के बीच का अंतर (₹ में) क्या है?

Solution :

Using the relationship between cost price:

$$A \times \frac{43}{100} = B \times \frac{57}{100}$$

$$A : B = 57 : 43$$

Total cost price = ₹624:

100 units = ₹624 ⇒ 1 unit = ₹6.24

Difference in cost price: 14 units = ₹14 × ₹6.24 = ₹87.36

67. Two articles are sold for ₹10384 each. On one, the seller gains 18% and on the other, he loses 12%. What is his overall gain or loss?

दो वस्तुएँ प्रत्येक ₹10384 में बेची गईं। एक पर, विक्रेता को 18% का लाभ होता है और दूसरे पर, उसे 12% की हानि होती है। उसका कुल लाभ या हानि क्या है?

Solution :

Using the given formula for profit/loss:

$$\text{Profit/Loss} = 10384 \times \frac{9}{59} - 10384 \times \frac{3}{22}$$

Simplifying:

Profit = 1584 − 1416 = 168 (Profit)

68. A shopkeeper marks his goods 30% higher than the cost price and allows a discount of 10% on the marked price. In order to earn 6.5% more profit, what discount percent should he allow on the marked price?

एक दुकानदार अपने सामान पर लागत मूल्य से 30% अधिक मूल्य अंकित करता है और अंकित मूल्य पर 10% की छूट देता है। 6.5% अधिक लाभ अर्जित करने के लिए उसे अंकित मूल्य पर कितने प्रतिशत की छूट देनी चाहिए?

Solution :

Let Marked Price (MP) = ₹130:

$$SP = 130 \times \frac{9}{10} = ₹117$$

New Selling Price (SP):

$$SP = ₹117 + ₹6.5 = ₹123.5$$

Discount:

$$\text{Discount \%} = \frac{6.5}{130} \times 100 = 5\%$$

69. A pair of jeans marked at ₹960 is offered in a departmental store for ₹816. What is the percentage of discount given by the shopkeeper?

₹960 अंकित एक जोड़ी जींस एक डिपार्टमेंटल स्टोर में ₹816 में उपलब्ध है। दुकानदार द्वारा दी गई छूट का प्रतिशत क्या है?

Solution :

Using the discount formula:

$$SP = MP \times \frac{100 - D\%}{100}$$

Substituting values:

$$816 = 960 \times \frac{100 - D\%}{100}$$

Solving: 85 = 100 − D%

D% = 100 − 85 = 15%

70. An article was sold for ₹512 after giving two equal successive discounts on the marked price of ₹800. The percentage discount offered is:

एक वस्तु को ₹800 के अंकित मूल्य पर लगातार दो समान छूट देने के बाद ₹512 में बेचा गया। प्रस्तावित प्रतिशत छूट है:

Solution :

Using the successive discount formula:

$$800 \times x^2 = 512$$

$$x^2 = \frac{512}{800} = \frac{64}{100} = \left(\frac{4}{5}\right)^2$$

$$x = \frac{4}{5}$$

Discount %:

$$D\% = 1 - \frac{4}{5} = \frac{1}{5} = 20\%$$

71. A ₹1000 box of cookies is offered at 10% discount, and a ₹400 bar of chocolate is offered at 8% discount. If we buy 2 boxes of cookies and 3 bars of chocolate, what is the effective discount we get (in %)?

कुकीज़ का ₹1000 बॉक्स 10% छूट पर और ₹400 चॉकलेट बार 8% छूट पर पेश किया जाता है। यदि हम कुकीज़ के 2 डिब्बे और चॉकलेट के 3 बार खरीदते हैं। हमें मिलने वाली प्रभावी छूट क्या है (% में)?

Solution :

Total cost of 2 cookies boxes and 3 chocolate bars:

$$2 \times 1000 + 3 \times 400 = ₹3200$$

$$\text{Total discount} = \frac{2 \times 1000 \times 10}{100} + \frac{3 \times 400 \times 8}{100}$$

$$= ₹200 + ₹96 = ₹296$$

$$\text{\% discount: } \frac{296}{3200} \times 100 = 9.25\%$$

72. A ₹750 tin of cheese is offered at 8% discount, and a ₹1250 tin of butter at 20% discount. If we buy 5 tins of cheese and 3 tins of butter, what is the effective discount we get (in %)?

₹750 के पनीर के टिन पर 8% की छूट और ₹1250 के मक्खन के टिन पर 20% की छूट दी जाती है। यदि हम 5 टिन पनीर और 3 टिन मक्खन खरीदते हैं, तो हमें मिलने वाली प्रभावी छूट (% में) क्या है?

Solution :

$750 \times 5 : 1250 \times 3$

1 : 1

8%　　20%

x

1　:　1

$$x = \frac{8+20}{2} = 14\%$$

73. If house tax is paid before the due date, one gets a reduction of 12% on the amount of the bill. By paying the tax before the due date, a person got a reduction of ₹2100. What is the total amount (in ₹) of house tax?

यदि देय तिथि से पहले गृह कर का भुगतान किया जाता है, तो बिल की राशि पर 12% की कटौती की जाती है। तय तारीख से पहले टैक्स चुकाने पर एक व्यक्ति को ₹2100 की छूट मिली। गृह कर की राशि (₹ में) क्या थी?

Solution :

Let the total amount of house tax be 100%:

12% = ₹2100

$$100\% = ₹\frac{2100 \times 100}{12} = ₹17500$$

74. A person marks his 40 items at 20% above the cost price. He sells 10 items at 10% discount and 20 items at 15% discount on the marked price. What is the maximum discount (rounded off) he can offer on the remaining items if he still gets some profit?

एक व्यक्ति अपनी 40 वस्तुओं का मूल्य लागत मूल्य से 20% अधिक पर अंकित करता है। वह अंकित मूल्य पर 10 वस्तुएँ 10% छूट पर और 20 वस्तुएँ 15% छूट पर बेचता है। यदि उसे अभी भी कुछ लाभ मिलता है तो वह शेष वस्तुओं पर अधिकतम कितनी छूट (पूर्णांकित) दे सकता है?

Solution :

Profit equation:

$$\frac{10 \times 6}{5} \times \frac{9}{10} + \frac{20 \times 6}{5} \times \frac{17}{20} + \frac{10 \times 6}{5} \times x > 40$$

Simplifying:

$$\frac{54}{5} + \frac{102}{5} + 12x > 40$$

$12x > 40 - 31.2 \quad \Rightarrow 12x > 8.8$

$x > \frac{88}{120} = \frac{11}{15} \quad \Rightarrow x = 26.66\%$ Maximum discount: 26.66%

75. A person buys an object at ₹16. If he has to buy a dozen objects, he pays a total amount of ₹160. How much is the discount in percent (rounded to the nearest integer)?

एक व्यक्ति एक वस्तु ₹16 में खरीदता है। यदि उसे एक दर्जन वस्तुएँ खरीदनी हैं तो वह कुल राशि ₹160 का भुगतान करता है। जब वह एक दर्जन वस्तुएँ (निकटतम पूर्णांक में) खरीदता है तो प्रतिशत में कितनी छूट मिलती है?

Solution :

Cost of one object = ₹16

Cost of a dozen objects without discount: $16 \times 12 =$ ₹192

Discounted cost: ₹160

Discount: ₹192 − ₹160 = ₹32

Discount percentage: $\frac{32}{192} \times 100 = 16.66\%$

Nearest discount: 17%

76. Marked price and cost price of an article are in the ratio 5 : 4. If the profit earned by selling the article is 12.5%, then what is the discount percentage?

एक वस्तु का अंकित मूल्य और लागत मूल्य 5 : 4 के अनुपात में है। यदि वस्तु को बेचने पर अर्जित लाभ 12.5% है, तो छूट प्रतिशत क्या है?

Solution :

According to the question:

$$4 \times \frac{9}{8} = 5 \times x$$

Solving for x:

$$x = \frac{9}{10} = 10\% \text{ discount}$$

77. A vegetable seller bought 10 dozens of potatoes for ₹120, another 5 dozens for ₹50, and another 5 dozens for ₹30. He sold the potatoes for ₹9 per dozen. How much discount did he overall offer?
एक सब्जी विक्रेता ने ₹120 में 10 दर्जन आलू खरीदे, ₹50 में 5 दर्जन और ₹30 में 5 दर्जन आलू खरीदे। उसने आलू ₹9 प्रति दर्जन में बेच दिए। उसने कुल मिलाकर कितनी छूट की पेशकश की?

Solution :

Cost price (CP) for vegetable seller:

10 dozen	=	120
5 dozen	=	50
5 dozen	=	30
Total dozen 20	=	₹ 200

Average CP for 1 dozen = ₹10

Selling price (SP) for 1 dozen: ₹9

Discount percentage:

$$\text{Discount }\% = \frac{10-9}{10} \times 100 = 10\%$$

78. What is the net discount (in %) for successive discounts of 30% and 50%?
30% और 50% की क्रमिक छूट के लिए शुद्ध छूट (% में) क्या है?

Solution :

$$30 + 50 - \frac{30 \times 50}{100}$$

$$30 + 50 - 15$$

$$= 80 - 15$$

$$= 65\%$$

79. Find a single discount percentage equivalent to successive discounts of 10%, 20%, and 25%.
10%, 20%, और 25% की क्रमिक छूट के बराबर एकल छूट प्रतिशत ज्ञात कीजिए।

Solution :

Calculating the final price after successive discounts:

$$100 \times \frac{9}{10} \times \frac{4}{5} \times \frac{3}{4} = 54$$

Single discount:

$100 - 54 = 46\%$ Discount

80. If two successive discounts, each of 20% on the marked price of an article, are equal to a single discount of ₹331.20, then what is the marked price (in ₹) of the article?
यदि किसी वस्तु के अंकित मूल्य पर दो क्रमिक छूट, प्रत्येक 20%, ₹331.20 की एकल छूट के बराबर है, तो वस्तु का अंकित मूल्य (₹ में) क्या है?

Solution :

Equivalent single discount for two successive 20% discounts:

$$\text{Discount }\% = -20 - 20 + \frac{20 \times 20}{100} = -36\%$$

According to the question:

$36\% = ₹331.20 \Rightarrow 100\% = \dfrac{331.20 \times 100}{36} = ₹920$

Alternatively, $\dfrac{4}{5} \times \dfrac{4}{5} = \dfrac{16}{25}$:

If 9 units = ₹331.20, then 25 units = ₹920

Marked Price: ₹920

81. After giving two successive discounts, each of $x\%$, on the marked price of an article, the total discount is ₹259.20. If the marked price of the article is ₹720, then the value of x is:
एक वस्तु के अंकित मूल्य पर प्रत्येक $x\%$ की दो क्रमिक छूट देने के बाद कुल छूट ₹259.20 है। यदि वस्तु का अंकित मूल्य ₹720 है, तो x का मान है:

Solution :

Marked price of the article = ₹720

Selling Price after two equal and successive discounts: $(x + x - \dfrac{x^2}{100})\% = \left(2x - \dfrac{x^2}{100}\right)\%$

According to the question:

$$720 \times \left(\frac{2x - \frac{x^2}{100}}{100}\right) = ₹259.20$$

$$2x - \frac{x^2}{100} = \frac{2592}{7200} \times 100 = 32$$

Simplifying:

$200x - x^2 = 3600 \Rightarrow x^2 - 200x + 3600 = 0$

Factoring:

$x(x-180) - 20(x-180) = 0 \Rightarrow (x-20)(x-180) = 0$

Since $x = 180$ is invalid, $x = 20$.

82. After allowing a discount of 15% on the marked price of an article, it was sold for ₹425. Had the discount not been given, the profit would have been 25%. What was the cost price (in ₹) of the article?
एक वस्तु के अंकित मूल्य पर 15% की छूट देने के बाद, इसे ₹425 में बेचा गया। यदि छूट न दी गई होती तो लाभ 25% होता। वस्तु का लागत मूल्य (₹ में) क्या था?

Solution :

Cost Price of the article:

$CP = ₹425 \times \frac{20}{17} \times \frac{4}{5} = ₹400$

83. Two successive discounts, each of $x\%$, on the marked price of an article, are equal to a single discount of ₹331.20. If the marked price of the article is ₹920, then the value of x is:
एक वस्तु के अंकित मूल्य पर दो क्रमिक छूट, प्रत्येक $x\%$, ₹331.20 की एक से एक छूट के बराबर है। यदि वस्तु का अंकित मूल्य ₹920 है, तो x का मान है:

Solution :

Marked Price: Selling Price = 920 : 588.8:

$9200 : 5888 \Rightarrow 400 : 256$

$\Rightarrow 25 : 16$

Since both discounts are equal:

Ratio of MP and SP after discount $= \sqrt{25} : \sqrt{16} = 5 : 4$

Discount percentage: $x = \frac{1}{5} \times 100 = 20\%$

84. A shopkeeper earns a profit of 17% on selling a book at 10% discount on the printed price. If the cost price is ₹500, then the printed price (in ₹) is:
एक दुकानदार मुद्रित मूल्य पर 10% छूट पर एक किताब बेचने पर 17% का लाभ कमाता है। यदि लागत मूल्य ₹500 है, तो मुद्रित मूल्य (₹ में) है:

Solution :

Given:

Cost price (CP) = ₹500, Profit = 17%,

Discount = 10%

Formula: $\frac{MP}{CP} = \frac{100 + P\%}{100 - D\%}$

Substituting values:

$\frac{MP}{500} = \frac{100+17}{100-10} \Rightarrow \frac{MP}{500} = \frac{117}{90}$

$MP = ₹500 \times \frac{117}{90} = ₹650$

85. The marked price of an object is 42% more than the cost price. If on selling the object a profit of 20.7% is obtained, then the percent discount obtained on the marked price is:
किसी वस्तु का अंकित मूल्य उसके लागत मूल्य से 42% अधिक है। यदि किसी वस्तु को बेचने पर 20.7% का लाभ प्राप्त होता है, तो वस्तु के अंकित मूल्य पर प्राप्त प्रतिशत छूट है:

Solution :

Let Cost Price (CP) = $100x$:

$MP = 142x, \quad SP = 120.7x$

Discount percentage:

$\text{Discount\%} = \frac{142x - 120.7x}{142x} \times 100 = \frac{21.3x}{142x} \times 100 = 15\%$

86. On a television of brand A, the discount is 25%, and on a television of brand B, the discount is 40%. The price of B after discount is ₹2, 250 greater than the price of A after discount. What is the marked price of A (in ₹) if the marked price of B is ₹35000?
ब्रांड A के टेलीविजन पर छूट 25% है और ब्रांड B के टेलीविजन पर छूट 40% है। छूट के बाद B की कीमत छूट के बाद A की कीमत से ₹2250 अधिक है। यदि B का अंकित मूल्य ₹35000 है तो A का अंकित मूल्य (₹ में) क्या है?

Solution :

Let the Marked Price of product A = x.

According to the question:

$$35000 \times \frac{60}{100} - x \times \frac{75}{100} = 2250$$

$$21000 - \frac{3x}{4} = 2250$$

$$3x = 84000 - 9000 \quad \Rightarrow \quad x = ₹25000$$

87. The marked price of an article is ₹800, and it is sold at a discount of 19%. If there is a gain of 8%, then by what percent above the cost price was the article marked?

एक वस्तु का अंकित मूल्य ₹800 है और इसे 19% की छूट पर बेचा जाता है। यदि 8% का लाभ हुआ, तो वस्तु पर लागत मूल्य से कितने प्रतिशत अधिक अंकित किया गया?

Solution :

$$\frac{\text{CP}}{\text{MP}} = \frac{100 - 19}{100 + 8} = \frac{81}{118}$$

Markup Percentage:

$$\frac{27 \times 100}{81} = 33\frac{1}{3}\%$$

88. A dealer allows 25% discount on the marked price of an article and gains 20%. If the cost price of the article increases by 20%, how much discount percentage should he allow on the marked price to earn the same percentage of profit as before?

एक डीलर एक वस्तु के अंकित मूल्य पर 25% की छूट देता है और 20% का लाभ प्राप्त करता है। यदि वस्तु का लागत मूल्य 20% बढ़ जाता है, तो उसे अंकित मूल्य पर कितने प्रतिशत की छूट देनी चाहिए ताकि पहले के समान लाभ अर्जित किया जा सके?

Solution :

Let the Marked Price of the article = ₹100.

According to the question:

MP	SP	CP
100	75	$75 \times \frac{100}{120}$

New Cost Price after 20% increase:

$$\frac{100}{120} \times 75 \times \frac{120}{100} = 75$$

New Selling Price:

$$75 \times \frac{120}{100} = 90$$

Discount:

$$100 - 90 = 10\%$$

89. An article was marked at Ex and sold at a discount of $(x - 40)\%$. If the customer paid $(x - 32)$, then find the marked price of the article.

एक वस्तु को Ex पर अंकित किया गया और $(x-40)\%$ की छूट पर बेचा गया। यदि ग्राहक ने $(x - 32)$ भुगतान किया है, तो वस्तु का अंकित मूल्य ज्ञात कीजिए।

Solution :

Given: MP = ₹x

$$MP \times \left(\frac{100 - \text{Discount\%}}{100}\right) = SP$$

Substituting values:

$$x \times \left(\frac{140 - x}{100}\right) = x - 32$$

Simplifying:

$$140x - x^2 = 100x - 3200$$

$$x^2 - 40x - 3200 = 0$$

$$\Rightarrow \quad x^2 - 80x + 40x - 3200 = 0$$

Factoring:

$$x(x - 80) + 40(x - 80) = 0$$

$$\Rightarrow \quad (x - 80)(x + 40) = 0$$

$$x = ₹80$$

90. A man purchases 100 copies of a book from the publisher and gets a discount of 25%. He buys 50 copies from a retailer at a discount of 10%. He got an overall discount of:

एक व्यक्ति प्रकाशक से एक पुस्तक की 100 प्रतियां खरीदता है और उसे 25% की छूट मिलती है। वह एक खुदरा विक्रेता से 10% की छूट पर 50 प्रतियां खरीदता है। उसे कुल मिलाकर छूट मिली:

Solution :

Suppose the Cost Price of 1 book = ₹1:

Discount on 100 books:

$\frac{100 \times 25}{100} = ₹25$

Discount on 50 books:

$\frac{50 \times 10}{100} = ₹5$

Total Discount:

₹30

Discount Percentage:

$\frac{30}{150} \times 100 = 20\%$

91. A shopkeeper usually gives a discount of 10% on the marked price of each item. During a sale season, he decides to give two more discounts - the first discount is at the rate of 50% of the original discount, and the second discount is at the rate of 40% of the first. What would be the equivalent rate of a single discount (correct to two places of decimal) of all these discounts?

एक दुकानदार आमतौर पर प्रत्येक वस्तु के अंकित मूल्य पर 10% की छूट देता है। एक बिक्री सीज़न के दौरान, वह दो और छूट देने का निर्णय लेता है - पहली छूट मूल छूट के 50% की दर से है और दूसरी छूट पहली छूट के 40% की दर से है। इन सभी छूटों में से एक छूट (दशमलव के दो स्थानों तक सही) की समतुल्य दर क्या होगी?

Solution :

Original discount = 10%

First discount $= 10 \times \frac{50}{100} = 5\%$

Second discount $= 5 \times \frac{40}{100} = 2\%$

Now, the equivalent discount percentage of three successive discounts of 10%, 5%, and 2% is:

$-10-5-2+\frac{10 \times 5}{100}+\frac{5 \times 2}{100}+\frac{10 \times 2}{100}-\frac{5 \times 10 \times 2}{10000}$

$-17 + 0.5 + 0.1 + 0.2 - 0.01 = -16.21\%$

Equivalent single discount percentage = 16.21%.

92. A shopkeeper bought 80 kg of rice at a discount of 10%. Besides, 1 kg rice was offered free to him on the purchase of every 20 kg rice. If he sells the rice at the marked price, what is his profit percentage?

एक दुकानदार ने 80 किलो चावल 10% की छूट पर खरीदा। इसके अलावा, 20 किलो चावल की खरीद पर उसे 1 किलो चावल मुफ्त दिया गया। यदि वह चावल अंकित मूल्य पर बेचता है, तो उसका लाभ प्रतिशत क्या होगा?

Solution :

Let CP of 1 kg rice = ₹1.

Cost Price of 80 kg rice:

$80 \times \frac{90}{100} = ₹72$ But he gets 4 kg rice free.

Hence, he gets 84 kg rice at ₹72.

Selling Price = ₹84.

Profit percentage:

$\frac{12}{72} \times 100 = 16\frac{2}{3}\%$.

93. A shopkeeper allows 28% discount on the marked price of an article and still makes a profit of 20%. If he gains ₹30.80 on the sale of one article, then what will be the cost price of the article?

एक दुकानदार किसी वस्तु के अंकित मूल्य पर 28% की छूट देता है और फिर भी 20% का लाभ कमाता है। यदि उसे एक वस्तु की बिक्री पर ₹30.80 का लाभ होता है, तो वस्तु का लागत मूल्य क्या होगा?

Solution :

Let the Marked Price of the article = ₹$100x$.

Selling Price $= 100x \times \frac{100-28}{100} = ₹72x$.

Cost Price $= 72x \times \frac{100}{120} = ₹60x$.

Profit:

$72x - 60x = 12x$ According to the question:

$12x = 30.80 \Rightarrow 60x = ₹154$.

94. By using faulty weight, a shopkeeper cheats to the extent of 6% while buying and selling rice. Find his gain percentage (rounded to two decimal places).
दोषपूर्ण वजन का उपयोग करके, एक दुकानदार चावल खरीदते और बेचते समय 6% की सीमा तक धोखाधड़ी करता है। उसका लाभ प्रतिशत ज्ञात करें (दशमलव के दो स्थानों तक पूर्णांकित)।

Solution :

$\frac{53}{50} \times \frac{50}{47} = \frac{53}{47}$ Simplify:
$\frac{600}{47} = 12.77\%$ Profit.

95. A dishonest dealer is selling sugar at ₹40 per kg and pretending that he is selling it at a loss of ₹5 per kg but he actually gives $800g$ instead of 1 kg using a faulty weight. What is his actual profit or loss percentage?
एक बेईमान व्यापारी चीनी ₹40 प्रति किलोग्राम पर बेच रहा है और दिखावा कर रहा है कि वह इसे ₹5 प्रति किलोग्राम के घाटे पर बेच रहा है। लेकिन वह वास्तव में 1 किलोग्राम के बजाय 800 ग्राम देता है। उसका वास्तविक लाभ या हानि प्रतिशत क्या है?

Solution :

$\frac{\text{Sp}}{\text{Cp}} = \frac{40}{45} \times \frac{1000}{800}$
$\frac{\text{Sp}}{\text{Cp}} = \frac{8}{9} \times \frac{5}{4} = \frac{10}{9}$
$\frac{1}{9} \times 100 = 11\frac{1}{9}\%$ Profit

96. Subir claimed to sell his items at only 5% above the cost of production, but used a weight that had 1 kg written on it, though it actually weighed 960 grams. What was the actual profit percentage earned by Subir?
सुबीर ने अपनी वस्तुओं को उत्पादन लागत से केवल 5% अधिक पर बेचने का दावा किया, लेकिन एक वजन का इस्तेमाल किया जिस पर 1 किलो लिखा था, हालांकि वास्तव में इसका वजन 960 ग्राम था। सुबीर द्वारा अर्जित वास्तविक लाभ प्रतिशत क्या था?

Solution :

$\frac{21}{20} \times \frac{1000}{960} = \frac{35}{32}\,\frac{300}{32} = 9.375\%$ Profit.

97. The cost price and selling price of rice are the same. Due to a faulty weighing machine, the seller earns a 15% profit. If ₹x is the cost price of 1000 gm rice and the machine is changed to show 1000 gm instead of 950 gm, what should be the selling price now to get the same percentage of profit?
चावल का लागत मूल्य और विक्रय मूल्य समान है। दोषपूर्ण वजन मशीन के कारण, विक्रेता 15% लाभ कमाता है। यदि ₹x, 1000 ग्राम चावल का लागत मूल्य है और मशीन बदल दी गई है जो 950 ग्राम के बजाय 1000 ग्राम दिखाती है, लाभ का समान प्रतिशत प्राप्त करने के लिए अब विक्रय मूल्य क्या होना चाहिए?

Solution :

$\frac{\text{SP}}{x} \times \frac{1000}{950} = \frac{23}{20}$ $\text{SP} = 1.0925x.$

98. A shopkeeper advertises for selling cloth at 7% loss. However, by using a false scale of length 1 metre, he actually gains 24%. What will be the actual length he uses instead of 1 metre?
एक दुकानदार 7% हानि पर कपड़ा बेचने का विज्ञापन करता है। हालाँकि, 1 मीटर लंबाई के झूठे पैमाने का उपयोग करने से उसे वास्तव में 24% का लाभ होता है। 1 मीटर के स्थान पर उसके द्वारा उपयोग की जाने वाली वास्तविक लंबाई क्या होगी?

Solution :

$\frac{93}{100} \times \frac{100}{x} = \frac{31}{25}$ $x = 75$ cm.

99. A shopkeeper normally makes a profit of 20% in a certain transaction; he weighs 900 g instead of 1 kg due to an issue with the weighing machine. If he charges 10% less than what he normally charges, what is his actual profit or loss percentage?
एक दुकानदार आम तौर पर एक निश्चित लेनदेन में 20% का लाभ कमाता है; वजन मापने वाली मशीन में समस्या के कारण उसका वजन 1 किलो के बजाय 900 ग्राम हो गया। यदि वह सामान्य शुल्क से 10% कम शुल्क लेता है, तो उसका वास्तविक लाभ या हानि प्रतिशत क्या है?

Solution :

$\frac{6}{5} \times \frac{1000}{900} \times \frac{9}{10}\,\frac{6}{5} = \frac{1}{5} = 20\%$ Profit.

100. A faulty weighing machine reads 1 kg when 900 g is actually weighed on it. The shopkeeper marked the price of his goods by 10%, but unfortunately, he was caught by the metrology department and then ordered to repair his weighing machine and sell the goods at a 10% discount on cost price for a month. If each customer is now paying ₹20 for 1 kg, then before the raid, what amount (rounded off to two decimal places) did they pay for the same quantity?
एक दोषपूर्ण वजन मशीन 1 किलोग्राम पढ़ती है जबकि वास्तव में उस पर 900 ग्राम वजन होता है। दुकानदार ने अपने सामान की कीमत 10% अंकित की, लेकिन दुर्भाग्य से उसे मेट्रोलॉजी

विभाग ने पकड़ लिया और फिर उन्होंने उसे अपनी वजन मापने वाली मशीन की मरम्मत करने और लागत मूल्य पर 10% छूट पर सामान बेचने का आदेश दिया। यदि प्रत्येक ग्राहक अब 1 किलो के लिए ₹20 का भुगतान कर रहा है, तो छापे से पहले समान मात्रा के लिए उन्होंने कितनी राशि का भुगतान किया? (दशमलव के दो स्थानों तक पूर्णांकित)

Solution :

$\frac{10}{9} \times \frac{11}{10} = \frac{11}{9}$ Discount of 10% on CP: $9 \times 0.9 = 8.1$ $8.1 \to ₹20$ $11 \to ₹27.16$

101. A vegetable vendor bought 100 kg of potatoes at the rate of ₹19 per kg and spent ₹100 as cartage. He sold 60 kg of potatoes with a 50% profit and half of the remaining stock with a 40% profit. He sold half of the still remaining potatoes with a 25% profit. What profit percentage should he aim for when selling the ultimate remaining potatoes to achieve an overall profit of 42%?

एक सब्जी विक्रेता ने ₹19 प्रति किलो की दर से 100 किलो आलू खरीदे और गाड़ी पर ₹100 खर्च किए। उसने 60 किलोग्राम आलू 50% लाभ पर और शेष स्टॉक का आधा 40% लाभ पर बेचा। उसने बचे हुए आधे आलू को 25% लाभ पर बेच दिया। 42% का समग्र लाभ प्राप्त करने के लिए शेष बचे आलू को बेचते समय उसे कितने प्रतिशत लाभ का लक्ष्य रखना चाहिए?

Solution :

$60 \times 50\% + 20 \times 40\% + 10 \times 25\% + 10 \times x\% =$

$100 \times 42\%$ $3000 + 800 + 250 + 10x = 4200$

$4050 + 10x = 4200$ $10x = 150 \implies x =$

15% Profit.

102. A shopkeeper sells an item for ₹953.7 after giving two successive discounts of 49% and 75% on its marked price. Had he not given any discount, he would have earned a profit of 10%. What is the cost price (in ₹) of the item?

एक दुकानदार एक वस्तु को उसके अंकित मूल्य पर 49% और 75% की दो क्रमिक छूट देने के बाद ₹953.7 में बेचता है। यदि उसने कोई छूट नहीं दी होती तो उसे 10% का लाभ होता। वस्तु का लागत मूल्य (₹ में) क्या है?

Solution :

$953.7 \times \frac{100}{51} \times \frac{4}{1} \times \frac{10}{11} = ₹6800.$

103. By selling an article at 6/14 of its actual selling price, Sharad incurs a loss of 12%. If he sells it at 90% of its actual selling price, then the profit percentage is:

एक वस्तु को उसके वास्तविक विक्रय मूल्य के 6/14 पर बेचने पर, शरद को 12% की हानि होती है। यदि वह इसे इसके वास्तविक विक्रय मूल्य के 90% पर बेचता है, तो लाभ प्रतिशत है।

Solution :

Let SP be the actual selling price.

Given: $SP \times \frac{6}{14} = CP \times \frac{88}{100}$. $\frac{CP}{SP} = \frac{75}{154}$.

Now, at 90%: $\frac{75}{154} \times \frac{9}{10} = \frac{125}{231}$.

Profit Percentage: $\frac{106}{125} = 84.8\%$.

104. A trader gives 5% discount on the marked price of an article and gives 4 articles free for buying 14 articles and gains 33% in the entire transaction. Assuming a customer pays for 14 articles, the marked price of an article is what percent more than its cost price?

एक व्यापारी एक वस्तु के अंकित मूल्य पर 5% की छूट देता है और 14 वस्तुएं खरीदने पर 4 वस्तुएं मुफ्त देता है और पूरे लेनदेन में 33% का लाभ प्राप्त करता है। यह मानते हुए कि एक ग्राहक 14 वस्तुओं के लिए भुगतान करता है, एक वस्तु का अंकित मूल्य उसके लागत मूल्य से कितने प्रतिशत अधिक है?

Solution :

$\frac{19}{20} \times \frac{14}{18} \times x = \frac{133}{100}$. $x = \frac{133 \times 18 \times 20}{100 \times 19 \times 14}$.

$x = \frac{18}{10} = 80\%$Markup.

105. The list price of a washing machine is ₹41000. It is sold to a retailer after two successive discounts of 12% and 30%. The retailer wants to earn a profit of 25% on its cost price after allowing a 50% discount (on its new list price) to the customer. At what price should he list the washing machine?

एक वॉशिंग मशीन की सूची कीमत ₹41000 है। इसे 12% और 30% की दो क्रमिक छूट के बाद एक खुदरा विक्रेता को बेचा जाता है। खुदरा विक्रेता ग्राहक को 50% छूट (नए सूची मूल्य पर) देने के बाद लागत मूल्य पर 25% का लाभ कमाना चाहता है। उसे वॉशिंग मशीन को किस कीमत पर सूचीबद्ध करना चाहिए?

Solution :

$41000 \times \frac{7}{10} \times \frac{22}{25} \times \frac{5}{4} \times \frac{2}{1} = ₹63140.$

106. Amit and Bilal buy goods for ₹1500 and ₹2000 respectively. Amit marks his goods up by $y\%$, while Bilal marks his goods up by $2y\%$ and offers a discount of $y\%$. If both make the same profit, then find the value of y.

अमित और बिलाल क्रमशः ₹1500 और ₹2000 का सामान खरीदते हैं। अमित अपने सामान पर $y\%$ अधिक मूल्य अंकित करता है, जबकि बिलाल अपने सामान पर $2y\%$ अधिक मूल्य अंकित करता है और $y\%$ की छूट प्रदान करता है। यदि दोनों समान लाभ कमाते हैं, तो y का मान ज्ञात करें।

Solution :

$20(1 + 2y)(1 - y) - 15(1 + y) = 5\ 4(1 + 2y)(1-y)-3(1+y) = 1\ 4+4y-8y^2 = 4+3y$

$8y^2 = y \implies 8y = 1 \implies y = \frac{1}{8} = 12.5\%$.

107. If a shopkeeper sells sugar at ₹44.8 per kg, he is able to make a 12% profit. Due to water seepage, $\frac{1}{5}$ of the sugar is damaged. What should now be the selling price per kg of the rest of the sugar to have a 5% profit?

यदि कोई दुकानदार ₹44.8 प्रति किलोग्राम पर चीनी बेचता है। वह 12% लाभ कमाने में सक्षम है। पानी के रिसाव के कारण चीनी का $\frac{1}{5}$ भाग खराब हो जाता है। अब 5% लाभ पाने के लिए शेष चीनी का प्रति किलोग्राम विक्रय मूल्य क्या होना चाहिए?

Solution :

$$\text{Cost Price (CP)} = \frac{44.8 \times 100}{112} = ₹40/\text{kg}.$$

$$\text{For 5\% Profit, SP} = 40 \times 1.05 = ₹42/\text{kg}.$$

$$\text{After damage: SP} = 42 \times \frac{5}{4} = ₹52.5/\text{kg}.$$

108. If Ram purchases a fan, he gets 8% discount. However, if he purchases three fans, he gets 6% on the first, 10% on the second, and 12% on the third. If the price paid by Ram for three fans is ₹6800, then what will be the marked price of each fan?

यदि राम एक पंखा खरीदता है, तो उसे 8% की छूट मिलती है। हालाँकि, यदि वह तीन पंखे खरीदता है, तो उसे पहले पर 6%, दूसरे पर 10% और तीसरे पर 12% की छूट मिलती है। यदि राम द्वारा तीन पंखों के लिए भुगतान की गई कीमत ₹6800 है, तो प्रत्येक पंखे का अंकित मूल्य क्या होगा?

Solution :

$$x \times 0.94 + x \times 0.90 + x \times 0.88 = 6800$$

$$2.72x = 6800 \implies x = ₹2500.$$

109. A shopkeeper has a fault of 100 g in a 3 kg weight. He sold 3 kg of potatoes to Charu using the same faulty weight for ₹46. The cost price of the potatoes for the seller was ₹12 per kg. How much profit (in ₹) did he earn on the potatoes?

एक दुकानदार के पास 3 kg वजन में 100 g का दोष है। उसने चारू को उसी दोषपूर्ण वजन का उपयोग करके ₹46 में 3 kg आलू बेच दिया। विक्रेता के लिए आलू का लागत मूल्य ₹12 प्रति किलोग्राम था। उसने आलू पर कितना लाभ (₹ में) कमाया?

Solution :

$\text{CP} = ₹12/\text{kg. Total CP} = 12 \times 3 \times \frac{2900}{3000} = ₹34.8.$

$\text{Profit} = \text{SP} - \text{CP} = 46 - 34.8 = ₹11.2.$

110. A shopkeeper sold 3 wristwatches for ₹2800 each. If he sold the watches with 40%, 25%, and 12% profit, respectively, then find his overall profit percentage.

एक दुकानदार ने 3 कलाई घड़ियाँ प्रत्येक ₹2800 में बेचीं। यदि उसने घड़ियाँ क्रमशः 40%, 25%, और 12% लाभ के साथ बेची हैं, तो उसका कुल लाभ प्रतिशत ज्ञात कीजिए।

Solution :

$$\text{Total CP} = 100 \times \frac{5}{7} + 100 \times \frac{4}{5} + 100 \times \frac{25}{28}.$$

$$\text{Total CP} = \frac{500}{7} + 80 + \frac{625}{7} = 80 + \frac{1125}{7}.$$

$$\text{Total CP} = \frac{1685}{7}$$

$$\text{Total SP} = 300.$$

$$\text{Profit Percentage: } \frac{83}{337} = 24.62\%.$$

3 Ratio and Proportion

Ratio / अनुपात

Ratio: The ratio of a to b is written as $a : b$ and read as "a is to b". Here, A is called the "Antecedent" and B is called the "Consequent".
अनुपात:- a से b का अनुपात $a : b$ के रूप में लिखा जाता है और "a is to b" के रूप में पढ़ा जाता है। जहां A को "पूर्ववर्ती" कहा जाता है और B को "परिणामी" कहा जाता है।

Types of Ratio/ अनुपात के प्रकार

1. Duplicate Ratio of $x : y$ is $x^2 : y^2$
2. Sub-Duplicate Ratio of $x : y$ is $\sqrt{x} : \sqrt{y}$
3. Triplicate Ratio of $x : y$ is $x^3 : y^3$
4. Sub-Triplicate Ratio of $x : y$ is $\sqrt[3]{x} : \sqrt[3]{y}$
5. Inverse Ratio of $x : y$ is $\frac{1}{x} : \frac{1}{y}$

Important Results महत्वपूर्ण परिणाम

1. If $\frac{a}{b} = \frac{c}{d}$
   Then
   (a) $\frac{a+b}{b} = \frac{c+d}{d}$
   (b) $\frac{a-b}{b} = \frac{c-d}{d}$
   (c) $\frac{a+b}{a-b} = \frac{c+d}{c-d}$

Proportion / समानुपात

Proportion: When two ratios are equal, such as $\frac{a}{b} = \frac{c}{d}$, then $a,\ b,\ c,\ d$ are said to be in proportion.

1. a, b and c are in proportion then $\frac{a}{b} = \frac{b}{c}$
2. a, b, c and d are in proportion then $\frac{a}{b} = \frac{c}{d}$

Mean Proportion

Mean Proportion: Let x be the mean proportion between a and b, then: $\frac{a}{x} = \frac{x}{b}$
or
$x = \sqrt{ab}$
माध्य अनुपात:- मान लीजिए x, a और b के बीच माध्य अनुपात है, तो: $\frac{a}{x} = \frac{x}{b}$
अथवा
$x = \sqrt{ab}$

Important Results/ महत्वपूर्ण परिणाम

(i) Let x be a number which is added to $a,\ b,\ c$ and d to make them in proportion then
मान लीजिए कि x एक संख्या है जिसे $a,\ b,\ c$ और d में जोड़ने पर वे अनुपात में आ जाते हैं

$$x = \frac{bc - ad}{(a+d) - (b+c)}$$

(ii) Let x be a number which is subtracted to $a,\ b,\ c$ and d to make them in proportion then
मान लीजिए कि x एक संख्या है जिसे $a,\ b,\ c$ और d में घटाकर उन्हें अनुपात में बनाया जाता है

$$x = \frac{ad - bc}{(a+d) - (b+c)}$$

Type - 1 Problems based on Basic Ratio and Proportion
मूल अनुपात एवं समानुपात पर आधारित समस्याएँ

1. If $A : B = 11 : 7$ and $B : C = 5 : 19$, then what is $A : B : C$? यदि : $= 11 : 7$ और : $= 5 : 19$, तो : : क्या है?

Solution :

$A : B = (11 : 7) \times 5 = 55 : 35$

$B : C = (5 : 19) \times 7 = 35 : 133$

$A : B : C = 55 : 35 : 133$

2. If $a : b = 3 : 5$, $b : c = 7 : 8$, and $c : d = 2 : 3$, then $2a : 3d$ is equal to?
यदि $a : b = 3 : 5$, $b : c = 7 : 8$ और $c : d = 2 : 3$, तो $2a : 3d$ इसके बराबर है:

Solution :

3 : 5
7 : 8
2 : 3
21 : 35 : 40 : 60
42 : 70 : 80 : 120

Now, $2a : 3d$:

42 : 180

7 : 30

3. If $x : y = 5 : 7$, then $(6x - y) : (5x + 3y)$ is equal to?
यदि $x : y = 5 : 7$, तो $(6x - y) : (5x + 3y)$ बराबर है:

Solution :

$x : y = 5 : 7$

$x = 5, \quad y = 7$

According to the question: $\frac{6x - y}{5x + 3y} = \frac{30 - 7}{25 + 21}$

$\frac{23}{46} = \frac{1}{2} = 1 : 2$

4. If $(5x + 2y) : (10x + 3y) = 5 : 9$, then $(2x^2 + 3y^2) : (4x^2 + 9y^2)$ is equal to?
यदि $(5x + 2y) : (10x + 3y) = 5 : 9$, तो $(2x^2 + 3y^2) : (4x^2 + 9y^2)$ का मान क्या होगा?

Solution :

$\frac{5x + 2y}{10x + 3y} = \frac{5}{9}$

$45x + 18y = 50x + 15y$

$5x = 3y,$

$\frac{x}{y} = \frac{3}{5}$

Hence, $(2x^2 + 3y^2) : (4x^2 + 9y^2) = (18 + 75) : (36 + 225) = 93 : 261 = 31 : 87$

5. If $(a + b) : (b + c) : (c + a) = 7 : 6 : 5$ and $a + b + c = 27$, then what will be the value of $\frac{1}{a} : \frac{1}{b} : \frac{1}{c}$?
यदि $(a + b) : (b + c) : (c + a) = 7 : 6 : 5$ और $a + b + c = 27$, तो $\frac{1}{a} : \frac{1}{b} : \frac{1}{c}$ का मान क्या होगा?

Solution :

Let, $(a + b) = 7k, \quad (b + c) = 6k, \quad (c + a) = 5k$

$\therefore a + b + c = \frac{18k}{2} = 9k$

$27 = 9k, \quad k = 3$

$\therefore a + b = 21, \quad b + c = 18, \quad c + a = 15$

$\therefore \frac{1}{a} : \frac{1}{b} : \frac{1}{c} = \frac{1}{9} : \frac{1}{12} : \frac{1}{6} = 4 : 3 : 6$

6. If $\frac{a}{b} = \frac{3}{4}$, $\frac{b}{c} = \frac{4}{5}$, and $\frac{c}{d} = \frac{5}{6}$, then the sum of the numerator and the denominator (which are co-primes) of $\left(\frac{a}{b}\right)^{10}$ is:
यदि $\frac{a}{b} = \frac{3}{4}$, $\frac{b}{c} = \frac{4}{5}$, और $\frac{c}{d} = \frac{5}{6}$, तो $\left(\frac{a}{b}\right)^{10}$ के अंश और हर (जो सह-अभाज्य हैं) का योग है:

Solution :

$\because \frac{a}{b} = \frac{3}{4}, \quad \frac{b}{c} = \frac{4}{5}, \quad \frac{c}{d} = \frac{5}{6},$

$\therefore a : b : c : d = 3 : 4 : 5 : 6$

$\frac{a}{d} = \frac{3}{6} = \frac{1}{2}$

$\left(\frac{a}{d}\right)^{10} = \left(\frac{1}{2}\right)^{10} = \frac{1}{1024}$

Then, $(a + d) = 1 + 1024 = 1025$

7. If x and y are two numbers having opposite signs, $\frac{x^2}{y^2} = \frac{16}{81}$, then what is the value of $\frac{3x + 4y}{2x - 3y}$?

यदि x और y विपरीत चिन्ह वाली दो संख्याएँ हैं, $\frac{x^2}{y^2} = \frac{16}{81}$ तो $\frac{3x+4y}{2x-3y}$ का मान क्या है?

Solution :

According to the question, let x is positive and y is negative. $\frac{x^2}{y^2} = \frac{16}{81} \Rightarrow \frac{x}{y} = \frac{4}{-9}$, $x = 4$, $y = -9$

$\therefore \frac{3x+4y}{2x-3y} = \frac{3\times4+4\times(-9)}{2\times4-3\times(-9)}$

$= \frac{12-36}{8+27} = \frac{-24}{35}$

8. The product of two positive numbers is 1344 and their ratio is $\frac{7}{12}$. The smaller of these numbers is:

दो धनात्मक संख्याओं का गुणनफल 1344 है और उनका अनुपात $\frac{7}{12}$ है। इनमें से छोटी संख्या है:

Solution :

$7x \times 12x = 1344$

$x^2 = 16 \Rightarrow x = 4$

$7x = 28$

9. Find the fourth proportional to 7, 13, and 21:

7, 13, और 21 का चौथा आनुपातिक ज्ञात कीजिए:

Solution :

$\frac{7}{13} = \frac{21}{x}$

$x = 39$

10. In a proportion, the product of the first and fourth terms is 70 and that of the second and third terms is $3.5y$. The value of y is:

एक अनुपात में, पहले और चौथे पद का गुणनफल 70 है और दूसरे और तीसरे पद का गुणनफल $3.5y$ है। y का मान है:

Solution :

$\frac{a}{b} = \frac{c}{d}$,

$ad = bc$

$70 = 3.5y$

$y = \frac{70}{3.5} = 20$

11. If p is the third proportional to 3 and 9, then what is the fourth proportional to 6, p, and 4?

यदि p, 3, 9 का तीसरा समानुपाती है, तो 6, p, 4 का चौथा समानुपाती क्या है?

Solution :

If a and b are numbers, then the third proportional is: $p = \frac{b^2}{a}$

$p = \frac{9^2}{3} = 27$

Let the fourth proportional be x: $\frac{6}{p} = \frac{4}{x}$

$\frac{6}{27} = \frac{4}{x}$

$x = 18$

12. Find the ratio between the fourth proportional of 12, 16, 6 and the third proportional of 4 and 6:

12, 16, 6 के चौथे अनुपात और 4, 6 के तीसरे अनुपात के बीच अनुपात ज्ञात कीजिए।

Solution :

Let the fourth proportional of a, b, and c be x:

$12 : 16 :: 6 : x$

$x = \frac{16\times6}{12} = 8$

The third proportional of a and b is: $\frac{b^2}{a} = \frac{6^2}{4} = 9$

The required ratio is: $8 : 9$

13. What is the difference in the mean proportional between 1.8 and 3.2 and the third proportional to 5 and 3?

1.8 और 3.2 के माध्यानुपाती तथा 5 और 3 के तृतीयानुपाती में क्या अंतर है?

Solution :

According to the question: $\left|\sqrt{1.8\times3.2} - \frac{3\times3}{5}\right| = \left|\sqrt{5.76} - \frac{9}{5}\right|$

$= |2.4 - 1.8| = 0.6$

Type - 2 Increase or Decrease in Ratio/Proportion
अनुपात/समानुपात में वृद्धि या कमी

14. Two numbers are in the ratio of 13 : 8. If each number is increased by 25, then the ratio becomes 44 : 29. The difference between the two numbers is:
दो संख्याएँ 13 : 8 के अनुपात में हैं। यदि प्रत्येक संख्या में 25 की वृद्धि की जाती है, तो अनुपात 44 : 29 हो जाता है। दोनों संख्याओं के बीच का अंतर है:

Solution :

Let first number = $13x$, second number = $8x$.

According to the question:

$$\frac{13x+25}{8x+25}=\frac{44}{29}$$

$$377x+725=352x+1100$$

$$377x-352x=1100-725$$

$$25x=375$$

$$x=15$$

Now, difference of both numbers:

$$13x-8x=5x=5\times15=75$$

15. Two numbers are in the ratio 4 : 5. If 2 is subtracted from the first number and 2 is added to the second number, then their ratio becomes 2 : 3. The difference between the two numbers is:
दो संख्याएँ 4 : 5 के अनुपात में हैं। यदि पहली संख्या में से 2 घटा दिया जाए और दूसरी संख्या में 2 जोड़ दिया जाए, तो उनका अनुपात 2 : 3 हो जाता है। दोनों संख्याओं के बीच का अंतर है:

Solution :

Ratio of numbers is 4 : 5.

Let, the numbers are $4x$ and $5x$ respectively.

If the first number is decreased by 2 and the second number is increased by 2, then the new ratio = 2 : 3.

As per the question:

$$\frac{4x-2}{5x+2}=\frac{2}{3}$$

$$12x-6=10x+4$$

$$12x-10x=4+6$$

$$2x=10\Rightarrow x=5$$

Difference between both numbers:

$$5x-4x=x=5$$

16. The ratio of two numbers is 3 : 5. If eight is added to the first, and seven to the second, then the ratio becomes 2 : 3. What will the ratio become if six is added to each?
दो संख्याओं का अनुपात 3 : 5 है। यदि पहले में आठ और दूसरे में सात जोड़ दिया जाए, तो अनुपात 2 : 3 हो जाता है। प्रत्येक में छह जोड़ने पर अनुपात क्या होगा?

Solution :

Let numbers are $3x$ and $5x$.

According to the question:

$$\frac{3x+8}{5x+7}=\frac{2}{3}$$

$$9x+24=10x+14$$

$$x=10$$

Required Ratio:

$$\frac{3x+6}{5x+6}=\frac{36}{56}=\frac{9}{14}$$

17. If x is subtracted from each of 52, 47, 20, and 19, the numbers so obtained in this order are in proportion. What is the mean proportional between $(x+4)$ and $(x-8)$?
यदि 52, 47, 20 और 19 में से प्रत्येक में से x घटा दिया जाए, तो इस क्रम में प्राप्त संख्याएँ अनुपात में होंगी। $(x+4)$ और $(x-8)$ के बीच माध्य आनुपातिक क्या है?

Solution :

$$(52-x)(20-19)=(20-x)(52-47)$$

$$1(52-x)=5(20-x)$$

$$52-x=100-5x$$

$$4x=48\Rightarrow x=12$$

Mean proportional between 16 and 4:

$$y=\sqrt{16\times4}=8$$

18. If x is added to each of 12, 28, 21, and 45, the numbers so obtained, in this order, are in proportion. What is the mean proportional between $(x+3)$ and $(4x+1)$?
यदि 12, 28, 21 और 45 में से प्रत्येक में x जोड़ा जाए, तो इस क्रम में प्राप्त संख्याएँ अनुपात में होंगी। $(x+3)$ और $(4x+1)$ के बीच माध्य आनुपातिक क्या है?

Solution :

$(12+x)(45-21) = (21+x)(28-12)$

$(12+x)24 = (21+x)16$

$36+3x = 42+2x$

$x = 6$

Mean proportional between 9 and 25:

$y = \sqrt{9 \times 25} = 15$

19. Before a battle, the ratio of captains to soldiers was 2 : 7. During the war, 25 captains and 100 soldiers were martyred. The new ratio of captains becomes 3 : 10. What is the number of soldiers after the war?
एक युद्ध से पहले कप्तानों और सैनिकों का अनुपात 2 : 7 था। युद्ध के दौरान 25 कप्तान और 100 सैनिक शहीद हुए। यहाँ कप्तानों का नया अनुपात 3 : 10 हो जाता है। युद्ध के बाद सैनिकों की संख्या क्या है?

Solution :

Let the number of captains and soldiers before the war be $2x$ and $7x$ respectively.

According to the question:

$\frac{2x-25}{7x-100} = \frac{3}{10}$

$20x - 250 = 21x - 300$

$x = 50$

Number of soldiers after the war:

$7x - 100 = 7 \times 50 - 100 = 250$

20. What is the number of candidates who had applied if the ratio of selected to unselected was 14 : 25? If 35 fewer had applied and 10 fewer were selected, the ratio of selected to unselected would have been 3 : 5.
यदि चयनित और अचयनित का अनुपात 14 : 25 था, तो आवेदन करने वाले उम्मीदवारों की संख्या क्या है। यदि 35 कम ने आवेदन किया था और 10 कम चयनित हुए थे, तो चयनित और अचयनित का अनुपात 3 : 5 होता?

Solution :

Let the selected and unselected candidates be $14x$ and $25x$ respectively.

Total applicants:

$14x + 25x = 39x$

According to the question:

Total new applicants $= 39x - 35$

If 10 candidates were less selected: $14x - 10$

Now:

$\frac{14x-10}{39x-35} = \frac{3}{8}$

$112x - 80 = 117x - 105$

$5x = 25 \Rightarrow x = 5$

Hence, total applicants:

$39x = 39 \times 5 = 195$

21. The salaries of Vipin and Dinesh are in the ratio 5 : 8. If the salary of each is increased by 4800, then the new ratio becomes 7 : 10. What is Vipin's salary?
विपिन और दिनेश का वेतन अनुपात 5 : 8 है। यदि प्रत्येक का वेतन 4800 बढ़ जाता है, तो नया अनुपात 7 : 10 हो जाता है। विपिन का वेतन क्या है?

Solution :

Let Vipin's salary $= 5x$ and Dinesh's salary $= 8x$. According to the question: $\frac{5x+4800}{8x+4800} = \frac{7}{10}$

$50x + 48000 = 56x + 33600$

$6x = 14400 \Rightarrow x = 2400$

Hence, Vipin's salary:

$5x = 5 \times 2400 =$ ₹12000

22. The monthly salaries of A and B are in the ratio 11 : 21. If both of them get a salary increment of ₹4000 per month, the new ratio becomes 3 : 5. What is the new monthly salary of A?
A और B का मासिक वेतन 11 : 21 के अनुपात में है। यदि उन दोनों को प्रति माह ₹4000 की वेतन वृद्धि मिलती है, तो नया अनुपात 3 : 5 हो जाता है। A का नया मासिक वेतन क्या है?

Solution :

Let salaries of A and B be $11x$ and $21x$ respectively. According to the question:

$$\frac{11x + 4000}{21x + 4000} = \frac{3}{5}$$

$$55x + 20000 = 63x + 12000$$

$$8x = 8000 \Rightarrow x = 1000$$

New monthly salary of A:

$$11x + 4000 = ₹15000$$

23. In a school there are 550 students. The ratio of the boys to that of the girls is 6 : 5. How many more girls should join the school so that the ratio becomes 5 : 6?

एक स्कूल में 550 छात्र हैं। लड़कों का लड़कियों से अनुपात 6 : 5 है। कितनी और लड़कियों को स्कूल में शामिल होना चाहिए ताकि अनुपात 5 : 6 हो जाए?

Solution :

Total Students
550

Boys: $550 \times \frac{6}{11} = 300$

Girls: $550 \times \frac{5}{11} = 250$

Let x new girls join the class.

Then:

$$\frac{300}{250 + x} = \frac{5}{6}$$

$$1250 + 5x = 1800$$

$$5x = 550 \Rightarrow x = 110$$

24. The ratio of boys and girls in a group is 7 : 6. If 4 more boys join the group and 3 girls leave the group, then the ratio of boys to girls becomes 4 : 3. What is the total number of boys and girls initially in the group?

एक समूह में लड़कों और लड़कियों का अनुपात 7 : 6 है। यदि 4 और लड़के समूह में शामिल हो जाते हैं और 3 लड़कियां समूह छोड़ देती हैं, तो लड़कों का लड़कियों से अनुपात 4 : 3 हो जाता है। प्रारंभ में लड़कों और लड़कियों की कुल संख्या क्या है?

Solution :

Let the number of boys and girls in the group be $7x$ and $6x$. According to the question:

$$\frac{7x + 4}{6x - 3} = \frac{4}{3}$$

$$21x + 12 = 24x - 12$$

$$3x = 24 \Rightarrow x = 8$$

Initially, total number of boys and girls:

$$13x = 13 \times 8 = 104$$

25. How many job applicants had applied if the ratio of selected to unselected was 19 : 17? If 1200 less had applied and 800 less were selected, then the ratio of selected to unselected would have been 1 : 1.

यदि चयनित और अचयनित का अनुपात 19 : 17 है तो कितने नौकरी आवेदकों ने आवेदन किया था। यदि 1200 कम ने आवेदन किया होता और 800 कम ने चयन किया होता, तो चयनित और अचयनित का अनुपात 1 : 1 होता।

Solution :

Let the number of selected applicants = $19x$,

and unselected applicants = $17x$.

If 1200 less had applied and 800 less were selected:

$$\frac{36x - 1200}{19x - 800} = \frac{2}{1}$$

$$18x - 600 = 19x - 800$$

$$x = 200$$

Hence, total number of applicants:

$$36x = 36 \times 200 = 7200$$

Type - 3 Problems based on Income and Expenditure
आय एवं व्यय पर आधारित समस्याएँ

26. The ratio of the monthly incomes of A and B is 11 : 13 and the ratio of their expenditures is 9 : 11. If both of them manage to save ₹4, 000 per month, then the difference in their incomes (in ₹) is:

A और B की मासिक आय का अनुपात 11 : 13 है और उनके व्यय का अनुपात 9 : 11 है। यदि वे दोनों प्रति माह ₹4, 000 बचाने में कामयाब होते हैं, तो उनकी आय में अंतर (₹ में) है।

Solution :

Income → 11 : 13

Expenditure → 9 : 11

Savings → 4000 4000

121 ∼ 117 units = 44000 ∼ 36000

4 units = 8000   1 unit = 2000

Difference in their incomes = 2 × 2000 = 4000

27. The salaries of P and Q together amount to ₹9, 60, 000. P and Q save 64% and 44%, respectively, of their individual salaries. If P's expenditure is 1.5 times Q's expenditure, then what is the ratio of P's salary to Q's salary?

P और Q का वेतन मिलाकर ₹9, 60, 000 है। P और Q अपने व्यक्तिगत वेतन का क्रमशः 64% और 44% बचाते हैं। यदि P का व्यय Q के व्यय का 1.5 गुना है, तो P के वेतन का Q के वेतन से अनुपात क्या है?

Solution :

P exp. $= 1.5 \times Q$ exp.

$$\frac{P \text{ exp.}}{Q \text{ exp.}} = \frac{3}{2}$$

For P:

P exp. = 3, savings = 64%

$$\Rightarrow \text{P inc.} = \frac{3 \times 100}{36}$$

For Q:

Q exp. = 2, savings = 44%

$$\Rightarrow \text{Q inc.} = \frac{2 \times 100}{56}$$

Income $P : Q$:

$$\frac{3}{36} : \frac{2}{56}$$

Income $P : Q = 7 : 3$

28. The incomes of A and B are in the ratio 3 : 4, and their expenditures are in the ratio 9 : 5. If the income of A is equal to three times the expenditure of B, then what is the ratio of the savings of A and B?

A और B की आय का अनुपात 3 : 4 है और उनके व्यय का अनुपात 9 : 5 है। यदि A की आय B के व्यय के तीन गुना के बराबर है, तो A की बचत का अनुपात B से क्या है?

Solution :

Income → 3 : 4

Expenditure → 9 : 5

Income of $A = 3x$,

Income of $B = 4x$

Expenditure of $A = 9k$,

Expenditure of $B = 5k$

According to the question:

$3x = 3 \times 5k$

$x = 5k$

$$\frac{\text{Saving of } A}{\text{Saving of } B} = \frac{3x - 9k}{4x - 5k} = \frac{15k - 9k}{20k - 5k} = \frac{6}{15} = \frac{2}{5}$$

Required ratio: 2 : 5

29. A sum of ₹11236 is divided among A, B, and C such that the ratio of the shares of A and B is 3 : 5, and the ratio of the shares of A and C is 4 : 7. The share of B is:

11236 की राशि को A, B, और C के बीच इस प्रकार विभाजित किया जाता है कि A और B के शेयरों का अनुपात 3 : 5 है और A और C के शेयरों का अनुपात 4 : 7 है। B का हिस्सा है:

Solution :

Given:

$A : B = (3 : 5) \times 4$,

$A : C = (4 : 7) \times 3$

$A : B : C = 12 : 20 : 21$

According to the question:

$$B's\ share = \frac{20}{12 + 20 + 21} \times 11236$$

$$= \frac{20}{53} \times 11236$$

= ₹4240

30. The ratio of the incomes of A and B is 3 : 5, whereas the ratio of their expenditures is 4 : 7 respectively. If A and B save ₹16, 000 and ₹26, 000, respectively, then what is the difference (in ₹) between their expenditures?

A और B की आय का अनुपात 3 : 5 है, जबकि उनके व्यय

का अनुपात क्रमश: 4 : 7 है। यदि A और B क्रमश: ₹16, 000 और ₹26, 000 बचाते हैं, तो उनके व्यय के बीच का अंतर (₹) क्या है?

Solution :

Let the ratio of income of A and B be $3a$ and $5a$, and the ratio of their expenditures be $4b$ and $7b$, respectively.

According to the question:

$3a - 4b = 16000 \quad (i)$

$5a - 7b = 26000 \quad (ii)$

On solving equations (i) and (ii), we get:

$a = 8000,$

$b = 2000$

Difference between their expenditures:

$7b - 4b = 3b = 3 \times 2000 =$ ₹6000

31. Between two consecutive years, my income is in the ratio 2 : 5 and expenses are in the ratio 4 : 7. If my income in the second year is ₹75,000 and my expenses in the first year are ₹20,000, then my total savings in the two years together are:

लगातार दो वर्षों के बीच, मेरी आय 2 : 5 के अनुपात में है और व्यय 4 : 7 के अनुपात में है। यदि दूसरे वर्ष में मेरी आय ₹75,000 है और पहले वर्ष में मेरा खर्च ₹20,000 है, तो दोनों वर्षों में मेरी कुल बचत है:

Solution :

Given:

Ratio of income = 2 : 5

Ratio of expenditure = 4 : 7

Expenditure in first year = ₹20,000 Income in second year = ₹75,000

Let the income and expenditure in the first and second year be X and Y, respectively.

According to the question:

$$\frac{x}{75000} = \frac{2}{5}$$

$$5x = 150000$$

$$x = 30000$$

$$\frac{20000}{y} = \frac{4}{7}$$

$$4y = 140000$$

$$y = 35000$$

Now, total saving in two years:

$(30000 - 20000) + (75000 - 35000) =$

$10000 + 40000 = 50,000$

32. The ratio of the sum of the salaries of A and B to the difference of their salaries is 11 : 1, and the ratio of the sum of the salaries of B and C to the difference of their salaries is also 11 : 1. If A's salary is the highest and C's is the lowest, then what is B's salary (in ₹) given the total of all their salaries is ₹1,82,000?

A और B के वेतन के योग का उनके वेतन के अंतर से अनुपात 11 : 1 है और B और C के वेतन के योग का उनके वेतन के अंतर से अनुपात भी 11 : 1 है। यदि A का वेतन सबसे अधिक है और C का सबसे कम है, तो B का वेतन (₹) क्या है, जबकि उनके सभी वेतन का कुल योग ₹1,82,000 है?

Solution :

$$\frac{\text{Sum of salary of } A \text{ and } B}{\text{Difference of salaries of } A \text{ and } B} = \frac{11}{1}$$

$$A + B + A - B = 2A = 12$$

$$A = 6, \quad B = 5$$

$$\frac{\text{Sum of salary of } B \text{ and } C}{\text{Difference of salaries of } B \text{ and } C} = \frac{11}{1}$$

$$B + C + B - C = 2B = 12$$

$$B = 6, \quad C = 5$$

According to the question, the ratio between A, B, C:

A	:	B	:	C
6	:	5	:	
	:	6	:	5
36	:	30	:	25

Salary of B:

$$\frac{30}{91} \times 182000 = 60000$$

33. The ratio of the income of A to that of B is 5 : 7. A and B save ₹4,000 and ₹5,000 respectively. If the expenditure of A is equal to $66\frac{2}{3}\%$ of the expenditure of B, then the total income of A and B is:

A की आय का B की आय से अनुपात 5 : 7 है। A और B क्रमशः ₹4,000 और ₹5,000 बचाते हैं। यदि A का व्यय ₹B के व्यय का 66% है, तो A और B की कुल आय:

Solution :

Let income of $A = 5x$, Income of $B = 7x$.

According to the question,

Expenditure of A = Expenditure of $B \times 66\frac{2}{3}\%$.

$5x - 4000 = (7x - 5000) \times \frac{2}{3}$

$15x - 12000 = 14x - 10000$

$x = 2000$

Total income of A and $B = 5x + 7x = 12x = 24,000$

34. The ratio of income of A and B last year was 4 : 3, respectively. The ratios of their individual incomes of the last year and the present year are 3 : 4 and 5 : 6, respectively. If their total income for the present year is ₹8.04 lakh, then the income of B last year was:

पिछले वर्ष A और B की आय का अनुपात क्रमशः 4 : 3 था। पिछले वर्ष और वर्तमान वर्ष की उनकी व्यक्तिगत आय का अनुपात क्रमशः 3 : 4 और 5 : 6 है। यदि वर्तमान वर्ष के लिए उनकी कुल आय ₹8.04 लाख है, तो पिछले वर्ष B की आय थी:

Solution :

$A : B$ (Last Year) $= 4 : 3 = 4x : 3x$.

Last : Present $(A) = 3 : 4$.

$3 = 4x,\ 4 = \frac{16x}{3}$.

Last : Present $(B) = 5 : 6$.

$5 = 3x,\ 6 = \frac{18x}{5}$.

Present Year $(A : B)$:

$\frac{16x}{3} : \frac{18x}{5} = 40 : 27$.

Total income for the present year:

67 units = 8.04 lakh.

27 units = 3.24 lakh.

6 = 3.24 lakh, 5 (Last Year) = 2.7 lakh.

35. A, B, and C spend 80%, 85%, and 75% of their incomes, respectively. If their savings are in the ratio 8 : 9 : 20 and the difference between the incomes of A and C is ₹18,000, then the income of B is:

A, B, और C अपनी आय का क्रमशः 80%, 85%, और 75% खर्च करते हैं। यदि उनकी बचत का अनुपात 8 : 9 : 20 है और A और C की आय के बीच का अंतर ₹18,000 है, तो B की आय:

Solution :

$A \sim 20\% = 8$, then $100\% = 40$.

$B \sim 15\% = 9$, then $100\% = 60$.

$C \sim 25\% = 20$, then $100\% = 80$.

$A : B : C = 2 : 3 : 4$.

2 units = 18000.

1 unit = 9000.

Income of B = 3 units = 27, 000.

Type - 4 Problem based on Tickets & Revenue
टिकट पर आधारित समस्या

36. If by increasing the price of a ticket in the ratio 8 : 11, the number of tickets sold fall in the ratio 23 : 21, then what is the increase (in ₹) in revenue if revenue before the increase in price of the ticket was ₹36,800?

यदि एक टिकट की कीमत 8 : 11 के अनुपात में बढ़ाने से, बेची गई टिकटों की संख्या 23 : 21 के अनुपात में गिर जाती है, तो राजस्व में वृद्धि (₹ में) क्या है, यदि टिकट की कीमत में वृद्धि से पहले राजस्व ₹36,800 था?

Solution :

Revenue = Price × No. of tickets

Old price		:	New price
Price	8	:	11
Ticket	23	:	21
	184	:	231

Total number of old tickets 184 = 36800

Increase in revenue $= (231 - 184) = 47$

Increase in revenue $= \frac{36800}{184} \times 47 = 9400$

37. When the ticket prices of a water park are increased in the ratio 11 : 12, the number of daily visitors to the park falls in the ratio 8 : 7. If the daily revenues before the increase in ticket price were ₹1,76,000, then find the daily revenues after the increase in ticket price.

जब एक वॉटर पार्क की टिकट की कीमतें 11 : 12 के अनुपात

में बढ़ाई जाती हैं तो पार्क में दैनिक आगंतुकों की संख्या 8 : 7 के अनुपात में गिर जाती है। यदि टिकट की कीमत में वृद्धि से पहले दैनिक राजस्व ₹1,76,000 था, तो टिकट की कीमत में वृद्धि के बाद दैनिक राजस्व ज्ञात कीजिए।

Solution :

Revenue = Number of tickets × Number of visitors

Price of ticket = 11 : 12

No. of daily visitors = 8 : 7

Daily revenue = 88 : 84

$88 = 176000, \quad 1 = 2000, \quad 84 = 168000$

Revenue after increment in price of ticket = 168000

38. The entry ticket at a fun park was increased in the ratio 7 : 9, due to which customers fell in the ratio 13 : 11. What is the new daily collection (in ₹), if the daily collection before the price hike was ₹2,27,500?
एक फन पार्क में प्रवेश टिकट 7 : 9 के अनुपात में बढ़ा दिया गया, जिसके कारण ग्राहकों का अनुपात 13 : 11 हो गया। यदि मूल्य वृद्धि से पहले दैनिक संग्रह ₹2,27,500 था, तो नया दैनिक संग्रह (₹ में) क्या है?

Solution :

Ratio of price of ticket = 7 : 9

Ratio of no. of visitors = 13 : 11

Daily collection = 91 : 99

$91 = 227500, \quad 1 = 2500, \quad 99 = 247500$

Type - 5 (Problems based on Coins Rupees)

39. Atul purchased bread costing ₹20 and gave a ₹100 note to the shopkeeper. The shopkeeper gave the balance money in coins of denomination ₹2, ₹5, and ₹10. If these coins are in the ratio 5 : 4 : 1, then how many ₹5 coins did the shopkeeper give?
अतुल ने ₹20 की रोटी खरीदी और दुकानदार को ₹100 का नोट दिया। दुकानदार ने शेष पैसे ₹2, ₹5 और ₹10 मूल्य के सिक्कों में दिए। यदि ये सिक्के 5 : 4 : 1 के अनुपात में हैं, तो दुकानदार ने कितने ₹5 के सिक्के दिए?

Solution :

Remaining Amount = 100 − 20 = 80

$5x \times 2 + 4x \times 5 + x \times 10 = 80$

$10x + 20x + 10x = 80$

$40x = 80$

$x = 2$

$= 4x = 8$ coins

40. A payment of ₹120 is made with ₹10, ₹5, and ₹2 coins. A total of 25 coins are used. Which of the following is the number of ₹10 coins used in the payment?
₹120 का भुगतान ₹10, ₹5 और ₹2 के सिक्कों से किया जाता है। कुल 25 सिक्कों का उपयोग किया जाता है। भुगतान में उपयोग किए गए ₹10 के सिक्कों की संख्या निम्नलिखित में से कौन सी है?

Solution :

$10x + 5y + 2(25 - x - y) = 120$

$8x + 3y = 70$

By options:

If $x = 10$, then $y =$ negative (not valid).

If $x = 8$, then $y = 2$ (valid).

If $x = 6$, then $y = \frac{22}{3}$ (not valid).

If $x = 4$, then $y = \frac{38}{3}$ (not valid).

Number of ₹10 coins = 8

41. In a wallet, there are notes of the denomination ₹10 and ₹50. The total number of notes is 12. The number of ₹10 and ₹50 notes are in the ratio 1 : 2. Total money in the wallet is:
एक बटुए में ₹10 और ₹50 मूल्यवर्ग के नोट हैं। नोटों की कुल संख्या 12 है। ₹10 और ₹50 के नोटों की संख्या 1 : 2 के अनुपात में है। बटुए में कुल पैसा है:

Solution :

Total number of notes = 12

Ratio = 1 : 2

3 = 12, 1 = 4 notes of ₹10, 2 = 8 notes of ₹50

Total money $= 10 \times 4 + 50 \times 8 = 40 + 400 = 440$

42. A bag contains ₹10, ₹5, and ₹2 notes in the ratio 10 : 5 : 2. The total value of ₹5 notes in the bag is ₹84 more than that of ₹2 notes. The total value of ₹10 notes in the bag (in ₹) is:

एक बैग में 10 : 5 : 2 के अनुपात में ₹10, ₹5 और ₹2 के नोट हैं। बैग में ₹5 के नोटों का कुल मूल्य ₹2 के नोटों से ₹84 अधिक है। बैग में ₹10 के नोटों का कुल मूल्य (₹ में) है:

Solution :

$5 \times 5x - 2 \times 2x = 84$

$25x - 4x = 84$

$21x = 84, \quad x = 4$

Total value of ₹10 notes $= 10 \times 10x = 10 \times 10 \times 4 = 400$

Type - 6 (Miscellaneous)

43. In a school, $\frac{3}{8}$ of the students are girls, and the rest are boys. One-third of the boys are below 10 years, and $\frac{2}{3}$ of the girls are also below 10 years. If the number of students aged 10 or more years is 260, then the number of boys in the school is:

एक स्कूल में, विद्यार्थियों की संख्या में से $\frac{3}{8}$ लड़कियाँ हैं और शेष लड़के हैं। लड़कों की संख्या में से एक तिहाई 10 वर्ष से कम हैं और लड़कियों की संख्या में से $\frac{2}{3}$ भी 10 वर्ष से कम हैं। यदि 10 वर्ष या उससे अधिक आयु के विद्यार्थियों की संख्या 260 है, तो विद्यालय में लड़कों की संख्या है:

Solution :

Let total students be $8x$.

Girls = $3x$, Boys = $5x$.

Boys (below 10 years) = $\frac{5x}{3}$, Girls (below 10 years) = $2x$.

Students aged 10+ years:

$\frac{10x}{3} + x = 260$

$\frac{13x}{3} = 260, \quad x = 60$

Total boys = $5x = 300$

44. By adding 3 and 5 to the numerator and denominator of a fraction, it becomes 3. If 1 is subtracted from the numerator and 3 is added to the denominator, it becomes $\frac{2}{5}$. Find the fraction.

किसी भिन्न के अंश और हर में 3 और 5 जोड़ने पर यह 3 हो जाता है। यदि अंश में से 1 घटाया जाए और हर में 3 जोड़ा जाए, तो यह $\frac{2}{5}$ हो जाता है। भिन्न ज्ञात कीजिए।

Solution :

Let the fraction be $\frac{x}{y}$.

$\frac{x+3}{y+5} = 3 \quad \Rightarrow \quad 3x + 9 = 2y + 10 \quad \Rightarrow$

$3x - 2y = 1 \quad ...\text{(Eq. 1)}$

$\frac{x-1}{y+3} = \frac{2}{5} \quad \Rightarrow \quad 5x - 5 = 2y + 6 \quad \Rightarrow$

$5x - 2y = 11 \quad ...\text{(Eq. 2)}$

Solve Eq. 1 and Eq. 2:

$x = 5, \quad y = 7$

Fraction $= \frac{5}{7}$

45. The ratio of boys and girls in a group is 7 : 6. If 4 more boys join the group and 3 girls leave, then the ratio of boys to girls becomes 4 : 3. What is the total number of boys and girls initially in the group?

एक समूह में लड़कों और लड़कियों का अनुपात 7 : 6 है। यदि 4 और लड़के समूह में शामिल हो जाते हैं और 3 लड़कियाँ समूह छोड़ देती हैं, तो लड़कों का लड़कियों से अनुपात 4 : 3 हो जाता है। प्रारंभ में समूह में लड़कों और लड़कियों की कुल संख्या कितनी है?

Solution :

$\frac{7x+4}{6x-3} = \frac{4}{3}$

$21x + 12 = 24x - 12$

$3x = 24 \quad \Rightarrow \quad x = 8$

Total boys and girls initially: $7x + 6x = 13x = 104$

46. A diamond is broken into three parts in the ratio 1 : 2 : 3. The price of a diamond is directly proportional to the square of its weight. If after breaking the diamond, the loss is ₹39,600 in the price, then find the price of the diamond before breaking it.

एक हीरे को 1 : 2 : 3 के अनुपात में तीन भागों में तोड़ा जाता है। हीरे की कीमत उसके वजन के वर्ग के समानुपाती होती है। यदि हीरे को तोड़ने के बाद कीमत में ₹39,600 का नुकसान होता है, तो तोड़ने से पहले हीरे की कीमत ज्ञात करें।

Solution :

Price $\sim$ weight^2

$1:2:3 \quad \Rightarrow \quad 6^2:(1^2+2^2+3^2)$

$36:(1+4+9)=36:14$

Ratio $=18:7$

11 units $=39600 \quad \Rightarrow \quad 1$ unit $=3600$

18 units $=\ 64800$

47. Two-thirds of the number of employees of a company are males, and the rest are females. If $\frac{3}{8}$ of the male employees and $\frac{2}{5}$ of the female employees are temporary employees, and the total number of permanent employees is 740, then $\frac{7}{15}$ of the total number of employees exceeds the number of temporary female employees by:

किसी कंपनी के कर्मचारियों की संख्या में दो-तिहाई पुरुष हैं और शेष महिलाएँ हैं। अगर पुरुष कर्मचारियों में से $\frac{3}{8}$ और महिला कर्मचारियों में से $\frac{2}{5}$ अस्थायी कर्मचारी हैं और स्थायी कर्मचारियों की कुल संख्या 740 है, तो कर्मचारियों की कुल संख्या में से $\frac{7}{15}$ अस्थायी महिला कर्मचारियों की संख्या से 15 अधिक है:

Solution :

Let total employees be $3x$.

Males $=2x$, Females $=x$.

Temporary employees:

Male $=\frac{3x}{8}$, Female $=\frac{2x}{5}$

Permanent employees:

$\frac{5x}{4}+\frac{3x}{5}=740$

$37x=20\times 740 \quad \Rightarrow \quad x=400$

Total employees $=3x=1200$

$\frac{7}{15}$ of $1200=560$

$560-160=400$

48. ₹600 has been divided among A, B, and C in such a way that ₹40 more than $\frac{2}{5}$ of A's share, ₹20 more than $\frac{2}{7}$ of B's share, and ₹10 more than $\frac{9}{17}$ of C's share are all equal. Find A's share.

₹600 रुपये को A, B और C के बीच इस प्रकार विभाजित किया गया है कि A के हिस्से के $\frac{2}{5}$ से ₹40 अधिक, B के हिस्से के $\frac{2}{7}$ से ₹20 अधिक और C के हिस्से के $\frac{9}{17}$ से ₹10 अधिक, सभी बराबर हैं। A का हिस्सा ज्ञात करें।

Solution :

$\frac{5}{2}(k-40)+\frac{7}{2}(k-20)+\frac{17}{9}(k-10)=600$

$108k-1800-1260+34k-340=10800$

$142k=14200$

$k=100$

$A=150$

49. On the whole surface of the earth, the ratio of land and water is $1:2$. The land and water in the northern hemisphere is in the ratio $2:3$. Then what is the ratio of land and water in the southern hemisphere?

पृथ्वी की सम्पूर्ण सतह पर भूमि एवं जल का अनुपात $1:2$ है। उत्तरी गोलार्ध में भूमि और जल का अनुपात $2:3$ है। तो दक्षिणी गोलार्ध में भूमि और जल का अनुपात क्या है?

Solution :

$(1:2)\times 2\times 5$

$(2:3)\times 3$

$10:20$

$6:9$

Southern hemisphere land : water ratio

$=4:11$

50. The railway fares of air-conditioned sleeper and ordinary sleeper classes are in the ratio $4:1$. The number of passengers traveling by air-conditioned sleeper and ordinary sleeper classes were in the ratio $3:25$. If the total collection was ₹37,000, how much did air-conditioned sleeper passengers pay?

वातानुकूलित शयनयान और साधारण शयनयान श्रेणी का रेलवे किराया $4:1$ के अनुपात में है। वातानुकूलित शयनयान और साधारण शयनयान श्रेणी में यात्रा करने वाले यात्रियों की संख्या का अनुपात $3:25$ था। यदि कुल संग्रह ₹37,000 है, तो वातानुकूलित शयनयान यात्रियों ने कितना भुगतान किया?

Solution :

Fare × Number of Passengers = Total Revenue

4 : 1 and 3 : 25

Combined ratio = 12 : 25

37 units = 37, 000

12 units = 12, 000

51. The ratio of successful and unsuccessful examinees in an examination in a school is 6 : 1. The ratio would have been 9 : 1 if 6 more examinees had been successful. The total number of examinees is:
एक स्कूल में एक परीक्षा में सफल और असफल परीक्षार्थियों का अनुपात 6 : 1 है। यदि 6 और परीक्षार्थी सफल होते तो अनुपात 9 : 1 होता। कुल परीक्षार्थियों की संख्या है:

Solution :

$$\frac{6x+6}{x-6} = \frac{9}{1}$$

$6x + 6 = 9x - 54$

$3x = 60$

$x = 20$

Total number of examinees $= 7x = 140$

4 Mixture and Alligation

Mixture
मिश्रण

Mixture is defined as a result of mixing two or more items or substances in a certain ratio or proportion.
मिश्रण को दो या दो से अधिक वस्तुओं या पदार्थों को एक निश्चित अनुपात या अनुपात में मिलाने के परिणामस्वरूप परिभाषित किया जाता है

Alligation
एलीगेशन

Alligation is a process or rule for the solution of problems concerning the compounding or mixing of ingredients differing in price or quality.
एलीगेशन कीमत या गुणवत्ता में भिन्न सामग्री के संयोजन या मिश्रण से संबंधित समस्याओं के समाधान के लिए एक प्रक्रिया या नियम है

Type 1 (Basic)

1. A mixture contains acid and water in the ratio of $6 : 1$. On adding 12 litres of water to the mixture, the ratio of acid to water becomes $3 : 2$. The quantity of water (in litres) in the original mixture was:
एक मिश्रण में अम्ल और पानी का अनुपात $6 : 1$ है। मिश्रण में 12 लीटर पानी मिलाने पर, अम्ल और पानी का अनुपात $3 : 2$ हो जाता है। मूल मिश्रण में पानी की मात्रा (लीटर में) थी:

Solution :

Make Acid same

$6 : 1$

$(3 : 2) \times 2$

$6 : 1$

$6 : 4$

$3 = 12$

$1 = 4$ litre

2. A vessel contains a solution of two liquids A and B in the ratio $5 : 3$. When 10 litres of the solution is taken out and replaced by the same quantity of B, the ratio of A and B in the vessel becomes $10 : 11$. The quantity (in litres) of the solution in the vessel was:
एक बर्तन में $5 : 3$ के अनुपात में दो तरल पदार्थ A और B का घोल है। जब 10 लीटर घोल निकाला जाता है और उसके स्थान पर उतनी ही मात्रा में B मिलाया जाता है, तो बर्तन में A और B का अनुपात $10 : 11$ हो जाता है। बर्तन में घोल की मात्रा (लीटर में) थी:

Solution :

$10 : 6$

$10 : 11$

$5 = 10$

$16 = 32$

$32 + 10 = 42$ Ltr

Or:

5/8   0

     10/21

10/21 : 25/168

16 : 25

$5 = 10$

$1 = 2$

$21 = 42$ Ltr Total Quantity

3. A container contains 20 L mixture in which there is 10% sulphuric acid. Find the quantity of sulphuric acid to be added in it to make the solution to contain 25% sulphuric acid.

एक कंटेनर में 20 लीटर मिश्रण है जिसमें 10% सल्फ्यूरिक एसिड है। घोल में 25% सल्फ्यूरिक एसिड बनाने के लिए इसमें मिलाई जाने वाली सल्फ्यूरिक एसिड की मात्रा ज्ञात कीजिए।

Solution :

10   100

    25

75   :   15

5   :   1

$5 = 20$

$1 = 4$ ltr

4. In a 56 liters mixture of milk and water, the ratio of milk to water is 5 : 2. In order to make the ratio of milk to water 7 : 2, some quantity of milk is to be added to the mixture. The quantity of the milk present in the new mixture will be:

दूध और पानी के 56 लीटर मिश्रण में, दूध और पानी का अनुपात 5 : 2 है। दूध और पानी का अनुपात 7 : 2 बनाने के लिए, मिश्रण में दूध की कुछ मात्रा मिलानी पड़ती है। नए मिश्रण में मौजूद दूध की मात्रा होगी:

Solution :

5 : 2

7 : 2

$7 = 56$ liters

5. A large container has a 50-litre mixture of juice and water in the ratio 3 : 2. In this mixture, a 60-litre juice and water mixture is added, that has a juice to water ratio of 2 : 1. After this, 11 litres of the solution is replaced with pure juice. What is the ratio of water to juice in the final mixture?

एक बड़े कंटेनर में जूस और पानी का 50-लीटर मिश्रण है जिसका अनुपात 3 : 2 है। इस मिश्रण में 60 लीटर रस और पानी का मिश्रण मिलाया जाता है, जिसमें रस और पानी का अनुपात 2 : 1 होता है। इसके बाद 11 लीटर घोल को शुद्ध जूस से बदल दिया जाता है। अंतिम मिश्रण में पानी और रस का अनुपात क्या है?

Solution :

Mixture (i) 50 Litres → Juice, Water

Mixture (i) 60 Litres → Juice, Water

Total juice $= 30 + 40 = 70$ litres

Total water $= 20 + 20 = 40$ litres

Total mixture $= 50 + 60 = 110$ litres

According to the question:

Water : Juice $= (110 - 11) \times \frac{4}{11} : (110 - 11) \times \frac{7}{11} + 11$

$= 99 \times 4 : 99 \times 7 + 11$

$= 396 : 704$

$= 18 : 37$

6. Two kinds of rice, the first costing 65 per kg and the second 95 per kg, are mixed together. Find the ratio in which the two types are mixed so that the mixture costs 83 per kg.

दो प्रकार के चावल, पहले की कीमत 65 प्रति किलोग्राम और दूसरे की कीमत 95 प्रति किलोग्राम, को एक साथ मिलाया जाता है। वह अनुपात ज्ञात कीजिए जिसमें दोनों प्रकार को मिलाया जाए ताकि मिश्रण की लागत 83 प्रति किलोग्राम हो।

Solution :

According to the question:

I   II

65   95

   83

12   18

Required ratio $= 95 - 83 : 83 - 65$

7. Alloy A contains copper and zinc in the ratio of 4 : 3 and alloy B contains copper and zinc in the ratio of 5 : 2. A and B are taken in the ratio of 5 : 6 and melted to form a new alloy. The percentage of zinc in the new alloy is closest to:

मिश्र धातु A में 4 : 3 के अनुपात में तांबा और जस्ता होता है और मिश्र धातु B में 5 : 2 के अनुपात में तांबा और जस्ता होता है। A और B को 5 : 6 के अनुपात में लिया जाता है और एक नया मिश्र धातु बनाने के लिए पिघलाया जाता है। नई मिश्रधातु में जिंक का प्रतिशत निकटतम है:

Solution :

Copper : Zinc

Alloy A → 4 : 3

Alloy B → 5 : 2

On mixing A and B in the ratio of 5 : 6:

New alloy composition $= (4 \times 5 + 5 \times 6)$:

$(3 \times 5 + 2 \times 6) = 50 : 27$

Zinc Percentage $= \frac{27}{77} \times 100 = 35.06\%$

8. How many litres of water should be added to a 7.5 litre mixture of acid and water in the ratio of 1 : 2 such that the resultant mixture has 20% acid in it?

एसिड और पानी के 7.5 लीटर मिश्रण में 1 : 2 के अनुपात में कितने लीटर पानी मिलाया जाना चाहिए ताकि परिणामी मिश्रण में 20% एसिड हो?

Solution :

Quantity of acid in mixture $= 7.5 \times \frac{1}{3}$ litre $=$

2.5 ltr

Let x litres of water be added:

$$\frac{2.5}{7.5 + x} \times 100 = 20$$

$12.5 = 7.5 + x \implies x = 5$ litres

9. An alloy contains copper and tin in the ratio 3 : 2. If 250 gm of copper is added to this alloy, then the copper in it becomes double the quantity of tin in it. What is the amount (in gm) of tin in the alloy?

एक मिश्रधातु में तांबा और टिन का अनुपात 3 : 2 है। यदि इस मिश्रधातु में 250 ग्राम तांबा मिलाया जाए तो इसमें मौजूद तांबा टिन की मात्रा से दोगुना हो जाता है। मिश्रधातु में टिन की मात्रा (ग्राम में) क्या है?

Solution :

Amount of copper in the alloy $= 3x$

Amount of tin in the alloy $= 2x$

According to the question:

$$\frac{3x + 250}{2x} = \frac{2}{1}$$

$3x + 250 = 4x$

$x = 250$

Hence, amount of tin in the alloy $= 2x =$

$2 \times 250 = 500$ gm.

10. How many kg of salt costing 28 per kg must be mixed with 39.6 kg of salt costing 16 per kg, so that selling the mixture at 29.90, there is a gain of 15%?

28 प्रति किलोग्राम मूल्य वाले कितने किलोग्राम नमक को 16 प्रति किलोग्राम मूल्य वाले 39.6 किलोग्राम नमक के साथ मिलाया जाना चाहिए, ताकि मिश्रण को 29.90 पर बेचने पर 15% का लाभ हो?

Solution :

Cost price of the mixture per kg $=$

$$\frac{29.90 \times 100}{115} = 26 \text{ Rs.}$$

From the Rule of Alligation:

16 Rs. kg   28 Rs. kg

26 Rs. kg

2 Rs. kg   10 Rs. kg

Ratio = 2 : 10 = 1 : 5

Amount of 28 Rs/kg salt:

$39.6 \times 5 = 198$ kg

11. In what ratio, sugar costing ₹60 per kg be mixed with sugar costing ₹42 per kg such that by selling the mixture at ₹56 per kg there is a gain of 12%?

₹60 प्रति किलोग्राम कीमत वाली चीनी को ₹42 प्रति किलोग्राम कीमत वाली चीनी के साथ किस अनुपात में मिलाया जाए ताकि मिश्रण को ₹56 प्रति किलोग्राम बेचने पर 12% का लाभ हो?

Solution :

Cost price of mixing sugar $= \frac{56 \times 100}{112} = 50$

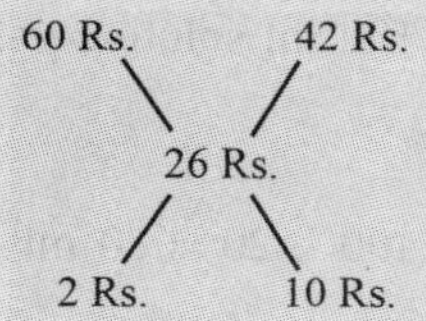

Intended ratio = = 8 : 10

= 4 : 5

12. 40 litres of 60% concentration of acid solution is added to 35 litres of 80% concentration of acid solution. What is the concentration of acid in the new solution?

60% सांद्रता वाले एसिड घोल के 40 लीटर को 80% सांद्रता वाले एसिड घोल के 35 लीटर में मिलाया जाता है। नये घोल में अम्ल की सांद्रता क्या है?

Solution :

Total solution = 40 + 35 = 75 litre

Amount of total acid $= (40 \times \frac{60}{100}) + (35 \times \frac{80}{100}) = 24 + 28 = 52$ litre

Concentration of acid in the new solution $= \frac{52}{75} \times 100 = 69\frac{1}{3}\%$

13. A mixture is composed of 11 parts of pure milk and 2 parts of water. If 35 litres of water were added to the mixture, then the new mixture will contain twice as much pure milk as water. How many litres of pure milk does the original mixture contain?

एक मिश्रण 11 भाग शुद्ध दूध और 2 भाग पानी से बना है। यदि मिश्रण में 35, लीटर पानी मिलाया जाए तो नए मिश्रण में पानी से दोगुना शुद्ध दूध होगा, तो मूल मिश्रण में कितने लीटर शुद्ध दूध है?

Solution :

Ratio of milk and water in the given mixture

$= 11x : 2x$

After adding 35 litres of water to the mixture,

the ratio of the mixture is 2 : 1:

$$\frac{11x}{2x + 35} = \frac{2}{1}$$

$x = 10$

Quantity of milk in the mixture $= 11x = 11 \times 10 = 110$ litres

14. Two bottles of the same capacity are 35% and $33\frac{1}{3}\%$ full of orange juice, respectively. They are filled up completely with apple juice, and then the contents of both bottles are emptied into another vessel. The percentage of apple juice in the mixture is:

समान क्षमता की दो बोतलें क्रमशः 35% और $33\frac{1}{3}\%$ संतरे के रस से भरी हैं। उन दोनों को सेब के रस से पूरी तरह भर दिया जाता है और फिर दोनों बोतलों की सामग्री को दूसरे बर्तन में खाली कर दिया जाता है। मिश्रण में सेब के रस का प्रतिशत है:

Solution :

Let each bottle's capacity = 300 litres

Quantity of apple juice in the mixture $= (300 \times \frac{65}{100}) + (300 \times \frac{66\frac{2}{3}}{100})$

$= 195 + 200 = 395$ litres

Total quantity = 600 litres

Apple juice percentage $= \frac{395}{600} \times 100 = 65\frac{5}{6}\%$

15. The volume of the water in two tanks, A and B, is in the ratio of 6 : 5. The volume of water in tank A is increased by 30%. By what percentage should the volume of water in tank B be increased so that both tanks have the same volume of water?

दो टैंक A और B में पानी की मात्रा 6 : 5 के अनुपात में है। टैंक A में पानी की मात्रा 30% बढ़ जाती है। टैंक B में पानी की मात्रा कितने प्रतिशत बढ़ाई जानी चाहिए ताकि दोनों टैंकों में पानी की मात्रा समान हो?

Solution :

Let the volume of water in tank A = $6x$

And the volume of water in tank B = $5x$

Again, let the tank B increase by $m\%$:

$$\frac{6x \times 130}{100} = \frac{5x \times (100 + m)}{100}$$

$780 = 500 + 5m$

16. A beaker contains acid and water in the ratio $1 : x$. When 300 ml of the mixture and 50 ml of water are mixed, the ratio of acid and water becomes $2 : 5$. What is the value of x?

एक बीकर में अम्ल और पानी का अनुपात $1 : x$ है। जब 300 मिलीलीटर मिश्रण और 50 मिलीलीटर पानी मिलाया जाता है, तो अम्ल और पानी का अनुपात $2 : 5$ हो जाता है। x का मान क्या है?

Solution :

Initial ratio: $2 : 2x$

New ratio: $2 : 5$

$$\frac{5-2x}{2+2x} = \frac{50}{300} = \frac{1}{6}$$

$$6(5-2x) = 2+2x$$

$$30 - 12x = 2 + 2x \implies 14x = 28 \implies$$

$$x = 2$$

17. A jar contains a mixture of fruit juice and water in the ratio $5 : x$. When 1 litre of water is added to 4 litres of the mixture, the ratio of fruit juice to water becomes $1 : 1$. What is the value of x?

एक जार में फलों के रस और पानी का मिश्रण $5 : x$ के अनुपात में है। जब 4 लीटर मिश्रण में 1 लीटर पानी मिलाया जाता है तो फलों के रस और पानी का अनुपात $1 : 1$ हो जाता है। x का मूल्य क्या है?

Solution :

Reverse Process:

Final ratio: 1 : 1 (for 5 litres)

2.5 litres fruit juice : 2.5 litres water

Before adding water: 2.5 litres fruit juice : (2.5 – 1) litres water

$2.5 : 1.5 = 5 : 3$

$x = 3$

18. If x beakers of 100 ml containing $1 : 4$ acid-water solution are mixed with y beakers of 200 ml containing $3 : 17$ acid-water solution, then the ratio of acid to water in the resulting mixture becomes $19 : 91$. Find $x : y$.

यदि $1 : 4$ एसिड-पानी के घोल वाले 100 मिलीलीटर के x बीकर को $3 : 17$ एसिड-पानी के घोल वाले 200 मिलीलीटर के y बीकर के साथ मिलाया जाता है तो परिणामी मिश्रण में एसिड और पानी का अनुपात $19 : 91$ हो जाता है। $x : y$ ज्ञात कीजिए।

Solution :

Amount of acid in 100 ml solution $= 100 \times \frac{1}{5} = 20$ ml

Amount of water $= 100 - 20 = 80$ ml

Amount of acid in 200 ml solution

$= 200 \times \frac{3}{20} = 30$ ml

Amount of water $= 200 - 30 = 170$ ml

According to the question:

$$\frac{20x + 30y}{80x + 170y} = \frac{19}{91}$$

$$\frac{2x + 3y}{8x + 17y} = \frac{19}{91}$$

$$182x + 273y = 152x + 323y$$

$$30x = 50y$$

$$\frac{x}{y} = \frac{5}{3} \quad x : y = 5 : 3$$

19. A milkman uses three containers for selling milk, their capacities being $40L$, $30L$ and $20L$ respectively. He fills respectively 87.5%, 80% and 90% of the containers with a mix of milk and water in the ratios $3 : 2$, $5 : 1$ and $7 : 2$ respectively. What is the ratio of the total quantity of milk to that of water carried by him?

एक दूधवाला दूध बेचने के लिए तीन कंटेनरों का उपयोग करता है, उनकी क्षमता क्रमश: $40L$, $30L$ और $20L$ है। वह क्रमश: 87.5%, 80% और 90% कंटेनरों को दूध और पानी के मिश्रण से क्रमश: $3 : 2$, $5 : 1$ और $7 : 2$ के अनुपात में भरता है। उसके द्वारा ले जाए गए दूध की कुल मात्रा और पानी की कुल मात्रा का अनुपात क्या है?

Solution :

$40 \times 87.5\% = 35$ litres (ratio $3 : 2 \implies 21L$ milk and $14L$ water)

$30 \times 80\% = 24$ litres (ratio $5 : 1 \implies 20L$ milk and $4L$ water)

$20 \times 90\% = 18$ litres (ratio $7 : 2 \implies 14L$ milk and $4L$ water)

Milk : Water $= 21 + 20 + 14 : 14 + 4 + 4$

$= 55 : 22$

$= 5 : 2$

20. Fresh fruit contains 68% water and dry fruit contains 20% water. How much dry fruit can be obtained from 100 kg of fresh fruit?
ताजे फल में 68% पानी होता है और सूखे फल में 20% पानी होता है। 100 किलोग्राम ताजे फल से कितना सूखा फल प्राप्त किया जा सकता है?

Solution :

32% of Fresh Fruit = 80% of Dry Fruit
Fresh : Dry = 5 : 2
5 = 100 kg
1 = 20 kg
2 = 40 kg

21. A vessel contains a mixture of milk and water in the ratio 14 : 3. 25.5 litres of the mixture is taken out from the vessel, and 2.5 litres of pure water and 5 litres of pure milk is added to the mixture. If the resultant mixture contains 20% water, what was the initial quantity of the mixture in the vessel before the replacement (in litres)?
एक बर्तन में दूध और पानी का अनुपात 14 : 3 है। बर्तन से 25.5 लीटर मिश्रण निकालकर बर्तन में 2.5 लीटर पानी और 5 लीटर दूध मिलाया गया। यदि अब नए मिश्रण में 20% पानी है तो प्रतिस्थापन से पहले प्रारंभ में बर्तन में कितना मिश्रण था?

Solution :

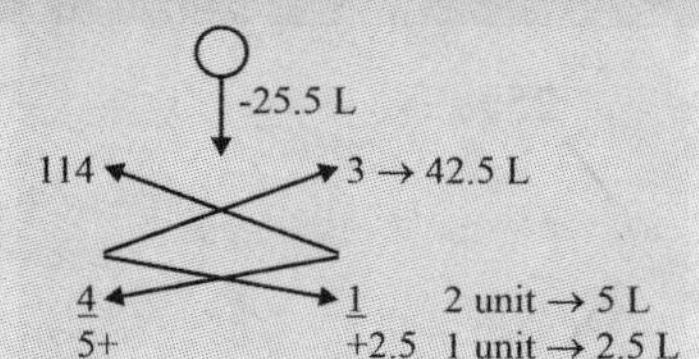

Initial mixture = 42.5 + 25.5

= 68 litres

17 units = 17 × 2.5 = 42.5 litres

22. A drum contains 80 litres of ethanol. 20 litres of this liquid is removed and replaced with water. 20 litres of this mixture is again removed and replaced with water. How much water (in litres) is present in this drum now?
एक ड्रम में 80 लीटर इथेनॉल होता है। इस तरल में से 20 लीटर निकाल लिया जाता है और उसके स्थान पर पानी डाल दिया जाता है। इस मिश्रण का 20 लीटर फिर से निकाला जाता है और पानी से बदल दिया जाता है। इस ड्रम में अब कितना पानी (लीटर में) मौजूद है?

Solution :

Formula Used:

Final quantity of ethanol = Initial quantity ×

$(1 - \frac{\text{quantity replaced}}{\text{Initial quantity}})^2$

$80 \times (1 - \frac{20}{80})^2 = 80 \times (\frac{3}{4})^2$

Water = 80 − 45 = 35 litres

23. A container has 30 litres of milk, from which 3 litres of milk is taken out and replaced with water. The process is done three times. What is the final ratio of the water and the milk in the container?
एक कंटेनर में 30 लीटर दूध है, जिसमें से 3 लीटर दूध निकाला जाता है और उसके स्थान पर पानी डाल दिया जाता है। यह प्रक्रिया तीन बार की जाती है। कंटेनर में पानी और दूध का अंतिम अनुपात क्या है?

Solution :

Remaining quantity of milk = Initial quantity

$(1 - \frac{\text{quantity taken out}}{\text{Initial quantity}})^3$

$= 30 \times (1 - \frac{3}{30})^3 = 30 \times (\frac{9}{10})^3$

$= 30 \times \frac{729}{1000} = \frac{2187}{100} = 21.87$ litres milk

Water = 30.00 − 21.87 = 8.13 litres

Required ratio:

Water : Milk = 813 : 2187 = 271 : 729

24. A vessel contains a mixture of O_2 and N_2. The quantity of O_2 is 36% of the total mixture. Some quantity of the mixture is taken out and replaced by N_2. This process is repeated once more. Now the quantity of O_2 is 16% of the total mixture. If the total quantity of the mixture was 18 litres, then find the quantity of the mixture which is taken out?
एक बर्तन में O_2 और N_2 का मिश्रण है। जिसमें O_2 की मात्रा कुल मिश्रण का 36% है। हमने मिश्रण की कुछ मात्रा निकालकर उतनी ही मात्रा में N_2 मिला दी। यह प्रक्रिया कुल दो बार की। अब उसमें बची हुई O_2 की मात्रा मिश्रण की कुल मात्रा का 16% है। यदि कुल मिश्रण 18 लीटर हो, तो बताइए कि कितनी मात्रा निकाली गई थी?

Solution :

$$\frac{\text{Remaining } O_2}{\text{Total mixture}} = \frac{16}{36} = \left(\frac{4}{6}\right)^2$$

$$\frac{4}{6} = \frac{2}{3}$$

$$\text{Replaced fraction} = \frac{1}{3} = \frac{6L}{18L}$$

25. A man has 10, 000 Rs. He lent a part of it at 15% simple interest and the remaining at 10% simple interest. The total interest he received after 5 years amounted to 6, 500 Rs. The difference between the parts of the amounts he lent is:

एक आदमी के पास 10, 000 रु हैं। उसने इसका एक भाग 15% साधारण ब्याज पर और शेष 10% साधारण ब्याज पर उधार दिया। 5 वर्ष बाद उसे कुल ब्याज 6, 500 रु प्राप्त हुआ। उसके द्वारा उधार दी गई राशि के भागों के बीच का अंतर है:

Solution :

Interest for 5 years = ₹6,500

$$\text{Interest for 1 year} = \frac{6500}{5} = 1,\ 300$$

$$\text{Overall Interest Rate} = \frac{1300}{10000} \times 100 = 13\%$$

15   10

    13

3   2

Difference between parts of amount:

$$10,\ 000 \times \frac{(15-10)}{(15+10)}$$

$$= 10,\ 000 \times \frac{5}{25} = 2,\ 000$$

26. A sum of ₹50,000 is lent partly at 4% and the remaining at 5% per annum. If the yearly simple interest on the average is 4.6%, the two parts are:

₹50,000 की राशि आंशिक रूप से 4% और शेष 5% प्रति वर्ष की दर पर उधार दी जाती है। यदि वार्षिक साधारण ब्याज औसतन 4.6% है, तो दो भाग हैं:

Solution :

From Alligation:

I: R → 4%   II: R → 5%

mix R% 4.6%

(5-4.6) 0.4   (4-6.4) 0.6

I : II = 0.4 : 0.6

I : II = 2 : 3

$$\text{then I} = 50000 \times \frac{2}{2+3} = 20000$$

$$\text{And II} = 50000 \times \frac{3}{5} = 30000$$

27. A dealer had 120 kg of wheat. A part of it was sold by him at 10% gain and the rest at 25% gain. Overall, he had a gain of 15%. How much of the wheat was sold at 10% gain?

एक डीलर के पास 120 किलो गेहूं था। इसका एक हिस्सा उसने 10% लाभ पर और शेष 25% लाभ पर बेच दिया। कुल मिलाकर उसे 15% का फ़ायदा हुआ। कितना गेहूं 10% लाभ पर बेचा गया?

Solution :

10%   25%

    15%

10   5

2 : 1

3 units = 120 kg

1 unit = 40 kg

2 units = 80 kg

Hence, 80 kg of wheat was sold at 10% gain.

28. Radhika saves 25% of her income. If her expenditure increases by 15% and her income increases by 35%, then her savings increased by:

राधिका अपनी आय का 25% बचाती है। यदि उसका व्यय 15% बढ़ जाता है और उसकी आय 35% बढ़ जाती है, तो उसकी बचत बढ़ जाती है?

Solution :

15 x

35

75 : 25

3 : 1

$15 \times 3 + 1 \times x = 4 \times 35$

$45 + x = 140$

$x = 95\%$

29. Three containers A, B, and C are having a mixture of milk and water in the ratio 1 : 3, 2 : 3, and 2 : 5 respectively. If the capacities of the containers are in the ratio 2 : 3 : 5, find the ratio of milk to water when the mixtures of all three containers are mixed together.

तीन बर्तन A, B और C में दूध और पानी का अनुपात क्रमशः 1 : 3, 2 : 3, और 2 : 5 है तथा इन बर्तनों के आयतन के अनुपात 2 : 3 : 5 है। यदि इन तीनों बर्तनों का मिश्रण को मिलाया जाए तो नए मिश्रण में दूध और पानी का अनुपात क्या होगा?

Solution :

Milk : Water (A) = $(1 : 3) \times 5 \times 7 \times 2$

Milk : Water (B) = $(2 : 3) \times 4 \times 7 \times 3$

Milk : Water (C) = $(2 : 5) \times 4 \times 5 \times 5$

Solution :

M : W = (35 + 84 + 100) :

$(105 + 126 + 250) \implies 219 : 481$

30. Three glasses of sizes $3L$, $4L$, and xL contain a mixture of spirit and water in the ratio 2 : 3, 3 : 7, and 4 : 11 respectively. All three glasses are poured into a single vessel. The ratio of spirit to water in the resultant mixture is 14 : 31. Then, the value of x is:

तीन गिलास जिनके आकार 3 ली, 4 ली, और x ली है, स्पिरिट और पानी के मिश्रण से भरे हुए हैं, जिनका अनुपात क्रमशः 2 : 3, 3 : 7, और 4 : 11 है। सभी तीनों गिलास की मात्रा को नए बर्तन में पलट दिया जाता है। मिश्रण के उस बर्तन में स्पिरिट और पानी का अनुपात 14 : 31 हो जाता है तो x का मान क्या है?

Solution :

Spirit : Water (A) = $(2 : 3) \times 5 \times 8 \times 3$

Spirit : Water (B) = $(3 : 7) \times 10 \times 3$

Spirit : Water (C) = $(4 : 11) \times 15 \times x$

Resultant Ratio: $\dfrac{36 + 4x}{69 + 11x} = \dfrac{14}{31}$

Cross-multiply:

$31(36 + 4x) = 14(69 + 11x)$

$1116 + 124x = 966 + 154x \implies 30x = 150 \implies x = 5$

5 Partnership

Partnership/ साझेदारी

Definition / परिभाषा

Partnership: Partners invest same/different capital for same/different time period.

साझेदारी: साझेदार समान/अलग-अलग समय अवधि के लिए समान/अलग-अलग पूंजी निवेश करते हैं।

Types of Partners / साझेदारों के प्रकार

(i) Active Partner: Invests money as well as takes part in business activity for which he is paid salary from the profit.

सक्रिय भागीदार: पैसा निवेश करता है और साथ ही व्यावसायिक गतिविधि में भाग लेता है जिसके लिए उसे लाभ से वेतन दिया जाता है।

(ii) Sleeping Partner: Only invests money and does not take part in business activities.

स्लीपिंग पार्टनर: जो केवल पैसा निवेश करता है और व्यावसायिक गतिविधियों में भाग नहीं लेता है।

$Profit = Capital \times time$

लाभ = पूंजी $\times$ समय

Type 1 (Basic)

1. A, B, and C started a business. A invested $33\frac{1}{3}\%$ of the total capital. B invested $33\frac{1}{3}\%$ of the remaining capital, and C invested the rest. If the total profit at the end of a year was ₹20,250, then how much does C's profit exceed B's profit?

A, B और C ने एक व्यवसाय शुरू किया। A ने कुल पूंजी का $33\frac{1}{3}\%$ निवेश किया। B शेष पूंजी का $33\frac{1}{3}\%$ निवेश करता है और C, शेष। यदि एक वर्ष के अंत में कुल लाभ ₹20, 250 था, तो C का लाभ B के लाभ से कितना अधिक है?

Solution :

Let Total Capital be 9 units

A : B : C = 3 : 2 : 4

9 = ₹20, 250

2 = ₹4, 500

2. A and B invest in a business in the ratio 3 : 7. The business makes a profit of ₹60,000 in one year. They decide to distribute the profit remaining after reinvesting 40% of the profit. How much will A get (in ₹)?

A और B एक व्यवसाय में 3 : 7 के अनुपात में निवेश करते हैं। व्यवसाय 1 वर्ष में ₹60, 000 का लाभ कमाता है। वे लाभ का 40% पुनः निवेश करने के बाद शेष लाभ को वितरित करने का निर्णय लेते हैं। A को कितना मिलेगा (₹ में)?

Solution :

Profit after reinvestment:

60,000×3/5 = 36,000

A = 36,000×3/10 = ₹10, 800

3. The profit earned by a company is to be divided among three friends who invested their money in the company, in the proportion of 6 : 9 : 11. If the share of the person who invested the most is ₹2,310, then what is the sum of the money (in ₹) received by the other two friends?

एक कंपनी द्वारा अर्जित लाभ को तीन दोस्तों के बीच 6 : 9 : 11 के अनुपात में विभाजित किया जाना है, जिन्होंने अपना पैसा कंपनी में निवेश किया था। यदि सबसे अधिक निवेश करने वाले व्यक्ति का हिस्सा (₹2, 310) है, तो इसका योग क्या है? अन्य दो दोस्तों द्वारा प्राप्त धन (₹ में)?

Solution :

Let I, II, III investments be $6x : 9x : 11x$:
Since Ratio of investment = Ratio of profit,
$11x = ₹2,310$
$x = \frac{2310}{11} = 210$
Money received by the 1st friend = $6x$ = $6 \times 210 = ₹1,260$
Money received by the 2nd friend = $9x$ = $9 \times 210 = ₹1,890$
Intended sum = $1,260 + 1,890 = ₹3,150$

4. A, B, and C together invest ₹53, 000 in a business. A invests ₹5, 000 more than B and B invests ₹6, 000 more than C. Out of a total profit of ₹31, 800, find the share of A.
A, B और C मिलकर एक व्यवसाय में (₹53, 000) का निवेश करते हैं। A, B (₹5, 000) अधिक निवेश करता है और B, C से (₹6, 000) अधिक निवेश करता है। (₹31, 800) के कुल लाभ में से, Aका हिस्सा ज्ञात कीजिए।

Solution :

Let C's investment = x ₹
B's investment = $(x + 6,000)$ ₹
and, A's investment = $(x+6,000)+5,000 = (x + 11,000)$ ₹
Hence $x + (x + 6,000) + (x + 11,000) = ₹53,000$
$3x = ₹53,000 - ₹17,000 = ₹36,000$
$x = ₹12,000$
Investment ratio:

A	:	B	:	C
23000	:	18000	:	12000
23	:	18	:	12

Share of A = $\frac{23}{53} \times ₹31,800 = ₹600 \times 23 = ₹13,800$

5. A and B invest in a business in the ratio 2 : 5. If 50% of the total profit goes to charity and A's share is ₹3.6 lakhs, what is the total profit (in ₹ lakhs)?
A और B एक व्यवसाय में 2 : 5 के अनुपात में निवेश करते हैं। यदि कुल लाभ का 50% दान में जाता है और A का हिस्सा (₹3.6 लाख) है, तो कुल लाभ (₹ लाख) में है।

Solution :

Investment Ratio: 2 : 5
Profit Ratio: $2x : 5x$
Profit after donating to charity:
$\frac{7x}{2}$ where $x = ₹3.6$ lakhs
Share of A:
$\frac{\frac{7x}{2} \times 2}{2+5} = ₹3.6$ lakhs
$x = ₹3.6$ lakhs
Hence, Total profit = $7x$
$= 7 \times ₹3.6$
$= ₹25.2$ lakhs

6. 'A' started a business with a capital of ₹54,000 and admitted 'B' and 'C' after 4 months and 6 months, respectively. At the end of the year, the profit was divided in the ratio 1 : 4 : 5. What is the difference between the capitals invested by 'B' and 'C'?
'A' ने ₹54, 000 की पूंजी के साथ एक व्यवसाय शुरू किया और क्रमश: 4 महीने और 6 महीने के बाद 'B' और 'C' को शामिल किया। वर्ष के अंत में, लाभ को 1 : 4 : 5 के अनुपात में विभाजित किया गया। B और C द्वारा निवेश की गई पूंजी के बीच का अंतर क्या है?

Solution :

Let the capital invested by B be x ₹
And the capital invested by C be y ₹
According to the question:
$54,000 \times 12 : x \times 8 : y \times 6 = 1 : 4 : 5$
$8x = 54,000 \times 12 \times 4 \implies x = \frac{54,000 \times 6}{1} = ₹324,000$
$6y = 54,000 \times 12 \times 5 \implies y = \frac{54,000 \times 10}{1} = ₹540,000$
$y - x = ₹540,000 - ₹324,000 = ₹216,000$
Hence, the difference between the capital invested by B and C is ₹216,000

7. A starts a cement trading business by investing ₹5 lakhs. After 2 months, B joins the business by investing ₹10 lakhs, and then after 4 months, C too joins them by investing ₹20 lakhs. 1 year after A started the business, they make ₹3,50,000 in profit. What is B's share of the profit (in ₹)?
A ने ₹5 लाख का निवेश करके एक सीमेंट व्यापार व्यवसाय शुरू किया। 2 महीने के बाद, B ₹10 लाख का निवेश करके व्यवसाय में शामिल हो जाता है और फिर 4 महीने के बाद C भी

₹20 लाख का निवेश करके उनके साथ जुड़ जाता है। A द्वारा व्यवसाय शुरू करने के 1 वर्ष बाद वे ₹3, 50, 000 का लाभ कमाते हैं। लाभ में B का हिस्सा (₹) क्या है?

Solution :

Ratio of shares:

$A : B : C = (5 \times 12) : (10 \times 10) : (20 \times 6) =$

$60 : 100 : 120 = 3 : 5 : 6$

Profit of B:

$\frac{5}{3+5+6} \times 3,50,000 = \frac{5}{14} \times 3,50,000 =$

₹1, 25, 000

8. A and B started a business by investing equal amounts. Four months later, C joins them by investing ₹3.5 lakhs. By withdrawing his investment in the business, B leaves the business 4 months after C joined. At the end of the year, the business makes ₹62,400 in profit, out of which A collects ₹24,000 as his share of profit. How much should be paid to C (in ₹) as his share of profit?

A और B ने समान राशि का निवेश करके एक व्यवसाय शुरू किया। चार महीने बाद, C ₹3.5 लाख का निवेश करके उनके साथ जुड़ जाता है। व्यवसाय में अपना निवेश निकालकर B, C के शामिल होने के 4 महीने बाद व्यवसाय छोड़ देता है। वर्ष के अंत में व्यवसाय से ₹62, 400 का लाभ होता है, जिसमें से A अपने लाभ के हिस्से के रूप में ₹24, 000 एकत्र करता है। C को (₹) के रूप में कितना भुगतान किया जाना चाहिए?

Solution :

Let A and B invest x lakh in equal amounts.

According to the question:

Ratio of shares of A, B, and C = $(x \times 12)$: $(x \times 8) : (3.5 \times 8) = 12x : 8x : 28 = 3x : 2x : 7$

$\frac{3x}{5x+7} = \frac{₹24,000}{₹62,400}$

$\frac{3x}{5x+7} = \frac{5}{13}$

$39x = 25x + 35$

$14x = 35$

$x = 2.5$ ₹ lakh

Hence, ratio of shares of A, B and C = 7.5 : 5.7

Ratio of shares of A, B, and C = 7.5 : 5 : 7

Profit of C:

$\frac{7}{19.5} \times$ ₹62, 400 = ₹22, 400

9. A sum (in ₹) is distributed between A, B, and C in the ratio 9 : 6 : 11. If A gives ₹500 from his share to C, the ratio of shares of A, B, and C becomes 4 : 3 : 6. What is the sum of shares (in ₹) of B and C, in the beginning?

एक राशि (₹ में) A, B और C के बीच 9 : 6 : 11 के अनुपात में वितरित की जाती है। यदि A अपने हिस्से से C को ₹500 देता है, तो A, B और C के शेयरों का अनुपात 4 : 3 : 6 हो जाता है। आरंभ में B और C के शेयरों का योग (₹) में क्या है?

Solution :

Make B the same:

9 : 6 : 11 and 8 : 6 : 12 1 = ₹500

17 = ₹8, 500

10. A, B, and C enter into a partnership with capitals in the ratio $\frac{2}{3} : \frac{3}{5} : \frac{5}{6}$. After 8 months, A increases his share of capital by 25%. If at the end of the year, the total profit earned is ₹5,820, then the share of C in the profit is:

A, B और C अनुपात $\frac{2}{3} : \frac{3}{5} : \frac{5}{6}$ में पूंजी के साथ साझेदारी में प्रवेश करते हैं। 8 महीने के बाद, A ने पूंजी में अपना हिस्सा 25% बढ़ा दिया। यदि वर्ष के अंत में अर्जित कुल लाभ ₹5, 820 है तो लाभ में C का हिस्सा है:

Solution :

Initial ratio:

$\frac{2}{3} : \frac{3}{5} : \frac{5}{6}$

$20 : 18 : 25$

$(20 \times 8 + 25 \times 4) : (18 \times 12) : (25 \times 12) =$

$65 : 54 : 75$

194 = ₹5, 820

$75 = \frac{75}{194} \times$ ₹5, 820 = ₹2, 250

11. A, B, and C are partners in a business. A, whose capital has been used for 5 months, claims $\frac{1}{7}$ of the profit. B, whose capital has been used for 7 months, claims $\frac{1}{2}$ of the profit. C has invested ₹4,600 for 9 months. How much capital did A contribute?

A, B और C एक व्यवसाय में भागीदार हैं। A, जिसकी पूंजी का उपयोग 5 महीने के लिए किया गया है, लाभ का $\frac{1}{7}$ दावा करता है। B, जिसकी पूंजी 7 महीने तक उपयोग की गई है, लाभ का $\frac{1}{2}$ दावा करता है। C ने ₹4,600 का निवेश 9 महीने के लिए किया है। A ने कितनी पूंजी का योगदान दिया?

Solution :

Let the total profit be 35:

Profit (A:B:C) = 5 : 7 : 23

$$\text{Investment} = \frac{\text{Profit}}{\text{Time}}:$$

$$\frac{5}{5} : \frac{7}{7} : \frac{23}{9} = 9 : 9 : 23$$

23 = ₹4, 600

Investment of A = 9 = ₹1, 800

12. A started a business with a capital of ₹1,12,000. After 2 months, B joined the business with a capital of ₹80,000, and after another 2 months, C joined the business with a capital of ₹72,000. After 10 months from the start of the business, B withdrew ₹8,000, and C also withdrew ₹8,000. If B received ₹9,800 as his share in the profit at the end of a year, then the total profit was:

A ने ₹1,12,000 की पूंजी के साथ एक व्यवसाय शुरू किया। 2 महीने के बाद, B ₹80,000 की पूंजी के साथ व्यवसाय में शामिल हो गया और अगले 2 महीने के बाद, C ₹72,000 की पूंजी के साथ व्यवसाय में शामिल हो गया। व्यवसाय शुरू होने के 10 महीने बाद, B ने ₹8,000 निकाल लिए और C ने भी ₹8,000 निकाल लिए। यदि B को एक वर्ष के अंत में लाभ में अपने हिस्से के रूप में ₹9,800 प्राप्त हुए, तो कुल लाभ था:

Solution :

Ratio of investments of A, B, and C:

$(112000 \times 12) : (80000 \times 8 + 72000 \times 2) : (72000 \times 6 + 64000 \times 2)$

$(112 \times 12) : (80 \times 8 + 72 \times 2) : (72 \times 6 + 64 \times 2) = 1344 : 784 : 560 = 12 : 7 : 5$

Since at the end of the year, B got a profit of ₹9,800:

$$\text{Total profit} = 9800 \times \frac{(12 + 7 + 5)}{7}$$

= ₹33, 600

13. A, B, and C are partners in a firm sharing profit in the ratio 3 : 4 : 5. If they set aside 4% of the profits as an emergency fund and shared the rest of the profit, and B gets his share of profit as ₹1,81,400, the amount of profit set aside for the emergency fund is:

A, B और C एक फर्म में साझेदार हैं और लाभ को 3 : 4 : 5 के अनुपात में बाँटते हैं। यदि वे लाभ का 4% आपातकालीन निधि के रूप में अलग रखते हैं और शेष लाभ साझा करते हैं और B को लाभ का हिस्सा ₹1,81,400 प्राप्त होता है, तो आपातकालीन निधि के लिए निर्धारित लाभ की राशि क्या है?

Solution :

Profit Ratio: 3 : 4 : 5

4 = 1, 81, 400

Total Profit = 12 = 5, 44, 200

96% = 5, 44, 200

4% = 22, 675

14. A, B, and C started a business by investing ₹27,500, ₹32,500, and ₹37,500 respectively. A is a working partner and gets 20% of the profit as a working allowance, and the remaining is distributed in proportion to their investments. If the money received by C is ₹13,500, what is the total profit?

A, B और C ने क्रमश: ₹27,500, ₹32,500 और ₹37,500 का निवेश करके एक व्यवसाय शुरू किया। A एक कामकाजी भागीदार है और उसे लाभ का 20% कामकाजी भत्ते के रूप में मिलता है और शेष को उनके निवेश के अनुपात में वितरित किया जाता है। यदि C द्वारा प्राप्त धन ₹13,500 है, तो कुल लाभ कितना है?

Solution :

27500 : 32500 : 37500

55 : 65 : 75

11 : 13 : 15

C = 15 = ₹13,500

Total 39 = ₹35,100

80% = ₹35,100

100% = ₹43,875

Total Profit = ₹43,875

15. Two businessmen A and B invest in a business in the ratio $5:8$. They decided to reinvest 30% of the profit they earned back into the business. The remaining profit they distributed amongst themselves. If A's share of the profit was ₹87,500, how much profit (in ₹) did the business make?

दो व्यवसायी A और B एक व्यवसाय में $5:8$ के अनुपात में निवेश करते हैं। उन्होंने अपने द्वारा अर्जित लाभ का 30% पुनः व्यवसाय में निवेश करने का निर्णय लिया। शेष लाभ उन्होंने आपस में बाँट लिया। यदि लाभ में A का हिस्सा ₹87,500 था, तो व्यवसाय ने कितना लाभ (₹ में) कमाया?

Solution :

Let us take the total profit = P

Since they decided to reinvest the 30% profit,

the remaining profit that is distributed = $0.7P$

A's share in the profit = ₹87,500

$\frac{5}{13}\times 0.7P = 87,500$

$\frac{7P}{26} = 87,500$

$P = 3,25,000$

The total profit is ₹3,25,000.

16. A and B entered into a partnership with investments in the ratio $3:5$. After a few months, A withdrew his investment. At the end of the year, they received profit in the ratio $2:5$. After how many months did A withdraw?

A और B ने $3:5$ के अनुपात में निवेश के साथ साझेदारी की। कुछ महीनों के बाद, A ने अपना पैसा वापस ले लिया। वर्ष के अंत में, उन्हें $2:5$ के अनुपात में लाभ प्राप्त हुआ। कितने महीनों के बाद A ने पैसा वापस ले लिया?

Solution :

$\frac{3x}{5\times 12} = \frac{2}{5}$

$3x = 24 \implies x = 8$ months

After 8 months, A withdrew his amount.

17. A invests ₹6,00,000 more than B in a business. B invests his capital for $7\frac{1}{2}$ months, while A invests his capital for $2\frac{1}{2}$ months more than B. Out of the total profit, the share of B is ₹2,48,000 less than the share of A. If the total profit is ₹12,40,000, what is the share of A?

A व्यवसाय में B से ₹6,00,000 अधिक निवेश करता है। B अपनी पूंजी $7\frac{1}{2}$ महीने के लिए निवेश करता है, जबकि A अपनी पूंजी B से $2\frac{1}{2}$ महीने अधिक के लिए निवेश करता है। यदि कुल लाभ ₹12,40,000 में से B का हिस्सा A के हिस्से से ₹2,48,000 कम है, तो A का हिस्सा क्या है?

Solution :

Total profit = ₹12,40,000

Total profit:

A + B = ₹12,40,000

A - B = ₹2,48,000

A = ₹7,44,000 and B = ₹4,96,000

Profit ratio: A : B = 93 : 62

Time ratio: $A:B = 10:\frac{15}{2} = 4:3$

Investment ratio: A : B = 9 : 8

1 = ₹6,00,000

8 = ₹48,00,000

18. A, B, and C started a business with their capitals in the ratio $2:3:5$. A increased his capital by 50% after 4 months, B increased his capital by $33\frac{1}{3}\%$ after 6 months, and C withdrew 50% of his capital after 8 months from the start of the business. If the total profit at the end of a year was ₹86,800, then the difference between the shares of A and C in the profit was:

A, B और C ने $2:3:5$ के अनुपात में अपनी पूंजी के साथ एक व्यवसाय शुरू किया। A ने 4 महीने के बाद अपनी पूंजी में 50% की वृद्धि की, B ने 6 महीने के बाद अपनी पूंजी में $33\frac{1}{3}\%$ की वृद्धि की और C ने अपनी पूंजी का 50% वापस ले लिया। व्यवसाय शुरू होने से 8 महीने। यदि एक वर्ष के अंत में कुल लाभ ₹86,800 था, तो लाभ में A और C के शेयरों के बीच का अंतर था:

Solution :

$2\times 4+3\times 8 : 3\times 6+4\times 6 : 5\times 8+5/2\times 4$

$8+24 : 18+24 : 40+10$

32 : 42 : 50

16 : 21 : 25

62 = ₹86,800

1 = ₹1,400

9 = ₹12,600

19. A, B, and C started a business with their capitals in the ratio 1 : 4 : 4. At the end of every 3 months, A doubles his capital, B halves his capital, and C leaves his capital unchanged. At the end of the year, if B's share in the profit was ₹4,50,000, then the total profit (in lakhs) was:

A, B और C ने 1 : 4 : 4 के अनुपात में अपनी पूंजी के साथ एक व्यवसाय शुरू किया। प्रत्येक 3 माह के अंत में. A अपनी पूंजी को दोगुना कर देता है, B अपनी पूंजी को आधा कर देता है और C अपनी पूंजी को अपरिवर्तित छोड़ देता है। वर्ष के अंत में, यदि लाभ में B का हिस्सा ₹4,50,000 था, तो कुल लाभ (लाख में) कितना था:

Solution :

$1 \times 3 + 2 \times 3 + 4 \times 3 + 8 \times 3 : 4 \times 3 + 2 \times 3 + 1 \times 3 + \frac{1}{2} \times 3 : 4 \times 12$

$45 : \frac{45}{2} : 48$

90 : 45 : 96

30 : 15 : 32

15x = ₹4,50,000

x = ₹30,000

Total profit = 30x + 15x + 32x = 77x

= 77 × ₹30,000 = ₹23,10,000 = 23.1 lakhs

20. A, B, and C started a business with their capitals in the ratio 4 : 2 : 9. At the end of every quarter, A halves his capital, B doubles his capital, and C leaves his capital unchanged. If at the end of a year, A's profit was ₹24,000, then what is the total profit (in ₹)?

A, B और C ने 4 : 2 : 9 के अनुपात में अपनी पूंजी के साथ एक व्यवसाय शुरू किया। प्रत्येक तिमाही के अंत में, A अपनी पूंजी को आधा कर देता है जबकि B अपनी पूंजी को दोगुना कर देता है और C अपनी पूंजी को अपरिवर्तित छोड़ देता है। यदि एक वर्ष के अंत में A का लाभ ₹24,000 था, तो कुल लाभ (₹ में) क्या है?

Solution :

Capital of A, B and C = 4 : 2 : 9

Total capital of A after end of the year = $4 \times 3 + \frac{4}{2} \times 3 + \frac{4}{4} \times 3 + \frac{4}{8} \times 3$

$A = \frac{45}{2}$

Total capital of B after end of the year = $2 \times 3 + 4 \times 3 + 8 \times 3 + 16 \times 3 = 6 + 12 + 24 + 48$

B = 90

Total capital of C after end of the year

= 9 × 12 = 108

Profit ratio of A, B and C = $\frac{45}{2} : 90 : 108$

= 5 : 20 : 24

5 unit = ₹24,000

1 unit = $\frac{24000}{5} = 4,800$

(5 + 20 + 24) unit = 49 unit = ₹4, 800×49

= 2, 35, 200

Total profit = ₹2,35,200

21. A, B, and C are partners in a firm sharing profit in the ratio 3 : 4 : 5. If they set aside 4% of the profits as an emergency fund and shared the rest of the profit, and B gets his share of profit as ₹1,81,400, the amount of profit set aside for the emergency fund is:

A, B और C एक फर्म में साझेदार हैं और लाभ को 3 : 4 : 5 के अनुपात में बाँटते हैं। यदि वे लाभ का 4% आपातकालीन निधि के रूप में अलग रखते हैं और शेष लाभ साझा करते हैं और B को लाभ का हिस्सा ₹1,81,400 प्राप्त होता है, तो आपातकालीन निधि के लिए निर्धारित लाभ की राशि क्या है?

Solution :

Let the total profit be P.

Profit remaining after emergency fund = 96% of P.

Total share = 3 + 4 + 5 = 12

Share of $B = \frac{4}{12} \times \frac{96}{100} \times P =$ ₹1, 81, 400

$P =$ ₹5, 66, 875

Emergency fund = $\frac{4}{100} \times 5,66,875 =$ ₹22, 675

22. A, B, and C invest in a business in the ratio 3 : 6 : 5. A and C are working partners. Only B is a sleeping partner; hence his share will be $\frac{3}{4}$ of what it would have been if he were a working partner. If they make ₹50,000 profit, half of which is reinvested in the business and the other half is distributed between the partners, then how much does C get (in ₹)?

A, B और C एक व्यवसाय में 3 : 6 : 5 के अनुपात में निवेश करते हैं। A और C कामकाजी भागीदार हैं। केवल B एक स्लीपिंग पार्टनर है इसलिए उसका हिस्सा उस राशि का $\frac{3}{4}$ होगा जो अगर वह एक कामकाजी पार्टनर होता तो होता। यदि वे ₹50,000 का लाभ कमाते हैं, जिसका आधा हिस्सा व्यवसाय में पुनः निवेश किया जाता है और दूसरा आधा भागीदारों के बीच वितरित किया जाता है, तो C को कितना मिलता है (₹ में)?

Solution :

Let the profit received by A, B, and C be $3x$, $6x$, and $5x$ if all were working partners.

Since B is a sleeping partner, his share $= \frac{3}{4} \times 6x = \frac{9x}{2}$.

Half of the profit $= 25,000$ is divided among A, B, and C.

$3x + \frac{9x}{2} + 5x = 25,000 \implies x = 2,000$

C's share from the profit = $5x = 5 \times 2,000$ = ₹10, 000.

23. A, B, and C invested ₹40,000, ₹48,000, and ₹80,000 respectively for a business at the start of a year. After six months, for the remaining time of the year, A added ₹4,000, B added ₹4,000, while C withdrew ₹4,000 every month. If the total profit is ₹6,72,000, then what is C's share (in ₹)?

A, B, और C ने एक वर्ष की शुरुआत में एक व्यवसाय के लिए क्रमशः ₹40,000, ₹48,000, और ₹80,000 का निवेश किया। छह महीने के बाद, वर्ष के शेष समय के लिए, A ने ₹4,000 जोड़े, B ने ₹4,000 जोड़े जबकि C ने हर महीने ₹4,000 निकाले। यदि कुल लाभ ₹6,72,000 है, तो C का हिस्सा (₹ में) क्या है?

Solution :

A, B, and C invested ₹40000, ₹48000, and ₹80000, respectively

Total profit = ₹ 672000

After six months, for the remaining time of the year, A added ₹4000, B added ₹4000 and C withdrew ₹4000 every month

So, (4000 × 6) + (4000 × 5) + (4000 × 4) + (4000 × 3) + (4000 × 2) + (4000 × 1)

= 4000 × 21 = 84000

A : B : C = [(40, 000 × 12) + 84, 000] : [(48, 000×12)+84, 000] : [(80, 000×12) − 84, 000]

= (4, 80, 000 + 84, 000) : (5, 76, 000 + 84, 000) : (9, 60, 000 − 84, 000)

= 5, 64, 000 : 6, 60, 000 : 8, 76, 000

= 564 : 660 : 876

The share of C = $\left(\frac{876}{564 + 660 + 876}\right) \times$ 6, 72, 000 = $\left(\frac{876}{2,100}\right) \times$ 6, 72, 000 = ₹2, 80, 320

C's share = ₹2,80,320.

24. A, B, and C entered into a partnership with investments of ₹24,000, ₹32,000, and ₹48,000 respectively. After one year, A withdrew ₹4,000, and B invested ₹8,000 more. After one more year, A invested ₹15,000 more, and C withdrew ₹8,000. At the end of three years, they earned a profit of ₹1,14,450. Find the share of A in the total profit.

A, B और C ने ₹24,000, ₹32,000 और ₹48,000 के निवेश के साथ साझेदारी की। एक वर्ष के बाद, A ने ₹4,000 निकाल लिए और B ने ₹8,000 अधिक निवेश किया। एक और वर्ष के बाद, A ने ₹15,000 अधिक निवेश किया और C ने ₹8,000 निकाल लिए। तीन साल के अंत में, उन्होंने ₹1,14,450 का लाभ कमाया। कुल लाभ में A का हिस्सा ज्ञात करें।

Solution :

$24k + 20k + 35k$: $32k + 40k + 40k$: $48k + 48k + 40k$

79 : 112 : 136

327 = ₹1,14,450

79 = ₹27,650

A's Profit = ₹27,650

25. Veer and Bhavya started a business by investing ₹45,000 and ₹50,000 respectively. At the end of the year, they decided to divide 50% of the total profit equally and the rest in the investment ratio. If they had divided the entire profit in the investment ratio, Bhavya got ₹1,500 more profit than that of actual. Find the total profit.

वीर और भव्या ने ₹45,000 और ₹50,000 का निवेश करके एक व्यवसाय शुरू किया। वर्ष के अंत में, उन्होंने कुल लाभ का 50% समान रूप से और शेष को निवेश अनुपात में विभाजित करने का निर्णय लिया। यदि उन्होंने पूरे लाभ को निवेश अनुपात में विभाजित किया होता, तो भव्या को वास्तविक से ₹1,500 अधिक लाभ होता। कुल लाभ ज्ञात करें।

Solution :

Veer : Bhavya = 9 : 10 = 36 : 40
1st Condition
18 + 19 : 19 + 20
37 : 39
2nd Condition
36 : 40
Bhavya Got ₹1,500 More
1 = 1, 500
76 = 1, 14, 000
Total Profit = ₹1,14,000

6 Simple Interest and Compound Interest

Simple Interest
साधारण ब्याज

- Principal - Borrowed money is called Principal and it is denoted by 'P'.
  मूलधन - उधार लिया गया धन मूलधन कहलाता है और इसे 'P' से दर्शाया जाता है।

- Time - Money is borrowed for certain time period, that time is called interest time and it is denoted by 'T' or 't'.
  समय - पैसा एक निश्चित समय अवधि के लिए उधार लिया जाता है, उस समय को ब्याज समय कहा जाता है और इसे 'T' या 't' द्वारा दर्शाया जाता है।

- Amount - The principal becomes Amount when interest is added to it Amount is represented as A.
  राशि - मूलधन राशि बन जाती है जब इसमें ब्याज जोड़ा जाता है राशि को A के रूप में दर्शाया जाता है।

- Interset - It is the amount to be paid on the borrowed money or the amount received on the money lent.
  इंटरसेट - यह उधार लिए गए पैसे पर भुगतान की जाने वाली राशि या उधार दिए गए पैसे पर प्राप्त राशि है

Formula to Calculate Simple Interest/ साधारण ब्याज की गणना करने का सूत्र

$$SI = \frac{PRT}{100}$$

Formula To Calculate Amount/ राशि की गणना करने का सूत्र

Amount "A" = P + SI

Compound Interest - The interest on a loan or deposit calculated based on both the intial principal and the accumulated interest from previous periods and it is denoted by 'CI'.

चक्रवृद्धि ब्याज - किसी ऋण या जमा पर ब्याज की गणना प्रारंभिक मूलधन और पिछली अवधि के संचित ब्याज दोनों के आधार पर की जाती है और इसे '(CI)' द्वारा दर्शाया जाता है।

Amount = $P\left(1 + \frac{r}{100}\right)^n$

(where n = Time & r = rate of interest)/(जहाँ n = समय & r = ब्याज दर)

Interest Rates to Remember in CI/ CI में याद रखने योग्य ब्याज दरें

Compound Interest

1st %	2nd %	3rd %	4th %
2	4.04	-	-
3	6.09	-	-
4	8.16	-	-
5	10.25	-	-
6	12.36	-	-
8	16.64	25.9712	-
10	21	33.1	46.41
12	25.44	40.49 28	-
15	32.25	52.0875	-
20	44	72.8	-

Type 1 (Problems based on Simple Interest)

1. What is the simple interest on a sum of ₹ 99000 at $12\frac{1}{2}\%$ per annum for a period of 9 months?

₹ 99000 की राशि पर 9 महीने की अवधि के लिए $12\frac{1}{2}\%$ प्रति वर्ष की दर से साधारण ब्याज कितना है?

Solution :

$$SI = \frac{P \times R \times T}{100}$$

$$SI = \frac{99000 \times 25 \times 9}{2 \times 12 \times 100}$$

$$SI = 9281.25$$

2. The difference between the simple interest on a fixed amount for 7 years and on the same amount for 12 years is ₹ 2,500. If the rate of interest is 10% p.a., then the amount is:

7 वर्षों के लिए एक निश्चित राशि पर और 12 वर्षों के लिए समान राशि पर साधारण ब्याज के बीच का अंतर ₹ 2,500 है। यदि ब्याज दर 10

Solution :

According to the question,

$$\Delta SI = \frac{P \times R \times (t_2 - t_1)}{100}$$

Δ = Difference

$$\frac{P \times 10 \times (12 - 7)}{100} = 2500$$

$$\frac{P \times 50}{100} = 2500$$

$$P = 5000$$

3. A sum of 25600 is invested on simple interest partly at 7% per annum and the remaining at 9% per annum. The total interest at the end of 3 years is 5832. How much money (in ₹) was invested at 9% per annum?

₹ 25600 की राशि को आंशिक रूप से 7% प्रति वर्ष की दर से और शेष को 9% प्रति वर्ष की दर से साधारण ब्याज पर निवेश किया जाता है। 3 वर्ष के अंत में कुल ब्याज ₹ 5832 है। प्रति वर्ष 9% पर कितना पैसा निवेश किया गया था?

Solution :

Let x was invested at 9% per annum.

$$SI = \frac{P \times R \times T}{100}$$

$$\frac{(25600 - x) \times 7 \times 3}{100} + \frac{x \times 9 \times 3}{100} = 5832$$

$$179200 - 7x + 9x = 194400$$

$$2x = 15200$$

$$x = 7600$$

4. If the simple interest for 9 years be equal to 45% of the principal, then the rate of interest per annum is equal to:

यदि 9 वर्षों के लिए साधारण ब्याज मूलधन के 45% के बराबर हो, तो प्रति वर्ष ब्याज की दर बराबर होगी:

Solution :

Let principal be ₹ x.

$$SI = \frac{P \times R \times T}{100}$$

$$\frac{45}{100} \times x = \frac{x \times R \times 9}{100}$$

$$R = \frac{45}{9} = 5\%$$

5. A person invested one-fourth of the sum of ₹ 25,000 at a certain rate of simple interest and the rest at 4% p.a. higher rate. If the total interest received for 2 years is ₹ 4,125, what is the rate at which the second sum was invested?

एक व्यक्ति ने ₹ 25,000 की राशि का एक-चौथाई साधारण ब्याज की एक निश्चित दर पर और शेष 4% प्रति वर्ष की दर पर निवेश किया। यदि 2 वर्षों के लिए प्राप्त कुल ब्याज ₹ 4,125 है, तो वह दर क्या है जिस पर दूसरी राशि का निवेश किया गया था?

Solution :

$$\frac{6250 \times r \times 2}{100} + \frac{18750 \times (r + 4) \times 2}{100}$$

$$= 4125$$

$$125r + 375(r + 4) = 4125$$

$$r + 3(r + 4) = 33$$

$$4r = 33 - 12 = 21$$

$$r = 5.25\%$$

Higher rate: $r + 4 = 9.25\%$

6. A sum of Rs. 36,000 is divided into two parts A and B such that the simple interest at the rate of 15% p.a. on A and B after two years and four years respectively is equal. The total interest (Rs. in) received from A is:

₹ 36,000 की राशि को दो भागों A और B में इस प्रकार बांटा गया है कि 15% प्रति वर्ष की दर से साधारण ब्याज प्राप्त हो। A और B पर क्रमशः दो साल और चार साल के बाद बराबर है। A से प्राप्त कुल ब्याज (₹ में) है:

Solution :

Let amount of A be x Rs.

$\therefore$ Amount of B = $(36000 - x)$ Rs.

According to the question,

$$S.I. = \frac{\text{Principal} \times \text{Rate} \times \text{Time}}{100}$$

$$\frac{x \times 15 \times 2}{100} = \frac{(36000 - x) \times 15 \times 4}{100}$$

$3x = 72000$

$x = 24000$ Rs.

Hence, Total interest received by A: $\frac{24000 \times 15 \times 2}{100} = 7200$ Rs.

7. If the annual rate of simple interest increases from 11% to 17% then a person's yearly income increases by ₹ 1,071.20. The simple interest (in ₹) on the same sum at 10% for 5 years is:

यदि साधारण ब्याज की वार्षिक दर 11% से बढ़कर 17% हो जाती है तो व्यक्ति की वार्षिक आय ₹ 1,071.20 बढ़ जाती है। उसी राशि पर 5 वर्षों के लिए 10% की दर से साधारण ब्याज (₹ में) है:

Solution :

Annual rate of S.I. increased from 11% to $17\frac{1}{2}\%$.

Rate increased by $= 6\frac{1}{2}\% = \frac{13}{2}\%$

$\frac{13}{2}\% = 1071.20$

$1\% = \frac{1071.20 \times 2}{13} = 164.8$

Simple interest on the same sum at 10% for 5 years:

$50\% = 164.8 \times 50$

$= 8240$

8. Sudeep invested $\frac{1}{8}$ of a certain sum at 5% p.a. for two years and $\frac{3}{5}$ of the sum at 6% p.a. for two years and the remaining at 10% p.a. for two years. If the total interest received is ₹ 1,674, then the total sum invested is:

सुदीप ने एक निश्चित राशि का $\frac{1}{8}$ भाग 5% प्रति वर्ष की दर से निवेश किया। दो वर्ष के लिए और राशि का $\frac{3}{5}$ भाग 6% प्रति वर्ष की दर से। दो वर्षों के लिए और शेष पर 10% वार्षिक दर से। दो साल के लिए। यदि प्राप्त कुल ब्याज ₹ 1,674 है, तो निवेश की गई कुल राशि है:

Solution :

Let total amount invested = $40x$.

$$\frac{5x \times 5 \times 2}{100} + \frac{24x \times 6 \times 2}{100} + \frac{11x \times 10 \times 2}{100} = 1674$$

$25x + 144x + 110x = 83700$

$279x = 83700$

$x = 300$

$\therefore$ Total amount = ₹ 12000

9. An amount of ₹ 12,000 was borrowed at simple interest, at an interest rate. After four months, ₹ 6,000 more were added to it, and the rate of interest on the total principal was doubled against the previous rate. ₹ 2,800 was paid as interest at the end of the year. Calculate the rate of interest initially charged?

₹ 12,000 की राशि साधारण ब्याज पर, ब्याज दर पर उधार ली गई थी। चार महीने के बाद, इसमें ₹ 6,000 और जोड़ दिए गए और कुल मूलधन पर ब्याज की दर पिछली दर के मुकाबले दोगुनी हो गई। अंत में ब्याज के रूप में ₹ 2,800 का भुगतान किया गया। वर्ष, प्रारंभ में लगाए गए ब्याज की दर की गणना करें:

Solution :

Let an amount of Rs. 12000 was borrowed at simple interest, at an interest rate on r%.

After four months the amount becomes (12000+6000) = 18000

According to the question,

The rate of interest on the total principal amount on 18000 was doubled = 2r%

Interest for four months $= 12000 \times r \times \frac{1}{3} \times \frac{1}{100}$

$\because$2800 was paid as interest at the end of the year.

So, $12000 \times r \times \frac{1}{3} \times \frac{1}{100} + 18000 \times 2r \times \frac{2}{3} \times \frac{1}{100} = 2800$

$280r = 2800$

$r = 10\%$

10. Suresh borrowed out a sum of money to Rakesh for 5 years at simple interest. At the end of 5 years, Rakesh paid $9/8$ of the sum of Suresh to clear out the amount. Find the rate of simple interest per annum.

सुरेश ने राकेश को 5 वर्ष के लिए साधारण ब्याज पर एक धनराशि उधार दी। 5 वर्ष के अंत में, राकेश ने राशि चुकाने के लिए सुरेश की राशि का $9/8$ भाग का भुगतान किया। प्रति वर्ष साधारण ब्याज की दर ज्ञात कीजिए:

Solution :

Let the principal amount x given to Rakesh by Suresh.

Rakesh was returned money to Suresh by the end of the year = $\frac{9}{8}x$

$\therefore$ Interest = $\frac{9}{8} \times x - x = \frac{x}{8}$

$$\text{S.I} = \frac{\text{Principal amount} \times \text{rate} \times \text{times}}{100}$$

$$\frac{x}{8} = \frac{x \times R \times 5}{100}$$

Or

$$R = \frac{x \times 100}{8 \times 5 \times x} = \frac{100}{40} = 2.5$$

So the rate of simple interest is 2.5

11. A sum of money amounts to ₹ 7,500 in 5 years, and to ₹ 8,500 in 7 years at simple interest at the same rate of interest. The rate of interest per annum is:

समान ब्याज दर पर साधारण ब्याज पर एक धनराशि 5 वर्षों में ₹ 7,500 और 7 वर्षों में ₹ 8,500 हो जाती है। प्रति वर्ष ब्याज दर है:

Solution :

5yrs (amt) $\sim$ 7500

7yrs (amt) $\sim$ 8500

2yrs (int) $\sim$ 1000

5yrs (int) $\sim$ 2500

P = 7500 - 2500 = 5000

$$1000 = \frac{5000 \times r \times 2}{100}$$

$r = 10\%$

12. A certain sum doubles itself on simple interest in 10 years. At the same rate of interest in how many years it will become 3 times of itself?

एक निश्चित राशि साधारण ब्याज पर 10 वर्षों में दोगुनी हो जाती है। समान ब्याज दर पर कितने वर्षों में यह स्वयं का 3 गुना हो जाएगा?

Solution :

2 times:

$$1 = \frac{1 \times r \times 10}{100}$$

$r = 10\%$

3 times:

$$2 = \frac{1 \times 10 \times t}{100}$$

$t = 20$ yrs

13. A sum of ₹ 8,400 amounts to ₹ 11,046 at 8.75 ₹ 8,400 की राशि 8.75

Solution :

$r_1 = 8.75\%,\ P_1 = 8400$ Rs.

$SI_1 = 11046 - 8400 = 2646$ Rs.

$$SI = \frac{P \times r \times t}{100}$$

$$2646 = \frac{8400 \times 8.75 \times t_1}{100}$$

$t_1 = 3.6$

$SI_2 = ?,\ r_2 = 8.75\%,\ t_2 = 3.6,\ P_2 = 9600$

$$SI_2 = \frac{P_2 \times r_2 \times t_2}{100} = \frac{9600 \times 8.75 \times 3.6}{100}$$

$SI_2 = 3024$

14. A sum of amounts to ₹ 14,395.20 at 9.25% p.a. simple interest in 5.4 years. What will be the simple interest on the same sum at 8.6 9.25% प्रति वर्ष की दर से कुल राशि ₹ 14,395.20 होती है। 5.4 वर्ष में साधारण ब्याज। उसी राशि पर 8.6% प्रति वर्ष की दर से साधारण ब्याज कितना होगा? 4.5 साल में?

Solution :

$\because$ Principal amount = $\dfrac{\text{Final amount}}{1+\dfrac{R\times T}{100}}$

Principal amount $=\dfrac{14395.20\times 100}{100+9.25\times 5.4}$

$=\dfrac{1439520}{100+49.95}=\dfrac{1439520}{149.95}=9600$ Rs.

$\therefore$ S.I. = $\dfrac{\text{Principal amount}\times\text{Rate}\times\text{Time}}{100}$

$=\dfrac{9600\times 8.6\times 4.5}{100}=3715.20$ Rs.

15. Two equal sums were lent on simple interest at 6% and 10% per annum respectively. The first sum was recovered two years later than the second sum and the amount in each case was ₹ 1105. What was the sum (in ₹) lent in each scheme?

दो समान राशियाँ क्रमशः 6% और 10% प्रति वर्ष साधारण ब्याज पर उधार दी गईं। पहली राशि दूसरी राशि की तुलना में दो साल बाद वसूल की गई और प्रत्येक मामले में राशि ₹ 1105 थी। प्रत्येक योजना में उधार दी गई राशि (₹ में) क्या थी?

Solution :

Let, Time = t years, Rate = $R\%$, Principal = P

$(t+2)\times 6\%=t\times 10\%$

$t\times 6\%+12\%=t\times 10\%$

$4\%\times t=12\%$

$t=3$ years

$\therefore$ Amount = $P+SI$

$1105=P+\dfrac{P\times R\times t}{100}$

$1105=P+\dfrac{P\times 6\times 5}{100}$

$1105=P+\dfrac{3P}{10}$

$1105=\dfrac{13P}{10}$

$P=\dfrac{11050}{13}=850$ Rs.

16. A sum of x amounts to ₹ 9246 in 4 years and to ₹ 11,298.75 in $7\frac{1}{2}$ years, at $y\%$ p.a. simple interest. The values of x and y are, respectively:

$y\%$ साधारण ब्याज प्रति वर्ष की दर से x की एक राशि 4 वर्षों में ₹ 9246 और $7\frac{1}{2}$ वर्षों में ₹ 11,298.75 हो जाती है। x और y के मान क्रमशः हैं:

Solution :

Let, Principal is x.

According to the question,

Sum of an amount to ₹ 9246 in 4 years and to ₹ 11,298.75 in 7.5 years.

3.5 years SI = 11298.75 − 9246 = 2052.75

Simple Interest for 1 year $=\dfrac{2052.75}{3.5}=586.5$

Simple Interest for 4 years $=4\times 586.5=2346$

Principal = 9246 − 2346 = 6900

$R\%=\dfrac{586.5}{6900}\times 100=8.5\%$

Hence, Principal and rate are ₹ 6900 and 8.5%, respectively.

17. A sum of ₹ 5,000 is divided into two parts such that the simple interest on the first part for $4\frac{1}{5}$ years at $6\frac{2}{3}\%$ p.a. is double the simple interest on the second part for $2\frac{2}{4}$ years at 4% p.a. What is the difference between the two parts?

₹ 5,000 की एक राशि को दो भागों में इस प्रकार विभाजित किया जाता है कि पहले भाग पर $4\frac{1}{5}$ वर्षों के लिए $6\frac{2}{3}\%$ प्रति वर्ष की दर से साधारण ब्याज प्राप्त हो। $2\frac{2}{4}$ वर्षों के लिए 4% प्रति वर्ष की दर से दूसरे भाग पर साधारण ब्याज का दोगुना है। दोनों भागों में क्या अंतर है?

Solution :

Let the first part = x Rs.

Second part = $(5000-x)$ Rs.

$\dfrac{x\times 20\times 21}{100\times 3\times 5}=\dfrac{(5000-x)\times 4\times 11}{4\times 100}$

$14x = 55000 - 11x$

$25x = 55000$

$x = 2200$

Difference between both parts: $5000 - 2x = 5000 - 4400 = 600$

18. Two equal sums are lent at 8% and 4% simple interest p.a, respectively, at the same time. The first sum is received 2 years earlier than the other, and the amount received in each case is ₹ 14,500. Each sum is:

दो समान राशियाँ क्रमश: 8% और 4% साधारण ब्याज प्रति वर्ष पर एक ही समय में उधार दी जाती हैं। पहली राशि दूसरे की तुलना में 2 वर्ष पहले प्राप्त होती है और प्रत्येक मामले में प्राप्त राशि ₹ 14,500 है। प्रत्येक योग है:

Solution :

Let two equal amounts = x.

Interest amount of both sums will be equal.

In case of 8% and 4%, amounts are borrowed for $t-2$ and t years respectively.

$$\frac{x \times 8 \times (t-2)}{100} = \frac{x \times 4 \times t}{100}$$

For $t - 2 = 2$,

$$A = P\left(1 + \frac{R \times T}{100}\right)$$

$$14500 = x\left(1 + \frac{8 \times 2}{100}\right)$$

$$x = \frac{14500 \times 25}{29} = 12500$$

19. A borrowed ₹ 3,000 from his friend B on 31 December, 2007 on the condition that he would return it after one year with a simple annual interest of 15%. But A is in a position to return the money only on 31 August, 2008. How much money does A have to return to B?

A ने 31 दिसंबर, 2007 को अपने मित्र B से ₹ 3,000 इस शर्त पर उधार लिया कि वह इसे एक वर्ष के बाद 15% के साधारण वार्षिक ब्याज के साथ वापस कर देगा। लेकिन A केवल 31 अगस्त, 2008 को पैसे लौटाने की स्थिति में है। A को B को कितना पैसा लौटाना है?

Solution :

Principal amount = ₹ 3000, Rate = 15%

Time = 8 months (31 December to 31 August)

Solution :

$$S.I = \frac{P \times R \times T}{100} = \frac{3000 \times 15 \times 8}{100 \times 12} = 300$$

Amount returned to B by A = 300 + 3000 = 3300

20. A man takes a loan of some amount at some rate of simple interest. After three years, the loan amount is doubled and the rate of interest is decreased by 2%. After 5 years, if the total interest paid on the whole is ₹ 13,600, which is equal to the same when the first amount was taken for $11\frac{1}{3}$ years, then the loan taken initially is:

एक व्यक्ति साधारण ब्याज की कुछ दर पर कुछ राशि का ऋण लेता है। तीन वर्षों के बाद, ऋण राशि दोगुनी हो जाती है और ब्याज दर 2% कम हो जाती है। 5 वर्षों के बाद, यदि कुल भुगतान किया गया ब्याज ₹ 13,600 है, जो कि $11\frac{1}{3}$ वर्षों के लिए पहली राशि लेने के बराबर है, तो शुरू में लिया गया ऋण है:

Solution :

Let amount be P and rate be $r\%$.

According to the question, $\frac{P \times r \times 3}{100} + \frac{2P \times (r-2) \times 5}{100} = 13600 \quad . . . (i)$

$$\frac{P \times r \times 11\frac{1}{3}}{100} = 13600 \quad . . . (ii)$$

$$\frac{P \times r \times 34}{3 \times 100} = 13600$$

$$P \times r = 120000$$

Substitute $Pr = 120000$ in equation (i):

$$\frac{3Pr}{100} + \frac{10Pr - 20P}{100} = 13600$$

$$\frac{13Pr - 20P}{100} = 13600$$

$$20P = 200000$$

$\therefore$ The initial loan amount is ₹ 10,000.

21. If the interest on an amount of ₹ 28,000 in three years is greater than the interest on ₹ 27,000 in three years by ₹ 225, then what is the interest on ₹ 35,500 in $2\frac{3}{5}$ years?

यदि तीन वर्षों में ₹ 28,000 की राशि पर ब्याज, तीन वर्षों में ₹ 27,000 की राशि से ₹ 225 अधिक है, तो $2\frac{3}{5}$ वर्षों में ₹ 35,500 पर ब्याज कितना होगा?

Solution :

$\because SI = \frac{P \times R \times T}{100}$

According to the question:

$225 = \frac{28000 \times R \times 3}{100} - \frac{27000 \times R \times 3}{100}$

$225 = \frac{3R}{100} \times (28000 - 27000)$

$225 = \frac{3R}{100} \times 1000$

$75 = 10R$

$R = 7.5\%$

Again,

$\text{SI} = \frac{P \times R \times T}{100} = \frac{35500 \times 7.5 \times 13}{100 \times 5}$

$\text{SI} = 71 \times 7.5 \times 13$

$\text{SI} = 6922.50$

22. A sum of ₹ 7500 amounts to ₹ 11,000 in $5\frac{1}{3}$ years at a certain rate percent per annum simple interest. The sum will amount to ₹ x in $2\frac{2}{3}$ years at the same rate of simple interest. The value of x is:

₹ 7500 की राशि $5\frac{1}{3}$ वर्ष में एक निश्चित दर प्रतिशत प्रति वर्ष साधारण ब्याज पर ₹ 11,000 हो जाती है। साधारण ब्याज की समान दर पर $2\frac{2}{3}$ वर्षों में राशि x होगी। x का मान है:

Solution :

Principal = ₹ 7500, Total Amount = ₹ 11,000

Time = $5\frac{1}{3}$ years

Interest = Amount − Principal = 11000 − 7500 = 3500

$SI = \frac{P \times R \times T}{100}$

$3500 = \frac{7500 \times R \times \frac{16}{3}}{100}$

$R = \frac{35}{4}\%$

According to the question: $SI = \frac{P \times R \times T}{100} = \frac{7500 \times \frac{35}{4} \times \frac{8}{3}}{100}$

Total Amount (x) = Interest + Principal = 1750 + 7500 = 9250

23. Ankit borrowed Rs. 6000 from bank and returned Rs.2000 after 6 years. After 10 years from starting he returned Rs. 5300 and settle his account. What was the rate of interest?

अंकित ₹ 6000 के साधारण ब्याज पर बैंक से उधार लेता है। 6 साल बाद ₹ 2000 रू चूका देता है। शुरुआत में 10 साल बाद ₹ 5300 रु देकर कर्ज से मुक्त हो जाता है। ब्याज दर ज्ञात करें?

Solution :

6000×R% × 6 +4000×R% × 4 = 1300 Rs→ SI

52000×R% = 1300

$40R\% = \frac{1}{40} = 2.5\%$

24. Pranav borrows a sum of ₹ 335180 at the rate of 13% per annum simple interest. At the end of the first year, he repays ₹ 23060 towards return of principal amount borrowed. If Pranav clears all pending dues at the end of the second year, including interest payment that accrued during the first year, how much does he pay (in ₹) at the end of the second year?

प्रणव ने 13% प्रति वर्ष साधारण ब्याज की दर से ₹ 335180 की राशि उधार ली। पहले वर्ष के अंत में, वह उधार ली गई मूल राशि की वापसी के लिए ₹ 23060 चुकाता है। यदि प्रणव दूसरे वर्ष के अंत में सभी लंबित बकाया राशि का भुगतान करता है, जिसमें पहले वर्ष के दौरान अर्जित ब्याज भुगतान भी शामिल है, तो वह दूसरे वर्ष के अंत में कितना भुगतान (₹ में) करेगा?

Solution :

1st Year Interest = 335180×13%

Principal at end of 1st Year = (335180-23060) = ₹ 312120

2nd Year Interest = ₹ 312120×13%

Solution :

Amount pay at the end of 2nd Year = P at 2nd year + 1st Year Int + 2nd Year Int

= 312120 + (335180+312120)×13%

= 312120 + 647300×13%

= 312120 + 84149

= ₹ 396269

25. If the simple interest on a certain sum for $2\frac{2}{3}$ years at 15% p.a. is Rs. 514.80 less than the simple interest on the same sum for $4\frac{1}{4}$ years at 12% p.a. then the sum is:

यदि किसी निश्चित राशि पर $2\frac{2}{3}$ वर्षों के लिए 15% प्रति वर्ष की दर से साधारण ब्याज रुपये है। उसी राशि पर $4\frac{1}{4}$ वर्षों के लिए 12% प्रति वर्ष की दर से साधारण ब्याज से ₹ 514.80 कम। तो योग है:

Solution :

Let principal be Rs. P

According to the question:

$$\frac{P \times \frac{17}{4} \times 12}{100} - \frac{P \times \frac{8}{3} \times 15}{100} = 514.80$$

$51P - 100 = 51480$

$11P = 51480$

$P = 4680$

26. A man invested an amount of Rs.105750 at simple interest in the name of his son, daughter and his wife in such a way that they get the same interest after 3, 4 and 5 years respectively. If the rate of interest is 5% per annum, then the amount invested for the wife is:

एक आदमी ने अपने बेटे, बेटी और अपनी पत्नी के नाम पर साधारण ब्याज पर 105750 रुपये की राशि इस तरह निवेश की कि उन्हें क्रमश: 3, 4 और 5 साल बाद समान ब्याज मिले। यदि ब्याज दर 5% प्रति वर्ष है, तो पत्नी के लिए निवेश की गई राशि है:

Solution :

Same interest after 3, 4, and 5 years.

Then Principal Ratio will be:

$\frac{1}{3} : \frac{1}{4} : \frac{1}{5} = 20 : 15 : 12$

$47 = 105750$

12 units = 27000

Amount invested for the wife is ₹ 27000.

Type 2 (Problems based on Compound Interest)

27. An amount of ₹ 3000 at the rate of 20% rate of compound interest (changed annually) is invested after 2 years. What is the compound interest?

2 वर्षों के बाद 20% चक्रवृद्धि ब्याज (वार्षिक परिवर्तन) की दर पर ₹ 3000 की राशि का निवेश किया जाता है। चक्रवृद्धि ब्याज क्या है?

Solution :

Method 1

20Effective Rate = $20\% + 20\% + \frac{20 \times 20}{100}$

Effective Rate = $20 + 20 + 4 = 44\%$

44% of ₹ 3000 = ₹ 1320

Method 2

$(\frac{6}{5})^2 = \frac{36}{25}$

$25 = 3000$

$11 = 1320$

Method 3 (Tree Method)

Principal = ₹ 3000

600 600
120

$600 + 600 + 120 = 1320$

Method 4

Ratio for 2 years = 2 : 1

600 : 120

$2 \times 600 + 1 \times 120 = 1320$

28. The compound interest on an amount of ₹ 5120 at the rate of 12.5% (compounded annually) for 3 years is:

₹ 5120 की राशि पर 3 वर्षों के लिए 12.5% (वार्षिक रूप से संयोजित) की दर से चक्रवृद्धि ब्याज है:

Solution :

$12.5\% = \frac{1}{8}$

$(\frac{9}{8})^3 = \frac{729}{512}$

$512 = 5120$

$1 = 10$

$217 = 2170$

CI = ₹ 2170

29. What is the compound interest on ₹ 20,000 for 9 months at the rate of 4% per annum, when interest is compounded quarterly?

₹ 20,000 पर 9 महीने के लिए 4% प्रति वर्ष की दर से चक्रवृद्धि ब्याज क्या है, जब ब्याज त्रैमासिक रूप से संयोजित होता है?

Solution :

Compound Quarterly:

$R = \frac{4}{4} = 1\%$

$Time = \frac{9 \times 4}{12} = 3$ quarters

By Ratio Method:

$3 : 3 : 1$

$200 : 2 : 0.02$

$3 \times 200 + 3 \times 2 + 1 \times 0.02 = 600 + 6 + 0.02 =$ 606.02

30. What is the compound interest (in ₹) on a sum of ₹ 46,000 for 2 years at 15% per annum, interest being compounded annually (nearest to ₹)?

₹ 46,000 की राशि पर 2 वर्षों के लिए 15% प्रति वर्ष की दर से चक्रवृद्धि ब्याज (₹ में) क्या है, ब्याज वार्षिक रूप से संयोजित किया जा रहा है (₹ के निकटतम)?

Solution :

$$A = P\left(1 + \frac{R}{100}\right)^n \left(1 + \frac{R \times T}{100}\right)$$

$$A = 46000\left(1 + \frac{15}{100}\right)^2 \left(1 + \frac{15 \times 2}{5 \times 100}\right)$$

$$A = 46000 \times \left(\frac{23}{20}\right)^2 \times \frac{53}{50}$$

$$A = 46 \times 23 \times 23 \times 53/20 = 64485.1$$

$CI = A - P = 64485.1 - 46000 =$ $18485.1 \approx 18485$

31. What will be the compound interest (in ₹) on a sum of ₹ 7,200 for 18 months at the rate of 20% per annum if the interest is compounded half-yearly (nearest to an integer)?

₹ 7,200 की राशि पर 18 महीने के लिए 20% प्रति वर्ष की दर से चक्रवृद्धि ब्याज (₹ में) क्या होगा? यदि ब्याज अर्ध-वार्षिक रूप से संयोजित किया जाता है (एक पूर्णांक के निकटतम)?

Solution :

$\text{Time} = \frac{18 \times 2}{12} = 3$ periods

By Ratio Method:

$3 : 3 : 1$

$720 : 72 : 7.2$

$2160 + 216 + 7.2 = 2383.2$ **Or by Effective Rate:**

10% for 3 times:

$R = 33.1\%$

33.1% of 7200 = 2383.2

32. What is the compound interest on a sum of ₹ 37,500 for $1\frac{1}{3}$ years at a rate of 12% p.a., if the interest is compounded 8-monthly?

12% प्रति वर्ष की दर से $1\frac{1}{3}$ वर्षों के लिए ₹ 37,500 की राशि पर चक्रवृद्धि ब्याज क्या है? यदि ब्याज 8-मासिक चक्रवृद्धि है?

Solution :

Rate = 12% annually = 8% (8-monthly)

Time = $1\frac{1}{3}$ years = 16 months = 2 (8-monthly periods)

$$CI = P\left(1+\frac{R}{100}\right)^{r} - P$$
$$= 37500\left(1+\frac{8}{100}\right)^{2} - 37500$$
$$= 37500 \times \frac{108}{100} \times \frac{108}{100} - 37500$$
$$= 43740 - 37500$$

Compound Interest = ₹ 6240

33. In what time will ₹ 3,90,625 amount to ₹ 4,56,976 at 8% per annum, interest being compounded half-yearly?

कितने समय में ₹ 3,90,625 की राशि 8% प्रति वर्ष की दर से ₹ 4,56,976 हो जाएगी, ब्याज अर्धवार्षिक रूप से संयोजित होगा?

Solution :

If the interest is compounded half-yearly:

$$R = \frac{8}{2}\% = 4\%$$
$$A = P\left(1+\frac{R}{100}\right)^{n}$$
$$\frac{456976}{390625} = \left(1+\frac{4}{100}\right)^{2t}$$
$$\frac{456976}{390625} = \left(\frac{26}{25}\right)^{2t}$$
$$\left(\frac{26}{25}\right)^{4} = \left(\frac{26}{25}\right)^{2t}$$
$$2t = 4$$
$$T = 2 \text{ years}$$

34. In what time will a sum of ₹ 1,25,000 amount to ₹ 1,48,877 at 12% per annum, if interest is being compounded half-yearly?

कितने समय में ₹ 1,25,000 की राशि 12% प्रति वर्ष की दर से ₹ 1,48,877 हो जाएगी। यदि ब्याज अर्धवार्षिक रूप से संयोजित किया जा रहा है?

Solution :

Given:

$P = 125000,\ A = 148877,\ R = \frac{12}{2} = 6\%$

$$148877 = 125000\left(1+\frac{6}{100}\right)^{2t}$$
$$\frac{148877}{125000} = \left(1+\frac{6}{100}\right)^{2t}$$
$$\frac{148877}{125000} = \left(\frac{53}{50}\right)^{2t}$$
$$\left(\frac{53}{50}\right)^{3} = \left(\frac{53}{50}\right)^{2t}$$
$$2t = 3$$
$$t = \frac{3}{2} = 1\frac{1}{2} \text{ years}$$

35. A sum amounts to ₹ 18,600 after 3 years and to ₹ 27,900 after 6 years, at a certain rate percent p.a., when the interest is compounded annually. The sum is:

एक राशि 3 साल के बाद ₹ 18,600 हो जाती है और 6 साल बाद एक निश्चित प्रतिशत प्रति वर्ष की दर से ₹ 27,900 हो जाती है, जब ब्याज वार्षिक रूप से संयोजित होता है। योग है:

Solution :

$$P \rightarrow 18600 \rightarrow 27900$$

The amounts form a geometric progression (GP):

$$(18600)^2 = 27900 \times P$$
$$P = \frac{18600^2}{27900} = 12400$$

36. At what rate percent per annum will a sum of ₹ 15,625 amount to ₹ 21,952 in three years, if the interest is compounded annually?

यदि ब्याज वार्षिक रूप से संयोजित किया जाता है, तो प्रति वर्ष किस दर से ₹ 15,625 की राशि तीन वर्षों में ₹ 21,952 हो जाएगी?

Solution :

$$\sqrt[3]{15625} : \sqrt[3]{21952} = 25 : 28$$
$$r\% = \frac{3}{25} \times 100 = 12\%$$

37. The compound interest on a certain sum at 10% p.a. for 2 years is ₹ 1,201.60, interest compounded yearly. The sum is:

एक निश्चित राशि पर 10% प्रति वर्ष की दर से चक्रवृद्धि ब्याज 2 वर्षों के लिए ₹ 1,201.60 है, ब्याज वार्षिक चक्रवृद्धि है। योग है:

Solution :

$10\%, 10\%, \frac{10}{3}\%$

10 : 11

10 : 11

30 : 31

3000 : 3751

751 = 1201.60

1 = 1.6

3000 = 4800

38. A and B together borrowed a sum of ₹ 451,750 at an interest rate of 7% p.a. compound interest in such a way that to settle the loan, A paid as much amount after three years as paid by B after 4 years from the day of borrowing. The sum (in ₹) borrowed by A was:

A और Bने मिलकर 7% प्रति वर्ष की ब्याज दर पर ₹ 451,750 की राशि उधार ली। चक्रवृद्धि ब्याज इस प्रकार है कि ऋण का निपटान करने के लिए, A ने तीन वर्षों के बाद उतनी ही राशि का भुगतान किया, जितनी B द्वारा उधार लेने के दिनों से 4 वर्षों के बाद भुगतान की गई थी। A द्वारा उधार ली गई राशि (₹ में) थी:

Solution :

Let A and B borrow amounts P_1 and P_2 respectively:

$P_1\left(1+\frac{7}{100}\right)^3 = P_2\left(1+\frac{7}{100}\right)^4$

$\frac{P_1}{P_2} = \frac{107}{100}$

$207 \to 451750$

$1 \to 250$

$107 \to 250 \times 107 = 26750$

Hence, A borrowed ₹ 26750.

39. In how many months will ₹ 8,000 yield ₹ 2,648 as compound interest at 20% per annum compounded semi-annually?

कितने महीने में ₹ 8,000 उपज ₹ 2,648 रु. 20% प्रति वर्ष की दर से अर्धवार्षिक रूप से संयोजित चक्रवृद्धि ब्याज के रूप में मिलेंगे?

Solution :

Given:

$P = 8000,\ r = 20\%$ per annum $=$ 10% half-yearly

$CI = 2648$

$CI = P\left[\left(1+\frac{r}{100}\right)^n - 1\right]$

$2648 = 8000\left[\left(1+\frac{10}{100}\right)^n - 1\right]$

$\frac{331}{1000} = \left(\frac{11}{10}\right)^n - 1$

$\left(\frac{11}{10}\right)^n = 1+\frac{331}{1000} = \frac{1331}{1000}$

$\left(\frac{11}{10}\right)^n = \left(\frac{11}{10}\right)^3$

$n = 3$ half-yearly periods $= 18$ months

40. The compound interest on ₹ 4,000 after 3 years is ₹ 630.50. Then the rate of interest compounded yearly is:

3 वर्ष के बाद ₹ 4,000 पर चक्रवृद्धि ब्याज ₹ 630.50 है। तो वार्षिक चक्रवृद्धि ब्याज दर है:

Solution :

Compound Interest = Amount (A) - Principal (P):

$630.50 = A - 4000$

$A = 4630.50$

$A = P\left(1+\frac{R}{100}\right)^n$

$\frac{4630.50}{4000} = \left(1+\frac{R}{100}\right)^3$

$\frac{9261}{8000} = \left(1+\frac{R}{100}\right)^3 = \left(\frac{21}{20}\right)^3$

$\frac{21}{20} = 1+\frac{R}{100}$

$210 = 200 + R$

$R = 10/2 = 5\%$

41. The compound interest and the amount obtained on a certain sum of money are ₹ 820 and ₹ 8,820 respectively after 2 years. If the rate of interest is compounded yearly, then the rate of interest is:

2 वर्षों के बाद एक निश्चित धनराशि पर चक्रवृद्धि ब्याज और प्राप्त राशि क्रमशः ₹ 820 और ₹ 8,820 है। यदि ब्याज की दर वार्षिक रूप से संयोजित होती है, तो ब्याज की दर है:

Solution :

Principal = Amount - Compound Interest

$P = 8820 - 820 = 8000$

$\frac{A}{P} = \frac{8820}{8000} = \frac{882}{800} = \frac{441}{400} = \left(\frac{21}{20}\right)^2$

$r = \frac{1}{20} = 5\%$

42. ₹ 4,000 is given at 5% per annum for one year and interest is compounded half-yearly. ₹ 2,000 is given at 40% per annum compounded quarterly for one year. The total interest received is nearest to:

₹ 4,000 रुपये एक वर्ष के लिए 5% प्रति वर्ष की दर से दिए जाते हैं और ब्याज अर्धवार्षिक रूप से संयोजित किया जाता है। ₹ 2,000 1 वर्ष के लिए 40% प्रति वर्ष त्रैमासिक चक्रवृद्धि दर पर दिए जाते हैं। प्राप्त कुल ब्याज इसके निकटतम है:

Solution :

For ₹ 4,000 compounded half-yearly:

$R\% = \frac{5}{2} = 2.5\%$ (half-yearly)

Amount/Principal $= \left(\frac{41}{40}\right)^2 = \frac{1681}{1600}$

Compound Interest/Principal $= \frac{81}{1600}$

Compound Interest $= \frac{81}{1600} \times 4000 = 202.5$

For ₹ 2,000 compounded quarterly:

$R = \frac{40}{4} = 10\%$ (quarterly)

Amount/Principal $= \left(\frac{11}{10}\right)^4 = \frac{14641}{10000}$

Compound Interest/Principal $= \frac{4641}{10000}$

Interest $= \frac{4641}{10000} \times 2000 = 928.2$

Total interest received:

$202.5 + 928.2 = 1130.70$

43. The difference between the compound interest on a sum of ₹ 8,000 for 1 year at the rate of 10% per annum, interest compounded yearly and half-yearly is:

₹ 8,000 की राशि पर 1 वर्ष के लिए 10% प्रति वर्ष की दर से चक्रवृद्धि ब्याज, वार्षिक और अर्धवार्षिक चक्रवृद्धि ब्याज के बीच का अंतर है:

Solution :

10% for 1 year (compounded yearly)

5% for 2 half-years

Effective rate = 10.25%

Difference = 0.25% of 8000 = 20

44. Any certain amount was invested at a certain compound interest rate. The compound amount after five years is 1.1881 times the amount obtained in three years. What is the percentage rate of interest?

किसी निश्चित राशि को एक निश्चित चक्रवृद्धि ब्याज दर पर निवेश किया गया था। पांच वर्षों के बाद का चक्रवृद्धि तीन वर्षों में प्राप्त राशि का 1.1881 गुना है। ब्याज की प्रतिशत दर क्या है?

Solution :

According to the question: $y = x \times 1.1881$

$\frac{y}{x} = \frac{11881}{10000}$ (for 2 years)

For 1 year:

$\sqrt{\frac{11881}{10000}} = \frac{\sqrt{11881}}{\sqrt{10000}} = \frac{109}{100}$

Required rate (r):

$r = \frac{9}{100} \times 100 = 9\%$

45. If a sum of money triples in four years at compound interest, then in how many years will that amount become 27 times at the same interest rate?

यदि कोई धनराशि चक्रवृद्धि ब्याज पर चार वर्षों में तीन गुना हो जाती है, तो उसी ब्याज दर पर कितने वर्षों में वह राशि 27 गुना हो जाएगी?

Solution :

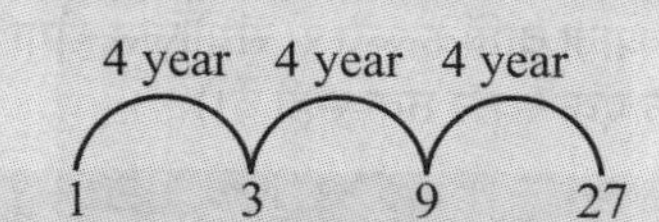

Required years = 4 + 4 + 4 = 12 years

46. A person borrowed a certain sum at 10% p.a. for three years, interest being compounded annually. At the end of two years, he repaid a sum of ₹ 6,634, and at the end of the third year, he cleared off the debt by paying ₹ 13,200. What was the sum borrowed by him?

एक व्यक्ति ने 10% वार्षिक दर पर एक निश्चित राशि उधार ली। तीन वर्षों के लिए, ब्याज वार्षिक चक्रवृद्धि होगी। दो साल के अंत में, उसने ₹ 6,634 का भुगतान किया और तीसरे वर्ष के अंत में, उसने ₹ 13,200 का भुगतान करके कर्ज चुकाया। उससे उधार ली गई राशि क्या थी?

Solution :

Reverse Process:

$13200 \times \frac{10}{11} = 12000$

$12000 + 6634 = 18634$

$18634 \times \frac{10}{11} \times \frac{10}{11} = 15400$

Principal borrowed by him = ₹ 15400

47. A bank offers 5.6% compound interest per annum calculated on a half-yearly basis. A customer deposits ₹ 7346 each on 1 January and 1 July of a year. At the end of the year, the amount he would have gained by way of interest is ______. [Give your answer correct to 2 decimal places.]

एक बैंक अर्ध-वार्षिक आधार पर गणना की गई प्रति वर्ष 5.6% चक्रवृद्धि ब्याज प्रदान करता है। एक ग्राहक एक वर्ष में 1 जनवरी और 1 जुलाई को ₹ 7346 जमा करता है। वर्ष के अंत में, ब्याज के माध्यम से उसे जो राशि प्राप्त होगी वह ______ है। [अपना उत्तर दशमलव 2 स्थानों तक सही दें।]

Solution :

$\frac{5.6}{2} = 2.8\%$

$\{7346 \times (1.028) + 7346\} \times (1.028) - 2 \times 7346$

$= 622.82$ Ans.

48. At what rate percent per annum will a certain sum of money multiply itself by 55 times in 2 years, the interest being compounded annually? (Give your answer correct to 1 decimal place).

किस प्रतिशत प्रति वर्ष की दर से एक निश्चित राशि 2 वर्षों में 55 गुना हो जाएगी, ब्याज वार्षिक रूप से संयोजित होगा? (अपना उत्तर दशमलव के एक स्थान तक सही दें।)

Solution :

$y^2 = 55$

$\cdot\left(1 + \frac{x}{100}\right)^2 = 55$

$1 + \frac{x}{100} = \sqrt{55} = 7.416$

$\frac{x}{100} = 6.416$

$x = 641.6\%$

49. If the amount received at the end of 2nd and 3rd year at compound interest on a certain principal is ₹ 9,600 and ₹ 10,272 respectively, what is the rate of interest (in यदि किसी निश्चित मूलधन पर चक्रवृद्धि ब्याज पर दूसरे और तीसरे वर्ष के अंत में प्राप्त राशि ₹ 9,600 और ₹ 10,272 है, ब्याज दर (% में) क्या है?

Solution :

Amount at the end of 2nd year = ₹ 9600

Amount at the end of 3rd year = ₹ 1027

Difference = ₹ 672

Note: The amount for the second year is the principal amount for the third year.

Rate (%) = $\frac{672}{9600} \times 100 = 7\%$

50. Rahul invested two equal sums of money at compound interest under two schemes A and B. Under scheme A, the interest rate was 10% per annum, and under scheme B, the interest rate was 12% p.a. Rahul got ₹ 1,050 as interest on the amount invested in scheme A after 2 years. How much interest will Rahul earn under scheme B after two years, if the interest is compounded annually in both schemes?

राहुल ने दो योजनाओं ए और बी के तहत चक्रवृद्धि ब्याज पर दो बराबर धनराशि का निवेश किया। योजना ए के तहत, ब्याज दर 10% प्रति वर्ष थी और योजना बी के तहत, ब्याज दर 12% प्रति वर्ष थी। राहुल को 2 साल बाद स्कीम ए में निवेश की गई राशि पर ब्याज के रूप में ₹ 1,050 मिले। यदि दोनों योजनाओं में ब्याज वार्षिक रूप से संयोजित किया जाता है, तो दो साल बाद राहुल को योजना बी के तहत कितना ब्याज मिलेगा?

Solution :

Let equal amount = ₹ 1000

For scheme A:

Compound interest of 1st year = ₹ 100

Compound interest of 2nd year = ₹ 100 + ₹ 10

Total interest = ₹ 210

$210 \rightarrow 1050$

$1 \rightarrow 5$

Amount = ₹ 5000

For scheme B:

Compound interest of 1st year = ₹ 600

Compound interest of 2nd year = ₹ 600 + ₹ 72

Total interest = ₹ 600 + ₹ 600 + ₹ 72 = ₹ 1272

51. A sum of ₹ 15,000 is lent at 16% p.a. compound interest. What is the difference between the compound interest for the second year and the third year?

₹ 15,000 की राशि 16% प्रति वर्ष की दर पर उधार दी जाती है। चक्रवृद्धि ब्याज। दूसरे वर्ष और तीसरे वर्ष के चक्रवृद्धि ब्याज के बीच क्या अंतर है?

Solution :

$$15000 \times \frac{4}{25} \times \frac{29}{25} \times \frac{4}{25} = 445.44$$

Type 3 (Misc. Problems)

52. The simple interest on a sum of ₹ 8,000 at a certain rate percent per annum for 3 years is ₹ 3,600. What will be the amount (in ₹) of the same sum after 2 years at the same rate, if the interest is compounded 8-monthly?

₹ 8,000 की राशि पर 3 वर्ष के लिए एक निश्चित प्रतिशत वार्षिक दर पर साधारण ब्याज ₹ 3,600 है। यदि ब्याज 8-मासिक रूप से संयोजित होता है, तो 2 वर्षों के बाद उसी दर पर उसी राशि की राशि (₹ में) क्या होगी?

Solution :

$$SI = \frac{PRT}{100}$$

$$3600 = \frac{8000 \times R \times 3}{100}$$

$$R = 15\%$$

According to the question:

12 months = 15%

$$8 \text{ months} = \frac{15}{12} \times 8 = 10\%$$

Time = 3 (8-month periods in 2 years):

$$A = 8000\left(1 + \frac{10}{100}\right)^3$$

$$= 8000 \times \frac{11 \times 11 \times 11}{1000} = 10,648$$

53. A borrowed a sum of ₹ 160,000 from B at 10% per annum simple interest. At the same time, he lent the same sum to C at the same rate on compound interest, compounded semi-annually for 2 years. Find the amount (in ₹) earned by A in the whole transaction.

A ने B से 10% प्रति वर्ष साधारण ब्याज पर ₹ 160,000 की राशि उधार ली। उसी समय उसने C को समान राशि चक्रवृद्धि ब्याज पर समान दर पर उधार दी, जो 2 वर्षों के लिए अर्ध-वार्षिक रूप से संयोजित होती है। पूरे लेनदेन में A द्वारा अर्जित राशि (₹ में) ज्ञात कीजिए।

Solution :

Amount earned by A = Difference between compound interest and simple interest in the given year.

$$= 160000\left[\left(1 + \frac{5}{100}\right)^4 - 1\right] - \frac{160000 \times 10 \times 2}{100}$$

$$= 160000\left(\frac{21^4 - 20^4}{20^4}\right) - 32000$$

$$= (194481 - 160000) - 32000$$

$$= 34481 - 32000 = 2481$$

54. A man invested a total of ₹ 12,050 in two parts, one at 10% p.a. simple interest for 2 years and the other at the same rate at compound interest, interest being compounded annually, for the same time. The amounts he received from both the parts are equal. The sum (in ₹) invested at the compound interest is:

एक आदमी ने दो भागों में कुल ₹ 12,050 का निवेश किया, एक 10% प्रति वर्ष की दर से 2 वर्षों के लिए साधारण ब्याज और दूसरा समान दर पर चक्रवृद्धि ब्याज, उसी समय के लिए ब्याज वार्षिक रूप से संयोजित किया जाता है। दोनों भागों से उसे प्राप्त रकम बराबर है। चक्रवृद्धि ब्याज पर निवेश की गई राशि (₹ में) है:

Solution :

Let the amount given on compound interest = x.

According to the question:

$$x \times \left(1 + \frac{10}{100}\right)^2 = (12050 - x) + \frac{(12050 - x) \times 10 \times 2}{100}$$

$$x \times \frac{11 \times 11}{10 \times 10} = (12050 - x) + \frac{241000 - 20x}{100}$$

$$\frac{121x}{100} + \frac{20x}{100} = 12050 - x + 2410$$

$$\frac{141x}{100} + x = 14460$$

$$241x = 1446000$$

$$x = 6000$$

55. The simple interest on a sum of ₹ 12,000 at the end of 5 years is ₹ 46,000. What would have been the compound interest on the same sum at the same rate for 3 years when compounded annually?

5 वर्ष के अंत में ₹ 12,000 की राशि पर साधारण ब्याज ₹ 46,000 है। वार्षिक रूप से संयोजित होने पर 3 वर्षों के लिए समान राशि पर समान दर पर चक्रवृद्धि ब्याज कितना होगा?

Solution :

$$SI = \frac{P \times R \times T}{100}$$

$$46000 = \frac{12000 \times R \times 5}{100}$$

$$R = 10\%$$

$$CI = P\left[\left(1 + \frac{R}{100}\right)^t - 1\right]$$

$$= 12000\left[\left(1 + \frac{10}{100}\right)^3 - 1\right]$$

$$= 12000\left[\frac{1331}{1000} - 1\right]$$

$$= 12000 \times \frac{331}{1000} = 3972$$

56. A man invests ₹ 7,000 at 8% per annum simple interest for 2 years and ₹ 10,000 at compound interest at the same rate for the same period, compounded annually. What will be the total interest and the total amount (in ₹) respectively, on maturity?

एक आदमी 2 साल के लिए 8% प्रति वर्ष साधारण ब्याज पर ₹ 7,000 निवेश करता है और समान अवधि के लिए समान दर पर चक्रवृद्धि ब्याज पर ₹ 10,000 निवेश करता है, जो वार्षिक रूप से संयोजित होता है। परिपक्वता पर कुल ब्याज और कुल राशि क्रमशः (₹ में) क्या होगी?

Solution :

$$SI = \frac{P \times R \times T}{100}$$

$$SI = \frac{7000 \times 8 \times 2}{100} = 1120$$

$$A = P\left(1 + \frac{R}{100}\right)^n$$

$$A = 10000 \times \left(1 + \frac{8}{100}\right)^2$$

$$A = 10000 \times \frac{54}{50} \times \frac{54}{50} = 11664$$

$$CI = A - P = 11664 - 10000 = 1664$$

$$\text{Total interest} = SI + CI = 1120 + 1664 = 2784$$

$$\text{Total amount} = 7000 + 10000 + 2784 = 19784$$

57. The simple interest on a certain sum for 3 years at 12% p.a. is ₹ 6,750. What is the compound interest (in ₹) on the same sum for 2 years at 20% p.a., if interest is compounded half-yearly? (rounded off to the nearest ₹)

एक निश्चित राशि पर 3 वर्ष के लिए 12% प्रति वर्ष की दर से साधारण ब्याज ₹ 6,750 है। यदि ब्याज अर्ध-वार्षिक रूप से संयोजित होता है, तो 2 वर्षों के लिए 20% प्रति वर्ष की दर से उसी राशि पर चक्रवृद्धि ब्याज (₹ में) क्या है? (निकटतम तक पूर्णांकित)

Solution :

$$SI = \frac{P \times R \times T}{100}$$

$$6750 = \frac{P \times 12 \times 3}{100}$$

$$P = \frac{6750 \times 100}{12 \times 3} = 18750$$

When interest is compounded half-yearly:

$$R = \frac{20}{2} = 10\%$$

$$CI = P\left[\left(1 + \frac{R}{100}\right)^T - 1\right]$$

$$= 18750\left[\left(1 + \frac{10}{100}\right)^4 - 1\right]$$

$$= 18750\left[\frac{14641 - 10000}{10000}\right]$$

$$= \frac{18750 \times 4641}{10000} = 8702$$

58. The difference in compound interest on a certain sum at 10% p.a. for one year, when the interest is compounded half-yearly and yearly, is ₹ 88.80. What is the simple interest on the same sum for $1\frac{2}{3}$ years at the same rate?

एक निश्चित राशि पर 10% प्रति वर्ष की दर से चक्रवृद्धि ब्याज का अंतर एक वर्ष के लिए, जब ब्याज अर्धवार्षिक और वार्षिक रूप से संयोजित होता है, तो ₹ 88.80 होता है। उसी धनराशि पर समान दर से $1\frac{2}{3}$ वर्षों तक साधारण ब्याज कितना होगा?

Solution :

When interest is compounded annually:

$R = 10\%$

When interest is compounded half-yearly:

Two cycles will be formed in 1 year.

5%	5%
6 month	6 month

Equivalent rate: $5 + 5 + \frac{25}{100} = 10.25\%$

Difference in rate: $10.25\% - 10\% = 0.25\%$

$P \times \frac{0.25}{100} = 88.80$

$P \times 0.25 = 8880$

$P = 35520$

$SI = \frac{P \times R \times T}{100} = \frac{35520 \times 10 \times 5}{100 \times 3} =$ 5920

59. The compound interest on a certain sum at the end of two years is ₹ 408. The simple interest on the same sum for the same time is ₹ 400. The rate of interest per annum is:

दो वर्ष के अंत में एक निश्चित राशि पर चक्रवृद्धि ब्याज ₹ 408 है। उसी राशि पर समान समय के लिए साधारण ब्याज ₹ 400 है। प्रति वर्ष ब्याज दर है:

Solution :

$SI = \frac{P \times R \times T}{100}$

$400 = \frac{P \times R \times 2}{100}$

$P \times R = 20000$ ....(i)

Difference between CI and SI for 2 years:

$P\left(\frac{R}{100}\right)^2 = 8$

$\frac{P \times R^2}{10000} = 8$

$R = \frac{8 \times 100 \times 100}{20000} = 4\%$

60. What is the rate of interest if simple interest earned on a certain sum for the 3rd year is ₹ 1750 and compound interest earned for 2 years is ₹ 3622.5?

यदि तीसरे वर्ष के लिए एक निश्चित राशि पर अर्जित साधारण ब्याज ₹ 1750 है और 2 वर्षों के लिए अर्जित चक्रवृद्धि ब्याज ₹ 3622.5 है तो ब्याज की दर क्या है?

Solution :

1750 1750

122.5

$1750 \times r\% = 122.5$

$r = \frac{122.5}{1750} = 7\%$

61. A sum of ₹ 9,000 amounts to ₹ 13,356 at a certain rate percent per annum in $4\frac{2}{5}$ years at simple interest. What will be the simple interest on the same sum at double the rate for $2\frac{1}{3}$ years?

₹ 9,000 की एक राशि साधारण ब्याज पर $4\frac{2}{5}$ वर्षों में एक निश्चित प्रतिशत प्रति वर्ष की दर से ₹ 13,356 हो जाती है। उसी राशि पर $2\frac{1}{3}$ वर्षों में दोगुनी दर पर साधारण ब्याज कितना होगा?

Solution :

Principal (P) = ₹ 9,000

Amount (A) = ₹ 13,356

Rate = $r\%$

Time (t) = $4\frac{2}{5} = \frac{22}{5}$ years

Simple Interest = Amount - Principal

$\frac{P \times r \times t}{100} = 13356 - 9000$

$4356 = \frac{9000 \times r \times \frac{22}{5}}{100}$

$r = 11\%$

According to the question:

$SI = \frac{P \times R \times T}{100}$

$= \frac{9000 \times 22 \times 7}{100 \times 3}$

$= 210 \times 22 =$ 4620

62. A sum of ₹ 6,600 amounts to ₹ 8,756 in $4\frac{2}{3}$ years at $x\%$ p.a. simple interest. What will the same sum amount to if invested at $(x+3)\%$ p.a. for $2\frac{1}{2}$ years, interest compounded yearly (nearest to ₹ 1)?

रुपये की राशि ₹ 6,600, $x\%$ प्रति वर्ष की दर से $4\frac{2}{3}$ वर्षों में ₹ 8,756 साधारण ब्याज। यदि $(x+3)\%$ प्रति वर्ष की दर से निवेश किया जाए तो समान राशि कितनी होगी? $2\frac{1}{2}$ वर्षों के लिए, ब्याज वार्षिक रूप से संयोजित (लगभग ₹ 1)?

Solution :

$$8756 - 6600 = \frac{6600 \times \frac{14}{3} \times x}{100}$$

$$2156 = \frac{2200 \times 14 \times x}{100}$$

$$x = 7\%$$

New rate:

$$x + 3 = 7 + 3 = 10\%$$

New amount for two years:

$$6600 \times \left(\frac{110}{100}\right)^2 = 7986$$

Interest for next $\frac{1}{2}$ year:

$$7986 \times \frac{10}{100} \times \frac{1}{2} = 399.3$$

Total amount:

$$7986 + 399.3 = 8385.3 \approx 8385$$

Type 4 (Instalment/ किश्त)

63. In a store, Arun has two options to buy a mobile phone, either he can pay the full amount of ₹ 2,500 or he may buy it on an installment scheme, i.e., ₹ 520 as a down payment and four equal installments, with the rate of interest being charged at 25% p.a. simple interest. Arun buys a mobile phone on the installment scheme. What is the monthly installment for Arun?

एक स्टोर में, अरुण के पास मोबाइल फोन खरीदने के लिए दो विकल्प हैं, या तो वह ₹ 2,500 की पूरी राशि का भुगतान कर सकता है या वह किस्त योजना में खरीद सकता है, यानी ₹ 520 का डाउन पेमेंट और चार समान किस्तों में, ब्याज दर 25% प्रति वर्ष की दर से शुल्क लिया जाता है। साधारण ब्याज। अरुण किस्त योजना पर एक मोबाइल फोन खरीदता है। अरुण की मासिक किस्त क्या है?

Solution :

$$2500 - 520 = 1980$$

$$1980 + \frac{1980 \times 25 \times 4}{1200}$$

$$= 4x + \frac{x \times 25 \times (0+1+2+3)}{1200}$$

$$1980 + 165 = 4x + \frac{x}{8}$$

$$\frac{33x}{8} = 2145$$

$$x = 520$$

64. P borrows ₹ 5,00,000 from the bank, which charges 18% simple interest per annum at the start of the year. He pays 11 equal monthly installments of ₹ 40,000 and his bike at the end of the year to pay the amount. What is the value of the bike given?

P उस बैंक से ₹ 5,00,000 उधार लेता है जो वर्ष की शुरुआत में प्रति वर्ष 18% साधारण ब्याज लेता है। वह ₹ 40,000 की 11 समान मासिक किश्तों का भुगतान करता है। शेष धनराशी का भुगतान करने के लिए वर्ष के अंत में अपनी बाइक को गिरवी रख देता है। गिरवी रखी गई बाइक का मूल्य क्या है?

Solution :

$$5,00,000 \times \frac{18 \times 12}{1200} = 90,000$$

Total Amount = 500, 000 + 90, 000 = 590, 000

Bike Value = 590, 000 − 11 × 40, 000 = 1, 50, 000

65. A man borrows ₹ 24,000 from a bank at 10% per annum simple interest and clears the debt in three years. If the installments paid at the end of the first and second years to clear the debt are ₹ 1,500 and ₹ 2,500 respectively, what amount (in ₹) should be paid at the end of the third year to clear the debt?

एक आदमी एक बैंक से 10% प्रति वर्ष साधारण ब्याज पर ₹ 24,000 उधार लेता है और तीन साल में कर्ज चुका देता है। यदि ऋण चुकाने के लिए पहले और दूसरे वर्ष के अंत में भुगतान की जाने वाली किश्तें क्रमशः ₹ 1,500 और ₹ 2,500 हैं, तो ऋण चुकाने के लिए तीसरे वर्ष के अंत में कितनी राशि (₹ में) का भुगतान किया जाना चाहिए?

Solution :

$$4000 \times 10\% + (4000 - 1500) \times 10\% + (2500 - 2500) \times 10\%$$

$$400 + 250 + 0 = 650$$

66. A mobile can be purchased on cash payment of ₹ 1,500. But the same mobile can also be purchased on a cash down payment of ₹ 350 and the rest can be paid in three equal annual installments of ₹ 400 for the next three months. Find the rate of simple interest.

एक मोबाइल ₹ 1,500 नगद में खरीदा जा सकता है अथवा वही मोबाइल ₹ 350 की अग्रिम भुगतान के साथ प्रति माह ₹ 400 की तीन बराबर किश्तों में खरीदा जा सकता है। साधारण ब्याज की दर क्या होगी?

Solution :

$1500 - 350 = 1150$

$1150 + \frac{1150 \times 3 \times r}{1200} = 3 \times 400 + \frac{400 \times r \times (0+1+2)}{1200}$

$1150 + \frac{23r}{8} = 1200 + r$

$\frac{15r}{8} = 50$

$r = \frac{80}{3} = 26\frac{2}{3}$

67. A mobile phone is available for ₹ 79,860 by cash payment or by paying ₹ 60,000 as a down payment and the remaining amount in three equal annual installments. If the shopkeeper charges interest at the rate of 10% per annum compounded annually, then the amount of each installment (in ₹) will be:

एक मोबाइल फोन ₹ 79,860 नकद भुगतान द्वारा या ₹ 60,000 नकद भुगतान करके और शेष राशि तीन समान वार्षिक किस्तों में। यदि दुकानदार 10% वार्षिक चक्रवृद्धि ब्याज की दर से ब्याज लेता है, तो प्रत्येक किस्त की राशि (₹ में) होगी:

Solution :

$79860 - 60000 = 19860$

$10 : 11) \times 11 \times 11$

$100 : 121) \times 11$

$1000 : 1331$

$1210 : 1331$

$1100 : 1331$

$1000 : 1331$

Sum $= 1210 + 1100 + 1000 = 3310,$ units

$3310 = 19860$

$1 = 6$

$1331 = 7986$

Each Installment: ₹ 7986

68. A sum of money is paid back in two annual installments of ₹ 17,640 each, allowing 5% compound interest compounded annually. The sum borrowed was:

कुछ धनराशि वार्षिक चक्रवृद्धि की दर से 5% चक्रवृद्धि ब्याज देकर प्रति वर्ष ₹ 17,640 की दो वार्षिक किस्तों में चुकाई जाती है। उधार ली गई राशि कितनी थी?

Solution :

$20 : 21 \times 21$

$400 : 441$

$420 : 441$

$400 : 441$

$441 = 17640$

$1 = 40$

$820 = 32800$

Sum = ₹ 32,800

69. A computer is available for ₹ 76,800 cash or ₹ 23,840 cash down payment and three equal half-yearly installments. If the interest charged is 20% p.a., compounded semi-annually, then the interest charged in the installment scheme is:

एक कंप्यूटर ₹ 76,800 नकद या ₹ 23,840 नकद अग्रिम भुगतान और तीन समान अर्ध-वार्षिक किस्तों पर उपलब्ध है। यदि लगाया गया ब्याज 20% प्रति वर्ष है, जो अर्ध-वार्षिक रूप से संयोजित होता है, तो किस्त योजना में लगाया गया ब्याज है:

Solution :

$76800 - 23840 = 52960$

$r = \frac{20}{2} = 10\%$

$10 : 11 \times 11^2$

$100 : 121 \times 11$

$1000 : 1331$

$1210 : 1331$

$1100 : 1331$

$1000 : 1331$

$3310 : 3993$(Int = 683 units)

$3310 = 52960$

$1 = 16$

$683 =$ ₹ 10,928

70. Calculate annual installment required to pay off the debt of ₹ 21,080 due in 5 years at 12% simple interest.

₹ 21,080 के ऋण का भुगतान करने के लिए आवश्यक वार्षिक किस्त की गणना करें, जो 5 वर्षों में 12% साधारण ब्याज

पर देय है।

Solution :

Let installment be 100 units: $100 + 112 + 124 + 136 + 148 = 620$ units

$620 \text{ units} = 21,080$

$1 \text{ unit} = 34$

$100 \text{ units} = 3,400$

Installment Amount: ₹ 3,400

71. A car with a price of ₹ 6,50,000 is bought by making some down payment. On the balance, a simple interest of 10% is charged in lump sum, and the money is to be paid in 20 equal annual installments of ₹ 25,000. How much is the down payment?

₹ 6,50,000 की कीमत वाली कार कुछ डाउन पेमेंट करके खरीदी जाती है। शेष राशि पर एकमुश्त 10% का साधारण ब्याज लगाया जाता है और ₹ 25,000 की 20 समान वार्षिक किस्तों में भुगतान किया जाना है। डाउन पेमेंट की राशि कितनी है?

Solution :

$25000 \times 20 = 5,00,000$

$\frac{5,00,000}{1.1} \approx 4,54,545.45$

$6,50,000 - 4,54,545.45 = 1,95,455$

Down Payment: ₹ 1,95,455

7 Time Work and Pipe Cistern

Total work = time × efficiency
कुल कार्य = समय × दक्षता

Efficiency $\propto$ total work (When time is constant)
दक्षता $\propto$ कुल कार्य (जब समय स्थिर हो)

$\frac{1}{\text{Time}}$ / Efficiency / क्षमता $\propto$ wages / मजदूरी $\propto$ $\frac{1}{\text{Time (समय)}}$

For same time period wages $\propto$ efficiency
समान समय अवधि के लिए मजदूरी $\propto$ क्षमता

For different time period wages $\propto$ work done
अलग-अलग समयावधि के लिए वेतन $\propto$ किया गया कार्य

If M_1 men finish W_1 work in D_1 days, working T_1 time each day and M_2 men finish W_2 work in D_2 days, working T_2 time each day then

यदि M_1 पुरुष W_1 कार्य को D_1 दिनों में पूरा करते हैं, प्रत्येक दिन T_1 समय कार्य करते हैं और M_2 पुरुष W_2 कार्य को D_2 दिनों में समाप्त करते हैं, प्रत्येक दिन T_2 समय कार्य करते हैं, तो

$$\frac{M_1D_1T_1}{W_1} = \frac{M_2D_2T_2}{W_2}$$

If A does the work in $(x+a)$ days, and B does in $(x+b)$ days while both can do the work in x days.

यदि A कार्य को $(x+a)$ दिनों में करता है, और B $(x+b)$ दिनों में करता है जबकि दोनों कार्य को x दिनों में कर सकते हैं।

$$x = \sqrt{ab}$$

Type 1 (Basic)

1. A, B and C can do a work in 5 days, 6 days and 10 days, respectively. Working together, in how many days will they finish the same work?

A, B और C एक काम को क्रमशः 5 दिन, 6 दिन और 10 दिन में कर सकते हैं। एक साथ काम करते हुए वे उसी काम को कितने दिनों में पूरा करेंगे?

Solution :

A→5 (efficiency 6)
B→6 (efficiency 5) → 30 total work
C→10 (efficiency 3)

Efficiency of A, B and C = 6:5:3

Total work = 30

If they work together, time taken by them =

$$\frac{30}{6+5+3} = \frac{30}{14} = 2\frac{1}{7}$$

2. A and B can do a work in $26\frac{2}{3}$ days, B and C together can complete the same work in 48 days, while A and C together can complete the same work in 30 days. How long (in days) will A alone take to complete 60% of the work?

A और B एक काम को $26\frac{2}{3}$ दिनों में पूरा कर सकते हैं, B और C एक साथ उसी काम को 48 दिनों में पूरा कर सकते हैं, जबकि A और C एक साथ उसी काम को 30 दिनों में पूरा कर सकते हैं। A अकेले 60% कार्य पूरा करने में कितना समय (दिनों में) लेगा?

Solution :

(A + B) ⟶ $\frac{80}{3}$ — 9

(B + C) ⟶ 48 — 5 ⟶ 240

(A + C) ⟶ 30 — 8

2(A + B + C) = 22

A + B + C = 11

Efficiency of A = 11 - 5 = 6

60% of total work = $\frac{240 \times 60}{100} = 144$

Time taken by A to complete the 60% of the work = $\frac{144}{6} = 24$

Hence, A will alone complete 60% of the work in 24 days.

3. A can complete a work in 15 days and B can complete the same work in 20 days. If they are working together, then in how many days will they complete 35% of the same work?

A एक कार्य को 15 दिनों में पूरा कर सकता है और B उसी कार्य को 20 दिनों में पूरा कर सकता है। यदि वे एक साथ काम कर रहे हैं तो वे उसी काम का 35% कितने दिनों में पूरा करेंगे?

Solution :

A can complete a work = 15 days = 15

B can complete a work = 20 days

(A + B) working together to complete the work

$= \frac{15 \times 20}{35} = \frac{300}{35} = \frac{60}{7}$ days

$\therefore$ 35% of total work $= \frac{60}{7} \times \frac{7}{20} = 3$ days

4. 40 persons take 6 days to complete a certain task, working 10 hours a day. How many hours a day will be sufficient for 30 persons to complete the same task in 10 days?

40 व्यक्ति प्रतिदिन 10 घंटे कार्य करके एक निश्चित कार्य को पूरा करने में 6 दिन लगाते हैं। उसी कार्य को 10 दिनों में पूरा करने के लिए 30 व्यक्तियों के लिए प्रतिदिन कितने घंटे पर्याप्त होंगे?

Solution :

By formula: $M_1D_1H_1 = M_2D_2H_2$

then, $40 \times 6 \times 10 = 30 \times 10 \times H_2$

$H_2 = \frac{40 \times 6}{30} = 8$ hours

5. Eighteen persons working 8 hours a day can complete 3 units of work in 10 days. How many days are required by 25 persons to complete 5 units of work working 6 hours a day?

अठारह व्यक्ति प्रतिदिन 8 घंटे काम करके 3 इकाइयों का काम 10 दिनों में पूरा कर सकते हैं। 25 व्यक्तियों को प्रतिदिन 6 घंटे कार्य करके 5 इकाई कार्य पूरा करने में कितने दिन लगेंगे?

Solution :

$$\frac{M_1D_1H_1}{W_1} = \frac{M_2D_2H_2}{W_2}$$

$$\frac{18 \times 10 \times 8}{3} = \frac{25 \times D_2 \times 6}{5}$$

$$D_2 = \frac{8 \times 6 \times 10}{30} = 16 \text{ days}$$

6. 10 men or 15 women can complete a work in 30 days. In how many days can 15 men and 27 women complete the work?

10 पुरुष या 15 महिलाएँ एक कार्य को 30 दिनों में पूरा कर सकते हैं। 15 पुरुष और 27 महिलाएँ उस कार्य को कितने दिनों में पूरा कर सकते हैं?

Solution :

10 Men = 15 Women

Men/Women = 3/2

Ratio of efficiencies of 1 man and 1 woman = 3:2

$\therefore M_1D_1 = M_2D_2$

$10 \times 3 \times 30 = (15 \times 3 + 27 \times 2) \times D_2$

$\frac{900}{99} = D_2$

$D_2 = 9\frac{1}{11}$ days

7. Suman and Lata working together can complete a task in 8 days. If Lata can complete the same task in 12 days, then how many days will Suman take to complete the same task?

सुमन और लता एक साथ काम करते हुए एक कार्य को 8 दिनों में पूरा कर सकते हैं। यदि लता उसी कार्य को 12 दिनों में पूरा कर सकती है, तो सुमन को उसी कार्य को पूरा करने में कितने दिन लगेंगे?

Solution :

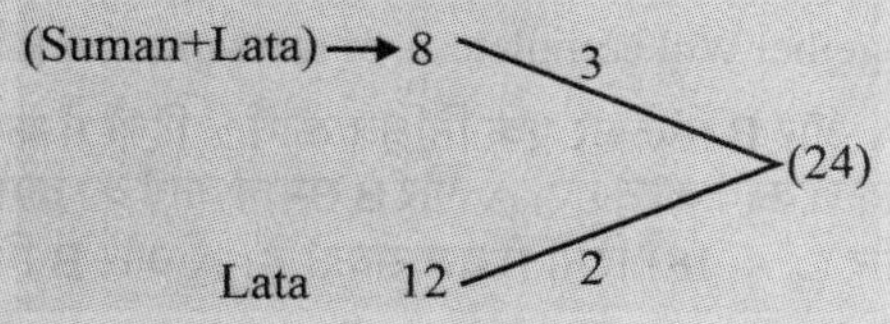

According to the question Time taken by Suman to complete the same task $= \frac{24}{1} =$ 24 days

8. A can complete one-sixth of a work in 5 days and B can complete one-fourth of the same work in 15 days. In how many days, will both working together complete the work?

A किसी काम का छठा हिस्सा 5 दिनों में पूरा कर सकता है और B उसी काम का एक-चौथाई 15 दिनों में पूरा कर सकता है। दोनों मिलकर काम कितने दिनों में पूरा करेंगे?

Solution :

A can complete one-sixth of a work in 5 days

A can complete total work in 5×6 days = 30 days B can complete total work in 15×4 days = 60 days

$\therefore$ A and B together can complete total work in $= \frac{30 \times 60}{30 + 60} = 20$ days

9. Amit and Sunil together can complete a work in 9 days, Sunil and Dinesh together can complete the same work in 12 days, and Amit and Dinesh together can complete the same work in 18 days. In how many days will they complete the work if Amit, Sunil and Dinesh work together?

अमित और सुनील एक साथ एक काम को 9 दिनों में पूरा कर सकते हैं, सुनील और दिनेश एक साथ उसी काम को 12 दिनों में पूरा कर सकते हैं, और अमित और दिनेश एक साथ उसी काम को 18 दिनों में पूरा कर सकते हैं। यदि अमित, सुनील और दिनेश एक साथ काम करते हैं तो वे कितने दिनों में काम पूरा करेंगे?

Solution :

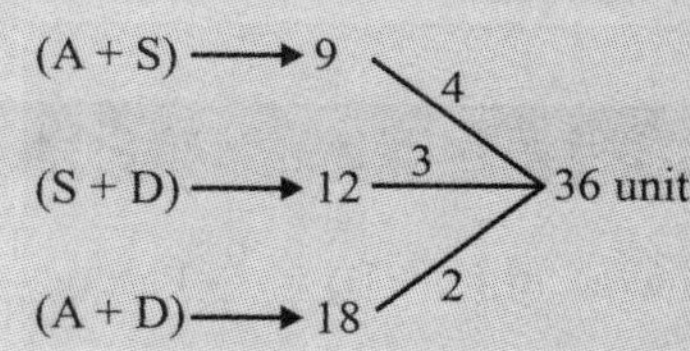

If Amit, Sunil and Dinesh work together, then their total efficiencies $= \frac{9}{2}$

Required time $= \frac{36 \times 2}{9} = 8$ days

10. A can do 50% of the job in 16 days, B can do $\frac{1}{4}$ of the job in 24 days. In how many days can they do $\frac{3}{4}$ of the job working together?

A 50% काम 16 दिनों में कर सकता है, B $\frac{1}{4}$ काम 24 दिनों में कर सकता है। वे एक साथ काम करके $\frac{3}{4}$ काम कितने दिनों में पूरा कर सकते हैं?

Solution :

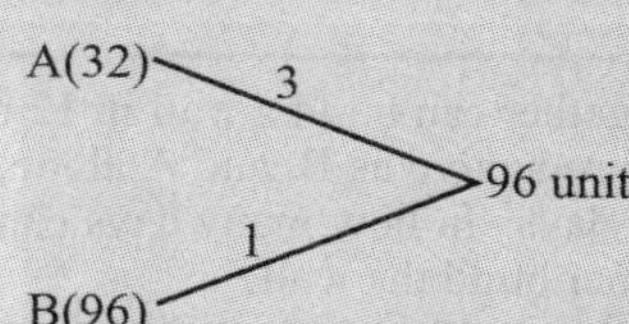

Time taken by A to do 50% of the work = 16 days = 16

$\therefore$ Time taken by A to complete the work = 32 days

Time taken by B to do $\frac{1}{4}$ part = 24 days

$\therefore$ Time taken by B to complete the work = $24 \times 4 = 96$ days

$\therefore$ Time taken by (A + B) to do $\frac{3}{4}$ part of the work $= \frac{96 \times \frac{3}{4}}{4} = 18$ days

11. A and B can together complete a task in 18 hours. After 6 hours, A leaves. B takes 36 hours to finish the rest of the task. How many hours would A have taken to do the task if he worked alone?

A और B मिलकर एक कार्य को 18 घंटे में पूरा कर सकते हैं। 6 घंटे के बाद A चला जाता है। शेष कार्य को पूरा करने में B

को 36 घंटे लगते हैं। यदि A अकेले कार्य करता तो उसे कार्य करने में कितने घंटे लगते?

Solution :

6 hour's work of (A + B) = $\frac{6}{18}$ Part or $\frac{1}{3}$ Part

Remaining work = $1 - \frac{1}{3} = \frac{2}{3}$

The remaining $\frac{2}{3}$ part of the work is done by B in 36 hours. Then B will complete the whole work in $36 \times \frac{3}{2}$ hours.

Let A will do the work in x hours. $\frac{1}{54} + \frac{1}{x} = \frac{1}{18}$

$\frac{1}{x} = \frac{1}{18} - \frac{1}{54}$

$\frac{1}{x} = \frac{1}{27}$

$x = 27$

Hence, A will complete the work in 27 hours.

12. A, B and C together can build a wall in 12 days. C is four times as productive as B, and A alone can build the wall in 48 days. In how many days can A and B working together build the wall?

A, B और C मिलकर 12 दिनों में एक दीवार बना सकते हैं। C, B से चार गुना अधिक उत्पादक है और A अकेले 48 दिनों में दीवार बना सकता है। A और B मिलकर कितने दिनों में दीवार बना सकते हैं?

Solution :

Let B and C build the wall in $4x$ and x days, respectively.

$\frac{1}{48} + \frac{1}{4x} + \frac{1}{x} = \frac{1}{12}$

$\frac{5}{4x} = \frac{1}{12} - \frac{1}{48}$

$\frac{5}{4x} = \frac{3}{48}$

$x = 20$

1 day's work of A and B = $\frac{1}{48} + \frac{1}{80} = \frac{8}{240} = \frac{1}{30}$

So, both will build the wall in 30 days.

13. A and B together can do a certain work in x days. Working alone, A and B can do the same work in $(x + 8)$ and $(x + 18)$ days, respectively. A and B together will complete $\frac{5}{6}$ of the same work in:

A और B मिलकर एक निश्चित कार्य x दिनों में कर सकते हैं। अकेले कार्य करते हुए, A और B समान कार्य को क्रमश: $(x+8)$ और $(x + 18)$ दिनों में कर सकते हैं। A और B मिलकर उसी कार्य का $\frac{5}{6}$ भाग कितने समय में पूरा करेंगे:

Solution :

A and B together complete a work in x days, whereas A and B separately complete the work in $(x+a)$ and $(x+b)$ days, respectively.

$x = \sqrt{ab}$

$x = \sqrt{8 \times 18} = 12$

A and B together will complete the work in 12 days

$\therefore$ The time taken to complete $\frac{5}{6}$ part of the same work $= 12 \times \frac{5}{6} = 10$ days

14. A is twice as efficient as B, and C is thrice as efficient as B. Working together, they can finish a certain work in 5 days. A and C worked together for 5 days. B alone would complete the remaining work in:

A, B से दोगुना कुशल है और C, B से तीन गुना कुशल है। एक साथ काम करते हुए, वे एक निश्चित कार्य को 5 दिनों में पूरा कर सकते हैं। A और C ने 5 दिनों तक एक साथ काम किया। B अकेले ही शेष कार्य पूरा करेगा:

Solution :

Ratio of work capacity of A, B, and C = 2:1:3

$\therefore$ Work done by (A + B + C) in 5 days = $6 \times 5 = 30$ units

Work done by (A + C) in 5 days = $5 \times 5 = 25$ units

Remaining work = $30 - 25 = 5$ units

Time taken by B to complete the remaining work = $\frac{5}{1} = 5$ days

15. Working together, A and B can complete a work in 12 days. They work together for 9 days, after which B leaves. If A finishes the remaining work in 5 days, then the number of days that B alone would take to complete the work is:

एक साथ काम करते हुए, A और B एक कार्य को 12 दिनों में पूरा कर सकते हैं। वे 9 दिनों तक एक साथ काम करते हैं जिसके बाद B चला जाता है। यदि A शेष कार्य को 5 दिनों में पूरा करता है, तो B को अकेले कार्य पूरा करने में कितने दिन लगेंगे:

Solution :

9 days' work of A and B together = $\frac{9}{12} = \frac{3}{4}$

Remaining work = $1 - \frac{3}{4} = \frac{1}{4}$

A finishes the remaining work in 5 days.

Time taken by A to complete the work $= 5 \times 4 = 20$ days.

$B = \frac{xy}{x-y} = \frac{20 \times 12}{20 - 12}$

Using the formula

$= \frac{20 \times 12}{8} = 30$ days.

16. X takes 10 days less than Y to complete a task. Both X and Y together can complete the work in 12 days. In how many days will Y alone complete the work?

X को एक कार्य पूरा करने में Y से 10 दिन कम लगते हैं। X और Y दोनों मिलकर कार्य को 12 दिनों में पूरा कर सकते हैं। Y अकेले कितने दिनों में काम पूरा करेगा?

Solution :

Let Y complete the work in x days. $\frac{1}{x-10} + \frac{1}{x} = \frac{1}{12}$

$\frac{x + x - 10}{x(x-10)} = \frac{1}{12}$

$x^2 - 10x = 24x - 120$

$x^2 - 34x + 120 = 0$

$(x - 30)(x - 4) = 0$

If $x - 30 = 0$, then $x = 30$

If $x - 4 = 0$, then $x = 4$

$\therefore$ Y alone can finish the work in 30 days.

17. A can complete a work in 20 days, while B can complete the same work in 25 days. Both worked together for 10 days, and then C alone completed the remaining work in 10 days. In how many days will A, B, and C together complete the same work?

A एक कार्य को 20 दिनों में पूरा कर सकता है, जबकि B उसी कार्य को 25 दिनों में पूरा कर सकता है। दोनों ने 10 दिनों तक एक साथ काम किया और फिर C ने अकेले शेष काम 10 दिनों में पूरा कर लिया। A, B और C मिलकर उसी कार्य को कितने दिनों में पूरा करेंगे?

Solution :

Let total work be 100 units. $A = \frac{100}{20} = 5$, $B = \frac{100}{25} = 4$

Work done by A and B together in 10 days $= 10 \times (5 + 4) = 90$

Remaining work = $100 - 90 = 10$

Work done by C in 10 days = 10

Efficiency of C = $\frac{10}{10} = 1$

Time taken by A, B, and C together to complete the work $= \frac{100}{5+4+1} = \frac{100}{10} =$ 10 days.

18. A and B can complete a work in 15 days and 10 days, respectively. They started doing the work together, but after 4 days, B had to leave. Then A, working with a new worker C, completed the remaining work in 3 days. If C works alone, in how many days can he do 40% of the same work?

A और B एक कार्य को क्रमशः 15 दिन और 10 दिन में पूरा कर सकते हैं। उन्होंने एक साथ काम करना शुरू किया लेकिन 4 दिन बाद B को काम छोड़ना पड़ा। फिर A ने एक नए कर्मचारी C के साथ काम करके शेष कार्य 3 दिनों में पूरा किया। यदि C अकेले कार्य करता है, तो वह उसी कार्य का 40% कितने दिनों में पूरा कर सकता है?

Solution :

Let total work be 30 units.

A's efficiency = $\frac{30}{15} = 2$, B's efficiency = $\frac{30}{10} = 3$

$4(a + b) + 3(a + c) = 30$

$4 \times 5 + 3 \times 2 + 3c = 30$

$3c = 30 - 20 - 6 = 4$

$c = \frac{4}{3}$

40% of 30 = 12 units

Time taken by C to complete 40% of the work: $12 \times \frac{3}{4} = 9$ days.

19. A, B, and C together can complete a work in x, 30, and 45 days, respectively. B and C worked together for 6 days. The remaining work was completed by A alone in 12 days. The value of x is:

A, B और C मिलकर एक काम को क्रमशः x, 30 और 45 दिनों में पूरा कर सकते हैं। B और C ने 6 दिनों तक एक साथ काम किया। शेष कार्य A द्वारा अकेले 12 दिनों में पूरा किया गया। x का मान है:

Solution :

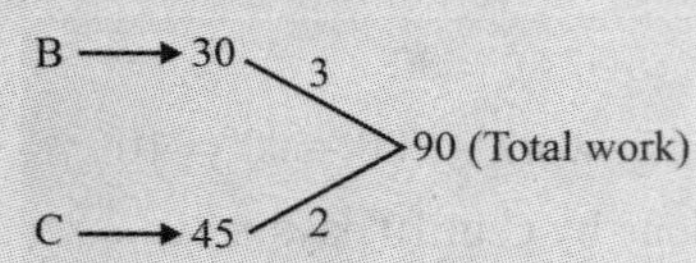

6 days' work of (B + C) $= (3+2)\times 6 =$ 30 units.

Remaining work $= 90 - 30 = 60$ units.

A completed the remaining work in 12 days.

A's efficiency $= \frac{60}{12} = 5$

Time taken by A to complete the total work $= \frac{90}{5} = 18$ days.

$\therefore x = 18$

20. A and B can complete a certain task in 24 days and 40 days, respectively. They worked together for 8 days. C completed the remaining task in 14 days. Working together, A and C will complete 75% of the same work in:

A और B एक निश्चित कार्य को क्रमशः 24 दिन और 40 दिन में पूरा कर सकते हैं। उन्होंने 8 दिनों तक एक साथ काम किया। C ने शेष कार्य 14 दिनों में पूरा किया। एक साथ काम करते हुए, A और C समान कार्य का 75% कार्य कितने समय में पूरा करेंगे:

Solution :

Let total work be 120 units.
A - 24 - 5
B - 40 - 3
8(a+b) + 14c = 120
8×8 + 14c = 120
14c = 120-64 = 56
c = 4
75% of 120 = 90 units

$\frac{90}{5+4}$

$\frac{90}{9}$

10 days

21. Y can complete half of the work in 20 days, and Y can do one-fifth of the same work in 10 days. X started the work and left after 8 days. Then Y took over to complete the remaining work. The total number of days taken by them to complete the work is:

Y काम का आधा हिस्सा 20 दिनों में पूरा कर सकता है और Y उसी काम का पांचवां हिस्सा 10 दिनों में पूरा कर सकता है। X ने काम शुरू किया और 8 दिन बाद छोड़ दिया। फिर शेष कार्य को पूरा करने के लिए Y ने कार्यभार संभाला। कार्य पूरा करने में उनके द्वारा लिए गए कुल दिनों की संख्या है:

Solution :

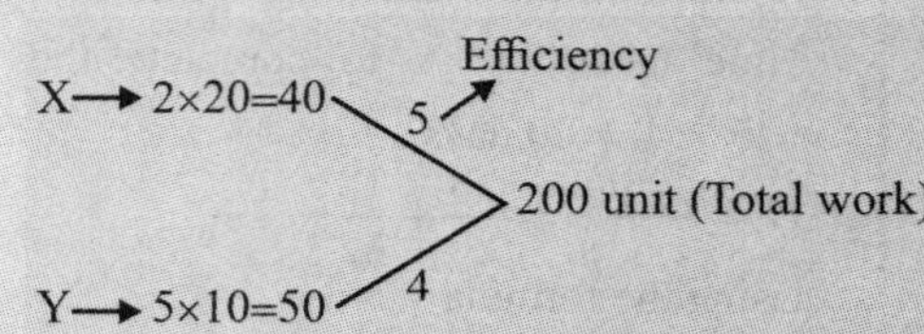

$\because$ X stopped the work after 8 days. Work done by X in 8 days: $8 \times 5 = 40$ units.

Remaining work is completed by Y.

Remaining work: $200 - 40 = 160$ units.

Time taken by Y: $\frac{160}{4} = 40$ days.

Total time taken by X and Y to complete the work: $8 + 40 = 48$ days.

22. The ratio of the efficiencies of A, B, and C is 7 : 5 : 4. Working together, they can finish a work in 35 days. A and B work together for 28 days. The remaining work will be completed (in days) by C alone:

A, B और C की दक्षताओं का अनुपात 7 : 5 : 4 है। एक साथ काम करते हुए, वे एक काम को 35 दिनों में पूरा कर सकते हैं। A और B 28 दिनों तक एक साथ काम करते हैं। शेष कार्य अकेले C द्वारा (दिनों में) पूरा किया जाएगा:

Solution :

Total work done by (A + B + C): $(7+5+4)\times 35 = 560$ units.

Work done by (A + B) in 28 days: $28 \times 12 = 336$ units.

Time taken by C to complete the remaining work: $\frac{560-336}{4} = \frac{224}{4} = 56$ days.

23. A and B, working together, can complete a work in d days. Working alone, A takes $(8+d)$ days and B takes

$(18+d)$ days to complete the same work. A works for 4 days. The remaining work will be completed by B alone, in:

A और B, एक साथ काम करते हुए, एक कार्य को d दिनों में पूरा कर सकते हैं। अकेले काम करते हुए, A को उसी काम को पूरा करने में $(8 + d)$ दिन लगते हैं और B को $(18 + d)$ दिन लगते हैं। A 4 दिनों तक काम करता है। शेष कार्य B अकेले कितने समय में पूरा करेगा:

Solution :

If A does the work in $(x + a)$ days, and B does it in $(x + b)$ days, while both can do the work in x days, then: $x = \sqrt{ab}$

$\therefore d = \sqrt{8 \times 18} = 12$ days.

A ⟶ 20 — 3

60 units

B ⟶ 30 — 2

Remaining work: $60 - 12 = 48$ units.

Time required by B: $\frac{48}{2} = 24$ days.

24. A is as efficient as B and C together. Working together, A and B can complete a work in 36 days, and C alone can complete it in 60 days. A and C work together for 10 days. B alone will complete the remaining work in:

A, B और C के समान ही कुशल है। A और B एक साथ काम करते हुए एक काम को 36 दिनों में पूरा कर सकते हैं और C अकेले इसे 60 दिनों में पूरा कर सकता है। A और C 10 दिनों तक काम करते हैं। B अकेले शेष कार्य को कितने समय में पूरा करेगा:

Solution :

Let total work be 180

A+B - 36 - 5

C - 60 - 3

A = B+C

A = B + 3

A+B = 5

A = 4 , B = 1 , C = 3

10(a+c) + b×n = 180

10(7) + 1×n = 180

n = 180-70 = 110 Days

25. A can complete a certain work in 35 days and B can complete the same work in 15 days. They worked together for 7 days, then B left the work. In how many days will A alone complete 60% of the remaining work?

A एक निश्चित कार्य को 35 दिनों में पूरा कर सकता है और B उसी कार्य को 15 दिनों में पूरा कर सकता है। उन्होंने 7 दिनों तक एक साथ काम किया, फिर B ने काम छोड़ दिया। A अकेले शेष कार्य का 60% कितने दिनों में पूरा करेगा?

Solution :

7 days' work of (A+B) = $7 \times (\frac{1}{35} + \frac{1}{15}) = \frac{70}{105} = \frac{2}{3}$

$\therefore$ Remaining work = $1 - \frac{2}{3} = \frac{1}{3}$

60% part of the remaining work = $\frac{1}{3} \times \frac{60}{100} = \frac{1}{5}$ part

$\therefore$ Time taken by A to complete $\frac{1}{5}$ part of the work: $35 \times \frac{1}{5} = 7$ days.

26. Smith and Ajit can complete a task in 12 days and 18 days, respectively. If they work together on the task for 4 days, then the fraction of the task that will be left is:

स्मिथ और अजीत एक कार्य को क्रमशः 12 दिन और 18 दिन में पूरा कर सकते हैं। यदि वे किसी कार्य पर 4 दिनों तक एक साथ काम करते हैं, तो शेष कार्य का अंश होगा:

Solution :

Total work = 36

Smith - 12 - 3

Ajit - 18 - 2

$4(3 + 2) = 20$ units done

Remaining work = $36 - 20 = 16$ units

Fraction of remaining work = $\frac{16}{36} = \frac{4}{9}$

27. A and B can do a work in 15 days and 10 days respectively. They begin the work together but B leaves after two days. Now A completes the remaining work. The total number of days needed for the completion of the work is:

A और B एक काम को क्रमशः 15 दिन और 10 दिन में कर सकते हैं। वे एक साथ काम शुरू करते हैं लेकिन B दो दिन बाद काम छोड़ देता है। अब A शेष कार्य पूरा करता है। कार्य पूरा करने के लिए आवश्यक दिनों की कुल संख्या है:

Solution :

A → 15 (2)
B → 10 (3)
→ 30 units

Work done in 2 days by both = $(2+3)\times 2 = 10$ units

Remaining work = $30 - 10 = 20$ units

Time taken by A to complete the remaining work: $\frac{20}{2} = 10$ days.

Total Time = $10 + 2 = 12$ days.

28. A can finish a piece of work in a certain number of days. B takes 45% more number of days to finish the same work independently. They worked together for 58 days and then the remaining work was done by B alone in 29 days. In how many days could A have completed the work, had he worked alone?

A किसी कार्य को निश्चित दिनों में पूरा कर सकता है। B को उसी कार्य को स्वतंत्र रूप से पूरा करने में 45% अधिक दिन लगते हैं। उन्होंने 58 दिनों तक एक साथ काम किया और फिर शेष काम B ने अकेले 29 दिनों में पूरा किया। यदि A अकेले कार्य करता तो वह कार्य कितने दिनों में पूरा कर सकता था?

Solution :

B:A = 29 : 20 (Time Ratio)

B:A = 20 : 29 (Efficiency Ratio)

Total work = $58(a + b) + 29b$

Tw = $58(20 + 29) + 29 \times 20$

A will complete Total work in: $\frac{58(20 + 29) + 29 \times 20}{29}$

$2(20 + 29) + 20 = 2 \times 49 + 20 = 98 + 20 =$ 118 days.

29. A can complete work in 25 days and B can complete the same work in 20 days. They started the work together but B left after 4 days and A continued to work. In how many days will the entire work be completed?

A कार्य को 25 दिनों में पूरा कर सकता है और B उसी कार्य को 20 दिनों में पूरा कर सकता है। उन्होंने एक साथ काम शुरू किया लेकिन B ने 4 दिन बाद काम छोड़ दिया और A ने काम करना जारी रखा। पूरा काम कितने दिनों में पूरा हो जाएगा?

Solution :

A → 25 (4)
B → 20 (5)
→ 100 units (Total work)

Work done in 4 days by (A + B) = (5 + 4) × 4 = 36 units

Remaining work = 100 - 36 = 64 units

Then, time taken by A to complete the remaining work = $64/4 = 16$ days

Hence, entire work be completed = $16 + 4 =$ 20 days

30. A, B, and C can do a work separately in 18, 36, and 54 days, respectively. They started the work together, but B and C left 5 days and 10 days, respectively, before the completion of the work. In how many days was the work finished?

A, B और C एक कार्य को अलग-अलग क्रमश: 18, 36 और 54 दिनों में कर सकते हैं। उन्होंने एक साथ काम शुरू किया, लेकिन B और C ने काम पूरा होने से क्रमश: 5 दिन और 10 दिन पहले छोड़ दिया। कार्य कितने दिनों में समाप्त हुआ?

Solution :

Let Tw = 108 units

A - 18 - 6

B - 36 - 3

C - 54 - 2

$108 + 5 \times (3) + 10 \times (2)$

$108 + 15 + 20$

$108 + 35$

143 units done by (A+B+C)

$\frac{143}{11} = 13$ days.

31. 25 persons can complete a work in 60 days. They started the work. 10 persons left the work after x days. If the whole work was completed in 80 days, then what is the value of x?

25 व्यक्ति एक कार्य को 60 दिनों में पूरा कर सकते हैं। उन्होंने काम शुरू कर दिया. x दिनों के बाद 10 व्यक्तियों ने काम छोड़

दिया। यदि पूरा कार्य 80 दिनों में पूरा हो गया, तो x का मान क्या है?

Solution :

We know that, Formula $M_1D_1 = M_2D_2 + M_3D_3$
$25 \times 60 = 25 \times x + 15 \times (80 - x)$
$5 \times 60 = 5x + 3(80 - x)$
$300 = 5x + 240 - 3x$
$2x = 60$
$x = 30$ days.

32. 12 men can complete a painting work in 8 days. However, 16 women can complete the same painting work in 12 days. 8 men started painting the house. After 6 days of painting, 2 men were replaced by 4 women. Now how many days will they take to complete the remaining painting?

12 आदमी एक पेंटिंग का काम 8 दिनों में पूरा कर सकते हैं। हालाँकि, 16 महिलाएँ उसी पेंटिंग कार्य को 12 दिनों में पूरा कर सकती हैं। 8 आदमी घर की पुताई करने लगे। 6 दिनों की पेंटिंग के बाद, 2 पुरुषों का स्थान 4 महिलाओं ने ले लिया। अब शेष पेंटिंग को पूरा करने में उन्हें कितने दिन लगेंगे?

Solution :

12 men can complete a work in 8 days = $12 \times 8 = 96$

16 women can complete the same work in 12 days = $16 \times 12 = 192$

The ratio of men and women = $96 : 192 = 1 : 2$

Total work = $12 \times 1 \times 8 = 96$ units

Now, 8 men started painting for 6 days = $8 \times 1 \times 6 = 48$ units

Remaining work = $96 - 48 = 48$ units

2 men were replaced by 4 women = $4 \times 2 = 8$ women

Time taken to complete the remaining work: $\frac{48}{8} = 6$ days.

33. A contract is to be completed in 75 days and 187 men are to work 15 hours per day. After 65 days, $\frac{3}{5}$ of the work is completed. How many additional men may be employed, so that the work may be completed in time, each man now working 17 hours per day?

एक अनुबंध को 75 दिनों में पूरा किया जाना है और 187 पुरुषों को प्रति दिन 15 घंटे काम करना है। 65 दिनों के बाद, कार्य का $\frac{3}{5}$ भाग पूरा हो जाता है। कितने अतिरिक्त आदमी नियोजित किए जा सकते हैं, ताकि काम समय पर पूरा हो सके, प्रत्येक आदमी अब प्रति दिन 17 घंटे काम कर रहा है?

Solution :

Let x more men should be employed.

$$\frac{M_1D_1H_1}{W_1} = \frac{(M_2D_2H_2)}{W_2}$$

$$\frac{187 \times 65 \times 15}{\frac{3}{4}} = \frac{(187 + x) \times 10 \times 17}{\frac{2}{5}}$$

$11 \times 65 = 187 + x$

$x = 715 - 187 = 528$

34. A certain number of persons can complete a task in 28 days. If there were 150 persons more, it would have taken 7 days less for the task to be completed. How many persons are there in the beginning?

व्यक्तियों की एक निश्चित संख्या किसी कार्य को 28 दिनों में पूरा कर सकती है। यदि 150 व्यक्ति अधिक होते, तो कार्य पूरा होने में 7 दिन कम लगते। आरंभ में कितने व्यक्ति हैं?

Solution :

Let the number of persons in the beginning $= n$
$n \times 28 = (n + 150) \times (28 - 7)$
$n \times 28 = (n + 150) \times 21$
$n \times 4 = (n + 150) \times 3$
$4n - 3n = 450$
$n = 450$

35. If 45 persons complete a work in 18 days, working 8 hours a day, then how many persons are required to complete two-thirds of the same work in 20 days, working 9 hours a day?

यदि 45 व्यक्ति प्रतिदिन 8 घंटे कार्य करके एक कार्य को 18 दिनों में पूरा करते हैं, तो कितने व्यक्तियों को उसी कार्य का दो-तिहाई भाग प्रतिदिन 9 घंटे कार्य करके 20 दिनों में पूरा करना होगा?

Solution :

Let the number of persons required = x

$$\frac{M_1D_1H_1}{W_1} = \frac{M_2D_2H_2}{W_2}$$

$$\frac{45 \times 18 \times 8}{1} = \frac{x \times 20 \times 9}{\frac{2}{3}}$$

$x = 24$

36. A certain number of men can do a piece of work in 54 days. If there were 30 more men, it would take 18 days less to complete the work. The number of men, initially is:

एक निश्चित संख्या में आदमी एक काम को 54 दिनों में कर सकते हैं। यदि 30 आदमी और हों तो काम पूरा करने में 18 दिन कम लगेंगे। प्रारंभ में पुरुषों की संख्या है:

Solution :

Let x persons can do the work. As per question: $x \times 54 = (x + 30) \times (54 - 18)$
$x \times 54 = (x + 30) \times 36$
$3x = 2x + 60$
$x = 60$

37. A and B working separately can do a piece of work in 21 and 24 days, respectively. If they work for a day alternately with A beginning the work, it would be completed in ______:

A और B अलग-अलग कार्य करते हुए एक कार्य को क्रमशः 21 और 24 दिनों में पूरा कर सकते हैं। यदि वे A के साथ काम शुरू करके एक दिन के लिए बारी-बारी से काम करते हैं, तो यह ______ में पूरा हो जाएगा।

Solution :

Let total work be 168 units.
A - 21 - 8
B - 24 - 7
$8 + 7 = 15$ units in 2 days.
165 units in 22 days
$\frac{3}{8}$ days for remaining 3 units.
$22\frac{3}{8}$ days.

38. A, B, and C complete a work in 5 days, 8 days, and 12 days, respectively. If they work together and get ₹ 9,800 for the work, then how much does A get?

A, B और C एक कार्य को क्रमशः 5 दिन, 8 दिन और 12 दिन में पूरा करते हैं। यदि वे एक साथ काम करते हैं और काम के लिए उन्हें ₹ 9,800 मिलते हैं, तो A को कितना मिलता है?

Solution :

The ratio of time of A, B, and C = 5 : 8 : 12
Ratio of efficiency = $\frac{1}{5} : \frac{1}{8} : \frac{1}{12} = 24 : 15 : 10$
∵ Money distribution of work efficiency:
Amount received by A = $\frac{9800}{24 + 15 + 10} \times 24$

$\frac{9800}{49} \times 24 = 200 \times 24 = 4800$

39. To do a certain work A and B work on alternate days, with B beginning the work on the first day. A can finish the work alone in 48 days. If the work gets completed in $1\frac{11}{3}$ days, then B alone can finish 4 times the same work in:

एक निश्चित कार्य करने के लिए A और B वैकल्पिक दिनों में कार्य करते हैं, B पहले दिन कार्य शुरू करता है। A अकेले कार्य को 48 दिनों में पूरा कर सकता है। यदि कार्य $1\frac{11}{3}$ दिन में पूरा हो जाता है तो B अकेले उसी कार्य को 4 गुना समय में पूरा कर सकता है:

Solution :

B works on odd days.
A works on even days.
So, B works for 6 days while A works for $5\frac{1}{3}$ days.
$5\frac{1}{3}$ days' work of A = $\frac{1}{48} \times \frac{16}{3} = \frac{1}{9}$
Remaining work = $1 - \frac{1}{9} = \frac{8}{9}$
B completes $\frac{8}{9}$ work in 6 days.
B will complete the whole work: $6 \times \frac{9}{8} = \frac{27}{4}$ days.
B will complete 4 times the work in 27 days.

40. A, B and C can do a piece of work in 20 days, 30 days and 60 days respectively. If A works daily and on every third day B and C also work with A, then in how many days, will the work be completed?

A, B और C एक काम को क्रमशः 20 दिन, 30 दिन और 60 दिन में कर सकते हैं। यदि A प्रतिदिन कार्य करता है और हर तीसरे दिन B और C भी A के साथ कार्य करते हैं, तो कार्य कितने दिनों में पूरा होगा?

Solution :

Total Work = 60 units
$A = 20$ -- 3
$B = 30$ -- 2
$C = 60$ -- 1
In 3 Days,
work = $(3 + 3 + 6) = 12$ units
$\frac{60}{12} = 5$
$5 \times 3 = 15$ Days

41. A and B are equally efficient, and each can individually complete a piece of work in 36 days, if none takes any holiday. A and B started working together on this piece of work, but A took a day off after every five days of work, while B took a day off after every seven days of work. If the duo had started work on 1 July 2021, on which date was the work completed?

A और बी समान रूप से कुशल हैं, और यदि कोई भी छुट्टी नहीं लेता है, तो प्रत्येक व्यक्तिगत रूप से 36 दिनों में एक काम पूरा कर सकता है। A और B ने इस काम पर एक साथ काम करना शुरू किया, लेकिन A ने हर पांच दिन के काम के बाद एक दिन की छुट्टी ली, जबकि B ने हर सात दिन के काम के बाद एक दिन की छुट्टी ली। यदि दोनों ने 1 जुलाई 2021 को काम शुरू किया था, तो काम किस तारीख को पूरा हुआ?

Solution :

Total Work = 36 Units
A takes holiday on 6th Day
B takes holiday on 8th Day
In 8 Days,
work done = 14 units
In 16 Days,
work done = 28 units
Remaining 8 units
17 Day – 2
18 Day – 1(Only B working)
19 Day – 2
20 Day – 2
21 Day – Remaining 1 unit will be done on 21st Day
On 21st July, work will be completed.

42. A can do 1/5 of a piece of work in 20 days, B can do 30% of the same work in 36 days, while C can do 80% of the same work in 160 days. B and C together started and worked for x days. After x days B left the work, and A joined C and both completed the remaining work in $(x - 41)$ days. If the ratio between the work done by (B+C) together to the work done by (A + C) together is 19:6, then what fraction of the same work can be completed by C alone in $2x$ days?

A किसी काम का $1/5$ भाग 20 दिनों में कर सकता है, ▯ उसी काम का 30% 36 दिनों में कर सकता है, जबकि C उसी काम का 80% 160 दिनों में कर सकता है। B और C ने एक साथ काम शुरू किया और x दिनों तक काम किया। x दिनों के बाद B ने काम छोड़ दिया, और A, C में शामिल हो गया और दोनों ने शेष कार्य $(x - 41)$ दिनों में पूरा किया। यदि (B+C) द्वारा एक साथ किए गए कार्य का (A + C) द्वारा एक साथ किए गए कार्य से अनुपात 19:6 है, तो उसी कार्य का कितना भाग C अकेले $2x$ दिनों में पूरा कर सकता है?

Solution :

600
A - 100-6
B - 120-5
C - 200-3

$$\frac{8x}{9(x-41)} = \frac{19}{6}$$

$$48x = 171x - 171 \times 41$$

$$x = \frac{171 \times 41}{123} = 57$$

$$\frac{3 \times 2 \times 57}{600} = \frac{57}{100}$$

43. There are two water taps in a tank which can fill the empty tank in 12 hours and 18 hours respectively. It is seen that there is a leakage point at the bottom of the tank which can empty the completely filled tank in 36 hours. If both the water taps are opened at the same time to fill the empty tank and the leakage point was repaired after 1 hour, then in how much time the empty tank will be completely filled?

एक टंकी में दो पानी के नल हैं जो खाली टंकी को क्रमशः 12 घंटे और 18 घंटे में भर सकते हैं। ऐसा देखा गया है कि टैंक के तल पर एक रिसाव बिंदु है जो 36 घंटे में पूरी तरह से भरे हुए टैंक को खाली कर सकता है। यदि खाली टंकी को भरने के लिए दोनों पानी के नल एक ही समय में खोले जाएं और रिसाव बिंदु की मरम्मत 1 घंटे के बाद की जाए, तो खाली टंकी कितने समय में पूरी भर जाएगी?

Solution :

Suppose, filling taps and leakage are respectively
Total work = 36 units
12-1
18-2
36- (-1)
Work done in first hour = $3 + 2 - 1 = 4$ units
Remaining work = $36 - 4 = 32$ units
Time required to fill the tank after the leakage is repaired: $\frac{32}{3+2} = 6\frac{2}{5}$ hours.

Total time to fill the tank = $1 + 6\frac{2}{5}$ hours
= 7 hours 24 minutes

44. Pipes A, B and C can fill a tank in 30h, 40h and 60h respectively. Pipes A, B and C are opened at 7 a.m., 8 a.m. and 10 a.m., respectively on the same day. When will the tank be full?

पाइप A, B और C एक टैंक को क्रमशः 30 घंटे, 40 घंटे और 60 घंटे में भर सकते हैं। पाइप A, B और C एक ही दिन में क्रमशः सुबह 7 बजे, सुबह 8 बजे और सुबह 10 बजे खोले जाते हैं। टंकी कब भरेगी?

Solution :

TW = 120
30-4
40-3
60-2
Let Pipe A can fill = In x hour
Then, Pipe B can fill = In $(x-1)$ hour
Pipe C can fill = $(x-3)$ hour
$4x+3(x-1)+2(x-3)=120$
$4x+3x-3+2x-6=120$
$9x=120+9$
$x=\frac{129}{9}=14\frac{1}{3}$
$=$ 14 hours $+ \frac{1}{3} \times$ 60 minutes $=$ 14 hours 20 minutes.
Hence, Tank will be full: 7 : 00 am + 14 hours 20 minutes = 9 : 20 pm.

45. Two pipes A and B can fill a tank in 12 hours and 18 hours, respectively. Both pipes are opened simultaneously. In how much time will the empty tank be filled completely?
दो पाइप A और B एक टैंक को क्रमशः 12 घंटे और 18 घंटे में भर सकते हैं। दोनों पाइप एक साथ खोले जाते हैं। खाली टंकी कितने समय में पूरी भर जाएगी?

Solution :

Time taken by pipe A and pipe B to fill the tank: $\frac{12\times18}{12+18}=\frac{36}{5}$ hours $=7\frac{1}{5}$ hours $=$ 7 hours 12 minutes.

46. Pipes A, B and C can fill a tank in 15, 30 and 40 hours, respectively. Pipes A, B and C are opened at 6 a.m., 8 a.m. and 10 a.m., respectively, on the same day. When will the tank be full?
पाइप A, B और C एक टैंक को क्रमशः 15, 30 और 40 घंटे में भर सकते हैं। पाइप A, B और C एक ही दिन क्रमशः सुबह 6 बजे, सुबह 8 बजे और 10 बजे खोले जाते हैं। टंकी कब भरेगी?

Solution :

Let tank will be filled after t hours from 6:00 am: $\frac{t}{15}+\frac{t-2}{30}+\frac{t-4}{40}=1$
$\frac{8t+4t-8+3t-12}{120}=1$
$15t-20=120$
$15t=140$
$t=\frac{28}{3}=9\frac{1}{3}$ hours $=$ 9 hours 20 minutes.
Hence, the tank will be filled at 3 : 20 PM.

47. Pipe A can fill a tank in 10 hours. Pipe B can fill the same tank in 12 hours. Pipe C can empty the full tank in 16 hours. All the pipes are opened at 8:00 AM. and Pipe A and B are closed at 10:00 A.M. After how much time from starting will the tank be empty?
पाइप A एक टंकी को 10 घंटे में भर सकता है। पाइप B उसी टंकी को 12 घंटे में भर सकता है। पाइप C पूरी टंकी को 16 घंटे में खाली कर सकता है। सुबह आठ बजे सभी पाइप खोल दिए जाते हैं। और पाइप A और B सुबह 10:00 बजे बंद कर दिए जाते हैं। प्रारंभ करने के कितने समय बाद टैंक खाली हो जाएगा?

Solution :

Pipe A and B fill the tank but pipe C empties the tank.

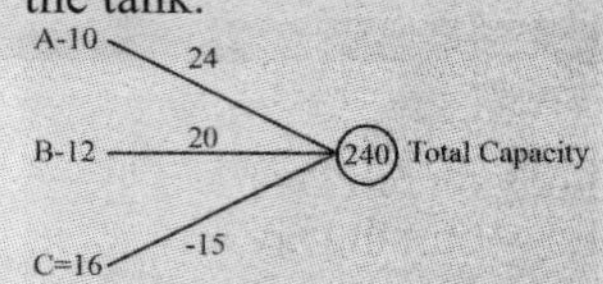

Part filled by (A+B-C) in 1 hour:
$\frac{24}{240}+\frac{20}{240}-\frac{15}{240}=\frac{29}{240}$
Part filled by (A + B) in 2 hours:
$\frac{29}{240}\times2=\frac{58}{240}$.
C can empty the tank: $\frac{\frac{58}{240}}{\frac{15}{240}}=\frac{58}{15}$ hours $=$ 3 hours 52 minutes.
Hence, time taken to empty the tank:
2 hours + 3 hours 52 minutes = 5 hours 52 minutes.

48. A pump can fill a tank with water in 1 hour. Because of a leak, it took $1\frac{1}{3}$ hours to fill the tank. In how many hours can the leak along drain all the water of the tank when it is full?
एक पंप एक टंकी को 1 घंटे में पानी से भर सकता है। रिसाव के कारण टैंक को भरने में $1\frac{1}{3}$ घंटे लगे। टंकी के भर जाने पर रिसाव कितने घंटों में सारा पानी निकाल सकता है?

Solution :

Let all water will drain, due to leak in 'T' hours.

According to the question: $\frac{1}{1} - \frac{1}{T} = \frac{1}{\frac{4}{3}}$

$\frac{1}{T} = 1 - \frac{3}{4}$

$\frac{1}{T} = \frac{1}{4}$

$T = 4$ hours.

49. Two pipes A and B can fill a cistern in $12\frac{1}{2}$ hours and 25 hours, respectively. The pipes are opened simultaneously and it is found that due to a leakage in the bottom, it took 1 hour 40 minutes more to fill the cistern. When the cistern is full, in how much time will the leak empty the cistern?

दो पाइप A और B एक टंकी को क्रमशः $12\frac{1}{2}$ घंटे और 25 घंटे में भर सकते हैं। पाइपों को एक साथ खोला गया और पाया गया कि तली में रिसाव के कारण टंकी को भरने में 1 घंटा 40 मिनट अधिक लगे। जब टंकी भर जाती है तो रिसाव से टंकी कितने समय में खाली हो जाएगी?

Solution :

Let, the cistern will be empty in 'T' hours.

According to the question:

$\frac{1}{\frac{25}{2}} + \frac{1}{25} - \frac{1}{T} = \frac{1}{10}$

$\frac{3}{25} - \frac{1}{T} = \frac{1}{10}$

$\frac{1}{T} = \frac{3}{25} - \frac{1}{10}$

$\frac{1}{T} = \frac{6-5}{50} = \frac{1}{50}$

$T = 50$ hours.

50. Pipes A, B and C can fill an empty tank in $\frac{30}{7}$ hours if all the three pipes are opened simultaneously. A and B are filling pipes, and C is an emptying pipe. Pipe A can fill the tank in 15 hours, and pipe C can empty it in 12 hours. In how long (in hours) can pipe B alone fill the empty tank?

यदि तीनों पाइप एक साथ खोले जाएं तो पाइप A, B और C एक खाली टैंक को $\frac{30}{7}$ घंटे में भर सकते हैं। A और B भरने वाले पाइप हैं और C खाली करने वाला पाइप है। पाइप A टंकी को 15 घंटे में भर सकता है और पाइप C इसे 12 घंटे में खाली कर सकता है। पाइप B अकेले खाली टैंक को कितने समय में (घंटों में) भर सकता है?

Solution :

When pipe A, B, and C are opened together, it takes $\frac{30}{7}$ hours to fill the tank.

$\frac{7}{30}$ part filled in 1 hour.

According to the question:

$\frac{1}{15} + B - \frac{1}{12} = \frac{7}{30}$

$B = \frac{7}{30} - \frac{1}{15} + \frac{1}{12} = \frac{14+4+5}{60} = \frac{1}{4}$

Part filled by B in 1 hour = $\frac{1}{4}$

Time taken by B to fill the empty tank: $B = \frac{1}{\frac{1}{4}} = 4$ hours.

51. Taps A and B can fill a tank in 15 minutes and 10 minutes, respectively while tap C can empty the full tank in x minutes. If all the three taps are opened together, the tank is filled completely in 8 minutes. Tap C alone will empty $\frac{3}{8}$th part of the tank in:

नल A और B एक टंकी को क्रमशः 15 मिनट और 10 मिनट में भर सकते हैं, जबकि नल C पूरी टंकी को x मिनट में खाली कर सकता है। यदि तीनों नल एक साथ खोल दिए जाएं तो टंकी 8 मिनट में पूरी भर जाती है। अकेले नल C कितने समय में टैंक का $\frac{3}{8}$वाँ भाग खाली कर देगा:

Solution :

According to the question, let the time taken by C to empty the tank be x minutes. When all three taps are opened together: $\frac{8}{15} + \frac{8}{10} - \frac{8}{x} = 1$

Solve the equation: $x = 24$ minutes.

Time taken to empty $\frac{3}{8}$ part: $24 \times \frac{3}{8} = 9$ minutes.

52. Pipe A and B can empty a full tank in 20 and 15 hours, respectively, whereas C can fill the tank alone in x hours. Three pipes are opened together at the same time, and to empty $\frac{1}{18}$th part of the tank it took 40 minutes. What is the value of x?

पाइप A और B एक भरे हुए टैंक को क्रमशः 20 और 15 घंटे में खाली कर सकते हैं, जबकि C अकेले टैंक को x घंटे में भर सकता है। तीन पाइप एक ही समय में एक साथ खोले जाते हैं और टैंक के $\frac{1}{18}$ हिस्से को खाली करने में 40 मिनट का समय लगता है। x का मूल्य क्या है?

Solution :

Time taken to empty $\frac{1}{18}$ part by all three pipes = 40 minutes

Time taken to empty the tank: $40 \times 18 =$ 720 minutes = 12 hours.

According to the question: $\frac{1}{x} = \frac{1}{20} + \frac{1}{15} - \frac{1}{12}$

$\frac{1}{x} = \frac{3+4-5}{60} = \frac{2}{60}$

$x = 30$ hours.

53. Pipes A and B can fill a tank in 12 hours respectively and pipe C can empty the full tank in 24 hours. All three pipes are opened together, but after 4 hours pipe B is closed. In how many hours, the empty tank will be completely filled?

पाइप A और B एक टंकी को क्रमशः 12 घंटे में भर सकते हैं और पाइप C पूरी टंकी को 24 घंटे में खाली कर सकता है। तीनों पाइप एक साथ खोले जाते हैं, लेकिन 4 घंटे बाद पाइप B बंद हो जाता है। खाली टंकी कितने घंटे में पूरी भर जाएगी?

Solution :

According to the question:

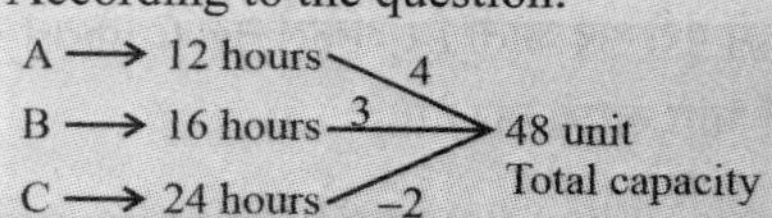

Part of tank filled by (A + B - C) in 4 hours:

$(4+3-2) \times 4 = 20$ units.

After 4 hours, pipe B is closed,

Time taken to fill the remaining part by (A - C): $\frac{48-20}{4-2} = \frac{28}{2} = 14$ hours.

Hence, total time: $4 + 14 = 18$ hours.

54. Pipes A and B can fill a tank in 16 hours and 24 hours, respectively, whereas pipe C can empty the full tank in 40 hours. All three pipes are opened together, but pipe C is closed after 10 hours. After how many hours will the remaining part of the tank be filled?

पाइप A और B एक टंकी को क्रमशः 16 घंटे और 24 घंटे में भर सकते हैं, जबकि पाइप C पूरी टंकी को 40 घंटे में खाली कर सकता है। तीनों पाइप एक साथ खोले जाते हैं, लेकिन पाइप C 10 घंटे बाद बंद हो जाता है। कितने घंटे बाद टंकी का शेष भाग भर जाएगा?

Solution :

Part filled by pipes A, B & C in 1 hour: $\frac{1}{16} + \frac{1}{24} - \frac{1}{40} = \frac{19}{240}$

Part filled by pipes A, B & C in 10 hours: $\frac{19}{240} \times 10 = \frac{19}{24}$

Remaining part: $1 - \frac{19}{24} = \frac{5}{24}$

Part filled by pipes A & B in 1 hour: $\frac{1}{16} + \frac{1}{24} = \frac{5}{48}$

Time taken by pipes (A + B) to fill $\frac{5}{24}$th part: $\frac{\frac{5}{24}}{\frac{5}{48}} = 2$ hours.

55. An inlet pipe A originating from a river can fill a reservoir in 30 days. An outlet pipe B, which is capable of emptying the completely filled reservoir in 50 days, drains water from the reservoir to an irrigation canal. The pipes are opened on alternate days, starting with A. On which day from the beginning will the reservoir get completely filled for the first time?

एक नदी से निकलने वाला इनलेट पाइप A एक जलाशय को 30 दिनों में भर सकता है। और एक आउटलेट पाइप B, जो 50 दिनों में पूरी तरह से भरे जलाशय को खाली करने में सक्षम है और जलाशय से पानी को सिंचाई नहर में निकाल देता है। पाइपों को A से शुरू करके वैकल्पिक दिनों में खोला जाता है। शुरुआत से किस दिन जलाशय पहली बार पूरा भर जाएगा?

Solution :

Part filled by pipe A in 1 day: $\frac{1}{30}$

Part emptied by pipe B in 1 day: $\frac{1}{50}$

Part filled by both pipes A & B in 2 days: $\frac{1}{30} - \frac{1}{50} = \frac{2}{150} = \frac{1}{75}$

Part filled in $73 \times 2 = 146$ days: $\frac{73}{75}$

Remaining part: $1 - \frac{73}{75} = \frac{2}{75}$

Pipe A will fill $\frac{2}{75}$ part in 1 day. Hence, total number of days: $146 + 1 = 147$ days.

Reservoir is completely filled on the 147th day.

56. Three pipes, A, B and C, can fill a cistern in 12, 18 and 24 minutes, respectively. If all the pipes are opened together for 7 minutes, what will be the volume of the water that overflows as the percentage of the total volume of the cistern?
तीन पाइप, A, B और C, एक टंकी को क्रमशः 12, 18 और 24 मिनट में भर सकते हैं। यदि सभी पाइपों को एक साथ 7 मिनट के लिए खोल दिया जाए, तो ओवरफ्लो होने वाले पानी की मात्रा टंकी की कुल मात्रा के प्रतिशत के रूप में क्या होगी?

Solution :

Part filled by pipes A, B & C in 7 minutes:

$\frac{7}{12}+\frac{7}{18}+\frac{7}{24}=\frac{42+28+21}{72}=\frac{91}{72}$ part.

Overflowed part: $\frac{91}{72}-1=\frac{19}{72}$ part.

Required percentage: $\frac{19}{72}\times 100=\frac{267}{18}\%$.

57. Pipes A and B can fill a tank in 16 hours and 24 hours, respectively, and pipe C alone can empty the full tank in x hours. All the pipes were opened together at 10:30 a.m., but C was closed at 2:30 p.m. If the tank was full at 8:30 p.m. on the same day, then what is the value of x?/
पाइप A और B एक टैंक को क्रमशः 16 घंटे और 24 घंटे में भर सकते हैं, और पाइप C अकेले पूरे टैंक को x घंटे में खाली कर सकता है। सुबह 10:30 बजे सभी पाइप एक साथ खोले गए, लेकिन C दोपहर 2:30 बजे बंद कर दिया गया। यदि टंकी रात 8:30 बजे भरी हुई थी। उसी दिन x का मान क्या है?

Solution :

Pipe A and B can fill a tank in 16 hrs and 24 hrs respectively. Pipe C alone can empty the full tank in x hrs.

Pipe A and Pipe B work for 10 : 30 AM to 8 : 30 PM = 10 hours.

Pipe C works for 10 : 30 AM to 2 : 30 PM = 4 hours.

According to the question:

$\frac{10}{16}+\frac{10}{24}-\frac{4}{x}=1$

$\frac{30+20}{48}=1+\frac{4}{x}$

$\frac{50}{48}=1+\frac{4}{x}$

$\frac{4}{x}=\frac{25}{24}-1=\frac{1}{24}$

$x=24\times 4=96$ hours.

58. Two taps can fill a tank in 20 min and 30 min respectively. There is an outlet tap at of that rectangular tank which can pump out 100 L of water per minute. 1000 L of water it takes 24 min to fill an empty tank. What is the volume of the tank? Exactly half level if the outlet tap is open, then:
दो नल एक टैंक को क्रमशः 20 मिनट और 30 मिनट में भर सकते हैं। उस आयताकार टैंक के किनारे एक आउटलेट नल है जो प्रति मिनट 100 लीटर पानी पंप कर सकता है। 1000 L पानी से एक खाली टैंक को भरने में 24 मिनट लगते हैं। टैंक का आयतन क्या है? ठीक आधे स्तर पर यदि आउटलेट नल खुला है, तो?

Solution :

A & B: Inlet, C: Outlet Tap.

Combined rate of A + B + C: $\frac{5}{3}$

Rate of individual taps: $A\rightarrow\frac{3}{1}$, $B\rightarrow\frac{2}{1}$, $C\rightarrow-\frac{10}{3}$.

Calculation:

$\frac{60}{5}=24$ min (time for combined inlet and outlet).

Volume: $10/3\rightarrow 100$ L.

$1\rightarrow 30$ L.

$60\rightarrow 30\times 60=1800$ L.

59. 23 Outlet fitted in a tank; some are inlet taps and some are outlet taps. Each inlet tap can fill the tank in 24 hours, and each outlet tap can empty the tank in 40 hours. If all the taps are open, the tank fills in 40 hours. Find the number of outlet taps:
एक टैंक में 23 लगे आउटलेट में से कुछ इनलेट नल हैं और कुछ आउटलेट नल हैं। प्रत्येक इनलेट नल 24 घंटे में टैंक को भर सकता है और प्रत्येक आउटलेट नल 40 घंटे में टैंक को खाली कर सकता है। यदि सभी नल खुले हैं तो टैंक 40 घंटे में भर जाता है, तो आउटलेट नल की संख्या ज्ञात कीजिए।
(a) 11 (b) 14 (c) 16 (d) 13

Solution :

TW = 120
Rate of inlets: $24\rightarrow 5$ units/hour
Rate of outlets: $40\rightarrow -3$ units/hour
$\frac{120}{4}=3$ unit
$5x-3(23-x)=3$
$5x-69+3x=3$
$8x=72$
$x=9$
$23-9=14$ Outlet taps.

60. A could fill a cistern in 18 hours, while Pipe B can drain a filled cistern in 30 hours. When the cistern is empty, Pipe A is turned on for an hour and then turned off. Now Pipe B is allowed to drain water from the cistern for an hour, and then it is turned off. The pipes are alternately opened until the cistern is full. How much time did it take to fill the cistern completely? पाइप A 18 घंटे में एक खाली टंकी भर सकता है जबकि पाइप B 30 घंटे में एक भरा हुआ टंकी खाली करता है। जब टंकी खाली होती है, तो पाइप A को एक घंटे के लिए चालू किया जाता है और फिर बंद कर दिया जाता है। अब पाइप B को एक घंटे के लिए चालू किया जाता है और फिर बंद कर दिया जाता है। प्रत्येक पाइप को बारी-बारी से एक घंटे के लिए तब तक खोला गया जब तक कि टंकी पूरी तरह से न भर जाए। टंकी को पूरी तरह से भरने में कितना समय लगा?

Solution :

18 hr → 5
30 hr → -3
→ 90

+5 -3 ↓ ×43 ↓
86 Hrs 86 Hrs

A → $\frac{4}{5}$ Hrs +4 unit
90

$86\frac{4}{5}$ Hr

61. Two pipes X and Y can fill an empty tank in 36 and 45 hours respectively. Ravi opens these two pipes simultaneously. After 6 hours 45 minutes, he comes back and sees that there is a leakage in the tank. He stops the leakage, and the tank took 15.5 hours to be filled completely after closing the leak. In what time will the leakage empty the filled tank? दो पाइप X और Y क्रमश: 36 और 45 घंटे में एक खाली टैंक भर सकते हैं। रवि इन दोनों पाइपों को एक साथ खोलता है। 6 घंटे 45 मिनट के बाद वह वापस आता है और देखता है कि टैंक में एक रिसाव है। वह रिसाव को रोकता है और इस तरह रिसाव बंद करने के बाद टैंक को पूरी तरह से भरने में 15.5 घंटे लगते हैं। भरे हुए टैंक को रिसाव अकेले कितने समय में खाली कर देगा?

Solution :

x → 36 → 5
y → 45 → 4
→ 180

Leek(z)=6hr 45min

$\frac{180}{9} = 20\ hr$

$z \times \frac{27}{4} = (x + y) \times \frac{9}{4}$

$z \times \frac{27}{4} = 9 \times \frac{9}{4}$

$x = 3$

Empty $= \frac{180}{3} = 60$

Extra time $22.15 - 20 = 2\text{hr}15\text{min} = \frac{9}{4}\text{hr}$

62. Two pumps, which can fill a tank in 16 hours and 20 hours while working alone, were opened by Gourav for 10 hours 12 minutes. It was realized that 413 liters of water spilled over the ground due to overflow in the tank. What is the total capacity of the tank? पंप अकेले काम करते हुए क्रमश: 16 घंटे और 20 घंटे में एक टैंक भर सकते हैं। गौरव ने दोनों पंपों को 10 घंटे 12 मिनट के लिए खोला और पाया कि टैंक में ओवरफ्लो होने के कारण 413 लीटर पानी जमीन पर गिर गया। टैंक की कुल क्षमता क्या है?

Solution :

A → 36 → 5
B → 45 → 4
→ 80

Leek(z)=6hr 45min

$\frac{180}{9} = 20\ hr$

filled in $10\frac{1}{5} = \frac{51}{5} \times 9 = \frac{459}{5}$

Overflow $= \frac{459}{5} - 80 = \frac{59}{5} \rightarrow 413L$

$1 \rightarrow 35L$

Capacity $= 80 \times 35 = 2800L$

63. Pipe A takes $\frac{4}{5}$ of the time required by pipe B to fill an empty tank individually. When an outlet pipe C is also opened simultaneously with pipes A and B, it takes $\frac{4}{5}$ more time to fill the empty tank than it takes when only pipe A and pipe B are opened together. If it takes 40 hours to fill the tank when all three pipes are opened simultaneously, in what time (in hours) will pipe C empty the full tank, operating alone?
पाइप A एक खाली टैंक को व्यक्तिगत रूप से भरने में पाइप B द्वारा आवश्यक समय का $\frac{4}{5}$ समय लेता है। जब एक आउटलेट पाइप C को पाइप A और B के साथ एक साथ खोला जाता है, तो खाली टैंक को भरने में केवल पाइप A और पाइप B को एक साथ खोलने की तुलना में $\frac{4}{5}$ अधिक समय लगता है। यदि तीनों पाइपों को एक साथ खोलने पर टैंक को भरने में 40 घंटे लगते हैं, तो पाइप C अकेले काम करते हुए कितने समय (घंटों) में पूरा टैंक खाली कर देगा?

Solution :

A : B (Time) = 4 : 5

A : B (Efficiency) = 5 : 4

A + B + C : A + B (Time) = 9 : 5

A + B + C : A + B (Efficiency) = 5 : 9

C (Efficiency) = 9 − 5 = 4

Time taken by C to empty the full tank

$= \frac{(9-4)\times 40}{4} = 50$ hours

64. A cistern can be filled by two pipes in 8 minutes and 10 minutes, separately. Both the pipes are opened together for a certain time, but due to an obstruction, only $\frac{5}{8}$ of the full quantity of water flowed through the former pipe and $\frac{3}{5}$ through the latter pipe. However, the obstruction was suddenly removed, and the cistern was filled in 3 minutes from that moment. How long did it take before the full flow began?

एक टंकी को दो पाइपों द्वारा अलग-अलग 8 मिनट और 10 मिनट में भरा जा सकता है। दोनों पाइपों को एक निश्चित समय के लिए एक साथ खोला जाता है, लेकिन एक रुकावट के कारण, पानी की पूरी मात्रा का केवल $\frac{5}{8}$ पहले वाले पाइप से और $\frac{3}{5}$ दूसरे पाइप से प्रवाहित होता है। हालाँकि, रुकावट अचानक हटा दी गई, और उस क्षण से 3 मिनट में टंकी भर गई। पूर्ण प्रवाह शुरू होने में कितना समय लगा?

Solution :

40 = Total volume of the cistern

Efficiencies of pipes: 8 → 5, 10 → 4

$$x\left(\frac{5\times 5}{8}+\frac{4\times 3}{5}\right)+(5+4)\times 3=40$$

$$x\left(\frac{25}{8}+\frac{12}{5}\right)+27=40$$

$$x\left(\frac{125+96}{40}\right)=13 \implies x=\frac{13\times 40}{221}=$$

$\frac{40}{17}$ minutes

$x = 2\frac{6}{17}$ minutes

65. Pipes M, N, and S can fill a tank in 25, 50, and 100 minutes, respectively. Initially, pipes N and S are kept open for 10 minutes, and then pipe N is shut while pipe M is opened. Pipe S is closed 15 minutes before the tank overflows. How much time (in minutes) will it take to fill the tank if the three pipes work in this pattern?

पाइप M, N और S एक टैंक को क्रमशः 25, 50 और 100 मिनट में भर सकते हैं। प्रारंभ में, पाइप N और S को 10 मिनट के लिए खुला रखा जाता है, और फिर पाइप N को बंद कर दिया जाता है जबकि पाइप M को खोल दिया जाता है। टैंक ओवरफ्लो होने से 15 मिनट पहले पाइप S को बंद कर दिया जाता है। यदि तीनों पाइप इस पैटर्न पर काम करें तो टैंक को भरने में कितना समय (मिनटों में) लगेगा?

Solution :

Efficiencies: $25 \to 4,\ \ 50 \to 2,\ \ 100 \to 1$

$$4(x-10)+2\times 10+1(x-15)=100$$

$$4x-40+20+x-15=100$$

$$5x=135 \implies x=27 \text{ minutes}$$

8 Speed, Time and Distance

Speed/ गति

Speed - Speed is defined as the distance travelled in unit time and it is denoted by 'S'
गति- गति को इकाई समय में तय की गई दूरी के रूप में परिभाषित किया गया है और इसे 'S' द्वारा दर्शाया जाता है।

Distance/ दूरी

Distance - The length of space between two points is considered the distance and it is denoted by 'D'
दूरी- दो बिंदुओं के बीच की दूरी को दूरी माना जाता है और इसे 'D' से दर्शाया जाता है।

Time/ समय

Time - Time is a fundamental physical quantity which defines the sequence and duration of events and it is denoted by 'T'
समय- समय एक मौलिक भौतिक मात्रा है जो घटनाओं के अनुक्रम और अवधि को परिभाषित करती है और इसे 'T' द्वारा दर्शाया जाता है।
Speed = Distance/Time
गति = दूरी/समय

Some Important Results
कुछ महत्वपूर्ण परिणाम

Speed = Distance/Time
गति = दूरी/समय
If A and B starts walking towards each other. After meeting each other. A covered his remaining distance in T^1 time and B covered his remaining distance in T^2 time. Then ratio of their speed is:
यदि A और B एक दूसरे की ओर चलना शुरू करते हैं। एक दूसरे से मिलने के बाद. A ने अपनी शेष दूरी T^1 समय में तय की और B ने अपनी शेष दूरी T^2 समय में तय की। तो उनकी गति का अनुपात है:

$$\frac{S^1}{S^2} = \sqrt{\frac{T^2}{T^1}}$$

Product of Time before Meeting = Product of Time after Meeting
मीटिंग से पहले के समय का गुणनफल = मीटिंग के बाद के समय का गुणनफल

1. When the two objects travel in same direction
   जब दो वस्तुएँ एक ही दिशा में यात्रा करती हैं
   Relative Speed = Diff of Speed
   सापेक्ष गति = गति का अंतर
2. When the two objects travel in opposite direction
   जब दो वस्तुएं विपरीत दिशा में यात्रा करती हैं
   Relative Speed = Sum of Speed
   सापेक्ष गति = गति का योग

When train passes a pole or stationary man
जब ट्रेन किसी खंभे या स्थिर आदमी से गुजरती है

Distance covered will be equal to length of train
तय की गई दूरी ट्रेन की लंबाई के बराबर होगी

$$T = \frac{L}{S}$$

T = Crossing time (पार करने का समय)
When train passes a bridge/ platform
जब ट्रेन पुल/प्लेटफार्म से गुजरती है

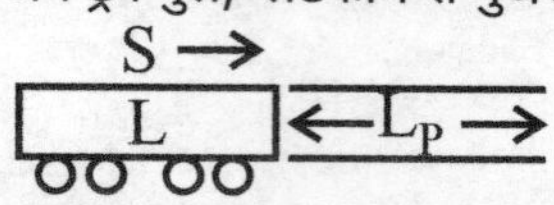

Distance covered will be equal to length of train + length of bridge/platform
तय की गई दूरी = ट्रेन की लंबाई + पुल/प्लेटफार्म की लंबाई के बराबर होगी

$$vT = \frac{L_T + L_P}{S}$$

T = Crossing time (पार करने का समय)

when a train passes another train in opposite direction
जब एक ट्रेन विपरीत दिशा में दूसरी ट्रेन को पार करती है

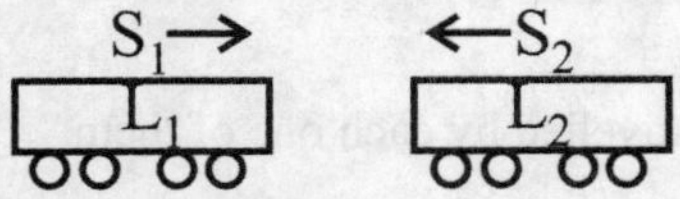

$$T = \frac{L_1 + L_2}{S_1 + S_2}$$

Distance covered = $L_1 + L_2$
तय की गई दूरी = $L_1 + L_2$

When a train passes another train in same direction.
जब एक ट्रेन उसी दिशा में दूसरी ट्रेन को पार करती है

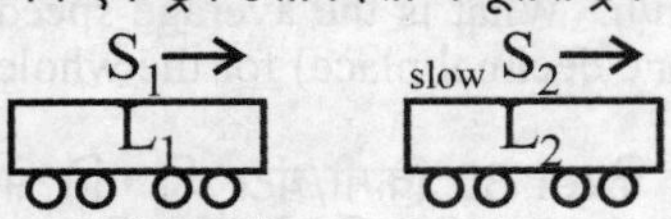

$$T = \frac{L_1 + L_2}{S_1 - S_2}$$

Distance covered = Relative speed
तय की गई दूरी = सापेक्ष गति

When a train passes a person sitting in another moving train.
जब एक ट्रेन दूसरी चलती ट्रेन में बैठे व्यक्ति को पार कर जाती है

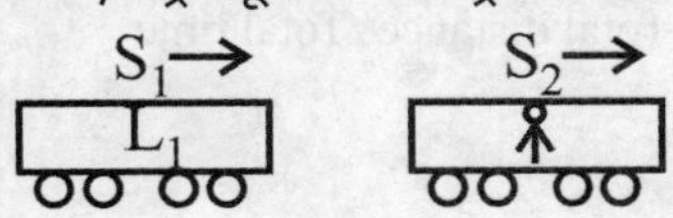

$$T = \frac{L_1}{S_1 - S_2}$$

Type 1

1. With an average speed of 60 km/h, a train reaches its destination on time. If it goes with an average speed of 45 km/h, it is late by 20 minutes. The distance covered by the train is

   60 किमी/घंटा की औसत गति से एक ट्रेन समय पर अपने गंतव्य तक पहुंचती है। यदि यह 45 किमी/घंटा की औसत गति से चलती है, तो 20 मिनट की देरी होती है। ट्रेन द्वारा तय की गई दूरी है

Solution :

Let the distance covered by the train = x km.

According to the question,

$$\frac{x}{45} - \frac{x}{60} = \frac{20}{60}$$

$$\frac{4x - 3x}{180} = \frac{1}{3}$$

$$\frac{x}{180} = \frac{1}{3}$$

$$x = 60 \text{ km}$$

2. Ram is travelling at a speed of 72 km/h while Shyam is travelling at a speed of 90 km/h. What is the difference in their speeds (in metres per second)?

   राम 72 किमी प्रति घंटे की गति से यात्रा कर रहा है जबकि श्याम 90 किमी प्रति घंटे की गति से यात्रा कर रहा है। उनकी गति में क्या अंतर है (मीटर प्रति सेकंड में)?

Solution :

According to the question,

Speed of Shyam (in m/s) - Speed of Ram (in m/s)

$$90 \times \frac{5}{18} - 72 \times \frac{5}{18} = 25 - 20 = 5 \text{ m/s}$$

3. A person travels $5x$ distance at a speed of 5 km/h, x distance at a speed of 5 km/h, and $4x$ distance at a speed of 6 km/h, and takes a total of 112 minutes. What is the total distance (in km) travelled by the person?

   एक व्यक्ति $5x$ दूरी 5 किमी/घंटा की गति से, x दूरी 5 किमी/घंटा की गति से और $4x$ दूरी 6 किमी/घंटा की गति से तय करता है और कुल 112 मिनट लेता है। व्यक्ति द्वारा तय की गई कुल दूरी (किमी में) क्या है?

Solution :

$$\text{Total time} = \frac{5x}{5} + \frac{x}{5} + \frac{4x}{6} = x + \frac{x}{5} + \frac{2x}{3}$$

$$\frac{112}{60} = \frac{28x}{15}$$

$$x = 1 \text{ km}$$

$$\text{Total distance} = 5x + x + 4x = 10x = 10 \text{ km}$$

4. A journey of 900 km is completed in 11 h. If two-fifth of the journey is completed at the speed of 60 km/h, at what speed (in km/h) is the remaining journey completed?

   900 किमी की यात्रा 11 घंटे में पूरी होती है। यदि यात्रा का दो-पाँचवाँ भाग 60 किमी/घंटा की गति से पूरा किया जाता है, तो शेष यात्रा किस गति (किमी/घंटा में) से पूरी की जाएगी?

Solution :

From the question:

$\frac{2}{5}$ of the journey $= \frac{2}{5} \times 900 = 360$ km

Time taken for 360 km $= \frac{360}{60} = 6$ hours

Remaining time $= 11 - 6 = 5$ hours

Remaining distance $= 900 - 360 = 540$ km

Required speed $= \frac{540}{5} = 108$ km/h

5. While covering a distance of 51 km, a man noticed that after walking for 1 hour and 40 minutes, the distance covered by him was $\frac{5}{12}$ of the remaining distance. What was his speed?

51 किमी की दूरी तय करते समय, एक व्यक्ति ने देखा कि 1 घंटा 40 मिनट तक चलने के बाद, उसके द्वारा तय की गई दूरी शेष दूरी का $\frac{5}{12}$ थी। उसकी गति क्या थी?

Solution :

Distance covered $= \frac{5}{17}$ of total distance

Time taken for $5x = 100$ minutes

$17x = 340$ minutes $= \frac{17}{3}$ hours

Speed $= \frac{51 \times 3}{17} = 9$ km/h

6. A and B travel the same distance at speeds of 8 km/h and 12 km/h, respectively. If B takes 30 minutes less than that taken by A, what in the distance (in km) travelled by each one of them?

A और B क्रमश: 8 किमी/घंटा और 12 किमी/घंटा की गति से समान दूरी तय करते हैं। यदि B, A द्वारा लिए गए समय से 30 मिनट कम लेता है, तो उनमें से प्रत्येक द्वारा तय की गई दूरी (किमी में) क्या है?

Solution :

Speed of A = 8 km/h

Speed of B = 12 km/h

Let Total distance is d.

According to the question,

$\frac{d}{8} - \frac{d}{12} = \frac{30}{60}$

$\frac{3d - 2d}{24} = \frac{1}{2}$

$d = 12$

Hence distance travelled by each one of them is 12 km.

7. One-fifth of a journey is covered at a speed of 30 km/h, one-fourth of the journey at a speed of 25 km/h and the rest at 60 km/h. What is the average speed (in km/h, correct to one decimal place) for the whole journey?

एक यात्रा का पांचवां हिस्सा 30 किमी/घंटा की गति से, एक-चौथाई यात्रा 25 किमी/घंटा की गति से और शेष 60 किमी/घंटा की गति से तय किया जाता है। पूरी यात्रा के लिए औसत गति (किमी/घंटा में, दशमलव स्थान तक सही) क्या है?

Solution :

Average speed = Total distance /Total time

$\frac{1}{5}, \frac{1}{4}$ and 1

(Let total distance = $5 \times 4 \times 1 = 20$ km)

4, 5 and 11 km respectively the distance covered by the speed of 30, 25 and 60 km/h.

According to the question,

Average speed $= \frac{20}{\frac{4}{30} + \frac{5}{25} + \frac{11}{60}}$

$= \frac{20}{\frac{2}{15} + \frac{1}{5} + \frac{11}{60}} = \frac{20}{\frac{8 + 12 + 11}{60}}$

$= \frac{20 \times 60}{31} = 38.70$ km/h.

8. A man is running at the speed of 20 km/hr. What is time (in seconds) taken by man to cover one round of a circular garden of radius 350 metres?

एक आदमी 20 किमी/घंटा की गति से दौड़ रहा है। 350 मीटर त्रिज्या वाले एक वृत्ताकार बगीचे का एक चक्कर लगाने में मनुष्य को कितना समय (सेकंड में) लगेगा?

Solution :

Given

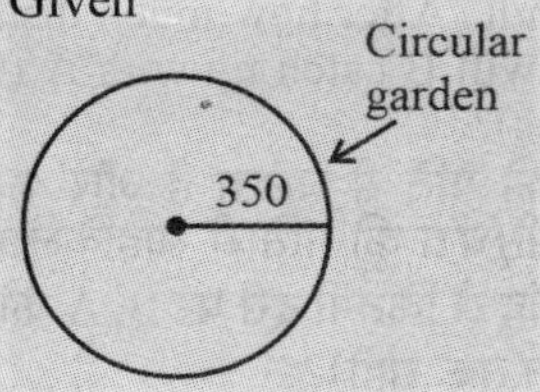

Speed of the man = 20 km/h, $r = 350$ m. = $20 \times \frac{5}{18}$ m/s

Perimeter (P) = $2\pi r$

$= 2 \times \frac{22}{7} \times 350$

$$\text{Time} = \frac{2 \times \frac{22}{7} \times 350}{20 \times \frac{5}{18}} = 396 \text{ seconds.}$$

9. B starts 4 minutes after A from the same point, for a place at a distance of 7 miles from the starting point. A on reaching the destination turns back and walks a mile where he meets B. If A's speed is a mile in 8 minutes then B's speed is a mile in minutes.

B, A के 4 मिनट बाद उसी बिंदु से आरंभ करता है, प्रारंभिक बिंदु से 7 मील की दूरी पर एक स्थान के लिए। गंतव्य पर पहुंचने पर A वापस मुड़ता है और एक मील चलता है जहां उसकी मुलाकात B से होती है। यदि A की गति 8 मिनट में एक मील है तो B की गति मिनटों में एक मील है।

Solution :

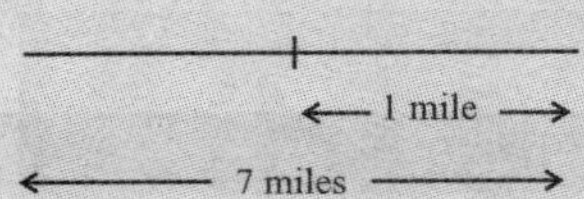

Total distance travelled by A = 7+1 = 8 miles

Total distance travelled by B = 7−1 = 6 miles

Let the speed of $B = x$ miles/min

According to the question-

Time taken by A = Time taken by B + 4 minutes.

$$\frac{8}{\frac{1}{8}} = \frac{6}{x} + 4 \text{ minutes.}$$

$$60 = \frac{6}{x}$$

$$x = \frac{1}{10}$$

Hence, Time taken by B to walk 1 mile = 10 minutes.

10. Ravi starts for his school from his house on his cycle at 8:20 a.m. If he runs his cycle at a speed of 10 km/h, he reaches his school 8 minutes late, and if he drives the cycle at a speed of 16 km/h, he reaches his school 10 minutes early.

रवि सुबह 8:20 बजे अपने घर से साइकिल पर स्कूल के लिए निकलता है। यदि वह अपनी साइकिल 10 किमी/घंटा की गति से चलाता है, तो वह अपने स्कूल में 8 मिनट देरी से पहुंचता है, और यदि वह साइकिल 16 किमी की गति से चलाता है, तो वह अपने स्कूल में 10 मिनट पहले पहुंचता है।

Solution :

Let the distance of school from house = x km

According to the question,

$$\frac{x}{10} - \frac{x}{16} = \frac{18}{60}$$

$$\frac{16x - 10x}{160} = \frac{3}{10}$$

$$6x = 48$$

$$x = 8 \text{ km.}$$

Time taken to reach school at the speed of 10 km/hr

$$\text{Time} = \frac{8}{10} \times 60 = 48 \text{ minutes.}$$

School starts at time = 8 : 20 + (48 − 8) minutes = 9 : 00 a.m.

11. Alok starts walking from P at the speed of 6 km/h towards Q. Raman starts at the same time from P towards Q at the speed of 9 km/h. Raman reaches Q and turns back and starts walking towards P. He meets Alok at R. If PQ is 15 km, then what is PR?

आलोक P से Q की ओर 6 किमी/घंटा की गति से चलना शुरू करता है। रमन उसी समय P से Q की ओर 9 किमी/घंटा की गति से चलना शुरू करता है। रमन Q पर पहुंचता है और वापस मुड़कर P की ओर चलना शुरू कर देता है। वह R पर आलोक से मिलता है। यदि PQ 15 किमी है, तो PR क्या है?

Solution :

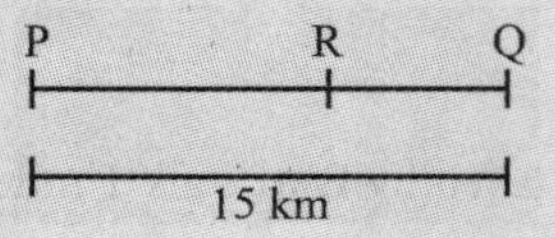

The time taken by Raman to cover a distance of 15 km $\frac{15}{9} = \frac{5}{3}$ hours.

The distance covered by Alok in $\frac{5}{3}$ hours $\frac{5}{3} \times 6 = 10$ km.

Hence, remaining distance $= 15 - 10 = 5$ km.

Relative speed $= 9 + 6 = 15$ km/h.

Time taken to cover a distance of 5 km $\frac{5}{15} = \frac{1}{3}$ hours.

Distance covered by Raman in $\frac{1}{3}$ hours $9 \times \frac{1}{3} = 3$ km.

Hence, the distance (PR) = PQ - RQ $15 - 3 = 12$ km.

12. A boy increases his speed to $\frac{9}{5}$ times of his original speed. By doing this, he reaches his school 40 minutes before the usual time. How much time (in minutes) does he take usually?

एक लड़का अपनी गति को अपनी मूल गति से $\frac{9}{5}$ गुना तक बढ़ा देता है। ऐसा करने पर वह सामान्य समय से 40 मिनट पहले अपने स्कूल पहुंच जाता है। वह आमतौर पर कितना समय (मिनटों में) लेता है?

Solution :

Ratio of the speed of boy = 5 : 9

लड़के की गति का अनुपात = 5 : 9

$\because$ Distance is constant

$\therefore$ Ratio of the time = 9 : 5

9 : 5

–4

$\because$ 4 units = 40 minutes

$\therefore$ 9 units = 90 minutes.

Hence, he takes 90 minutes usually.

13. A starts walking at 4 km/h and after 4 hours. B starts cycling from the same point as that of A in the same direction at 10 km/h. After how much distance from the starting point will B catch up with A (Correct to two decimal places)?

A 4 किमी/घंटा की गति से चलता है और 4 घंटे बाद B उसी बिंदु से 10 किमी/घंटा की गति से साइकिल चलाना शुरू करता है। प्रारंभिक बिंदु से कितनी दूरी पर B, A को पकड़ लेगा (दशमलव के दो स्थान तक सही)?

Solution :

Let B will catch A after t hours.

In this case, the distance travelled by both of them will be equal.

$4(4 + t) = 10 \times t$

$16 + 4t = 10t$

$6t = 16$

$t = \frac{8}{3}$ hours.

Required Distance $= 10t = 10 \times \frac{8}{3} =$ 26.67 km.

14. Train A takes 45 minutes more than train B to travel a distance of 450 km. Due to engine trouble speed of train B falls by a quarter, so it takes 30 minutes more than Train A to complete the same journey. What is the speed of Train A (in km/h)?

ट्रेन A को 450 किमी की दूरी तय करने में ट्रेन B से 45 मिनट अधिक लगते हैं। इंजन में खराबी के कारण ट्रेन B की गति एक चौथाई कम हो जाती है, इसलिए उसी यात्रा को पूरा करने में उसे ट्रेन A से 30 मिनट अधिक लगते हैं। ट्रेन A की गति (किमी/घंटा में) क्या है?

Solution :

Ratio of speed of train B $= 1 : (1 - \frac{1}{4}) = 4 : 3$

The ratio of the time taken by train B = 3 : 4

Let the time taken by B is $3x$ and $4x$ respectively. According to the question-

$4x - 3x = 45 + 30 = 75$ minutes.

Initially, the time taken by train B $= 3x =$ 225 minutes.

Initially, the time taken by train A $= 225 + 45 = 270$ minutes.

Speed of train A $= \frac{450}{4.5} = 100$ km/h.

15. Two cars A and B travel from one city to another, at the speeds of 72 km/h and 90 km/h respectively. If car B takes 1 hour lesser than car A for the journey, then what is the distance (in km) between the two cities?
दो कारें A और B क्रमशः 72 किमी/घंटा और 90 किमी/घंटा की गति से एक शहर से दूसरे शहर तक यात्रा करती हैं। यदि कार B यात्रा के लिए कार A से 1 घंटा कम लेती है, तो दोनों शहरों के बीच की दूरी (किमी में) क्या है?

Solution :

Let the distance between the two cities = x km

And the time taken by car A = t hours

According to the question-

$72t = 90(t-1)$

$72t = 90t - 90$

$18t = 90$

$t = 5$ hours.

$x = 72 \times 5 = 360$ km.

16. A and B are travelling towards each other from the points P and Q respectively. After crossing each other, A and B take $6\frac{1}{8}$ hours and 8 hours, respectively, to reach their destinations Q and P. If the speed of B is 16.8 km/h, then the speed (in km/h) of A is:
A और B क्रमशः बिंदु P और Q से एक दूसरे की ओर यात्रा कर रहे हैं। एक दूसरे को पार करने के बाद, A और B को अपने गंतव्य Q और P तक पहुँचने में क्रमशः $6\frac{1}{8}$ घंटे और 8 घंटे लगते हैं। यदि B की गति 16.8 किमी/घंटा है, तो A की गति (किमी/घंटा में) है:

Solution :

$\frac{V_A}{V_B} = \sqrt{\frac{t_B}{t_A}}$ (By formula)

$\frac{V_A}{16.8} = \sqrt{\frac{8 \times 8}{49}}$

$\frac{V_A}{16.8} = \frac{8}{7}$

$V_A = 19.2$ km/h.

17. If Abhi travels a certain distance at 6 km/h, he reaches his destination 12 minutes early, but if he travels at 4 km/h, he reaches his destination 10 minutes late. The speed (in km/h) at which he should travel to reach his destination on time is:
यदि अभि एक निश्चित दूरी 6 किमी/घंटा की गति से तय करता है, तो वह अपने गंतव्य पर 12 मिनट पहले पहुंचता है, लेकिन यदि वह 4 किमी/घंटा की गति से यात्रा करता है, तो वह अपने गंतव्य पर 10 मिनट की देरी से पहुंचता है। वह गति (किमी/घंटा में) जिस पर उसे समय पर अपने गंतव्य तक पहुंचने के लिए यात्रा करनी चाहिए:

Solution :

Speed ratio = $6 : 4 = 3 : 2$

Time ratio = $2 : 3$

1 unit = 22 minutes, hence 2 units = 44 minutes.

Distance = $\frac{6 \times 44}{60} = 4.4$ km.

Speed = $\frac{4.4}{56 \text{ minutes}} = \frac{33}{7} = 4\frac{5}{7}$ km/h.

18. A train covers 360 km at a uniform speed. If the speed had been 10 km/h more, it would have taken 3 hours less for the same journey. What is the speed of the train (in km/h)?
एक ट्रेन एक समान गति से 360 किमी की दूरी तय करती है। यदि गति 10 किमी/घंटा अधिक होती, तो उसी यात्रा में 3 घंटे कम लगते। ट्रेन की गति (किमी/घंटा में) क्या है?

Solution :

Let the speed of the train = x km/h.

According to the question,

Time difference = $\frac{360}{x} - \frac{360}{x+10} = 3$

$\frac{360x + 3600 - 360x}{x^2 + 10x} = 3$

$3x^2 + 30x - 3600 = 0$

$x^2 + 10x - 1200 = 0$

$(x+40)(x-30) = 0$

$x = 30$ km/h.

19. A car covers 25 km at a uniform speed. If the speed had been 8 km/h more, it would have taken 10 hours less for the same journey. What is the speed of the car (in km/h)?
एक कार एक समान गति से 25 किमी की दूरी तय करती है। यदि गति 8 किमी/घंटा अधिक होती, तो उसी यात्रा में 10 घंटे कम लगते। कार की गति (किमी/घंटा में) क्या है?

Solution :

Let the speed of the car = s km/h.

$$25 = \frac{s(s+8)}{8 \times 10}$$

$$5 = \frac{s(s+8)}{4}$$

$$s(s+8) = 20$$

$$s = 2 \text{ km/h.}$$

20. A man on a tour travels first 360 km by train at 72 km/h, the next 160 km on a motorcycle at 12.8 km/h, and the last 200 km by bicycle at 16 km/h. Ignoring the buffer times between the different modes of travel, what is the average speed (in m/s) for his tour?

यात्रा पर निकला एक आदमी पहले 360 किमी की दूरी ट्रेन से 72 किमी/घंटा की गति से तय करता है, अगले 160 किमी की दूरी मोटर साइकिल पर 12.80 किमी/घंटा की गति से और अंतिम 200 किमी की यात्रा साइकिल से 16 किमी/घंटा की गति से तय करता है। यात्रा के विभिन्न तरीकों के बीच बफर समय को नजरअंदाज करते हुए, उसके दौरे के लिए औसत गति (मीटर/सेकेंड में) क्या है?

Solution :

$$\text{Average speed} = \frac{\text{Total distance}}{\text{Total time}}$$

$$= \frac{360+160+200}{\frac{360}{72}+\frac{160}{12.8}+\frac{200}{16}}$$

$$= \frac{720}{5+12.5+12.5} = \frac{720}{30}$$

$$= 24 \text{ km/h.}$$

$$= 24 \times \frac{5}{18} = \frac{20}{3} \text{ m/s.}$$

$$= 6.67 \text{ m/s.}$$

21. A person goes from point A to B and comes back. His average speed for the whole journey is 30 km/hr. If his speed while going from A to B is 24 km/hr, then what will the speed of the person (in km/hr) while coming back from B to A?

एक व्यक्ति बिंदु A से B तक जाता है और वापस आ जाता है। पूरी यात्रा के दौरान उसकी औसत गति 30 किमी/घंटा है। यदि A से B तक जाते समय उसकी गति 24 किमी/घंटा है, तो B से A तक वापस आते समय व्यक्ति की गति (किमी/घंटा में) क्या होगी?

Solution :

$$\text{Average speed} = \frac{2xy}{x+y}$$

$$30 = \frac{2 \times 24 \times y}{24+y}$$

$$30(24+y) = 48y$$

$$720 + 30y = 48y$$

$$48y - 30y = 720$$

$$18y = 720$$

$$y = \frac{720}{18} = 40$$

Hence the speed of return from B to A = 40

km/hr.

22. A person covered 50% of a certain distance at 2 km/h, 25% of the distance at 4 km/h, and the remaining distance at 8 km/h. His average speed (in km/h) for the entire journey was:

एक व्यक्ति ने एक निश्चित दूरी का 50% 2 किमी/घंटा, 25% दूरी 4 किमी/घंटा और शेष दूरी 8 किमी/घंटा की गति से तय की, पूरी यात्रा के लिए उसकी औसत गति (किमी/घंटा में) थी:

Solution :

Let total distance = d km.

According to the question:

$$\text{Average speed} = \frac{\text{Total distance}}{\text{Total time}}$$

$$= \frac{d}{\frac{d}{2\times 2}+\frac{d}{4\times 4}+\frac{d}{4\times 8}}$$

$$= \frac{d \times 32}{8d+2d+d} = \frac{32}{11} \text{ km/h.}$$

23. A truck covers a distance of 384 km at a certain speed. If the speed is decreased by 16 km/h, it will take 2 hours more to cover the same distance. 75% of its original speed (in km/h) is:

एक ट्रक एक निश्चित गति से 384 किमी की दूरी तय करता है। यदि गति 16 किमी/घंटा कम कर दी जाए तो समान दूरी तय करने में 2 घंटे अधिक लगेंगे, इसकी मूल गति का 75% (किमी/घंटा में) है:

Solution :

Let the original speed = x km/h.

$$\frac{384}{x-16} - \frac{384}{x} = 2$$

$$384\left[\frac{1}{x-16} - \frac{1}{x}\right] = 2$$

$$384\frac{(x-(x-16))}{x(x-16)} = 2$$

$$x^2 - 16x = 3072$$

$$x^2 - 16x - 3072 = 0$$

$$(x-64)(x+48) = 0$$

$$x = -48(\text{Invalid})$$

$$x = 64$$

$$x \times \frac{75}{100} = 64 \times \frac{3}{4} = 48 \text{ km/h}$$

OR Trick:

$$\text{Distance} = \frac{S_1 \times S_2}{S_1 - S_2} \times (t_1 - t_2)$$

Let the original speed of the track = s km/h

$$384 = \frac{s(s-16)}{16 \times 2}$$

s(s - 16) = 384 × 8

s(s - 16) = 64 × 48

$\therefore s = 64$ km/h

Hence $64 \times \frac{75}{100} = 48$ km/h.

24. A train without stoppage travels with an average speed of 70 km/h, and with stoppage, it travels with an average speed of 56 km/h. How many minutes does the train stop on an average per hour?
एक ट्रेन बिना रुके 70 किमी/घंटा की औसत गति से चलती है और रुकने पर यह 56 किमी/घंटा की औसत गति से चलती है। ट्रेन प्रति घंटे औसतन कितने मिनट रुकती है?

Solution :

The time of stop of the train per hour = $\frac{(x-y)}{x} \times 60$ min.

Where x = Speed of the train without stoppage

y = Speed of the train with stoppage.

Hence, the time of stop of the train per hour

$$= \frac{70-56}{70} \times 60$$

$$= \frac{14}{70} \times 60 = 12 \text{ minutes.}$$

25. A takes 2 hours 30 minutes more than B to walk 40 km. If A doubles his speed, then he can make it in 1 hour less than B. How much time (in hours) does A require for walking a 40 km distance?
A को 40 किमी चलने में B से 2 घंटे 30 मिनट अधिक लगते हैं। यदि A अपनी गति दोगुनी कर देता है, तो वह इसे B से 1 घंटा कम में पूरा कर सकता है। A को 40 किमी की दूरी तय करने में कितना समय (घंटे में) लगेगा?

Solution :

Ratio of speed of A = 1 : 2

Ratio of the time taken by A = 2 : 1

$1 \rightarrow 3\frac{1}{2}$

$2 \rightarrow 7$ hours

Hence, the time taken by A to cover the distance of 40 km = 7 hours

26. A man leaves from P at 6 am and reaches Q at 2 pm on the same day. Another man leaves Q at 8 am and reaches P at 3 pm on the same day. At what time do they meet?
एक आदमी सुबह 6 बजे P से निकलता है और उसी दिन दोपहर 2 बजे Q पहुंचता है। एक अन्य व्यक्ति Q को सुबह 8 बजे छोड़ता है और उसी दिन दोपहर 3 बजे P पर पहुंचता है। वे किस समय मिलते हैं?

Solution :

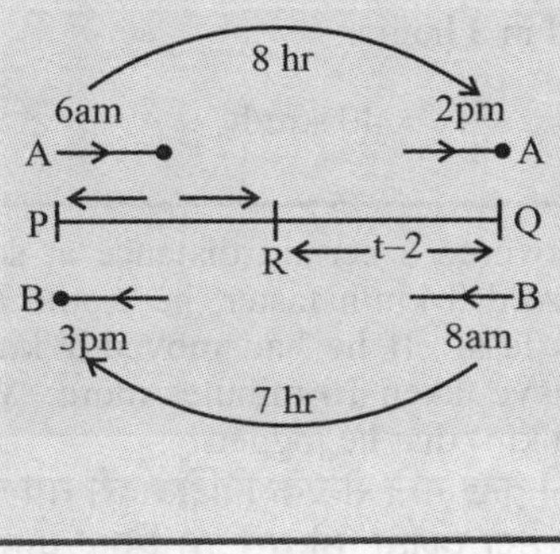

Let both persons meet at place R. From 6 am, person A reaches R after t hours, then from 8 am, person B will reach R after $(t-2)$ hours.

Then,

$$\frac{t}{8}+\frac{(t-2)}{7}=1$$

$$\frac{7t+8(t-2)}{56}=1$$

$$7t+8t-16=56$$

$$15t=72$$

$$t=\frac{72}{15}=4\ \frac{12}{15}\ \text{hr}$$

$$t=4\ \text{hr}\ 12/15\times 60\ \text{min}=4\ \text{hr}\ 48\ \text{min}$$

Time of meeting $=$ 6 am $+$ 4 hr 48 min $=$ 10 : 48 am

27. Travelling at 60 km/h, a person reaches his destination in a certain time. He covers 60% of his journey in $\frac{2}{5}$ of the time. At what speed (in km/h) should he travel to cover the remaining journey so that he reaches the destination on time?

60 किमी/घंटा की गति से यात्रा करते हुए एक व्यक्ति एक निश्चित समय में अपने गंतव्य तक पहुँच जाता है। वह अपनी यात्रा का 60% भाग $\frac{2}{5}$ समय में तय करता है। शेष यात्रा तय करने के लिए उसे किस गति (किमी/घंटा में) से यात्रा करनी चाहिए ताकि वह सही समय पर गंतव्य तक पहुंच सके?

Solution :

Let the distance be 300 km.

Time taken $=\frac{300}{60}=5$ hours

60% of journey $=180$ km in 2 hours

Remaining distance of 120 km will be covered in 3 hours.

Speed $=\frac{120}{3}=40$ km/h

28. A jogger covered a certain distance at some speed. Had he moved 3 km/h faster, he would have taken 20 minutes less. If he had moved 1 km/h slower, he would have taken 10 minutes more. What is the distance (in km) that he jogged?

एक जॉगर ने कुछ गति से एक निश्चित दूरी तय की। यदि वह 3 किमी/घंटा तेज चलता, तो उसे 20 मिनट कम लगते। यदि वह 1 किमी/घंटा धीमी गति से चलता, तो उसे 10 मिनट अधिक लगते। उसने कितनी दूरी (किमी में) जॉगिंग की?

Solution :

Equating distance in both cases,

$$x(x+3)\times\frac{20}{3}=x(x-1)\times 10$$

$$2x+6=3x-3$$

$$x=9\ \text{km/h}$$

$$\text{Distance}=9\times\frac{12\times 20}{3\times 60}=12\ \text{km}$$

29. A man travels 450 km to his home partly by train and partly by car. He takes 8 hours 40 minutes if he travels 240 km by train and the rest by car. He takes 20 minutes more if he travels 180 km by train and the rest by car. The speed of the car in km/hr is:

कोई व्यक्ति 450 किमी अपने घर जाने के लिए आंशिक रूप से रेलगाड़ी से और आंशिक रूप से कार से यात्रा करता है यदि वह 240 किमी रेलगाड़ी से और शेष यात्रा कार से करें तो उसे 8 घंटे 40 मिनट का समय लगता है। यदि वह 180 किलोमीटर रेलगाड़ी से और शेष यात्रा कार से करे तो 20 मिनट का समय अधिक लगता है। कार की किमी./घंटा में चाल बताइए?

Solution :

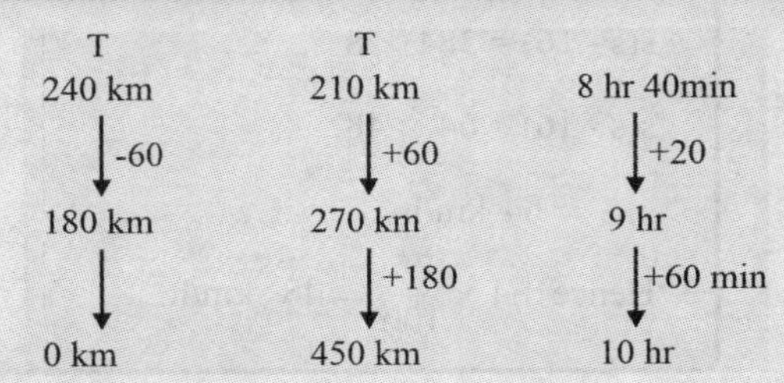

$$\therefore \text{speed of car}=\frac{450}{10}=45\ \text{km/hr}$$

30. A train has to cover a distance of 900 km in 25 hours. What should be its average speed in meters/second?

एक ट्रेन को 25 घंटे में 900 किमी की दूरी तय करनी है। इसकी औसत गति मीटर/सेकंड में क्या होनी चाहिए?

Solution :

Distance = 900 km

Time = 25 hours

$$\text{Speed}=\frac{900}{25}=36\ \text{km/hr}$$

$$\text{Speed (m/sec)}=36\times\frac{5}{18}=10\ \text{m/sec}$$

31. A 600 metres long train can cross a 1200 meters long platform in 36 seconds. In how much time (in seconds) can it cross a bridge 2200 metres long?
600 मीटर लंबी ट्रेन 1200 मीटर लंबे प्लेटफॉर्म को 36 सेकंड में पार कर सकती है। यह 2200 मीटर लंबे पुल को कितने समय (सेकंड में) में पार कर सकती है?

Solution :

Speed of Train $= \frac{1200+600}{36} = 50$ m/sec

Time taken to cross bridge: $\frac{2200+600}{50} =$ 56 seconds

32. A train crosses a pole in 12 sec and a bridge of length 170 m in 36 sec. Then the speed of the train is:
एक ट्रेन एक खंभे को 12 सेकंड में और 170 मीटर लंबे पुल को 36 सेकंड में पार करती है। तो ट्रेन की गति है:

Solution :

$L = 12 \times S \quad ...\ (1)$

$L + 170 = 36S \quad ...\ (2)$

From (1) and (2), $\frac{L}{L+170} = \frac{1}{3}$

$3L - L = 170$

$2L = 170$

$L = 85$ m

$S = \frac{85}{12} \times \frac{18}{5} = 25.5$ km/h

33. Two trains running in opposite directions on parallel tracks at speeds of 42 km/h and 57 km/h take 18 seconds to cross each other. If the length of one train is 270 m, then the length of the other train is:
42 किमी/घंटा और 57 किमी/घंटा की गति से समानांतर पटरियों पर विपरीत दिशाओं में चलने वाली दो ट्रेनें एक-दूसरे को पार करने में 18 सेकंड लेती हैं। यदि एक ट्रेन की लंबाई 270 मीटर है तो दूसरी ट्रेन की लंबाई है:

Solution :

Relative speed $= \frac{\text{distance}}{\text{time}}$

$(42+57) \times \frac{5}{18} = \frac{270+D}{18}$

$99 \times 5 = 270 + D$

$495 = 270 + D$

$D = 225$ m

34. Train A and B start at the same time. Train A travels at 55 km/hr from station X to station Y and train B travels at 80 km/hr from station Y to station X. They cross each other after 1 hour 36 minutes. What is the distance (in km) between stations X and Y? / ट्रेन A और B एक ही समय पर शुरू होती हैं। ट्रेन A स्टेशन X से स्टेशन Y तक 55 किमी/घंटा की गति से यात्रा करती है और ट्रेन B स्टेशन Y से स्टेशन X और Y के बीच की दूरी (में) क्या है?

Solution :

Time $= 1 + \frac{36}{60} = 1 + \frac{3}{5} = \frac{8}{5}$ hours

Distance $= (80+55) \times \frac{8}{5}$

Distance $= 135 \times \frac{8}{5} = 27 \times 8 = 216$ km

35. The distance between two railway stations is 1176 km. To cover this distance, an express train takes 5 hours less than a passenger train while the average speed of the passenger train is 70 km/h less than that of the express train. The time taken by the passenger train to complete the travel is:
दो रेलवे स्टेशनों के बीच की दूरी 1176 किमी है। इस दूरी को तय करने में एक एक्सप्रेस ट्रेन को पैसेंजर ट्रेन की तुलना में 5 घंटे कम समय लगता है जबकि पैसेंजर ट्रेन की औसत गति एक्सप्रेस ट्रेन की औसत गति 70 किमी/घंटा कम है। यात्री ट्रेन द्वारा यात्रा पूरी करने में लगा समय है:

Solution :

Let speed of express train be x km/h, then speed of passenger train will be $(x - 70)$ km/h.

According to the question: $\frac{1176}{x} + 5 = \frac{1176}{x-70}$

$1176\left(\frac{1}{x-70} - \frac{1}{x}\right) = 5$

$1176 \times 70 = 5x(x-70)$

$$x(x-70) = 16464$$

$$x = 168 \text{ km/h} \quad \text{(Speed of Express Train)}$$

Speed of Passenger Train = 98 km/h

$$\text{Time} = \frac{1176}{98} = 12 \text{ hours}$$

36. The lengths of train A and B are in the ratio of 2 : 3. They pass a station pole in 4 seconds and 5 seconds, respectively. In what time (in seconds) will they cross each other when they move in the same direction?

ट्रेन A और B की लंबाई 2 : 3 के अनुपात में है। वे एक स्टेशन के खंभे को क्रमश: 4 सेकंड और 5 सेकंड में पार करती हैं। जब वे एक ही दिशा में चलते हैं तो वे कितने समय में (सेकंड में) एक दूसरे को पार कर लेंगे?

Solution :

Let, $l_1 = 20$ m, $t_1 = 4$ sec, and $l_2 = 30$ m, $t_2 = 5$ sec.

Speed of Train A: $V_1 = \frac{l_1}{t_1} = \frac{20}{4} = 5$ m/sec

Speed of Train B: $V_2 = \frac{l_2}{t_2} = \frac{30}{5} = 6$ m/sec

Relative speed when moving in the same direction: $V_2 - V_1 = \frac{l_1 + l_2}{T}$

$$6 - 5 = \frac{20 + 30}{T}$$

$$T = \frac{50}{1} = 50 \text{ seconds}$$

37. A train X travelling at 60 km/h overtakes another train Y 225 m long and completely passes it in 72 seconds. If the trains had been going in opposite directions, they would have passed each other in 18 seconds. The length (in m) of X and the speed (in km/h) of Y are, respectively:

60 किमी/घंटा की गति से यात्रा करने वाली एक ट्रेन यदि रेलगाड़ियाँ विपरीत दिशा में जा रही होतीं, तो वे 18 सेकंड में एक-दूसरे को पार कर जातीं। X की लंबाई (मीटर में) और Y की गति (किमी/घंटा में) क्रमशः हैं:

Solution :

Equate the distance:
$$(60 - x) \times 72 = (60 + x) \times 18$$
$$240 - 4x = 60 + x$$
$$5x = 180$$
$$x = 36 \text{ km/h}$$

$$L_1 + 225 = \left(\frac{60 - 36}{\frac{18}{5}}\right) \times 72$$

$$L_1 + 225 = 24 \times 5 \times 4 = 480$$

$$L_1 = 255 \text{ m}$$

38. Places A and B are 396 km apart. Train X leaves from A for B and train Y leaves from B for A at the same time on the same day on parallel tracks. Both trains meet after $5\frac{1}{2}$ hours. The speed of Y is 10 km/h more than that of X. What is the speed (in km/h) of Y?

स्थान A और B 396 किमी दूर हैं। ट्रेन X, A से B के लिए रवाना होती है और ट्रेन Y, B से A के लिए एक ही दिन में समानांतर पटरियों पर रवाना होती है। दोनों ट्रेनें $5\frac{1}{2}$ घंटे बाद मिलती हैं। Y की गति X की गति से 10 किमी/घंटा अधिक है। Y की गति (/घंटा में) क्या है?

Solution :

Distance between A & B = 396 km

Let the speed of $X = a$ km/h and speed of $Y = (a + 10)$ km/h.

From the rule of relative speed:

$$\text{Relative speed of } (X + Y) = \frac{396}{\frac{11}{2}}$$

$$= \frac{396 \times 2}{11} = 72 \text{ km/h}$$

$$a + a + 10 = 72$$

$$2a = 62$$

$$a = 31 \text{ km/h}$$

Speed of $Y = a + 10 = 31 + 10 = 41$ km/h

39. Two trains each having a length of 160 m moving in opposite directions crossed each other in 9 seconds. If one train crossed a 200 m-long platform in 18 seconds, then the ratio of their speeds is:

160 मीटर लंबी दो रेलगाड़ियाँ विपरीत दिशा में चलते हुए 9 सेकंड में एक-दूसरे को पार करती हैं। यदि एक ट्रेन 200 मीटर लंबे प्लेटफार्म को 18 सेकंड में पार करती है, तो उनकी गति का अनुपात है:

Solution :

Let the speeds of the trains be S_1 and S_2, respectively.

Then, $S_1 + S_2 = \frac{160+160}{9} = \frac{320}{9}$ ...(i)

And, $S_1 = \frac{160+200}{18} = \frac{360}{18} =$ 20 m/sec ...(ii)

From eq. (i) and (ii): $20 + S_2 = \frac{320}{9}$

$S_2 = \frac{320}{9} - 20 = \frac{320-180}{9} = \frac{140}{9}$

Required ratio: $S_1 : S_2 = 20 : \frac{140}{9} = \frac{20 \times 9}{140} = \frac{180}{140} = 9 : 7$

40. An electric engine can travel 36 km/h without the train wagons. The reduction in speed of the engine is directly proportional to the square root of the number of the train wagons. With 9 wagons, the speed of the engine is 300 km/h. Find the number of maximum wagons which can be pulled by the engine:

एक इलेक्ट्रिक इंजन ट्रेन, बिना बोगियों के 36 किमी/घंटा की चाल से यात्रा कर सकता है। इंजन की गति में कमी रेल बोगियों की संख्या के वर्गमूल के सीधे आनुपातिक है। 9 बोगियों के साथ इंजन की गति 300 किमी/घंटा है। इंजन द्वारा खींचे जा सकने वाले अधिकतम बोगियों की संख्या ज्ञात करें:

Solution :

W=0, 8=36	W=9, 8=30	Speed=0
$\Delta S \propto \sqrt{w}$	$\therefore \Delta S = 2\sqrt{w}$	$\Delta S = 36$
$\Delta S = k\sqrt{w}$	$36 = 2\sqrt{w}$	Max. wagons to be pulled = 324 - 1
$\Delta S = k\sqrt{q}$	$18 = \sqrt{w}$	323
$K = 2$	$w = 324$	

41. A man walks to a viewpoint and returns to the starting point by his car maintaining constant speed and thus takes a total time of 8 hours 30 minutes. He would have gained 5 hours by driving both ways. How long would it have taken for him to walk both ways with the same walking speed?

एक व्यक्ति एक दृश्य बिंदु तक चलता है और अपनी कार द्वारा स्थिर गति बनाए रखते हुए प्रारंभिक बिंदु पर लौटता है और इस प्रकार उसे कुल 8 घंटे 30 मिनट का समय लगता है। दोनों तरफ गाड़ी चलाने से उसे 5 घंटे का लाभ होता। उसे समान गति से दोनों ओर चलने में कितना समय लगा होगा?

Solution :

$$\frac{D}{x} + \frac{D}{y} = \frac{17}{2}$$

$$\frac{D}{y} + \frac{D}{y} = \frac{17}{2} - 5 = \frac{7}{2}$$

$$\frac{D}{y} = \frac{7}{4}$$

$$\frac{D}{x} + \frac{7}{4} = \frac{17}{2}$$

$$\frac{D}{x} = \frac{27}{4}$$

$$2 \times \frac{D}{x} = \frac{27}{2} \text{ hr} = 13 \text{ hr } 30 \text{ min}$$

42. Two guns are fired from the same place at an interval of 6 minutes. A person approaching the place observes that 5 minutes 52 seconds have elapsed between the hearing of the sound of the two guns. If the velocity of sound is 330 m/sec, then the man was approaching that place at what speed (in km/h):

6 मिनट के अंतराल पर एक ही जगह से दो बंदूकें दागी जाती हैं। उस जगह की ओर आते हुए एक व्यक्ति देखता है कि दो बंदूकों की आवाज के बीच 5 मिनट 52 सेकंड बीत चुके हैं। यदि ध्वनि का वेग 330 मी/सेकंड है, तो मनुष्य उस स्थान पर किस गति से (किमी/घंटा में) आ रहा था:

Solution :

Time interval between firing of two guns = 6 minutes = 360 seconds.

The person hears the second gunshot after 5 minutes 52 seconds = 352 seconds.

Velocity of sound = 330 m/s.

The difference in time is due to the person's approach. The time difference between the hearing of the two sounds is:

Time difference = 360 − 352 = 8 seconds.

$S \sim \frac{1}{T}$

Time Ratio $= \frac{8}{352} = \frac{1}{44}$

Speed Ratio = 44 : 1

44 = 330

$1 = \frac{330}{44} = 7.5$ m/sec

Speed of Man $= \frac{7.5 \times 18}{5} = 27$ Km/hr

9 Race

Linear Race/ रेखीय दौड़

1. In an 800 m race, the ratio of the speeds of two contestants Alka and Mira is 4 : 7. If Alka has a start of 520 m, then Alka will win by how much distance?
800 मीटर की दौड़ में, दो प्रतियोगियों अलका और मीरा की गति का अनुपात 4 : 7 है। यदि अलका की शुरुआत 520 मीटर है, तो अलका किस अंतर से जीतेगी?

Solution :

Remaining Distance Alka has to cover =

$800 - 520 = 280$ m.

Speed Ratio $= 4 : 7$

Distance Ratio $= 4 : 7$

4 units $= 280$ m, 7 units $= 490$ m.

Alka wins by $800 - 490 = 310$ m.

2. In a 100 m race, Kamal defeats Bimal by 5 seconds. If the speed of Kamal is 18 km/h, then what is the speed of Bimal?
100 मीटर दौड़ में कमल ने बिमल को 5 सेकंड से हरा दिया। यदि कमल की गति 18 किमी/घंटा है, तो बिमल की गति है?

Solution :

Speed of Kamal $= 18 \times \frac{5}{18} = 5$ m/s.

$$\frac{100}{x} - \frac{100}{5} = 5$$

$$\frac{100}{x} = 25$$

$x = 4$ m/s.

Speed of Bimal $= 4 \times \frac{18}{5} = 14.4$ km/h.

3. A can run 250 m in 25 seconds and B in 30 seconds. How many metres start can A give to B in a 1 km race so that the race may end in a dead-heat?
A 25 सेकंड में 250 मीटर और B 30 सेकंड में दौड़ सकता है। एक किलोमीटर की दौड़ में A, B को कितने मीटर की शुरुआत दे सकता है ताकि दौड़ बराबरी पर समाप्त हो सके?

Solution :

In 250 m, B takes 5 seconds extra.

In 1000 m, B will take 20 seconds extra.

30 seconds = 250 m.

B will take $\frac{500}{3} = 166.67$ m.

Start given by A $= 166.67$ m.

4. Geeta runs $\frac{5}{2}$ times as fast as Babita. In a race, if Geeta gives a lead of 40 m to Babita, find the distance from the starting point where both of them will meet (correct up to two decimal places).
गीता, बबीता से $\frac{5}{2}$ गुना तेज दौड़ती है। एक दौड़ में, यदि गीता बबीता को 40 मीटर की बढ़त देती है, तो शुरुआती बिंदु से दूरी ज्ञात करें जहां वे दोनों मिलेंगे (दशमलव के दो स्थानों तक सही)।

Solution :

Speed Ratio $= 5 : 2$

Difference of ratio $= 3$ units $= 40$ m.

5 units $= \frac{200}{3} = 66.67$ m.

Hence, they will meet 66.67 m from the starting point.

5. In a 400 m race, A runs at a speed of 16 m/s. If A gives B a start of 15 m and still beats him by 10 seconds, then what is the speed of B?

400 मीटर की दौड़ में, A 16 मीटर/सेकंड की गति से दौड़ता है। यदि A, B को 15 मीटर की शुरुआत देता है और फिर भी उसे 10 सेकंड से हरा देता है, तो B की गति क्या है?

Solution :

$$\frac{385}{x} - \frac{400}{16} = 10$$
$$\frac{385}{x} - 25 = 10$$
$$x = \frac{385}{35} = 11 \text{ m/s.}$$

6. In a race of 1500 m, if vehicle P leads vehicle Q by 200 m, then vehicle P wins the race by 8 seconds. Conversely, if vehicle P gives a lead of 400 m to vehicle Q, the race ends in a draw. What is the time taken by vehicle P to cover 1500 m?
1500 मीटर की दौड़ में, यदि वाहन P, वाहन Q से 200 मीटर आगे है, तो वाहन P 8 सेकंड से दौड़ जीत जाता है। इसके विपरीत, यदि वाहन P वाहन Q को 400 मीटर की बढ़त देता है, तो दौड़ बराबरी पर समाप्त होती है। वाहन P को 1500 मीटर की दूरी तय करने में कितना समय लगेगा?

Solution :

Speed Ratio $P : Q = 1500 : (1500 - 400) = 15 : 11$

$$\frac{1300}{11x} - \frac{1500}{15x} = 8$$
$$\frac{1300}{11} - 100 = 8x$$
$$8x = \frac{200}{11}$$
$$x = \frac{25}{11}$$

Time taken by P: $\frac{1500 \times 11}{15 \times 25} =$ 44 seconds.

7. A takes 3 minutes 40 seconds while B takes 4 minutes to complete a race. A beats B by 60 m in the race if both start simultaneously from the starting point. Find the length of the course.
A को एक दौड़ पूरी करने में 3 मिनट 40 सेकंड लगते हैं जबकि B को 4 मिनट लगते हैं। यदि दोनों प्रारंभिक बिंदु से एक साथ दौड़ शुरू करते हैं तो A, B को 60 मीटर से हरा देता है। पाठ्यक्रम की लंबाई ज्ञात कीजिए।

Solution :

Time Ratio $A : B = 220 : 240 = 11 : 12$

Solution :

Speed Ratio $= 12 : 11$

Distance Ratio $= 12 : 11$

Difference of distance (1 unit) $= 60$ m.

Total distance (12 units) $= 720$ m.

Length of Course $= 720$ m.

8. In a race of one kilometre, A gives B a start of 100 m and still wins by 20 seconds. But if A gives B a start of 25 seconds, B wins by 50 m. The time taken by A to run one kilometre is?
एक किलोमीटर की दौड़ में, A, B को 100 मीटर की शुरुआत देता है और फिर भी 20 सेकंड से जीत जाता है। लेकिन यदि A, B को 25 सेकंड की शुरुआत देता है, तो B 50 मीटर से जीत जाता है। A को एक किलोमीटर दौड़ने में कितना समय लगेगा?

Solution :

$\frac{900}{B} - \frac{1000}{A} = 20$ (Eq 1) $\frac{1000}{B} - \frac{950}{A} = 25$ (Eq 2)

Solving both equations: $9000/B - 10000/A = 200$ and $9000/B - 8550/A = 225$

$1450/A = 25$

$A = 58$ m/s.

Time taken by A: $\frac{1000}{58} = \frac{500}{29}$ seconds.

9. In a race of 200 metres, B can give a start of 10 metres to A, and C can give a start of 20 metres to B. The start that C can give to A, in the same race, is?
200 मीटर की दौड़ में, B, A को 10 मीटर की शुरुआत दे सकता है, और C, B को 20 मीटर की शुरुआत दे सकता है। उसी दौड़ में, C, A को कितनी शुरुआत दे सकता है?

Solution :

$B : A = 200 : 190 = 20 : 19$

$C : A = 200 : 180 = 10 : 9$

$A : B : C = 171 : 190 : 200$

Start given by C to A: $200 - 171 = 29$ m.

10. Arun and Bhaskar run a race of 3 km. First, Arun gives Bhaskar a head start of 400 m and beats him by 30 sec. While coming back, Arun gives Bhaskar a lead of 2.5 min and gets beaten by 500 m. What is the difference between the times in minutes in which Arun and Bhaskar can run the race for one side separately?

अरुण और भास्कर 3 किमी की दौड़ लगाते हैं। पहले अरुण ने भास्कर को 400 मीटर की बढ़त दिलाई और उसे 30 सेकंड से हरा दिया। वापस आते समय, अरुण ने भास्कर को 2.5 मिनट की बढ़त दी और 500 मीटर से हरा दिया। मिनटों में उस समय के बीच क्या अंतर है जिसमें अरुण और भास्कर एक पक्ष के लिए अलग-अलग दौड़ लगा सकते हैं?

Solution :

3000 2600 30

2500 3000 150

500 $-$ 400 $-$ 120

3000 $-$ 2400 $-$ 720

5000 in 750 sec

3000 in 450 sec (Bhaskar).

2500 in $450 - 150 = 300$ sec

3000 in 360 sec (Arun).

Difference $= 450 - 360 = 90$ sec $= 1.5$ min.

11. In a race of 800 m, A can beat B by 40 m. In a race of 500 m, B can beat C by 5 m. In a race of 200 m, A will beat C by?

800 मीटर की दौड़ में A, B को 40 मीटर से हरा सकता है। 500 मीटर की दौड़ में, B, C को 5 मीटर से हरा सकता है। 200 मीटर की दौड़ में A, C को कितने से हरा देगा?

Solution :

$A : B = 800 : 760 = 20 : 19$

$B : C = 500 : 495 = 100 : 99$

$A : C = 20 \times 100 : 99 \times 19$

$A : C = 2000 : 1881$

$A : C = 200 : 188.1$

Difference $= 200 - 188.1 = 11.9$ m.

Circular Race/ गोलाकार दौड़

12. On a circular path of 693 m, Sujata and Anjali start from the same point but in opposite directions with speeds of 2.85 km/h and 1.5 m/sec respectively. When will they meet for the first time?

693 मीटर के वृत्ताकार पथ पर, सुजाता और अंजलि एक ही बिंदु से लेकिन विपरीत दिशाओं में क्रमशः 2.85 किमी/घंटा और 1.5 मीटर/सेकंड की गति से शुरू होती हैं। वे पहली बार कब मिलेंगे?

Solution :

Anjali's speed in km/h $= 1.5 \times \frac{18}{5} = 5.4$ km/h

Time of first meeting $= \frac{\text{Track length}}{\text{Relative speed}}$

Relative speed in opposite directions: $2.85 + 5.4 = 8.25$ km/h

$= \frac{0.693 \times 60}{8.25} = 5.04$ minutes.

13. On a circular path of 120 m, Radhika and Deepika start from the same point but in the same direction with speeds of 12 m/min and 20 m/min respectively. When will they meet for the first time?

120 मीटर के वृत्ताकार पथ पर, राधिका और दीपिका एक ही बिंदु से लेकिन समान दिशा में क्रमशः 12 मीटर/मिनट और 20 मीटर/मिनट की गति से शुरू करती हैं। वे पहली बार कब मिलेंगे?

Solution :

Relative speed in the same direction: $20 - 12 = 8$ m/min

Time of first meeting $= \frac{\text{Track length}}{\text{Relative speed}} = \frac{120}{8} = 15$ minutes.

14. A, B, and C run simultaneously, starting from a point, around a circular track of length 1200 m, at respective speeds of 2 m/s, 4 m/s, and 6 m/s. A and B run in the same direction, while C runs in the opposite direction to the other two. After how much time will they meet for the first time?

A, B और C, एक बिंदु से शुरू करके, 1200 मीटर लंबे गोलाकार ट्रैक के चारों ओर, 2 मीटर/सेकंड, 4 मीटर/सेकंड और 6 मीटर/सेकंड की गति से एक साथ दौड़ते हैं। A और B एक ही दिशा में दौड़ते हैं, जबकि C अन्य दो के विपरीत दिशा में दौड़ता है। वे कितने समय बाद पहली बार मिलेंगे?

Solution :

Time for A and B to meet: $\frac{1200}{4-2} =$ 600 seconds.

Time for A and C to meet: $\frac{1200}{2+6} =$ 150 seconds.

LCM of 600 and 150 = 600 seconds or 10 minutes.

15. Two runners, Sony and Mony, start running on a circular track of length 200 m at speeds of 18 km/h and 24 km/h, respectively, in the same direction. After how much time from the start will they meet again at the starting point?

दो धावक, सोनी और मोनी, 200 मीटर लंबे एक गोलाकार ट्रैक पर क्रमश: 18 किमी/घंटा और 24 किमी/घंटा की गति से एक ही दिशा में दौड़ना शुरू करते हैं। प्रारंभ से कितने समय बाद वे पुन: प्रारंभिक बिंदु पर मिलेंगे?

Solution :

Speed of Sony: 5 m/sec, Speed of Mony: $\frac{20}{3}$ m/sec

Time for Sony to complete one lap: $\frac{200}{5} = 40$ seconds.

Time for Mony to complete one lap: $\frac{200 \times 3}{20} = 30$ seconds.

LCM of 40 and 30 = 120 seconds.

16. Racer A and racer B run a race of 18 km on a circular track of length 800 m. Both complete one round in 200 sec and 250 sec, respectively. After how much time from the start will the faster person meet the slower person for the last time?

रेसर A और रेसर B 800 मीटर लंबे गोलाकार ट्रैक पर 18 किमी की दौड़ लगाते हैं। दोनों क्रमश: 200 सेकंड और 250 सेकंड में एक चक्कर पूरा करते हैं। प्रारंभ से कितने समय बाद तेज व्यक्ति धीमे व्यक्ति से आखिरी बार मिलेगा?

Solution :

LCM of (200, 250) = 1000 seconds.
In 200 seconds, A covers 800 m.

In 1000 seconds, A covers 4 km.

1st meetup: 4 km.

4th and last meetup: 16 km.

Time $= 1000 \times 4 = 4000$ seconds.

17. Having started from the same point and at the same time, two runners - P and Q - are running around a circular track of length 500 m in opposite directions with the speeds of 6 m/s and 10 m/s, respectively. If they exchange their speeds after meeting for the first time, who will reach the starting point first?

एक ही बिंदु से और एक ही समय पर चलना शुरू करते हुए, दो धावक - P और Q - क्रमश: 6 मीटर/सेकंड और 10 मीटर/सेकंड की गति के साथ विपरीत दिशाओं में 500 मीटर लंबे एक गोलाकार ट्रैक के चारों ओर दौड़ रहे हैं। यदि वे पहली बार मिलने के बाद अपनी गति बदल लेते हैं, तो शुरुआती बिंदु पर सबसे पहले कौन पहुंचेगा?

Solution :

Both P and Q will reach at the same time
No change due to exchange their Speeds after meeting.

18. Ram and Shyam are racing along a circular track. The speed of Ram is thrice the speed of Shyam. The length of the circular track is 1440 m. After the start of the race from the same point simultaneously, Ram meets Shyam for the first time at the end of the 8th minute. If Ram and Shyam start the race again from the same starting point simultaneously, then the time taken by Shyam to finish the race is: (given that the length of the race is the same as the length of the track).

राम और श्याम एक गोलाकार ट्रैक पर दौड़ रहे हैं। राम की गति श्याम की गति से तीन गुना है। वृत्ताकार ट्रैक की लंबाई 1440 मीटर है। एक ही बिंदु से एक साथ दौड़ शुरू करने के बाद, राम 8वें मिनट के अंत में पहली बार श्याम से मिलता है। यदि राम और श्याम एक ही प्रारंभिक बिंदु से एक साथ फिर से दौड़ शुरू करते हैं, तो श्याम द्वारा दौड़ पूरी करने में लिया गया समय है: (यह देखते हुए कि दौड़ की लंबाई ट्रैक की लंबाई के समान है)

Solution :

3:1
2---1440
1---720
720 in 8 min
1440 in 16 min

19. Two persons started running on a circular track simultaneously with speeds of 20 m/s and 30 m/s in opposite directions. If the circumference of the circular track is 100 m, then find at how many distinct points they will cross each other.

दो व्यक्तियों ने एक गोलाकार ट्रैक पर एक साथ विपरीत दिशाओं में 20 मीटर/सेकंड और 30 मीटर/सेकंड की गति से दौड़ना शुरू किया। यदि वृत्ताकार पथ की परिधि 100 मीटर है, तो ज्ञात कीजिए कि वे कितने अलग-अलग बिंदुओं पर एक-दूसरे को पार करेंगे?

Solution :

To find distinct points

Take the ratio of speed: $20 : 30 = 2 : 3$.

For opposite directions: $2 + 3 =$ 5 meeting points.

20. Seema walks around a circular field at the rate of 3 rounds/hour, while Priya runs around it at the rate of 8 rounds/hour. They start in the same direction from the same point at 11 : 54 p.m.. They shall first cross each other at?

सीमा एक गोलाकार मैदान के चारों ओर 3 चक्कर/घंटा की दर से चलती है, जबकि प्रिया इसके चारों ओर 8 चक्कर/घंटा की दर से दौड़ती है। वे रात 11 : 54 बजे एक ही बिंदु से एक ही दिशा में चलना शुरू करते हैं। वे पहले एक-दूसरे को पार करेंगे?

Solution :

Let Dis be 1 km.

Speed of A = 3 km/hr.

Speed of B = 8 km/hr.

$1/(8 - 3) = 1/5$ hr = 12 min.

11 : 54 + 12 min = 12 : 06 a.m.

21. M and N walk along a circular track. They start at 5 : 00 a.m. from the same point in opposite directions. M and N walk at a speed of 5 rounds/hour and 2 rounds/hour, respectively. How many times will they cross each other before 6 : 30 a.m. on the same day?

M और N एक वृत्ताकार पथ पर चलते हैं। वे सुबह 5 : 00 बजे एक ही बिंदु से विपरीत दिशाओं में चलना शुरू करते हैं। M और N क्रमशः 5 चक्कर/घंटा और 2 चक्कर/घंटा की गति से चलते हैं। एक ही दिन प्रातः 6 : 30 बजे से पहले वे कितनी बार एक दूसरे को पार करेंगे?

Solution :

Let Distance be 10 km.

Speed $A = 50$ km/hr.

Speed $B = 20$ km/hr.

$10/70 = 1/7$ hr or $60/7$ min.

$90/(60/7) \approx 10.5$.

10 times, they will cross each other.

22. The friends P and Q simultaneously start running from the same point around a circular track. They run in the same direction. P runs at 6 m/s and Q runs at 2 m/s, then how many distinct points they will cross each other?

वे मित्र P और Q एक साथ एक गोलाकार ट्रैक के चारों ओर एक ही बिंदु से दौड़ना शुरू करते हैं। वे एक ही दिशा में दौड़ते हैं। P 6 मीटर/सेकंड की गति से दौड़ता है और Q 2 मीटर/सेकंड की गति से दौड़ता है, तो वे एक दूसरे को कितने अलग-अलग बिंदुओं को पार करेंगे?

Solution :

To find distinct points

Take the ratio of speed: $6 : 2 = 3 : 1$.

For the same direction: $3 - 1 =$ 2 meeting points.

10 Boats and Streams

1. If a boat goes upstream at a speed of 24 km/h and comes back the same distance at 40 km/h. What is the average speed (in km/h) for the total journey?

यदि एक नाव 24 किमी/घंटा की गति से धारा के प्रतिकूल जाती है और 40 किमी/घंटा की गति से समान दूरी तक वापस आती है। कुल यात्रा के लिए औसत गति (किमी/घंटा में) क्या है?

Solution :

Average speed $= \dfrac{2V_1V_2}{V_1+V_2}$

$\dfrac{2\times 24\times 40}{24+40} = \dfrac{1920}{64} = 30$ km/h.

2. The speed of a boat downstream is 150% more than its speed upstream. If the time taken by the boat for going 80 km downstream and 50 km upstream is 8.2 hours, then what is the speed (in km/h) of the boat downstream?

एक नाव की गति धारा के विपरीत दिशा में उसकी गति से 150% अधिक है। यदि नाव को धारा के अनुकूल 80 किमी और धारा के प्रतिकूल 50 किमी जाने में 8.2 घंटे का समय लगता है, तो धारा के अनुकूल नाव की गति (किमी/घंटा में) क्या है?

Solution :

Let upstream speed $= x$ km/h.

Downstream speed $= \dfrac{5x}{2}$ km/h.

According to the question: $\dfrac{80}{\frac{5x}{2}} + \dfrac{50}{x} = 8.2$

$\dfrac{160}{5x} + \dfrac{50}{x} = 8.2$

$\dfrac{160+250}{5x} = 8.2$

$x = \dfrac{410}{41} = 10$ km/h.

Hence downstream speed $= \dfrac{5x}{2} = \dfrac{5\times 10}{2} =$ 25 km/h.

3. The time taken by a boat to travel 13 km downstream is the same as the time taken by it to travel 7 km upstream. If the speed of the stream is 3 km/h, then how much time (in hours) will it take to travel a distance of 44.8 km in still water?

एक नाव द्वारा धारा के अनुकूल 13 किमी की यात्रा करने में लिया गया समय, धारा के प्रतिकूल 7 किमी की यात्रा करने में लिया गया समय के समान है। यदि धारा की गति 3 किमी/घंटा है, तो शांत पानी में 44.8 किमी की दूरी तय करने में कितना समय (घंटे में) लगेगा?

Solution :

Let the speed of the boat in still water $=$ x km/h.

According to the question: $\dfrac{13}{x+3} = \dfrac{7}{x-3}$

$13x - 39 = 7x + 21$

$6x = 60 \implies x = 10$ km/h.

Required time $= \dfrac{44.8}{10} = \dfrac{112}{25}$ hr.

4. A boat can go 3.6 km upstream and 5.4 km downstream in 54 minutes, while it can go 5.4 km upstream and 3.6 km downstream in 58.5 minutes. The time (in minutes) taken by the boat in going 10 km downstream is: / एक नाव 54 मिनट में धारा के प्रतिकूल 3.6 किमी और धारा के अनुकूल 5.4 किमी जा सकती है, जबकि 58.5 मिनट में धारा के प्रतिकूल 5.4 किमी और धारा के अनुकूल 3.6 किमी जा सकती है। धारा के अनुकूल 10 किमी जाने में नाव द्वारा लिया गया समय (मिनटों में) है:

Solution :

Let the speed of the boat downstream = x km/min.

Speed of the boat upstream = y km/min.

$\frac{3.6}{y} + \frac{5.4}{x} = 54 \implies \frac{0.2}{y} + \frac{0.3}{x} = 3$ (i).

$\frac{5.4}{y} + \frac{3.6}{x} = 58.5 \implies \frac{0.6}{y} + \frac{0.4}{x} = 6.5$ (ii).

On solving equations (i) and (ii): $0.4/x - 0.3/x = 6.5 - 3$

$0.5/x = 2.5 \implies x = \frac{1}{5}$.

Required time $= \frac{10}{\frac{1}{5}} = 50$ minutes.

5. In a stream running at 3 km/h, a motorboat goes 12 km upstream and back to the starting point in 60 min. Find the speed of the motorboat in still water. (in km/h)
3 किमी/घंटा की गति से चलने वाली धारा में, एक मोटरबोट 12 किमी धारा के प्रतिकूल जाती है और 60 मिनट में प्रारंभिक बिंदु पर वापस आती है। शांत जल में मोटरबोट की गति ज्ञात कीजिए। (किमी/घंटा में)

Solution :

Let the speed of the boat in still water be x km/h.

$\frac{12}{x-3} + \frac{12}{x+3} = 1 \quad [\because 60 \text{ min} = 1 \text{ hour}]$

$12\left(\frac{x+3+x-3}{x^2-9}\right) = 1$

$24x = x^2 - 9$

$x^2 - 24x - 9 = 0$

$x = \frac{24 \pm \sqrt{576+36}}{2} = \frac{24 \pm \sqrt{612}}{2} = 12 \pm 3\sqrt{17}$

Taking the positive value, $x = 3(4 + \sqrt{17})$ km/h.

6. The speed of a boat in still water is 6 km/h. Time taken by the boat to cover a certain distance upstream is 3 hours more than the time taken to cover the same distance downstream. If the speed of the stream is 2 km/h, then what is the total distance, upstream and downstream, covered by the boat?
शांत जल में एक नाव की गति 6 किमी/घंटा है। नाव द्वारा धारा के प्रतिकूल एक निश्चित दूरी तय करने में लिया गया समय धारा के अनुकूल समान दूरी तय करने में लिए गए समय से 3 घंटे अधिक है। यदि धारा की गति 2 किमी/घंटा है, तो नाव द्वारा धारा के प्रतिकूल और धारा के अनुकूल कुल कितनी दूरी तय की गई?

Solution :

Let the distance covered by the boat in each direction (downstream and upstream) be x km.

Speed of boat downstream $= 6 + 2 = 8$ km/h.

Speed of boat upstream $= 6 - 2 = 4$ km/h.

$\frac{x}{4} - \frac{x}{8} = 3$

$\frac{x}{8} = 3$

$x = 24$

$\therefore$ Total covered distance upstream and downstream by boat $= x + x = 24 + 24 = 48$ km.

7. A boat covered a distance of 15 km upstream in 5 hours and a distance of 42 km downstream in 6 hours. The speed of the stream (in km/h) is:
एक नाव धारा के प्रतिकूल 15 किमी की दूरी 5 घंटे में और धारा के अनुकूल 42 किमी की दूरी 6 घंटे में तय करती है। धारा की गति किमी/घंटा में है:

Solution :

Speed of boat downstream $= \frac{42}{6} = 7$ km/h.

Speed of boat upstream $= \frac{15}{5} = 3$ km/h.

Speed of the stream $= \frac{7-3}{2} = 2$ km/h.

8. Ravi can row a boat in still water at the speed of 14 km/h. If a river is flowing at the speed of 2 km/h and Ravi takes 3 hours to cover a certain distance upstream, then how much time will he take to cover the same distance downstream?
रवि शांत पानी में 14 किमी/घंटा की गति से नाव चला सकता है। यदि एक नदी 2 किमी/घंटा की गति से बह रही है और रावी को धारा के प्रतिकूल एक निश्चित दूरी तय करने में 3 घंटे लगते हैं, तो धारा के अनुकूल समान दूरी तय करने में उसे कितना समय लगेगा?

Solution :

In upstream:

$$V_R - V_S = \frac{\text{Total distance (D)}}{3}$$

$$14 - 2 = \frac{D}{3}$$

$$D = 36 \text{ km}$$

Time taken by Ravi to cover the distance D in downstream

$$\text{Time} = \frac{D}{V_R + V_S} = \frac{36}{14+2} = \frac{36}{16} = \frac{9}{4} = 2\frac{1}{4} = 2 \text{ hours } 15 \text{ minutes.}$$

9. A person can row a distance of 4 km upstream in one hour 20 minutes and can row back to the starting point in just 24 minutes. How much time (in hours) will he take to row 13 km in still water?
एक व्यक्ति एक घंटे 20 मिनट में धारा के विपरीत 4 किमी की दूरी तय कर सकता है और केवल 24 मिनट में शुरुआती बिंदु पर वापस आ सकता है। शांत पानी में 13 किमी की दूरी तय करने में उसे कितना समय (घंटों में) लगेगा?

Solution :

Let, Speed of boat in still water = x km/hr

Speed of stream = y km/hr

Time taken to cover 4 km distance in upstream:

$$\frac{4}{x-y} = 1\frac{20}{60} = \frac{4}{3}$$

$$x - y = 3 \text{ ...(i)}$$

Time taken to return:

$$\frac{4}{x+y} = \frac{24}{60} = \frac{2}{5}$$

$$x + y = 10 \text{ ...(ii)}$$

From equations (i) and (ii):

$$x + y = 10, \quad x - y = 3$$

$$x = \frac{13}{2}$$

Time taken to cover a distance of 13 km in still water:

$$\frac{13}{\frac{13}{2}} = 2 \text{ hours}$$

10. A boat goes 30 km upstream in 3 hours and downstream in 1 hour. How much time (in hours) will this boat take to cover 60 km in still water?
एक नाव धारा के विपरीत 30 किमी की दूरी 3 घंटे में और धारा की दिशा में 1 घंटे में तय करती है। शांत पानी में 60 किमी की दूरी तय करने में इस नाव को कितना समय (घंटों में) लगेगा?

Solution :

Speed of boat in upstream $= \frac{30}{3} = 10$ km/h

Speed of boat in downstream $= \frac{30}{1} = 30$ km/h

Speed of boat in still water $= \frac{10+30}{2} = 20$ km/h

Required time:

$$\frac{60}{20} = 3 \text{ hours}$$

11. The speed of a boat in still water is 18 km/h and the speed of the current is 6 km/h. In how much time (in hours) will the boat travel a distance of 90 km upstream and the same distance downstream?
शांत पानी में एक नाव की गति 18 किमी/घंटा है और धारा की गति 6 किमी/घंटा है। नाव कितने समय (घंटों में) में धारा के प्रतिकूल 90 किमी की दूरी और धारा के अनुकूल समान दूरी तय करेगी?

Solution :

Let the speed of boat in still water = x km/h

and speed of stream = y km/h

Speed of boat in upstream = $x - y$

Speed of boat in downstream = $x + y$

Total time taken:

$$\frac{90}{x+y} + \frac{90}{x-y} = \frac{90}{24} + \frac{90}{12}$$

$$= 90\left(\frac{1+2}{24}\right)$$

$$= \frac{90}{8} = 11\frac{1}{4} \text{ hours}$$

12. The speed of a boat in still water is 15 km/h. The speed of the current is 3 km/h. In how much time (in hours) will the boat travel a distance of 54 km upstream and the same distance downstream?
शांत जल में एक नाव की गति 15 किमी/घंटा है। धारा की गति 3 किमी/घंटा है। नाव कितने समय (घंटों में) में धारा के प्रतिकूल 54 किमी की दूरी और धारा के अनुकूल समान दूरी तय करेगी?

Solution :

Speed of boat in upstream = 15 − 3 = 12 km/h

Speed of boat in downstream = 15 + 3 = 18 km/h

Time taken to cover the distance in downstream:

$\frac{54}{18} = 3$ hours

Time taken to cover the distance in upstream:

$\frac{54}{12} = 4\frac{1}{2}$ hours

Total time taken:

$4\frac{1}{2} + 3 = 7\frac{1}{2}$ hours

13. Anup can row 33 km downstream and 35 km upstream in 8 hours. He can also row 44 km downstream and 28 km upstream in the same time. How much time (in hours) will he take to row 55 km downstream and 14 km upstream?

अनुप 8 घंटे में धारा के अनुकूल 33 किमी और धारा के प्रतिकूल 35 किमी नाव चला सकता है। वह समान समय में धारा के अनुकूल 44 किमी और धारा के विपरीत 28 किमी तक नाव चला सकता है। धारा के अनुकूल 55 किमी और धारा के प्रतिकूल 14 किमी चलने में उसे कितना समय (घंटों में) लगेगा?

Solution :

Let speed of downstream = a km/hr

Speed of upstream = b km/hr

According to the question:

$\frac{33}{a} + \frac{35}{b} = 8$ ...(i)

$\frac{44}{a} + \frac{28}{b} = 8$ ...(ii)

From equations (i) and (ii):

$\frac{33}{a} + \frac{35}{b} = \frac{44}{a} + \frac{28}{b}$

$\frac{44}{a} - \frac{33}{a} = \frac{35}{b} - \frac{28}{b}$

$\frac{11}{a} = \frac{7}{b}$

$a : b = 11 : 7$

Time taken:

$\frac{55}{11} + \frac{14}{7} = 5 + 2 = 7$ hours

Hence, option (d) is correct.

14. A boat covers a round trip journey between two points A and B in a river in T hours. If its speed in still water becomes 2 times, it would take $\frac{80}{161}T$ hours for the same journey. Find the ratio of its speed in still water to the speed of the river?

एक नाव एक नदी में दो बिंदु A और B की एक गोल यात्रा T घंटे में तय करती है। यदि शांत पानी में इसकी गति 2 गुना हो जाए, तो उसी यात्रा में इसे $\frac{80}{161}T$ घंटे लगेंगे। शांत जल में इसकी गति और नदी की गति का अनुपात ज्ञात कीजिये?

Solution :

Let boat's speed in still water = x km/h

Speed of the stream = y km/h

Let total distance = k km

$\frac{k}{x+y} + \frac{k}{x-y} = T$ i

$\frac{k}{2x+y} + \frac{k}{2x-y} = \frac{80}{161}T$ ii

$\frac{1}{2x+y} + \frac{1}{2x-y} = \frac{80}{161} \times \frac{T}{k}$

$\frac{1}{2x+y} + \frac{1}{2x-y} = \frac{80}{161} \times \left(\frac{1}{x+y} + \frac{1}{x-y}\right)$ [From Eq. (i)]

$\frac{4x}{4x^2 - y^2} = \frac{80}{161} \times \frac{2x}{x^2 - y^2}$

$160x^2 - 40y^2 = 161x^2 - 161y^2$

$x^2 = 121y^2$

$\frac{x}{y} = \frac{11}{1} = 11 : 1$

15. A boat goes 27 km upstream and 33 km downstream in 6 hours. In the same time, it can go 36 km upstream and 22 km downstream. How much time will it take to go 36 km upstream and 44 km downstream?

एक नाव 6 घंटे में धारा के विपरीत 27 किमी और धारा के अनुकूल 33 किमी चलती है। एक ही समय में यह धारा के विपरीत 36 किमी और धारा के अनुकूल 22 किमी जा सकता है। धारा के विपरीत 36 किमी और धारा के अनुकूल 44 किमी जाने में कितना समय लगेगा?

Solution :

Let the speed of boat = a km/hr

And the speed of stream = b km/hr

According to the question,

$$\frac{33}{a+b} + \frac{27}{a-b} = \frac{22}{a+b} + \frac{36}{a-b}$$

$$\frac{11}{a+b} = \frac{9}{a-b}$$

$$11a + 11b = 9a + 9b$$

$$a = 10b$$

then,

$$\frac{33}{10b+b} + \frac{27}{10b-b} = 6$$

$$\frac{3}{b} + \frac{3}{b} = 6$$

Speed of downstream = $10 + 1 = 11$ km/hr

Speed of upstream = $10 - 1 = 9$ km/hr

$$\frac{44}{11} + \frac{36}{9} = 4 + 4 = 8 \text{ hours}$$

$\therefore$ It will take 8 hours to go 36 km upstream and 44 km downstream

11 Average

Average / औसत

Average is defined as the mean value which is equal to the sum of the number of a given set of values divided by the total number of values present in the set.

औसत को माध्य मान के रूप में परिभाषित किया जाता है जो दिए गए मानों के सेट की संख्या के योग को सेट में मौजूद मानों की कुल संख्या से विभाजित करने के बराबर होता है।

1. Sum of the observation Average total number of the observation
   प्रेक्षणों का योग - औसत प्रेक्षणों की कुल संख्या
2. Average of two or more numbers/quantities is called the mean of these numbers, which is given by
   दो या दो से अधिक संख्याओं/मात्राओं का औसत इन संख्याओं का माध्य कहलाता है, जो निम्न द्वारा दिया जाता है
   $$\text{Average (A)} = \frac{\text{Sum of all observation (सभी प्रेक्षणों का योग)}}{\text{Total no. of all observation (प्रेक्षणों की कुल संख्या)}}$$
3. If the given observation (x) are occurring with certain frequency (A) then,
   यदि दिए गए अवलोकन (x) निश्चित आवृत्ति (A) के साथ हो रहे हैं, तो,
   $$\text{Average} = \frac{(A_1x_1 + A_2x_2 + \cdots + A_nx_n)}{(x_1 + x_2 + \cdots + x_n)}$$
   जहां $A_1, A_2, A_3, \ldots, A_n$ आवृत्तियाँ हैं।
4. If the average of 'n1' numbers is a2, and the average of 'n2' numbers is a2, then average of total numbers n1, and n2, is
   यदि 'n1' संख्याओं का औसत a2 है, और 'n2' संख्याओं का औसत a2 है, तो कुल संख्याओं n1 और n2 का औसत:
   $$\text{Average} = \frac{(n_1a_2 + n_2a_2)}{(n_1 + n_2)}$$

Average of Consecutive Numbers

1. The average of 'n' consecutive natural numbers starting from 1 i.e. average of 1, 2, 3, ..., n
   1 से शुरू होने वाली 'n' क्रमागत प्राकृतिक संख्याओं का औसत यानी 1, 2, 3 का औसत,...n
   $$\text{Average} = \frac{(n+1)}{2}$$
2. The average of squares of 'n' consecutive natural numbers starting from 1:
   1 से शुरू होने वाले 'n' क्रमागत प्राकृतिक संख्याओं के वर्गों का औसत:
   $$\text{Average} = \frac{(n+1)(2n+1)}{6}$$
3. The average of cubes of first 'n' consecutive natural numbers:
   प्रथम 'n' क्रमागत प्राकृतिक संख्याओं के घनों का औसत:
   $$\text{Average} = \frac{(n(n+1)^2)}{4}$$
4. The average of first 'n' consecutive even natural numbers:
   प्रथम 'n' क्रमागत सम प्राकृत संख्याओं का औसत:
   $\text{Average} = n + 1$
5. The average of first 'n' consecutive odd natural numbers:
   प्रथम 'n' क्रमागत विषम प्राकृत संख्याओं का औसत:
   $\text{Average} = n$
6. The average of certain consecutive numbers $a, b, c, \ldots, n$ is
   कुछ क्रमागत संख्याओं $a, b, c, \ldots, n$ का औसत:
   $$\text{Average} = \frac{\text{First No. + Last No.}}{2} = \frac{(a+n)}{2}$$

   Ex. Find average of 4, 5, 6, ..., 20
   4, 5, 6, ..., 20 का औसत ज्ञात कीजिए:
   $$\text{Average} = \frac{4+20}{2} = \frac{24}{2} = 12$$

7. The average of 1st 'n' multiples of certain numbers
x के पहले 'n' गुणकों का (कुछ संख्याओं का औसत):
Average $= \frac{x(n+1)}{2}$

Ex. Find the average of the first 10 multiples of 3
3 के पहले 10 गुणकों का औसत ज्ञात कीजिए:

First 10 multiples of 3 = 3, 6, 9, . . . , 30

$$\text{Average} = \frac{3(1+10)}{2} = \frac{33}{2} = 16.5$$

8. Average of squares of the first n even numbers
पहली n सम संख्या के वर्गों का औसत:
Average $= \frac{2(n+1)(2n+1)}{3}$

9. Average of cubes of the first n even numbers
पहली n सम संख्या के घनों का औसत:
Average $= 2n(n+1)^2$

10. Average of squares of the first n odd numbers
पहली n विषम संख्या के वर्गों का औसत:
Average $= \frac{(2n+1)(2n-1)}{3}$

11. Average of cubes of the first n odd numbers
पहली n विषम संख्या के घनों का औसत:
Average $= n(2n^2-1)$

12. Average of the first n odd numbers
पहली n विषम संख्याओं का औसत:
Average $= \frac{\text{Last odd no.} + 1}{2}$

13. Average of the first n even numbers
पहली n सम संख्याओं का औसत:
Average $= \frac{\text{Last even no.} + 2}{2}$

Average Speed:

1. If A goes from P to Q with a speed of x km/h and returns from Q to P with a speed of y km/h, then the average speed of the total journey is
यदि A, P से Q तक x किमी/घंटा की गति से जाता है और Q से P तक y किमी/घंटा की गति से वापस आता है, तो कुल यात्रा की औसत गति है: Average Speed $= \frac{2xy}{x+y}$ or Average Speed $= \frac{\text{Total Distance}}{\text{Total Time Taken}}$

2. If a distance is traveled with three different speeds a km/h, b km/h, and c km/h, then the average speed of the total journey is
यदि एक दूरी तीन अलग-अलग गति a किमी/घंटा, b किमी/घंटा और c किमी/घंटा से तय की जाती है, तो कुल यात्रा की औसत गति: Average Speed $= \frac{3abc}{ab+bc+ca}$ km/h

distance in travelled with 1m (hg 3 1m (hy and 4 lum (h Find og need of the who)
1 मी (एचजी 3 1 मी (एचजी और 4 एलयूएम (एच) के साथ तय की गई दूरी और कौन की आवश्यकता का पता लगाएं)

Average Speed

1. The average of all even numbers between 104 and 148 is
104 और 148 के बीच सभी सम संख्याओं का औसत है

Solution :

Intended average $= \frac{a+l}{2} = \frac{104+148}{2} =$ 126

2. What is the average of all numbers between 100 and 200 which are divisible by 13?
100 और 200 के बीच 13 से विभाज्य सभी संख्याओं का औसत क्या है?

Solution :

Numbers between 100 to 200 which are divisible by 13 are 104, 117, and 195.

Average $= \frac{\text{First term + Last term}}{2} = \frac{104+195}{2} = 149.5$

3. The average of the first 10 multiples of 5 is:
5 के प्रथम 10 गुणजों का औसत है:

Solution :

First 10 multiples of 5 = 5, 10, 15, 20, 25, 30, 35, 40, 45, and 50.

Average $= \frac{5+10+15+20+25+30+35+40+45+50}{10} = \frac{275}{10} = 27.5$

4. The average of the square of the first seven natural numbers is:
प्रथम सात प्राकृतिक संख्याओं के वर्ग का औसत है:

Solution :

Average of squares of first n natural numbers $= \dfrac{n(n+1)(2n+1)}{6}$

Hence, required average

$$= \frac{(7+1)(7\times 2+1)}{6} = \frac{8\times 15}{6} = 20$$

5. Of the three numbers whose average is 22, the first is $\frac{3}{8}$ times the sum of the other two. What is the first number?
तीन संख्याओं में से जिनका औसत 22 है, पहली अन्य दो के योग का $\frac{3}{8}$ गुना है। पहला नंबर क्या है?

Solution :

Let the first, second, and third numbers be x, y, and z respectively.

As per the question,

$$x = \frac{3}{8}(y+z) \implies 8x = 3(y+z)$$

$$\frac{x+y+z}{3} = 22 \implies x+y+z = 66$$

$$x + \frac{8x}{3} = 66 \implies 11x = 198 \implies x = 18$$

6. The average of 9 numbers is 19. If the average of the first four numbers is 14, then the average of the last 5 numbers is:
9 संख्याओं का औसत 19 है। यदि पहली चार संख्याओं का औसत 14 है, तो अंतिम 5 संख्याओं का औसत है:

Solution :

According to the question,

Average of the last 5 numbers $= \dfrac{9\times 19 - 4\times 14}{5} = \dfrac{171-56}{5} = \dfrac{115}{5} = 23$

7. The average of the squares of four consecutive odd natural numbers is 201. The average of 7 times of the largest number and 3 times of the smallest number is:
चार क्रमागत विषम प्राकृत संख्याओं के वर्गों का औसत 201 है। सबसे बड़ी संख्या का 7 गुना और सबसे छोटी संख्या का 3 गुना का औसत है:

Solution :

Let the consecutive odd numbers be x, $x+2$, $x+4$, and $x+6$.

According to the question,

$(x)^2+(x+2)^2+(x+4)^2+(x+6)^2 = 201\times 4$

$x^2+x^2+4+4x+x^2+16+8x+x^2+36+12x = 804$

$4x^2+24x+56 = 804 \implies 4x^2+24x-748 = 0$

Factoring: $4x^2+12x+374 = 0 \implies 2x^2+12x-374 = 0$

$2x^2+34x-22x-374 = 0 \implies 2x(x+17)-22(x+17) = 0$

$(x+17)(2x-22) = 0$

$x = 11$ (as $x = -17$ is not possible).

Again, according to the question: Average $= \dfrac{7\times 17+3\times 11}{2} = \dfrac{152}{2} = 76$

8. The average of nine 2-digit numbers is decreased by 6 when the digits of one of the 2-digit numbers are interchanged. Find the difference between the digits of that number.
जब दो अंकों वाली संख्याओं में से एक के अंकों को आपस में बदल दिया जाता है, तो नौ दो अंकों वाली संख्याओं का औसत 6 कम हो जाता है। उस संख्या के अंकों के बीच अंतर ज्ञात कीजिए।

Solution :

Let the two-digit number be $10x+y$.

According to the question,

$$\frac{10x+y}{9} - \frac{10y+x}{9} = 6$$

$$9x - 9y = 6\times 9$$

$$x - y = 6$$

Thus, the difference between the digits of that number is $x - y = 6$.

9. The average weight of 25 boxes in a boat is increased by 2 kg when one of the boxes weighing 68 kg is replaced by a new box. The weight (in kg) of the new box is:
एक नाव में 25 बक्सों का औसत वजन 2 किलोग्राम बढ़ जाता है जब 68 किलोग्राम वजन वाले बक्सों में से एक को एक नए बक्से से बदल दिया जाता है। नये डिब्बे का वजन (किलो में) है:

Solution :

Let the weight of the new box be y kg.

Average of 25 boxes = x.

According to the question,

$25x - 68 + y = (x + 2) \times 25$

$25x - 68 + y = 25x + 50$

$-68 + y = 50$

$y = 50 + 68 = 118$ kg.

10. The average of five numbers, given in a particular order, is 32. The average of the first three numbers is 28, while that of the last three numbers is 34. What is the average of the first two numbers?
एक विशेष क्रम में दी गई पाँच संख्याओं का औसत 32 है। पहली तीन संख्याओं का औसत 28 है, जबकि अंतिम तीन संख्याओं का औसत 34 है। पहली दो संख्याओं का औसत क्या है?

Solution :

Let the five numbers be X_1, X_2, X_3, X_4, X_5.

Sum of the first three numbers: $X_1 + X_2 + X_3 = 28 \times 3 = 84$

Sum of the last three numbers: $X_3 + X_4 + X_5 = 34 \times 3 = 102$ Sum of all five numbers: $X_1 + X_2 + X_3 + X_4 + X_5 = 32 \times 5 = 160$

From the above: $X_1 + X_2 = 160 - 102 = 58$

Average of the first two numbers: $\frac{X_1 + X_2}{2} = \frac{58}{2} = 29.$

11. The average of x occurring 5 times and y occurring 7 times is 37. Also, the average of x occurring 7 times and y occurring 5 times is 35. The value of y is:/ x के 5 बार और y के 7 बार होने का औसत 37 है। साथ ही, x के 7 बार और y के 5 बार आने का औसत 35 है। y का मान है:

Solution :

According to the question:

$5x + 7y = 37 \times 12$ (i)

$7x + 5y = 35 \times 12$ (ii)

Adding equations (i) and (ii):

$12(x + y) = 72 \times 12$

$x + y = 72$ (iii).

Subtracting equation (ii) from (i):

$x - y = -12$ (iv).

Solving (iii) and (iv):

$x = 30, \quad y = 42.$

12. The average of 46 numbers is 50.5. The average of the first 25 numbers is 45, and that of the last 18 numbers is 56. The 28th number is 67. If the 26th and 27th numbers are excluded, then what is the average of the remaining numbers?
46 संख्याओं का औसत 50.5 है। पहली 25 संख्याओं का औसत 45 है और अंतिम 18 संख्याओं का औसत 56 है। 28वीं संख्या 67 है। यदि 26वीं और 27वीं संख्याओं को हटा दिया जाए, तो शेष संख्याओं का औसत क्या है?

Solution :

Sum of all 46 numbers: $46 \times 50.5 = 2323$

Sum of the first 25 numbers: $25 \times 45 = 1125$

Sum of the last 18 numbers: $18 \times 56 = 1008$

Sum of the 26th, 27th, and 28th numbers:

$2323 - 1125 - 1008 = 190.$

According to the question: $26th + 27th + 67 = 190$

$26th + 27th = 123.$

Average of the remaining numbers: $\frac{2323 - 123}{44} = 50.$

13. The average of 33 numbers is 74. The average of the first 17 numbers is 72.8 and that of the last 17 numbers is 77.2. If the 17th number is excluded, then what will be the average of the remaining numbers (correct to one decimal place)?

33 संख्याओं का औसत 74 है। पहली 17 संख्याओं का औसत 72.8 है और अंतिम 17 संख्याओं का औसत 77.2 है। यदि 17वीं संख्या को हटा दिया जाए, तो शेष संख्याओं का औसत (दशमलव के एक स्थान तक सही) क्या होगा?

Solution :

Sum of 33 numbers $= 74 \times 33 = 2442$

17th number $=$ [Sum of first 17 numbers $+$ Sum of last 17 numbers] $-$ Sum of 33 numbers

$= (17 \times 72.8 + 17 \times 77.2) - 2442$

$= 2550 - 2442 = 108$

After removing the 17th number, the average of the remaining numbers is: $\frac{2442 - 108}{32} = \frac{2334}{32} = 72.93$

14. The average weight of a certain number of students in a group is 72 kg. If 10 students having an average weight of 78 kg leave and 4 students having an average weight of 80 kg join the group, the average weight of the students in the group decreases by 0.7 kg. The number of students initially in the group is:

एक समूह में छात्रों की एक निश्चित संख्या का औसत वजन 72 किलोग्राम है। यदि 78 किलोग्राम औसत वजन वाले 10 छात्र समूह छोड़ देते हैं और 80 किलोग्राम औसत वजन वाले 4 छात्र समूह में शामिल हो जाते हैं, तो समूह में छात्रों का औसत वजन 0.7 किलोग्राम कम हो जाता है। प्रारंभ में समूह में विद्यार्थियों की संख्या है:

Solution :

Let the total number of students be x.

Total weight $= 72x$

According to the question: Total weight $= 71.3 \times (x - 6)$

$72x - 780 + 320 = 71.3x - 427.8$

$0.7x = 460 - 427.8 = 32.2$

$x = 46$

15. The average weight of some students in a class was 60.5 kg. When 8 students, whose average weight was 65 kg, joined the class, then the average weight of all the students increased by 0.9 kg. The number of students in the class initially was:

एक कक्षा में कुछ छात्रों का औसत वजन 60.5 किलोग्राम था। जब 8 छात्र, जिनका औसत वजन 65 किग्रा था, कक्षा में शामिल हुए, तो सभी छात्रों का औसत वजन 0.9 किग्रा बढ़ गया। प्रारंभ में कक्षा में विद्यार्थियों की संख्या थी:

Solution :

Let the number of students initially be x.

According to the question: $(x + 8) \times (60.5 + 0.9) = x \times 60.5 + 8 \times 65$

$(x + 8) \times 61.4 = 60.5x + 520$

$61.4x + 491.2 = 60.5x + 520$

$0.9x = 520 - 491.2$

$x = \frac{28.8}{0.9} = 32$

16. The average of ten numbers is 32.5. The average of the first four numbers is 25.6, and that of the last three numbers is 38.2. The 5th number is 50% more than the 6th number and 8 less than the 7th number. What is the average of the 5th and 7th numbers?

दस संख्याओं का औसत 32.5 है। पहली चार संख्याओं का औसत 25.6 है और अंतिम तीन संख्याओं का औसत 38.2 है। 5वीं संख्या 6ठी संख्या से 50% अधिक है और 7वीं संख्या से 8 कम है। 5वीं और 7वीं संख्या का औसत क्या है?

Solution :

Sum of total numbers $= 32.5 \times 10 = 325$

Sum of first four numbers $= 4 \times 25.6$

$= 102.4$

Sum of last three numbers $= 38.2 \times 3$

$= 114.6$

Remaining numbers (5th, 6th, and 7th): $325 - (102.4 + 114.6) = 108$

Let the 5th, 6th, and 7th numbers be $3x, 2x, 3x + 8$: $3x + 2x + 3x + 8 = 108$

$8x = 100$

$x = 12.5$

Average of the 5th and 7th numbers: $\frac{3x + (3x + 8)}{2} = \frac{6x + 8}{2} = \frac{6 \times 12.5 + 8}{2} = \frac{83}{2} = 41.5$

17. The average weight of 30 persons of group A is 3 kg more than the average weight of 25 persons of group B. The average weight of 25 persons of group B is 2.5 kg more than the average weight of 20 persons of group C. If the total weight of 30 persons of group A is 1725 kg, then what will be the average weight of the persons of group A and group C taken together (in kg)?

समूह A के 30 व्यक्तियों का औसत वजन समूह B के 25 व्यक्तियों के औसत वजन से 3 किलोग्राम अधिक है। समूह B के 25 व्यक्तियों का औसत वजन समूह C के 20 व्यक्तियों के औसत वजन से 2.5 किलोग्राम अधिक है। यदि समूह A के 30 व्यक्तियों का कुल वजन 1725 किलोग्राम है, तो समूह A और समूह C के व्यक्तियों का मिलाकर औसत वजन (किलो में) क्या होगा?

Solution :

Total weight of 30 persons of group A = 1725 kg

Average weight of 30 persons of group A = $\frac{1725}{30} = 57.5$ kg

Average weight of 25 persons of group B = 54.5 kg

Average weight of 20 persons of group C = 52 kg

Average weight of persons in groups A and C:

$$\frac{1725 + 20 \times 52}{30 + 20} = \frac{1725 + 1040}{50} = \frac{2765}{50} = 55.3 \text{ kg.}$$

18. In a class of 60 students, 40% are girls. The average weight of the whole class is 59.2 kg and the average weight of the girls is 55 kg. What is the average weight of the boys?

60 छात्रों की एक कक्षा में, 40% लड़कियाँ हैं। पूरी कक्षा का औसत वजन 59.2 किलोग्राम है और लड़कियों का औसत वजन 55 किलोग्राम है। लड़कों का औसत वजन कितना है?

Solution :

Let the average weight of boys = x kg

Number of girls = $\frac{60}{100} \times 40 = 24$

Thus, number of boys = 60 − 24 = 36

Average of all class:

$$\frac{\text{Weight of all boys} + \text{Weight of all girls}}{\text{Number of students}}$$

$$59.2 = \frac{36 \times x + 24 \times 55}{60}$$

$$59.2 \times 60 = 36 \times x + 24 \times 55$$

$$36 \times x = 3552 - 1320$$

$$x = \frac{2232}{36}$$

$$x = 62 \text{ kg}$$

Thus, the average weight of boys in class will be 62 kg.

19. The average age of 40 students of a class is 16 years. After admission of 10 new students to the class the average becomes 15 years. If the average age of 5 of the new students is 11 years, then the average age (in years) of the remaining 5 new students is:

एक कक्षा के 40 विद्यार्थियों की औसत आयु 16 वर्ष है। कक्षा में 10 नए विद्यार्थियों के प्रवेश के बाद औसत 15 वर्ष हो जाता है। यदि 5 नये विद्यार्थियों की औसत आयु 11 वर्ष है, तो शेष 5 नये विद्यार्थियों की औसत आयु (वर्षों में) है:

Solution :

Total age of 40 students = 40 × 16 = 640 years

(40 + 10 new students) Total age = 50 × 15 = 750 years

Total age of 10 new students = 750 − 640 = 110 years

Total age of 5 new students = 11 × 5 = 55 years

Thus, the average age of 5 new students = $\frac{55}{5} = 11$ years.

20. In a company with 600 employees, the average age of the male employees is 42 years and that of the female employees is 41 years. If the average age of all the employees in the company is 41 years 9 months, then the number of female employees is

600 कर्मचारियों वाली एक कंपनी में पुरुष कर्मचारियों की औसत आयु 42 वर्ष और महिला कर्मचारियों की औसत आयु 41 वर्ष है। यदि कंपनी में सभी कर्मचारियों की औसत आयु 41 वर्ष 9 महीने है, तो महिला कर्मचारियों की संख्या है

Solution :

Let the number of female employees $= x$

Number of male employees $= 600 - x$

Total sum $=$ Average $\times$ No. of terms

$(600 - x) \times 42 + 41x = 600 \times \left(41 + \dfrac{9}{12}\right)$

$600 \times 42 - 42x + 41x = 600 \times \dfrac{167}{4}$

$x = \dfrac{600 \times 42 - 600 \times \dfrac{167}{4}}{42 - 41}$

$x = \dfrac{600 \times \dfrac{1}{4}}{1} = 150$

21. In a group of 150 people, $\frac{2}{5}$ are men, $\frac{1}{3}$ are women and the rest are children. The average age of the women is $\frac{4}{5}$ of the average age of the men. The average age of the children is $\frac{1}{5}$ of the average age of the men. If the average age of the men is 50 years, then the average age of all the people in the group is:

150 लोगों के समूह में $\frac{2}{5}$ पुरुष, $\frac{1}{3}$ महिलाएं और शेष बच्चे हैं। महिलाओं की औसत आयु पुरुषों की औसत आयु का $\frac{4}{5}$ है। बच्चों की औसत आयु पुरुषों की औसत आयु का $\frac{1}{5}$ है। यदि पुरुषों की औसत आयु 50 वर्ष है, तो समूह के सभी लोगों की औसत आयु है:

Solution :

No. of men $= 150 \times \dfrac{2}{5} = 60$

No. of women $= 150 \times \dfrac{1}{3} = 50$

No. of children $= 150 - (60 + 50) = 40$

Average age ratio of men, women and children $=$ 50 : 40 : 10

Average age of all persons $= \dfrac{(60 \times 50) + (50 \times 40) + (40 \times 10)}{150}$

$= \dfrac{5400}{150} = 36$ years

22. The average of the weights of boys in a class is 69.3 kg and that of girls in the same class is 59.4 kg. If the average of the weights of all the boys and girls in the class is 63.8 kg then the percentage of the number of girls in the class is:

एक कक्षा में लड़कों का औसत वजन 69.3 किलोग्राम है और उसी कक्षा में लड़कियों का औसत वजन 59.4 किलोग्राम है। यदि कक्षा में सभी लड़कों और लड़कियों के वजन का औसत 63.8 किलोग्राम है तो कक्षा में लड़कियों की संख्या का प्रतिशत है:

Solution :

B = Boys, G = Girls

B ↓ 69.3 G ↓ 59.4

63.8

4.4 5.5

Boys : Girls $= B : G = 44 : 55 = 4 : 5$

Percentage of girls in the class $= \dfrac{5}{9} \times 100 = 55\dfrac{5}{9}\%$

23. The number of students in a class is 75, out of which $33\frac{1}{3}\%$ are boys and the rest are girls. The average score in mathematics of the boys is $66\frac{2}{3}\%$ more than that of the girls. If the average score of all the students is 66, then the average score of the girls is:

एक कक्षा में विद्यार्थियों की संख्या 75 है, जिनमें से $33\frac{1}{3}\%$ लड़के हैं और शेष लड़कियाँ हैं। लड़कों का गणित में औसत अंक लड़कियों की तुलना में $66\frac{2}{3}\%$ अधिक है। यदि सभी विद्यार्थियों का औसत अंक 66 है, तो लड़कियों का औसत अंक है:

Solution :

Total number of students = 75

Number of boys $= 75 \times 33\frac{1}{3}\% = 75 \times \dfrac{100}{3 \times 100} = 25$

Let the average score of girls $= x$

(Since $66\frac{2}{3}\% = \frac{2}{3}$)

Average score of boys $= x + \dfrac{2x}{3} = \dfrac{5x}{3}$

$25 \times \dfrac{5x}{3} + 50 \times x = 66 \times 75$

$$\frac{125x}{3} + 50x = 4950$$

$125x + 150x = 14850$ (Multiply by 3)

$275x = 14850$

$x = 54$

24. Several students have taken an exam. There was an error in the answer key which affected the marks of 48 students and their average marks reduced from 78 to 66. The average of remaining students increased by 3.5 marks. This resulted in the reduction of the average of all students by 4.5 marks. The number of students that attended the exam is:

कई छात्रों ने परीक्षा दी है. उत्तर कुंजी में एक त्रुटि थी जिससे 48 छात्रों के अंक प्रभावित हुए और उनके औसत अंक 78 से घटकर 66 हो गए। शेष छात्रों के औसत अंक 3.5 अंक बढ़ गए। इसके परिणामस्वरूप सभी छात्रों के औसत में 4.5 अंक की कमी आई। परीक्षा में शामिल होने वाले छात्रों की संख्या है:

Solution :

Let the number of remaining students = x

According to the question:

$48 \times 66 + x \times 81.5 = (48 + x) \times 73.5$

$3168 + 81.5x = 3528 + 73.5x$

$81.5x - 73.5x = 3528 - 3168$

$8x = 360$

$x = \frac{360}{8} = 45$

Total number of students = 48 + 45 = 93

25. The average marks of 42 students in an examination was 64. Later it was found that the marks of three students were misread as 72, 58, and 44 instead of the actual marks 64, 42, and 26, respectively. What is the correct average marks?

एक परीक्षा में 42 छात्रों के औसत अंक 64 थे। बाद में यह पाया गया कि तीन छात्रों के अंक वास्तविक अंक 64, 42 और 26 के बजाय क्रमश: 72, 58 और 44 के रूप में पढ़े गए थे। सही औसत अंक क्या हैं?

Solution :

The sum of the marks of 42 students = $64 \times 42 = 2688$

Marks read incorrectly = $72 + 58 + 44 = 174$

Correct marks = $64 + 42 + 26 = 132$

Error in marks = $174 - 132 = 42$

Actual marks of 42 students = $2688 - 42 = 2646$

Correct average of students = $\frac{2646}{42} = 63$

26. A batsman scores 92 runs in the 15th inning and thus increases his average by 4. What is his average after the 15th inning?

एक बल्लेबाज ने 15वीं पारी में 92 रन बनाए और इस प्रकार उसका औसत 4 बढ़ गया। 15वीं पारी के बाद उसका औसत क्या है?

Solution :

Let the average after the 14th inning = x

According to the question:

$14x + 92 = 15(x + 4)$

$14x + 92 = 15x + 60$

$x = 32$

Average after the 15th inning

$= x + 4 = 32 + 4 = 36$

27. x, y, z are three positive numbers such that y is $\frac{4}{5}$ times x, and z is $\frac{5}{8}$ times y. If the average of the reciprocals of x, y, z is $\frac{17}{240}$, then the average of $3x$ and $5y$ will be:

x, y और z तीन धनात्मक संख्याएँ हैं जैसे कि y, x का $\frac{4}{5}$ गुना है और z, y का $\frac{5}{8}$ गुना है। यदि संख्याओं x, y, z के व्युत्क्रमों का औसत $\frac{17}{240}$ है तो x का 3 गुना और y का 5 गुना का औसत होगा:

Solution :

$x : y : z = 10 : 8 : 5$

Sum of reciprocals: $\frac{1}{10k} + \frac{1}{8k} + \frac{1}{5k} = \frac{17}{80}$

$k = 2$

Average of $3x$ and $5y$:

$$\frac{(3 \times 2 \times 10) + (5 \times 2 \times 8)}{2} = \frac{60 + 80}{2} = \frac{140}{2} = 70$$

28. Three numbers are such that if the average of any two of them is added to the third number, the sums obtained are 168, 174, and 180 respectively. What is the average of the original three numbers?

तीन संख्याएँ ऐसी हैं कि यदि उनमें से किन्हीं दो का औसत तीसरी संख्या में जोड़ा जाए, तो प्राप्त योग क्रमश: 168, 174 और 180 है। मूल तीन संख्याओं का औसत क्या है?

Solution :

Let the numbers be $x,\ y$, and z:

$\frac{x+y}{2}+z=168 \implies x+y+2z=336$ (i)

$\frac{y+z}{2}+x=174 \implies y+z+2x=348$ (ii)

$\frac{z+x}{2}+y=180 \implies z+x+2y=360$ (iii)

Adding equations (i), (ii), and (iii):

$x+y+2z+y+z+2x+z+x+2y=336+348+360$

$4(x+y+z)=1044$

$x+y+z=\frac{1044}{4}=261$

Average of the three numbers:

$\frac{x+y+z}{3}=\frac{261}{3}=87$

29. The average weight of a group of 3 people A, B and C is 70 kg. When D joins this group, the average becomes 60 kg. A and E, whose weight is 5 kg more than that of D, replaces A, and the average weight of B, C, D and E now becomes 59 kg. What is the average weight (in kg) of A, D and E? (correct to the nearest integer)

3 लोगों A, B और C के समूह का औसत वजन 70 किलोग्राम है। जब D इस समूह में शामिल होता है, तो औसत 60 किग्रा हो जाता है। A और E, जिनका वजन D से 5 किलोग्राम अधिक है, A का स्थान लेते हैं और B, C, D और E का औसत वजन अब 59 किलोग्राम हो जाता है। A, D और E का औसत वजन (किलो में) क्या है? (निकटतम पूर्णांक तक सही)

Solution :

According to the question:

$A+B+C=210$ (1)

Including D in this group: $\frac{A+B+C+D}{4}=60$

$A+B+C+D=240$ (2)

Average weight: $\frac{E+B+C+D}{4}=59$

$E+B+C+D=236$ (3)

From equations (2) and (3):

$A-E=4$

From equation (2): $D=30$

From equation (3): $E=30+5=35$

According to the question:

$A-35=4 \implies A=39$

Hence, the average weight of $A,\ D,\ E$:

$\frac{39+30+35}{3}=\frac{104}{3}=34.66\approx 35$ kg

30. In a class with a certain number of students, if one student weighing 30 kg is added, then the average weight of the class increases by 1 kg. If one more student weighing 30 kg is added, then the average weight of the class increases by 1.5 kg over the original average. What is the original average weight (in kg) of the class?

छात्रों की एक निश्चित संख्या वाली कक्षा में, यदि 30 किलोग्राम वजन वाला एक छात्र जोड़ा जाता है, तो कक्षा का औसत वजन 1 किलोग्राम बढ़ जाता है। यदि 30 किलोग्राम वजन वाला एक और छात्र जोड़ा जाता है, तो कक्षा का औसत वजन मूल औसत से 1.5 किलोग्राम बढ़ जाता है। कक्षा का मूल औसत वजन (किलो में) क्या है?

Solution :

Let the number of students be n and the original average be x:

$n\cdot x+30=(n+1)(x+1)$

$n\cdot x+30=n\cdot x+n+x+1$

$n+x=29\ n\cdot x+60=(n+2)(x+1.5)$

$n\cdot x+60=n\cdot x+1.5n+2x+3$

$1.5n+2x=57$

From $n+x=29$: $n=29-x$:

$1.5(29-x)+2x=57$

$43.5-1.5x+2x=57$

$0.5x=13.5 \implies x=27$

31. A person is told to guess his weight. He has estimated that his weight is more than 70 kg but less than 80 kg. His friend has estimated his weight to be more than 74 kg but less than 84 kg. His mother told that his weight is less than 78 kg. Assuming that the weight is an integer and all three of them have given correct estimates, what is the average weight of the estimates together?

एक व्यक्ति को अपने वजन का अनुमान लगाने के लिए कहा जाता है। उनका अनुमान है कि उनका वज़न 70 किलो से ज़्यादा लेकिन 80 किलो से कम है। उनके दोस्त ने अनुमान लगाया है कि उनका वजन 74 किलोग्राम से अधिक लेकिन 84 किलोग्राम से कम है। उनकी मां ने बताया कि उनका वजन 78 से कम है। मान लिया जाए कि वजन एक पूर्णांक है और उन तीनों ने सही अनुमान दिया है तो अनुमान का औसत वजन कितना है?

Solution :

Weight according to the person = More than 70 kg and less than 80 kg.

Weight according to the friend = More than 74 kg and less than 84 kg. Weight according to the mother = Less than 78 kg.

Thus, the possible weights are: 75, 76, 77 kg.

Average:

$$\frac{75 + 76 + 77}{3} = 76 \text{ kg}$$

12 Ages

1. Three years ago, the ratio of the age of father to that of his son was $8 : 3$. After 4 years, their ages will be in the ratio $11 : 5$. What is the present age (in years) of the father?

तीन वर्ष पहले, पिता की आयु का उसके पुत्र की आयु से अनुपात $8 : 3$ था। 4 वर्ष बाद, उनकी आयु का अनुपात $11 : 5$ होगा, पिता की वर्तमान आयु (वर्षों में) क्या है?

Solution :

Let the ages of the father and son 3 years ago be $8x$ years and $3x$ years, respectively.

According to the question:

$$\frac{8x+3+4}{3x+3+4} = \frac{11}{5}$$

$$\frac{8x+7}{3x+7} = \frac{11}{5}$$

$$40x + 35 = 33x + 77$$

$$7x = 42$$

$$x = 6$$

Age of father 3 years ago $= 8 \times 6 = 48$ years

Age of son 3 years ago $= 3 \times 6 = 18$ years

Present age of father $= 48 + 3 = 51$ years

Hence, the correct answer is: 51 years.

2. The average age of Akhil and Sunita is 51 years, of Sunita and Veena is 31 years, and that of Veena and Akhil is 33 years. What is the age of Veena?

अखिल और सुनीता की औसत आयु 51 वर्ष है, सुनीता और वीना की 31 वर्ष है, और वीना और अखिल की औसत आयु 33 वर्ष है। वीणा की आयु है:

Solution :

Akhil + Sunita $= 51 \times 2 = 102$ (i)

Sunita + Veena $= 31 \times 2 = 62$ (ii)

Veena + Akhil $= 33 \times 2 = 66$ (iii)

Adding equations (i), (ii), and (iii):

2(Akhil + Sunita + Veena) $= 102 + 62 + 66$

Akhil + Sunita + Veena $= \frac{230}{2} = 115$

Age of Veena $= 115 -$ (Akhil + Sunita) $=$ $115 - 102 = 13$ years.

3. Nalini's younger brother is 12 years old. If the ratio of Nalini's age to that of her brother is $7 : 6$, then what will be the ratio of their ages 6 years hence?

नलिनी का छोटा भाई 12 साल का है। यदि नलिनी की आयु का उसके भाई की आयु से अनुपात $7 : 6$ है, तो 6 वर्ष बाद उनकी आयु का अनुपात क्या होगा?

Solution :

Nalini's age $= \frac{12}{6} \times 7 = 14$ years

Required ratio $= (14 + 6) : (12 + 6) = 20 : 18 = 10 : 9$

4. The ratio of the present ages of A and B is $8 : 9$. After 9 years, this ratio will become $19 : 21$. C is 3 years younger than B. What is the present age (in years) of C?

A और B की वर्तमान आयु का अनुपात $8 : 9$ है। 9 वर्ष बाद यह अनुपात $19 : 21$ हो जाएगा। C, B से 3 वर्ष छोटा है। C की वर्तमान आयु (वर्षों में) क्या है?

Solution :

Let the present ages of A and B be $8x$ and $9x$ years, respectively. From the question:

$$\frac{8x+9}{9x+9} = \frac{19}{21}$$

$168x + 189 = 171x + 171$

$3x = 18$

$x = 6$

Present age of $C = 9x - 3 = 54 - 3 = 51$ years.

5. The ratio of the present ages of A and B is $8 : 15$. Eight years ago, the ratio of their ages was $6 : 13$. What will be the ratio of their ages 8 years from now?

A और B की वर्तमान आयु का अनुपात $8 : 15$ है। आठ वर्ष पहले, उनकी आयु का अनुपात $6 : 13$ था। अब से 8 वर्ष बाद A और B की आयु का अनुपात क्या होगा?

Solution :

Let the present age of $A = 8x$

Present age of $B = 15x$

According to the question:

$$\frac{8x-8}{15x-8} = \frac{6}{13}$$

$104x - 104 = 90x - 48$

$14x = 56$

$x = 4$

Present age of $A = 32$ years

Present age of $B = 60$ years

Ratio of their ages after 8 years:

$$\frac{32+8}{60+8} = \frac{40}{68} = \frac{10}{17} = 10 : 17$$

6. Five years ago, the ratio of the age of A to that of B was $4 : 5$. Five years hence, the ratio of the age of A to that of B will be $6 : 7$. If, at present, the age of C is 10 years younger than B, then what will be the ratio of the present age of A to that of C?

पाँच वर्ष पहले, A की आयु का B की आयु से अनुपात $4 : 5$ था। पाँच वर्ष बाद, A की आयु का B की आयु से अनुपात $6 : 7$ होगा। यदि वर्तमान में C की आयु B से 10 वर्ष कम है, तो A की वर्तमान आयु का C की वर्तमान आयु से अनुपात क्या होगा?

Solution :

5 years ago:

Let the age of $A = 4x$

Age of $B = 5x$

Present age of $A = (4x + 5)$

Present age of $B = (5x + 5)$

According to the question:

$$\frac{(4x+5)+5}{(5x+5)+5} = \frac{6}{7}$$

$$\frac{4x+10}{5x+10} = \frac{6}{7}$$

$28x + 70 = 30x + 60$

$2x = 10$

$x = 5$

Present age of $A = (4x + 5) = 25$ years

Present age of $B = (5x+5) = (5 \times 5+5) = 30$ years

Present age of $C = B - 10 = 30 - 10 = 20$ years

The ratio of the present age of A to that of $C = 25 : 20 = 5 : 4$

7. Seven years ago, the ratio of the ages of A and B was $4 : 5$. Eight years hence, the ratio of the ages of A and B will be $9 : 10$. What is the difference between their present ages in years?

सात वर्ष पहले, A और B की आयु का अनुपात $4 : 5$ था। आठ वर्ष बाद, A और B की आयु का अनुपात $9 : 10$ होगा। उनकी वर्तमान आयु के बीच वर्षों में क्या अंतर है?

Solution :

Let 7 years ago, the age of $A = 4x$

Age of $B = 5x$

Present age of $A = (4x + 7)$ years

Present age of $B = (5x + 7)$ years

According to the question:

$$\frac{(4x+7)+8}{(5x+7)+8} = \frac{9}{10}$$

$40x + 150 = 45x + 135$

$5x = 15$

$x = 3$

Present age of $A = 4 \times 3 + 7 = 19$ years

Present age of $B = 5 \times 3 + 7 = 22$ years

Difference between their present ages $= 22 - 19 = 3$ years.

8. The present ages of A and B are in the ratio $3 : 4$. Twelve years ago, their ages were in the ratio $2 : 3$. The sum of the present ages of A and B (in years) is:
A और B की वर्तमान आयु का अनुपात $3 : 4$ है। बारह वर्ष पहले, उनकी आयु का अनुपात $2 : 3$ था। A और B की वर्तमान आयु का योग (वर्षों में) है:

Solution :

Let the present ages of A and B be $3x$ and $4x$ years, respectively.

$$\frac{3x - 12}{4x - 12} = \frac{2}{3}$$

$9x - 36 = 8x - 24$

$x = 12$

The sum of the present ages of A and $B = 3x + 4x = 7x = 7 \times 12 = 84$ years.

9. The ages of Fatima and Ahmed are in the ratio $3 : 8$. The sum of their present ages is 44 years. The difference in their ages is:
फातिमा और अहमद की उम्र का अनुपात $3 : 8$ है। उनकी वर्तमान उम्र का योग 44 वर्ष है। उनकी उम्र का अंतर है:

Solution :

Let the age of Fatima $= 3x$ years

And the age of Ahmed $= 8x$ years

Sum of their ages $= 3x + 8x = 44$

$11x = 44$

$x = 4$

Age of Fatima $= 3x = 3 \times 4 = 12$ years

Age of Ahmed $= 8x = 8 \times 4 = 32$ years

Difference in their ages $= 32 - 12 = 20$ years.

10. Six years ago, the average of the ages of Ravi, Mohan, and Govind was 32 years. If Shyam joins them now, the average of the ages of all four of them is 36 years. The present age of Shyam is:
छह वर्ष पहले, रवि, मोहन और गोविंद की औसत आयु 32 वर्ष थी। यदि श्याम अब उनके साथ जुड़ जाए, तो उन चारों की आयु का औसत 36 वर्ष है। श्याम की वर्तमान आयु है:

Solution :

Total sum of the ages of Ravi, Mohan, and Govind at present:

$(32 \times 3) + (6 \times 3) = 96 + 18 = 114$ years

Total ages of all four of them:

$36 \times 4 = 144$ years

Present age of Shyam $= 144 - 114 = 30$ years

11. The ratio of a man's age to his father's age is $4 : 5$, and the ratio of his age to his son's age is $6 : 1$. Four years ago, these ratios were $11 : 14$ and $11 : 1$, respectively. The ratio of the age of the grandfather to that of the grandson 12 years from now will be:
एक आदमी की उम्र का उसके पिता की उम्र से अनुपात $4 : 5$ है, और उसकी उम्र का उसके बेटे की उम्र से अनुपात $6 : 1$ है। चार साल पहले ये अनुपात क्रमशः $11 : 14$ और $11 : 1$ था। अब से 12 वर्ष बाद दादा की आयु का पोते की आयु से अनुपात होगा:

Solution :

Let the present age of the man and his father be $4x$ and $5x$ years, respectively.

And, let the present age of the man and his son be $6y$ and y years, respectively.

Four years ago, the ratio of the man's age to his father's age $= 11 : 14$:

$$\frac{4x - 4}{5x - 4} = \frac{11}{14}$$

$56x - 56 = 55x - 44$

$x = 12$

Present age of the man $= 4 \times 12 = 48$ years

Present age of the father $= 5 \times 12 = 60$ years

The ratio of the man's age to his son's age $= 6y : y$:

$6y = 48$

$y = 8$

Present age of the son $= y = 8$ years

12 years after, the ratio of the grandfather's age to the grandson's age:

$$\frac{60+12}{8+12} = \frac{72}{20} = \frac{18}{5}$$

12. Ramesh is three times elder than Suresh. After 2 years, Ramesh will be two times elder than Suresh. What is the present age of Ramesh (in years)?
रमेश, सुरेश से तीन गुना बड़ा है। 2 वर्ष बाद रमेश, सुरेश से दो गुना बड़ा होगा। रमेश की वर्तमान आयु (वर्षों में) क्या है?

Solution :

Let the age of Suresh $= x$ years

Then, the age of Ramesh $= 3x$ years

After two years:

Age of Suresh $= x + 2$,

Age of Ramesh $= 3x + 2$

According to the question:

$3x + 2 = 2(x + 2)$

$3x + 2 = 2x + 4$

$x = 2$

Hence, the present age of Ramesh $= 3 \times 2 =$ 6 years

13. Ravi's age is $\frac{3}{5}$ of Shyam's age. After x years, the ratio of the ages of Ravi and Shyam becomes 5 : 7. If initially, the sum of their ages is 32, then what is the value of x?
रवि की उम्र श्याम की उम्र का $\frac{3}{5}$ है। x वर्ष बाद रवि और श्याम की आयु का अनुपात 5 : 7 हो जाता है। यदि प्रारंभ में उनकी आयु का योग 32 है, तो x का मान क्या है?

Solution :

Let the ages of Ravi and Shyam be $3x$ and $5x$, respectively.

$3x + 5x = 32$

$8x = 32$

$x = 4$

Age of Ravi $= 3 \times 4 = 12$ years

Age of Shyam $= 5 \times 4 = 20$ years

According to the question, after x years:

$$\frac{12+x}{20+x} = \frac{5}{7}$$

$84 + 7x = 100 + 5x$

$2x = 16$

$x = 8$ years

14. The sum of the present ages of a father and his son is 78 years. After five years, the ratio of their ages becomes 7 : 4. Find the present age of the father.
एक पिता और उसके पुत्र की वर्तमान आयु का योग 78 वर्ष है। पाँच वर्ष बाद, उनकी आयु का अनुपात 7 : 4 हो जाता है। पिता की वर्तमान आयु ज्ञात कीजिए।

Solution :

Let the present age of the father $= x$ years

Then, the present age of the son $= 78 - x$ years.

According to the question:

$$\frac{x+5}{78-x+5} = \frac{7}{4}$$

$$\frac{x+5}{83-x} = \frac{7}{4}$$

$4(x + 5) = 7(83 - x)$

$4x + 20 = 581 - 7x$

$11x = 561$

$x = 51$

Hence, the present age of the father $=$ 51 years

15. The present age of Manoj is twice the sum of the ages of his two children. After 20 years, the age of Manoj will become equal to the sum of the ages of his two children. What is the present age of Manoj?
मनोज की वर्तमान आयु उसके दो बच्चों की आयु के योग से दोगुनी है। 20 वर्ष बाद, मनोज की उम्र उसके दो बच्चों की उम्र के योग के बराबर हो जाएगी। मनोज की वर्तमान आयु क्या है?

Solution :

Let the ages of the children be a and b, and let the age of Manoj $= m$.

According to the question:

$m = 2(a+b)$

After 20 years:

$m + 20 = (a+20) + (b+20)$

$m + 20 = (a+b) + 40$

$m = (a+b) + 20$ Substituting $m = 2(a+b)$:

$2(a+b) = (a+b) + 20$

$a + b = 20$

Substitute $a+b = 20$ back into $m = 2(a+b)$:

$m = 2 \times 20 = 40$

Hence, the present age of Manoj $= 40$ years

16. Five years ago, the average of the ages of 4 persons was 40 years. If a new person joins the group now, then the average of the age of all five persons is 46 years. The age of the fifth person (in years) is
पाँच वर्ष पहले, 4 व्यक्तियों की औसत आयु 40 वर्ष थी। यदि अब एक नया व्यक्ति समूह में शामिल होता है, तो सभी पांच व्यक्तियों की औसत आयु 46 वर्ष है। पांचवें व्यक्ति की आयु (वर्षों में) है।

Solution :

5 years ago, total age of 4 persons $= 4 \times 40 =$ 160 years

Their present age $= 160 + 20 = 180$ years

Total age of 5 persons after adding one person to the group $= 5 \times 46 = 230$ years

Age of the 5th person $= 230 - 180 = 50$ years

17. Four years ago, the ratio of the ages (in years) of A and B was 8 : 9. The ratio of their ages, 8 years from now, will be 11 : 12. What is the sum of their present ages (in years)?
चार वर्ष पहले, A और B की आयु का अनुपात (वर्षों में) 8 : 9 था। अब से 8 वर्ष बाद उनकी आयु का अनुपात 11 : 12 होगा। वर्षों में उनकी वर्तमान आयु का योग क्या है?

Solution :

Let, 4 years ago the ages of A and B were $8x$ and $9x$ respectively.

According to the question:

$$\frac{8x+4+8}{9x+4+8} = \frac{11}{12}$$

$$\frac{8x+12}{9x+12} = \frac{11}{12}$$

$96x + 144 = 99x + 132$

$3x = 12$

$x = 4$

Hence, sum of present ages of A and B:

$8x + 4 + 9x + 4 = 17x + 8 = 17 \times 4 + 8 =$ $68 + 8 = 76$ years

18. A father said to his son, "I was as old as you are at the present at the time of your birth." If the father's age is 38 years now, the son's age five years back was?
एक पिता ने अपने बेटे से कहा, "तुम्हारे जन्म के समय मैं उतना ही बूढ़ा था जितना तुम वर्तमान में हो"। यदि पिता की आयु अभी 38 वर्ष है, तो पाँच वर्ष पहले पुत्र की आयु थी?

Solution :

Let the son's present age be x years.

Then, $38 - x = x$

$2x = 38$

$x = 19$

Son's age 5 years back $= 19 - 5 = 14$ years

19. When Rajesh was born, his father's age was 29 years older than his brother, and his mother was 25 years older than his sister. If his brother is 2 years elder than his sister, after 6 years the average age of the family is 20 years. Then what was the age of the mother when Rajesh was born?

जब राजेश का जन्म हुआ, तो उनके पिता की उम्र उनके भाई से 29 वर्ष बड़ी थी और उनकी माँ की उम्र उनकी बहन से 25 वर्ष बड़ी थी। यदि उसका भाई अपनी बहन से 2 वर्ष बड़ा है। 6 वर्ष बाद परिवार की औसत आयु 20 वर्ष है। तो जब राजेश का जन्म हुआ तो माँ की आयु क्या थी?

Solution :

Let the age of the sister $= x$, brother $= x + 2$,

father $= 29 + x + 2$, and mother $= 25 + x$.

Present age $= 4x + 58$

After 6 years:

$$4x + 58 + 30 = 4x + 88$$

$$\frac{4x + 88}{5} = 20$$

$$4x + 88 = 100$$

$$4x = 12$$

$$x = 3$$

Mother's age $= 25 + x = 25 + 3 = 28$ years

20. Six years ago Manisha's age was equal to the sum of the present ages of her son and daughter. Four years hence, the ratio of the ages of her son and daughter at that time will be 7 : 6. Manisha is 6 years younger than her husband. Manisha's present age is 2.5 times the present age of her daughter. What is the age of Manisha's husband?

छह वर्ष पहले मनीषा की उम्र उसके बेटे और बेटी की वर्तमान उम्र के योग के बराबर थी। चार वर्ष बाद, उस समय उसके बेटे और बेटी की उम्र का अनुपात 7 : 6 होगा। मनीषा अपने पति से 6 साल छोटी हैं। मनीषा की वर्तमान आयु उसकी बेटी की वर्तमान आयु से 2.5 गुना है। तो फिर मनीषा के पति की उम्र क्या है?

Solution :

$$M - 6 = S + D$$

$$M = 2.5D$$

$$\frac{S + 4}{D + 4} = \frac{7}{6}$$

$$D = 20, \quad M = 50, \quad H = 56$$

13 Number System

Natural Numbers: Set of counting numbers is called natural numbers. It is denoted by N, where,
प्राकृतिक संख्याएँ: गिनती की संख्याओं के समूह को प्राकृतिक संख्याएँ कहते हैं। इसे N द्वारा निरूपित किया जाता है, जहाँ,
$N = \{1, 2, 3, \ldots \infty\}$

Even Numbers: The set of all natural numbers which are divisible by 2 are called even numbers. It is denoted by E, where,
सम संख्याएँ: सभी प्राकृतिक संख्याओं का समुच्चय जो 2 से विभाज्य हो, सम संख्याएँ कहलाती हैं। इसे E से दर्शाया जाता है। $E = \{2, 4, 6, 8, 10, \ldots \infty\}$

Odd Numbers: The set of all natural numbers which are not divisible by 2 are called odd numbers. In other words, the natural numbers which are not even numbers, are odd numbers. i.e.,
विषम संख्याएँ: सभी प्राकृतिक संख्याओं का समूह जो 2 से विभाज्य नहीं होती, विषम संख्याएँ कहलाती हैं। दूसरे शब्दों में, वे प्राकृतिक संख्याएँ जो सम संख्याएँ नहीं हैं, विषम संख्याएँ हैं। अर्थात,
$O = \{1, 3, 5, 7, \ldots \infty\}$

Whole Numbers: When zero is included in the set of natural numbers, then it forms the set of whole numbers. It is denoted by W, where,
पूर्ण संख्याएँ: जब प्राकृतिक संख्याओं के समुच्चय में शून्य को शामिल कर लिया जाता है, तो यह पूर्ण संख्याओं का समुच्चय बन जाता है। इसे W द्वारा निरूपित किया जाता है, जहाँ,
$W = \{0, 1, 2, 3, \ldots \infty\}$

Integers: When, in the set of whole numbers, natural numbers with a negative sign are included, then it becomes the set of integers. It is denoted by I or Z.
पूर्णांक: जब पूर्ण संख्याओं के समुच्चय में ऋणात्मक चिन्ह वाली प्राकृत संख्याओं को शामिल किया जाता है तो वह पूर्णांकों का समुच्चय बन जाता है। इसे I या Z से दर्शाया जाता है।
$Z = \{\ldots, -4, -3, -2, -1, 0, 1, 2, 3, 4, \ldots \infty\}$

Integers can further be classified into negative or positive integers. Negative integers are denoted by Z^- and positive integers are denoted by Z^+.
पूर्णांकों को आगे ऋणात्मक या धनात्मक पूर्णांकों में वर्गीकृत किया जा सकता है। ऋणात्मक पूर्णांकों को Z^- द्वारा और धनात्मक पूर्णांकों को Z^+ द्वारा निरूपित किया जाता है।
$Z^- = \{\ldots, -3, -2, -1\} and$
$Z^+ = \{1, 2, 3, \ldots \infty\}$

Further, 0 is neither a negative nor a positive integer.
इसके अलावा 0 न तो ऋणात्मक और न ही धनात्मक पूर्णांक है।

Prime Numbers: The natural numbers which have no factors other than 1 and itself are called prime numbers.
अभाज्य संख्याएँ: वे प्राकृतिक संख्याएँ जिनका 1 और स्वयं के अलावा कोई गुणनखंड नहीं होता, अभाज्य संख्याएँ कहलाती हैं।

Note that:

(i) In other words, they can be divided only by themselves or 1. As, 2, 3, 5, 7, 11, etc.
दूसरे शब्दों में, उन्हें केवल स्वयं या केवल 1 के द्वारा विभाजित किया जा सकता है। जैसे, 2, 3, 5, 7, 11 आदि।

(ii) All prime numbers other than 2 are odd numbers, but all odd numbers are not prime numbers. 2 is the only even prime number.
2 के अलावा अन्य सभी अभाज्य संख्याएँ विषम संख्याएँ हैं लेकिन सभी विषम संख्याएँ अभाज्य संख्याएँ नहीं हैं। 2 एकमात्र सम अभाज्य संख्या है।

Co-Prime Numbers: Two numbers which have no common factor except 1, are called co-prime numbers. Such as, 9 and 16, 4 and 17, 80 and 81, etc.
It is not necessary that two co-prime numbers always. They may or may not be prime numbers.
सह-अभाज्य संख्याएँ: दो संख्याएँ जिनका 1 को छोड़कर अन्य कोई सामान्य गुणनखंड नहीं होता, सह-अभाज्य संख्याएँ कहलाती हैं। जैसे, 9 और 16, 4 और 17, 80 और 81 आदि।
यह आवश्यक नहीं है कि दो सह-अभाज्य संख्याएँ हमेशा अभाज्य हों। वे अभाज्य संख्याएँ हो भी सकती हैं और नहीं भी

1. Rational Numbers: The numbers which can be expressed in the form of $\frac{p}{q}$, where p and q are integers and coprime, and $q \neq 0$, are called rational numbers. It is denoted by Q. These may be positive or negative. Examples: $\frac{4}{5}, \frac{5}{1}, \frac{-1}{2}$, etc. are rational numbers.
परिमेय संख्याएँ: वे संख्याएँ जिन्हें $\frac{p}{q}$ के रूप में व्यक्त किया

जा सकता है जहाँ p और q पूर्णांक हैं और सह-अभाज्य हैं तथा $q \neq 0$ होता है। इसे Q द्वारा दर्शाया जाता है। ये सकारात्मक या नकारात्मक हो सकती हैं।

$\frac{4}{5}, \frac{5}{1}, \frac{-1}{2}$, जैसे आदि परिमेय संख्याएँ हैं।

Irrational Numbers: The numbers which are not rational numbers are called irrational numbers. Examples: $\sqrt{2} = 1.414213562, \pi = 3.141592653$.

अपरिमेय संख्याएँ: वे संख्याएँ जो परिमेय संख्याएँ नहीं हैं, अपरिमेय संख्याएँ कहलाती हैं। जैसे $\sqrt{2} = 1.414213562$, $\pi = 3.141592653$।

Real Numbers: The set of all rational numbers as well as irrational numbers is called real numbers. The square of all of them is positive.

वास्तविक संख्याएँ: सभी परिमेय संख्याओं के साथ-साथ अपरिमेय संख्याओं के समुच्चय को वास्तविक संख्याएँ कहा जाता है। इन सभी का वर्ग धनात्मक होता है।

4/5, 5/1,-1/2 जैसे आदि परिमेय संख्याएँ हैं।

अपरिमेय संख्याएँ: वे संख्याएँ जो परिमेय संख्याएँ नहीं हैं, अपरिमेय संख्याएँ कहलाती हैं। जैसे कि √2 = 1.414213562. π = 3.141592653

वास्तविक संख्याएँ: सभी परिमेय संख्याओं के साथ-साथ अपरिमेय संख्याओं के समुच्चय को वास्तविक संख्याएँ कहा जाता है। अल का वर्ग

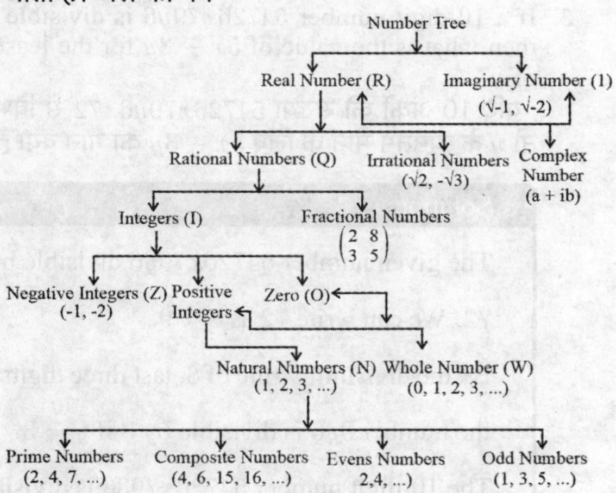

Divisibility Rules/ विभाज्यता नियम

2 - Last digit should be divisible by 2
अंतिम अंक 2 से विभाज्य होना चाहिए

3 - Sum of digits should be divisible by 3
अंकों का योग 3 से विभाज्य होना चाहिए

4 - Last 2 digit should be divisible by 4
अंतिम 2 अंक 4 से विभाज्य होने चाहिए

5 - Last digit should be 0 or 5
अंतिम अंक 0 या 5 होना चाहिए

6 - $6 = 2\times3$ (co-prime factors)
$6 = 2\times3$ (सह-अभाज्य गुणनखंड)

If a number is divisible by 2 & 3 both, that number will also be divisible by 6
यदि कोई संख्या 2 और 3 दोनों से विभाज्य है, तो वह संख्या 6 से भी विभाज्य होगी

8 - Last 3 digit should be divisible by 8
अंतिम 3 अंक 8 से विभाज्य होना चाहिए

9 - Sum of digits should be divisible by 9
अंकों का योग 9 से विभाज्य होना चाहिए

10 - Last Digit should be 0
अंतिम अंक 0 होना चाहिए

16 - Last 4 digit should be divisible by 16
अंतिम 4 अंक 16 से विभाज्य होने चाहिए

1. Divisibility rule of 7
7 का विभाज्यता नियम

Ex. Num = 861
$86 - 1\times2 = 84$
$8 - 2\times4 = 0$

Divisible by 7 (R = 0)
7 से विभाज्य (R = 0)

2. Divisibility Rule of 11 - If the difference between the sum of the digits at odd places and sum of the digits at even places is zero or multiple of 11.
11 का विभाज्यता नियम - यदि विषम स्थानों पर अंकों के योग और सम स्थानों पर अंकों के योग के बीच का अंतर शून्य या 11 का गुणक है।

3. Divisibility rule of 13
13 का विभाज्यता नियम
Ex. Num = 975
$97 + 5\times4 = 117$
$11 + 7\times4 = 39$

39 divisible by 13

Some more Divisibility Rules

- ABAB - divisible by 101
- ABCABC - divisible by 1001
- $a^n + b^n$ divisible by $(a + b)$
  जब n विषम हो
- $a^n - b^n$ divisible by $(a - b)$
  जब n विषम हो
- $a^b - b^n$ divisible by $(a + b)$ & $(a - b)$
  जब n सम हो

Fermat's Theorem
फर्मेट की प्रमेय

Fermat's Theorem $\rightarrow \frac{a^{P-1}}{P} = 1$ (Remainder)

फर्मेट की प्रमेय $\rightarrow \frac{a^{P-1}}{P} = 1$ (शेष)

$p = \frac{\text{prime number}}{\text{अभाज्य संख्या}}$

$a, p \rightarrow \frac{\text{co-prime}}{\text{सह अभाज्य}}$

Ex. $\frac{50^{16}}{17} \rightarrow R = 1$

$\frac{20^{48}}{13} = \frac{(20^4)^{12}}{13} \rightarrow R = 1$

Wilson's Theorem (विल्सन की प्रमेय)

P = any prime number / P = कोई भी अभाज्य संख्या

$\frac{(P-1)!}{P} \rightarrow \text{Remainder } = -1 \text{ or } (P-1)$

$\frac{(ax+k)^n}{a} \rightarrow \text{Remainder } = K^n$

$\frac{(ax+1)^n}{a} \rightarrow \text{Remainder } = 1^n = 1$

$\frac{(ax-k)^n}{a} \rightarrow \text{Remainder } = (-k)^n$

$\frac{(ax-1)^n}{a} \rightarrow \text{Remainder } = (-1)^n$

n = even → 1

n = odd → –1

Problem based on Divisibility/ विभाज्यता पर आधारित समस्या

1. $5^{71} + 5^{72} + 5^{73} + 5^{74} + 5^{75}$ is divisible by which of the following number:
निम्नलिखित में से किस संख्या से विभाज्य है?

(a) 71 (c) 89
(b) 69 (d) 73

Solution:

$5^{71} + 5^{72} + 5^{73} + 5^{74} + 5^{75}$

$5^{71}(1 + 5 + 5^2 + 5^3 + 5^4)$

$5^{71}(1 + 5 + 25 + 125 + 625)$

$5^{71}(781)$

$5^{71}(71 \times 11)$

Hence it is divisible by 71.

2. If a 10-digit number $75462A97B6$ is divisible by 72, then the value of $\sqrt{8A - 4B}$ is:
यदि 10 अंकों की संख्या $75462A97B6$, 72 से विभाज्य है, तो $\sqrt{8A - 4B}$ का मान है:

Solution:

A number which is divisible by 72, then it must be divisible by 8 and 9.

Divisibility Rule of 8: Last three digits of the number should be perfectly divisible by 8.

Divisibility Rule of 9: Sum of digits of the number must be divisible by 9.

For the 10-digit number $75462A97B6$, $B = 3$, because 736 is divisible by 8.

Now, sum of digits $= 7 + 5 + 4 + 6 + 2 + A + 9 + 7 + 3 + 6 = 49 + A$. Then, $A = 5$ [54 is divisible by 9].

$A = 5, B = 3$

Now, $\sqrt{8A - 4B}$

$\sqrt{40 - 12}$

$\sqrt{28}$.

3. If a 10-digit number $54726x79y6$ is divisible by 72, then what is the value of $5x - 3y$ for the least value of y?
यदि 10 अंकों की संख्या $54726x79y6$, 72 से विभाज्य है, तो y के न्यूनतम मान के लिए $5x - 3y$ का मान क्या है?

Solution:

The given number $54726x79y6$ divisible by 72. We can write 72 as 8×9.

By the divisibility rule of 8, last three digits of the number $9y6$ is divisible by 8 if $y = 3$.

The 10-digit number $54726x7936$ is divisible by 9 if the sum of the digit is divisible by 9, so $5+4+7+2+6+x+7+9+3+6 = 49+x$.

For $x = 5$, we get the resultant number divisible by 9.

The value of $5x - 3y = 5 \times 5 - 3 \times 3 = 25 - 9 = 16$.

4. How many numbers are there from 200 to 800 which are neither divisible by 5 nor by 7?/ 200 से 800 तक ऐसी कितनी संख्याएँ हैं जो न तो 5 से और न ही 7 से विभाज्य हैं?

Solution:

Number divisible by 5 between 200 to 800:

205, 210, 215, . . . , 795

$$n = \frac{795 - 205}{5} + 1 = 119$$

Numbers divisible by 7 between 200 to 800:

203, 210, 217, . . . , 798

$$n = \frac{798 - 203}{7} + 1 = 86$$

Numbers divisible by 35 between 200 to 800:

210, 245, 280, . . . , 770

$$n = \frac{770 - 210}{35} + 1 = 17$$

Hence, on removing numbers which are divisible by both 5 and 7:

$119 + 86 - 17 = 188.$

Required number = 599 − 188 = 411.

5. If the nine-digit number $9m2365n48$ is completely divisible by 88, what is the value of $m^2 \times n^2$, for the smallest value of n, where m and n are natural numbers?
यदि नौ अंकों की संख्या $9m2365n48$, 88 से पूर्णतः विभाज्य है, तो n के सबसे छोटे मान के लिए $m^2 \times n^2$ का मान क्या है, जहाँ m और n प्राकृतिक संख्याएँ हैं?

Solution:

The given number $9m2365n48$ is divisible by 88. We can write 88 as 8 × 11.

By the divisibility rule of 8, we have $n48$ is divisible by 8.

For $n = 2$ (smallest value of n), we get the resultant number divisible by 8.

And for $n = 2$, we will check the divisibility of 11.

$= (9 + 2 + 6 + 2 + 8) - (m + 3 + 5 + 4)$

$= 27 - (m + 12)$

$= 15 - m$

For $m = 4$, we get the resultant number divisible by 11.

The value of $m^2 \times n^2 = 4^2 \times 2^2 = 16 \times 4 = 64$.

6. The smallest six-digit number that is exactly divisible by 53 is:
53 से पूर्णतः विभाज्य छह अंकों की सबसे छोटी संख्या है:

Solution :

The smallest 6-digit number = 100000

The smallest 6-digit number is divisible by 53.

```
      1886
 53)100000
    53
     470
     424
      460
      424
       360
       318
       x42
```

Required 6-digit smallest number = 100000 + (53 − 42) = 100000 + 11 = 100011

7. If a nine-digit number $1263487xy$ is divisible by both 8 and 5, then the greatest possible values of x and y, respectively, are:
यदि नौ अंकों की संख्या $1263487xy$, 8 और 5 दोनों से विभाज्य है, तो क्रमशः x और y के अधिकतम संभव मान हैं:

(a) 6 and 0 (c) 6 and 5
(b) 2 and 5 (d) 2 and 0

Solution :

The given 9-digit number $1263487xy$ is divisible by both 8 and 5.

By divisibility rule of 5, the last digit will be 0 or 5.

If the last digit is 5, the number will not be divisible by 8.

$y = 0$

By divisibility rule of 8:

1263487×0

$= \frac{7 \times 0}{8}$

$= x = 6$ (Greatest value)

So, the greatest value of x and y is 6 and 0.

8. The largest six-digit number exactly divisible by 243 is:

243 से पूर्णतः विभाज्य छह अंकों की सबसे बड़ी संख्या है:

Solution :

Largest 6-digit number = 999999

The largest 6-digit number is exactly divisible by 243.

$$\begin{array}{r} 4115 \\ 243\overline{)999999} \\ \underline{972} \\ \times 279 \\ \underline{243} \\ \times 369 \\ \underline{243} \\ 1269 \\ \underline{1215} \\ \times\times 54 \end{array}$$

So, the required 6-digit largest number = $999999 - 54 = 999945$

9. If the number $645A2879B8$ is divisible by both 8 and 9, then the smallest possible values of A and B will be:
यदि संख्या $645A2879B8$, 8 और 9 दोनों से विभाज्य है, तो A और B का न्यूनतम संभव मान होगा:
(a) $A = 3,\ B = 4$ (b) $A = 4,\ B = 3$
(c) $A = 2,\ B = 3$ (d) $A = 3,\ B = 2$

Solution :

The given number $645A2879B8$ is divisible by both 8 and 9.

By divisibility rule of 8:

$9B8/8$, $B = 2$ (Smallest value)

By divisibility rule of 9:

$645A287928/9$

$(6+4+5+A+2+8+7+9+2+8)/9$

$(51+A)/9$

For $A = 3$, 54 is divisible by 9.

$\therefore A = 3$ (Smallest value)

So, the smallest possible value of A and B will be 3 and 2 respectively.

Hence, option (d) is correct.

10. If the nine-digit number $8475639AB$ is divisible by 99, then find the value of A and B?
यदि नौ अंकों की संख्या $8475639AB$, 99 से विभाज्य है, तो A और B का मान ज्ञात कीजिए?
(a) $A = 5,\ B = 3$ (b) $A = 4,\ B = 8$
(c) $A = 4,\ B = 6$ (d) $A = 3,\ B = 9$

Solution :

The given 9-digit number $8475639AB$ is divisible by 99, i.e., also divisible by both 9 and 11.

By divisibility rule of 9:

$8+4+7+5+6+3+9+A+B = 42+A+B$

$A + B$ can be 3 or 12.

From options, only option (b) and (d) satisfy $A + B = 12$.

From option (d):

$(8+7+6+9+9) - (4+5+3+3) = 39 - 15 = 24$ (not a multiple of 11)

From option (b):

$(8+7+6+9+8) - (4+5+3+4) = 38 - 16 = 22$ (multiple of 11)

$\therefore$ Required answer will be option (b).

11. Find the lowest positive value of $(c–b)$ such that the 7-digit number $1738b9c$ is divisible by 12:
$(c–b)$ का न्यूनतम धनात्मक मान ज्ञात कीजिए जिससे कि 7 अंकों की संख्या $1738b9c$ 12 से विभाज्य हो?

Solution :

$1738b9c$

By divisibility rule of 4:

Last two digits $9c$ should be divisible by 4.

$c = 2$ or 6.

By divisibility rule of 3:

Sum of digits $= 28 + b + c$.

For $c = 2$, $b = 0,\ 3,\ 6,\ 9$.

For $c = 6$, $b = 2,\ 5,\ 8$.

Lowest positive value of $(c - b) = 6 - 5 = 1$.

12. If the six-digit number $15x1y2$ is divisible by 44, then $(x+y)$ is equal to:
यदि छह अंकों की संख्या $15x1y2$, 44 से विभाज्य है, तो $(x+y)$ बराबर है:

Solution :

Since the given number is divisible by 44

Hence the number will be divisible by both 11 and 4.

Divisibility rule of 11–The difference between the sum of digits present at odd places and the sum of digits present at even places will be zero or multiple of 11.

Now,

$(2+1+5)-(y+x+1)=0$

$8-(x+y)=0$

$(x+y)=7.$

13. When 200 is divided by a positive integer x, the remainder is 8. How many values of x are there?
जब 200 को एक धनात्मक पूर्णांक x से विभाजित किया जाता है, तो शेषफल 8 होता है। x के कितने मान हैं?

Solution :

Required number which is divisible by $x = 200-8 = 192 = 2^6 \times 3.$

The value of x will always be greater than 8.

Hence, value of x can be-

12, 16, 24, 32, 48, 64, 96 and 192.

Hence, the number of values of $x = 8$.

14. What is the sum of all natural numbers between 100 and 400 which are divisible by 13?
100 और 400 के बीच की सभी प्राकृतिक संख्याओं का योग क्या है जो 13 से विभाज्य हैं?

Solution :

Numbers divisible by 13 between 100 and 400 are 104, 117, 130, . . . , 390.

$$T=\left(\frac{l-a}{d}+1\right)n$$

$$=\left(\frac{390-104}{13}+1\right)$$

$$=\frac{286}{13}+1=22+1$$

$$n=23$$

$$S_n=\frac{n}{2}\ [2a+(n-1)d]$$

$$=\frac{23}{2}\ [2\times 104+(23-1)d]$$

$$=\frac{23}{2}\ [2\times 104+(23-1)\times 13]$$

$$=\frac{23}{2}\times[208+286]=\frac{23}{2}\times 494=5681$$

15. a, b, and c are three fractions such that $a<b<c$. If c is divided by a, the result is $9/2$, which exceeds b by $23/6$. The sum of a, b, and c is $19/12$. What is the value of $(2a+b-c)$?
a, b और c तीन भिन्न हैं जैसे कि $a<b<c$। यदि c को a से विभाजित किया जाता है, तो परिणाम $9/2$ होता है जो b से $23/6$ अधिक होता है। a, b और c का योग $19/12$ होता है। $(2a+b-c)$ का मान क्या है?

Solution :

According to the question,

$$\frac{c}{a}=\frac{9}{2},\quad c=b+\frac{23}{6}$$

$$b=\frac{9}{2}-\frac{23}{6}=\frac{27}{6}-\frac{23}{6}=\frac{4}{6}=\frac{2}{3}$$

Now, $c=9x,\ a=2x$ (say)

$$a+b+c=\frac{19}{12}$$

$$11x+\frac{2}{3}=\frac{19}{12}$$

$$11x=\frac{19}{12}-\frac{8}{12}=\frac{11}{12}$$

$$x=\frac{1}{12}$$

$$\therefore c=\frac{3}{4},\quad a=\frac{1}{6}$$

$$2a+b-c=2\times\frac{1}{6}+\frac{2}{3}-\frac{3}{4}$$

$$=\frac{2}{6}+\frac{4}{6}-\frac{9}{12}=\frac{6}{6}-\frac{3}{12}=\frac{1}{4}$$

16. If $N=9^9$, then N is divisible by how many positive perfect cubes?
यदि $N=9^9$ है, तो N कितने धनात्मक पूर्ण घनों से विभाज्य है?

Solution :

$N = 9^9 = (3^2)^9 = 3^{18}$.

The number of positive factors of N that are cubes is $\left(\frac{18}{3} + 1\right) = 7$.

Hence, N is divisible by 7 positive perfect cubes:

$(3^3, 3^6, 3^9, 3^{12}, 3^{15}, 3^{18})$.

17. Given that $2^{20} + 1$ is completely divisible by a whole number, which of the following is completely divisible by the same number? यह देखते हुए कि $2^{20}+1$ एक पूर्ण संख्या से पूरी तरह से विभाज्य है, निम्नलिखित में से कौन सा एक ही संख्या से पूरी तरह से विभाज्य है?

(a) $2^{15}+1$ (b) $2^{90}+1$
(c) 5×2^{30} (d) $2^{60}+1$

Solution :

From option (d): $\frac{2^{60}+1}{2^{20}+1} = \frac{(2^{20})^3+1^3}{2^{20}+1} =$
$(2^{20}+1)(2^{40}-2^{20}+1)/(2^{20}+1)$

Hence, $2^{60}+1$ will be exactly divisible by the same number as $2^{20}+1$.

18. How many natural numbers less than 1000 are divisible by 5 or 7 but NOT by 35? 1000 से कम कितनी प्राकृतिक संख्याएँ 5 या 7 से विभाज्य हैं लेकिन 35 से नहीं?

Solution :

Total natural numbers below 1000 = 999

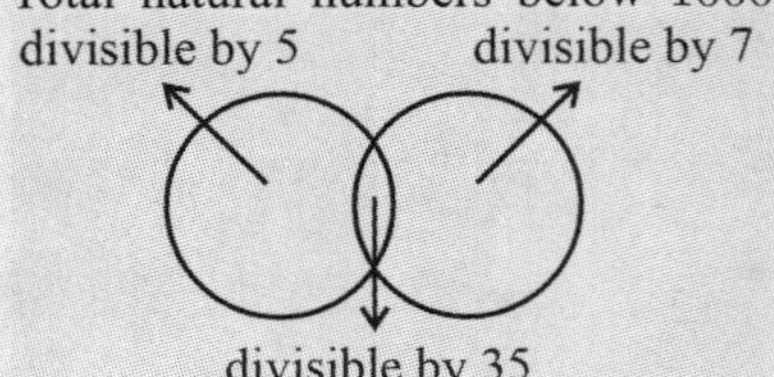

Total numbers divisible by 5 $= \frac{999}{5} = 199$

Total numbers divisible by 7 $= \frac{999}{7} = 142$

Total numbers divisible by 35 $= \frac{999}{35} = 28$

Total numbers till 999 divisible by 5 and 7 but not divisible by 35: $199+142-2\times 28 = 285$

19. How many multiples of 7 are there from 1 to 100 which are not multiples of 4? 1 से 100 तक 7 के ऐसे कितने गुणज हैं जो 4 के गुणज नहीं हैं?

Solution :

Total multiples of 7 from 1 to 100:
$\frac{100}{7} = 14$

LCM of 7 and 4: 28

Numbers divisible by both 7 and 4: $\frac{100}{28} = 3$

Numbers divisible by 7 but not 4: $14 - 3 = 11$

Type 2 (Problems based on Factors) टाइप 2 (कारकों पर आधारित समस्याएं)

20. The number of factors of 3600 is: 3600 के गुणनखंडों की संख्या है:

Solution :

$3600 \rightarrow 2^4 \times 3^2 \times 5^2$

No. of factors (NOF) $\rightarrow (4+1)\times(2+1)\times(2+1) = 5\times 3\times 3 = 45$

21. Find the number of prime factors in the product $(30)^5 \times (24)^5$. गुणनफल $(30)^5 \times (24)^5$ में अभाज्य गुणनखंडों की संख्या ज्ञात कीजिए।

Solution :

$30^5 \times 24^5 = (2\times 3\times 5)^5 \times (2^3 \times 3)^5$

$= 2^{20} \times 3^{10} \times 5^5$

Number of prime factors $= 20+10+5 = 35$

22. How many positive factors of 333 are there? 333 के कितने सकारात्मक गुणनखंड हैं?

Solution :

$333 = 3^2 \times 37^1$

Positive factors $= (2+1)\times(1+1)$

$= 3\times 2 = 6$

23. How many factors of 420 are prime numbers? 420 के कितने गुणनखंड अभाज्य संख्याएँ हैं?

Solution :

$420 = 2^2 \times 3 \times 5 \times 7$

Prime factors = 2, 3, 5, 7

Total prime factors = 4

24. Sum of all factors of 19600 is: संख्या 19600 के सभी गुणनखंड का योग क्या है?

Solution :

$19600 \rightarrow 2^4 \times 7^2 \times 5^2$

No. of factors (NOF) $= (4+1) \times (2+1) \times (2+1) = 45$

Sum of factors (SOF) $= \frac{2^5-1}{2-1} \times \frac{7^3-1}{7-1} \times \frac{5^3-1}{5-1}$

$= 31 \times 57 \times 31 = 54777$

25. Sum of all even factors of 64800: संख्या 64800 के सभी सम गुणनखंड का योग क्या है?

Solution :

$64800 = 2^5 \times 3^4 \times 5^2 \rightarrow 2 \times (2^4 \times 3^4 \times 5^2)$

No. of even factors $= (5+1-1) \times (4+1) \times (2+1) = 75$

Sum of even factors $= (2^1 + 2^2 + 2^3 + 2^4 + 2^5) \times \frac{3^5-1}{3-1} \times \frac{5^3-1}{5-1}$

$= 62 \times 121 \times 31 = 236562$

26. Sum of all the odd factors of 64800 is: संख्या 64800 के सभी विषम गुणनखंड का योग क्या है?

Solution :

$64800 \rightarrow 2^5 \times 3^4 \times 5^2$

No. of odd factors $= (4+1) \times (2+1) = 15$

Sum of odd factors $= \frac{3^5-1}{3-1} \times \frac{5^3-1}{5-1}$

$= 121 \times 31 = 3751$

27. Which of the following statement(s) is/are TRUE?
I. The total number of positive factors of 72 is 12
II. The sum of first 20 odd numbers is 400
III. Largest two-digit prime number is 97
निम्नलिखित में से कौन सा कथन सत्य है/हैं?
प्रथम. 72 के सकारात्मक कारकों की कुल संख्या 12 है
द्वितीय. प्रथम 20 विषम संख्याओं का योग 400 है
तृतीय. दो अंकों की सबसे बड़ी अभाज्य संख्या 97 है

(a) Only I and III
(b) Only II and III
(c) Only I and III
(d) All are true

Solution :

I. $72 = 8 \times 9 = 2^3 \times 3^2$

Number of factors $= (3+1)(2+1) = 12$

It is correct.

II. Sum of first n odd numbers = n^2 $[n = 20]$

Sum of first 20 odd numbers $= 20^2 = 400$

It is also correct.

III. Largest two-digit prime number = 97 (correct).

Thus, all the statements are true.

28. One of the factors of $(8^{2k} + 5^{2k})$, where k is an odd number, is: $(8^{2k} + 5^{2k})$ का एक गुणनखंड, जहां k एक विषम संख्या है, है:
(a) 88 (b) 84 (c) 89 (d) 86

Solution :

(c) Given k is an odd number. Let $k = 1$:

$8^2 + 5^2 = 64 + 25 = 89$

Hence, it is divisible by 89.

Formula:

(i) $(a^n + b^n)$, will always be divisible by $(a+b)$ where n is an odd number.

(ii) $(a^n - b^n)$, will always be divisible by $(a-b)$ where n is an odd number.

(iii) $(a^2 - b^2)$, will always be divisible by $(a+b)$ and $(a-b)$ where n is even.

Now, $(8^{2k}+5^{2k}) = (8^2)^k + (5^2)^k = (64^k + 25^k)$

Since k is an odd number, it is clear from formula (i) that the number will always be divisible by $(64+25) = 89$.

29. If $N = 4^{11}+4^{12}+4^{13}+4^{14}$, then how many positive factors of N are there?
यदि $N = 4^{11}+4^{12}+4^{13}+4^{14}$ है, तो N के कितने धनात्मक गुणनखंड हैं?

Solution :

$N = 4^{11}+4^{12}+4^{13}+4^{14}$

$N = 4^{11}(1+4+4^2+4^3)$

$= 4^{11} \times 85$

$= 2^{22} \times 5 \times 17$

Positive factors of $N = (22+1)(1+1)(1+1) = 23 \times 2 \times 2 = 92$

30. If $N = 3^{14}+3^{13}-12$, then what is the largest prime factor of N?
यदि $N = 3^{14}+3^{13}-12$, तो N का सबसे बड़ा अभाज्य गुणनखंड क्या है?

Solution :

$N = 3^{14}+3^{13}-12$

$N = 3^{12} \times (3^2+3) - 12$

$N = 3^{12} \times 12 - 12$

$= 12(3^{12}-1)$

$= 12 \times (729+1)(729-1)$

$= 12 \times 730 \times 728$

The largest prime factor = 73.

Type 3 (Problem based on Ramainder)
प्रकार 3 (विभाज्यता पर आधारित समस्या)

31. Given n an integer, what is the remainder when $(6n+3)^2$ is divided by 9?
n एक पूर्णांक दिया गया है जब $(6n+3)^2$ को 9 से विभाजित किया जाता है तो शेषफल क्या होता है?

Solution :

$$\frac{(6n+3)^2}{9} = \frac{36n^2+9+36n}{9}$$
$$= \frac{36n^2}{9} + \frac{9}{9} + \frac{36n}{9} = 4n^2+1+4n$$

In $(6n+3)^2$, all the three terms are divisible by 9. Hence, the remainder will be zero.

32. If $71^{83}+73^{83}$ is divided by 36, the remainder is:
यदि $71^{83}+73^{83}$ को 36 से विभाजित किया जाए, तो शेषफल क्या है?

Solution :

On $a^m + b^m$, if m is an odd number, then the sum of $(a+b)$ is divisible by 36. $\frac{71+73}{36} = \frac{144}{36} = 4$

Then, remainder = 0.

33. If $31^{47}+43^{47}$ is divided by 37, the remainder is:
यदि $31^{47}+43^{47}$ को 37 से विभाजित किया जाए, तो शेषफल क्या है?

Solution :

(a^n+b^n) is divisible by $(a+b)$ if n is odd.

$31^{47}+43^{47}$ is divisible by $(31+43) = 74$.

Since 74 is a multiple of 37, $31^{47}+43^{47}$ is exactly divisible by 37. Remainder = 0.

34. What is the remainder when we divide $4^{50}+7^{50}$ by 65?
जब हम $4^{50}+7^{50}$ को 65 से विभाजित करते हैं तो शेषफल क्या होता है?

Solution :

$$\frac{4^{50}+7^{50}}{65} = \frac{(4^2)^{25}+(7^2)^{25}}{65}$$
$$= \frac{16^{25}+49^{25}}{65}$$

Since $(a^n + b^n)$ is always divisible by $(a+b)$ where n is odd: $16^{25}+49^{25}$ is always divisible by $(16+49) = 65$.

Hence, remainder = 0.

35. What is the remainder when $(127^{97} + 97^{97})$ is divided by 32?
जब $(127^{97} + 97^{97})$ को 32 से विभाजित किया जाता है तो शेषफल क्या होता है?

Solution :

$$\frac{(127^{97} + 97^{97})}{32}$$
$$= \frac{(128-1)^{97}}{32} + \frac{(96+1)^{97}}{32}$$
$$= (-1)^{97} + (1)^{97} = -1 + 1 = 0$$

36. If $14331433 \times 1422 \times 1425$ is divided by 10, then what is the remainder?
यदि $14331433 \times 1422 \times 1425$ को 10 से विभाजित किया जाए, तो शेषफल क्या होगा?

Solution :

Given: Number $= 14331433 \times 1422 \times 1425$

Last digit of number $= 3 \times 2 \times 5 = 30$

The last digit is 0. Hence, on dividing by 10, the remainder obtained will be zero.

37. The remainder when $72 \times 73 \times 78 \times 76$ is divided by 35 is:
$72 \times 73 \times 78 \times 76$ को 35 से विभाजित करने पर शेषफल क्या है?

Solution :

By remainder theorem,

$$+2 \times +3 \times +6 \times +8 = +1 \times +8$$
$$\frac{72 \times 73 \times 76 \times 78}{35} = \frac{36x}{35} = 1 \times 8 = 8$$

38. If 2941 + 3741 is divided by 33, then the remainder is:
यदि 2941+3741 को 33 से विभाजित किया जाए, तो शेषफल क्या है?

Solution :

$(a^n + b^n)$ is divisible by $(a+b)$ when n is odd.

$29^{41} + 37^{41}$ is divisible by $(29 + 37 = 66)$.

Since 66 is divisible by 33, remainder = 0.

39. If a number is divided by 899, then the remainder is 63. If the same number is divided by 29, then what will be the remainder?
यदि किसी संख्या को 899 से विभाजित किया जाता है, तो शेषफल 63 होता है। यदि उसी संख्या को 29 से विभाजित किया जाता है, तो शेषफल क्या होगा?

Solution :

899 is divisible by 29.

$$29\overline{)63}\;(2$$
$$\underline{58}$$
$$5$$

40. What is the remainder when 622 is divided by 13?
जब 622 को 13 से विभाजित किया जाता है तो शेषफल क्या होता है?

Solution :

$$13\overline{)622}\;(47$$
$$\underline{52}$$
$$102$$
$$\underline{91}$$
$$11$$

41. What will be the remainder when $(631 \times 632 \times 633)$ is divided by 5?
जब $(631 \times 632 \times 633)$ को 5 से विभाजित किया जाए तो शेषफल क्या होगा?

Solution :

$$\frac{631 \times 632 \times 633}{5}$$
$$= \frac{1 \times 2 \times 3}{5}$$
$$= \frac{6}{5}$$

42. What will be the remainder when 421 is divided by 11?
जब 421 को 11 से विभाजित किया जाए तो शेषफल क्या होगा?

Solution :

$$11\overline{)421}\;(38$$
$$\underline{33}$$
$$91$$
$$\underline{88}$$
$$3$$

43. What will be the remainder when $(401 + 402 + 403 + 404)$ is divided by 4?

जब (401 + 402 + 403 + 404) को 4 से विभाजित किया जाए तो शेषफल क्या होगा?

Solution :

$401 + 402 + 403 + 404 = 1610$

On dividing 1610 by 4:

$$\begin{array}{r} 402 \\ 4\overline{)1610} \\ \underline{16} \\ 10 \\ \underline{8} \\ 2 \end{array}$$

44. What is the remainder when we divide $5^{70} + 7^{70}$ by 74?
जब हम $5^{70} + 7^{70}$ को 74 से विभाजित करते हैं तो शेषफल क्या होता है?

Solution :

$$\frac{5^{70} + 7^{70}}{74} = \frac{(5^2)^{35} + (7^2)^{35}}{74}$$
$$= \frac{25^{35} + 49^{35}}{74}$$

$a^n + b^n$ is divisible by $(a + b)$ when n is odd number as 35 is odd.

Hence$25^{35} + 49^{35}$ is divisible by $25 + 49 = 74$

remainder = 0.

45. When $(2^{36} - 1)$ is divided by 9, the remainder is:
जब $(2^{36} - 1)$ को 9 से विभाजित किया जाता है, तो शेषफल होता है:

Solution :

$$\frac{(2^{36} - 1)}{9} = \frac{((2^3)^{12} - 1)}{9}$$
$$= \frac{(8^{12} - 1)}{9}$$
$$= \frac{((9-1)^{12} - 1)}{9} = \frac{(1-1)}{9} = 0$$

46. If a positive integer n is divided by 7, the remainder is 2. Which of the numbers in the options yields a remainder of 0 when it is divided by 7?
यदि एक धनात्मक पूर्णांक n को 7 से विभाजित किया जाता है, तो शेषफल 2 होता है। विकल्पों में से किस संख्या को 7 से विभाजित करने पर 0 शेषफल प्राप्त होता है?

(a) $n + 5$ (c)$n + 3$
(b) $n + 2$ (d) $n + 1$

Solution :

According to question,

$\frac{n}{7} \Rightarrow$ remainder $= 2$

From option (a),

$$\frac{n+5}{7} = \frac{2+5}{7} = 1$$

Hence remainder = zero.

Hence option (a) is correct.

47. When a number is divided by 14, the remainder is 9. If the square of the same number is divided by 14, then the remainder will be:
जब किसी संख्या को 14 से विभाजित किया जाता है, तो शेषफल 9 होता है। यदि उसी संख्या के वर्ग को 14 से विभाजित किया जाता है, तो शेषफल होगा:

Solution :

Let the number be 23. On dividing 23 by 14 the remainder is 9.

Again, on dividing $(23)^2 = 529$ by 14, the remainder is 11.

48. If a number is divided by 3, the remainder will be 2. If the number is added by 5 and then divided by 3, then what will be the remainder?
यदि किसी संख्या को 3 से विभाजित किया जाए, तो शेषफल 2 होगा। यदि संख्या को 5 से जोड़ा जाए और फिर 3 से विभाजित किया जाए, तो शेषफल क्या होगा?

Solution :

Let quotient be x:

Dividend = Divisor × Quotient + Remainder $= 3x + 2$

According to question,

$$\frac{(3x+2)+5}{3} = \frac{3x+7}{3} = \frac{(3x+6)+1}{3}$$

Hence remainder = 1.

49. When a number M is divided by 7, the remainder is 6. What is the remainder if the square of M is divided by 7?
जब एक संख्या M को 7 से विभाजित किया जाता है, तो शेषफल 6 होता है। यदि M के वर्ग को 7 से विभाजित किया जाता है, तो शेषफल क्या होगा?

Solution :

If M is divided by 7, the remainder is 6, that can be written as:

$M = 7x + 6$ for integer x.

Then, M^2 is divided by 7, that can be written as:

$M^2 = (7x + 6)^2 = 49x^2 + 84x + 36 = 49x^2 + 84x + 35 + 1$

So, $(49x^2 + 84x + 35)$ is divisible by 7 and leaves remainder 1.

Hence remainder = 1.

50. If a number is divisible by 624, the remainder will be 53. If the same number is divisible by 16, then the remainder will be:
यदि कोई संख्या 624 से विभाज्य है, तो शेषफल 53 होगा। यदि वही संख्या 16 से विभाज्य है, तो शेषफल होगा:

Solution :

Formula:

Dividend = Divisor × Quotient + Remainder

Let the number that divides 624 be N:

$\frac{624}{N} \Rightarrow \text{Remainder} = 53$

When it is divisible by 16, it can be written as:

$\frac{(39 \times 16 + 53)}{N} = \frac{(39 \times 16 + 16 \times 3 + 5)}{N}$

Required remainder = 5.

51. When an integer n is divided by 6, the remainder is 5. What is the remainder if $9n$ is divided by 6?
जब एक पूर्णांक n को 6 से विभाजित किया जाता है, तो शेषफल 5 होता है। यदि $9n$ को 6 से विभाजित किया जाता है, तो शेषफल क्या होता है?

Solution :

When an integer n is divided by 6, the remainder is 5.

Number = Divisor × Quotient + Remainder

$n = 6q + 5$

According to the question,

If $9n$ is divided by 6,

$9n = 9(6q + 5)$

$9n = 54q + 45$

$9n = 54q + 42 + 3$

$9n = 6(9q + 7) + 3$

So, 3 is remainder if $9n$ is divided by 6.

52. If 7 divided a positive integer n, the remainder is 2. Which of the following numbers gives a remainder of 0 when divided by 7?
यदि 7 ने एक धनात्मक पूर्णांक n को विभाजित किया है, तो शेषफल 2 है। निम्नलिखित में से कौन सी संख्या 7 से विभाजित करने पर 0 का अनुस्मारक देती है?

(a) $n + 5$ (b) $n - 5$
(c) $n + 2$ (d) $n + 1$

Solution :

Let $n = 9$.

On dividing by 7, remainder found = 2.

From option,

$n + 5 = 14$ is exactly divisible by 7.

53. If 5 divided the integer n, the remainder is 2. What will be the remainder if $7n$ is divided by 5?
यदि 5 पूर्णांक n को विभाजित करता है, तो शेषफल 2 होता है। यदि $7n$ को 5 से विभाजित किया जाता है, तो शेषफल क्या होगा?

Solution :

Let integer $n = 7$.

Remainder left on dividing 7 by 5 = 2.

$7n = 49$.

Remainder left on dividing $7n$ by 5 = 4.

54. N is the largest two-digit number, which when divided by 3, 4, and 6 leaves the remainder 1, 2, and 4 respectively. What is the remainder when N is divided by 5?
N दो अंकों की सबसे बड़ी संख्या है, जिसे 3, 4 और 6 से विभाजित करने पर क्रमश: 1, 2 और 4 शेष बचता है। जब N को 5 से विभाजित किया जाता है तो शेषफल क्या होता है?

Solution :

L.C.M of 3, 4, and 6 = 12.

$[3-1-4-2=5-3=2]$

Required number = $12k-2$.

On putting $k=8$, $N=94$.

Remainder on dividing 94 by 5 = 4.

55. If a number is divided by 899, then remainder is 63. If the same number is divided by 29, then remainder will be:

यदि किसी संख्या को 899 से विभाजित किया जाता है, तो शेषफल 63 होता है। यदि उसी संख्या को 29 से विभाजित किया जाता है, तो शेषफल होगा:

Solution :

899 is divisible by 29.

$$29\overline{)63}\ (2,\quad 63-58=5$$

56. When an integer n is divided by 5, then remainder is 3. What is the remainder if $8n$ is divided by 5?

जब एक पूर्णांक n को 5 से विभाजित किया जाता है, तो शेषफल 3 होता है। यदि $8n$ को 5 से विभाजित किया जाता है, तो शेषफल क्या होता है?

Solution :

According to question,

Number in first condition (n) = $5K+3$.

(K = quotient)

Number in 2nd condition ($8n$) = $40K+24$.

Required remainder = 24/5 = 4.

57. If a number is divided by 30 then it leaves 17 as a remainder. What will be the remainder when the same number is divided by 10?

यदि किसी संख्या को 30 से विभाजित किया जाए तो शेषफल 17 बचता है। जब उसी संख्या को 10 से विभाजित किया जाए तो शेषफल क्या होगा?

Solution :

Let number be x and quotient be P:

dividend = divisor × quotient + remainder

$x=30P+17$

On dividing by 10:

$$\frac{30P+17}{10}=7 \quad \text{(remainder)}$$

58. A number when divided by 36 leaves remainder 25. What is the remainder when the same number is divided by 6?

एक संख्या को जब 36 से विभाजित किया जाता है तो शेषफल 25 बचता है। जब उसी संख्या को 6 से विभाजित किया जाता है तो शेषफल क्या होता है?

Solution :

Let assume that on dividing number by 36, K is obtained as the quotient and 25 is the remainder.

Dividend = $36\times K+25$

$=6\times 6K+6\times 4+1$

$=6(6K+4)+1$

Hence on dividing by 6, 1 is left as remainder.

59. What is the largest two-digit number which when divided by 6 and 7 gives remainder 3 and 4 respectively?

दो अंकों की सबसे बड़ी संख्या कौन सी है जिसे 6 और 7 से विभाजित करने पर क्रमश: 3 और 4 शेष बचता है?

Solution :

L.C.M of 6 and 7 = 42.

required number = $42k-3$

For $k=2$:

Number = $42\times 2-3=81$

60. When an integer n is divided by 8, the remainder is 3. What will be the remainder if $6n-1$ is divided by 8?

जब एक पूर्णांक n को 8 से विभाजित किया जाता है, तो शेषफल 3 होता है। यदि $6n-1$ को 8 से विभाजित किया जाता है, तो शेषफल क्या होगा?

Solution :

On dividing n by 8, the remainder is 3.

We can take $n = 8 + 3 = 11$.

On putting $n = 11$ in $6n - 1$:

$\frac{65}{8} = 1$ (remainder)

61. When a positive integer is divided by d, then remainder is 15. When ten times of the same number is divided by d, then remainder is 6. The least possible value of d is:

जब एक धनात्मक पूर्णांक को d से विभाजित किया जाता है, तो शेषफल 15 होता है। जब उसी संख्या के दस गुना को d से विभाजित किया जाता है, तो शेषफल 6 होता है। d का न्यूनतम संभव मान है:

Solution :

Let positive integer = x:

$\frac{x}{d} = 15$ (remainder)

$\frac{10x}{d} = 150 = (144 + 6)$

So, d will be a factor of 144 and will be greater than 15. Minimum value of $d = 16$

62. When a certain number is divided by 52, the remainder is 49. When the same number is divided by 13, the remainder is x. What is the value of $\sqrt{5x - 1}$?

जब एक निश्चित संख्या को 52 से विभाजित किया जाता है, तो शेषफल 49 होता है। जब उसी संख्या को 13 से विभाजित किया जाता है, तो शेषफल x होता है। $\sqrt{5x - 1}$ का मान क्या है?

Solution :

Dividend = Divisor × Quotient + Remainder

$D = Q_1 \times 52 + 49$

$D = Q_2 \times 13 + x$

52 is divisible by 13, so we can suppose that D is divisible by 13, and this gives 49 as remainder.

But again, 49 can be divided by 13 (49 > 13). This gives 10 as remainder which is equal to x.

$\sqrt{5x - 1} = \sqrt{5 \times 10 - 1} = \sqrt{49} = 7$

63. If x is the remainder when 361284 is divided by 5 and y is the remainder when 4^n is divided by 6, then what is the value of $(2x - y)$?

यदि 361284 को 5 से विभाजित करने पर x शेषफल है और 4^n को 6 से विभाजित करने पर y शेषफल है, तो $(2x - y)$ का मान क्या है?

Solution :

$\frac{3^{61284}}{5} = \frac{(3^4)^{15321}}{5} = \frac{(80+1)^{15321}}{5} =$ 1 (remainder)

$x = 1$

Remainder obtained on dividing 4^n by $6 = 4$

(where n is a natural number).

$\frac{496}{6} = 4$ (remainder)

$y = 4$

$2x - y = 2 \times 1 - 4 = -2$

64. The ratio of divisor and remainder is 3:2 and ratio of divisor and quotient is 7:12. If remainder is 14 then what is the remainder obtained when the dividend is divided by 9?

भाजक और शेषफल का अनुपात 3:2 है और भाजक और भागफल का अनुपात 7:12 है। यदि शेषफल 14 है तो लाभांश को 9 से विभाजित करने पर क्या शेषफल प्राप्त होता है?

Solution :

The ratio of divisor and remainder = 3 : 2 = $3x, 2x$

And ratio of divisor and quotient = 7 : 12 = $7y, 12y$

If remainder = 14 then $2x = 14 \Rightarrow x = 7$

Then, divisor $3x = 3 \times 7 = 21$, but divisor is $7y$ then $7y = 21$, $y = 3$

Then quotient $12y = 12 \times 3 = 36$

Now, dividend = divisor × quotient + remainder

dividend = $21 \times 36 + 14 = 770$

Then the remainder obtained on dividing the dividend by 9

```
   85
9)770
  72
   50
   45
    5
```

remainder = 5

65. M is the largest three-digit number which when divided by 6 and 5 leaves remainder 5 and 3 respectively. What will be the remainder when M is divided by 11?

M तीन अंकों की सबसे बड़ी संख्या है जिसे 6 और 5 से विभाजित करने पर क्रमश: 5 और 3 शेष बचता है। जब M को 11 से विभाजित किया जाए तो शेषफल क्या होगा?

Solution :

The largest 3-digit number which when divided by 6 and 5 leaves remainder as 5 and 3: 998, 993, 988, 983

Hence 983 is that number which when divided by 6 and 5 leaves remainder 5 and 3 respectively: $\Rightarrow M = 983$

$M \div 11 = 983 \div 11$

remainder = 4

66. How many natural numbers are there between 1000 to 2000, which when divided by 341 leaves remainder 5?

1000 से 2000 के बीच ऐसी कितनी प्राकृत संख्याएँ हैं जिन्हें 341 से विभाजित करने पर 5 शेष बचता है?

Solution :

Natural numbers between 1000 and 2000:

$1000 < 341K + 5 < 2000$

Where K is taken as 3, 4, 5:

$K = 3$ then $341K + 5 = 341 \times 3 + 5 = 1028$

$K = 4$ then $341K + 5 = 341 \times 4 + 5 = 1369$

$K = 5$ then $341K + 5 = 341 \times 5 + 5 = 1710$

Hence, the numbers of natural numbers between 1000 and 2000 are three which when divided by 341 leaves 5 as a remainder.

67. M is the largest 4-digit number, which when divided by 4, 5, 6, and 7 leaves remainder as 2, 3, 4, and 5 respectively. What will be the remainder when M is divided by 9?

M 4 अंकों की सबसे बड़ी संख्या है, जिसे 4, 5, 6 और 7 से विभाजित करने पर क्रमश: 2, 3, 4 और 5 शेष बचता है। जब M को 9 से विभाजित किया जाए तो शेषफल क्या होगा?

Solution :

Largest four-digit number = 9999

L.C.M of 4, 5, 6, 7 = 420

The difference between divisor and remainder is same: $(4 - 2 = 2, 5 - 3 = 2, 6 - 4 = 2, 7 - 5 = 2)$

Required number $= 420K - 2$

$(K = 1, 2, 3 \ldots)$:

On putting $K = 23$:

$420 \times 23 - 2 = 9658$

According to question: $9658 \div 9$

Remainder = 1

68. N is the smallest three-digit prime number. When N is divided by 13, then what will be the remainder?

N तीन अंकों की सबसे छोटी अभाज्य संख्या है। जब N को 13 से विभाजित किया जाए तो अनुस्मारक क्या होगा?

Solution :

The smallest three-digit prime number (N) = 101

Remainder on dividing 101 by 13 = 10

69. The sum of the digits of a two-digit number is $\frac{1}{7}$ of the number. The unit digit is 4 less than the tens digit. If the number obtained on reversing its digits is divided by 7, then the remainder will be:

दो अंकों की संख्या के अंकों का योग संख्या का $\frac{1}{7}$ है। इकाई का अंक दहाई के अंक से 4 कम है। यदि उसके अंकों को उलटने पर प्राप्त संख्या को 7 से विभाजित किया जाए, तो शेषफल होगा:

Solution :

Let number $= 10x + y$

According to question:

$10x + y = 7(x + y) \quad \ldots (i)$

$x = y + 4 \quad \ldots (ii)$

From equations (i) and (ii):

$10(y + 4) + y = 7(y + 4 + y)$

$10y + 40 + y = 14y + 28$

$3y = 12$

$y = 4$

$x = 4 + 4 = 8$

Hence, number = 84

Reversing the digits and dividing by 7:

$\frac{48}{7} = (6 \times 7 + 6) \div 7$

Remainder = 6

70. When a number is successively divided by 3, 4, and 7, the remainders obtained are 2, 3, and 5, respectively. What will be the remainder when 84 divides the same number?
जब किसी संख्या को क्रमिक रूप से 3, 4 और 7 से विभाजित किया जाता है, तो शेषफल क्रमशः 2, 3 और 5 प्राप्त होता है। जब उसी संख्या को 84 से विभाजित करने पर शेषफल क्या होगा?

Solution :

When the number is divided by 7, the remainder obtained = 5:

Number $= 7x + 5$

On dividing the same number by 4, remainder = 3:

Number $= 4(7x + 5) + 3$

On dividing the same number by 3, remainder = 2:

Number $= 3[4(7x + 5) + 3] + 2$

Number $= 3[28x + 23] + 2$

Number $= 84x + 69 + 2$

Number $= 84x + 71$

Hence, on dividing the number by 84, the remainder is 71.

71. When 2388, 4309, and 8151 are divided by a certain 3-digit number, then the remainder in each case is the same. The remainder is:
जब 2388, 4309 और 8151 को एक निश्चित 3-अंकीय संख्या से विभाजित किया जाता है, तो प्रत्येक स्थिति में शेषफल समान होता है। शेष है:

Solution :

Let the remainder be r.

Factor of $1921 = 113 \times 17$:

2388-r   4309-r   8151-r

1921   3842

1921

The three-digit number = 113:

$\frac{2388}{113} \Rightarrow$ remainder = 15

72. When a certain number is divided by 65, then the remainder is 56. When the same number is divided by 13, then the remainder is x. What is the value of $\sqrt{5x-2}$?
जब एक निश्चित संख्या को 65 से विभाजित किया जाता है, तो शेषफल 56 होता है। जब उसी संख्या को 13 से विभाजित किया जाता है, तो शेषफल x होता है। $\sqrt{5x-2}$ का मान क्या है?

Solution :

Remainder obtained on dividing 56 by 13 = 4:

$x = 4$

$\sqrt{5x-2} = \sqrt{5 \times 4 - 2} = \sqrt{20-2} = \sqrt{18} = 3\sqrt{2}$

73. If r is the remainder when each of 4749, 5601, and 7092 is divided by the greatest possible number ($d > 1$), then the value of $(d + r)$ will be:
यदि 4749, 5601 और 7092 में से प्रत्येक को अधिकतम संभव संख्या ($d > 1$) से विभाजित करने पर शेषफल r है, तो $(d + r)$ का मान होगा:

Solution :

(4749-r)   (5601-r)   (7092-r)

difference=852   difference=1451

difference=2343

H.C.F of 852, 1491, and 2343 = 213:

$213 \div 63$

The remainder left on dividing each number by 213 = 63:

$d = 213,\ r = 63$

Hence, $d + r = 276$

Type 4 (Unit Digit)

74. What is the unit digit of $1^5 + 2^5 + 3^5 + \ldots + 20^5$?
$1^5 + 2^5 + 3^5 + \ldots + 20^5$ का इकाई अंक क्या है?

Solution :

To obtain the unit digit, divide the power by 4, and the last remainder is taken as the power of the base number.

So,

$1^1 + 2^1 + 3^1 + 4^1 + 5^1 + 6^1 + \ldots + 20^1$

$S_{20} = \frac{20(20+1)}{2}$

$S_{20} = 210$

Hence, unit digit = 0.

75. What is the unit digit of $(217)^{413} \times (819)^{547} \times (414)^{624} \times (342)^{812}$?
$(217)^{413} \times (819)^{547} \times (414)^{624} \times (342)^{812}$ का इकाई अंक क्या है?

Solution :

$(217)^{413} \times (819)^{547} \times (414)^{624} \times (342)^{812}$

↓ ↓ ↓ ↓

$7^1 \times 9^3 \times 4^4 \times 2^4$

↓ ↓ ↓ ↓

$7 \times 9 \times 6 \times 6$

6③ × 3⑥

3×6=18

Hence, unit digit = 8.

76. What percentage of the numbers from 101 to 1000 have 9 in the unit's digit?
101 से 1000 तक कितने प्रतिशत संख्याओं का इकाई अंक 9 है?

Solution :

Required numbers from 101 to 200 = 10.

Hence, numbers from 101 to 1000 in which the unit digit is 9 = 10 × 9 = 90.

Total numbers from 101 to 1000 = 900.

Required percentage $= \frac{90}{900} \times 100 = 10\%$.

77. What is the unit digit of 5124×124^3?
5124×124^3 का इकाई अंक क्या है?

Solution :

When any even number is multiplied by 5, the unit digit always becomes zero (whatever the power is).

For example:

$5 \times 4 = 20$,

$125 \times 166 = \ldots 0$,

$12244 \times 12135 = \ldots 0$,

$5124 \times 1245 = \ldots 0$.

78. What is the digit at the unit's place in the product $(653 \times 308 \times 402 \times 413)$?
गुणनफल $(653 \times 308 \times 402 \times 413)$ में इकाई स्थान पर अंक क्या है?

Solution :

$(653 \times 308 \times 402 \times 413)$:

Unit's digit $= 3 \times 8 \times 2 \times 3 = 144$.

Unit digit = 4.

79. If $x = (164)^{169} + (333)^{357} - (727)^{726}$, then what is the unit digit of x?
यदि $x = (164)^{169} + (333)^{357} - (727)^{726}$, तो x का इकाई अंक क्या है?

Solution :

$x = (164)^{169} + (333)^{357} - (727)^{726}$

For the unit digit, divide the power of the base by 4 and place the remainder as the power of the unit digit of the base number.

$x = (4)^1 + (3)^1 - (7)^2 = 4 + 3 - 49$

On writing the unit digit:

$x = 4 + 3 - 9 = -2$

$x = 10 - 2 = 8$

Note: 10 is added to the unit digit to make it positive.

80. Let $x = (633)^{24} - (277)^{38} + (266)^{54}$. What is the unit digit of x?
माना $x = (633)^{24} - (277)^{38} + (266)^{54}$। x का इकाई अंक क्या है?

Solution :

$x = (633)^{24} - (277)^{38} + (266)^{54}$:

$x = 3^{24} - 7^{38} + 6^{54}$

$x = 4 - 2 + 6$

$x = -2$

$x = -2 + 10 = 8$

Unit digits are always positive; 10 is added to make it positive.

81. What is the unit digit of the sum of the first 111 whole numbers?
प्रथम 111 पूर्ण संख्याओं के योग का इकाई अंक क्या है?

Solution :

Whole numbers = 0, 1, 2, . . . , 110:

$\text{Sum} = \frac{110 \times 111}{2} = 55 \times 111$

Unit digit = 5.

Type 5 (Misc.)

82. Which of the following is not a prime number?
निम्नलिखित में से कौन सी अभाज्य संख्या नहीं है?
(a)41 (b)47 (c)91 (d)73

Solution :

Prime Number: The number which has only two factors, i.e., 1 and the number itself. For example: 2, 3, 5, 7, 31, 37, ...

In the given options, 91 has more than two factors (1, 13, 7, 91). Hence, 91 is not a prime number.

83. x, y, and z are prime numbers and $x + y + z = 38$. What is the maximum value of x?
x, y और z अभाज्य संख्याएँ हैं और $x + y + z = 38$। x का अधिकतम मान क्या है?

Solution :

$x + y + z = 38$

For x to be maximum, the values of y and z must be 2 and 5, respectively.

$x + 2 + 5 = 38$

$x = 31$

84. How many two-digit prime numbers are there between 10 and 100 which remain prime numbers when the order of their digits is reversed?
10 से 100 के बीच ऐसी कितनी दो अंकीय अभाज्य संख्याएँ हैं जिनके अंकों का क्रम उलटने पर भी अभाज्य संख्याएँ बनी रहती हैं?

Solution :

Number of prime numbers between 1 and 100 = 25

Prime numbers between 1 and 10 = 4

Prime numbers between 10 and 100 = 21

The required numbers are as follows:

11, 13, 17, 19, 23, 29, 37, 41, 43, 47, 53, 59, 61, 67, 71, 73, 79, 83, 89, 97

Hence, there are 9 prime numbers which remain prime after reversing their digits.

85. Evaluate: $\frac{1}{15} + \frac{1}{35} + \frac{1}{63} + \frac{1}{99} + \frac{1}{143}$
मूल्यांकन करें: $\frac{1}{15} + \frac{1}{35} + \frac{1}{63} + \frac{1}{99} + \frac{1}{143}$

Solution :

$$\frac{1}{15} + \frac{1}{35} + \frac{1}{63} + \frac{1}{99} + \frac{1}{143} = \frac{1}{2}\left[\frac{1}{3} - \frac{1}{5} + \frac{1}{5} - \frac{1}{7} + \frac{1}{7} - \frac{1}{9} + \frac{1}{11} - \frac{1}{13}\right]$$

$$= \frac{1}{2}\left[\frac{1}{3} - \frac{1}{13}\right] = \frac{1}{2} \times \frac{10}{39} = \frac{5}{39}$$

86. What is the sum of $1\frac{1}{2} + 4\frac{1}{6} + 7\frac{1}{12} + 10\frac{1}{20} + . . .$ up to 20 terms? / $1\frac{1}{2} + 4\frac{1}{6} + 7\frac{1}{12} + 10\frac{1}{20} + . . .$ का योग क्या है 20 पदों तक?

Solution :

$1\frac{1}{2}+4\frac{1}{6}+7\frac{1}{12}+10\frac{1}{20}+...$ up to 20 terms

$= \left(\frac{1}{2}+\frac{1}{6}+\frac{1}{12}+...\text{ up to 20 terms}\right) -$

$(1+4+7+...\text{ up to 20 terms})$

$= \left(1-\frac{1}{2}+\frac{1}{2}-\frac{1}{3}+\frac{1}{3}-\frac{1}{4}+..+\frac{1}{20}-\frac{1}{21}\right)$

$= 20/2\ [2+19\times 3] + \left(1-\frac{1}{21}\right)$

$= 10\times 59 + \frac{20}{21} = 590 + \frac{20}{21} = \frac{12410}{21}$

87. If $\frac{1}{21}+\frac{1}{22}+\frac{1}{23}+...+\frac{1}{210}=\frac{1}{k}$, then what is the value of k?

यदि $\frac{1}{21}+\frac{1}{22}+\frac{1}{23}+...+\frac{1}{210}=\frac{1}{k}$, तो k का मान क्या है?

Solution :

$\frac{1}{2^1}+\frac{1}{2^2}+\frac{1}{2^3}+...+\frac{1}{2^{20}}=\frac{1}{k}$

$\because 1+\frac{1}{r}+\frac{1}{r^2}+...+\frac{1}{r^n}=\frac{1-r^n}{1-r}$

$\frac{1}{2}\left(1-\frac{1}{2^{10}}\right)\div\left(1-\frac{1}{2}\right)=\frac{1}{k}$

$1-\frac{1}{2^{10}}=\frac{1}{k}$

$\frac{1023}{1024}=\frac{1}{k}$

$k=\frac{1024}{1023}$

88. What is the value of $16^2+17^2+18^2+...+25^2$?
$16^2+17^2+18^2+...+25^2$ का मान क्या है?

Solution :

Sum of square of n numbers $= \frac{n(n+1)(2n+1)}{6}$

Required sum

$= \left(\frac{25(25+1)(2\times 25+1)}{6}\right) - \left(\frac{15(15+1)(2\times 15+1)}{6}\right)$

$= \frac{25\times 26\times 51}{6} - \frac{15\times 16\times 31}{6}$

$= 5525 - 1240 = 4285$

89. Find the sum of $6+8+10+12+14+...+40$.
$6+8+10+12+14+...+40$ का योग ज्ञात कीजिये

Solution :

According to the question:

$6+(n-1)\times 2 = 40$

$2n-2=34$

$n=18$

Sum of AP $= \frac{n}{2}[a+l]$

$= \frac{18}{2}(6+40)$

$= 9\times 46 = 414$

Trick: $6+8+10+12+14+...+40$

$= 2(3+4+5+6+7+...+20)$

$= 2\left(\frac{20\times 21}{2}-3\right)$

$= 2(210-3) = 2\times 207 = 414$

90. The value of $1+3+5+7+...+(2n-1)$ is:
$1+3+5+7+...+(2n-1)$ का मान है:

Solution :

$1+3+5+7+...+(2n-1)$

The given series is in AP:

$d=3-1=2, \quad a=1$

Sum $= \frac{n}{2}[2a+(n-1)d]$

$= \frac{n}{2}[2\times 1+(n-1)\times 2]$

$= \frac{2n}{2}[1+n-1] = n\times n = n^2$

91. How many numbers less than 350 are multiples of both 4 and 3?
350 से कम कितनी संख्याएँ 4 और 3 दोनों के गुणज हैं?

Solution :

L.C.M of 3 and 4 = 12

Numbers less than 350 which are multiples of 3 and 4

12, 24, 36, 48, . . . , 348

In AP: $a=12, \quad d=12, \quad T_n=348$

$a+(n-1)d=T_n$

$12+(n-1)\times 12=348$

$12n=348$

$n=29$

Hence, the number of such terms = 29.

Trick:

L.C.M of 4 and 3 = 12

$\frac{350}{12}$ = quotient = 29

Hence, the required number of numbers = 29.

92. What is the sum of all 3-digit positive integers divisible by 11?

11 से विभाज्य सभी 3 अंकीय धनात्मक पूर्णांकों का योग क्या है?

Solution :

Three-digit numbers divisible by 11:

110, 121, 132, . . . , 990

$a=110,\ d=11,\ l=990$

From the formula of AP: $l=a+(n-1)d$

$990=110+(n-1)\times 11$

$880\div 11=n-1$

$n=81$

Sum $S=\frac{n}{2}[a+l]$

$S=\frac{81}{2}[110+990]$

$S=\frac{81}{2}\times 1100$

$S=44550$

93. In a two-digit number, its unit digit exceeds its tens digit by 2, and the product of the given number and the sum of its digits is equal to 460. The number is:

दो अंकों की संख्या में, इसका इकाई अंक इसके दहाई के अंक से 2 अधिक है और दी गई संख्या का गुणनफल और इसके अंकों का योग 460 के बराबर है। संख्या है:

Solution :

Let the number be $10x+y$:

$y-x=2$

$(10x+y)(x+y)=460$

Solve the equations:

$x=4,\ \ y=6$

Number = 46

94. If $A=0.\overline{abcabc}$, then by what number A should be multiplied to get an integral value?

यदि $A=0.\overline{abcabc}$, तो पूर्णांक मान प्राप्त करने के लिए A को किस संख्या से गुणा किया जाना चाहिए?

Solution :

$A=0.\overline{abc}$

$A=\frac{abc}{999}$

Multiply 999 or a multiple of 999 to A to make it an integer:

$2997=999\times 3,\quad 1998=999\times 2$

95. A and B are positive integers. If $A+B+AB=65$, then what is the difference between A and B ($A,\ B\leq 15$)?

A और B धनात्मक पूर्णांक हैं। यदि $A+B+AB=65$ है, तो A और B ($A,\ B\leq 15$) के बीच क्या अंतर है?

Solution :

$A+B+AB=65$

On putting $A=10,\ B=5$:

$10+5+10\times 5=65$

$65=65$

Hence, $A-B=10-5=5$

96. When a two-digit number is multiplied by the sum of its digits, the product is 424. When the number obtained by interchanging its digits is multiplied by the sum of its digits, the result is 280. The sum of the digits of the given number is:

जब दो अंकों की संख्या को उसके अंकों के योग से गुणा किया जाता है, तो गुणनफल 424 होता है। जब उसके अंकों को आपस में बदलने पर प्राप्त संख्या को अंकों के योग से गुणा किया जाता है, तो परिणाम 280 होता है। दिए गए अंकों का योग वह संख्या है:

Solution :

Let the two-digit number be $10x + y$.

According to the first condition:

$10x + y = \frac{424}{x+y} \quad \ldots (i)$

According to the second condition: $10y+x = \frac{280}{x+y} \quad \ldots (ii)$

By adding Equation (i) and Equation (ii):

$11x + 11y = \frac{704}{x+y}$

$x + y = \sqrt{64}$

$x + y = 8$

97. The sum of two numbers is 59, and their product is 840. Find the sum of their squares.
दो संख्याओं का योग 59 है और उनका गुणनफल 840 है। उनके वर्गों का योग ज्ञात कीजिए।

Solution :

Let the two numbers be a and b:

$a + b = 59, \quad ab = 840$

$(a+b)^2 = a^2 + b^2 + 2ab$

$59^2 = a^2 + b^2 + 2 \times 840$

$3481 = a^2 + b^2 + 1680$

$a^2 + b^2 = 3481 - 1680$

Hence, $a^2 + b^2 = 1801$

98. If the difference between two numbers is 6, and the difference between their squares is 60, what is the sum of their cubes?
यदि दो संख्याओं के बीच का अंतर 6 है और उनके वर्गों के बीच का अंतर 60 है, तो उनके घनों का योग क्या है?

Solution :

Let the two numbers be a and b:

$a - b = 6 \quad \ldots (i)$

$a^2 - b^2 = 60$

$a^2 - b^2 = (a-b)(a+b)$

$60 = 6(a+b)$

$a + b = 10 \quad \ldots (ii)$

From equations (i) and (ii):

$a = 8, \ b = 2$

$(a+b)^3 = a^3 + b^3 + 3ab(a+b)$

$10^3 = a^3 + b^3 + 3 \times 8 \times 2 \times 10$

$1000 = a^3 + b^3 + 480$

$a^3 + b^3 = 1000 - 480$

$a^3 + b^3 = 520$

Hence the sum of cubes of numbers = 520

99. How many zeroes are there in the product 46×1510?
गुणनफल 46×1510 में कितने शून्य होते हैं?

Solution :

The product of $4^6 \times 15^10$:

Number of 2's in $4^6 = 12$,

Number of 5's in $15^{10} = 10$

One pair of 2 and 5 combines to form a zero.

The possible number of pairs of 2 and 5 in

$46 \times 15^{10} = 10$.

Hence, number of zeroes $= 10$

100. A fraction is such that the numerator is five less than the denominator. Also, four times the numerator is one more than the denominator. The fraction is:
एक भिन्न ऐसी होती है जिसका अंश हर से पांच कम होता है। इसके अलावा अंश का चार गुना हर से एक अधिक है। अंश है:

Solution :

Let the fraction be $\frac{x}{y}$:

According to the question: $x = y - 5 \quad \ldots (i)$

$4x = y + 1 \quad \ldots (ii)$

From equations (i) and (ii):

$4x = x + 5 + 1$

$3x = 6$

$x = 2$

From equation (i):

$y = 7$

Hence, the fraction $= \frac{x}{y} = \frac{2}{7}$

101. By adding 3 and 5 to the numerator and denominator of a fraction, it becomes $\frac{2}{3}$. If 1 and 3 are subtracted from the numerator and denominator, it becomes $\frac{2}{5}$. Find the fraction.

किसी भिन्न के अंश में 3 और 5 तथा हर में 2 जोड़ने पर यह $\frac{2}{3}$ बन जाती है। यदि अंश और हर में से क्रमशः 1 और 3 घटाया जाए, तो यह $\frac{2}{5}$ हो जाती है। भिन्न ज्ञात कीजिए।

Solution :

Let the fraction be $\frac{a}{b}$:

$\frac{a+3}{b+5} = \frac{2}{3}$

$3a - 2b = 1 \quad . . . (i)$

$\frac{a-1}{b+3} = \frac{2}{5}$

$5a - 2b = 11 \quad . . . (ii)$

From equations (i) and (ii):

$a = 5, \quad b = 7$

Fraction $= \frac{a}{b} = \frac{5}{7}$

102. Three fractions, x, y, and z, are such that $x > y > z$. When the smallest of them is divided by the greatest, the result is $\frac{9}{16}$, which exceeds y by 0.0625. If $x + y + z = 1\frac{13}{24}$, then what is the value of $x + z$?

तीन भिन्न, x, y और z, ऐसे हैं कि $x > y > z$. जब उनमें से सबसे छोटे को सबसे बड़े से विभाजित किया जाता है, तो परिणाम $\frac{9}{16}$ होता है जो y से 0.0625 अधिक है। यदि $x + y + z = 1\frac{13}{24}$, तो $x + z$ का मान क्या है?

Solution :

Greatest number $= x$, Smallest number $= z$

According to the question: $\frac{z}{x} = y + 0.0625 = \frac{9}{16}$

$y = 0.5625 - 0.0625 = 0.5 = \frac{1}{2}$

$x + y + z = 1\frac{13}{24} = \frac{37}{24}$

$x + z = \frac{37}{24} - \frac{1}{2} = \frac{37 - 12}{24} = \frac{25}{24}$

103. Which of the following statement(s) is/are TRUE?
निम्नलिखित में से कौन सा कथन सत्य है/हैं?

I. $\frac{0.03}{0.2} + \frac{0.003}{0.02} + \frac{0.0003}{0.002} + \frac{0.00003}{0.0002} = 0.6$

II. $(0.01) + (0.01)^2 + (0.001)^2 = 0.010101$

(a) Only I
(b) Only II
(c) Neither I nor II
(d) Both I and II

Solution :

(I) $\frac{0.03}{0.2} + \frac{0.003}{0.02} + \frac{0.0003}{0.002} + \frac{0.00003}{0.0002} = 0.6$

$\frac{3}{2} \times 10^{-1} + \frac{3}{2} \times 10^{-1} + \frac{3}{2} \times 10^{-1} + \frac{3}{2} \times 10^{-1} = 0.6$

$4 \times \frac{3}{2} \times 10^{-1} = 0.6$

$0.6 = 0.6$

(II) $(0.01) + (0.01)^2 + (0.001)^2 = 0.010101$

$0.01 + 0.0001 + 0.000001 = 0.010101$

$0.010101 = 0.010101$

Hence, both statements I and II are true.

104. If $P = \frac{96}{95 \times 97}$, $Q = \frac{97}{96 \times 98}$, and $R = \frac{1}{97}$, then which of the following is TRUE?

यदि $P = \frac{96}{95 \times 97}$, $Q = \frac{97}{96 \times 98}$, और $R = \frac{1}{97}$ है, तो निम्नलिखित में से कौन सा सत्य है?

(a) P < Q < R (b) R < Q < P
(c) Q < P < R (d) R < P < Q

Solution :

Let $95 = 1,\ 96 = 2,\ 97 = 3,\ 98 = 4$:

$P = \frac{2}{1 \times 3} = 0.67$, $Q = \frac{3}{2 \times 4} = 0.375$, $R = \frac{1}{3} = 0.333$

Hence, $P > Q > R$ is right

105. The numerator of a fraction is less than 6 its denominator. If 1 is subtracted from its numerator and 5 is added to its denominator, then its denominator becomes 4 times its numerator. Find the fraction.

किसी भिन्न का अंश उसके हर 6 से कम है। यदि इसके अंश में से 1 घटा दिया जाए और इसके हर में 5 जोड़ दिया जाए, तो इसका हर इसके अंश का 4 गुना हो जाता है। भिन्न ज्ञात कीजिए।

Solution :

Let the denominator $= x$, numerator $= x - 6$:

According to the question: $4[(x-6)-1] = x + 5$

$4(x-7) = x + 5$

$4x - 28 = x + 5$

$3x = 33$

$x = 11$

Hence, numerator $= x - 6 = 11 - 6 = 5$

Fraction $= \frac{5}{11}$

14

LCM & HCF

LCM (Least Common Multiple)
The least common multiple (LCM) is defined as the smallest multiple that two or more numbers have in common.
एलसीएम (न्यूनतम समापवर्त्य) - लघुत्तम समापवर्त्य (एलसीएम) को सबसे छोटे गुणज के रूप में परिभाषित किया गया है जो दो या दो से अधिक संख्याओं में समान है।

HCF (Highest Common Factor)
HCF is used to find the highest common factor of any two or more given integers.
एचसीएफ (उच्चतम सामान्य गुणनखंड) - एचसीएफ का उपयोग किसी भी दो या अधिक दिए गए पूर्णांकों के उच्चतम सामान्य गुणनखंड को खोजने के लिए किया जाता है।

Type 1 (Problem based on LCM)

1. What is the LCM of 3.6, 1.8, and 0.144?
3.6, 1.8 और 0.144 का एलसीएम क्या है?

Solution:

LCM of 3.6, 1.8, and 0.144:

$$\text{LCM} = \frac{\text{LCM}(3600,\ 1800,\ 144)}{1000}$$

LCM of 3600, 1800, and 144 = 3600

$$\text{LCM of } 3.6,\ 1.8, \text{ and } 0.144 = \frac{3600}{1000} = 3.6$$

2. The sum of two numbers is 50, and their product is 525. The LCM of the two numbers is:
दो संख्याओं का योग 50 है और उनका गुणनफल 525 है। दोनों संख्याओं का एलसीएम है:

Solution:

Let the numbers be x and y respectively.

According to the question

$x + y = 50 \quad . . . (i)$

$xy = 525$

$(x+y)^2 - 4xy = (x-y)^2$

$50^2 - 4 \times 525 = (x-y)^2$

$2500 - 2100 = (x-y)^2$

$(x-y)^2 = 400 \Rightarrow x - y = 20 \quad . . . (ii)$

From equations (i) and (ii):

$x + y = 50,$

$x - y = 20$

$2x = 70$

$x = 35$

On putting the $x = 35$ value of x in equation (i):

$y = 50 - 35$

$y = 15$

LCM of 15 and 35

3	15,35
5	5,35
7	1,7
	1,1

$= 3 \times 5 \times 7 = 105$

3. What is the smallest integer that is divisible by 3, 7, and 18?
वह सबसे छोटा पूर्णांक कौन सा है जो 3, 7 और 18 से विभाज्य है?

Solution:

LCM of $3, 7,$ and $18 = 126$

126 is the smallest integer that is divisible by 3, 7, and 18.

4. Five bells ring together at intervals of 3, 5, 8, 9, and 10 seconds. All the bells ring simultaneously at the same time. They will ring again simultaneously after:
पाँच घंटियाँ 3, 5, 8, 9 और 10 सेकंड के अंतराल पर एक साथ बजती हैं। सभी घंटियाँ एक ही समय पर एक साथ बजती हैं। वे इसके बाद फिर से एक साथ बजेंगी:

Solution:

Again, the bells will ring simultaneously after =

LCM of $3, 5, 8, 9,$ and 10

LCM $= 360$ seconds

$= \frac{360}{60}$ minutes

$= 6$ minutes.

5. The least number which is exactly divisible by 4, 5, 8, 10, and 12 is:
सबसे छोटी संख्या जो 4, 5, 8, 10 और 12 से पूर्णतया विभाज्य हो वह है:

Solution:

LCM of $4, 5, 8, 10,$ and 12

$4 = 2 \times 2$

$8 = 2 \times 2 \times 2$

$10 = 2 \times 5$

$12 = 2 \times 2 \times 3$

LCM $= 2 \times 2 \times 2 \times 3 \times 5 = 120$

Hence it is clear that 120 is the least number which is exactly divisible by 4, 5, 8, 10 and 12.

6. The proportion among three numbers is 3 : 4 : 5, and their LCM is 1800. Then the second number is:
तीन संख्याओं का अनुपात 3 : 4 : 5 है और उनका LCM 1800 है। तो दूसरी संख्या है:

Solution:

Let the three numbers be $3x$, $4x$, and $5x$:

LCM of all three numbers $= 3 \times 4 \times 5 \times x = 60x$

LCM $= 1800$

$60x = 1800$

$x = 30$

Second number $= 4x = 4 \times 30 = 120$

7. The largest three-digit number that is exactly divisible by 6, 7, and 8 is:
तीन अंकों की सबसे बड़ी संख्या जो 6, 7 और 8 से पूर्णतः विभाज्य है, वह है:

Solution:

LCM of $6, 7,$ and $8 = 168$

The largest three-digit number $= 999$

$$168\overline{)999}\ (5,\quad \underline{840},\quad 159 \qquad = 999 - 159 = 840$$

The number 840 is completely divisible by 168.

8. Let x be the least 4-digit number which, when divided by 2, 3, 4, 5, 6, and 7, leaves a remainder of 1 in each case. If x lies between 2800 and 3000, then what is the sum of the digits of x?
मान लीजिए कि x सबसे छोटी 4 अंकों की संख्या है जिसे 2, 3, 4, 5, 6 और 7 से विभाजित करने पर प्रत्येक स्थिति में 1 शेष बचता है। यदि x 2800 और 3000 के बीच है, तो x के अंकों का योग क्या है?

Solution:

LCM of 2, 3, 4, 5, 6, and 7 = 420

Hence

$x = 420k + 1$

$x = 420 \times 7 + 1 = 2941$ (On putting $k = 7$)

Hence the sum of the digits of $x = 2+9+4+1 = 16$

9. Suppose that x is the largest four-digit number which, when divided by 7, 8, and 11, leaves 4, 5, and 8 as remainders respectively. When x is divided by $(7 + 8 + 11)$, what is the remainder?

मान लीजिए कि x चार अंकों की सबसे बड़ी संख्या है जिसे 7, 8, और 11 से विभाजित करने पर क्रमश: 4, 5, और 8 शेष बचता है। जब x को $(7 + 8 + 11)$ से विभाजित किया जाता है तो शेषफल क्या होता है?

Solution:

$(7 - 4) = (8 - 5) = (11 - 8) = 3$

LCM(7, 8, 11)$k - 3$

$x = 616k - 3$

For $k = 16$

largest four-digit number: $x = 9853$

$$\frac{9853}{7+8+11} = \frac{9853}{26}$$

Remainder = 25

10. Which is the largest six-digit number, which when divided by 12, 15, 20, 24, and 30, leaves the remainders 8, 11, 16, 20, and 26 respectively?

छह अंकों की सबसे बड़ी संख्या कौन सी है, जिसे 12, 15, 20, 24, और 30 से विभाजित करने पर क्रमश: 8, 11, 16, 20, और 26 शेष बचता है:

Solution:

12-84,15-11 = 4, 2016 = 4, 24-20 = 4,

30-26=4

LCM of 12, 15, 20, 24, and 30 = 120

Required number = $120k - 4$

For the largest six-digit number

$= 120 \times 8333 - 4$

$= 999956$

11. When 12, 16, 18, 20, and 25 divide the least number x, the remainder in each case is 4, but x is divisible by 7. What is the digit at the thousands' place in x?

जब 12, 16, 18, 20, और 25 सबसे छोटी संख्या x को विभाजित करते हैं, तो प्रत्येक स्थिति में शेषफल 4 होता है लेकिन x 7 से विभाज्य है। x में हजार के स्थान पर अंक क्या है?

Solution:

LCM of 12, 16, 18, 20, and 25 = 3600

$x = 3600k + 4$

For $k = 5$, the number divisible by 7

$x = 18004$

The digit at the thousands' place is 8.

12. What is the sum of the digits of the least number which, when divided by 15, 18, and 36, leaves the same remainder 9 in each case and is divisible by 11?

सबसे छोटी संख्या के अंकों का योग क्या है जिसे 15, 18, और 36 से विभाजित करने पर प्रत्येक स्थिति में समान शेषफल 9 बचता है और 11 से विभाज्य होता है?

Solution:

The least number

= (LCM of 15, 18, and 36)$k + 9$

$= 180k + 9$

$180k + 9$ is divisible by 11.

For $k = 6$, Number $= 180 \times 6 + 9 = 1089$

Sum of the digits $= 1 + 0 + 8 + 9 = 18$

13. What is the least number which, when divided by 15, 18, and 36, leaves the same remainder 9 in each case and is divisible by 11?

वह सबसे छोटी संख्या कौन सी है जिसे 15, 18 और 36 से

विभाजित करने पर प्रत्येक स्थिति में समान शेषफल 9 बचता है और 11 से विभाज्य हो जाता है?

Solution:

L.C.M of 15, 18, and 36 :

$15 = 3 \times 5$

$18 = 2 \times 3 \times 3,$

$36 = 2 \times 2 \times 3 \times 3$

LCM = 180

Required number $= 180k + 9$

$180 \times 6 + 9 = 1089$

Hence, on putting k = 6, the number is divisible by 11.

14. Let x be the least number divisible by 13, such that when x is divided by 4, 5, 6, 7, 8, and 12, the remainder in each case is 2. The sum of the digits of x is:
मान लीजिए कि x 13 से विभाज्य सबसे छोटी संख्या है, जैसे कि जब x को 4, 5, 6, 7, 8, और 12 से विभाजित किया जाता है, तो प्रत्येक स्थिति में शेषफल 2 होता है। x के अंकों का योग है:

Solution:

L.C.M of 4, 5, 6, 7, 8, and 12 = 840

$x = 840k + 2$

The number is exactly divisible by 13.

On putting $k = 3$,

the required number $= 840 \times 3 + 2 = 2522$

Sum of digits $= 2 + 5 + 2 + 2 = 11$

15. What is the least number which, when divided by 12, 18, 24, and 28, we obtain 5 as the remainder in each case?
वह सबसे छोटी संख्या कौन सी है जिसे 12, 18, 24 और 28 से विभाजित करने पर प्रत्येक स्थिति में शेषफल 5 प्राप्त होता है?

Solution:

On taking L.C.M of 12, 18, 24, and 28 :
$\text{LCM} = 2 \times 2 \times 2 \times 3 \times 3 \times 7$
$\text{LCM} = 8 \times 9 \times 7 = 72 \times 7 = 504$
The Required number $= 504 + 5 = 509$

Type 2 (Problem Based on HCF)

16. Find the greatest number that divides 556, 763, and 349 and leaves 4 as remainders respectively.
वह अधिकतम संख्या ज्ञात करें जिससे 556, 763 तथा 349 में भाग देने पर प्रत्येक स्थिति में 4 शेष बचता है?

Solution:

Greatest number = HCF(552, 759, 345)

207

↓

69×3 HCF = 69

OR

only even nos will give even HCF

$\therefore$ only option (a) Possible. 69

17. A farmer has 945 cows and 2475 sheep. He forms them into flocks, keeping cows and sheep separate and having the same number of animals in each flock. If these flocks are as large as possible, then the maximum number of animals in each flock and total number of flocks required for the purpose are:
एक किसान के पास 945 गायें तथा 2475 भेड़े हैं। वह उन्हें झुंडों में इस तरह बाँटता है कि प्रत्येक झुंड में बराबर पशु हो और गाय तथा भेंड़े अलग-अलग हों। यदि ये झुंड अधिक बड़ा हो, तो एक झुंड में कितने पशु हैं तथा झुंडों की संख्या कितनी है?

Solution:

HCF ⟶ 15x $\quad$ 9×5 absent

46x $\quad$ 45 ⟶ 9 & 5 both present

Maximum number of animals in each flock

= HCF of 945 and 2475

HCF = 45

Total number of flocks $= \frac{945}{45} + \frac{2475}{45} =$

$21 + 55 = 76$

18. In finding the HCF of two numbers by the division method, four successive quotients are 4, 3, 6, and 5 respectively, and the final divisor is 12. What are the two numbers?
विभाजन विधि द्वारा दो संख्याओं के HCF ज्ञात करने में चार क्रमिक भागफल क्रमश: 4, 3, 6 और 5 हैं। और अंतिम भाजक 12 है। दो संख्याएँ क्या हैं?

Solution:

Two numbers = 1176, 5076

1176)5076(4

372)1176(3

60)372(6

12)60(5

x

19. Two numbers are in the ratio 6:11. If their HCF is 28, then the sum of these two numbers will be
दो संख्याएं 6: 11 के अनुपात में है। यदि उनकी HCF 28 है, तो इन दो संख्याओं का योग होगा ।

Solution:

Sum → 28(6 + 17)→28×17→476

20. The product of two numbers is 11830 and their HCF is 13. How many pairs of such numbers can be formed?
दो संख्यओं का गुणनफल 11830 है और HCF 13 है। ऐसी संख्याओं के कितने युग्म संभव है?

Solution:

Nos $13x$, $13y$ $\quad xy = 70$

$13x \times 13y = 11890$

1×70, 2×35, 5×14, 7×10

4 pairs

21. When 7897, 8110 and 8536 are divided by the greatest number x, then the remainder in each case is the same. The sum of the digits of x is:
जब 7897, 8110 और 8536 को सबसे बड़ी संख्या x से विभाजित किया जाता है, तो प्रत्येक स्थिति में शेष समान होता है। तो x के अंकों का योग है:

Solution:

Take diff → (213, 426, 630, 639)→HCF

HCF = 213

$2 + 1 + 3 = 6$

22. If r is the remainder when each of 6454, 7306 and 8797 is divided by the greatest number $d(d > 1)$, then $(d - r)$ is equal to:
यदि r वह शेष है, जो 6454, 7306 और 8797 को बड़ी से बड़ी संख्या $d(d > 1)$ से विभाजित करने पर प्राप्त होता है, तो $(d - r)$ का मान ज्ञात कीजिए।

Solution:

Greatest no HCF(852, 1491, 2343)

HCF = 213

$\frac{6454}{213} = R = 64$

$213 - 64 = 149$

23. A fruit vendor brings 1092 apples and 3432 oranges to a market. He arranges them in heaps of equal number of oranges as well as apples such that every heap consists of the maximum possible number of the fruits. What is this number?
एक फल विक्रेता बाजार में 1092 सेब और 3432 संतरे लाता है। वह उन्हें समान संख्या में संतरे और सेब के ढेर में व्यवस्थित करता है ताकि प्रत्येक ढेर में अधिकतम संभव संख्या में फल हों। यह संख्या क्या है?

Solution:

HCF of 1092 Apples and 3432 Oranges:

1092)3432(7
3276
x156)1092(7
1092
xxx

Hence, the maximum possible number of fruits is 156.

Type 3 (Misc)

24. LCM and HCF of two numbers are 90 and 15, respectively. If the sum of the two numbers is 75, then find the greater number.
दो संख्याओं का LCM और HCF क्रमश: 90 और 15 हैं। यदि दोनों संख्याओं का योग 75 है, तो बड़ी संख्या ज्ञात कीजिए।

Solution:

Let the number be x and y.

According to the question,

Product of the number = LCM × HCF

$xy = 90 \times 15 = 1350$ (Given)

And

$x + y = 75$ (Given) . . . (i)

$x - y = \sqrt{(x+y)^2 - 4xy}$

$x - y = \sqrt{(75)^2 - 4 \times 1350}$

$x - y = \sqrt{5625 - 5400}$

$x - y = \sqrt{225}$

From equation (i) and (ii),

$x + y = 75, \quad x - y = 15$

$x = 45, \quad y = 30$

Hence, the greater number is 45.

25. Three numbers are in the proportion of 3 : 8 : 15 and their LCM is 8280. What is their HCF?
तीन संख्याएँ 3 : 8 : 15 के अनुपात में हैं और उनका LCM 8280 है। उनका HCF क्या है?

Solution:

Let the numbers be $3x$, $8x$, and $15x$ respectively.

Then their LCM $= 120x$

$120x = 8280$

$x = \frac{8280}{120} = 69$

Numbers are (3×69), (8×69), and (15×69)

Then HCF $= 69$

26. LCM of two numbers is 22 times their HCF. If one of the numbers is 132 and the sum of LCM and HCF is 276, then what is the other number?
दो संख्याओं का LCM उनके HCF का 22 गुना है। यदि एक संख्या 132 है और एलसीएम और एचसीएफ का योग 276 है, तो दूसरी संख्या क्या है?

Solution:

Let, second number = b.

Given that LCM $= 22 \times$ HCF . . . (i)

First number (a) = 132

LCM + HCF = 276 . . . (ii)

According to the question,

22HCF + HCF = 276

23HCF = 276

HCF = 12

LCM = 22 × 12

Product of two numbers = LCM × HCF

$a \times b = 22 \times 12 \times 12$

$132 \times b = 22 \times 12 \times 12$

$b = 24$

Hence, Option (a) is correct.

27. The LCM of two numbers x and y is 204 times their HCF. If their HCF is 12 and the difference between the numbers is 60, then $x + y = ?$
दो संख्याओं x और y का LCM उनके HCF का 204 गुना है। यदि उनका HCF 12 है और संख्याओं के बीच का अंतर 60 है, तो $x + y = ?$

Solution:

LCM of x and $y = 204 \times 12$

LCM $\times$ HCF = Product of the numbers

$204 \times 12 \times 12 = xy$

$x - y = 60$

$(x+y)^2 = (x-y)^2 + 4xy$

$= 3600 + 4 \times 204 \times 144$

$= 3600 + 117504 = 121104$

$x + y = \sqrt{121104} = 348$

28. The HCF of two numbers is 21 and their LCM is 221 times the HCF. If one of the numbers lies between 200 and 300, then the sum of the digits of the other number is:
दो संख्याओं का HCF 21 है और उनका LCM, HCF का 221 गुना है। यदि कोई एक संख्या 200 और 300 के बीच है, तो दूसरी संख्या के अंकों का योग है:

Solution:

Let those two numbers be $a = 21x$ and $b =$

$21y$.

$\text{LCM} = 21 \times 221$

$21xy = 21 \times 221$

$xy = 221 \quad x = 13, \quad y = 17$

$a = 21 \times 13 = 273, \quad b = 21 \times 17 = 357$

Sum of the digits of $b = 3 + 5 + 7 = 15$

29. The difference of two numbers is 15 and HCF and LCM of two numbers are 3 and 108 respectively. Find the difference between their reciprocals.
दो संख्याओं का अंतर 15 और HCF और LCM है। दो संख्याएँ क्रमश: 3 और 108 हैं। उनके व्युत्क्रमों के बीच अंतर ज्ञात कीजिए।

Solution:

Let the numbers be x and y respectively.

According to the question,

$x - y = 15 \quad ...(i)$

$\text{LCM} \times \text{HCF} = x \times y$

$108 \times 3 = x \times y \quad ...(ii)$

By dividing equation (i) from equation (ii)

$= \frac{x-y}{xy} = \frac{15}{108 \times 3}$

$\frac{1}{y} - \frac{1}{x} = \frac{5}{108}$

30. Suppose that x and y are such three-digit numbers such that their difference is 729 and HCF is 81. Then what is the value of $x + y$?
मान लीजिए कि x और y तीन अंकों की ऐसी संख्याएँ हैं जिनका अंतर 729 और HCF है। 81 है तो $x + y$ का मान क्या है?

Solution:

$\text{HCF} = 81$

$x = 81A, \quad y = 81B$

$x - y = 729$

$81A - 81B = 729$

$A - B = 9$

$A - 9 = B$ (A greater than 9)

81A & 81B (Three Digit number less than 1000)

$1000/81 \sim 12.xyz$

A will be less than 12 & more than 9

At $A = 11$, B will be 2

$x + y = 81(a+b) = 81 \times 13 = 1053$

15 Algebra

Square Formulae / वर्गाकार सूत्र

- $a^2 - b^2 = (a+b)(a-b)$
- $(a+b)^2 = a^2 + b^2 + 2ab$
- $(a-b)^2 = a^2 + b^2 - 2ab$
- $ab = \dfrac{(a+b)^2 - (a^2+b^2)}{2}$
- $2(a^2+b^2) = (a+b)^2 + (a-b)^2$
- $4ab = (a+b)^2 - (a-b)^2$
- $(a+b)^2 = (a-b)^2 + 4ab$
- $(a-b)^2 = (a+b)^2 - 4ab$
- $a^2 + b^2 = (a+b)^2 - 2ab = (a-b)^2 + 2ab$
- $(a^2-b^2)^2 = a^4 + b^4 - 2a^2b^2$
- $a^4 + b^4 = (a^2+b^2)^2 - 2a^2b^2$
- $a^4 + b^4 = (a^2-b^2)^2 + 2a^2b^2$
- $(ax+by)^2 + (ay-bx)^2 = (a^2+b^2)(x^2+y^2)$
- $\dfrac{(a+b)}{(a-b)} + \dfrac{(a-b)}{(a+b)} = \dfrac{2(a^2+b^2)}{(a^2-b^2)}$
- $\dfrac{(a+b)}{(a-b)} + \dfrac{(a-b)}{(a+b)} = \dfrac{4ab}{(a^2-b^2)}$

Cube Formulae / घन सूत्र

- $(a+b)^3 = a^3 + b^3 + 3a^2b + 3ab^2$
- $(a+b)^3 = a^3 + b^3 + 3ab(a+b)$
- $a^3 + b^3 = (a+b)^3 - 3ab(a+b)$
- $a^3 + b^3 = (a+b)[(a+b)^2 - 3ab]$
- $a^3 + b^3 = (a+b)(a^2+b^2-ab)$
- $a^2 + b^2 - ab = \dfrac{(a^3+b^3)}{(a+b)}$
- $a + b = \dfrac{(a^3+b^3)}{(a^2+b^2-ab)}$
- $(a-b)^3 = a^3 - b^3 - 3a^2b + 3ab^2$
- $(a-b)^3 = a^3 - b^3 - 3ab(a-b)$
- $a^3 - b^3 = (a-b)^3 + 3ab(a-b)$
- $a^3 - b^3 = (a-b)(a^2+b^2+ab)$
- $a^3 - b^3 = (a-b)[(a-b)^2 + 3ab]$
- $a^2 + b^2 + ab = \dfrac{(a^3-b^3)}{(a-b)}$
- $a - b = \dfrac{(a^3-b^3)}{(a^2+b^2+ab)}$

Some Results of $x + \frac{1}{x}$ / $x + \frac{1}{x}$ के कुछ परिणाम

- If $x + \dfrac{1}{x} = k$, then $\sqrt{x} + \dfrac{1}{\sqrt{x}} = \pm\sqrt{k+2}$
- If $x + \dfrac{1}{x} = k$, then $\sqrt{x} - \dfrac{1}{\sqrt{x}} = \pm\sqrt{k-2}$
- If $x + \dfrac{1}{x} = k$, then $x^2 + \dfrac{1}{x^2} = k^2 - 2$
- If $x - \dfrac{1}{x} = k$, then $x^2 + \dfrac{1}{x^2} = k^2 + 2$
- If $x + \dfrac{1}{x} = k$, then $x - \dfrac{1}{x} = \pm\sqrt{k^2-4}$

- If $x - \frac{1}{x} = k$, then $x + \frac{1}{x} = \pm\sqrt{k^2 + 4}$
- If $x + \frac{1}{x} = k$, then $x^3 + \frac{1}{x^3} = k^3 - 3k$
- If $x - \frac{1}{x} = k$, then $x^3 - \frac{1}{x^3} = k^3 + 3k$

Some Higher Power Results / कुछ उच्च शक्ति परिणाम

- $x^4 + \frac{1}{x^4} = \left(x^2 + \frac{1}{x^2}\right)^2 - 2$
- $x^4 - \frac{1}{x^4} = \left(x^2 + \frac{1}{x^2}\right)\left(x + \frac{1}{x}\right)\left(x - \frac{1}{x}\right)$
- $x^5 + \frac{1}{x^5} = \left(x^3 + \frac{1}{x^3}\right)\left(x^2 + \frac{1}{x^2}\right) - \left(x + \frac{1}{x}\right)$
- $x^5 - \frac{1}{x^5} = \left(x^3 - \frac{1}{x^3}\right)\left(x^2 + \frac{1}{x^2}\right) - \left(x - \frac{1}{x}\right)$
- $x^7 + \frac{1}{x^7} = \left(x^4 + \frac{1}{x^4}\right)\left(x^3 + \frac{1}{x^3}\right) - \left(x + \frac{1}{x}\right)$
- $x^7 - \frac{1}{x^7} = \left(x^4 + \frac{1}{x^4}\right)\left(x^3 - \frac{1}{x^3}\right) + \left(x - \frac{1}{x}\right)$
- $x^8 - \frac{1}{x^8} = \left(x^4 + \frac{1}{x^4}\right)\left(x^2 + \frac{1}{x^2}\right)\left(x + \frac{1}{x}\right)\left(x - \frac{1}{x}\right)$

Some More Results / कुछ और परिणाम

- $(x^2 + y^2 + xy)(x^2 + y^2 - xy) = x^4 + y^4 + x^2y^2$
- $(x^2 + x + 1)(x^2 - x + 1) = x^4 + x^2 + 1$
- $(x + a)(x + b) = x^2 + ax + bx + ab$
- $1 + A + B + AB = (1 + A)(1 + B)$
- $(1+a)(1+b)(1+c) = 1+a+b+c+ab+bc+ca+abc$
- If $x^2 + y^2 + z^2 = 0$, then $x = 0,\ y = 0,\ z = 0$
- $(a + b + c)^2 = a^2 + b^2 + c^2 + 2(ab + bc + ca)$
  $a^2 + b^2 + c^2 = (a + b + c)^2 - 2(ab + bc + ca)$
  $ab + bc + ca = \frac{(a + b + c)^2 - (a^2 + b^2 + c^2)}{2}$
  $(a - b + c)^2 = a^2 + b^2 + c^2 - 2ab - 2bc + 2ca$
  $(a + b - c)^2 = a^2 + b^2 + c^2 + 2ab - 2bc - 2ca$

- $a^2 + b^2 + c^2 - ab - bc - ca = \frac{1}{2}[(a - b)^2 + (b - c)^2 + (c - a)^2]$ $\frac{1}{2}[2a^2 + 2b^2 + 2c^2 - 2ab - 2bc - 2ca]$
  $\frac{1}{2}[a^2 + b^2 - 2ab + b^2 + c^2 - 2bc + a^2 + c^2 - 2ca]$
  $\frac{1}{2}[(a - b)^2 + (b - c)^2 + (c - a)^2]$

- $a^3 + b^3 + c^3 - 3abc = (a+b+c)(a^2 + b^2 + c^2 - ab - bc - ca)$ $\frac{1}{2}(a + b + c)[(a - b)^2 + (b - c)^2 + (c - a)^2]$
  $\frac{a^3 + b^3 + c^3 - 3abc}{(a - b)^2 + (b - c)^2 + (c - a)^2} = \frac{a + b + c}{2}$
  $a^3 + b^3 + c^3 - 3abc = (a + b + c)[(a + b + c)^2 - 3(ab + bc + ca)]$
  $a^3 + b^3 + c^3 - 3abc = \frac{1}{2}(a+b+c)[3(a^2 + b^2 + c^2) - (a + b + c)^2]$

- $a+b+c-abc = (a+b+c)[(a+b+c)^2 - 3(ab+bc+ca)]$
  If $a^3 + b^3 + c^3 - 3abc = 0$ then $a^3 + b^3 + c^3 = 3abc = 0$
  OR $a^3 + b^3 + c^3 = 3abc = 0$

- $a^3 + b^3 + c^3 - 3abc = \frac{1}{2}(a + b + c)[(a - b)^2 + (b - c)^2 + (c - a)^2]$
  If $a^3 + b^3 + c^3 - 3abc = 0$
  OR $a^3 + b^3 + c^3 = 3abc$
  ↙ ↘
  a + b + c = 0   a = b = c

- $(a+b+c)(ab+bc+ca) - abc = (a+b)(b+c)(c+a)$
- $(a + b + c)(ab + bc + ca) = a^2b + b^2a + b^2c + bc^2 + a^2c + ac^2 + 3abc$
- $(a+b+c)(ab+bc+ca) - 3abc = a^2(b+c) + b^2(c+a) + c^2(a + b)$
- $(a + b + c)^3 = a^3 + b^3 + c^3 + 3(a + b)(b + c)(c + a)$
  $(a + b + c)^3 = a^3 + b^3 + c^3 + 3[a^2(b + c) + b^2(c + a) + c^2(a + b)] + 2abc = (a + b + c)^3 = a^3 + b^3 + c^3 + 3(a + b)(b + c)(c + a)$
- If $x + \frac{1}{y} = a,\ \ y + \frac{1}{z} = b,\ \ z + \frac{1}{x} = c$, then:
  $xyz + \frac{1}{xyz} = abc - (a + b + c)$

- If $x - \frac{1}{y} = a, \ y - \frac{1}{z} = b, \ z - \frac{1}{x} = c,$ then:
  $xyz - \frac{1}{xyz} = abc + (a + b + c)$

- If $a + \frac{1}{b} = b + \frac{1}{c} = c + \frac{1}{a}$ (where $a \neq b \neq c$), then: $abc = \pm 1$

Type 1

1. If $x + \frac{16}{x} = 8$, then the value of $x^2 + \frac{32}{x^2}$ is

यदि $x + \frac{16}{x} = 8$, तो $x^2 + \frac{32}{x^2}$ का मान है

Solution :

$x + \frac{16}{x} = 8$ (Given)

$\therefore x^2 - 8x + 16 = 0$

$(x - 4)^2 = 0$

$x = 4$

Hence,

$x^2 + \frac{32}{x^2} = 4^2 + \frac{32}{4^2}$

$= 16 + \frac{32}{16}$

$= 16 + 2 = 18$

2. If $(x - \frac{1}{x}) = \frac{11}{5}$, what is the value of $(x^3 + \frac{1}{x^3})$?

यदि $(x - \frac{1}{x}) = \frac{11}{5}$, तो $(x^3 + \frac{1}{x^3})$ का मान क्या है?

Solution :

Given, $(x - \frac{1}{x}) = \frac{11}{5}$

We know that:

If $x + \frac{1}{x} = a$, then $x^3 + \frac{1}{x^3} = a^3 - 3a$

$a = \frac{11}{5}$

$x^3 + \frac{1}{x^3} = \left(\frac{11}{5}\right)^3 - 3 \times \frac{11}{5}$

$= \frac{1331}{125} - \frac{33}{5}$

$= \frac{1331 - 825}{125}$

$= \frac{506}{125} = 4\frac{6}{125}$

3. If $x + \frac{1}{x} = -2\sqrt{3}$, what is the value of $x^5 + \frac{1}{x^5}$?

यदि $x + \frac{1}{x} = -2\sqrt{3}$, तो $x^5 + \frac{1}{x^5}$ का मान क्या है?

Solution :

Given, $x + \frac{1}{x} = -2\sqrt{3}$ ......(i)

$x^5 + \frac{1}{x^5} = ?$

On cubing both sides of equation (i), we get:

$x^3 + \frac{1}{x^3} + 3 \times x \times \frac{1}{x} \times \left(x + \frac{1}{x}\right) = -8 \times \sqrt{3}$

$x^3 + \frac{1}{x^3} - 6\sqrt{3} = -24\sqrt{3}$ ......(ii)

Again, on squaring both sides of equation (i), we get:

$x^2 + \frac{1}{x^2} + 2 = 12$

$\therefore x^2 + \frac{1}{x^2} = 10$ ......(iii)

We know that:

$x^5 + \frac{1}{x^5} = \left(x^3 + \frac{1}{x^3}\right) \times \left(x^2 + \frac{1}{x^2}\right) - \left(x + \frac{1}{x}\right)$

From equations (i), (ii), and (iii), we get:

$x^5 + \frac{1}{x^5} = \left(-18\sqrt{3}\right) \times 10 - (-2\sqrt{3})$

$= -180\sqrt{3} + 2\sqrt{3}$

$= -182\sqrt{3}$

4. If $(a + \frac{1}{a} + 3) = 6$ where a is a non-zero real number, then find the value of $a^2 + \frac{1}{a^2}$.

यदि $(a + \frac{1}{a} + 3) = 6$ जहां a एक गैर-शून्य वास्तविक संख्या है, तो $a^2 + \frac{1}{a^2}$ का मान ज्ञात करें।

Solution :

$(a + \frac{1}{a} + 3) = 6$

$\therefore a + \frac{1}{a} = 3$

On squaring both sides:

$a^2 + \frac{1}{a^2} + 2 \times a \times \frac{1}{a} = 3^2$

$a^2 + \frac{1}{a^2} + 2 = 9$

$\therefore a^2 + \frac{1}{a^2} = 9 - 2 = 7$

5. If $x^4 - \frac{1}{x^4} = 194$ and $x > 0$, then the value of $(x-2)^2$ is:

यदि $x^4 - \frac{1}{x^4} = 194$ और $x > 0$, तो $(x-2)^2$ का मान है:

Solution :

$x^4 - \frac{1}{x^4} = 194$

Adding 2 to both sides:

$x^4 + \frac{1}{x^4} + 2 = 194 + 2$

$(x^2 + \frac{1}{x^2})^2 = 196$

$\therefore x^2 + \frac{1}{x^2} = 14$

Adding 2 to both sides:

$x^2 + \frac{1}{x^2} + 2 = 14 + 2$

$(x + \frac{1}{x})^2 = 4^2$

$\therefore x + \frac{1}{x} = 4$

$\therefore x^2 + 1 = 4x$

$\therefore x^2 - 4x + 1 = 0$

$\therefore x^2 - 4x + 4 = 3$

$(x-2)^2 = 3$

6. If $5x - \frac{1}{4x} = 6$, $x > 0$, then find the value of $25x^2 - \frac{1}{16x^2}$.

Solution :

Given: $5x - \frac{1}{4x} = 6$ -----(i)

From the formula $(a+b)^2 = (a-b)^2 + 4ab$:

$(5x + \frac{1}{4x})^2 = (5x - \frac{1}{4x})^2 + 4 \times 5x \times \frac{1}{4x}$

$\Rightarrow (5x + \frac{1}{4x})^2 = (5x - \frac{1}{4x})^2 + 5$

$(5x + \frac{1}{4x})^2 = (6)^2 + 5$

$5x + \frac{1}{4x} = \sqrt{41}$ -----(ii)

$\therefore 25x^2 - \frac{1}{16x^2} = (5x - \frac{1}{4x})(5x + \frac{1}{4x})$

$= 6\sqrt{41}$ (from eqs. (i) and (ii))

7. If $\sqrt{x} - \frac{1}{\sqrt{x}} = \sqrt{3}$, then what is the value of $x^4 + \frac{1}{x^4}$?

Solution :

From the question: $\sqrt{x} - \frac{1}{\sqrt{x}} = \sqrt{3}$

On squaring both sides:

$(\sqrt{x} - \frac{1}{\sqrt{x}})^2 = (\sqrt{3})^2$

$x + \frac{1}{x} - 2 \times \sqrt{x} \times \frac{1}{\sqrt{x}} = 3$

$x + \frac{1}{x} = 3 + 2$

$x + \frac{1}{x} = 5$

$x^2 + \frac{1}{x^2} = 23$

$x^4 + \frac{1}{x^4} = 527$

8. If $(x^2 + \frac{1}{x^2}) = 23$, $x > 0$, what is the value of $(x^3 + \frac{1}{x^3})$?

Solution :

$x^2 + \frac{1}{x^2} = 23$

$x + \frac{1}{x} = 5$

$x^3 + \frac{1}{x^3} = 5^3 - 3 \times 5 = 110$

9. If $x + \frac{1}{x} = 3$, $x \neq 0$, then the value of $x^7 + \frac{1}{x^7}$ is:

Solution :

$x+\frac{1}{x}=3$

$x^3-\frac{1}{x^3}=3^3-3\times 3=18$

Again:

$x+\frac{1}{x}=3$

On squaring both sides:

$x^2+\frac{1}{x^2}=7$

On squaring both sides again:

$x^4+\frac{1}{x^4}=47$

$x^7+\frac{1}{x^7}=(x^4+\frac{1}{x^4})(x^3+\frac{1}{x^3})-(x+\frac{1}{x})$

$=47\times 18-3$

$=843$

10. If $x=4+\sqrt{15}$, what is the value of $(x^2+\frac{1}{x^2})$?

Solution :

Given: $x=4+\sqrt{15}$

$\frac{1}{x}=\frac{1}{4+\sqrt{15}}=\frac{1}{4+\sqrt{15}}\times\frac{4-\sqrt{15}}{4-\sqrt{15}}=4-\sqrt{15}$

$x+\frac{1}{x}=4+\sqrt{15}+4-\sqrt{15}$

$x+\frac{1}{x}=8$

On squaring both sides:

$x^2+\frac{1}{x^2}+2\times x\times\frac{1}{x}=64$

$x^2+\frac{1}{x^2}=62$

11. If $x^4+\frac{1}{x^4}=727,\ x>1$, then what is the value of $x-\frac{1}{x}$?

Solution :

Given that $x^4+\frac{1}{x^4}=727,\ x>1$

Adding 2 on both sides:

$x^4+\frac{1}{x^4}+2=727+2$

$(x^2+\frac{1}{x^2})^2=27^2$

$x^2+\frac{1}{x^2}=27$

Subtracting 2 from both sides:

$x^2+\frac{1}{x^2}-2=27-2$

$x^2+\frac{1}{x^2}-2\times x\times\frac{1}{x}=25$

$(x-\frac{1}{x})^2=5^2$

Taking the square root of both sides:

$x-\frac{1}{x}=5$

Hence, option (b) is correct.

12. If $x-\frac{1}{x}=1$, then what is the value of $x^8+\frac{1}{x^8}$?

Solution :

From the question: $x-\frac{1}{x}=1$

On squaring both sides:

$(x-\frac{1}{x})^2=1^2$

$x^2+\frac{1}{x^2}-1=1$

$x^2+\frac{1}{x^2}=3$

Again, squaring both sides:

$(x^2+\frac{1}{x^2})^2=3^2$

$x^4+\frac{1}{x^4}+2=9$

$x^4+\frac{1}{x^4}=7$

Again, squaring both sides:

$(x^4+\frac{1}{x^4})^2=7^2$

$x^8+\frac{1}{x^8}=47$

13. If $x-\frac{1}{x}=\sqrt{77}$, then one of the values of $x^3+\frac{1}{x^3}$ is:

Solution :

Given: $x - \frac{1}{x} = \sqrt{77}$

From the formula $(a-b)^2 = (a+b)^2 - 4ab$:

$(\sqrt{77})^2 = (x + \frac{1}{x})^2 - 4$

$(x + \frac{1}{x})^2 = 77 + 4 = 81$

$x + \frac{1}{x} = 9$ -----(ii)

On cubing both sides of equation (ii):

$(x + \frac{1}{x})^3 = 9^3$

$x^3 + \frac{1}{x^3} = 9^3 - 3 \times 9 = 702$

14. If $\sqrt{x} + \frac{1}{\sqrt{x}} = \sqrt{6}$, then the value of $x^6 + \frac{1}{x^6}$ will be:

Solution :

$\sqrt{x} + \frac{1}{\sqrt{x}} = \sqrt{6}$

On squaring both sides:

$x + \frac{1}{x} + 2 = 6$

$x + \frac{1}{x} = 4$

$x^2 + \frac{1}{x^2} = 14$

Again, cubing both sides:

$x^6 + \frac{1}{x^6} = (14)^3 - 3 \times 14$

$= 2744 - 42 = 2702$

15. If $x^4 + \frac{1}{x^4} = 3842$, then the positive value of $x + \frac{1}{x}$ will be:

Solution :

$x^4 + \frac{1}{x^4} = 3842$

$x^2 + \frac{1}{x^2} = \sqrt{3842 + 2} = \sqrt{3844} = 62$

$x + \frac{1}{x} = \sqrt{62 + 2} = \sqrt{64} = 8$

Hence, $x + \frac{1}{x} = 8$

16. If $x^4 + \frac{1}{x^4} = 6887$, then the positive value of $x - \frac{1}{x}$ is:

Solution :

Given: $x^4 + \frac{1}{x^4} = 6887$

Adding 2 to both sides:

$x^4 + \frac{1}{x^4} + 2 = 6887 + 2 = 6889$

$(x^2 + \frac{1}{x^2})^2 = 83^2$

$x^2 + \frac{1}{x^2} = 83$

Subtracting 2 from both sides:

$x^2 + \frac{1}{x^2} - 2 = 83 - 2 = 81$

$(x - \frac{1}{x})^2 = 9^2$

$x - \frac{1}{x} = 9$

17. If $x^4 + \frac{16}{x^4} = 27217,\ x > 0$, then the value of $x + \frac{2}{x}$ is:

Solution :

Given: $x^4 + \frac{16}{x^4} = 27217$

On adding 8 to both sides:

$(x^2)^2 + \left(\frac{4}{x^2}\right)^2 + 8 = 27217 + 8$

$(x^2 + \frac{4}{x^2})^2 = 27225$

$x^2 + \frac{4}{x^2} = \sqrt{27225} = 165$

Again, adding 4 to both sides:

$(x^2)^2 + \left(\frac{2}{x}\right)^2 + 4 = 165 + 4 = 169$

$(x + \frac{2}{x})^2 = 169$

$x + \frac{2}{x} = \sqrt{169} = 13$

Hence, $x + \frac{2}{x} = 13$

18. If $x - \frac{1}{2x} = 4$, then the value of $8x^3 - \frac{1}{x^3}$ will be:

Solution :

Given: $x - \frac{1}{2x} = 4$

On multiplying by 2 on both sides:

$2x - \frac{1}{x} = 8$ ----(i)

On cubing both sides:

$8x^3 - \frac{1}{x^3} - 3 \times 2x \times \frac{1}{x}(2x - \frac{1}{x}) = 512$

$8x^3 - \frac{1}{x^3} = 512 + 48$

$8x^3 - \frac{1}{x^3} = 560$

19. If $x(3 - \frac{2}{x}) = \frac{3}{x}$, then the value of $x^3 - \frac{1}{x^3}$ is:

Solution :

$x(3 - \frac{2}{x}) = \frac{3}{x}$

$3x - 2 = \frac{3}{x}$

$x - \frac{1}{x} = \frac{2}{3}$

From the formula, if $x - \frac{1}{x} = K$:

$x^3 - \frac{1}{x^3} = K^3 + 3K$

Hence:

$x^3 - \frac{1}{x^3} = \left(\frac{2}{3}\right)^3 + 3 \times \frac{2}{3}$

$= \frac{8}{27} + 2 = \frac{62}{27}$

20. If $\sqrt{x} + \frac{1}{\sqrt{x}} = 3$, then the value of $x^3 + \frac{1}{x^3}$ is:

Solution :

Let $\sqrt{x} + \frac{1}{\sqrt{x}} = 3 = K$

$x + \frac{1}{x} = K^2 - 2$

$x + \frac{1}{x} = 3^2 - 2 = 7$

$x^3 + \frac{1}{x^3} = K^3 - 3K$

$= 7^3 - 3 \times 7$

$= 343 - 21$

$= 322$

21. If $a - \frac{1}{a} = 3$, then the value of $a^6 + \frac{1}{a^6}$ is:

Solution :

Given: $a - \frac{1}{a} = 3$ ----(i)

By squaring equation (i):

$a^2 + \frac{1}{a^2} - 2 = 9$

$a^2 + \frac{1}{a^2} = 11$ ----(ii)

By cubing equation (ii):

$a^6 + \frac{1}{a^6} + 3(a^2 + \frac{1}{a^2}) = 1331$

$a^6 + \frac{1}{a^6} = 1331 - 33 = 1298$

22. If $x^4 + \frac{1}{x^4} = \frac{257}{16}$, then find $\frac{8}{13}(x^3 + \frac{1}{x^3})$, where $x > 0$:

Solution :

Given: $x^4 + \frac{1}{x^4} = \frac{257}{16}$

$(x^2 + \frac{1}{x^2})^2 = \frac{257}{16} + 2 = \frac{257 + 32}{16} = \frac{289}{16}$

$x^2 + \frac{1}{x^2} = \frac{17}{4}$

$(x + \frac{1}{x})^2 = \frac{17}{4} + 2 = \frac{25}{4}$

$x + \frac{1}{x} = \frac{5}{2}$

$\therefore x^3 + \frac{1}{x^3} = \left(\frac{5}{2}\right)^3 - 3 \times \frac{5}{2}$

$= \frac{125}{8} - \frac{15}{2} = \frac{65}{8}$

$\therefore \frac{8}{13}(x^3 + \frac{1}{x^3}) = \frac{8}{13} \times \frac{65}{8} = 5$

23. If $2x - y = 2$ and $xy = \frac{3}{2}$, then what is the value of $\frac{x^3 - y^3}{8}$?

Solution :

Given: $2x - y = 2$ ----(i)

$xy = \frac{3}{2}, \frac{x^3 - y^3}{8} = ?$

Dividing equation (i) by 2 and cubing both sides:

$\left(x - \frac{y}{2}\right)^3 = 1^3$

$\frac{x^3 - y^3}{8} - 3 \times \frac{xy}{2} \times 1 = 1$

$\frac{x^3 - y^3}{8} = \frac{13}{4}$

24. If $4x - 3y = 12$ and $xy = 5$, then find the value of $\frac{16x^2 + 9y^2}{8}$:

Solution :

Given: $4x - 3y = 12$

Squaring both sides:

$16x^2 + 9y^2 - 24xy = 144$

$16x^2 + 9y^2 = 144 + 24xy = 144 + 120$

Dividing both sides by 8:

$\frac{16x^2 + 9y^2}{8} = 3$

25. If $a^2 + b^2 = 65$ and $ab = 8$, then find the value of $a^2 - b^2$:

Solution :

Given: $a^2 + b^2 = 65,\ ab = 8$

Assuming $a = 8,\ b = 1$:

$a^2 - b^2 = 8^2 - 1^2$

$= 64 - 1 = 63$

26. If $x^4 - 79x^2 + 1 = 0$, then the value of $x + \frac{1}{x}$ can be:

Solution :

$x^4 - 79x^2 + 1 = 0$

Dividing by x^2 on both sides:

$x^2 - 79 + \frac{1}{x^2} = 0$

$x^2 + \frac{1}{x^2} = 79$

Adding 2 to both sides:

$x^2 + \frac{1}{x^2} + 2 = 79 + 2$

$(x + \frac{1}{x})^2 = 81$

$x + \frac{1}{x} = 9$

27. If $a^3 + 3a^2 + 9a = 1$, then what is the value of $a^3 + \frac{3}{a}$?

Solution :

Given: $a^3 + 3a^2 + 9a = 1$

$\therefore a^2 + 3a + \frac{9}{a} = \frac{1}{a}$

Multiplying both sides by $(a - 3)$:

$(a - 3)(a^2 + 3a + 9) = \frac{1}{a}(a - 3)$

$a^3 - 3^3 = \frac{a}{a} - \frac{3}{a}$

$a^3 + \frac{3}{a} = 1 + 27 = 28$

28. If $x + y + 3 = 0$, then find the value of $x^3 + y^3 - 9xy + 9$:

Solution :

Given: $x + y + 3 = 0$

$\therefore x + y = -3$

Cubing both sides:

$(x + y)^3 = (-3)^3$

$x^3 + y^3 + 3xy(x + y) = -27$

$x^3 + y^3 + 3xy(-3) = -27$

$x^3 + y^3 - 9xy = -27$

$x^3 + y^3 - 9xy + 9 = -27 + 9 = -18$

29. If $x^2 - 3x + 1 = 0$, then the value of $x^3 + y^3 - 9xy + 9$:

Solution :

Given: $x + y + 3 = 0$

$\therefore x + y = -3$

Cubing both sides:
$(x+y)^3 = (-3)^3$
$x^3 + y^3 + 3xy(x+y) = -27$
$x^3 + y^3 + 3xy(-3) = -27$
$x^3 + y^3 - 9xy = -27$
$x^3 + y^3 - 9xy + 9 = -27 + 9 = -18$

30. If $x^2 - 3x + 1 = 0$, then the value of $\dfrac{x^4 + \frac{1}{x^2}}{x^2 + 5x + 1}$:

Solution :

$x^2 - 3x + 1 = 0$

$\therefore x(x - 3 + \frac{1}{x}) = 0$

$x + \frac{1}{x} = 3$

Cubing both sides:

$x^3 + \frac{1}{x^3} + 3 \times 3 = 27$

$x^3 + \frac{1}{x^3} = 27 - 9 = 18$

$$\frac{x^4 + \frac{1}{x^2}}{x^2 + 5x + 1} = \frac{x(x^3 + \frac{1}{x^3})}{x(x + \frac{1}{x}) + 5}$$

$\frac{18}{3+5} = \frac{9}{4}$

31. If $(3x+2y)^3 + (3x-2y)^3 = 3kx(3x^2+4y^2)$, then the value of k will be:

Solution :

$(3x+2y)^3 + (3x-2y)^3 = 3kx(3x^2+4y^2)$

Using the identity $a^3 + b^3 = (a+b)(a^2 - ab + b^2)$:

$(3x+2y+3x-2y)[(3x+2y)^2 + (3x-2y)^2 - (3x+2y)(3x-2y)]$

$= 3kx(3x^2+4y^2)$

$6x[(9x^2 + 4y^2) \times 2 - (9x^2 - 4y^2)] = 3kx(3x^2+4y^2)$

$6x[18x^2 + 8y^2 - 9x^2 + 4y^2] = 3kx(3x^2+4y^2)$

$6x[9x^2 + 12y^2] = 3kx(3x^2+4y^2)$

$3 \times 6x(3x^2+4y^2) = 3kx(3x^2+4y^2)$

$6x = kx$

$k = 6$

32. If $2x^2 - 7x + 5 = 0$, then what is the value of $x^3 + \frac{125}{8x^3}$?

Solution :

Given: $2x^2 - 7x + 5 = 0$

Dividing by $2x$ on both sides:

$x - \frac{7}{2} + \frac{5}{2x} = 0$

$x + \frac{5}{2x} = \frac{7}{2}$

Cubing both sides:

$x^3 + \frac{125}{8x^3} + 3 \times x \times \frac{5}{2x}(x + \frac{5}{2x}) = \frac{343}{8}$

$x^3 + \frac{125}{8x^3} + \frac{15}{2} \times \frac{7}{2} = \frac{343}{8}$

$x^3 + \frac{125}{8x^3} = \frac{343}{8} - \frac{105}{4}$

$= \frac{343}{8} - \frac{210}{8} = \frac{133}{8}$

$= 16\frac{5}{8}$

33. If $4x^4 - 37x^2 + 9 = 0$, $x > \sqrt{\frac{3}{2}}$, then what is the value of $8x^3 - \frac{27}{x^3}$?

Solution :

$4x^4 - 37x^2 + 9 = 0$

Let us consider $x^2 = a$:

$4a^2 - 37a + 9 = 0$

$4a^2 - 36a - a + 9 = 0$

$4a(a-9) - 1(a-9) = 0$

$(4a-1)(a-9) = 0$

$a = \frac{1}{4}$ or $a = 9$

On putting the value of $a = x^2$:

$x^2 = \frac{1}{4}$ (not acceptable as $x > \sqrt{\frac{3}{2}}$)

$x^2 = 9 \implies x = 3$ or $x = -3$

$(x = -3$ not valid as $x > \sqrt{\frac{3}{2}})$

Then put $x = 3$:

$8x^3 - \frac{27}{x^3} = ?$

$? = 8 \times (3)^3 - \frac{27}{(3)^3}$

$? = 8 \times 27 - \frac{27}{27}$

$? = 216 - 1$

$? = 215$

34. If $x^3 + y^3 = 468$ and $x + y = 12$, then the value of $x^4 + y^4$ will be:

Solution :

Given: $x^3 + y^3 = 468$

$(x + y)(x^2 + y^2 - xy) = 468$

$x^2 + y^2 - xy = \frac{468}{12}$ [on putting the value of $x + y = 12$]

$x^2 + y^2 - xy = 39$ ----(i)

And $x + y = 12$

Squaring both sides:

$x^2 + y^2 + 2xy = 144$ ----(ii)

Subtracting equation (i) from equation (ii):

$xy = 144 - 39 = 105 \div 3 = 35$ ----(iii)

On putting the value of xy in equation (ii):

$x^2 + y^2 + 2 \times 35 = 144$

$x^2 + y^2 = 144 - 70 = 74$

Again, squaring both sides:

$x^4 + y^4 + 2x^2y^2 = (74)^2$

On putting the value of $xy = 35$:

$x^4 + y^4 + 2 \times (35)^2 = 5476$

$x^4 + y^4 = 5476 - 2450 = 3026$

Hence, $x^4 + y^4 = 3026$

35. If $x^2 - 3\sqrt{2}x + 1 = 0$, then what is the value of $x^3 + \frac{1}{x^2}$?

Solution :

$x^2 - 3\sqrt{2}x + 1 = 0$

$x^2 + 1 = 3\sqrt{2}x$

Dividing both sides by x:

$x + \frac{1}{x} = 3\sqrt{2}$

Cubing both sides:

$x^3 + \frac{1}{x^3} + 3(x + \frac{1}{x}) = 54\sqrt{2}$

$x^3 + \frac{1}{x^3} + 3 \times 3\sqrt{2} = 54\sqrt{2}$

$x^3 + \frac{1}{x^3} = 54\sqrt{2} - 9\sqrt{2}$

$x^3 + \frac{1}{x^3} = 45\sqrt{2}$

36. If $x^2 - 3x + 1 = 0$, then the value of $2(x^8 + \frac{1}{x^8}) - 5(x^2 + \frac{1}{x^2})$ is:

Solution :

Given: $x^2 - 3x + 1 = 0$

$x + \frac{1}{x} = 3$

Squaring both sides:

$x^2 + \frac{1}{x^2} = 9 - 2 = 7$ ----(ii)

Again, squaring both sides:

$x^8 + \frac{1}{x^8} = 2209 - 2 = 2207$ ----(iii)

From equations (ii) and (iii):

$2(x^8 + \frac{1}{x^8}) - 5(x^2 + \frac{1}{x^2}) = 4414 - 35 = 4379$

Hence, $2(x^8 + \frac{1}{x^8}) - 5(x^2 + \frac{1}{x^2}) = 4379$

37. If $a^4 + 1 = \frac{a^2}{b^2}(4b^2 - b^4 - 1)$, then what is the value of $a + b^2$?

Solution :

$a^4+1=\frac{a^2}{b^2}(4b^2-b^4-1)$

$a^2+\frac{1}{a^2}=4-b^2-\frac{1}{b^2}$

$a^2+\frac{1}{a^2}-2+b^2+\frac{1}{b^2}-2=0$

$(a-\frac{1}{a})^2+(b-\frac{1}{b})^2=0$

$a^2-1=0,\ b^2-1=0$

$a=1,\ b=1$

According to the question:

$a^2+b^2=1+1=2$

38. If $x^8-2599x^4+1=0$, then the positive value of $x-\frac{1}{x}$ will be:

Solution :

Given, $x^8-2599x^4+1=0$

$x^8+1=2599x^4$

On multiplying by $\frac{1}{x^4}$ both sides:

$x^4+\frac{1}{x^4}=2599$

On adding 2 to both sides:

$x^4+\frac{1}{x^4}+2=2599+2$

$(x^2+\frac{1}{x^2})^2=51^2$

$x^2+\frac{1}{x^2}=51$

On subtracting 2 from both sides:

$x^2+\frac{1}{x^2}-2=51-2=49$

$(x-\frac{1}{x})^2=7^2$

$x-\frac{1}{x}=7$

39. If $x+y=1$ and $xy(xy-2)=12$, then the value of x^4+y^4 is:

Solution :

$x+y=1$

Squaring both sides:

$x^2+y^2+2xy=1$

$x^2+y^2=1-2xy$

Again, squaring both sides:

$x^4+y^4+2x^2y^2=1+4x^2y^2-4xy$

$x^4+y^4=1+2xy(xy-2)$ $[\because xy(xy-2)=12]$

$=1+2\times 12$

$=25$

40. If $\frac{6x}{(2x^2+5x-2)}=1,\ x>0$, then the value of $x^3+\frac{1}{x^3}$ is:

Solution :

Given: $\frac{6x}{(2x^2+5x-2)}=1$

$\frac{6}{(2x+5-\frac{2}{x})}=1$

$2(x-\frac{1}{x})=1$

$x-\frac{1}{x}=\frac{1}{2}$

$\therefore x+\frac{1}{x}=\sqrt{(x-\frac{1}{x})^2+4}$

$=\sqrt{\left(\frac{1}{2}\right)^2+4}=\frac{\sqrt{17}}{2}$

$\therefore x^3+\frac{1}{x^3}=(x+\frac{1}{x})[x^2+\frac{1}{x^2}-1]$

$=(x+\frac{1}{x})[(x+\frac{1}{x})^2-3]$

$=\frac{\sqrt{17}}{2}\times\left(\frac{17-12}{4}\right)$

$=\frac{5\sqrt{17}}{8}$

41. If $x^4-6x^2-1=0$, then the value of $x^6-5x^2+\frac{5}{x^2}-\frac{1}{x^6}+5$ is:

Solution :

$x^6 - \frac{1}{x^6} - 5(x^2 - \frac{1}{x^2}) + 5$ ----(A)

$\because x^4 - 6x^2 - 1 = 0$

$x^4 - 1 = 6x^2$

$x^2 - \frac{1}{x^2} = 6$ ----(B)

$(x^2 - \frac{1}{x^2})^3 = 6^3$

$x^6 - \frac{1}{x^6} - 3(x^2 - \frac{1}{x^2}) = 216$

$x^6 - \frac{1}{x^6} - 3 \times 6 = 216$

$x^6 - \frac{1}{x^6} = 216 + 18$

$x^6 - \frac{1}{x^6} = 234$ ----(C)

By substituting the values from equations (B) and (C) into equation (A):

$= 234 - 5(6) + 5$

$= 234 - 30 + 5$

$= 209$

42. If $x^2 - 2\sqrt{5}x + 1 = 0$, then what is the value of $x^5 + \frac{1}{x^5}$?

Solution :

$x^2 - 2\sqrt{5}x + 1 = 0$

$x + \frac{1}{x} = 2\sqrt{5}$

$\therefore x^5 + \frac{1}{x^5} = (x^2 + \frac{1}{x^2})(x^3 + \frac{1}{x^3}) - (x + \frac{1}{x})$

$x^2 + \frac{1}{x^2} = (2\sqrt{5})^2 - 2 = 18$

$x^3 + \frac{1}{x^3} = (2\sqrt{5})^2 = 34\sqrt{5}$

$(x^2 + \frac{1}{x^2})(x^3 + \frac{1}{x^3}) = 18 \times 34\sqrt{5}$

$\therefore x^5 + \frac{1}{x^5} = 612\sqrt{5} - 2\sqrt{5}$

$= 610\sqrt{5}$

43. $A = \frac{x^8 - 1}{x^4 + 1}$ and $B = \frac{y^4 - 1}{y^4 + 1}$. If $x = 2$ and $y = 9$, then what is the value of $A^2 + 2AB + AB^2$?

Solution :

Given:

$A = \frac{x^8 - 1}{x^4 + 1}$ & $B = \frac{y^4 - 1}{y^2 + 1}$

From formula: $(a - b)(a + b) = a^2 - b^2$

By solving the equations for A and B:

$A = \frac{(x^4 - 1)(x^4 + 1)}{(x^4 + 1)}$ & $B = \frac{(y^2 - 1)(y^2 + 1)}{(y^2 + 1)}$

$A = (x^4 - 1)$ & $B = (y^2 - 1)$

If $x = 2$ and $y = 9$:

$A = 2^4 - 1 = 15$

$B = 9^2 - 1 = 80$

By putting the values of A and B:

$A^2 + AB^2 + 2AB$

$\Rightarrow (15)^2 + 15 \times (80)^2 + 2 \times 15 \times 80$

$= 225 + 96000 + 2400 = 98625$

44. If $x^2 - 16x + 59 = 0$, then what is the value of $(x - 6)^2 + \frac{1}{(x - 6)^2}$?

Solution :

Given:

$x^2 - 16x + 59 = 0$

$x^2 - 16x + 64 - 5 = 0$

$x^2 - 16x + 64 = 5$

$(x - 8)^2 = 5$

$x - 8 = \sqrt{5}$

$(x - 6) = \sqrt{5} + 2$

$\frac{1}{x - 6} = \sqrt{5} - 2$

$(x-6)^2 + \frac{1}{(x - 6)^2} = (\sqrt{5}+2)^2 + (\sqrt{5}-2)^2$

$= 5 + 4 + 4\sqrt{5} + 5 + 4 - 4\sqrt{5}$

$= 18$

45. If $x^2 + 1 = 3x$, then the value of $\dfrac{x^4 + x^2}{x^2 + 5x + 1}$ is:

Solution :

$x^2 + 1 = 3x$

Dividing by x both sides:

$x + \dfrac{1}{x} = 3$

On cubing both sides:

$x^3 + \dfrac{1}{x^3} + 3 \times x \times \dfrac{1}{x}(x + \dfrac{1}{x}) = 27$

$x^3 + \dfrac{1}{x^3} = 27 - 9$

$x^3 + \dfrac{1}{x^3} = 18$

$\dfrac{x^4 + x^2}{x^2 + 5x + 1} = \dfrac{x(x^3 + \dfrac{1}{x^3})}{x(x + \dfrac{1}{x} + 5)}$

$= \dfrac{18}{3 + 5} = \dfrac{18}{8} = \dfrac{9}{4}$

$= 2\dfrac{1}{4}$

46. If $(a + b + 4)\{(ab + 4(a + b)\} - 4ab = 0$, and $a \neq -4$, $b \neq -4$, then the value of $\dfrac{1}{(a + b + 4)^{117}} - 2^{-234}$ is:

Solution :

$(a + b + 4)\{(ab + 4(a + b)\} - 4ab = 0$

Let $a = 1$ and $b = -1$:

On putting the values $a = 1$ and $b = -1$:

$(1 + (-1) + 4)\{1 \times (-1) + 4(1 + (-1))\} - 4 \times 1 \times (-1) = 0$

$4 \times (-1) + 0 + 4 = 0$

$0 = 0$

$\therefore \quad \dfrac{1}{(a + b + c)^{117}} - 2^{-234} = \dfrac{1}{(1 + (-1) + 4)^{117}} - 2^{-234}$

$= \dfrac{1}{4^{117}} - 2^{-234}$

$= 2^{-234} - 2^{-234}$

$= 0$

47. If $a = \sqrt{8} - \sqrt{7}$ and $a = \dfrac{1}{b}$, then which of the value $\dfrac{a^2 + b^2 - 3ab}{a^2 + ab + b^2}$ is equal to?

Solution :

$a = \sqrt{8} - \sqrt{7}$

$b = \dfrac{1}{a}$

$b = \dfrac{1}{\sqrt{8} - \sqrt{7}} \times \dfrac{\sqrt{8} + \sqrt{7}}{\sqrt{8} + \sqrt{7}}$

$b = \sqrt{8} + \sqrt{7}$

$\dfrac{a^2 + b^2 - 3ab}{a^2 + ab + b^2} = \dfrac{(a + b)^2 - ab}{(a + b)^2 - ab}$

$\dfrac{(-\sqrt{7} - \sqrt{7})^2 - (\sqrt{8} - \sqrt{7})(\sqrt{8} + \sqrt{7})}{(\sqrt{8} + \sqrt{8})^2 - (\sqrt{8} - \sqrt{7})(\sqrt{8} + \sqrt{7})}$

$= \dfrac{(-2\sqrt{7})^2 - 1}{(2\sqrt{8})^2 - 1}$

$= \dfrac{28 - 1}{28 - 1} = \dfrac{27}{31}$

48. If $1 + 4x^2 + 16x^4 = 512$, and $1 - 2x + 4x^2 = 64$, then the value of $1 - 2x + 4x^2$ is:

Solution :

Given:

$1 + 4x^2 + 16x^4 = 512$ ----(i)

$1 - 2x + 4x^2 = 64$ ----(ii)

On dividing equation (i) by equation (ii):

$\dfrac{1 + 4x^2 + 16x^4}{1 - 2x + 4x^2} = \dfrac{512}{64}$

$(1+2x+4x^2)(1-2x+4x^2)/(1-2x+4x^2) = 8$

$1 + 2x + 4x^2 = 8$

49. If $x^4 + x^2y^2 + y^4 = 21$ and $x^2 + xy + y^2 = 7$, then the value of $\dfrac{1}{x^2} + \dfrac{1}{y^2}$ is:

Solution :

$x^2 + xy + y^2 = 7$ ----(i)

$\because (x^2-xy+y^2)(x^2+xy+y^2) = x^4+x^2y^2+y^4$

$x^2 - xy + y^2 = \dfrac{x^4 + x^2y^2 + y^4}{x^2 + xy + y^2}$

$= \dfrac{21}{7} = 3$ ----(ii)

From equations (i) and (ii):

$(x^2 + y^2) = 5$ and $xy = 2$

$\therefore \dfrac{1}{x^2} + \dfrac{1}{y^2} = \dfrac{x^2 + y^2}{x^2y^2} = \dfrac{5}{4}$

50. If $x^4 + x^2y^2 + y^4 = 273$ and $x^2 - xy + y^2 = 13$, then the value of xy is:

Solution :

$x^2 + xy + y^2 = \dfrac{x^4 + x^2y^2 + y^4}{x^2 - xy + y^2}$

$x^2 + xy + y^2 = \dfrac{273}{13} = 21$ ----(i)

$x^2 - xy + y^2 = 13$ ----(ii)

From equations (i) and (ii):

$2xy = 8$

$xy = 4$

51. If $1 + 9r^2 + 81r^4 = 256$ and $1 + 3r + 9r^2 = 32$, then find the value of $1 - 3r + 9r^2$:

Solution :

$x^4+y^4+x^2y^2 = (x^2+y^2-xy)(x^2+y^2+xy)$

According to the formula:

$1 + 9r^2 + 81r^4 = 256$

$(1 + 3r + 9r^2)(1 - 3r + 9r^2) = 256$

$1 - 3r + 9r^2 = \dfrac{256}{32} = 8$

52. If $x + y + z = 2,\ xy + yz + zx = -11,$ and $xyz = -12$, then what is the value of $x^3 + y^3 + z^3$:

Solution :

Given:

$x + y + z = 2,\ xy + yz + zx = -11,\ xyz = -12$

$x^3 + y^3 + z^3 - 3xyz = (x + y + z)[(x + y + z)^2 - 3(xy + yz + zx)]$

$x^3 + y^3 + z^3 - 3(-12) = 2[4 + 33]$

$x^3 + y^3 + z^3 = 74 - 36$

$x^3 + y^3 + z^3 = 38$

53. If $a + b + c = 11$ and $ab + bc + ca = 28$, then find the value of $a^3 + b^3 + c^3 - 3abc$:

Solution :

$a + b + c = 11,\ ab + bc + ca = 28$

$a^3 + b^3 + c^3 - 3abc = (a + b + c)[(a + b + c)^2 - 3(ab + bc + ca)]$

$a^3 + b^3 + c^3 - 3abc = 11(121 - 84)$

$a^3 + b^3 + c^3 - 3abc = 11 \times 37$

$a^3 + b^3 + c^3 - 3abc = 407$

54. If $a^2 + b^2 + 49c^2 + 18 = 2(b + 28c - a)$, then the value of $2a - b + 7c$ is:

Solution :

$a^2 + b^2 + 49c^2 + 18 = 2(b + 28c - a)$

$a^2 + b^2 + 49c^2 + 18 = 2b + 56c - 2a$

$(a^2 + 2a + 1) + (b^2 - 2b + 1) + (49c^2 - 56c + 16) = 0$

$(a + 1)^2 + (b - 1)^2 + (7c - 4)^2 = 0$

$a + 1 = 0,\ b - 1 = 0,\ 7c - 4 = 0$

$a = -1,\ b = 1,\ c = \dfrac{4}{7}$

$2a - b + 7c = 2(-1) - 1 + 7 \times \dfrac{4}{7}$

$= -2 - 1 + 4 = 1$

55. If $x+y+z = 18,\ xyz = 81,$ and $xy+yz+zx = 90$, then the value of $x^3 + y^3 + z^3 + xyz$ is:

Solution :

$x+y+z = 18,\ xyz = 81,\ xy+yz+zx = 90$

From the formula:

$x^3+y^3+z^3-3xyz = (x+y+z)[(x+y+z)^2-3(xy+yz+zx)]$

$x^3+y^3+z^3-3xyz = 18[(18)^2-3(90)]$

$x^3+y^3+z^3+xyz = 18(324-270)+4xyz$

$= 18 \times 54 + 4 \times 81$

$= 972 + 324$

$x^3+y^3+z^3+xyz = 1296$

56. If $a^2+b^2+c^2 = 6.25$ and $(ab+bc+ca) = 0.52$, what is the value of $(a+b+c)$, if $(a+b+c) < 0$?

Solution :

Given,

$a^2+b^2+c^2 = 6.25$

$ab+bc+ca = 0.52$

Let $a+b+c = y$.

Squaring both sides:

$a^2+b^2+c^2+2(ab+bc+ca) = y^2$

$6.25 + 2 \times 0.52 = y^2$

$6.25 + 1.04 = y^2$

$y^2 = 7.29$

$y = \pm\sqrt{7.29}$

If $a+b+c < 0$:

$a+b+c = -2.7$

57. If $x+y+z = 13$, $x^2+y^2+z^2 = 91$ and $xz = y^2$, then the difference between z and x is:

Solution :

Given,

$x+y+z = 13$, $x^2+y^2+z^2 = 91$, and $xz = y^2$.

$x+y+z = 13$

Squaring both sides:

$x^2+y^2+z^2+2(xy+yz+zx) = 169$

$91+2(xy+yz+zx) = 169$

$2(xy+yz+zx) = 169-91$

$xy+yz+zx = 39/13 = 3$

$x+z = 13-3$

$x+z = 10$

$xz = 9$

Solving: $x+\dfrac{9}{x} = 10$

$x(10-x) = 9$

$x = 1$

$z = 9-1 = 8$

58. If $x+y+z = 5$, $\dfrac{1}{x}+\dfrac{1}{y}+\dfrac{1}{z} = 0$, $xyz = 12$, and $x^3+y^3+z^3 = 151$, then the value of $x^2+y^2+z^2$ is:

Solution :

Given: $x+y+z = 5$, $\dfrac{1}{x}+\dfrac{1}{y}+\dfrac{1}{z} = 0$, $xyz = 12$

$x^2+y^2+z^3 = 151$

$\because \dfrac{1}{x}+\dfrac{1}{y}+\dfrac{1}{z} = 0$:

$xy+yz+zx = 0$

$x^3+y^3+z^3-3xyz = (x+y+z)(x^2+y^2+z^2-xy-yz-zx)$

$151-3\times 12 = 5(x^2+y^2+z^2-0)$

$115 = 5(x^2+y^2+z^2)$

$x^2+y^2+z^2 = 115/5 = 23$

59. If $x+y+z = 7$, $x^2+y^2+z^2 = 85$, and $x^3+y^3+z^3 = 913$, then the value of $\sqrt[3]{xyz}$ is:

Solution :

Given:

$x+y+z = 7$, $x^2+y^2+z^2 = 85$, and

$x^3+y^3+z^3 = 913$

$x^3+y^3+z^3-3xyz = (x+y+z)[x^2+y^2+z^2-xy-yz-zx]$

$(x+y+z)^2 = x^2+y^2+z^2+2(xy+yz+zx)$

$$x^2 + y^2 + z^2 + 2(xy + yz + zx) = 49$$
$$xy + yz + zx = -36$$
$$(xy + yz + zx) = -18$$
$$x^3 + y^3 + z^3 - 3xyz = 7[85 + 18]$$
$$3xyz = 913 - 721 = 192$$
$$xyz = 64$$
$$\sqrt[3]{xyz} = \sqrt[3]{64} = 4$$

60. If $a^2 + b^2 + c^2 + 48 = 8(a + b + c)$, then what is the value of $\sqrt[3]{a^3 - b^3 + c^3}$?

Solution :

$$a^2 + b^2 + c^2 + 48 = 8(a + b + c)$$
$$a^2 - 8a + 16 + b^2 - 8b + 16 + c^2 - 8c + 16 = 0$$
$$(a - 4)^2 + (b - 4)^2 + (c - 4)^2 = 0$$
$$(a - 4)^2 = 0 \implies a = 4$$
$$(b - 4)^2 = 0 \implies b = 4$$
$$(c - 4)^2 = 0 \implies c = 4$$
$$\sqrt[3]{a^3 - b^3 + c^3} = \sqrt[3]{4^3 - 4^3 + 4^3} = 4$$

61. If $a^3 + b^3 + c^3 - 3abc = 250$ and $a + b + c = 10$, then what will be the value of $(ab + bc + ca)$?

Solution :

Given: $a^3 + b^3 + c^3 - 3abc = 250$, and $a + b + c = 10$

From formula:

$$a^3 + b^3 + c^3 - 3abc = (a + b + c)[(a + b + c)^2 - 3(ab + bc + ca)]$$
$$250 = 10[(10)^2 - 3(ab + bc + ca)]$$
$$100 - 3(ab + bc + ca) = 250/10 = 25$$
$$3(ab + bc + ca) = 100 - 25 = 75$$
$$ab + bc + ca = 75/3 = 25$$
$$\frac{1}{5}(ab + bc + ca) = \frac{1}{5} \times 25 = 5$$

62. If $x = 555$, $y = 556$, and $z = 557$, then find the value of $x^3 + y^3 + z^3 - 3xyz$:

Solution :

From formula:

$$x^3 + y^3 + z^3 - 3xyz = \frac{1}{2}(x + y + z)[(x - y)^2 + (y - z)^2 + (z - x)^2]$$
$$x^3 + y^3 + z^3 - 3xyz = \frac{1}{2}(555 + 556 + 557)[(555 - 556)^2 + (556 - 557)^2 + (557 - 555)^2]$$
$$= \frac{1}{2} \times 1668 \times (1 + 1 + 4)$$
$$= \frac{1}{2} \times 1668 \times 6 = 5004$$

Hence, $x^3 + y^3 + z^3 - 3xyz = 5004$

63. If $a^2 + b^2 + 64c^2 + 16c + 3 = 2(a + b)$, then the value of $4a^7 + b^7 + 8c^2$ is:

Solution :

$$a^2 + b^2 + 64c^2 + 16c + 1 + 1 + 1 - 2a - 2b = 0$$
$$a^2 - 2a + 1 + b^2 - 2b + 1 + 64c^2 + 16c + 1 = 0$$
$$(a - 1)^2 + (b - 1)^2 + (8c + 1)^2 = 0$$

$(a - 1)^2 = 0 \implies a = 1$ $\quad (b - 1)^2 = 0 \implies b = 1$ $\quad (8c + 1)^2 = 0 \implies 8c = -1 \implies c = -\frac{1}{8}$

$$4a^7 + b^7 + 8c^2 = 4 \times 1 + 1 + 8 \times \frac{1}{64}$$
$$= 4 + 1 + \frac{1}{8} = 5\frac{1}{8}$$

64. If $a + b + c = 9$, $ab + bc + ca = 26$, $a^3 + b^3 + c^3 = 72$, and $c^3 + a^3 = 35$, then what is the value of abc?

Solution :

$a^3 + b^3 = 91$ ----(i)

$b^3 + c^3 = 72$ ----(ii)

$c^3 + a^3 = 35$ ----(iii)

Adding equations (i), (ii), and (iii):

$2(a^3+b^3+c^3)=198$

$a^3+b^3+c^3=99$

Using the formula:

$a^3+b^3+c^3-3abc=(a+b+c)[(a+b+c)^2-3(ab+bc+ca)]$

$99-3abc=9(81-78)$

$99-3abc=27$

$3abc=99-27=72$

$abc=24$

65. If $x+y+z=22$ and $xy+yz+zx=35$, then what is the value of $(x-y)^2+(y-z)^2+(z-x)^2$?

Solution :

$(x-y)^2+(y-z)^2+(z-x)^2$

$=[x^2+y^2-2xy+y^2+z^2-2yz+z^2+x^2-2zx]$

$=2(x^2+y^2+z^2-xy-yz-zx)$

$=2[(x+y+z)^2-3(xy+yz+zx)]$

$=2[(22)^2-3\times 35]$

$=2[484-105]$

$=2\times 379=758$

66. If $x=2+\sqrt{3}$, $y=2-\sqrt{3}$, and $z=1$, then what is the value of $\frac{x}{yz}+\frac{y}{xz}+\frac{z}{xy}+2\left(\frac{1}{x}+\frac{1}{y}+\frac{1}{z}\right)$?

Solution :

Given:

$x=2+\sqrt{3}, y=2-\sqrt{3}, z=1$

$x\times y\times z=(2+\sqrt{3})(2-\sqrt{3})\times 1=1$

$x+y+z=2+\sqrt{3}+2-\sqrt{3}+1=5$

$x^2+y^2+z^2=(2+\sqrt{3})^2+(2-\sqrt{3})^2+1$

$=7+4\sqrt{3}+7-4\sqrt{3}+1$

$=7+7+1=15$

Using: $(x+y+z)^2=x^2+y^2+z^2+2(xy+yz+zx)$:

$5^2=15+2(xy+yz+zx)$

$25=15+2(xy+yz+zx)$

$xy+yz+zx=\frac{10}{2}=5$

$\frac{x}{yz}+\frac{y}{xz}+\frac{z}{xy}+2\left[\frac{1}{x}+\frac{1}{y}+\frac{1}{z}\right]=\frac{x^2+y^2+z^2}{xyz}+2\left[\frac{xy+yz+zx}{xyz}\right]$

Substituting the values:

$\frac{15}{1}+2\left[\frac{5}{1}\right]=15+10=25$

67. If $x^3+y^3+z^3=3(1+xyz)$, $P=y+z-x$, $Q=z+x-y$, and $R=x+y-z$, then what is the value of $P^3+Q^3+R^3-3PQR$?

Solution :

Substituting:

$y=z=0, x^3+0=3(1+0)$

$x^3=3$

$P=0+0-x=-x, Q=0+x-0=x,$

$R=x+0-0=x$

$P^3+Q^3+R^3-3PQR=(-x)^3+x^3+x^3-3(-x)x^2$

$=-x^3+x^3+x^3+3x^3=4x^3$

$=4\times 3=12$

68. If $(2x+3y+4)(2x+3y-5)$ is equivalent to $(ax^2+by^2+2hxy+2gx+2fy+c)$, then what is the value of $\frac{g+f-c}{abh}$?

Solution :

$(2x+3y+4)(2x+3y-5)$

$=4x^2+6xy-10x+6xy+9y^2-15y+8x+12y-20$

$= 4x^2+9y^2+12xy-2x-3y-20$ Comparing with $ax^2+by^2+2hxy+2gx+2fy+c$:

$a = 4,\ b = 9,\ h = 6,\ g = -1,\ f = -\frac{3}{2}$, $c = -20$

$$\frac{g+f-c}{abh} = \frac{-1+\left(-\frac{3}{2}\right)-(-20)}{4\times 9\times 6}$$

$$= \frac{-\frac{5}{2}+20}{216} = \frac{35}{432}$$

69. If $a^{1/3}+b^{1/3}+c^{1/3}=0$, then the value of $(a+b+c)^6$ is equal to:

Solution :

$a^{1/3}+b^{1/3}+c^{1/3}=0$

If $x+y+z=0 \implies x^3+y^3+z^3=3xyz$

Hence, $a+b+c = a^{1/3}\times b^{1/3}\times c^{1/3}$

$(a+b+c) = 3(abc)^{1/3}$

$(a+b+c)^6 = 729a^2b^2c^2$

70. If $(x+y)^{1/3}+(y+z)^{1/3} = -(z+x)^{1/3}$, then the value of $(x^3+y^3+z^3)$ can be expressed as:

Solution :

$(x+y)^{1/3}+(y+z)^{1/3} = -(z+x)^{1/3}$

$(x+y)^{1/3}+(y+z)^{1/3}+(z+x)^{1/3} = 0$

If $a+b+c=0 \implies a^3+b^3+c^3=3abc$

$\implies (x+y)+(y+z)+(z+x) = 3(x+y)^{1/3}(y+z)^{1/3}(z+x)^{1/3}$

$$= \frac{2(x+y+z)}{3} = (x+y)^{1/3}(y+z)^{1/3}(z+x)^{1/3}$$

Again, on cubing both sides:

$x^3+y^3+z^3+3(x+y)(y+z)(z+x) = \frac{27}{8}(x+y)(y+z)(z+x)$

$$x^3+y^3+z^3 = (x+y)(y+z)(z+x)\left[\frac{27}{8}-3\right]$$

71. Given that $x,\ y,\ z$ are positive real numbers, if $(x+y)^2-z^2=8$, $(y+z)^2-x^2=10$, and $(x+z)^2-y^2=7$, then what is the value of $x+y+z$?

Solution :

$(x+y)^2-z^2=8$

$(x+y+z)(x+y-z)=8$ ----(i)

$(y+z)^2-x^2=10$

$(y+z+x)(y+z-x)=10$ ----(ii)

$(x+z)^2-y^2=7$

$(x+z+y)(x+z-y)=7$ ----(iii)

Adding equations (i), (ii), and (iii):

$(x+y+z)(x+y-z+y+z-x+x+z-y) = 8+10+7$

$(x+y+z)(x+y+z)=25$

$(x+y+z)^2=25$

$x+y+z=5$

72. The value of $(a^{2/3}+2a^{1/2}+3a^{1/3}+2a^{1/6}+1)(a^{1/3}-2a^{1/6}+1)-a^{1/2}(a^{1/2}-2)$ when $a=7$ is:

Solution :

Let $a^{1/6}=x$, then $a^{1/3}=x^2$, $a^{1/2}=x^3$, $a^{2/3}=x^4$.

According to the question:

$(x^4+2x^3+3x^2+2x+1)(x^2-2x+1) - x^3(x^3-2)$

Expanding:

$x^6-2x^5+x^4+2x^5-4x^4+2x^3+3x^4-6x^3+3x^2+2x^3-2x^2+2x+x^2-2x+1-x^6+2x^3$

$= 4x^4-4x^4-6x^3+6x^3-4x^2+4x^2+1$

73. If $x+y=4$, $xy=2$, $y+z=5$, $yz=3$, $z+x=6$, and $zx=4$, then find the value of $x^3+y^2+z^3-3xyz$:

Solution :

$2(x+y+z)=15$

$x+y+z=\frac{15}{2}$

$x+y=4 \implies x^2+y^2+2xy=16$

$y+z=5 \implies y^2+z^2+2yz=25$

$z+x=6 \implies z^2+x^2+2zx=36$

$2(x^2+y^2+z^2)+2(xy+yz+zx)=77$

Substituting $xy+yz+zx=9$ and $x+y+z=\frac{15}{2}$:

$2(x^2+y^2+z^2)+2\times 9=77$

$2(x^2+y^2+z^2)=77-18=59$

$x^2+y^2+z^2=\frac{59}{2}$

$x^2+y+z^3-3xyz=(x+y+z)(x^2+y^2+z^2-xy-yz-zx)$

$=\frac{15}{2}\times\left(\frac{59}{2}-9\right)=\frac{15}{2}\times\frac{41}{2}=\frac{615}{4}=153.75$

74. If $(5\sqrt{5}x^3-81\sqrt{3}y^3)+(\sqrt{5}x-3\sqrt{3}y)=(Ax^2+By^2+Cxy)$, then the value of $6A+B-\sqrt{15}C$ is:

Solution :

$5\sqrt{5}x^3-81\sqrt{3}y^3/\sqrt{5}x-3\sqrt{3}y=Ax^2+By^2+Cxy$

$a^3-b^3=(a-b)(a^2+b^2+ab)$

$\frac{(\sqrt{5}x-3\sqrt{3}y)(5x^2+27y^2+3\sqrt{15}xy)}{(\sqrt{5}x-3\sqrt{3}y)}=Ax^2+By^2+Cxy$

$5x^2+27y^2+3\sqrt{15}xy=Ax^2+By^2+Cxy$

By comparing the coefficient of x^2, y^2, and xy, we get:

$A=5, B=27, C=3\sqrt{15}$

$6A+B-\sqrt{15}C=30+27-\sqrt{15}\times\sqrt{15}\times 3$

$=57-45=12$

75. If $2\sqrt{2}x^2-3\sqrt{3}y^3=(\sqrt{2}x-\sqrt{3}y)(Ax^2-Bxy+Cy^2)$, then the value of $A^2+B^2+C^2$ is:

Solution :

$2\sqrt{2}x^2-3\sqrt{3}y^3=(\sqrt{2}x-\sqrt{3}y)(Ax^2-Bxy+Cy^2)$

$(\sqrt{2}x-\sqrt{3}y)(2x^2+\sqrt{6}xy+3y^2)=(\sqrt{2}x-\sqrt{3}y)(Ax^2-Bxy+Cy^2)$

$Ax^2-Bxy+Cy^2=(2x^2+\sqrt{6}xy+3y^2)$

$A=2,\ B=-\sqrt{6},\ C=3$

$A^2+B^2+C^2=2^2+6+3^2$

$=4+6+9=19$

76. If $(4x+2y)^3+(4x-2y)^3=16(Ax^3+Bxy^2)$, then what is the value of $\frac{1}{2}\sqrt{A^2+B^2}$?

Solution :

$(4x+2y)^3+(4x-2y)^3=16(Ax^3+Bxy^2)$

$64x^3+8y^3+24xy(4x+2y)+64x^3-8y^3-24xy(4x-2y)=16(Ax^3+Bxy^2)$

$128x^3+96xy^2=16(Ax^3+Bxy^2)$

$16(8x^3+6xy^2)=16(Ax^3+Bxy^2)$

$A=8,\ B=6$

$\frac{1}{2}\sqrt{A^2+B^2}=\frac{1}{2}\sqrt{8^2+6^2}$

$=\frac{1}{2}\sqrt{64+36}$

$=\frac{1}{2}\sqrt{100}=5$

77. If $27x^3-64y^3=(Ax+By)(Cx^2+Dy^2-Exy)$, then the value of $A-B+C-D+E$ will be:

Solution :

$27x^3-64y^3=(Ax+By)(Cx^2+Dy^2-Exy)$ (Given)

$(3x-4y)(9x^2+16y^2+12xy)=(Ax+By)(Cx^2+Dy^2-Exy)$

By comparing both sides, we get:

$A=3,\ B=-4,\ C=9,\ D=16,\ E=-12$

$A-B+C-D+E=3+4+9-16-12$

$=16-16-12=-12$

78. If $(x+y)^3+27(x-y)^3=(Ax-2y)(Bx^2+Cxy+13y^2)$, then the value of $A-B-C$ is:

Solution :

Given that, $(x+y)^3+27(x-y)^3=(Ax-2y)(Bx^2+Cxy+13y^2)$

$LHS=(x+y)^3+(3x-3y)^3$

$(x+y+3x-3y)[(x+y)^2+(3x-3y)^2-(x+y)(3x-3y)]$

Using $a^3+b^3=(a+b)(a^2+b^2-ab)$:

$=(4x-2y)[10x^2+10y^2-16xy-3(x^2-y^2)]$

$=(4x-2y)[7x^2+13y^2-16xy]$

After comparing:

$A=4,\ B=7,\ C=-16$

$A-B-C=4-7+16=13$

79. If $(27x^3-343y^3)\div(3x-7y)=Ax^2+By^2+7Cxy$, then the value of $4A-B+5C$ is:

Solution :

$(27x^3-343y^3)\div(3x-7y)=Ax^2+By^2+7Cxy$

Using $a^3-b^3=(a-b)(a^2+b^2+ab)$:

$$\frac{(3x-7y)(9x^2+49y^2+21xy)}{(3x-7y)}=Ax^2+By^2+7Cxy$$

$9x^2+49y^2+21xy=Ax^2+By^2+7Cxy$

Comparing coefficients:

$A=9,\ B=49,\ C=3$

$4A-B+5C=4\times9-49+5\times3$

$=36-49+15=2$

80. If $[8(x+y)^2-27(x-y)^2]+(5y-x)=Ax^2+Bxy+Cy^2$, then the value of $A+B+C$ is:

Solution :

$[8(x+y)^2-27(x-y)^2]+(5y-x)=Ax^2+Bxy+Cy^2$

Expanding and simplifying:

$4(x+y)^2+9(x-y)^2+6(x^2-y^2)+8xy-18xy=Ax^2+Bxy+Cy^2$

$19x^2+7y^2-10xy=Ax^2+Bxy+Cy^2$

Comparing coefficients:

$A=19,\ B=-10,\ C=7$

$A+B+C=19-10+7=16$

81. If $(135-\sqrt{5}x^3-2\sqrt{2}y^2)\div(3\sqrt{5}x-\sqrt{2}y)=Ax^2+By^2+\sqrt{10}Cxy$, then the value of $A+B-9C$ is:

Solution :

$(135-\sqrt{5}x^3-2\sqrt{2}y^2)\div(3\sqrt{5}x-\sqrt{2}y)=Ax^2+By^2+\sqrt{10}Cxy$

Using $a^3-b^3=(a-b)(a^2+b^2+ab)$:

$$\frac{(3\sqrt{5}x-\sqrt{2}y)(45x^2+2y^2+3\sqrt{10}xy)}{(3\sqrt{5}x-\sqrt{2}y)}=Ax^2+By^2+\sqrt{10}Cxy$$

$45x^2+2y^2+3\sqrt{10}xy=Ax^2+By^2+\sqrt{10}Cxy$

Comparing coefficients:

$A=45,\ B=2,\ C=3$

$A+B-9C=45+2-9\times3$

$=47-27=20$

Type 2 (Problems based on Equations in Two/ Three Variables)

82. If $2x + 3y - 5z = 18$, $3x + 2y + z = 29$, and $x+y+3z = 17$, then what is the value of $xy+yz+zx$?

Solution :

Given, $2x + 3y - 5z = 18$ -------(i)

$3x + 2y + z = 29$ -------(ii)

$x + y + 3z = 17$ -------(iii)

Multiplying by 5 in equation (ii) and adding it to equation (i): $2x + 3y - 5z = 18$

$15x + 10y + 5z = 145$

$17x + 13y = 163$ -------(iv)

Again, on multiplying by 3 in equation (ii) and subtracting it from equation (iii): $8x + 5y = 70$ -------(v)

By solving equations (iv) and (v): $x = 5,\ y = 6$

Substituting $x = 5$ and $y = 6$ into equation (ii): $15 + 12 + z = 29$

$z = 2$

$xy+yz+zx = (5\times 6)+(6\times 2)+(2\times 5) = 30 + 12 + 10 = 52$

83. If $3x + by + 9z = \frac{20}{3}$, $6x + 9y + 3z = \frac{17}{3}$, and $18x + 27y - z = \frac{113}{9}$, then what is the value of $75x + 113y$?

Solution :

$3x + 6y + 9z = \frac{20}{3}$ ....(i)

$6x + 9y + 3z = \frac{17}{3}$ ....(ii)

$18x + 27y - z = \frac{113}{9}$ ....(iii)

(eq1 + eq2) $\times \frac{1}{3}$ + (eq3) $\times$ 4: $(9x + 15y + 12z) \times \frac{1}{3} + (18x + 27y - z) \times 4$

$3x + 5y + 4z + 72x + 108y - 4z$

$75x + 113y = \left(\frac{20}{3} + \frac{17}{3}\right) \times \frac{1}{3} + 4 \times \frac{113}{9}$

$\frac{37}{9} + \frac{452}{9}$

$\frac{489}{9} = \frac{163}{3}$

84. If $a + b + c = \frac{7}{12}$, $3a - 4b + 5c = \frac{3}{4}$, and $7a - 11b - 13c = -\frac{7}{12}$, then what is the value of $a + c$?

Solution :

$a + b + c = \frac{7}{12}$ -------(i)

$3a - 4b + 5c = \frac{3}{4}$ -------(ii)

$7a - 11b - 13c = -\frac{7}{12}$ -------(iii)

Multiplying by 4 in equation (i) and adding it to equation (ii): $4a+4b+4c+3a-4b+5c = \frac{7}{3} + \frac{3}{4}$

$7a + 9c = \frac{37}{12}$ -------(iv)

Multiplying by 11 in equation (i) and adding it to equation (iii): $11a + 11b + 11c + 7a - 11b - 13c = \frac{77}{12} - \frac{7}{12}$

$18a - 2c = \frac{35}{6}$

$9a - c = \frac{35}{12}$ -------(v)

Multiplying equation (v) by 9 and adding it to equation (iv): $81a-9c+7a+9c = \frac{315}{12}+\frac{37}{12}$

$88a = \frac{352}{12}$

$a = \frac{1}{3}$

From equation (v): $3 - c = \frac{35}{12}$

$c = \frac{1}{12}$

$a + c = \frac{1}{3} + \frac{1}{12} = \frac{4+1}{12} = \frac{5}{12}$

85. The cost of 8 pencils, 5 pens, and 3 erasers is Rs. 111. The cost of 9 pencils, 6 pens, and 5 erasers is Rs. 130. The cost of 16 pencils, 11 pens, and 3 erasers is Rs. 221. What is the cost (in Rs.) of 39 pencils, 26 pens, and 13 erasers?

Solution :

Let the price of 1 pencil, 1 pen, and 1 eraser be x, y, and z respectively. As per the question:

$8x + 5y + 3z = 111$ -------(i)

$9x + 6y + 5z = 130$ -------(ii)

$16x + 11y + 3z = 221$ -------(iii)

Adding equations (i), (ii), and (iii):

$33x + 22y + 11z = 462$ -------(iv)

Dividing by 11: $3x + 2y + z = 42$ -------(v)

From equation (v) multiplied by 13: $39x + 26y + 13z = 546$

86. If $3x + 4y - 2z + 9 = 17$, $7x + 2y + 11z + 8 = 23$, and $5x + 9y + 6z - 4 = 18$, then what is the value of $x + y + z - 34$?

Solution :

$3x + 4y - 2z + 9 = 17$ -------(i)

$7x + 2y + 11z + 8 = 23$ -------(ii)

$5x + 9y + 6z - 4 = 18$ -------(iii)

By adding equations (i), (ii), and (iii): $15x + 15y + 15z = 45$

$x + y + z = 3$

$x + y + z - 34 = 3 - 34 = -31$

87. If $\frac{x + 3y - 2z}{4} = 6$, $x + \frac{2}{3}(2y + 3z) = 33$, and $\frac{1}{7}(x + y + z) + 2z = 9$, then what is the value of $46x + 131y$?

Solution :

$4x + 12y - 2z = 24$ ....(i)

$3x + 4y + 6z = 99$ ....(ii)

$x + y + 15z = 63$ ....(iii)

$(\text{eq1} \times \frac{21}{2}) + \text{eq2} + \text{eq3}$: $42x + 126y - 21z + 3x + 4y + 6z + x + y + 15z = 252 + 99 + 63$

$46x + 131y = 414$

88. If $3x + 4y - 11 = 18$ and $8x - 6y + 12 = 6$, then what is the value of $5x - 3y - 9$?

Solution :

$3x + 4y = 29$ -------(i)

$8x - 6y = -6$ -------(ii)

$4x - 3y = -3$ -------(iii)

On solving equations (i) and (iii): $x = 3$, $y = 5$

Substituting values: $5x - 3y - 9 = 15 - 15 - 9 = -9$

89. If the system of the following equations has the value of the variables as three consecutive integers, then the value of a is ______. $x - y + z = 2a$

$x + 4y - 2z = 3(4 - a)$

$2x - 3y + 4z = 6 - 2a$

Solution :

From the condition $x + z = 2y$: $x - y + z = 2a$

$y = 2a$

Substituting into equation (ii): $x + 2(x + z) - 2z = 3(4 - a)$

$x = 4 - a$

Substituting into equation (iii): $z = 2(2a) - 4 + a = 5a - 4$

In equation (iii): $2(4 - a) - 3(2a) + 4(5a - 4) = 6 - 2a$

$8 - 2a - 6a + 20a - 16 = 6 - 2a$

$14a = 14$

$a = 1$

Type 3 (Problems based on Quadratic Equations)

90. If α and β are the roots of the quadratic equation, and $\alpha+\beta=8$ and $a-\beta=2\sqrt{5}$, then what will be the equation which has roots α^4 and β^4?

Solution :

Given: $\alpha+\beta=8$ ....(i) and $a-\beta=2\sqrt{5}$ ....(ii)

Adding the squares of equations (i) and (ii):

$\alpha^2+\beta^2+2\alpha\beta=64$ ....(iii)

$\alpha^2+\beta^2-2\alpha\beta=20$

$2(\alpha^2+\beta^2)=84$

$\alpha^2+\beta^2=42$ ....(iv)

From equations (iii) and (iv): $42+2\alpha\beta=64$

$\alpha\beta=11$ ....(v)

$\alpha^4+\beta^4=(\alpha^2+\beta^2)^2-2\alpha^2\beta^2$

$=(42)^2-2\times(11)^2=1764-242=1522$

$\alpha^4\beta^4=(\alpha\beta)^4=11^4=14641$

Hence, the quadratic equation is: $x^2-(\alpha^4+\beta^4)x+\alpha^4\beta^4=0$

$x^2-1522x+14641=0$

91. If A and B are the roots of the equation $Ax^2-A^2x+AB=0$, then what is the value of A and B, respectively?

Solution :

Given: Quadratic equation $Ax^2-A^2x+AB=0$, where A and B are roots of the equation. Sum of roots: $A+B=-\frac{-A^2}{A}=A$

$A+B=A,\quad AB=0$

Product of roots: $A\times B=\frac{AB}{A}=B$

$A=1$

Hence, the values of A and B are 1 and 0, respectively.

92. If α and β are the roots of the quadratic equation $x^2-x-1=0$, what is the value of $\alpha^8+\beta^8$?

Solution :

Given: Quadratic equation $x^2-x-1=0$, where α and β are roots. $\alpha+\beta=1$ ....(i)

$\alpha\times\beta=-1$ ....(ii)

$\alpha^2+\beta^2=(\alpha+\beta)^2-2\alpha\beta=1^2-2(-1)=3$

$\alpha^4+\beta^4=(\alpha^2+\beta^2)^2-2(\alpha\beta)^2=9-2(1)=7$

$\alpha^8+\beta^8=(\alpha^4+\beta^4)^2-2(\alpha^4\beta^4)=7^2-2(1)=49-2=47$

93. If α and β are the roots of the equation $x^2-2x+4=0$, then what is the equation whose roots are $\frac{\alpha^3}{\beta^2}$ and $\frac{\beta^3}{\alpha^2}$?

Solution :

Given: α and β are the roots of $x^2-2x+4=0$. $\alpha+\beta=2,\quad \alpha\times\beta=4$

Sum of roots: $\frac{\alpha^3}{\beta^2}+\frac{\beta^3}{\alpha^2}=\frac{\alpha^5+\beta^5}{\alpha^2\beta^2}$

$\alpha^3+\beta^3=(\alpha+\beta)^3-3\alpha\beta(\alpha+\beta)=2^3-3(4)(2)=-16$

$\alpha^5+\beta^5=(\alpha^3+\beta^3)(\alpha^2+\beta^2)-\alpha^2\beta^2(\alpha+\beta)$

$=(-16)(-4)+16(2)=64$

Hence, the quadratic equation is: $x^2-\left(\frac{\alpha^5+\beta^5}{\alpha^2\beta^2}\right)x+\alpha\beta=0$

$x^2-2x+4=0$

94. If α and β are the roots of the equation $x^2-x+1=0$, then which equation will have roots α^3 and β^3?

Solution :

Given: $\alpha+\beta=1,\quad \alpha\times\beta=1$ $\alpha^3+\beta^3=(\alpha+\beta)^3-3\alpha\beta(\alpha+\beta)=1^3-3(1)(1)=-2$

$\alpha^3\beta^3=(\alpha\beta)^3=1^3=1$

Required equation: $x^2-(\alpha^3+\beta^3)x+\alpha^3\beta^3 = 0$

$x^2+2x+1=0$

95. If α and β are two roots of the quadratic equation $ax^2-bx+c=0$, where a, b, and c are constants and $a \neq 0$, then the value of $\frac{1}{\alpha}+\frac{1}{\beta}$ is:

Solution :

Given: $\alpha+\beta=\frac{b}{a}$, $\quad \alpha\times\beta=\frac{c}{a}\frac{1}{\alpha}+\frac{1}{\beta}=$

$\frac{\alpha+\beta}{\alpha\times\beta}=\frac{\frac{b}{a}}{\frac{c}{a}}=\frac{b}{c}$

Type 4 (Misc Problems)

96. The coefficient of x^3y in $(x-2y)\times(5x+y)^3$ is:

Solution :

$x^2y=(x-2y)(5x+y)^3$

$=(x-2y)\left[125x^3+y^3+3\times5xy(5x+y)\right]$

$=(x-2y)\left(125x^3+y^3+75x^2y+15xy^2\right)$

$=125x^4+xy^3+75x^3y+15x^2y^2-250x^2y-2y^4-150x^2y^2-30xy^3$

Required coefficient: $75-250=-175$

97. What is the coefficient of x in the expansion of $(3x-4)^3$?

Solution :

$(3x-4)^3=27x^3-64-3\times3x\times4(3x-4)$

$=27x^3-64-108x^2+144x$

Hence, the coefficient of x is 144.

98. If kx^3+4x^2+3x-4 and x^3-4x+k leave the same remainder when divided by $(x-3)$, then the value of k is:

Solution :

Dividing the given polynomials kx^3+4x^2+3x-4 and x^3-4x+k by $(x-3)$ leaves the same remainder. $x-3=0$ or $x=3$

By substituting $x=3$ in both equations:

$k\times3^3+4\times3^2+3\times3-4=3^3-4\times3+k$

$27k+36+9-4=27-12+k$

$27k+41=15+k$

$26k=-26$

$k=-1$

99. If $(2x+3y+4)(2x+3y-5)$ is equal to $(ax^2+by^2+2hxy+2gx+2fy+c)$, then what is the value of $\frac{3(g-f-c)}{ab}$?

Solution :

$(2x+3y+4)(2x+3y-5)$

$=4x^2+9y^2+12xy-2x-3y-20$

$=ax^2+by^2+2hxy+2gx+2fy+c$ (Given)

On comparing both sides: $a=4$, $b=9$, $h=6$, $g=-1$, $f=-\frac{3}{2}$, $c=-20$

$\frac{3(g-f-c)}{ab}=\frac{3(-1+\frac{3}{2}+20)}{4\times9}=\frac{\frac{41}{2}}{36}=\frac{41}{24}$

100. If $3a=27b=81c$ and $abc=144$, then the value of $12\left(\frac{1}{a}+\frac{1}{2b}+\frac{1}{5c}\right)$ is:

Solution :

$3a=27b=81c$

$3a=3^3b=3^4c$

$a=3b=4c$

$a : b : c = 1 : \frac{1}{3} : \frac{1}{4} = 12 : 4 : 3$

Let $a = 12k, b = 4k, c = 3k$: $abc = 144$

$12k \times 4k \times 3k = 144$

$144k^3 = 144$

$k^3 = 1$

$k = 1$

$a = 12k = 12 \times 1 = 12, \quad b = 4k = 4 \times 1 = 4, \quad c = 3k = 3 \times 1 = 3$

$12\left(\frac{1}{a} + \frac{1}{2b} + \frac{1}{5c}\right) =$

$12\left(\frac{1}{12} + \frac{1}{8} + \frac{1}{15}\right)$

$= 12\left(\frac{10 + 15 + 8}{120}\right) = 12 \times \frac{33}{120} = \frac{33}{10}$

101. If $\sqrt[3]{\frac{1-a}{a}} + 9 = 19 - \sqrt[3]{\frac{a}{1-a}}$, then what is the value of a?

Solution :

$\sqrt[3]{\frac{1-a}{a}} + 9 = 19 - 3\sqrt[3]{\frac{a}{1-a}}$

$3\left(\sqrt{\frac{1-a}{a}} + \sqrt{\frac{a}{1-a}}\right) = 10$

$\sqrt{\frac{1-a}{a}} + \sqrt{\frac{a}{1-a}} = \frac{10}{3}$

By squaring both sides: $\frac{1-a}{a} + \frac{a}{1-a} = \frac{100}{9}$

$\frac{1 + a^2 - 2a + a^2 + 2a - 2a^2}{a(1-a)} = \frac{100}{9}$

$100a - 100a^2 = 9$

$100a^2 - 100a + 9 = 0$

$a = \frac{90}{100}, \frac{10}{100}$

$a = \frac{9}{10}, \frac{1}{10}$

102. If $a + b = 10$ and $\sqrt{\frac{a}{b}} - 13 = -\sqrt{\frac{b}{a}} - 11$, then what is the value of $3ab + 4a^2 + b^2$?

Solution :

$a + b = 10$ -------(i)

$\sqrt{\frac{a}{b}} + \sqrt{\frac{b}{a}} = 2$

By squaring both sides: $\frac{a}{b} + \frac{b}{a} + 2 = 4$

$\frac{a^2 + b^2}{ab} = 2$

$a^2 + b^2 = 2ab$

$(a - b)^2 = 0$

$a = b$

$a = b = 5$

$3 \times 5 \times 5 + 4 \times 25 + 5 \times 25 = 75 + 100 + 125 = 300$

103. $p^3 + q^3 + r^3 - 3pqr = 4$. If $a = q + r, b = r + p$, and $c = p + q$, then what is the value of $a^3 + b^3 + c^3 - 3abc$?

Solution :

$a + b + c = q + r + r + p + p + q = 2(p + q + r)$

$a^3 + b^3 + c^3 - 3abc = \frac{1}{2}(a + b + c)\left[(a - b)^2 + (b - c)^2 + (c - a)^2\right]$

$= (p + q + r)$

$\left[(q - p)^2 + (r - q)^2 + (p - r)^2\right]$

$= 2(p^3 + q^3 + r^3 - 3pqr)$

$= 2 \times 4 = 8$

Method II: By value substitution, equations = 4, variables = 2. If $q = r = 0$: $p^3 = 4, \quad a = 0, \quad b = p, \quad c = p$

$a^3 + b^3 + c^3 - 3abc = 0 + p^3 + p^3 - 0 = 2p^3 = 2 \times 4 = 8$

104. If $x_1x_2x_3 = 4(4 + x_1 + x_2 + x_3)$, then what is the value of $\frac{1}{2+x_1} + \frac{1}{2+x_2} + \frac{1}{2+x_3}$?

Solution :

$x_1x_2x_3 = 4(4 + x_1 + x_2 + x_3)$

Substituting $x_1 = x_2 = x_3 = 4$: $4 \times 4 \times 4 = 4(4 + 4 + 4 + 4)$

$64 = 64$

$\frac{1}{2+x_1} + \frac{1}{2+x_2} + \frac{1}{2+x_3} = \frac{1}{2+4} + \frac{1}{2+4} + \frac{1}{2+4}$

$= \frac{1}{6} + \frac{1}{6} + \frac{1}{6} = \frac{3}{6} = \frac{1}{2}$

105. If $a + b + c + d = 2$, then the maximum value of $(1 + a)(1 + b)(1 + c)(1 + d)$ is:

Solution :

$a + b + c + d = 2$ --------(Given)

For the maximum value, take $a = b = c = d$:

$a = b = c = d = \frac{1}{2}$

$(1+a)(1+b)(1+c)(1+d) = \frac{3}{2} \times \frac{3}{2} \times \frac{3}{2} \times \frac{3}{2} = \frac{81}{16}$

106. If x and y are real numbers, then the least possible value of $4(x - 2)^2 + (y - 3)^2 - 2(x - 3)^2$ is:

Solution :

$4(x - 2)^2 + (y - 3)^2 - 2(x - 3)^2$

$= 4(x - 2)^2 - 2(x - 3)^2 + (y - 3)^2$

$= 4(x^2 + 4 - 4x) - 2(x^2 + 9 - 6x) + (y - 3)^2$

$= (2x^2 - 4x - 2) + (y - 3)^2$

$= 2(x^2 - 2x - 1) + (y - 3)^2$

$= 2\{(x - 1)^2 - 2\} + (y - 3)^2$

$= 2(x - 1)^2 - 4 + (y - 3)^2$

$= 2(x - 1)^2 + (y - 3)^2 - 4$

The minimum possible value of the given expression is: -4

107. If x, y, and z are positive numbers and $x + y + z = 1$, then the least value of $\frac{1}{x} + \frac{1}{y} + \frac{1}{z}$ is:

Solution :

$AM \geq HM$

$\frac{x+y+z}{3} \geq \frac{3}{\frac{1}{x} + \frac{1}{y} + \frac{1}{z}}$

$\frac{1}{3} \geq \frac{3}{\frac{1}{x} + \frac{1}{y} + \frac{1}{z}}$

$\frac{1}{x} + \frac{1}{y} + \frac{1}{z} \geq 9$

The least value is 9.

108. A function of two variables varies directly with x and inversely with y. Determine the function when $x = 5$ and $y = 3$. Given that, for $x = 0$ and $y = 1$, $f = 15$, and for $x = 1$ and $y = 15$, $f = 2$.

Solution :

$F = \frac{ax + b}{y}$

At $x = 0$ and $y = 1$, $F = 15$: $15 = \frac{a \times 0 + b}{1}$

$b = 15$

At $x = 1$ and $y = 15$, $F = 2$: $2 = \frac{a(1) + 15}{15}$

$a = 2$

For $x = 5$ and $y = 3$: $F = \frac{2 \times 5 + 15}{3}$

$F = \frac{10 + 15}{3}$

$F = 15$

16

Surds and Indices

INDICES/ सूचकांक

In seventeenth century a French mathematician Reni Dakata' multiplied a number several times and showed the obtained product by a special rule, which called 'indices' and the converse of indices is called surds.

सत्रहवीं शताब्दी में एक फ्रांसीसी गणितज्ञ रेनी डकाटा ने किसी संख्या को कई बार गुणा किया और प्राप्त उत्पाद को एक विशेष नियम द्वारा दिखाया, जिसे 'सूचकांक' कहा जाता था और सूचकांकों के विपरीत को सर्ड कहा जाता था।

Rule 1: If any number is multiplied by the same the number 'n' times, then,

नियम 1: यदि किसी संख्या को उसी संख्या से 'n' बार गुणा किया जाए, तो,

$a \times a \times a \times a \cdot \cdot \cdot \times a$ (n times) $= a^n$

(i) where n and a are real numbers. (including fractions)
जहां n और a वास्तविक संख्याएं हैं। (अंश सहित)

(ii) a is called base.
a को आधार कहा जाता है।

(iii) n is called indices.
n को सूचकांक कहा जाता है।

Rule 2: $a^m \times a^n = a^{m+n}$ and $a^m \times a^n \times a^p = a^{m+n+p}$
While multiplying, if base is same then powers get added.
गुणा करते समय. यदि आधार समान है तो शक्तियाँ जुड़ जाती हैं।

Rule 3: While multiplying, if bases are different but powers are same then,
गुणा करने पर, यदि आधार अलग-अलग हैं लेकिन घातें समान हैं, तो,
$a^x \times b^x \times c^x = (abc)^x$

Rule 4: While dividing, if base is same then powers get subtracted, as
विभाजित करते समय यदि आधार समान हो तो शक्तियाँ कम हो जाती हैं, जैसे

$$\frac{a^m}{a^n} = a^{m-n}$$

Rule 5: If there is negative indices on a number, then
यदि किसी संख्या पर ऋणात्मक सूचकांक है, तो

$$a^{-m} = \frac{1}{a^m} \quad \text{or} \quad a^m = \frac{1}{a^{-m}}$$

Rule 6: If there are indices on indices, then indices are multiplied, as:
यदि सूचकांकों पर सूचकांक हैं, तो सूचकांकों को गुणा किया जाता है। जैसा-

$(a^m)^n = a^{mn}$, $(a^m)^{1/n} = a^{m/n}$ and $\{(a^m)^n\}^p = a^{mnp}$

Rule 7:

(i) $a^{m^n} \neq (a^m)^n$

(ii) $a^{m^{1/n}} \neq (a^m)^{1/n}$ **(NOTE)**

(iii) $a^{m^{n^p}} \neq \{(a^m)^n\}^p$ **(NOTE)**

Rule 8: Indices as fraction.
भिन्न के रूप में सूचकांक.

(i) $\left(\frac{a}{b}\right)^m = \frac{a^m}{b^m}$

(ii) $\left(\frac{a}{b}\right)^{-m} = \left(\frac{b}{a}\right)^m$

Rule 9: If $a^x = a^y$, then $x = y$, and if $x^n = y^n$, then $x = y$.

Laws of Surds/ घातांक के नियम

- $\sqrt[n]{a} = a^{1/n}$
- $\sqrt[n]{ab} = \sqrt[n]{a} \times \sqrt[n]{b} = a^{1/n} \times b^{1/n} = (ab)^{1/n}$
- $\sqrt[n]{\frac{a}{b}} = \frac{\sqrt[n]{a}}{\sqrt[n]{b}} = \left(\frac{a}{b}\right)^{1/n}$

- $\sqrt[n]{a}$ → Radicand; n → order of surd
- $(\sqrt[n]{a})^m = a^{m/n} = \sqrt[n]{a^m}$
- $(\sqrt[n]{a})^n = a^{n/n} = a$
- $\sqrt[n]{\sqrt[m]{a}} = \sqrt[n]{a^{1/m}} = a^{1/mn}$
- $\sqrt[z]{\sqrt[y]{\sqrt[x]{a}}} = a^{mno/xyz}$
- $\sqrt{a\sqrt{a\sqrt{a\sqrt{a \ldots \infty}}}} = a$

$\sqrt{a\sqrt{a\sqrt{a \ldots \text{n times}}}} = a^{1-\frac{1}{2^n}} = a^{\frac{2^n-1}{2^n}}$

$\sqrt[3]{a\sqrt[3]{a\sqrt[3]{a \ldots \infty}}} = \sqrt{a}$

$\sqrt[n]{a\sqrt[n]{a\sqrt[n]{a \ldots \infty}}} = \sqrt[n-1]{a}$

$\sqrt{a \div \sqrt{a \div \sqrt{a \ldots \infty}}} = \sqrt[3]{a}$

- $\sqrt{a + \sqrt{a + \sqrt{a + \ldots \infty}}} = \frac{\sqrt{4a+1}+1}{2} = x$

$\sqrt{a - \sqrt{a - \sqrt{a - \ldots \infty}}} = \frac{\sqrt{4a+1}-1}{2} = y$

$x - y = 1$

$xy = a$

- $x = \sqrt{a + \sqrt{b \times a + \sqrt{b \times a + \ldots \infty}}} = \frac{\sqrt{4a+b^2}+b}{2}$

$y = \sqrt{a - \sqrt{b \times a - \sqrt{b \times a - \ldots \infty}}} = \frac{\sqrt{4a+b^2}-b}{2}$

$x - y = b$

$xy = a$

- $x = \sqrt{a\sqrt{a - \sqrt{a + \ldots \infty}}} = \frac{\sqrt{4a-3}+1}{2}$

$y = \sqrt{a - \sqrt{a + \sqrt{a - \ldots \infty}}} = \frac{\sqrt{4a-3}-1}{2}$

$x - y = 1$

$xy = 2(a-1)$

- $\sqrt{a + \sqrt{b \times a - \sqrt{b \times a + \ldots \infty}}} = \frac{\sqrt{4a-3b^2}+b}{2}$

$\sqrt{a - \sqrt{b \times a + \sqrt{b \times a - \ldots \infty}}} = \frac{\sqrt{4a-3b^2}-b}{2}$

Type 1 (Problem based on LCM)

1. Find $\sqrt{3\sqrt{3\sqrt{3 \ldots}}}$ is equal to:

Solution:

Let $x = \sqrt{3\sqrt{3\sqrt{3 \ldots}}}$

$x = \sqrt{3x}$

$x^2 = 3x$

$x = 3$

2. Find $\sqrt{7\sqrt{7\sqrt{7 \ldots}}}$ (10 terms):

Solution:

Let $x = \sqrt{7\sqrt{7\sqrt{7 \ldots}}}$ (10 terms)

$x = 7^{1/2} \times 7^{1/4} \times 7^{1/8} \ldots$ (10 terms)

$x = 7^{[1/2+1/4+\ldots(10\text{ terms})]}$

The G.P. is $1/2,\ 1/4,\ \ldots$ with $a = 1/2$ and $r = 1/2$.

Sum = $\frac{1-(1/2)^{10}}{1-1/2} = \frac{1023}{1024}$

$x = 7^{1023/1024}$

3. The value of $\sqrt{2\sqrt[3]{4\sqrt{2\sqrt[3]{4 \ldots}}}}$ is:

Solution:

Let $x = \sqrt{2\sqrt[3]{4x}}$

$x^2 = 2\sqrt[3]{4x}$

$x^6 = 8 \times 4x$

$x^5 = 32$

$x = 2$

4. Find the value of $\sqrt{12 + \sqrt{12 + \sqrt{12 + \ldots}}}$:

Solution:

Let $x = \sqrt{12 + x}$

$x^2 = 12 + x$

$x^2 - x - 12 = 0$

$x = 4$

5. Find the value of $\sqrt{7 + \sqrt{7 + \sqrt{7 + \ldots \infty}}}$:

Solution:

Let $x = \sqrt{7 + x}$

$x^2 = 7 + x$

$x^2 - x - 7 = 0$

$$x = \frac{1 \pm \sqrt{1 + 28}}{2}$$

$$x = \frac{1 + \sqrt{29}}{2}$$

6. Find the value of $\sqrt{72 - \sqrt{72 - \sqrt{72 \ldots \infty}}}$:

Solution:

Let $x = \sqrt{72 - x}$

$x^2 = 72 - x$

$x^2 + x - 72 = 0$

$x = 8$

7. Find the value of $\sqrt{17 - \sqrt{17 - \sqrt{17 \ldots \infty}}}$:

Solution:

Let $x = \sqrt{17 - x}$

$x^2 = 17 - x$

$x^2 + x - 17 = 0$

$$x = \frac{-1 \pm \sqrt{1 + 68}}{2} = \frac{-1 + \sqrt{69}}{2}$$

8. Find $\sqrt{35 + 2\sqrt{35 + 2\sqrt{35 + 2 \ldots \infty}}}$:

Solution:

Let $x = \sqrt{35 + 2x}$

$x^2 = 35 + 2x$

$x^2 - 2x - 35 = 0$

$x = 7$

9. Find $\sqrt{54 - 3\sqrt{54 - 3\sqrt{54 \ldots \infty}}}$:

Solution:

Let $x = \sqrt{54 - 3x}$

$x^2 = 54 - 3x$

$x^2 + 3x - 54 = 0$

$x = 6$

Formula:

If $x = \sqrt{a - b\sqrt{a + b\sqrt{a - b\sqrt{a \ldots \infty}}}}$,

$$x = \frac{\sqrt{4a - 3b^2} - b}{2}$$

If $x = \sqrt{a + b\sqrt{a - b\sqrt{a + b\sqrt{a - b \ldots \infty}}}}$,

$$x = \frac{\sqrt{4a - 3b^2} + b}{2}$$

10. Find $\sqrt{3 - \sqrt{3 + \sqrt{3 - \sqrt{3 + \ldots \infty}}}}$:

Solution:

Using the formula $x = \dfrac{\sqrt{4a - 3b^2} - b}{2}$ where

$a = 3,\ b = 1$:

$$x = \frac{\sqrt{12 - 3} - 1}{2}$$

$$x = \frac{3 - 1}{2} = 1$$

11. Find $\sqrt{4 + \sqrt{4 - \sqrt{4 + \ldots \infty}}}$:

Solution:

$$x = \frac{\sqrt{4a - 3b^2} + b}{2} \quad (a = 4,\ b = 1)$$
$$x = \frac{\sqrt{16 - 3} + 1}{2}$$
$$x = \frac{\sqrt{13} + 1}{2}$$

12. Find $\sqrt{12 + 2\sqrt{12 - 2\sqrt{12 + 2\sqrt{12 - 2 \ldots \infty}}}}$:

Solution:

$$x = \frac{\sqrt{4a - 3b^2} + b}{2} \quad (a = 12,\ b = 2)$$
$$x = \frac{\sqrt{48 - 12} + 2}{2} = \frac{6 + 2}{2}$$
$$x = 4$$

13. Find $\sqrt{12 - 2\sqrt{12 + 2\sqrt{12 - 2\sqrt{12 + 2 \ldots \infty}}}}$:

Solution:

$$x = \frac{\sqrt{4a - 3b^2} - b}{2} \quad (a = 12,\ b = 2)$$
$$x = \frac{\sqrt{48 - 12} - 2}{2} = \frac{6 - 2}{2}$$
$$x = \frac{4}{2} = 2$$

14. If $\sqrt{x}^{\sqrt{x}^{\sqrt{x}^{\cdots\infty}}} = 4$, then find x:

Solution:

$$(\sqrt{x})^4 = 4$$
$$((x)^{1/2})^4 = 4$$
$$x^2 = 4$$
$$x = 2$$

15. Find $\sqrt{1 + 2\sqrt{1 + 3\sqrt{1 + 4\sqrt{1 + 5 \ldots \infty}}}}$:

Solution:

$$x = 3 = \sqrt{9} = \sqrt{1 + 8}$$
$$= \sqrt{1 + 2 + 4}$$
$$= \sqrt{1 + 2\sqrt{16}}$$
$$= \sqrt{1 + 2\sqrt{1 + 3\sqrt{1 + 4}}}$$

16. If $x = \sqrt{-\sqrt{3} + \sqrt{3 + 8\sqrt{7 + 4\sqrt{3}}}}$ where $x > 0$, find the value of x:

Solution:

$$x = \sqrt{-\sqrt{3} + \sqrt{3 + 8\sqrt{(2 + \sqrt{3})^2}}}$$
$$= \sqrt{-\sqrt{3} + \sqrt{3 + 16 + 8\sqrt{3}}}$$
$$= \sqrt{-\sqrt{3} + 4 + \sqrt{3}}$$
$$= \sqrt{4} = 2$$

17. What is the value of $\sqrt{18 - 2\sqrt{77}}$:

Solution:

$$\sqrt{18 - 2\sqrt{77}}$$
$$\Rightarrow \sqrt{11 + 7 - 2\sqrt{11} \times \sqrt{7}}$$
$$\Rightarrow \sqrt{(\sqrt{11})^2 + (\sqrt{7})^2 - 2\sqrt{11} \times \sqrt{7}}$$
$$\Rightarrow (\sqrt{11} - \sqrt{7})^2$$
$$= \sqrt{11} - \sqrt{7}$$

18. The value of $\frac{1}{\sqrt{17 + 12\sqrt{2}}}$ is closest to:

Solution:

$$\frac{1}{\sqrt{17 + 12\sqrt{2}}} = \frac{1}{\sqrt{9 + 8 + 2 \times 3 \times 2\sqrt{2}}}$$

$$\frac{1}{\sqrt{(3+2\sqrt{2})^2}} = \frac{1}{3+2\sqrt{2}}$$
$$\frac{(3-2\sqrt{2})}{(3+2\sqrt{2})(3-2\sqrt{2})}$$
$$\frac{(3-2\times 1.414)}{9-8} = \frac{(3-2.828)}{1} = 0.17$$

19. If $\sqrt{1+\frac{\sqrt{3}}{2}}-\sqrt{1-\frac{\sqrt{3}}{2}} = C$, then the value of C is:

Solution:

$$\sqrt{\frac{2+\sqrt{3}}{2}} - \sqrt{\frac{2-\sqrt{3}}{2}} = C$$

On squaring both sides:

$$2\left(\frac{2+\sqrt{3}}{2}\right)+\sqrt{2-\sqrt{3}}-\sqrt{2+\sqrt{3}} = C^2$$
$$2-1 = C^2$$
$$C^2 = 1$$
$$C = \pm 1$$

Taking the positive value, $C = 1$

20. If $\frac{8+2\sqrt{3}}{3\sqrt{3}+5} = a\sqrt{3}-b$, find $a+b$:

Solution:

$$\frac{8+2\sqrt{3}}{3\sqrt{3}+5}\times\frac{3\sqrt{3}-5}{3\sqrt{3}-5} = a\sqrt{3}-b$$
$$= \frac{24\sqrt{3}-40+6\times 3-10\sqrt{3}}{27-15\sqrt{3}+15\sqrt{3}-25}$$
$$= \frac{24\sqrt{3}-40+18-10\sqrt{3}}{2} =$$
$$\frac{14\sqrt{3}-22}{2} = 7\sqrt{3}-11$$

Comparing both sides, $a = 7,\ b = 11$

$$a+b = 7+11 = 18$$

21. What is the value of $\sqrt{4600+\sqrt{540+\sqrt{1280+\sqrt{250+\sqrt{36}}}}}$?

Solution:

$$\sqrt{4600+\sqrt{540+\sqrt{1280+\sqrt{250+\sqrt{36}}}}}$$
$$= \sqrt{4600+\sqrt{540+\sqrt{1280+\sqrt{256}}}}$$
$$= \sqrt{4600+\sqrt{540+\sqrt{1280+16}}}$$
$$= \sqrt{4600+\sqrt{540+36}}$$
$$= \sqrt{4600+24}$$
$$= \sqrt{4624}$$
$$= 68$$

22. If $P = 7+4\sqrt{3}$ and $PQ = 1$, then what is the value of $\frac{1}{P^2}+\frac{1}{Q^2}$?

Solution:

$$P = 7+4\sqrt{3}$$
$$\frac{1}{P} = \frac{1}{7+4\sqrt{3}}\times\frac{7-4\sqrt{3}}{7-4\sqrt{3}} = 7-4\sqrt{3}$$

And $PQ = 1$, so $P = \frac{1}{Q}$

$$\therefore \frac{1}{P^2}+\frac{1}{Q^2} = \left(\frac{1}{P}\right)^2 + P^2$$
$$= (7-4\sqrt{3})^2+(7+4\sqrt{3})^2$$
$$= 49+48-56\sqrt{3}+49+48+56\sqrt{3}$$
$$= 194$$

23. The value of $\sqrt{28+10\sqrt{3}}-\sqrt{7-4\sqrt{3}}$ is closest to:

Solution:

$$\sqrt{28+10\sqrt{3}}-\sqrt{7-4\sqrt{3}}$$
$$= \sqrt{25+3+10\sqrt{3}}-\sqrt{4+3-4\sqrt{3}}$$
$$= \sqrt{(5+\sqrt{3})^2}-\sqrt{(2-\sqrt{3})^2}$$
$$= 5+\sqrt{3}-(2-\sqrt{3})$$

$$= 5 + \sqrt{3} - 2 + \sqrt{3}$$
$$= 3 + 2\sqrt{3}$$
$$= 3 + 2 \times 1.732 = 6.464 \approx 6.5$$

24. The expression $\sqrt{10 + 2(\sqrt{6} - 15 - \sqrt{10})}$ is equal to:

Solution:

$$\sqrt{10 + 2(\sqrt{6} - \sqrt{15} - \sqrt{10})}$$
$$= \sqrt{10 + 2\sqrt{6} - 2\sqrt{15} - 2\sqrt{10}}$$
$$= \sqrt{(\sqrt{3})^2 + (\sqrt{2})^2 + (\sqrt{5})^2 + 2 \times \sqrt{3} \times \sqrt{2}}$$
$$\sqrt{-2 \times \sqrt{3} \times \sqrt{5} - 2 \times \sqrt{2} \times \sqrt{5}}$$
$$= \sqrt{(\sqrt{3} + \sqrt{2} - \sqrt{5})^2}$$
$$= (\sqrt{3} + \sqrt{2} - \sqrt{5})$$

25. If $\sqrt{10 - 2\sqrt{21}} + \sqrt{8 + 2\sqrt{15}} = \sqrt{a} + \sqrt{b}$, where a and b are positive integers, then the value of $\sqrt{ab}$ is closest to:

Solution:

$$\sqrt{10 - 2\sqrt{21}} + \sqrt{8 + 2\sqrt{15}} = \sqrt{a} + \sqrt{b}$$
$$\sqrt{(\sqrt{7} - \sqrt{3})^2} + \sqrt{(\sqrt{5} + \sqrt{3})^2} = \sqrt{a} + \sqrt{b}$$
$$(\sqrt{7} - \sqrt{3}) + (\sqrt{5} + \sqrt{3}) = \sqrt{a} + \sqrt{b}$$
$$\sqrt{7} + \sqrt{5} = \sqrt{a} + \sqrt{b}$$

On comparison: $a = 7,\ b = 5$

$$\sqrt{ab} = \sqrt{7 \times 5} = \sqrt{35}$$
$$\sqrt{ab} = \sqrt{35} \approx 5.9$$

26. If $(\sqrt{2} + \sqrt{5} - \sqrt{3}) \times k = -12$, then what is the value of k?

Solution:

$$(\sqrt{2} + \sqrt{5} - \sqrt{3}) \times k = -12$$
$$k = \frac{-12}{\sqrt{2} + \sqrt{5} - \sqrt{3}} \times \frac{\sqrt{2} + \sqrt{5} + \sqrt{3}}{\sqrt{2} + \sqrt{5} + \sqrt{3}}$$
$$= \frac{-12(\sqrt{2} + \sqrt{5} + \sqrt{3})}{(\sqrt{2} + \sqrt{5})^2 - (\sqrt{3})^2}$$
$$= \frac{-12(\sqrt{2} + \sqrt{5} + \sqrt{3})}{2 + 5 + 2\sqrt{10} - 3}$$
$$= (\sqrt{2} + \sqrt{5} + \sqrt{3})(2 - \sqrt{10})$$

27. If $x = \sqrt{1 + \frac{\sqrt{3}}{2}} - \sqrt{1 - \frac{\sqrt{3}}{2}}$, then the value of $\frac{\sqrt{2} - x}{\sqrt{2} + x}$ will be closest to:

Solution:

$$x = \sqrt{1 + \frac{\sqrt{3}}{2}} - \sqrt{1 - \frac{\sqrt{3}}{2}}$$

On squaring both sides:

$$x^2 = 1 + \frac{\sqrt{3}}{2} + 1 - \frac{\sqrt{3}}{2} - 2 \times \sqrt{(1 + \frac{\sqrt{3}}{2})(1 - \frac{\sqrt{3}}{2})}$$
$$= 2 - 2 \times \sqrt{1^2 - \left(\frac{\sqrt{3}}{2}\right)^2}$$
$$= 2 - 2 \times 1 = 1$$

Hence $x = 1$

$$\therefore \frac{\sqrt{2} - 1}{\sqrt{2} + 1} = \frac{1.414 - 1}{1.414 + 1} = \frac{0.414}{2.414} = 0.17$$

28. If $\sqrt{86 - 60\sqrt{2}} = a - b\sqrt{2}$, then what will be the value of $\sqrt{a^2 + b^2}$ correct to one decimal place?

Solution:

$$\sqrt{86 - 60\sqrt{2}} = \sqrt{36 + 50 - 60\sqrt{2}}$$
$$= \sqrt{6^2 + (5\sqrt{2})^2 - 2 \times 6 \times 5\sqrt{2}}$$
$$= \sqrt{(6 - 5\sqrt{2})^2} = 6 - 5\sqrt{2} = a - b\sqrt{2}$$

On comparing both sides: $a = 6,\ b = 5$

$$\sqrt{a^2 + b^2} = \sqrt{36 + 25} = \sqrt{61} \approx 7.8$$

29. The value of $\sqrt{9-2\sqrt{11-6\sqrt{2}}}$ is closest to:

Solution:

$\sqrt{9-2\sqrt{11-6\sqrt{2}}}$

$=\sqrt{9-2\sqrt{9+2-6\sqrt{2}}}$

$=\sqrt{9-2\sqrt{(3-\sqrt{2})^2}}$

$=\sqrt{9-2(3-\sqrt{2})}$

$=\sqrt{3+2\sqrt{2}}$

$=\sqrt{2+1+2\sqrt{2}}$

$=\sqrt{(\sqrt{2}+1)^2}$

$=\sqrt{2}+1$

$=1.414+1=2.414\approx 2.4$

30. The value of $5\sqrt{3}+7\sqrt{2}-\sqrt{6}-\dfrac{23}{\sqrt{2}+\sqrt{3}+\sqrt{6}}$ is:

Solution:

$5\sqrt{3}+7\sqrt{2}-\sqrt{6}-\dfrac{23}{(\sqrt{2}+\sqrt{3})+\sqrt{6}}\times\dfrac{(\sqrt{2}+\sqrt{3})-\sqrt{6}}{(\sqrt{2}+\sqrt{3})-\sqrt{6}}$

$=5\sqrt{3}+7\sqrt{2}-\sqrt{6}-\dfrac{23(\sqrt{2}+\sqrt{3}-\sqrt{6})}{(\sqrt{2}+\sqrt{3})^2-6}$

$=5\sqrt{3}+7\sqrt{2}-\sqrt{6}-\dfrac{23(4\sqrt{3}+6\sqrt{2}-12+\sqrt{2}+\sqrt{3}-\sqrt{6})}{23}$

$=5\sqrt{3}+7\sqrt{2}-\sqrt{6}-5\sqrt{3}-7\sqrt{2}+12+\sqrt{6}$

$=12$

31. If $\dfrac{4}{1+\sqrt{2}+\sqrt{3}}=a+b\sqrt{2}+c\sqrt{3}-d\sqrt{6}$, where a, b, c, d are natural numbers, then the value of $a+b+c+d$ is:

Solution:

$\dfrac{4}{1+\sqrt{2}+\sqrt{3}}=a+b\sqrt{2}+c\sqrt{3}-d\sqrt{6}$

$=\dfrac{4(1-\sqrt{2}-\sqrt{3})}{(1+\sqrt{2}+\sqrt{3})^2}$

$=\dfrac{4(1-\sqrt{2}-\sqrt{3})(4-2\sqrt{6})}{16-24}$

$=2-2\sqrt{2}-2\sqrt{3}-\sqrt{6}+2\sqrt{3}+3\sqrt{2}$

$=2+\sqrt{2}-\sqrt{6}$

Comparing both sides: $a=2$, $b=1$, $c=0$, $d=1$

$a+b+c+d=2+1+0+1=4$

32. If the value of $\dfrac{3x\sqrt{y}+2y\sqrt{x}}{3x\sqrt{y}-2y\sqrt{x}}-\dfrac{3x\sqrt{y}-2y\sqrt{x}}{3x\sqrt{y}+2y\sqrt{x}}$ is the same as $\sqrt{x}\sqrt{y}$, then which of the following relations between x and y is correct?

Solution:

$\dfrac{a+b}{a-b}-\dfrac{a-b}{a+b}=\dfrac{4ab}{a^2-b^2}$

$\dfrac{3x\sqrt{y}+2y\sqrt{x}}{3x\sqrt{y}-2y\sqrt{x}}-\dfrac{3x\sqrt{y}-2y\sqrt{x}}{3x\sqrt{y}+2y\sqrt{x}}=\dfrac{4\times 3x\sqrt{y}\times 2y\sqrt{x}}{(3x\sqrt{y})^2-(2y\sqrt{x})^2}$

According to the question:

$\dfrac{24xy\sqrt{x}\sqrt{y}}{9x^2y-4y^2x}=\sqrt{x}\sqrt{y}$

$\dfrac{24xy}{xy(9x-4y)}=1$

$\therefore 24=9x-4y$

$9x-4y=24$

33. Which of the following statement(s) is/are TRUE?

I. $\sqrt{5}+\sqrt{5}>\sqrt{7}+\sqrt{3}$

II. $\sqrt{6}+\sqrt{7}-\sqrt{8}+\sqrt{5}$

III. $\sqrt{3}+\sqrt{9}>\sqrt{6}+\sqrt{6}$

(a) Only I (b) Only I and II

(c) Only II and III (d) Only I and III

Solution:

From statement (I):

$\sqrt{5}+\sqrt{5}>\sqrt{7}+\sqrt{3}$

On squaring both sides:

$(\sqrt{5}+\sqrt{5})^2 > (\sqrt{7}+\sqrt{3})^2$

$= 10 + 2\sqrt{25} > 10 + 2\sqrt{21}$ (True)

From statement (II):

$\sqrt{6}+\sqrt{7} > \sqrt{8}+\sqrt{5}$

Squaring both sides:

$(\sqrt{6}+\sqrt{7})^2 > (\sqrt{8}+\sqrt{5})^2$

$= 13 + 2\sqrt{42} > 13 + 2\sqrt{40}$ (True)

From statement (III):

$(\sqrt{3}+\sqrt{9}) > (\sqrt{6}+\sqrt{6})$

Squaring both sides:

$(\sqrt{3}+\sqrt{9})^2 > (\sqrt{6}+\sqrt{6})^2$

$= 12 + 2\sqrt{27} > 12 + 2\sqrt{36}$ (False)

Hence, only statements I and II are true.

34. Which of the following statement(s) is/are TRUE?
I. $\sqrt{64}+\sqrt{0.0064}+\sqrt{0.81}+\sqrt{0.0081} = 9.07$
II. $\sqrt{0.010201}+\sqrt{98.01}+\sqrt{0.25} = 11.51$
(a) Only I (b) Only II
(c) Both I and II (d) Neither I nor II

Solution:

From statement (I):

L.H.S. $= \sqrt{64}+\sqrt{0.0064}+\sqrt{0.81}+\sqrt{0.0081}$

$= 8 + 0.08 + 0.9 + 0.09$

$= 9.07 =$ R.H.S. (True)

From statement (II):

L.H.S. $= \sqrt{0.010201}+\sqrt{98.01}+\sqrt{0.25}$

$= 0.101 + 9.9 + 0.5$

$= 10.501 \neq$ R.H.S. (False)

Hence, only statement I is correct.

35. Which of the following statement(s) is/are TRUE?
I. $\sqrt{12} > \sqrt[3]{16} > \sqrt[4]{24}$
II. $\sqrt[3]{25} > \sqrt[4]{32} > \sqrt[6]{48}$
III. $\sqrt[4]{9} > \sqrt[3]{15} > \sqrt[6]{24}$
(a) Only I and II (b) Only I and III
(c) Only I (d) All are true

Solution:

From statement I:

$\sqrt{12} > \sqrt[3]{16} > \sqrt[4]{24}$

Or $12^{1/2} > 16^{1/3} > 24^{1/4}$

L.C.M of 2, 3, 4 = 12:

$12^6 > 16^4 > 24^3$

$144 \times 144 \times 144 > 256 \times 256 > 24 \times 24 \times 24$

So, it is clear that statement I is true.

From statement II:

$\sqrt[3]{25} > \sqrt[4]{32} > \sqrt[6]{48}$

Or $25^{1/3} > 32^{1/4} > 48^{1/6}$

L.C.M of 3, 4, 6 = 12:

$25^4 > 32^3 > 48^2$

So, it is clear that statement II is true.

From statement III:

$\sqrt[4]{9} > \sqrt[3]{15} > \sqrt[6]{24}$

Or $9^{1/4} > 15^{1/3} > 24^{1/6}$

L.C.M of 3, 4, 6 = 12:

$9^3 > 15^4 > 24^2$

Statement III is false.

Hence, only statements I and II are true.

36. Which of the following statement(s) is/are TRUE?
I. $\sqrt{625}+\sqrt[4]{1296}+\sqrt{1024} > 90$

II. $\sqrt[3]{\sqrt{729}}+\sqrt[4]{\sqrt{256}} = 5$

(a) Only I (b) Only II
(c) Neither I nor II (d) Both I and II

Solution:

From statement (I):

$\sqrt{625} + \sqrt[4]{1296} + \sqrt{1024}$

$= 25 + 6 + 32 = 63 \not> 90$ (False)

From statement (II):

$\sqrt[3]{\sqrt{729}} + \sqrt[4]{\sqrt{256}}$

$\sqrt[3]{27} + \sqrt[4]{16}$

$= 3 + 2 = 5$ (True)

Hence, it is clear that only statement II is true.

37. Which one among the following is the smallest?
(a) $\sqrt{301} - \sqrt{299}$ (b) $\sqrt{401} - \sqrt{399}$
(c) $\sqrt{101} - \sqrt{99}$ (d) $\sqrt{201} - \sqrt{199}$

Solution:

$\sqrt{301} - \sqrt{299} = 17.349 - 17.291 = 0.058$

$\sqrt{401} - \sqrt{399} = 20.024 - 19.974 = 0.05$

$\sqrt{101} - \sqrt{99} = 10.049 - 9.949 = 0.1$

$\sqrt{201} - \sqrt{199} = 14.177 - 14.106 = 0.071$

So, the smallest number is $\sqrt{401} - \sqrt{399}$.

38. Which of the following statement(s) is/are TRUE?
I. $\frac{2}{3\sqrt{5}} < \frac{3}{2\sqrt{5}} < \frac{5}{4\sqrt{3}}$
II. $\frac{3}{2\sqrt{5}} < \frac{2}{3\sqrt{3}} < \frac{7}{4\sqrt{5}}$
(a) Only I (b) Only II
(c) Both I and II (d) Neither I nor II

Solution:

From statement I:
$\frac{2}{3\sqrt{5}} < \frac{3}{2\sqrt{5}} < \frac{5}{4\sqrt{3}}$
L.C.M of $3\sqrt{5},\ 2\sqrt{5},\ 4\sqrt{3} = 12\sqrt{15}$:

$8\sqrt{3} < 18\sqrt{3} < 15\sqrt{5}$ (True)

From statement II:
$\frac{3}{2\sqrt{5}} < \frac{2}{3\sqrt{3}} < \frac{7}{4\sqrt{5}}$
L.C.M of $2\sqrt{5},\ 3\sqrt{3},\ 4\sqrt{5} = 12\sqrt{15}$:

$18\sqrt{3} < 8\sqrt{5} < 21\sqrt{3}$ (False)

Hence, only statement I is correct.

39. Which of the following is TRUE?
I. $3\sqrt[3]{11} > \sqrt{7} > 4\sqrt[4]{5}$
II. $\sqrt{7} > 3\sqrt[3]{11} > 4\sqrt[4]{5}$
III. $\sqrt{7} > 4\sqrt[4]{45} > 3\sqrt[3]{11}$
IV. $4\sqrt[4]{45} > \sqrt{7} > 3\sqrt[3]{11}$
(a) Only I (b) Only II
(c) Only III (d) Only IV

Solution:

Given: $3\sqrt[3]{11},\ \sqrt{7},\ 4\sqrt[4]{45}$

L.C.M of $3,\ 2,\ 4 = 12$:

$11^{1/(3\times12)},\ 7^{1/(2\times12)},\ 45^{1/(4\times12)}$

$= 114,\ 76,\ 453$

$= 14641,\ 117649,\ 91125$

Hence, $\sqrt{7} > 4\sqrt[4]{45} > 3\sqrt[3]{11}$.

17 Simplification

An expression must be simplified by following a defined order/sequence known as VBODMAS, which is given by:
नियम: एक अभिव्यक्ति को निम्नलिखित परिभाषित क्रम/अनुक्रम द्वारा सरलीकृत किया जाना चाहिए जिसे VBODMAS के रूप में जाना जाता है, जो कि दिया गया है:

V - Vineculum (line brackets)/Bar	वी - विनेकुलम (लाइन ब्रैकेट)/बार
B - Brackets	बी - कोष्ठक
O - Of	ओ - का
D - Division	डी - प्रभाग
M - Multiplication	एम - गुणन
A - Addition	ए - जोड़
S - Subtraction	एस - घटाव

There are four types of brackets given below:
नीचे चार प्रकार के ब्रैकेट दिए गए हैं।

1. $-\rightarrow$ Line/Bar लाइन/बार
2. () $\rightarrow$ Simple or Small Bracket/open brackets सरल या छोटे ब्रैकेट/खुले ब्रैकेट
3. {} $\rightarrow$ Curly Brackets/Braces घुंघराले ब्रैकेट/ब्रेसिज़
4. [] $\rightarrow$ Square Brackets/Closed brackets वर्गाकार कोष्ठक/बंद कोष्ठक

These brackets must be solved in the given order only.
इन कोष्ठकों को दिए गए क्रम में ही हल करना चाहिए।

1. The value of $3.8 - \left(\frac{4.2}{0.7} \times 3\right) + 5 \times 2 \div 0.5$ is:

Solution:

$$3.8 - \left(\frac{4.2}{0.7} \times 3\right) + 5 \times 2 \div 0.5$$
$$= 3.8 - (6 \times 3) + 5 \times 4$$
$$= 3.8 - 18 + 20$$
$$= 5.8$$

2. Find the value of the following expression:
$$\frac{(7.03)^3 - 0.027)}{(7.03)^2 + 2.109 + (0.3)^2}$$

Solution:

$$\frac{(7.03)^3 - 0.027}{(7.03)^2 + 2.109 + (0.3)^2}$$
$$= \frac{(7.03)^3 - (0.3)^3}{(7.03)^2 + 2.109 + (0.3)^2}$$
$$= \frac{(7.03 - 0.3)\ ((7.03)^2 + 2.109 + (0.3)^2)}{(7.03)^2 + 2.109 + (0.3)^2}$$
$$= 6.73$$

3. Find the value of the following expression:
$$\frac{\left(1\frac{1}{2} \div \frac{5}{6} \times 6 + \frac{4}{5} \times \frac{1}{2} + \frac{2}{3}\right)}{\left(2 - \left[1\frac{1}{3} \times \left(-\frac{3}{5}\right) - 6\left\{\frac{3}{5} - \left(3 - \frac{3}{10}\right)\right\}\right]\right)}$$

Solution:

$$\frac{\left(\frac{5}{6}\times\frac{6}{5}\times 6+\frac{2}{5}+\frac{2}{3}\right)}{\left(2-\left[\frac{4}{3}\times\left(-\frac{3}{5}\right)-6\left\{\frac{3}{5}-\frac{27}{10}\right\}\right]\right)}$$

$$=\frac{\frac{196}{15}}{\left(2-\left(-\frac{4}{5}+6\times-\frac{21}{10}\right)\right)}$$

$$=\frac{\frac{196}{15}}{-\frac{49}{5}}=-\frac{4}{3}$$

4. The value of:

$$\frac{2}{7}-\frac{3}{8}-\left[\frac{2\frac{1}{4}}{3\frac{1}{2}}\text{ of }1\frac{1}{3}+\left\{1\frac{17}{40}-\left(3-1\frac{1}{5}-\frac{3}{8}\right)\right\}\right]$$

Solution:

$$\frac{2}{7}-\frac{3}{8}-$$

$$\left[\frac{9}{4}\div\frac{7}{2}\times\frac{4}{3}+\left\{\frac{57}{40}-\left(\frac{120-48-15}{40}\right)\right\}\right]$$

$$=\frac{2}{7}-\frac{3}{8}-\left[\frac{27}{56}+\left\{\frac{57}{40}-\frac{57}{40}\right\}\right]$$

$$=\frac{2}{7}-\frac{3}{8}-\frac{27}{56}+0$$

$$=\frac{16-21-27}{56}=-\frac{32}{56}=-\frac{4}{7}$$

Hence, option (b) is correct.

5. The simplified value of:

$$\frac{119-\left[48\div 6-7\left\{5\times\frac{12}{3}-(15-(3-8))\right\}\right]}{24\div 3-9\text{ of }3+(52-(8-4))\div 6}$$

Solution:

$$\frac{119-[8-7\{20-(15+5)\}]}{8-27+\frac{52-4}{6}}$$

$$=\frac{119-[8-7\{20-20\}]}{8-27+8}$$

$$=\frac{119-(8-0)}{-11}$$

$$=\frac{111}{-11}=-11$$

6. If '+' means '-', '-' means '+', '×' means '÷', and '÷' means '÷', then the value of $(42-12\times 3+8+2+15)\div(8\times 2-4+9\div 3)$ is:

Solution:

$(42-12\times3+8+2+15)\div(8\times2-4+9\div3)$

As per the question, changing the symbols:

$=(42+12\div3-8\times2-15)\div(8\div2+4-9\times3)$

$=(46-31)\div(8-27)$

$=-15/-19$

7. If $M=\left(\frac{3}{7}\div\frac{6}{5}\times\frac{2}{3}\right)+\left(\frac{1}{5}\times\frac{3}{2}\right)$ and $N=\left(\frac{2}{5}\times\frac{5}{6}+\frac{1}{3}+\frac{3}{5}\times\frac{2}{3}\div\frac{3}{5}\right)$, then what is the value of M/N?

Solution:

$$M=\frac{3}{7}\times\frac{5}{6}\times\frac{2}{3}+\frac{3}{10}$$

$$=\frac{5}{21}+\frac{3}{10}=\frac{113}{210}$$

$$N=\frac{2}{5}\times\frac{5}{6}\div\frac{1}{3}+\frac{3}{5}\times\frac{2}{3}\div\frac{3}{5}$$

$$=\frac{2}{5}\times\frac{5}{6}\times\frac{3}{1}+\frac{3}{5}\times\frac{2}{3}\times\frac{5}{3}$$

$$=1+\frac{2}{3}$$

$$=\frac{5}{3}$$

According to the question:

$$\frac{M}{N}=\frac{\frac{113}{210}}{\frac{5}{3}}$$

$$=\frac{113\times 3}{210\times 5}=\frac{113}{350}$$

8. What is the value of $(0.08\% \text{ of } 0.008\% \text{ of } 8)^{1/9}$?

Solution:

$(8 \text{ of } 0.008\% \text{ of } 0.08\%)^{1/9}$

$= \left(8 \times \frac{0.008}{100} \times \frac{0.08}{100}\right)^{1/9}$

$= (8 \times 8 \times 8 \times 10^{-9})^{1/9}$

$= (2^9 \times 10^{-9})^{1/9}$

$= 2 \times 10^{-1}$

$= 0.2$

9. If x is the square of the number when $\frac{2}{5}$ of $\frac{25}{4} \div \frac{3}{7}$ of $\frac{9}{7}$ is divided by $11\frac{1}{4}$, then the value of $81x$ is:

Solution:

$\frac{2}{5}$ of $\frac{25}{4} \div \frac{3}{7}$ of $\frac{9}{7}$

$= \frac{2}{5} \times \frac{25}{4} \times \frac{7}{3}$

$= \frac{5}{2} \times 3 = \frac{15}{2}$

According to the question:

$\frac{\frac{15}{2}}{\frac{45}{4}} = \frac{60}{90} = \frac{2}{3}$

$\therefore x = \left(\frac{2}{3}\right)^2 = \frac{4}{9}$

(According to the question, x is squared)

$\therefore 81x = 81 \times \frac{4}{9} = 9 \times 4 = 36$

10. What is the value of the square root of:

$\sqrt{[\{(100 \text{ of } 0.9 \times 0.8 - 7 \times 1.2 \div 0.2 + 5 \times 4 - 3 \times 2)\} \div 10 + 1.85]} = ?$

Solution:

$\sqrt{[\{(100 \text{ of } 0.9 \times 0.8 - 7 \times 1.2 \div 0.2 + 5 \times 4 - 3 \times 2)\} \div 10 + 1.85]}$

$= \sqrt{[(90 \times 0.8 - 7 \times 6 + 20 - 6) \div 10 + 1.85]}$

$= \sqrt{[(72 - 42 + 20 - 6) \div 10 + 1.85]}$

$= \sqrt{[(72 - 42 + 14) \div 10 + 1.85]}$

$= \sqrt{[44 \div 10 + 1.85]}$

$= \sqrt{4.4 + 1.85}$

$= \sqrt{6.25}$

$= 2.5$

11. The value of $11.\overline{4} + 22.\overline{567} - 33.\overline{59}$ is:

Solution:

$11.\overline{4} + 22.\overline{567} - 33.\overline{59}$

$= 11 + 0.\overline{4} + 22 + 0.\overline{567} - 33 - 0.\overline{59}$

$= \frac{4}{9} + \frac{562}{990} - \frac{54}{90}$

$= \frac{440 + 562 - 594}{990}$

$= \frac{408}{990}$

$= 0.4\overline{12}$

12. The value of $\frac{(2.53)^3 + (2.47)^3}{25.3 \times 25.3 - 624.91 + 24.7 \times 24.7}$ is:

Solution:

$\frac{(2.53)^3 + (2.47)^3}{25.3 \times 25.3 - 25.3 \times 24.7 + 24.7 \times 24.7}$

$= 5 \times 10^k$

Let $a = 2.53,\ b = 2.47$:

$\frac{a^3 + b^3}{100(a^2 - ab + b^2)} = 5 \times 10^k$

$\frac{(a+b)(a^2 + b^2 - ab)}{100(a^2 - ab + b^2)} = 5 \times 10^k$

$\frac{2.53 + 2.47}{100} = 5 \times 10^k$

$\frac{5.00}{100} = 5 \times 10^k$

$5 \times 10^{-2} = 5 \times 10^k$

On comparing powers, $k = -2$.

13. The value of $0.\overline{46} + 0.\overline{723} - 0.\overline{39} \times 0.\overline{7}$ is:

Solution:

$0.\overline{46} + 0.\overline{723} - 0.\overline{39} \times 0.\overline{7}$

$\frac{46-4}{90} + \frac{723-7}{990} - \frac{39-3}{90} \times \frac{7}{9}$

$= \frac{42}{90} + \frac{716}{990} - \frac{36}{90} \times \frac{7}{9}$

$= \frac{42}{90} + \frac{716}{990} - \frac{28}{90}$

$\frac{462 + 716 - 308}{990}$

$= \frac{1178 - 308}{990}$

$= \frac{870}{990}$

$= 0.\overline{87}$

14. $(775 \times 775 \times 775 + 225 \times 225 \times 225) \div (77.5 \times 77.5 + 22.5 \times 22.5 - 77.5 \times 22.5)$ is equal to:

Solution:

$(775 \times 775 \times 775 + 225 \times 225 \times 225) \div (77.5 \times 77.5 + 22.5 \times 22.5 - 77.5 \times 22.5)$

From the formula: $a^3 + b^3 = (a+b)(a^2 + b^2 - ab)$

$\frac{a^3 + b^3}{a^2 + b^2 - ab} = (a+b)$

$= \frac{(775)^3 + (225)^3}{\frac{1}{100}\left[(775)^2 + (225)^2 - 775 \times 225\right]}$

$\Rightarrow 100\left[\frac{(775)^3 + (225)^3}{(775)^2 + (225)^2 - 775 \times 225}\right]$

$= 100(775 + 225)$

$= 100 \times 1000$

$= 100000$

15. What is the value of: $(3.6 \times 1.62 + 0.48 \times 3.6) \div (1.8 \times 0.8 + 10.8 \times 0.3 - 2.16)$?

Solution:

$(3.6 \times 1.62 + 0.48 \times 3.6) \div (1.8 \times 0.8 + 10.8 \times 0.3 - 2.16)$

Using $a^3 + b^3 = (a+b)(a^2 + b^2 - ab)$:

$= \frac{(1.8)^3 + (1.2)^3}{(1.2)^2 + (1.8)^2 - 1.2 \times 1.8}$

$= 1.8 + 1.2$

$= 3$

16. If $(1+\frac{1}{2})(1+\frac{1}{4})(1+\frac{1}{6})(1+\frac{1}{8})(1-\frac{1}{3})(1-\frac{1}{5})(1-\frac{1}{7}) = 1 + \frac{1}{x}$, then what is the value of x?

Solution:

$(1+\frac{1}{2})(1+\frac{1}{4})(1+\frac{1}{6})(1+\frac{1}{8})(1-\frac{1}{3})(1-\frac{1}{5})(1-\frac{1}{7}) = 1 + \frac{1}{x}$

$= \frac{3}{2} \times \frac{5}{4} \times \frac{7}{6} \times \frac{9}{8} \times \frac{2}{3} \times \frac{4}{5} \times \frac{6}{7} = 1 + \frac{1}{x}$

$\frac{9}{8} = 1 + \frac{1}{x}$

$1 + \frac{1}{8} = 1 + \frac{1}{x}$

On comparing:

$x = 8$

17. What is the value of: $\frac{1}{(0.1)^2} + \frac{1}{(0.01)^2} + \frac{1}{(0.5)^2} + \frac{1}{(0.05)^2}$?

Solution:

$$\frac{1}{(0.1)^2}+\frac{1}{(0.01)^2}+\frac{1}{(0.5)^2}+\frac{1}{(0.05)^2}$$

$$=100+10000+\frac{100}{25}+\frac{1000}{25}$$

$$=100+10000+4+400$$

$$=10504$$

18. What is the value of: $(2+\sqrt{2})+\frac{1}{(2+\sqrt{2})}+\frac{1}{(2-\sqrt{2})}(2+\sqrt{2})$?

Solution:

$$(2+\sqrt{2})+\frac{1}{(2+\sqrt{2})}+\frac{1}{(2-\sqrt{2})}(2+\sqrt{2})$$

$$=4+\frac{(2-\sqrt{2}+2+\sqrt{2})}{(2+\sqrt{2})(2+\sqrt{2})}$$

$$=4+\frac{4}{2}$$

$$=6$$

19. What is the value of $1006^2-1007\times1005+1008\times1004-1009\times1003$?

Solution:

$1006^2-1007\times1005+1008\times1004-1009\times1003$

Let $1003=x$:

$(x+3)^2-(x+4)(x+2)+(x+5)(x+1)-(x+6)x$

$=x^2+9+6x-x^2-6x-8+x^2+6x+5-x^2-6x$

$=6$

20. What is the value of $\frac{(1.2)^3+(0.8)^3+(0.7)^3-3\times1.2\times0.8\times0.7}{(1.35)[(1.2)^2+(0.8)^2+(0.7)^2-0.96-0.84-0.56]}$?

Solution:

$$\frac{(1.2)^3+(0.8)^3+(0.7)^3-3\times1.2\times0.8\times0.7}{(1.35)[(1.2)^2+(0.8)^2+(0.7)^2-0.96-0.84-0.56]}$$

$\therefore a^3+b^3+c^3-3abc=(a+b+c)(a^2+b^2+c^2-ab-bc-ca)$

$$=\frac{(1.2+0.8+0.7)[(1.2)^2+(0.8)^2+(0.7)^2-0.96-0.84-0.56]}{(1.35)[(1.2)^2+(0.8)^2+(0.7)^2-0.96-0.84-0.56]}$$

$$=\frac{-2.7}{1.35}$$

$$=2$$

21. If $(320+342+530+915)\div(20+22-x+18)=43$, then the value of x is:

Solution:

$(320+342+530+915)\div(20+22-x+18)=43$

$$\frac{2107}{60-x}=43$$

$$60-x=\frac{2107}{43}$$

$60-x=49$

$x=60-49=11$

22. A student was asked to find the value of:

$$\frac{\left(\frac{2\frac{1}{3}+2\frac{1}{2}-\frac{1}{6}}{\frac{7}{3}}\times\frac{17}{3}\div\frac{5}{3}\text{ of }\frac{17}{4}\right)}{\left(\frac{16}{5}\div\frac{9}{2}\text{ of }\frac{16}{3}+\frac{16}{3}\times\frac{3}{4}\div\frac{8}{3}\right)}$$

His answer was 6/7. What is the difference between the correct answer and his answer?

Solution:

$$\frac{\left(2\frac{1}{3}+2\frac{1}{2}-\frac{1}{6}\right)\div2\frac{1}{3}\times5\frac{2}{3}\div1\frac{2}{3}\text{ of }4\frac{1}{4}}{\left(3\frac{1}{5}\div4\frac{1}{2}\text{ of }5\frac{1}{3}+5\frac{1}{3}\times\frac{3}{4}\div2\frac{2}{3}\right)}$$

Solution:

$$\frac{\left(\frac{28}{6} \div \frac{7}{3} \times \frac{17}{3} \div \frac{12}{85}\right)}{\left(\frac{16}{5} \times \frac{1}{24} + \frac{16}{3} \times \frac{3}{4} \div \frac{8}{3}\right)}$$

$$= \frac{\frac{8}{5}}{\frac{2}{15} + \frac{3}{2}} = \frac{\frac{8}{5}}{\frac{4+5}{30}} = \frac{8}{5} \times \frac{30}{49}$$

$$= \frac{48}{49}$$

Intended difference: $\frac{48}{49} - \frac{6}{7} = \frac{6}{49}$

23. What is the simplified value of:
$\frac{1}{8}\left\{(x + \frac{1}{y})^2 - (x - \frac{1}{y})^2\right\}$?

Solution:

$$\frac{1}{8}\left\{(x + \frac{1}{y})^2 - (x - \frac{1}{y})^2\right\}$$

$$\frac{1}{8}\left\{x^2 + \frac{1}{y^2} + \frac{2x}{y} - x^2 - \frac{1}{y^2} + \frac{2x}{y}\right\}$$

$$= \frac{1}{8} \times \frac{4x}{y} = \frac{x}{2y}$$

24. The simplified value of: $\frac{1.0025 + 6.25 \times 10^{-6}}{0.0025 + 0.95}$ is:

Solution:

$$\frac{1.0025 + 6.25 \times 10^{-6}}{0.0025 + 0.95}$$

$$\frac{(1.0025 + 0.00000625)}{0.9525}$$

$$= \frac{1.00250625}{0.9525} = 1.0525$$

25. What is the value of: $\frac{(0.56 \times 0.36 + 0.42 \times 0.32)}{(0.8 \times 0.21)}$?

Solution:

$$\frac{(0.56 \times 0.36 + 0.42 \times 0.32)}{(0.8 \times 0.21)}$$

$$\frac{(0.2016 + 0.1344)}{(0.8 \times 0.21)}$$

$$= \frac{0.3360}{(0.8 \times 0.21)}$$

$$= \frac{336}{8 \times 21}$$

$$= \frac{42}{21} = 2$$

18 Trigonometry

Measurement of angles/ कोणों का मापन

Systems of Measurement of Angles: There are three systems for measuring angles:
(i) Sexagesimal or English System
(ii) Centesimal or French System
(iii) Circular System

कोणों को मापने की प्रणालियाँ: कोणों को मापने की तीन प्रणालियाँ हैं:
(i) सेक्सजेसिमल या अंग्रेजी प्रणाली
(ii) सेंटीसिमल या फ्रेंच प्रणाली
(iii) वृत्ताकार प्रणाली

Measure of an angle: The measure of an angle is the amount of rotation from the initial side to the terminal side.

कोण का माप: कोण का माप प्रारंभिक पक्ष से अंतिम पक्ष तक घूमने की मात्रा है।

Right angle: If the revolving ray starting from its initial position to final position describes one quarter of a circle, then we say that the angle formed is a right angle.

समकोण: यदि घूमने वाली किरण अपनी प्रारंभिक स्थिति से शुरू होकर अंतिम स्थिति तक एक वृत्त के एक चौथाई भाग का वर्णन करती है, तो हम कहते हैं कि बना हुआ कोण समकोण है।

Sexagesimal System: In this system a right angle is divided into 90 equal parts, called degrees. The symbol 1° is used to denote one degree. Thus, one degree is one-ninetieth part of a right angle. Each degree is divided into 60 equal parts, called minutes. The symbol $1'$ is used to denote one minute. And each minute is divided into 60 equal parts, called seconds. The symbol $1''$ is used to denote one second. Thus, one right angle = 90 degrees (90°), $1^\circ =$ 60 minutes (60'), $1' =$ 60 seconds (60")

सेक्सजेसिमल प्रणाली: इस प्रणाली में एक समकोण को 90 बराबर भागों में विभाजित किया जाता है, जिन्हें डिग्री कहा जाता है। प्रतीक 1° का प्रयोग एक डिग्री को दर्शाने के लिए किया जाता है। इस प्रकार, एक डिग्री समकोण का एक-उन्नीसवाँ भाग है। प्रत्येक डिग्री को 60 बराबर भागों में विभाजित किया जाता है, जिन्हें मिनट कहा जाता है। प्रतीक $1'$ का प्रयोग एक मिनट को दर्शाने के लिए किया जाता है। और प्रत्येक मिनट को 60 बराबर भागों में विभाजित किया जाता है, जिन्हें सेकंड कहा जाता है। प्रतीक $1''$ का प्रयोग एक सेकंड को दर्शाने के लिए किया जाता है। इस प्रकार,एक समकोण = 90 डिग्री (90°), 1° = 60 मिनट (60'), 1' = 60 सेकंड (60'')

Centesimal System: In this system a right angle is divided into 100 equal parts, called grades. Each grade is subdivided into 100 minutes, and each minute into 100 seconds. The symbols 15, $1'$, and $1''$ are used to denote a grade, a minute, and a second respectively. Thus, one right angle = 100 grades (100g), 1 grade = 100 minutes (100'), 1 minute = 100 seconds (100").

सेंटीसिमल प्रणाली: इस प्रणाली में एक समकोण को 100 बराबर भागों में विभाजित किया जाता है, जिन्हें ग्रेड कहा जाता है, प्रत्येक ग्रेड को 100 मिनट में और प्रत्येक मिनट को 100 सेकंड में विभाजित किया जाता है।
प्रतीक 15, $1'$, और $1''$ का उपयोग क्रमशः एक ग्रेड, एक मिनट और एक सेकंड को दर्शाने के लिए किया जाता है। इस प्रकार, एक समकोण = 100 ग्रेड (100g), 1 ग्रेड = 100 मिनट (100', 1 मिनट = 100 सेकंड (100'')

Circular System: One radian, written as 1^0, is the measure of an angle subtended at the centre of a circle by an arc of length equal to the radius of the circle. Radian is a constant angle.

वृत्ताकार प्रणाली: एक रेडियन, जिसे 1^0 लिखा जाता है, वृत्त की त्रिज्या के बराबर लंबाई के एक चाप द्वारा वृत्त के केंद्र पर बनाए गए कोण का माप है। रेडियन एक स्थिर कोण है।

The number of radians in an angle subtended by an arc of a circle at the centre is equal to arc/radius, i.e., $\theta = \dfrac{S}{R}$

किसी वृत्त के चाप द्वारा केंद्र पर अंतरित कोण में रेडियन की संख्या चाप/त्रिज्या के बराबर होती है, यानी $\theta = \dfrac{S}{R}$

Since $180^\circ = \pi$ radians $\quad \therefore 1^\circ = \dfrac{\pi}{180}$ radians

Degree	30°	45°	60°	90°	120°	135°
Radian	$\frac{\pi}{6}$	$\frac{\pi}{4}$	$\frac{\pi}{3}$	$\frac{\pi}{2}$	$\frac{2\pi}{3}$	$\frac{3\pi}{4}$

150°	180°	270°	360°
$\frac{5\pi}{6}$	π	$\frac{3\pi}{2}$	2π

Some Useful Points/ कुछ उपयोगी बिंदु

- The angle between two consecutive digits in a clock is $30^\circ (= \pi/6$ radians).
  एक घड़ी में दो क्रमागत अंकों के बीच का कोण $30^\circ (=$ 6 रेडियन) होता है।
- The hour hand rotates through an angle of 30° in one hour, i.e., $(1/2)^\circ$ in one minute.
  घंटे की सुई एक घंटे में 30° यानी एक मिनट में $1/2^\circ$ के कोण पर घूमती है।
- The minute hand rotates through an angle of 6° in one minute.
  मिनट की सुई एक मिनट में 6° के कोण पर घूमती है।

Trigonometric Functions

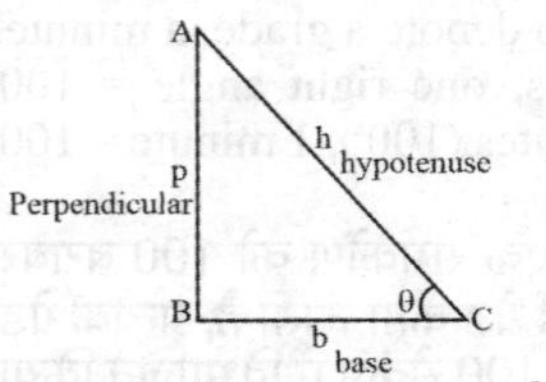

$\sin\theta = \frac{P}{h}$, $\cos\theta = \frac{b}{h}$, $\tan\theta = \frac{P}{b}$, $\cot\theta = \frac{b}{P}$

$\sec\theta = \frac{h}{b}$, $\operatorname{cosec}\theta = \frac{h}{P}$

$-1 \le \sin\theta$ or $\cos\theta \le 1$

$-\infty \le \tan\theta$ or $\cot\theta \le \infty$

$\sec\theta$ or $\operatorname{cosec}\theta \ge 1$ or $\sec\theta$ or $\operatorname{cosec}\theta \le -1$

$\sin\theta \cdot \operatorname{cosec}\theta = 1$, $\cos\theta \cdot \sec\theta = 1$, $\tan\theta \cdot \cot\theta = 1$

$\sin\theta = \frac{1}{\operatorname{cosec}\theta}$, $\operatorname{cosec}\theta = \frac{1}{\sin\theta}$

$\cos\theta = \frac{1}{\sec\theta}$, $\sec\theta = \frac{1}{\cos\theta}$

$\tan\theta = \frac{1}{\cot\theta}$, $\cot\theta = \frac{1}{\tan\theta}$

$\tan\theta = \frac{\sin\theta}{\cos\theta}$, $\cot\theta = \frac{\cos\theta}{\sin\theta}$

Angle and its measurement

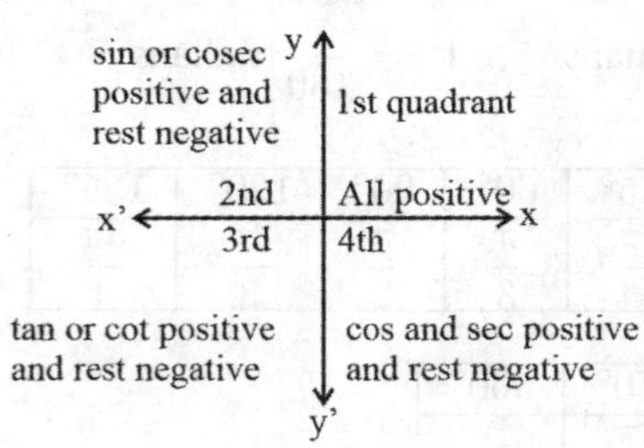

Some Special Formulae/ कुछ विशेष सूत्र

I. $\sin(-\theta) = -\sin\theta$
$\operatorname{cosec}(180^\circ + \theta) = -\operatorname{cosec}\theta$
$\cos(-\theta) = \cos\theta$
$\sec(-\theta) = \sec\theta$
$\tan(-\theta) = -\tan\theta$
$\cot(-\theta) = -\cot\theta$

II. $\sin(90^\circ - \theta) = \cos\theta$
$\cos(90^\circ - \theta) = \sin\theta$
$\tan(90^\circ - \theta) = \cot\theta$
$\cot(90^\circ - \theta) = \tan\theta$
$\sec(90^\circ - \theta) = \operatorname{cosec}\theta$
$\operatorname{cosec}(90^\circ - \theta) = \sec\theta$

III. $\sin(90^\circ + \theta) = \cos\theta$
$\cos(90^\circ + \theta) = -\sin\theta$
$\tan(90^\circ + \theta) = -\cot\theta$
$\cot(90^\circ + \theta) = \tan\theta$
$\sec(90^\circ + \theta) = -\operatorname{cosec}\theta$
$\operatorname{cosec}(90^\circ + \theta) = \sec\theta$

IV. $\sin(180^\circ - \theta) = \sin\theta$
$\cos(180^\circ - \theta) = -\cos\theta$
$\tan(180^\circ - \theta) = \tan\theta$
$\cot(180^\circ - \theta) = -\cot\theta$
$\sec(180^\circ - \theta) = -\sec\theta$
$\operatorname{cosec}(180^\circ - \theta) = \operatorname{cosec}\theta$

V. $\sin(180^\circ + \theta) = -\sin\theta$
$\cos(180^\circ + \theta) = -\cos\theta$
$\sec(180^\circ + \theta) = -\sec\theta$
$\operatorname{cosec}(180^\circ + \theta) = -\operatorname{cosec}\theta$
$\cot(180^\circ + \theta) = \cot\theta$
$\tan(180^\circ + \theta) = \tan\theta$

VI. $\sin(270^\circ - \theta) = -\cos\theta$
$\cos(270^\circ - \theta) = -\sin\theta$
$\tan(270^\circ - \theta) = \cot\theta$
$\cot(270^\circ - \theta) = \tan\theta$
$\operatorname{cosec}(270^\circ - \theta) = -\sec\theta$
$\sec(270^\circ - \theta) = -\operatorname{cosec}\theta$

VII. $\sin(270^\circ + \theta) = -\cos\theta$
$\cos(270^\circ + \theta) = \sin\theta$
$\tan(270^\circ + \theta) = -\cot\theta$
$\cot(270^\circ + \theta) = -\tan\theta$
$\operatorname{cosec}(270^\circ + \theta) = -\sec\theta$
$\sec(270^\circ + \theta) = \cos\theta$

VIII. $\sin(360^\circ - \theta) = -\sin\theta$
$\cos(360^\circ - \theta) = \cos\theta$
$\tan(360^\circ - \theta) = -\tan\theta$
$\operatorname{cosec}(360^\circ - \theta) = -\operatorname{cosec}\theta$
$\sec(360^\circ - \theta) = \sec\theta$
$\cot(360^\circ - \theta) = -\cot\theta$

IX. $\sin(360^\circ + \theta) = \sin\theta$
$\cos(360^\circ + \theta) = \cos\theta$
$\tan(360^\circ + \theta) = \tan\theta$
$\cot(360^\circ + \theta) = \cot\theta$
$\sec(360^\circ + \theta) = \sec\theta$
$\operatorname{cosec}(360^\circ + \theta) = \operatorname{cosec}\theta$

θ	$0°$	$30°$	$45°$	$60°$	$90°$
$\sin\theta$	0	$\frac{1}{2}$	$\frac{1}{\sqrt{2}}$	$\frac{\sqrt{3}}{2}$	1
$\cos\theta$	1	$\frac{\sqrt{3}}{2}$	$\frac{1}{\sqrt{2}}$	$\frac{1}{2}$	0
$\tan\theta$	0	$\frac{1}{\sqrt{3}}$	1	$\sqrt{3}$	∞
$\cot\theta$	∞	$\sqrt{3}$	1	$\frac{1}{\sqrt{3}}$	0
$\sec\theta$	1	$\frac{2}{\sqrt{3}}$	$\sqrt{2}$	2	∞
$\text{cosec}\,\theta$	∞	2	$\sqrt{2}$	$\frac{2}{\sqrt{3}}$	1

Few Results/ कुछ परिणाम:

$\tan 1° \cdot \tan 2° \cdots\cdots \tan 89° = 1$
$\cot 1° \cdot \cot 2° \cdots\cdots \cot 89° = 1$
$\cos 1° \cdot \cos 2° \cdots\cdots \cos 90° = 0 \quad [\cos 90° = 0]$
$\cos 1° \cdot \cos 2° \cdots\cdots \cos(\text{more than } 90°) = 0$
$\sin 1° \cdot \sin 2° \cdot \sin 3° \cdots \quad [\sin 180° = 0]$
$\sin 1° \cdot \sin 2° \cdot \sin 3° \cdots \quad [\sin(\text{more than } 180°) = 0]$

Trigonometric Identities/ त्रिकोणमितीय पहचान

- $\sin^2\theta + \cos^2\theta = 1$
- $\sec^2\theta - \tan^2\theta = 1$
- $\text{cosec}^2\theta - \cot^2\theta = 1$

I. $\sin(A+B) = \sin A\cos B + \cos A\sin B$
$\sin(A-B) = \sin A\cos B - \cos A\sin B$
$\cos(A+B) = \cos A\cos B - \sin A\sin B$
$\cos(A-B) = \cos A\cos B + \sin A\sin B$

II. $2\sin A\cos B = \sin(A+B) + \sin(A-B)$
$2\cos A\sin B = \sin(A+B) - \sin(A-B)$
$2\cos A\cos B = \cos(A+B) + \cos(A-B)$
$2\sin A\sin B = \cos(A-B) - \cos(A+B)$

III. $\sin(A+B)\sin(A-B) = \sin^2 A\cos^2 B - \cos^2 A\sin^2 B$
$\cos(A+B)\cos(A-B) = \cos^2 A - \sin^2 B = \cos^2 B - \sin^2 A$

IV. $\tan(A+B) = \dfrac{\tan A + \tan B}{1 - \tan A\tan B}$

$\tan(A-B) = \dfrac{\tan A - \tan B}{1 + \tan A\tan B}$

$\cot(A+B) = \dfrac{\cot A\cot B - 1}{\cot B + \cot A}$

$\cot(A-B) = \dfrac{\cot A\cot B + 1}{\cot B - \cot A}$

Sum Formulae/ योग सूत्र:

$\sin C + \sin D = 2\sin\dfrac{C+D}{2}\cos\dfrac{C-D}{2}$

$\sin C - \sin D = 2\sin\dfrac{C-D}{2}\cos\dfrac{C+D}{2}$

$\cos C + \cos D = 2\cos\dfrac{C+D}{2}\cos\dfrac{C-D}{2}$

$\cos C - \cos D = -2\sin\dfrac{C+D}{2}\sin\dfrac{C-D}{2}$

Double Angle Formulae (Multiple Angles) द्विकोण सूत्र: (एकाधिक कोण)

$\sin 2A = 2\sin A\cos A = \dfrac{2\tan A}{1+\tan^2 A}$

$\cos 2A = \cos^2 A - \sin^2 A = 2\cos^2 A - 1$

$1 + \tan^2 A = \dfrac{1-\tan^2 A}{1+\tan^2 A}$

$2A = 1 + \cos 2A \tan^2 A = \dfrac{1-\cos 2A}{1+\cos 2A}$

$\tan 2A = \dfrac{2\tan A}{1-\tan^2 A}$

Thrice Angle Formulae/ त्रिकोण सूत्र:

$\sin 3A = 3\sin A - 4\sin^3 A$
$\cos 3A = 4\cos^3 A - 3\cos A$

$\tan 3A = \dfrac{3\tan A - \tan^3 A}{1 - 3\tan^2 A}$

$\sin\theta\sin(60°-\theta)\sin(60°+\theta) = \dfrac{1}{4}\sin 3\theta$

$\cos\theta\cos(60°-\theta)\cos(60°+\theta) = \dfrac{1}{4}\cos 3\theta$

$\tan\theta\tan(60°-\theta)\tan(60°+\theta) = \tan 3\theta$
$\tan\theta + \tan(60°+\theta) + \tan(120°+\theta) = 3\tan 3\theta$
$\cos\theta + \cos(120°-\theta) + \cos(120°+\theta) = 0$

$\cos^2\theta + \cos^2(120°-\theta) + \cos^2(120°+\theta) = \dfrac{3}{2}$

Some Standard Angles/ कुछ मानक कोण

$\sin 18° = \cos 72° = \dfrac{\sqrt{5}-1}{4}$

$\sin 54° = \cos 36° = \dfrac{\sqrt{5}+1}{4}$

$\sin 72° = \cos 18° = \dfrac{\sqrt{10+2\sqrt{5}}}{4}$

$\sin 36° = \cos 54° = \dfrac{\sqrt{10-2\sqrt{5}}}{4}$

$\sin 15° = \cos 75° = \frac{\sqrt{3}-1}{2\sqrt{2}}$

$\sin 75° = \cos 15° = \frac{\sqrt{3}+1}{2\sqrt{2}}$

$\tan 15° = \cot 75° = 2-\sqrt{3}$

$\tan 75° = \cot 15° = 2+\sqrt{3}$

$\tan 22.5° = \cot 67.5° = \sqrt{2}-1$

$\tan 67.5° = \cot 22.5° = \sqrt{2}+1$

If $A+B+C = 90°$:

$\tan A \tan B + \tan B \tan C + \tan C \tan A = 1$

$\cot A + \cot B + \cot C = \cot A \cdot \cot B \cdot \cot C$

If $A+B+C = 180°$:

$\tan A + \tan B + \tan C = \tan A \cdot \tan B \cdot \tan C$

$\cot A \cdot \cot B + \cot B \cdot \cot C + \cot C \cdot \cot A = 1$

$\sin A + \sin B + \sin C = 4 \cos \frac{A}{2} \cos \frac{B}{2} \sin \frac{C}{2}$

$\sin A + \sin B - \sin C = 4 \sin \frac{A}{2} \sin \frac{B}{2} \sin \frac{C}{2}$

$\cos A + \cos B + \cos C = 1 + 4 \sin \frac{A}{2} \sin \frac{B}{2} \sin \frac{C}{2}$

$\sin 2A + \sin 2B - \sin 2C = 4 \cos A \cos B \sin C$

$\sin 2A + \sin 2B + \sin 2C = 4 \sin A \sin B \sin C$

$\cos 2A + \cos 2B + \cos 2C = 1 - 4 \cos A \cos B \cos C$

$\cos 2A + \cos 2B - \cos 2C = 1 - 4 \sin A \sin B \cos C$

Maxima and Minima

1. $-1 \le \sin \theta \le 1$
   $-1 \le \cos \theta \le 1$
   $-\infty \le \tan \theta \le +\infty$
   $-\infty \le \cot \theta \le +\infty$
   $\infty \le \sec \theta, \operatorname{cosec} \theta \le +\infty$
   (But not between -1 and 1).
2. $a \sin \theta + b \cos \theta \to \text{Max} = +\sqrt{a^2+b^2}$, $\text{Min} = -\sqrt{a^2+b^2}$
   $a \sin^2 X + b \cos^2 X$
   If $a > b$, Maximum value $= a$, Minimum value $= b$
   If $a < b$, Maximum value $= b$, Minimum value $= a$
   $a \sin^2 A + b \operatorname{cosec}^2 A$
   $a \tan^2 A + b \cot^2 A$
   $a \cos^2 A + b \sec^2 A$
   Minimum value $= 2\sqrt{ab}$
   The maximum value can go up to infinity.
   $a \operatorname{cosec}^2 X + b \sec^2 X$
   Minimum value $= (a+b)^2$
   Maximum value goes to infinity.

Type - 1 Questions based on Circular Measure of angles (Radian & Degree Measure)
कोणों के वृत्ताकार माप (रेडियन) पर आधारित प्रश्न

1. In circular measure, the value of the angle $11°15'$ is
वृत्ताकार माप में कोण का मान $11°15'$ होता है

Solution:

$11°15'$

$11° + \frac{15}{60}°$

$11° + \frac{1}{4}°$

$\frac{45}{4}°$

We know that

$180° = \pi$

$\frac{45}{4}° = \frac{\pi}{16}$

2. In a triangle ABC, $\angle ABC = 75°$ and $\angle ACB = \frac{\pi}{4}$. The circular measure of $\angle BAC$ is एक त्रिभुज ABC में, $\angle ABC = 75°$ और $\angle ACB = \frac{\pi}{4}$। $\angle BAC$ का वृत्ताकार माप है

Solution:

$75° = \frac{\pi}{180} \times 75 = \frac{5\pi}{12}$ radian

$\angle ACB = \pi - \frac{5\pi}{12} - \frac{\pi}{4}$

$\angle ACB = \frac{\pi}{3}$ radian

3. The degree measure of 1 radian (taking $a = \frac{22}{7}$) is
1 रेडियन की डिग्री माप ($a = \frac{22}{7}$ लेते हुए) है

Solution:

π radian = 180°

1 radian = $\frac{180}{\pi}$

$= \frac{180 \times 7}{22} = \frac{630}{11}$

$= (57\frac{3}{11})°$

$= 57° + \frac{3}{11} \times 60'$

$= 57° + \frac{180'}{11}$

$= 57°16' + \frac{4}{11} \times 60''$

$= 57°16'22''$ Approx

4. If the sum of two angles is $135° + \frac{\pi}{12}$ and their difference is then the circular measure of the greater angle is

यदि दो कोणों का योग $135° + \frac{\pi}{12}$ है और उनका अंतर है तो बड़े कोण का वृत्ताकार माप है

Solution:

$A + B = 135°$

$A - B = \frac{\pi}{12} = 15°$

$A = 75°$ and $B = 60°$

Greater Angle $= 75° = \frac{75\pi}{180}$

$= \frac{5\pi}{12}$

5. If θ is an acute angle and $\sin(\theta + 18°) = \frac{1}{2}$, then the value of θ in circular measure is:

यदि θ एक न्यून कोण है और $\sin(\theta + 18°) = \frac{1}{2}$ है तो वृत्ताकार माप का मान है:

Solution:

$\sin(\theta + 18°) = \frac{1}{2}$

$\sin(\theta + 18°) = \sin 30°$

$\theta + 18° = 30°$

$\theta = 12°$

$\theta = \frac{12\pi}{180} = \frac{\pi}{15}$ radian

Type 2 Problem Based on Angular Values
कोणीय मानों पर आधारित समस्या

6. Find the value of $\frac{\tan 60° - \tan 15°}{1 + \tan 60° \tan 15°}$

$\frac{\tan 60° - \tan 15°}{1 + \tan 60° \tan 15°}$ का मान ज्ञात कीजिए

Solution:

$\frac{\tan 60° - \tan 15°}{1 + \tan 60° \tan 15°}$

Compare from formula.

$\frac{\tan 60° - \tan 15°}{1 + \tan 60° \tan 15°} = \tan(60° - 15°)$

$= \tan 45° = 1$

7. What is the value of $\sin^2 60° + \tan^2 45° + \sec^2 45° - \text{cosec}^2 30°$?

$\sin^2 60° + \tan^2 45° + \sec^2 45° - \text{cosec}^2 30°$ का मान क्या है?

Solution:

$\sin^2 60° + \tan^2 45° + \sec^2 45° - \text{cosec}^2 30°$

$= (\frac{\sqrt{3}}{2})^2 + (1)^2 + (\sqrt{2})^2 - (2)$

$= \frac{3}{4} + 1 + 2 - 4$

$= \frac{3}{4} - 1 = -\frac{1}{4}$

8. What number should be subtracted from $4(\sin^4 60° + \cos^4 30°) - (\tan^2 45° - \cot^2 30°) + \cos^2 45° - \text{cosec}^2 45° + \sec^2 60°$ to get 2?

2 प्राप्त करने के लिए $4(\sin^4 60° + \cos^4 30°) - (\tan^2 45° - \cot^2 30°) + \cos^2 45° - \text{cosec}^2 45° + \sec^2 60°$ में से कौन सी संख्या घटाई जानी चाहिए?

Solution:

$4(\sin^4 60° + \cos^4 30°) - (\tan^2 45° - \cot^2 30°) + \cos^2 45° - \text{cosec}^2 45° + \sec^2 60° - ? = 2$

$? = 4(\sin^4 60° + \cos^4 30°) - (\tan^2 45° - \cot^2 30°) + \cos^2 45° - \text{cosec}^2 45° + \sec^2 60° - 2$

$? = 4[(\frac{\sqrt{3}}{2})^4 + (\frac{\sqrt{3}}{2})^4] - [(1)^2 - (\sqrt{3})^2] + (\frac{1}{\sqrt{2}})^2 - (\sqrt{2}) + (2)^2 - 2$

$? = 4[\frac{9}{16} + \frac{9}{16}] - [1 - 3] + \frac{1}{2} - 2 + 4 - 2$

$? = 4[\frac{18}{16}] + 2 + \frac{1}{2}$

$? = \frac{9}{2} + \frac{5}{2}$

$? = \frac{14}{2}$

$? = 7$

9. If A = 60°, what is the value of:
यदि $A = 60°$, तो इसका मान क्या है:

$$\frac{10 \sin \frac{A}{2} + 8 \cos A}{7 \sin \frac{3A}{2} - 12 \cos A}$$

Solution:

Given, A = 60°

The given expression is:

$$\frac{10 \sin \frac{A}{2} + 8 \cos A}{7 \sin \frac{3A}{2} - 12 \cos A}$$

$$= \frac{10 \sin \frac{60}{2} + 8 \cos 60°}{7 - 12 \cos 60°}$$

$$= \frac{10 \sin 30° + 8 \cos 60°}{7 - 12 \cos 60°}$$

$$= \frac{10 \times \frac{1}{2} + 8 \times \frac{1}{2}}{7 - 12 \times \frac{1}{2}}$$

$= 9$

10. If $2k \sin 30° \cos 30° \cot 60° = \frac{\cot^2 30° \sec 60° \tan 45°}{\text{cosec}^2 45° \text{ cosec } 30°}$, then find the value of k.

Solution:

From the question:

$$2k \sin 30° \cos 30° \cot 60° = \frac{\cot^2 30° \sec 60° \tan 45°}{\text{cosec}^2 45° \text{ cosec } 30°}$$

$$2k \times \frac{1}{2} \times \frac{\sqrt{3}}{2} \times \frac{1}{\sqrt{3}} = \frac{(\sqrt{3})^2 \times 2 \times 1}{(\sqrt{2})^2 \times 2}$$

$$\frac{k}{2} = \frac{3}{2}$$

$\therefore k = 3$

Hence, option (b) is correct

Type - 3 Problem based on A + B = 90°
A + B = 90° पर आधारित समस्या

11. Find the value of $\sin(60° + \theta) - \cos(30° - \theta)$.

Solution:

$\sin(60° + \theta) - \cos(30° - \theta)$

$= \sin(60° + \theta) - \sin(60° + \theta)$

$= 0$

12. The value of $\frac{\cos 8° \cos 24° \cos 60° \cos 66° \cos 82°}{\sin 82° \sin 66° \sin 60° \sin 8° \sin 24°}$ is:

Solution:

$$\frac{\cos 8° \cos 24° \cos 60° \cos 66° \cos 82°}{\sin 82° \sin 66° \sin 60° \sin 8° \sin 24°}$$

$$= \frac{\cos 8°}{\cos 8°} \times \frac{\cos 24°}{\cos 24°} \times \frac{\cos 60°}{\cos 60°} \times \frac{\cos 66°}{\cos 66°} \times \frac{\cos 82°}{\cos 82°}$$

$(\because \sin(90° - \theta) = \cos \theta)$

$$= \frac{\cos 60°}{\sin 60°} = \frac{\frac{1}{2}}{\frac{\sqrt{3}}{2}} = \frac{1}{\sqrt{3}}$$

13. The value of $\text{cosec}(58° + \theta) - \sec(32° - \theta) + \sin 15° \sin 35° \sec 55° \sin 30° \sec 75°$ is:

Solution:

$\text{cosec}(58° + \theta) - \sec(32° - \theta) + \sin 15° \sin 35° \sec 55° \sin 30° \sec 75°$

$$= \text{cosec}(58° + \theta) - \sec(32° - \theta) + \frac{\sin 15°}{\sin 15°} \times \frac{\sin 30°}{\sin 35°} \times \sin 30°$$

$$= \text{cosec}(58° + \theta) - \sec(32° - \theta) + \frac{1}{2}$$

$$= \frac{1}{\sin(58° + \theta)} - \frac{1}{\cos(32° - \theta)} + \frac{1}{2}$$

$$= \frac{1}{\sin(58° + \theta)} - \frac{1}{\sin(58° - \theta)} + \frac{1}{2}$$

$= \frac{1}{2}$

14. Find the value of
cot 25° cot 35° cot 45° cot 55° cot 65°.

Solution:

cot 25° cot 35° cot 45° cot 55° cot 65°

= cot 25°·cot 65°·cot 45°·cot 35°·cot 55°

(cot 45° = 1, and cot $A \cdot$ cot $B = 1$ where $A + B = 90°$)

$= 1 \times 1 \times 1 = 1$

15. If $4(\text{cosec}^2\ 66° - \tan^2\ 24°) + \frac{1}{2}\sin 90° - 4\tan^2\ 66° \cdot y \cdot \tan^2\ 24° = \frac{y}{2}$ then the value of y is:

Solution:

$$4\left(\frac{1}{\sin^2 66°} - \frac{\sin^2 24°}{\cos^2 24°}\right) + \frac{1}{2} - 4y = \frac{y}{2}$$
$$4\left(\frac{1 - \sin^2 24°}{\cos^2 24°}\right) + \frac{1}{2} - 4y = \frac{y}{2}$$
$$4\left(\frac{\cos^2 24°}{\cos^2 24°}\right) + \frac{1}{2} - 4y = \frac{y}{2}$$
$$4 + \frac{1}{2} - 4y = \frac{y}{2}$$
$$y = 1$$

16. If $\theta = 9°$, then find the value of:
$\cot\theta \cdot \cot 2\theta \cdot \cot 3\theta \cdot \cot 4\theta \cdot \cot 5\theta \cdot \cot 6\theta \cdot \cot 7\theta \cdot \cot 8\theta \cdot \cot 9\theta$

Solution:

$\theta = 9°$

= cot 9°·cot 18°·cot 27°·cot 36°·cot 45°· cot 54° · cot 63° · cot 72° · cot 81°

= (cot 9° · cot 81°) · (cot 18° · cot 72°) · (cot 27° · cot 63°) · (cot 36° · cot 54°) · cot 45°

= (tan 81° · cot 81°) · (cot 72° · tan 72°) · (cot 63° · tan 63°) · (cot 54° · tan 54°) · 1

= 1

17. The value of $\frac{(\cos 9° + \sin 81°)(\sec 9° + \text{cosec } 81°)}{\sin 56° \cdot \sec 34° + \cos 25° \cdot \text{cosec } 65°}$ is:

Solution:

$$\frac{(\cos 9° + \sin 81°)(\sec 9° + \text{cosec } 81°)}{\sin 56° \cdot \sec 34° + \cos 25° \cdot \text{cosec } 65°}$$
$$= \frac{(\cos(90° - 81°) + \sin 81°)(\sec(90° - 81°) + \text{cosec } 81°)}{\sin 56° \cdot \text{cosec } 56° + \sin 65° \cdot \text{cosec } 65°}$$
$$= \frac{(2\sin 81°)(2\text{ cosec } 81°)}{\sin 56° \cdot \frac{1}{\sin 56°} + \sin 65° \cdot \frac{1}{\sin 65°}}$$
$$= \frac{4}{2} = 2$$

18. The value of $(\tan 29° \cdot \cot 61° - \text{cosec}^2\ 61°) + \cot^2 54° - \sec^2 36° + (\sin^2 1° + \sin^2 3° + \cdots + \sin^2 89°)$ is:

Solution:

$(\tan 29° \cdot \cot 61° - \text{cosec}^2\ 61°) + \cot^2 54° - \sec^2 36° + (\sin^2 1° + \sin^2 3° + \cdots + \sin^2 89°)$

$= -1 - 1 = -2$

And $(\sin^2 1° + \sin^2 3° + \cdots + \sin^2 89°) = 45$

Required value: $-2 + 45 = 43$

19. If cosec 39° = x, then the value of $\frac{1}{\text{cosec}^2\ 51°} + \sin^2 39° + \tan^2 51° - \frac{1}{\sin^2 51° \cdot \sec^2 39°}$ is:

Solution:

$$\frac{1}{\text{cosec}^2\ 51°} + \sin^2 39° + \tan^2 51° - \frac{1}{\sin^2 51° \cdot \sec^2 39°}$$
$$= \sin^2 51° + \cos^2 51° + \cot^2 39° - \frac{1}{\sin^2 51° \cdot \text{cosec}^2\ 51°}$$
$$= 1 + \cot^2 39° - 1$$
$$= \cot^2 39° = \text{cosec}^2\ 39° - 1 = x^2 - 1$$

Type 4 (Find the Value of Theta)

20. If $\frac{\cos A}{\operatorname{cosec} A+1}+\frac{\cos A}{\operatorname{cosec} A-1}=2, \quad 0° \leq A \leq 90°$,
then A is equal to:

Solution:

From the question:

$$\frac{\cos A}{\operatorname{cosec} A+1}+\frac{\cos A}{\operatorname{cosec} A-1}=2$$

$$=\frac{\cos A(\operatorname{cosec} A-1)+\cos A(\operatorname{cosec} A+1)}{\operatorname{cosec}^2 A-1}=2$$

$$=\frac{2 \cos A \operatorname{cosec} A}{\cot^2 A}=2$$

$$\therefore \tan A=1 \implies A=45°$$

21. If $\sec^2 \alpha+4 \cos^2 \alpha=4, \ 0° \leq \alpha \leq 90°$, then find the value of α:

Solution:

$$\sec^2 \alpha+4 \cos^2 \alpha=4$$

$$\frac{1}{\cos^2 \alpha}+4 \cos^2 \alpha=4$$

Let $x=\cos^2 \alpha$: $\frac{1}{x}+4x=4$

$$4x^2-4x+1=0$$

$$(2x-1)^2=0 \implies x=\frac{1}{2}$$

$$\cos \alpha=\frac{1}{\sqrt{2}} \implies \alpha=45°$$

22. Solve for A (in degrees) if $\tan^2 A+5 \sec A=13$, where $0°<A<90°$:

Solution:

$$\tan^2 A+5 \sec A=13, \quad 0°<A<90°$$

$$\frac{\sin^2 A}{\cos^2 A}+\frac{5}{\cos A}=13$$

$$\frac{\sin^2 A+5 \cos A}{\cos^2 A}=13$$

$$\sin^2 A+5 \cos A=13 \cos^2 A$$

$$(1-\cos^2 A)+5 \cos A=13 \cos^2 A$$

$$-14 \cos^2 A+5 \cos A+1=0$$

$$14 \cos^2 A-5 \cos A-1=0$$

$$14 \cos^2 A-(7-2) \cos A-1=0$$

$$14 \cos^2 A-7 \cos A+2 \cos A-1=0$$

$$7 \cos A(2 \cos A-1)+1(2 \cos A-1)=0$$

$$(2 \cos A-1)(7 \cos A+1)=0$$

$$\therefore 2 \cos A-1=0$$

$$2 \cos A=1$$

$$\cos A=\frac{1}{2}$$

$$\cos A=\cos 60°$$

$$A=60°$$

23. If θ is a positive acute angle and $\tan 2\theta \cdot \tan 3\theta=1$, then the value of θ is:

Solution:

$$\tan 2\theta \cdot \tan 3\theta=1$$

$$\tan 2\theta=\frac{1}{\tan 3\theta}$$

$\tan 2\theta=\cot 3\theta$ [since $\tan(90°-\theta)=\cot \theta$]

$$2\theta+3\theta=90°$$

$$5\theta=90°$$

$$\theta=18°$$

24. यदि θ $(0° \leq \theta \leq 90°)$ के लिए $\sqrt{3} \cos \theta+\sin \theta=1$, तो θ क्या है?

Solution:

$$\sqrt{3} \cos \theta+\sin \theta=1$$

$$\sqrt{\sqrt{3}^2+1^2}=2$$

By dividing by 2 on both sides: $\frac{\sqrt{3}}{2} \cos \theta+\frac{1}{2} \sin \theta=\frac{1}{2}$

$$\cos \theta \cos 30°+\sin \theta \sin 30°=\frac{1}{2}$$

$$\cos(\theta-30°)=\cos 60°$$

$$\theta-30°=60°$$

$$\theta=90°$$

25. If $2\sin^2\theta + 3\sin\theta - 2 = 0$, $(0° < \theta < 90°)$, then find the value of θ:

Solution:

$2\sin^2\theta + 3\sin\theta - 2 = 0, \quad (0° < \theta < 90°)$

$\sin\theta = \dfrac{-3 \pm \sqrt{3^2 - 4\cdot 2\cdot(-2)}}{2\cdot 2}$

$\sin\theta = \dfrac{-3 \pm \sqrt{25}}{4}$

$\sin\theta = \dfrac{-3+5}{4} = \dfrac{1}{2}$

$\sin\theta = \sin 30°$

$\theta = 30°$

26. If $\sin(A-B) = \dfrac{1}{2}$ and $\cos(A+B) = \dfrac{1}{2}$, where $A > B > 0°$ and $A+B$ is an acute angle, then the value of A is:

Solution:

Given: $\sin(A-B) = \dfrac{1}{2}$

$\sin(A-B) = \sin 30°$

$A - B = 30°$ ---(i)

$\cos(A+B) = \cos 60°$

$A + B = 60°$ ---(ii)

On adding the equations (i) and (ii):

$A - B = 30°$

$A + B = 60°$

$2A = 90°$

$A = 45°$

Type - 5 (Problem based on Trigonometric Ratios)

27. If $\tan^2 A + 2\tan A - 63 = 0$ and $0 < A < \dfrac{\pi}{2}$, what is the value of $2\sin A + 5\cos A$?

Solution:

$\tan^2 A + 2\tan A - 63 = 0$

$\tan^2 A + 9\tan A - 7\tan A - 63 = 0$

$\tan A(\tan A + 9) - 7(\tan A + 9) = 0$

$(\tan A + 9)(\tan A - 7) = 0$

$\tan A = 7$

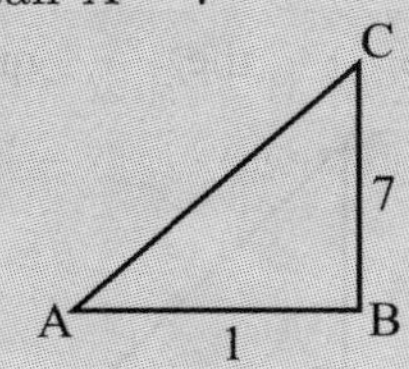

Value of $AC = \sqrt{7^2 + 1^2} = \sqrt{50}$: $2\sin A + 5\cos A$

$= 2\cdot\dfrac{7}{\sqrt{50}} + 5\cdot\dfrac{1}{\sqrt{50}}$

$= \dfrac{19}{\sqrt{50}}$

28. If $\cos\theta = \dfrac{2}{3}$, then $2\sec^2\theta + 2\tan^2\theta - 6$ equals:

Solution:

Given, $\cos\theta = \dfrac{\text{Base}}{\text{Hypotenuse}} = \dfrac{2}{3}$

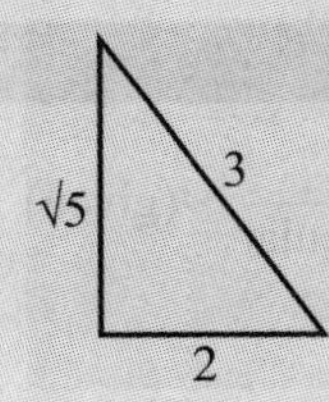

By Pythagoras theorem: $(\text{Hypotenuse})^2 = (\text{Base})^2 + (\text{Perpendicular})^2$

$(\text{Perpendicular})^2 = 3^2 - 2^2$

$\text{Perpendicular} = \sqrt{9-4} = \sqrt{5}$

According to the question: $2\sec^2\theta + 2\tan^2\theta - 6$

$= 2\left(\dfrac{3}{2}\right)^2 + 2\left(\dfrac{\sqrt{5}}{2}\right)^2 - 6$

$= 2\cdot\dfrac{9}{4} + 2\cdot\dfrac{5}{4} - 6$

$= \dfrac{9}{2} + \dfrac{5}{2} - 6$

$= 7 - 6 = 1$

29. If $\sin A = \dfrac{5}{13}$ and $7 \cot B = 24$, then the value of $(\sec A \cos B)(\text{cosec } B \tan A)$ is:

Solution:

According to the question:

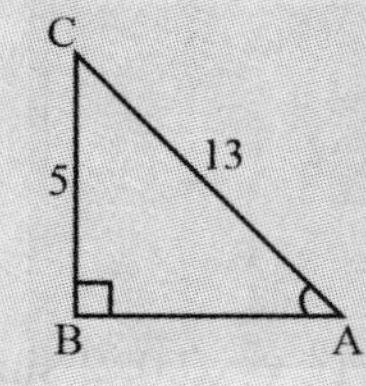

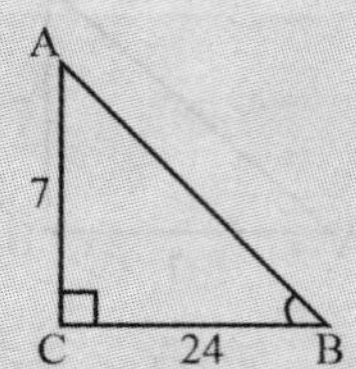

$AB = \sqrt{13^2 - 5^2}$ and $AB = \sqrt{24^2 + 7^2}$

$AB = \sqrt{144}$ and $AB = \sqrt{625}$

$AB = 12$ and $AB = 25$

$(\sec A \cos B)(\text{cosec } B \tan A)$

$= \left(\dfrac{13}{12} \cdot \dfrac{24}{25}\right)\left(\dfrac{25}{7} \cdot \dfrac{5}{12}\right)$

$= \dfrac{65}{42}$

30. In $\triangle ABC$, right-angled at B, if $\cot A = \dfrac{1}{2}$, the value of $\dfrac{\sin A(\cos C + \cos A)}{\cos C(\sin C - \sin A)}$ is:

Solution:

$\cot A = \dfrac{1}{2} = \dfrac{\text{Base}}{\text{Perpendicular}}$ (given)

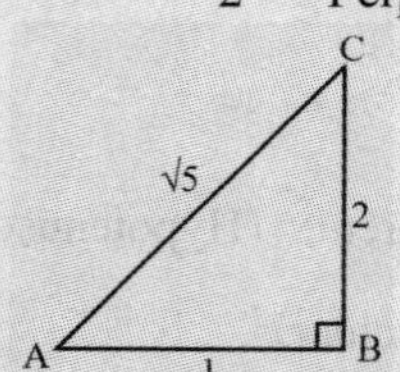

By Pythagoras theorem:

Hypotenuse $= \sqrt{(1)^2 + (2)^2} = \sqrt{5}$

$\dfrac{\sin A(\cos C + \cos A)}{\cos C(\sin C - \sin A)} = \dfrac{\dfrac{2}{\sqrt{5}}\left(\dfrac{2}{\sqrt{5}} + \dfrac{1}{\sqrt{5}}\right)}{\dfrac{2}{\sqrt{5}}\left(\dfrac{1}{\sqrt{5}} - \dfrac{2}{\sqrt{5}}\right)}$

$= \dfrac{\dfrac{3}{\sqrt{5}}}{-\dfrac{1}{\sqrt{5}}} = -3$

31. If $\cos \theta = \dfrac{4x}{1 + 4x^2}$, then what is the value of $\sin \theta$?

Solution:

$\cos \theta = \dfrac{4x}{1 + 4x^2}$

$\cos \theta = \dfrac{\text{Base}}{\text{Hypotenuse}}$

$(\text{Hypotenuse})^2 = (\text{Perpendicular})^2 + (\text{Base})^2$

$(1 + 4x^2)^2 = (\text{Perpendicular})^2 + 16x^2$

$(\text{Perpendicular})^2 = 1 + 16x^4 - 8x^2$

$\text{Perpendicular} = (1 - 4x^2)$

$\sin \theta = \dfrac{\text{Perpendicular}}{\text{Hypotenuse}} = \dfrac{1 - 4x^2}{1 + 4x^2}$

32. If $\cot \theta = \sqrt{2} + 1$, then $\text{cosec } \theta \sec \theta = ?$:

Solution:

$\cot \theta = \dfrac{\sqrt{2} + 1}{1} = \dfrac{\text{Base}}{\text{Perpendicular}}$

$\text{Hypotenuse} = \sqrt{(\sqrt{2} + 1)^2 + 1^2}$

$= \sqrt{3 + 2\sqrt{2} + 1} = \sqrt{4 + 2\sqrt{2}}$

$\text{cosec } \theta \sec \theta = \dfrac{\sqrt{4 + 2\sqrt{2}}}{1} \cdot \dfrac{\sqrt{4 + 2\sqrt{2}}}{\sqrt{2} + 1}$

$= \dfrac{4 + 2\sqrt{2}}{\sqrt{2} + 1} = 2\sqrt{2}$

33. If $3 \cos^2 \theta - 4 \sin \theta + 1 = 0$, $0° < \theta < 90°$, then the value of $3\theta + 5\theta$ will be:

Solution:

$3 \cos^2 \theta - 4 \sin \theta + 1 = 0$

$3(1 - \sin^2 \theta) - 4 \sin \theta + 1 = 0$

$3 - 3 \sin^2 \theta - 4 \sin \theta + 1 = 0$

$3 \sin^2 \theta + 4 \sin \theta - 4 = 0$

$(\sin \theta + 2)(3 \sin \theta - 2) = 0$

$\sin \theta = \dfrac{2}{3}$, $\sin \theta = -2$ (invalid)

$$\cos\theta = \sqrt{1-\sin^2\theta} = \sqrt{1-\frac{4}{9}} = \frac{\sqrt{5}}{3}$$

$$3\cos^2\theta + 5\tan^2\theta - \frac{3\cdot 5}{9} + \frac{5\cdot\left(\frac{2}{3}\right)^2}{\left(\frac{\sqrt{5}}{3}\right)^2}$$

$$= \frac{5}{3} + 5\cdot\frac{4}{5}$$

$$= \frac{17}{3} = 5\frac{2}{3}$$

34. If A lies in the first quadrant and $6\tan A = 5$, then the value of $\dfrac{8\sin A - 4\cos A}{\cos A + 2\sin A}$ is:

Solution:

$$6\tan A = 5 \implies \tan A = \frac{5}{6}$$

$$\frac{8\sin A - 4\cos A}{\cos A + 2\sin A}$$

On dividing by $\cos A$ in each term:

$$\frac{8\tan A - 4}{1 + 2\tan A} = \frac{8\cdot\frac{5}{6} - 4}{1 + 2\cdot\frac{5}{6}} = \frac{16}{16} = 1$$

35. If $\dfrac{\sin A + \cos A}{\cos A} = \dfrac{17}{12}$ then the value of $\dfrac{1-\cos A}{\sin A}$ is:

Solution:

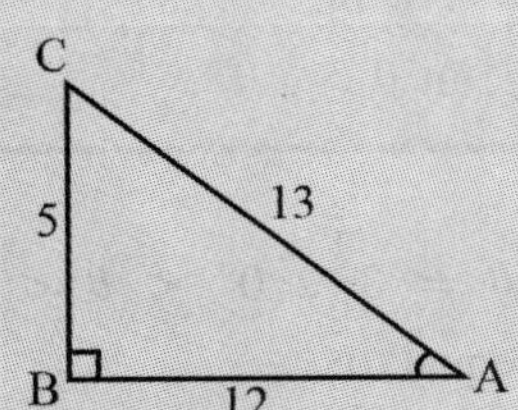

$$\frac{\sin A + \cos A}{\cos A} = \frac{17}{12}$$

$$\tan A + 1 = \frac{17}{12}$$

$$\tan A = \frac{5}{12}$$

$$\frac{1-\cos A}{\sin A} = \frac{1-\frac{12}{13}}{\frac{5}{13}} = \frac{\frac{1}{13}}{\frac{5}{13}} = \frac{1}{5}$$

36. If $\dfrac{\cos\theta + \sin\theta}{\cos\theta - \sin\theta} = \dfrac{\sqrt{3}+1}{\sqrt{3}-1}$, $0° < \theta < 90°$ then what is the value of $\sec\theta$?

Solution:

$$\frac{\cos\theta + \sin\theta}{\cos\theta - \sin\theta} = \frac{\sqrt{3}+1}{\sqrt{3}-1}$$

$$\frac{\cos\theta}{\sin\theta} = \sqrt{3} \quad \text{or} \quad \tan\theta = \frac{1}{\sqrt{3}}$$

$$\tan^2\theta = \frac{1}{3}$$

$$\sec^2\theta - 1 = \frac{1}{3}$$

$$\sec^2\theta = \frac{1}{3} + 1 = \frac{4}{3}$$

$$\sec\theta = \frac{2}{\sqrt{3}} = \frac{2\sqrt{3}}{3}$$

37. If $\dfrac{\sec\theta + \tan\theta}{\sec\theta - \tan\theta} = 5$ and θ is an acute angle, then the value of $\dfrac{3\cos^2\theta + 1}{3\cos^2\theta - 1}$ is:

Solution:

$$\frac{\sec\theta + \tan\theta}{\sec\theta - \tan\theta} = 5$$

$$\frac{\sec\theta + \tan\theta + \sec\theta - \tan\theta}{\sec\theta + \tan\theta - \sec\theta + \tan\theta} = \frac{5+1}{5-1}$$

$$\frac{2\sec\theta}{2\tan\theta} = \frac{6}{4}$$

$$\frac{1}{\cos\theta}\cdot\frac{\cos\theta}{\sin\theta} = \frac{2}{3}$$

$$\sin\theta = \frac{2}{3}$$

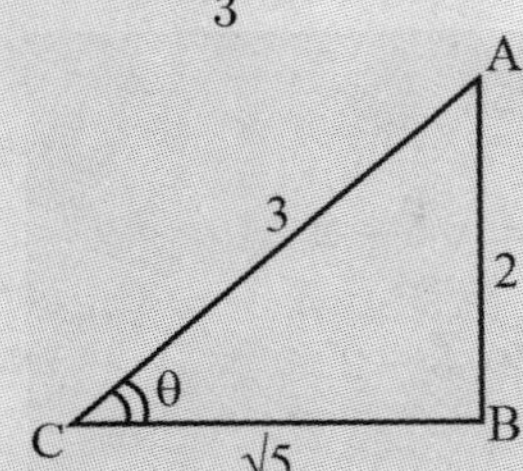

$$AC^2 = AB^2 + BC^2$$

$$3^2 = 2^2 + BC^2$$

$$9 = 4 + BC^2$$

$$BC = \sqrt{5}$$

Then:

$$\frac{3\cos^2\theta+1}{3\cos^2\theta-1}=\frac{3\cdot\left(\frac{\sqrt{5}}{3}\right)^2+1}{3\cdot\left(\frac{\sqrt{5}}{3}\right)^2-1}$$

$$=\frac{\frac{5}{3}+1}{\frac{5}{3}-1}=\frac{\frac{8}{3}}{\frac{2}{3}}=4$$

38. If $\sec\theta+\tan\theta=2+\sqrt{5}$ and θ is an acute angle, then the value of $\sin\theta$ is:

Solution:

Given: $\sec\theta+\tan\theta=2+\sqrt{5}$ ----(i)

Then $\sin\theta=?$: $\sec^2\theta-\tan^2\theta=1$

$(\sec\theta+\tan\theta)(\sec\theta-\tan\theta)=1$

$$\sec\theta-\tan\theta=\frac{1}{2+\sqrt{5}}\cdot\frac{2-\sqrt{5}}{2-\sqrt{5}}$$

$$=\frac{2-\sqrt{5}}{-1}=\sqrt{5}-2 \quad \text{----(ii)}$$

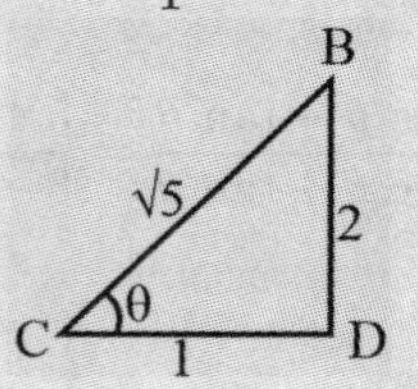

On adding equations (i) and (ii): $\sec\theta+\tan\theta=2+\sqrt{5}$

$\sec\theta-\tan\theta=\sqrt{5}-2$

$2\sec\theta=2+2\sqrt{5}-2$

$\sec\theta=\sqrt{5}$

$$\cos\theta=\frac{1}{\sqrt{5}}$$

$$\sin\theta=\frac{2}{\sqrt{5}}\cdot\frac{\sqrt{5}}{\sqrt{5}}$$

$$\sin\theta=\frac{2\sqrt{5}}{5}$$

Type 6 (Problems based on Trigonometric Identities)

39. Simplify the following expression: $\frac{\cos A}{1-\tan A}+\frac{\sin A}{1-\cot A}-\sin A$

Solution:

$$\frac{\cos A}{1-\tan A}+\frac{\sin A}{1-\cot A}-\sin A$$

$$=\frac{\cos^2 A}{\cos A-\sin A}+\frac{\sin^2 A}{\sin A-\cos A}-\sin A$$

$$=\frac{\cos^2 A-\sin^2 A}{\cos A-\sin A}-\sin A$$

$$=\cos A+\sin A-\sin A=\cos A$$

40. The value of $2-\sqrt{\frac{\cot\theta+\cos\theta}{\cot\theta-\cos\theta}}$, when $0°<\theta<90°$ is equal to:

Solution:

$$2-\sqrt{\frac{\cot\theta+\cos\theta}{\cot\theta-\cos\theta}}=2-\sqrt{\frac{(\cot\theta+\cos\theta)^2}{\cot^2\theta-\cos^2\theta}}$$

$$=2-\frac{\cot\theta+\cos\theta}{\sqrt{\frac{\cos^2\theta}{\sin^2\theta}-\cos^2\theta}}$$

$$=2-\frac{\frac{\cos\theta}{\sin\theta}(1+\sin\theta)}{\frac{\cos\theta}{\sin\theta}\cdot\cos\theta}$$

$$=2-\sec\theta+\tan\theta$$

41. If $\sec^2\theta+\tan^2\theta=3\frac{1}{2}$, $0°<\theta<90°$, then $(\cos\theta+\sin\theta)$ is equal to:

Solution:

$$\sec^2\theta+\tan^2\theta=\frac{7}{2}$$

$$\frac{\sec^2\theta-\tan^2\theta}{2\sec^2\theta}=\frac{1}{\frac{9}{2}} \implies \sec\theta=\frac{3}{2},\ \cos\theta=\frac{2}{3}$$

$$\sin\theta = \sqrt{1-\cos^2\theta} = \sqrt{1-\frac{4}{9}} = \frac{\sqrt{5}}{3}$$

$$\cos\theta + \sin\theta = \frac{2}{3} + \frac{\sqrt{5}}{3} = \frac{2+\sqrt{5}}{3}$$

42. Simplify: $\sec^2\alpha\left(1+\frac{1}{\text{cosec }\alpha}\right)\left(1-\frac{1}{\text{cosec }\alpha}\right)$

Solution:

$$\sec^2\alpha\left(1+\frac{1}{\text{cosec }\alpha}\right)\left(1-\frac{1}{\text{cosec }\alpha}\right)$$

$$= \sec^2\alpha\cdot(1+\sin\alpha)(1-\sin\alpha)$$

$$= \sec^2\alpha\cdot(1-\sin^2\alpha)$$

$$= \sec^2\alpha\cdot\cos^2\alpha = \frac{1}{\cos^2\alpha}\cdot\cos^2\alpha = 1$$

43. Simplify: $\frac{(1+\cos\theta)(\text{cosec }\theta-\cot\theta)\sec\theta}{\sin\theta(1-\sin\theta)(\sec\theta+\tan\theta)}$

Solution:

$$\frac{(1+\cos\theta)(\text{cosec }0.\theta-\cot\theta)\sec\theta}{\sin\theta(1-\sin\theta)(\sec\theta+\tan\theta)}$$

$$= \frac{(1+\cos\theta)\left(\frac{1-\cos\theta}{\sin\theta}\right)\sec\theta}{\sin\theta(1-\sin\theta)\left(\frac{1+\sin\theta}{\cos\theta}\right)}$$

$$= \frac{\sin^2\theta}{\cos^2\theta}\cdot\frac{1}{\sin^2\theta}$$

$$= \cos^2\theta$$

44. If $(\text{cosec }\theta+\cot\theta)/(\text{cosec }\theta+\cot\theta) = 7$, then the value of $\frac{4\sin^2\theta+5}{4\sin^2\theta-1}$ is:

Solution:

$$\frac{\text{cosec }\theta+\cot\theta}{\text{cosec }\theta+\cot\theta} = \frac{7}{1}$$

By componendo and dividendo: $\frac{2\text{ cosec }\theta}{2\cot\theta} = \frac{7+1}{7-1}$

$$\frac{1/\sin\theta}{\cos\theta/\sin\theta} = \frac{8}{6}$$

$$\frac{1}{\cos\theta} = \frac{4}{3} \implies \cos\theta = \frac{3}{4}$$

$$\sin^2\theta = 1-\cos^2\theta = 1-\frac{9}{16} = \frac{7}{16}$$

$$\frac{4\sin^2\theta+5}{4\sin^2\theta-1} = \frac{\frac{7}{4}+5}{\frac{7}{4}-1} = \frac{27}{3} = 9$$

45. $\frac{2+\tan^2\theta+\cot^2\theta}{\sec\theta\text{ cosec }\theta}$ is equal to:

Solution:

$$\frac{2+\tan^2\theta+\cot^2\theta}{\sec\theta\text{ cosec }\theta}$$

$$= \frac{1+\tan^2\theta+1+\cot^2\theta}{\sec\theta\text{ cosec }\theta}$$

$$= \frac{\sec^2\theta+\text{cosec}^2\theta}{\sec\theta\text{ cosec }\theta}$$

$$= \frac{\sec\theta}{\text{cosec }\theta}+\frac{\text{cosec }\theta}{\sec\theta}$$

$$= \frac{\sin\theta}{\cos\theta}+\frac{\cos\theta}{\sin\theta}$$

$$= \frac{\sin^2\theta+\cos^2\theta}{\sin\theta\cos\theta}$$

$$= \frac{1}{\sin\theta\cos\theta}$$

$$= \sec\theta\text{ cosec }\theta$$

46. If $\frac{(\tan\theta-\sec\theta+1)}{(\tan\theta+\sec\theta-1)}\sec\theta = \frac{1}{k}$, then $k =$ _____:

Solution:

$$\frac{(\tan\theta-\sec\theta+1)}{(\tan\theta+\sec\theta-1)}\sec\theta = \frac{1}{k}$$

$$\because \sec^2\theta-\tan^2\theta = 1$$

$$\frac{(\tan\theta-\sec\theta+\sec^2\theta-\tan^2\theta)}{(\tan\theta+\sec\theta-1)}\sec\theta = \frac{1}{k}$$

$$\frac{(\sec\theta-\tan\theta)(-1+\sec\theta+\tan\theta)}{(\tan\theta+\sec\theta-1)}\sec\theta = \frac{1}{k}$$

$$\frac{\frac{1}{\cos\theta}-\frac{\sin\theta}{\cos\theta}}{\frac{1}{\cos\theta}}-\frac{1}{k}$$
$$\frac{1-\sin\theta}{\cos^2\theta}-\frac{1}{k}$$
$$\frac{1-\sin\theta}{1-\sin^2\theta}=\frac{1}{k}$$
$$\frac{1}{1-\sin\theta}=\frac{1}{k}$$
$$k=1+\sin\theta$$

47. What is the value of $\left[(\sec 2\theta+1)\sqrt{\sec^2\theta-1}\right]\cdot\frac{1}{2}(\cot\theta-\tan\theta)$?:

Solution:

$$\left[(\sec 2\theta+1)\sqrt{\sec^2\theta-1}\right]\cdot\frac{1}{2}(\cot\theta-\tan\theta)$$
$$=\left[(\sec 2\theta+1)\tan\theta\cdot\frac{1}{2}\right]\left[\frac{\cos\theta}{\sin\theta}-\frac{\sin\theta}{\cos\theta}\right]$$
$$=\left[\frac{1}{\cos 2\theta}+1\right]\tan\theta\cdot\frac{\cos^2\theta-\sin^2\theta}{\sin\theta\cos\theta}$$
$$=\left[\frac{1}{2\cos^2\theta-1}+1\right]\tan\theta\cdot\frac{\cos 2\theta}{\sin 2\theta}$$
$$=\frac{\sin 2\theta}{\cos 2\theta}\cdot\frac{\cos 2\theta}{\sin 2\theta}=1$$

48. The value of $\frac{(\sin\theta+\cos\theta-1)}{(\sin\theta-\cos\theta+1)}\cdot\frac{\tan^2\theta(\csc^2\theta-1)}{(\sec\theta-\tan\theta)}$ is:

Solution:

$$\frac{(\sin\theta+\cos\theta-1)}{(\sin\theta-\cos\theta+1)}\cdot\frac{\tan^2\theta(\csc^2\theta-1)}{(\sec\theta-\tan\theta)}$$
$$=\frac{(\sin\theta+\cos\theta-1)}{(1+\sin\theta-\cos\theta)}\cdot\frac{\tan^2\theta\cot^2\theta\cos\theta}{(1-\sin\theta)}$$
$$=\frac{(\sin\theta\cos\theta+\cos^2\theta-\cos\theta)}{(1-\sin^2\theta-\cos\theta+\cos\theta\sin\theta)}$$
$$=\frac{(\sin\theta\cos\theta+\cos^2\theta-\cos\theta)}{(\cos^2\theta-\cos\theta+\cos\theta\sin\theta)}=1$$

49. The value of $(1+\cot\theta-\csc\theta)(1+\cos\theta+\sin\theta)\sec\theta$ is:

Solution:

$$(1+\cot\theta-\csc\theta)(1+\cos\theta+\sin\theta)\sec\theta$$
$$=\left(1+\frac{\cos\theta}{\sin\theta}-\frac{1}{\sin\theta}\right)\cdot(1+\cos\theta+\sin\theta)\cdot\frac{1}{\cos\theta}$$
$$=\frac{(\sin\theta+\cos\theta-1)(\sin\theta+\cos\theta+1)}{\sin\theta\cos\theta}$$
$$=\frac{2\sin\theta\cos\theta}{\sin\theta\cos\theta}=2$$

50. The value of $\frac{\sec\phi(1-\sin\phi)(\sin\phi+\cos\phi)}{\sin\phi(1+\tan\phi)+\cos\phi(1+\cot\phi)}\frac{(\sec\phi+\tan\phi)}{\sin\phi(1+\tan\phi)+\cos\phi(1+\cot\phi)}$ is:

Solution:

$$\frac{\sec\phi(1-\sin\phi)(\sin\phi+\cos\phi)(\sec\phi+\tan\phi)}{\sin\phi(1+\tan\phi)+\cos\phi(1+\cot\phi)}$$
$$=\frac{(\sec\phi-\tan\phi)(\sin\phi+\cos\phi)}{\sin\phi\left(\frac{\cos\phi+\sin\phi}{\cos\phi}\right)+\cos\phi\left(\frac{\sin\phi+\cos\phi}{\sin\phi}\right)}\cdot\frac{(\sec\phi+\tan\phi)}{\sin\phi\left(\frac{\cos\phi+\sin\phi}{\cos\phi}\right)+\cos\phi\left(\frac{\sin\phi+\cos\phi}{\sin\phi}\right)}$$
$$=\frac{(\sec^2\phi-\tan^2\phi)(\sin\phi+\cos\phi)}{(\sin\phi+\cos\phi)(\tan\phi+\cot\phi)}$$
$$=\frac{1}{\frac{\sin\phi}{\cos\phi}+\frac{\cos\phi}{\sin\phi}}=\frac{\sin\phi\cos\phi}{\sin^2\phi+\cos^2\phi}=\sin\phi\cos\phi$$

51. If $\frac{1}{\sec\theta-\tan\theta}-\frac{1}{\cos\theta}=\sec\theta\cdot k,\ 0°<\theta<90°$, then k is equal to:

Solution:

$$\frac{1}{\sec\theta-\tan\theta}-\frac{1}{\cos\theta}=\sec\theta\cdot k$$
$$\frac{1}{\frac{1}{\cos\theta}-\frac{\sin\theta}{\cos\theta}}-\frac{1}{\cos\theta}=\sec\theta\cdot k$$
$$\frac{\cos\theta}{1-\sin\theta}-\frac{1}{\cos\theta}=\sec\theta\cdot k$$
$$\frac{\cos^2\theta-(1-\sin\theta)}{\cos\theta(1-\sin\theta)}=\sec\theta\cdot k$$

Solution:

$$\frac{-\sin^2\theta+\sin\theta}{\cos\theta(1-\sin\theta)}=\sec\theta\cdot k$$
$$\frac{\sin\theta(-\sin\theta+1)}{\cos\theta(1-\sin\theta)}=\sec\theta\cdot k$$
$$\frac{\sin\theta}{\cos\theta}=k$$
$$k=\sin\theta$$

52. Simplify: $\dfrac{(\sec\theta+\tan\theta)(1-\sin\theta)}{\operatorname{cosec}\theta(1+\cos\theta)(\operatorname{cosec}\theta-\cot\theta)}$:

Solution:

$$\frac{(\sec\theta+\tan\theta)(1-\sin\theta)}{\operatorname{cosec}\theta(1+\cos\theta)(\operatorname{cosec}\theta-\cot\theta)}$$
$$=\frac{\left(\frac{1}{\cos\theta}+\frac{\sin\theta}{\cos\theta}\right)(1-\sin\theta)}{\frac{1}{\sin\theta}(1+\cos\theta)(\operatorname{cosec}\theta-\cot\theta)}$$
$$=\frac{\frac{1+\sin\theta}{\cos\theta}(1-\sin\theta)}{(\operatorname{cosec}\theta+\cot\theta)(\operatorname{cosec}\theta-\cot\theta)}$$
$$=\frac{\frac{1-\sin^2\theta}{\cos\theta}}{\operatorname{cosec}^2\theta-\cot^2\theta}$$
$$=\frac{\cos^2\theta}{\cos\theta}=\cos\theta$$

53. If $\dfrac{(\sin\theta-\operatorname{cosec}\theta)(\cos\theta-\sec\theta)}{\tan^2\theta-\sin^2\theta}=r^3$, then the value of r is:

Solution:

$$r^3=\frac{(\sin\theta-\operatorname{cosec}\theta)(\cos\theta-\sec\theta)}{\tan^2\theta-\sin^2\theta}$$
$$=\frac{\left(\sin\theta-\frac{1}{\sin\theta}\right)\left(\cos\theta-\frac{1}{\cos\theta}\right)}{\frac{\sin^2\theta}{\cos^2\theta}-\sin^2\theta}$$
$$=\frac{\frac{\sin^2\theta-1}{\sin\theta}\cdot\frac{\cos^2\theta-1}{\cos\theta}}{\sin^2\theta\left(\frac{1}{\cos^2\theta}-1\right)}$$
$$=\frac{(-\cos^2\theta)(-\sin^2\theta)}{\sin\theta\cos\theta\cdot\sin^2\theta\cdot\frac{1-\cos^2\theta}{\cos^2\theta}}$$
$$=\left(\frac{\cos\theta}{\sin\theta}\right)^3,\quad r=\cot\theta$$

54. If $12\cot^2\theta-31\operatorname{cosec}\theta+32=0,\ 0°<\theta<90°$, then the value of $\sin\theta$ is:

Solution:

$$12\cot^2\theta-31\operatorname{cosec}\theta+32=0$$
$$12(\operatorname{cosec}^2\theta-1)-31\operatorname{cosec}\theta+32=0$$
$$12\operatorname{cosec}^2\theta-12-31\operatorname{cosec}\theta+32=0$$
$$12\operatorname{cosec}^2\theta-31\operatorname{cosec}\theta+20=0$$
$$12\operatorname{cosec}^2\theta-16\operatorname{cosec}\theta-15\operatorname{cosec}\theta+20=0$$
$$4\operatorname{cosec}\theta(3\operatorname{cosec}\theta-4)-5(3\operatorname{cosec}\theta-4)=0$$
$$(3\operatorname{cosec}\theta-4)(4\operatorname{cosec}\theta-5)=0$$
$$\operatorname{cosec}\theta=\frac{4}{3}\quad\text{and}\quad\operatorname{cosec}\theta=\frac{5}{4}$$
$$\sin\theta=\frac{3}{4}\quad\text{and}\quad\sin\theta=\frac{4}{5}$$
$$\sin\theta=\frac{4}{5},\ \frac{3}{4}$$

55. If $\dfrac{\cos\alpha}{\cos\beta+\sin\alpha}+\dfrac{\cos\beta}{\sin\beta-\cos\alpha}=\dfrac{x}{\sin\alpha-\cos\beta}+\dfrac{\cos\beta}{\sin\beta+\cos\alpha}$, then x is equal to:

Solution:

$$\frac{\cos\alpha}{\cos\beta+\sin\alpha}+\frac{\cos\beta}{\sin\beta-\cos\alpha}=\frac{x}{\sin\alpha-\cos\beta}+\frac{\cos\beta}{\sin\beta+\cos\alpha}$$

Let $\alpha+\beta=180° \implies \beta=180°-\alpha$:

$$\cos\alpha=-\cos\beta,\quad \cos\beta=-\cos\alpha$$
$$\frac{\cos\alpha}{\sin\alpha+\cos(180°-\alpha)}+\frac{\cos\beta}{\sin(180°-\alpha)-\cos\alpha}=\frac{x}{\sin\alpha-\cos(180°-\alpha)}+\frac{\cos\beta}{\sin(180°-\alpha)+\cos\alpha}$$
$$=\frac{\cos\alpha}{\sin\alpha-\cos\alpha}-\frac{\cos\alpha}{\sin\alpha-\cos\alpha}$$
$$x=\cos\alpha$$

56. If $\tan\alpha=\dfrac{2}{\sqrt{13}}$, then the value of $\dfrac{\operatorname{cosec}^2\alpha+2\sec^2\alpha}{\operatorname{cosec}^2\alpha-3\sec^2\alpha}$ is:

Solution:

$\because \tan\alpha = \frac{2}{\sqrt{13}} \implies \cot\alpha = \frac{\sqrt{13}}{2}$

Then: $\frac{\text{cosec}^2\alpha + 2\sec^2\alpha}{\text{cosec}^2\alpha - 3\sec^2\alpha} =$

$$\frac{1+\cot^2\alpha + 2 + 2\tan^2\alpha}{1+\cot^2\alpha - 3 - 3\tan^2\alpha}$$
$$= \frac{3+\cot^2\alpha + 2\tan^2\alpha}{-2+\cot^2\alpha - 3\tan^2\alpha}$$
$$= \frac{3+\frac{13}{4}+2\cdot\frac{2\cdot 4}{13}}{-2+\frac{13}{4}-3\cdot\frac{4}{13}}$$
$$= \frac{3\cdot 52+169+32}{-2\cdot 53+169-48}$$
$$= \frac{156+201}{-104+121} = \frac{357}{17} = 21$$

57. The value of $\frac{\sec^2\theta}{\text{cosec}^2\theta} + \frac{\text{cosec}^2\theta}{\theta} - (\sec^2\theta + \text{cosec}^2\theta)$ is:

Solution:

$$\frac{\sec^2\theta}{\text{cosec}^2\theta} + \frac{\text{cosec}^2\theta}{\theta} - (\sec^2\theta + \text{cosec}^2\theta)$$
$$= \frac{\sin^2\theta}{\cos^2\theta} + \frac{\cos^2\theta}{\sin^2\theta}\left(\frac{1}{\cos^2\theta} + \frac{1}{\sin^2\theta}\right)$$
$$= \frac{\sin^4\theta + \cos^4\theta - (\sin^2\theta + \cos^2\theta)}{\cos^2\theta\sin^2\theta}$$
$$= \frac{1 - 2\sin^2\theta\cos^2\theta - 1}{\cos^2\theta\sin^2\theta}$$
$$= -2$$

58. The value of $\frac{2(\sin^6\theta + \cos^6\theta) - 3(\sin^4\theta + \cos^4\theta)}{\cos^4\theta - \sin^4\theta - 2\cos^2\theta}$ is:

Solution:

$$\frac{2(\sin^6\theta + \cos^6\theta) - 3(\sin^4\theta + \cos^4\theta)}{\cos^4\theta - \sin^4\theta - 2\cos^2\theta}$$
$$= \frac{2(\sin^4\theta + \cos^4\theta - \sin^2\theta\cos^2\theta) -}{\cos^2\theta - \sin^2\theta - 2\cos^2\theta}$$
$$= \frac{3(\sin^4\theta + \cos^4\theta)}{\cos^2\theta - \sin^2\theta - 2\cos^2\theta}$$

$$= \frac{-\sin^4\theta - \cos^4\theta - 2\sin^2\theta\cos^2\theta}{-(\cos^2\theta + \sin^2\theta)}$$
$$= \frac{-(\sin^2\theta + \cos^2\theta)^2}{-(\cos^2\theta + \sin^2\theta)}$$
$$= \frac{1}{1} = 1$$

59. The value of $4\left[\frac{(1-\sec A)^2 + (1+\sec A)^2}{1+\sec^2 A}\right]$ is:

Solution:

$$4\left[\frac{(1-\sec A)^2 + (1+\sec A)^2}{1+\sec^2 A}\right]$$
$$= 4\left[\frac{2(1+\sec^2 A)}{1+\sec^2 A}\right]$$
$$= 8$$

60. The value of $\sqrt{\sec^2\theta + \text{cosec}^2\theta}\cdot\sqrt{\tan^2\theta - \sin^2\theta}$ is equal to:

Solution:

$$\sqrt{\sec^2\theta + \text{cosec}^2\theta}\cdot\sqrt{\tan^2\theta - \sin^2\theta}$$
$$= \sqrt{\frac{1}{\cos^2\theta} + \frac{1}{\sin^2\theta}}\cdot\sqrt{\frac{\sin^2\theta}{\cos^2\theta} - \sin^2\theta}$$
$$= \sqrt{\frac{\sin^2\theta + \cos^2\theta}{\cos^2\theta\sin^2\theta}}\cdot\sqrt{\left(\frac{1}{\cos^2\theta} - 1\right)}$$
$$= \frac{1}{\cos\theta\sin\theta}\cdot\sqrt{\frac{\sin^2\theta\cdot\sin^2\theta}{\cos^2\theta}}$$
$$= \frac{1}{\cos\theta\sin\theta}\cdot\frac{\sin^2\theta}{\cos\theta}$$
$$= \sin\theta\sec^2\theta$$

61. The expression $(\cos^6 x + \sin^6 x - 1)(\tan^2 x + \cot^2 x + 2) + 3$ is equal to:

Solution:

$$(\cos^6 x + \sin^6 x - 1)(\tan^2 x + \cot^2 x + 2) + 3$$
$$= [1(\cos^2 x + \sin^2 x)^2 - 2\cos^2 x\sin^2 x - \cos^2 x\sin^2 x - 1]\left(\frac{\sin^2 x + \cos^2 x}{\sin x\cos x}\right)^2 + 3$$
$$= (1 - 3\cos^2 x\sin^2 x - 1)\cdot\frac{1}{\sin^2 x\cos^2 x} + 3$$
$$= -3\cos^2 x\sin^2 x\cdot\frac{1}{\sin^2 x\cos^2 x} + 3$$
$$= -3 + 3 = 0$$

62. If $117\cos^2 A + 129\sin^2 A = 120$ and $170\cos^2 B + 158\sin^2 B = 161$, then the value of $\text{cosec}^2 A \sec^2 B$ is:

Solution:

$117\cos^2 A + 129\sin^2 A = 120$

$117\cos^2 A + 117\sin^2 A + 12\sin^2 A = 120$

$117(\cos^2 A + \sin^2 A) + 12\sin^2 A = 120$

$12\sin^2 A = 120 - 117$

$\sin^2 A = \frac{3}{12}$

$\sin^2 A = \frac{1}{4}, \quad \text{cosec}^2 A = 4$ ----(i)

And: $170\cos^2 B + 158\sin^2 B = 161$

$158\cos^2 B + 12\cos^2 B + 158\sin^2 B = 161$

$12\cos^2 B + 158(\sin^2 B + \cos^2 B) = 161$

$12\cos^2 B = 161 - 158$

$\cos^2 B = \frac{3}{12}$

$\cos^2 B = \frac{1}{4}, \quad \sec^2 B = 4$ ----(ii)

Multiplying equations (i) and (ii):

$\text{cosec}^2 A \sec^2 B = 4 \cdot 4 = 16$

63. The value of $\left[\frac{\sqrt{3}+2\sin P}{1-2\cos P}\right]^3 + \left[\frac{1+2\cos P}{\sqrt{3}-2\sin P}\right]^{-3}$ is:

Solution:

$\left[\frac{\sqrt{3}+2\sin P}{1-2\cos P}\right]^3 + \left[\frac{1+2\cos P}{\sqrt{3}-2\sin P}\right]^{-3}$

Let $P = 90°$: $\left[\frac{\sqrt{3}+2\sin 90°}{1-2\cos 90°}\right]^3 + \left[\frac{1+2\cos 90°}{\sqrt{3}-2\sin 90°}\right]^{-3}$

$\left[\frac{\sqrt{3}+2\cdot 1^{-3}}{1-0}\right]^{-3} + \left[\frac{1+0}{\sqrt{3}-2\cdot 1}\right]^3$

$(\sqrt{3}+2)^3 + \left(\frac{\sqrt{3}+2}{\sqrt{3}-2}\cdot\frac{\sqrt{3}+2}{\sqrt{3}+2}\right)^3$

$(\sqrt{3}+2)^3 - (\sqrt{3}+2)^3 = 0$

64. If $\cos\theta = \frac{2P}{P^2+1}$, $(P \neq \pm 1)$, then $\text{cosec}\,\theta$ is equal to:

Solution:

$\cos\theta = \frac{2P}{P^2+1}$

$\sqrt{1-\sin^2\theta} = \frac{2P}{P^2+1}$ $(\cos^2\theta = 1 - \sin^2\theta)$

$1-\sin^2\theta = \left(\frac{2P}{P^2+1}\right)^2$

$\sin^2\theta = 1 - \frac{4P^2}{(P^2+1)^2}$

$\sin^2\theta = \frac{(P^2+1)^2 - 4P^2}{(P^2+1)^2}$

$\sin^2\theta = \frac{P^4+1+2P^2-4P^2}{(P^2+1)^2}$

$\sin^2\theta = \frac{(P^2-1)^2}{(P^2+1)^2}$

$\sin\theta = \frac{P^2-1}{P^2+1}$

$\text{cosec}\,\theta = \frac{P^2+1}{P^2-1}$

65. If $4 - 2\sin^2\theta - 5\cos\theta = 0$, $0° < \theta < 90°$, then the value of $\sin\theta + \tan\theta$ is:

Solution:

$4 - 2\sin^2\theta - 5\cos\theta = 0$

$4 - 2(1-\cos^2\theta) - 5\cos\theta = 0$

$4 - 2 + 2\cos^2\theta - 5\cos\theta = 0$

$2\cos^2\theta - 5\cos\theta + 2 = 0$

$2\cos^2\theta - 4\cos\theta - \cos\theta + 2 = 0$

$2\cos\theta(\cos\theta - 2) - 1(\cos\theta - 2) = 0$

$(\cos\theta - 2)(2\cos\theta - 1) = 0$

$\cos\theta = 2$ (Does not exist), $\cos\theta = \frac{1}{2} \implies \theta = 60°$

Then: $\sin\theta + \tan\theta = \sin 60° + \tan 60°$

$= \frac{\sqrt{3}}{2} + \sqrt{3}$

$= \frac{3\sqrt{3}}{2}$

66. What is the value of $\dfrac{2\ \cot(\pi - A)/2}{1 + \tan^2(2\pi - A)/2}$?

Solution:

$$\frac{2\ \cot\left(\dfrac{\pi - A}{2}\right)}{1 + \tan^2\left(\dfrac{2\pi - A}{2}\right)}$$

$$= \frac{2\ \cot\left(\dfrac{\pi}{2} - \dfrac{A}{2}\right)}{1 + \tan^2\left(\pi - \dfrac{A}{2}\right)} = \frac{2\ \tan\dfrac{A}{2}}{1 + \tan^2\dfrac{A}{2}} =$$

$\sin A$ (from the formula)

67. The value of $\dfrac{\sin 4\theta}{(1 - \cos 4\theta)}$ is:

Solution:

$$\frac{\sin 4\theta}{(1 - \cos 4\theta)}$$

$\because\ \sin 2\theta = 2\ \sin\theta\ \cos\theta = 2\ \sin^2\theta =$

$1 - \cos 2\theta$

$$\frac{\sin 2(2\theta)}{1 - \cos 2(2\theta)} = \frac{2\ \sin 2\theta\ \cos 2\theta}{2\ \sin^2 2\theta} =$$

$$\frac{\cos 2\theta}{\sin 2\theta} = \cot 2\theta$$

68. $\dfrac{(1 + \cos\theta)^2 + \sin^2\theta}{(\text{cosec}^2\ \theta - 1)\ \sin^2\theta} =?$

Solution:

$$\frac{(1 + \cos\theta)^2 + \sin^2\theta}{(\text{cosec}^2\ \theta - 1)\ \sin^2\theta}$$

$$= \frac{1 + \cos^2\theta + 2\ \cos\theta + \sin^2\theta}{\text{cosec}^2\ \theta\ \sin^2\theta - \sin^2\theta}$$

$$= \frac{1 + 2\ \cos\theta + 1}{1 - \sin^2\theta}$$

$$= 2\left(\frac{1 + \cos\theta}{\cos^2\theta}\right)$$

$$= 2(\sec^2\theta + \sec\theta)$$

$$= 2\ \sec\theta(1 + \sec\theta)$$

69. If $\sec\theta = 4x$ and $\tan\theta = \dfrac{4}{x}$, $(x \neq 0)$, then the value of $8(x^2 - \dfrac{1}{x^2})$ is:

Solution:

$\sec\theta = 4x$ and $\tan\theta = \dfrac{4}{x}$

$\therefore \sec^2\theta = (4x)^2$

$\sec^2\theta = 16x^2$ ----(i)

$\tan^2\theta = \dfrac{16}{x^2}$ ----(ii)

By subtracting equation (ii) from equation (i):

$\sec^2\theta - \tan^2\theta = 16x^2 - \dfrac{16}{x^2}$

$1 = 16\left(x^2 - \dfrac{1}{x^2}\right)$

$\dfrac{1}{2} = 8\left(x^2 - \dfrac{1}{x^2}\right)$

Hence: $8\left(x^2 - \dfrac{1}{x^2}\right) = \dfrac{1}{2}$

70. If $\left(\dfrac{1}{1 + \text{cosec}\ \theta} - \dfrac{1}{1 - \text{cosec}\ \theta}\right)\cos\theta = 2$, $0° < \theta < 90°$, then the value of $\sin^2\theta + \cot^2\theta + \sec^2\theta$ is:

Solution:

$$\left(\frac{1}{1 + \text{cosec}\ \theta} - \frac{1}{1 - \text{cosec}\ \theta}\right)\cos\theta = 2$$

$$= \frac{(1 - \text{cosec}\ \theta) - (1 + \text{cosec}\ \theta)}{(1 + \text{cosec}\ \theta)(1 - \text{cosec}\ \theta)}\cos\theta =$$

2

$$= \frac{1 - \text{cosec}\ \theta - 1 - \text{cosec}\ \theta}{1 - \text{cosec}^2\ \theta}\cos\theta = 2$$

$$= \frac{-2\ \text{cosec}\ \theta}{-\cot^2\theta}\cos\theta = 2$$

$$= \frac{2}{\sin\theta}\cdot\frac{\sin^2\theta}{\cos^2\theta}\cos\theta = 2$$

$2\tan\theta = 2 \implies \tan\theta = 1 \implies \theta = 45°$

$\therefore \sin^2\theta + \cot^2\theta + \sec^2\theta = \left(\dfrac{1}{\sqrt{2}}\right)^2 +$

$(1)^2 + (\sqrt{2})^2$

$= \dfrac{1}{2} + 1 + 2 = 3\dfrac{1}{2}$

71. If $A = 2(\sin^6\theta + \cos^6\theta) - 3(\sin^4\theta + \cos^4\theta)$, then the value of 3α such that $\cos\alpha = \sqrt{\dfrac{3+A}{5+A}}$ is:

Solution:

$$A = 2(\sin^6\theta + \cos^6\theta) - 3(\sin^4\theta + \cos^4\theta)$$
$$A = 2[(\sin^2\theta)^3 + (\cos^2\theta)^3] - 3(\sin^4\theta + \cos^4\theta)$$
$$A = 2[(\sin^2\theta + \cos^2\theta)((\sin^4\theta + \cos^4\theta) - \sin^2\theta\cos^2\theta)] - 3(\sin^4\theta + \cos^4\theta)$$
$$A = 2\sin^4\theta + 2\cos^4\theta - 2\sin^2\theta\cos^2\theta - 3\sin^4\theta - 3\cos^4\theta$$
$$A = -(\sin^2\theta + \cos^2\theta)$$
$$A = -1$$
$$\cos\alpha = \sqrt{\frac{3+A}{5+A}} = \frac{1}{\sqrt{2}}$$
$$\cos\alpha = \cos 45^\circ \implies \alpha = 45^\circ$$
$$3\alpha = 3\cdot 45^\circ = 135^\circ$$

72. What is the value of $\dfrac{\cos 7A + \cos 5A}{\sin 7A - \sin 5A}$?

Solution:

$$\frac{\cos 7A + \cos 5A}{\sin 7A - \sin 5A}$$
$$= \frac{2\cos\dfrac{7A+5A}{2}\cos\dfrac{7A-5A}{2}}{2\cos\dfrac{7A+5A}{2}\sin\dfrac{7A-5A}{2}}$$
$$= \cot A$$

73. What is the value of $\dfrac{2\sin(45^\circ+\theta)\sin(45^\circ-\theta)}{\cos 2\theta}$?

Solution:

$$\frac{2\sin(45^\circ+\theta)\sin(45^\circ-\theta)}{\cos 2\theta}$$
$$\because 2\sin A\sin B = \cos(A-B) - \cos(A+B)$$
$$= \frac{\cos(45^\circ+\theta-45^\circ+\theta) - \cos(45^\circ+\theta+45^\circ-\theta)}{\cos 2\theta}$$
$$= \frac{\cos 2\theta - \cos 90^\circ}{\cos 2\theta}$$
$$= \frac{\cos 2\theta}{\cos 2\theta} = 1$$

74. What is the value of $\dfrac{\sin(90^\circ-A) + \cos(180^\circ-2A)}{\cos(90^\circ-2A) + \sin(180^\circ-A)}$?

Solution:

$$\frac{\sin(90^\circ-A) + \cos(180^\circ-2A)}{\cos(90^\circ-2A) + \sin(180^\circ-A)} = \frac{\cos A - \cos 2A}{\sin 2A + \sin A}$$
$$= \frac{\cos A - \cos 2A}{\sin 2A + \sin A}$$
$$= \frac{2\sin\left(\dfrac{A+2A}{2}\right)\sin\left(\dfrac{2A-A}{2}\right)}{2\sin\left(\dfrac{2A+A}{2}\right)\cos\left(\dfrac{2A-A}{2}\right)}$$
$$= \frac{2\sin\left(\dfrac{3A}{2}\right)\sin\left(\dfrac{A}{2}\right)}{2\sin\left(\dfrac{3A}{2}\right)\cos\left(\dfrac{A}{2}\right)}$$
$$= \frac{\sin\left(\dfrac{A}{2}\right)}{\cos\left(\dfrac{A}{2}\right)} = \tan\left(\frac{A}{2}\right)$$

75. What is the value of $\dfrac{\sin(90^\circ-10\theta)\cos(\pi-6\theta)}{\cos\left(\dfrac{\pi}{2}-10\theta\right) - \sin(\pi-6\theta)}$?

Solution:

$$\frac{\sin(90^\circ-10\theta) - \cos(\pi-6\theta)}{\cos(90^\circ-10\theta) - \sin(\pi-6\theta)}$$
$$= \frac{\cos 10\theta + \cos 6\theta}{\sin 10\theta - \sin 6\theta}$$
$$= \frac{2\cos\left(\dfrac{10\theta+6\theta}{2}\right)\cos\left(\dfrac{10\theta-6\theta}{2}\right)}{2\cos\left(\dfrac{10\theta+6\theta}{2}\right)\sin\left(\dfrac{10\theta-6\theta}{2}\right)}$$
$$= \frac{\cos 2\theta}{\sin 2\theta} = \cot 2\theta$$

76. What is the value of $\dfrac{\cos 40° - \cos 140°}{\sin 80° + \sin 20°}$?

Solution:

$$\frac{\cos 40° - \cos 140°}{\sin 80° + \sin 20°}$$

$$= \frac{2 \sin\left(\dfrac{40° + 140°}{2}\right) \sin\left(\dfrac{140° - 40°}{2}\right)}{2 \sin\left(\dfrac{80° + 20°}{2}\right) \cos\left(\dfrac{80° - 20°}{2}\right)}$$

$$= \frac{2 \sin 90° \sin 50°}{2 \sin 50° \cos 30°}$$

$$= \frac{\sin 90°}{\cos 30°} = \frac{1}{\dfrac{\sqrt{3}}{2}} = \frac{2}{\sqrt{3}}$$

77. What is the value of $\dfrac{1 - \tan(90° - \theta) + \sec(90° - \theta)}{\tan(90° - \theta) + \sec(90° - \theta) + 1}$?

Solution:

$$\frac{1 - \tan(90° - \theta) + \sec(90° - \theta)}{\tan(90° - \theta) + \sec(90° - \theta) + 1}$$

$$= \frac{1 - \cot\theta + \operatorname{cosec}\theta}{\cot\theta + \operatorname{cosec}\theta + 1}$$

$$= \frac{1 - \dfrac{\cos\theta}{\sin\theta} + \dfrac{1}{\sin\theta}}{\dfrac{\cos\theta}{\sin\theta} + \dfrac{1}{\sin\theta} + 1}$$

$$= \frac{\sin\theta - \cos\theta + 1}{\cos\theta + \sin\theta + 1}$$

$$= \frac{2 \sin\left(\dfrac{\theta}{2}\right) \cos\left(\dfrac{\theta}{2}\right)}{2 \cos\left(\dfrac{\theta}{2}\right) \sin\left(\dfrac{\theta}{2}\right)}$$

$$= \frac{\sin\left(\dfrac{\theta}{2}\right)}{\cos\left(\dfrac{\theta}{2}\right)} = \tan\left(\frac{\theta}{2}\right)$$

78. What is the value of $\sin(90° - x) \cos[\pi - (x - y)] + \cos(90° - x) \sin[\pi - (x - y)]$?

Solution:

$\sin(90° - x) \cos[\pi - (x - y)] + \cos(90° - x) \sin[\pi - (x - y)]$

$$= -\cos x \cos(x - y) + \sin x \sin(x - y)$$

$$= -\cos x \cos(x - y) - \sin x \sin(x - y)$$

$$= -[\cos x \cos(x - y) + \sin x \sin(x - y)]$$

$$= -\cos[x - (x - y)]$$

$$= -\cos[x - x + y]$$

$$= -\cos y$$

79. What is the value of $\dfrac{32 \cos^6 x - 48 \cos^4 x + 18 \cos^2 x - 1}{4 \sin x \cos x \sin(60° - x) \cos(60° - x) \sin(60° + x) \cos(60° + x)}$?

Solution:

$$\frac{32 \cos^6 x - 48 \cos^4 x + 18 \cos^2 x - 1}{4 \sin x \cos x \sin(60° - x) \cos(60° - x) \sin(60° + x) \cos(60° + x)}$$

$\because \sin x \sin(60° - x) \sin(60° + x) = \dfrac{1}{4} \sin 3x$

and $\cos x \cos(60° - x) \cos(60° + x) = \dfrac{1}{4} \cos 3x$

$$= \frac{2\,(16 \cos^6 x - 24 \cos^4 x + 9 \cos^2 x - 1)}{\dfrac{4}{4} \sin 3x \cdot \dfrac{1}{4} \cos 3x}$$

$$= \frac{2[(4 \cos^2 x)^2 + (3 \cos x)^2 - 2 \cdot 4 \cos^2 x \cdot 3 \cos x] - 1}{\dfrac{1}{8} \cdot 2 \sin 3x \cos 3x}$$

$$= \frac{2(4 \cos^2 x - 3 \cos x)^2 - 1}{\dfrac{1}{8} \sin 6x}$$

$$= \frac{8 \cos 6x}{\sin 6x} = 8 \cot 6x$$

80. What is the value of $\dfrac{\dfrac{[4 \cos(90° - A) \sin^3(90° + A)]}{\cos\left(\dfrac{180° + 8A}{2}\right)}}{\dfrac{[4 \sin(90° + A) \cos^3(90° - A)]}{\cos\left(\dfrac{180° + 8A}{2}\right)}}$?

Solution:

$$\frac{\dfrac{[4 \sin(90° + A) \cos^3(90° - A)]}{\cos\left(\dfrac{180° + 8A}{2}\right)}}{\dfrac{[4 \sin(90° + A) \cos^3(90° - A)]}{\cos\left(\dfrac{180° + 8A}{2}\right)}}$$

$= \frac{4 \sin A \cos^3 A - 4 \cos A \sin^3 A}{\cos(90° + 4A)}$

$\because \cos(90° - \theta) = \sin\theta, \cos(90° + \theta) = -\sin\theta, \sin(90° - \theta) = \cos\theta, \sin(90° + \theta) = \cos\theta$

$= \frac{4 \sin A \cos A(\cos^2 A - \sin^2 A)}{-\sin 4A}$

$= \frac{2 \cdot 2 \sin A \cos A \cos 2A}{-\sin 4A}$ $(\because \sin 2A = 2 \sin A \cos A)$

$= \frac{2 \sin 2A \cos 2A}{-\sin 4A}$

$= \frac{\sin 4A}{-\sin 4A} = -1$

81. If $P + Q + R = 60°$, then what is the value of $\cos Q \cos R(\cos P - \sin P) + \sin Q \sin R(\sin P - \cos P)$?

Solution:

Given: $P + Q + R = 60°$ ----(i)

Assume $P = Q = 0°$ and $R = 60°$

$\cos Q \cos R(\cos P - \sin P) + \sin Q \sin R(\sin P - \cos P)$

$= 1 \cdot \frac{1}{2}(1 - 0) + 0$

$= \frac{1}{2}$

82. What is the value of $(90° - B) \sin(C - A) + \sin(90° + A) \cos(B + C) - \sin(90° - C) \cos(A + B)$?

Solution:

$\cos(90° - B) \sin(C - A) + \sin(90° + A) \cos(B + C) - \sin(90° - C) \cos(A + B)$

$= \sin B \sin(C - A) + \cos A \cos(B + C) - \cos C \cos(A + B)$

$= \sin B[\sin C \cos A - \cos C \sin A] + \cos A[\cos B \cos C - \sin B \sin C] - \cos C[\cos A \cos B - \sin A \sin B]$

$= 0$

83. If $A + B + C = 90°$, then what is the value of $\sin\left(\frac{A}{2}\right) \sin\left(\frac{180° - B - C}{2}\right) + \cos\left(\frac{A}{2}\right) \sin\left(\frac{B + C}{2}\right)$?

Solution:

$\sin\left(\frac{A}{2}\right) \sin\left(\frac{180° - B - C}{2}\right) + \cos\left(\frac{A}{2}\right) \sin\left(\frac{B + C}{2}\right)$

$= \sin\left(\frac{A}{2}\right) \sin\left(90° - \frac{B + C}{2}\right) + \cos\left(\frac{A}{2}\right) \sin\left(\frac{B + C}{2}\right)$

$= \sin\left(\frac{A}{2}\right) \cos\left(\frac{B + C}{2}\right) + \cos\left(\frac{A}{2}\right) \sin\left(\frac{B + C}{2}\right)$

$= \sin\left(\frac{A}{2} + \frac{B + C}{2}\right)$

$= \sin\left(\frac{90°}{2}\right) = \frac{1}{\sqrt{2}}$

84. What is the value of $\frac{(\sin 4x + \sin 4y) \tan(2x - 2y)}{\sin 4x - \sin 4y}$?

Solution:

$\frac{(\sin 4x + \sin 4y) \tan(2x - 2y)}{\sin 4x - \sin 4y}$

$= \frac{2 \sin\left(\frac{4x + 4y}{2}\right) \cos\left(\frac{4x - 4y}{2}\right)}{2 \cos\left(\frac{4x + 4y}{2}\right) \sin\left(\frac{4x - 4y}{2}\right)} \cdot \frac{\sin(2x - 2y)}{\cos(2x - 2y)}$

$= \frac{\sin(2x + 2y) \cos(2x - 2y)}{\cos(2x + 2y) \sin(2x - 2y)} \cdot \frac{\sin(2x - 2y)}{\cos(2x - 2y)}$

$= \tan(2x + 2y)$

85. What is the value of $\dfrac{\cos 3\theta + 2\cos 5\theta + \cos 7\theta}{\cos\theta + 2\cos 3\theta + \cos 5\theta} + \sin 2\theta \tan 3\theta$?

Solution:

$$\frac{\cos 3\theta + 2\cos 5\theta + \cos 7\theta}{\cos\theta + 2\cos 3\theta + \cos 5\theta} + \sin 2\theta \tan 3\theta$$
$$= \frac{\cos 3\theta + \cos 7\theta + 2\cos 5\theta}{\cos\theta + \cos 5\theta + 2\cos 3\theta} + \sin 2\theta \cdot \frac{\sin 3\theta}{\cos 3\theta}$$
$$= \frac{2\cos 5\theta \cos 2\theta + 2\cos 5\theta}{2\cos 3\theta \cos 2\theta + 2\cos 3\theta} + \frac{\sin 2\theta \sin 3\theta}{\cos 3\theta}$$
$$= \frac{2\cos 5\theta(\cos 2\theta + 1)}{2\cos 3\theta(\cos 2\theta + 1)} + \frac{\sin 2\theta \sin 3\theta}{\cos 3\theta}$$
$$= \frac{\cos 5\theta + \sin 2\theta \sin 3\theta}{\cos 3\theta}$$
$$= \frac{\cos(2\theta + 3\theta) + \sin 2\theta \sin 3\theta}{\cos 3\theta}$$
$$= \frac{\cos 2\theta \cos 3\theta}{\cos 3\theta} = \cos 2\theta$$

86. What is the value of $\dfrac{\tan 5\theta + \tan 3\theta}{4\cos 4\theta(\tan 5\theta - \tan 3\theta)}$?

Solution:

$$\frac{\tan 5\theta + \tan 3\theta}{4\cos 4\theta(\tan 5\theta - \tan 3\theta)} = \frac{\frac{\sin 5\theta}{\cos 5\theta} + \frac{\sin 3\theta}{\cos 3\theta}}{4\cos 4\theta\left(\frac{\sin 5\theta}{\cos 5\theta} - \frac{\sin 3\theta}{\cos 3\theta}\right)}$$
$$= \frac{\frac{\sin 5\theta \cos 3\theta + \cos 5\theta \sin 3\theta}{\cos 5\theta \cos 3\theta}}{\frac{\cos 4\theta \sin(5\theta - 3\theta)}{\cos 5\theta \cos 3\theta}}$$
$$= \frac{\sin(5\theta + 3\theta)}{4\cos 4\theta \sin(5\theta - 3\theta)}$$
$$= \frac{\sin 8\theta}{4\cos 4\theta \sin 2\theta}$$
$$= \frac{2\sin 4\theta \cos 4\theta}{4\cos 4\theta \sin 2\theta}$$
$$= \frac{2\sin 2\theta \cos 2\theta}{2\sin 2\theta} = \cos 2\theta$$

87. What is the value of $\dfrac{\sin 7x - \sin 5x}{\cos 7x + \cos 5x} - \dfrac{\cos 6x - \cos 4x}{\sin 6x + \sin 4x}$?

Solution:

$$\frac{\sin 7x - \sin 5x}{\cos 7x + \cos 5x} - \frac{\cos 6x - \cos 4x}{\sin 6x + \sin 4x}$$
$$= \frac{2\cos\left(\frac{7x+5x}{2}\right)\sin\left(\frac{7x-5x}{2}\right)}{2\cos\left(\frac{7x+5x}{2}\right)\cos\left(\frac{7x-5x}{2}\right)} - \frac{\sin\left(\frac{6x+4x}{2}\right)\sin\left(\frac{6x-4x}{2}\right)}{2\sin\left(\frac{6x+4x}{2}\right)\cos\left(\frac{6x-4x}{2}\right)}$$
$$= \tan x + \tan x = 2\tan x$$

88. What is the value of $\dfrac{\sin(y-z) + \sin(y+z) + 2\sin y}{\sin(x-z) + \sin(x+z) + 2\sin x}$?

Solution:

$$\frac{\sin(y-z) + \sin(y+z) + 2\sin y}{\sin(x-z) + \sin(x+z) + 2\sin x}$$
$$= \frac{2\sin y\cos(-z) + 2\sin y}{2\sin x\cos(-z) + 2\sin x}$$
$$= \frac{\sin y[\cos(-z) + 1]}{\sin x[\cos(-z) + 1]} = \frac{\sin y}{\sin x}$$

89. What is the value of $\dfrac{\sin(x+y) - 2\sin x + \sin(x-y)}{\cos(x-y) + \cos(x+y) - 2\cos x} \times \dfrac{\sin 10x - \sin 8x}{\cos 10x + \cos 8x}$?

Solution:

$$\frac{\sin(x+y) - 2\sin x + \sin(x-y)}{\cos(x-y) + \cos(x+y) - 2\cos x} \times \frac{\sin 10x - \sin 8x}{\cos 10x + \cos 8x}$$
$$= \frac{2\sin x\cos y - 2\sin x}{2\cos x\cos y - 2\cos x} \times \frac{2\cos\left(\frac{10x+8x}{2}\right)\sin\left(\frac{10x-8x}{2}\right)}{2\cos\left(\frac{10x+8x}{2}\right)\cos\left(\frac{10x-8x}{2}\right)}$$
$$= \frac{2\sin x(\cos y - 1)}{2\cos x(\cos y - 1)} \times \frac{2\cos 9x \sin x}{2\cos 9x \cos x}$$
$$= \frac{x}{x} = x$$

90. If $\tan 17° \tan 32° + \tan 32° \tan 41° + \tan 41° \tan 17° =?$,

Solution:

If $A+B+C=90°$
Then $\tan A \tan B + \tan B \tan C + \tan C \tan A = 1$
$A+B+C = 17° + 32° + 41° = 90°$
Ans = 1

91. If $\tan(A+8) = \frac{1}{2}$, $\tan(A-B) = \frac{1}{3}$, then find the value of $\tan 2A$:
(a) 5 (b) 7 (c) 1 (d) 3

Solution:

$\tan 2A = \tan(A+B+A-B)$

$= \frac{\frac{1}{2}+\frac{1}{3}}{1-\frac{1}{2}\times\frac{1}{3}} = \frac{\frac{5}{6}}{\frac{5}{6}} = 1$

92. If $\tan\alpha = \frac{1}{7}$, $\tan\beta = \frac{1}{3}$, then $\cos 2\alpha =?$

Solution:

$\cos 2\alpha = \frac{1-\tan^2\alpha}{1+\tan^2\alpha} = \frac{1-\frac{1}{49}}{1+\frac{1}{49}} = \frac{48}{50} = \frac{24}{25}$

$\sin 2\beta = \frac{2\times\frac{1}{3}}{1+\frac{1}{9}} = \frac{3}{5}$

93. If $\cos(\alpha+\beta) = \frac{4}{5}$, $\sin(\alpha-\beta) = \frac{5}{13}$, and $\alpha, \beta \in (0, \frac{\pi}{4})$, find $\tan 2\alpha$:

Solution:

$\tan 2\alpha = \tan[(\alpha+\beta)+(\alpha-\beta)]$

$= \frac{\tan(\alpha+\beta)+\tan(\alpha-\beta)}{1-\tan(\alpha+\beta)\tan(\alpha-\beta)}$

$= \frac{\frac{3}{4}+\frac{5}{12}}{1-\frac{3}{4}\cdot\frac{5}{12}} = \frac{\frac{9+5}{12}}{\frac{1-5}{16}} = \frac{\frac{14}{12}\cdot\frac{16}{11}}{} = \frac{56}{33}$

94. $\tan 13x - \tan 9x - \tan 4x =?$

Solution:

$\tan 13x = \tan(9x+4x)$

$\tan 13x = \frac{\tan 9x + \tan 4x}{1-\tan 9x \tan 4x}$

$\tan 13x - \tan 13x \tan 9x \tan 8x + \tan 4x$

$\tan 13x - \tan 9x \tan 4x = \tan 13x \tan 9x \tan 4x$

Type 7 (Maxima/Minima)

95. What is the maximum value of $3\sin x - 4$? $3\sin x - 4$?

Solution:

$3\sin x - 4$

For maximum value, $\sin x$ should be maximum: $3(1) - 4 = 3 - 4$

Ans = -1

96. Maximum value of $2\sin\theta + 3\cos\theta$ is:

Solution:

$a\sin x + b\cos x$

Maximum value: $\sqrt{a^2+b^2} = \sqrt{2^2+3^2} = \sqrt{4+9} = \sqrt{13}$

97. The minimum value of $2\sin^2\theta + 3\cos^2\theta$ is:

Solution:

$a\sin^2 x + b\cos^2 x$

If $a > b$ (max $= a$, min $= b$) If $a < b$ (max $= b$, min $= a$) Minimum value = 2

98. The minimum value of $4\tan^2\theta + 9\cot^2\theta$ is equal to:

Solution:

$a\tan^2 x + b\cot^2 x$

Minimum value: $2\sqrt{ab} = 2\sqrt{4\times 9} = 2\sqrt{36} = 2\cdot 6 = 12$

99. The minimum value of $\sin^2\theta + \cos^2\theta + \sec^2\theta + \operatorname{cosec}^2\theta + \tan^2\theta + \cot^2\theta$ is:

Solution:

$\sin^2 x + \cos^2 x + \sec^2 x + \operatorname{cosec}^2 x + \tan^2 x + \cot^2 x$

$1 + 1 + 2\tan^2 x + 1 + 2\cot^2 x$

$3 + 2(\tan^2 x + \cot^2 x)$

Minimum value of $\tan^2 x + \cot^2 x = 2$:

$3 + 2(2) = 3 + 4 = 7$

100. If x is a real quantity, what is the minimum value of $25\cos^2 x + 9\sec^2 x$?

Solution:

Minimum value of $a\cos^2 x + b\sec^2 x = 2\sqrt{ab}$

$2\sqrt{25 \cdot 9} = 2 \cdot 5 \cdot 3 = 30$

101. The least value of $8\operatorname{cosec}^2\theta + 25\sin^2\theta$ is:

Solution:

The minimum value of $a\operatorname{cosec}^2\theta + b\sin^2\theta$

$= 2\sqrt{a \times b} = 2\sqrt{25 \times 8} = 20\sqrt{2}$

Note:

For expressions of the form:

$a\sin^2\theta + b\operatorname{cosec}^2\theta$, $a\cos^2\theta + b\sec^2\theta$, $a\tan^2\theta + b\cot^2\theta$

Minimum value = $2\sqrt{ab}$

102. Find the minimum value of $32\sec^2\theta + 72\operatorname{cosec}^2\theta$:

Solution:

Formula: $a\sec^2 x + b\operatorname{cosec}^2 x = (\sqrt{a} + \sqrt{b})^2$

$(\sqrt{32} + \sqrt{72})^2 = (4\sqrt{2} + 6\sqrt{2})^2 = (10\sqrt{2})^2 = 200$

19 Height and Distance

1. A ladder 18m long rests against a wall so that the angle between the ladder and the wall is 30°. How far (in m) is the base of the ladder from the wall?
18 मीटर लंबी एक सीढ़ी एक दीवार पर इस प्रकार टिकी हुई है कि सीढ़ी और दीवार के बीच का कोण 30° है। सीढ़ी का आधार दीवार से कितनी दूर (मीटर में) है?

Solution:

Let the base of the ladder from the wall = x m.

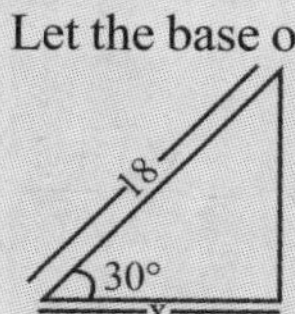

$$\cos 30^\circ = \frac{x}{18}$$

$$\frac{\sqrt{3}}{2} = \frac{x}{18}$$

$$x = 9\sqrt{3} \text{ m}$$

2. The length of the shadow of a vertical tower on level ground increases by 8.4 cm when the altitude of the sun changes from 45° to 30°. What is the height of the tower (in m)?
जब सूर्य की ऊंचाई 45° से 30° हो जाती है तो समतल जमीन पर एक ऊर्ध्वाधर टॉवर की छाया की लंबाई 8.4 सेमी बढ़ जाती है। टावर की ऊंचाई (मीटर में) क्या है?

Solution:

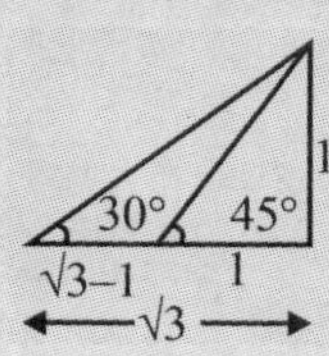

Given,

$$\sqrt{3} - 1 \text{ unit} = 8.4$$

$$1 \text{ unit} = \frac{8.4 \times (\sqrt{3}+1)}{2}$$

$$1 \text{ unit} = 4.2(\sqrt{3}+1)$$

Hence, the height of the tower is $4.2(\sqrt{3} + 1)$ m.

3. On walking 100 metres towards a building in a horizontal line, the angle of elevation of its top changes from 45° to 60°. What will be the height (in metres) of the building?
किसी भवन की ओर क्षैतिज रेखा में 100 मीटर चलने पर उसके शीर्ष का उन्नयन कोण 45° से 60° तक बदल जाता है। भवन की ऊंचाई (मीटर में) क्या होगी?

Solution:

Let the height of the building = h m.

In $\triangle DBC$:

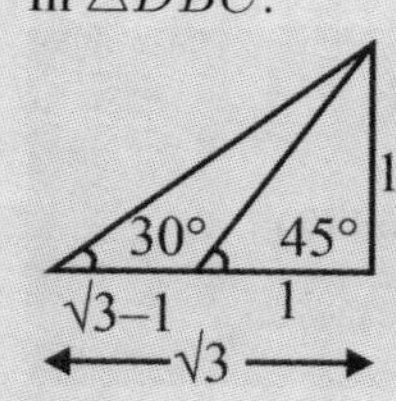

$$\tan 60^\circ = \frac{h}{x}$$

$$\sqrt{3} = \frac{h}{x}$$

$$x = \frac{h}{\sqrt{3}} \qquad (i)$$

In $\triangle DAC$:

$$\tan 45^\circ = \frac{h}{100+x}$$

$$1 = \frac{h}{100+x}$$

$$h = 100 + x$$

Substitute $x = \frac{h}{\sqrt{3}}$ from equation (1):

$$h = 100 + \frac{h}{\sqrt{3}}$$

$$h(\sqrt{3}-1)/\sqrt{3} = 100$$

$$h = \frac{100\sqrt{3}}{\sqrt{3}-1}$$

$$h = 50(3+\sqrt{3}) \text{ m}$$

4. A pole stands vertically on a road, which goes in the north-south direction. P, Q are two points towards the north of the pole, such that $PQ = b$, and the angles of elevation of the top of the pole at P, Q are α, β respectively. Then the height of the pole is:

एक सड़क पर एक खंभा लंबवत खड़ा है, जो उत्तर-दक्षिण दिशा में जाती है। P, Q खंभे के उत्तर की ओर दो बिंदु हैं, जैसे कि $PQ = b$, और P, Q पर खंभे के शीर्ष के उन्नयन कोण क्रमशः α, β हैं। तो खंभे की ऊंचाई है:

Solution:

$$\tan \beta = \frac{SR}{b+QR} \qquad (i)$$

$$\tan \alpha = \frac{SR}{QR} \qquad (ii)$$

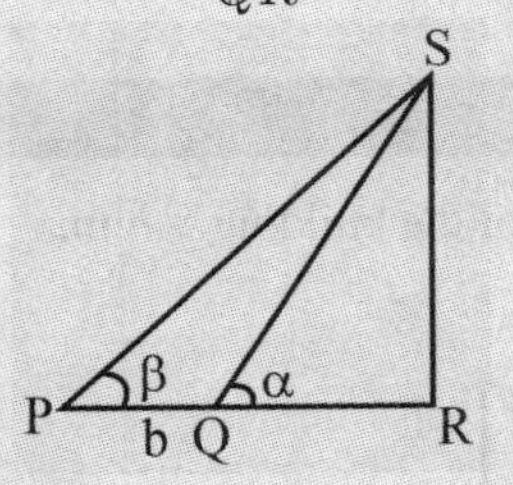

From equations (1) and (2):

$$\tan \beta = \frac{SR}{b + \frac{SR}{\tan \alpha}}$$

$$\frac{1}{\cot \beta} = \frac{SR}{b + SR \times \cot \alpha}$$

$$b = (\cot \beta - \cot \alpha)SR$$

$$SR = \frac{b}{\cot \beta - \cot \alpha}$$

5. A ladder is placed against a wall such that it just reaches the top of the wall. The foot of the ladder is at a distance of 5 metres from the wall. The angle of elevation of the top of the wall from the base of the ladder is 15°. What is the length (in metres) of the ladder?

एक सीढ़ी को एक दीवार के सामने इस प्रकार रखा गया है कि वह दीवार के शीर्ष तक पहुँचती है। सीढ़ी का पाया दीवार से 5 मीटर की दूरी पर है। सीढ़ी के आधार से दीवार के शीर्ष का उन्नयन कोण 15° है। सीढ़ी की लंबाई (मीटर में) क्या है?

Solution:

Let the length of the ladder $(AC) = l$ m.

In $\triangle ABC$:

$$\cos 15^\circ = \frac{5}{l}$$

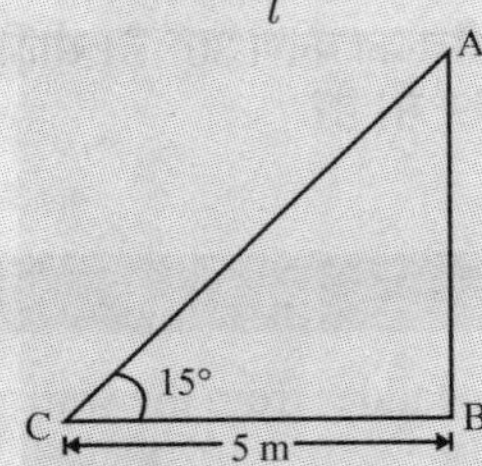

$$\frac{\sqrt{3}+1}{2\sqrt{2}} = \frac{5}{l} \qquad \left(\cos 15^\circ = \frac{\sqrt{3}+1}{2\sqrt{2}}\right)$$

$$l = \frac{10\sqrt{2}}{\sqrt{3}+1} \times \frac{\sqrt{3}-1}{\sqrt{3}-1}$$

$$l = \frac{10\sqrt{6}-10\sqrt{2}}{2} = 5\sqrt{6} - 5\sqrt{2} \text{ m}$$

6. From a point exactly midway between the foot of two towers P and Q, the angles of elevation of their tops are 30° and 60°, respectively. The ratio of the height of P to that of Q is:

दो टावरों P और Q के आधार के ठीक बीच में एक बिंदु से, उनके शीर्षों का उन्नयन कोण क्रमशः 30° और 60° है। P की ऊंचाई का Q की ऊंचाई से अनुपात है:

Solution:

In $\triangle$,

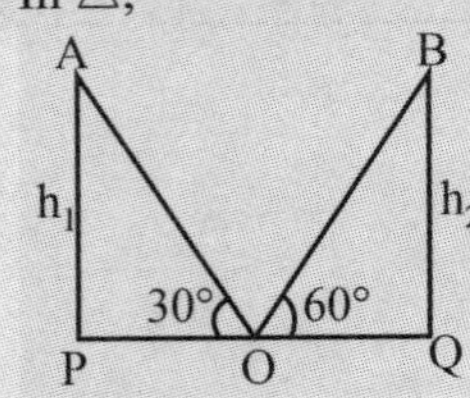

$$\tan 30° = \frac{h_1}{OP}$$

$$h_1 = OP \times \frac{1}{\sqrt{3}} \quad (i)$$

In $\triangle AQOB$,

$$\tan 60° = \frac{h_2}{OQ}$$

$$h_2 = OQ \times \sqrt{3} \quad (ii)$$

From equations (i) and (ii):

$$\frac{h_1}{h_2} = \frac{OP \times \frac{1}{\sqrt{3}}}{OQ \times \sqrt{3}} \quad (\because OP = OQ)$$

$$\frac{h_1}{h_2} = \frac{1}{3}$$

$$h_1 : h_2 = 1 : 3$$

7. Let A and B be two towers with the same base. From the midpoint of the line joining their feet, the angles of elevation of the tops of A and B are $30°$ and $45°$ respectively. The ratio of the heights of A and B is:

माना A और B एक ही आधार वाले दो टावर हैं। उनके पैरों को जोड़ने वाली रेखा के मध्य बिंदु से, A और B के शीर्षों का उन्नयन कोण क्रमशः $30°$ और $45°$ है। A और B की ऊंचाई का अनुपात है:

Solution:

Given,

Length of tower $A = PR = h_1$

Length of tower $B = QS = h_2$

$\angle ROP = 30°, \quad \angle SOQ = 45°$

and $OP = OQ$

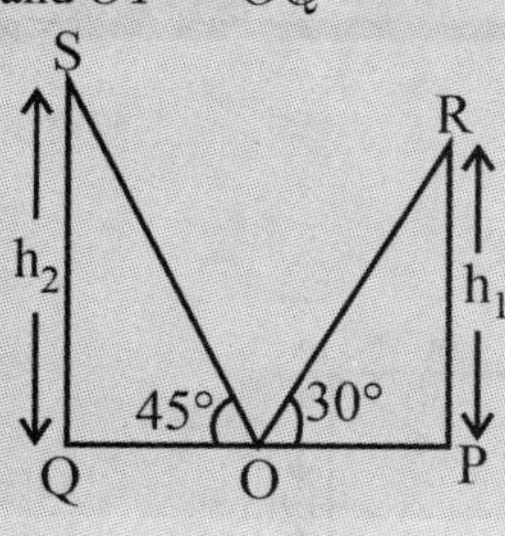

In $\triangle RPO$,

$$\tan 30° = \frac{PR}{OP}$$

$$\frac{1}{\sqrt{3}} = \frac{h_1}{OP}$$

$$OP = \sqrt{3}h_1 \quad (i)$$

In $\triangle SQO$,

$$\tan 45° = \frac{SQ}{OQ}$$

$$1 = \frac{h_2}{OQ} \quad (ii)$$

From equations (i) and (ii):

$$\sqrt{3}h_1 = h_2 \quad (\because OP = OQ)$$

$$h_1 : h_2 = 1 : \sqrt{3}$$

8. Two trees are standing along the opposite sides of a road. The distance between the two trees is 400 metres. There is a point on the road between the trees. The angle of depressions of the point from the tops of the trees are $45°$ and $60°$. If the height of the tree which makes $45°$ angle is 200 metres, then what will be the height (in metres) of the other tree?

एक सड़क के विपरीत किनारों पर दो पेड़ खड़े हैं। दोनों पेड़ों के बीच की दूरी 400 मीटर है। सड़क पर पेड़ों के बीच एक बिंदु है। पेड़ों के शीर्ष से बिंदु के अवनमन कोण $45°$ और $60°$ हैं। यदि $45°$ कोण बनाने वाले पेड़ की ऊंचाई 200 मीटर है, तो दूसरे पेड़ की ऊंचाई (मीटर में) क्या होगी?

Solution:

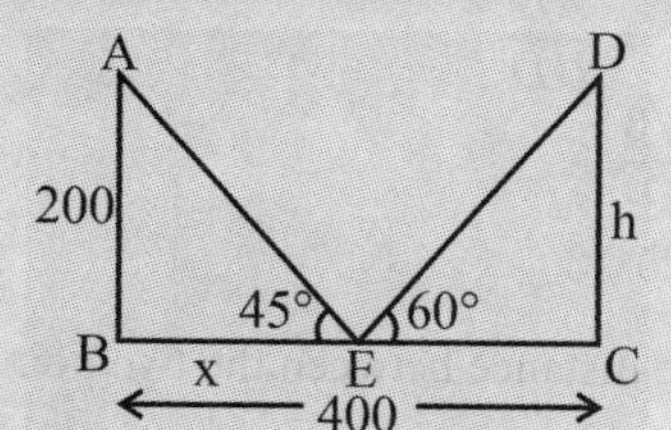

In $\triangle ABE$,

$$\tan 45° = \frac{200}{x}$$

$$x = 200 \text{ m}$$

In $\triangle DCE$,

$$\tan 60° = \frac{h}{400 - 200}$$

$$h = 200\sqrt{3} \text{ m}$$

9. The angle of elevation of the top of a tower $25\sqrt{3}$ m high from two points on the level ground on its opposite sides are $45°$ and $60°$. What is the distance (in m) between the two points (correct to one decimal place)?

$25\sqrt{3}$ मीटर ऊंचे एक टावर के शीर्ष का समतल जमीन पर विपरीत दिशा के दो बिंदुओं से उन्नयन कोण 45° और 60° है। दो बिंदुओं (दशमलव के एक स्थान तक सही) के बीच की दूरी (मीटर में) क्या है?

Solution:

Height of tower $(AB) = 25\sqrt{3}$

In $\triangle ABC$:

$$\tan 60^\circ = \frac{AB}{BC}$$

$$\sqrt{3} = \frac{25\sqrt{3}}{BC}$$

$$BC = 25 \text{ m}$$

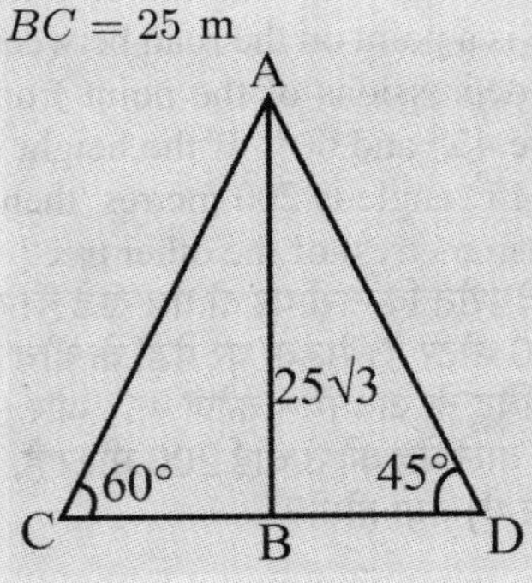

In $\triangle ABD$:

$$\tan 45^\circ = \frac{AB}{BD}$$

$$1 = \frac{25\sqrt{3}}{BD}$$

$$BD = 25\sqrt{3}$$

Now, the distance between the two points,

$$CD = BC + BD = 25 + 25\sqrt{3} \approx 68.3 \text{ m}$$

10. The angle of elevation of the top of a tree from a point on the ground which is 300 m away from the tree is 30°. When the tree grew up, its angle of elevation of the top of it became 60° from the same point. How much did the tree grow? (nearest to an integer)
पेड़ से 300 मीटर दूर जमीन पर एक बिंदु से पेड़ के शीर्ष का उन्नयन कोण 30° है। जब पेड़ बड़ा हुआ तो उसके शीर्ष का उन्नयन कोण उसी बिंदु से 60° हो गया। पेड़ कितना बड़ा हुआ? (एक पूर्णांक के निकटतम)

Solution:

Let the initial height of the tree = BD.

In $\triangle BDC$:

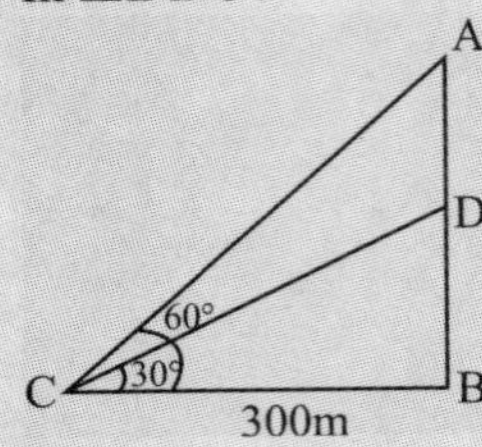

$$\tan 30^\circ = \frac{BD}{300}$$

$$BD = 100\sqrt{3} \text{ m}$$

Total height of the tree = AB.

In $\triangle ABC$:

$$\tan 60^\circ = \frac{AB}{300}$$

$$AB = 300\sqrt{3}$$

$$AD + BD = 300\sqrt{3}$$

$$AD = 300\sqrt{3} - 100\sqrt{3}$$

$$AD = 200\sqrt{3} \approx 346 \text{ m}$$

Hence, the increase in the height of the tree = 346 m.

11. From a point 12 m above the water level, the angle of elevation of the top of a hill is 60° and the angle of depression of the base of the hill is 30°. What is the height (in m) of the hill?
जल स्तर से 12 मीटर ऊपर एक बिंदु से, एक पहाड़ी की चोटी का उन्नयन कोण 60° है और पहाड़ी के आधार का अवनमन कोण 30° है। पहाड़ी की ऊंचाई (मीटर में) क्या है?

Solution:

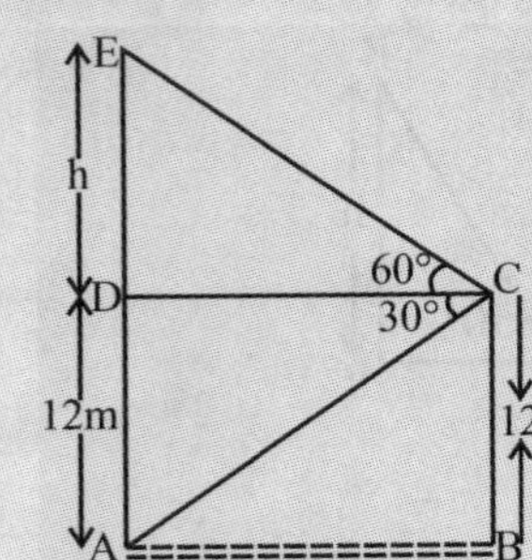

Let the angle of elevation and depression of the top and the base of the hill from point

C 12 m above the water level AB be 60° and 30° respectively.

मान लीजिए जल स्तर AB से 12 मीटर ऊपर बिंदु C से पहाड़ी की चोटी और आधार का उन्नयन और अवनमन कोण क्रमशः 60° और 30° है।

Height of the hill $(AE) = (h + 12)$ m

In $\triangle ADC$:

$$\tan 30° = \frac{AD}{CD} = \frac{12}{CD}$$

$$CD = 12\sqrt{3} \text{ m}$$

Again, in $\triangle CDE$:

$$\tan 60° = \frac{ED}{CD} = \frac{h}{12\sqrt{3}}$$

$$\sqrt{3} = \frac{h}{12\sqrt{3}}$$

$$h = 36 \text{ m}$$

Since, height of the hill $= (h+12) = (36+12)$

$= 48$ m.

12. A straight vertical pole was broken during a cyclone in such a way that its top touched the ground at a distance of $6\sqrt{3}$ m from the bottom of the pole and made an angle of 30° with the horizontal. What was the height (in m) of the pole?

चक्रवात के दौरान एक सीधा ऊर्ध्वाधर खंभा इस प्रकार टूट गया कि उसका शीर्ष खंभे के नीचे से $6\sqrt{3}$ मीटर की दूरी पर जमीन को छू गया और क्षैतिज के साथ 30° का कोण बना गया। खम्भे की ऊंचाई (मीटर में) क्या थी?

Solution:

Let, the height of pole is H m. In $\triangle ABC$:

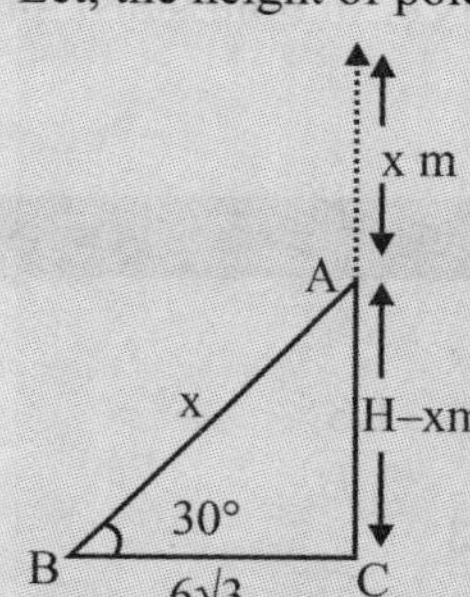

$$\sin 30° = \frac{(H-x)}{x}$$

$$\frac{1}{2} = \frac{(H-x)}{x}$$

$$3x = 2H$$

$$H = \frac{3x}{2}$$

Now, $\cos 30° = \frac{6\sqrt{3}}{x}$:

$$\frac{\sqrt{3}}{2} = \frac{6\sqrt{3}}{x}$$

$$x = 12 \text{ m}$$

Height of the pole: $H = \frac{3x}{2} = \frac{3 \times 12}{2} =$ 18 m.

13. The angles of elevation of the top of a tower 72 meters high from the top and bottom of a building are 30° and 60° respectively. What is the height (in m) of the building?

एक इमारत के शीर्ष और तल से 72 मीटर ऊंचे टावर के शीर्ष का उन्नयन कोण क्रमशः 30° और 60° है। इमारत की ऊंचाई (मीटर में) क्या है?

Solution:

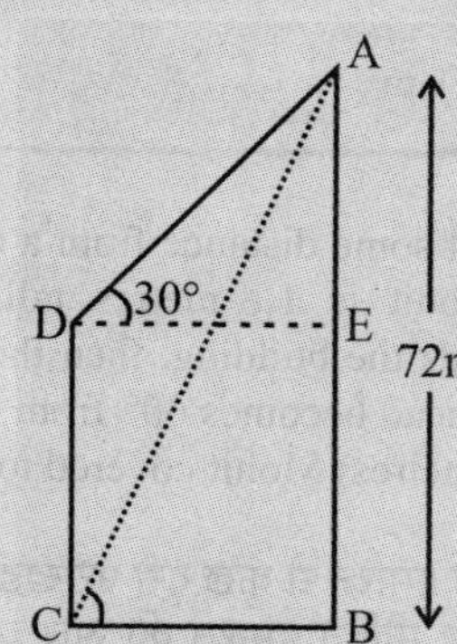

In $\triangle ABC$:

$$\tan 60° = \frac{72}{BC}$$

$$BC = \frac{72}{\sqrt{3}}$$

In $\triangle AED$:

$$\tan 30° = \frac{AE}{DE} \text{ (where } DE = BC)$$

$$\frac{1}{\sqrt{3}} = \frac{AE}{\frac{72}{\sqrt{3}}}$$

$$AE = 24 \text{ m}$$

Height of the building (DC):

$$DC = AB - AE = 72 - 24 = 48 \text{ m}$$

14. A tower stands on the top of a building which is 40 metres high. The angle of depression of a point situated on the ground from the top and bottom of the tower are found to be 60° and 45° respectively. What is the height (in m) of tower?

एक इमारत के शीर्ष पर एक टावर खड़ा है जो 40 मीटर ऊंचा है। टावर के ऊपर और नीचे से जमीन पर स्थित एक बिंदु का अवनमन कोण क्रमश: 60° और 45° पाया जाता है। टावर की ऊंचाई (मीटर में) क्या है?

Solution:

Let, height of tower = x.

$$\tan 45^\circ = \frac{40}{d}$$

$$d = 40 \text{ m}$$

$$\tan 60^\circ = \frac{x+40}{d}$$

$$\sqrt{3} = \frac{x+40}{40}$$

$$x = 40(\sqrt{3}-1) \text{ m}$$

15. Mohit is standing at some distance from a 60 metres tall building. Mohit is 1.8 metres tall. When Mohit walks towards the building, then the angle of elevation from his head becomes 60° from 45°. How much distance (in metres) Mohit covered towards the building?

मोहित 60 मीटर ऊंची इमारत से कुछ दूरी पर खड़ा है। मोहित 1.8 मीटर लंबा है। जब मोहित इमारत की ओर चलता है, तो उसके सिर से ऊंचाई का कोण 45° से 60° हो जाता है। मोहित ने इमारत की ओर कितनी दूरी (मीटर में) तय की?

Solution:

In $\triangle AGC$:

$$\tan 45^\circ = \frac{58.2}{CG}$$

$$1 = \frac{58.2}{CG}$$

$$CG = 58.2 \text{ m}$$

In $\triangle AGE$:

$$\tan 60^\circ = \frac{58.2}{EG}$$

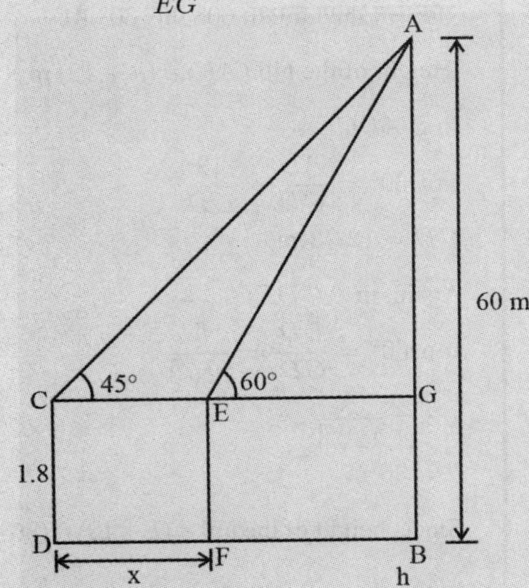

$$EG = \frac{58.2}{\sqrt{3}} \times \frac{\sqrt{3}}{3} = 19.4\sqrt{3} \text{ m}$$

$$x = 58.2 - 19.4\sqrt{3}$$

$$x = 19.4(3-\sqrt{3}) \text{ m}$$

16. An aeroplane is flying horizontally at a height of 1.8 km above the ground. The angle of elevation of the plane from point X is 60° and after 20 seconds, its angle of elevation from X becomes 30°. If point X is on the ground, then what is the speed (in km/hr) of the aeroplane?

एक हवाई जहाज़ ज़मीन से 1.8 किमी की ऊँचाई पर क्षैतिज रूप से उड़ रहा है। बिंदु X से विमान का उन्नयन कोण 60° है और 20 सेकंड के बाद, X से इसका उन्नयन कोण 30° हो जाता है। यदि बिंदु X ज़मीन पर है, तो हवाई जहाज की गति (/) में क्या है?

Solution:

In $\triangle ABX$:

$$\tan 60^\circ = \frac{AB}{XB} = \frac{1.8}{XB}$$

$$XB = \frac{1.8}{\sqrt{3}} \times \frac{\sqrt{3}}{3}$$

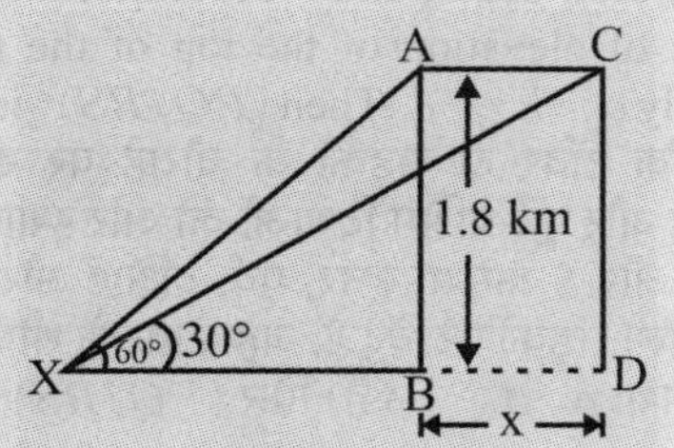

In $\triangle CDX$:

$$\tan 30^\circ = \frac{1.8}{0.6\sqrt{3}+x}$$

$0.6\sqrt{3}+x = 1.8\sqrt{3}$

$x = 1.2\sqrt{3}$ km

Speed of the aeroplane:

$$\frac{1.2\sqrt{3}}{20} \times 3600 = 216\sqrt{3} \text{ km/h}$$

17. Asha and Suman's mud forts have heights 9 cm and 16 cm. If the fort tops are 25 cm apart from each other, then the distance (in cm) between two forts is:

आशा और सुमन के मिट्टी के किलों की ऊंचाई 9 सेमी और 16 सेमी है। यदि किले के शीर्ष एक दूसरे से 25 सेमी की दूरी पर हैं, तो दो किलों के बीच की दूरी () में है:

Solution:

The heights of both the houses are 9 cm and 16 cm, respectively, and the tops of the houses are 25 cm away from each other.

दोनों घरों की ऊंचाई क्रमश: 9 सेमी और 16 सेमी है और घरों के शीर्ष एक दूसरे से 25 सेमी दूर हैं।

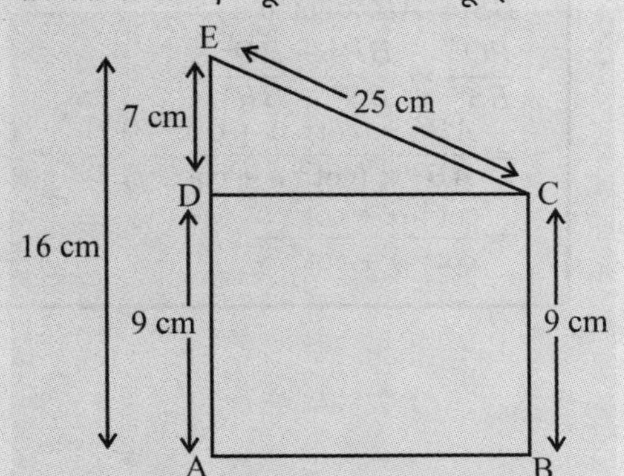

In $\triangle DEC$:

From Pythagoras theorem: $EC^2 = DE^2 + DC^2$

$625 = 49 + DC^2$

$DC^2 = 625 - 49$

$DC^2 = 576$

$DC = 24$ cm

Hence, the distance between two forts: 24 cm.

18. A person was standing on a road near a mall. He was 1425 m away from the mall and able to see the top of the mall from the road in such a way that the top of a tree, which is in between him and the mall, was exactly in line of sight with the top of the mall. The tree height is 10 m and it is 30 m away from him. How tall (in m) is the mall?

एक व्यक्ति एक मॉल के पास सड़क पर खड़ा था। वह मॉल से 1425 मीटर दूर था और सड़क से मॉल के शीर्ष को इस प्रकार देख पा रहा था कि एक पेड़ का शीर्ष, जो उसके और मॉल के बीच में है, बिल्कुल शीर्ष के दृश्य रेखा में था। पेड़ की ऊंचाई 10 मीटर है और यह उससे 30 मीटर दूर है। मॉल कितना लंबा () में है?

Solution:

Let the height of the mall = x m.

According to the question:

$$\frac{DE}{BC} = \frac{AE}{AC} \quad (\text{AADE} \sim \triangle ABC)$$

$$\frac{10}{x} = \frac{30}{1425}$$

$x = 475$ m

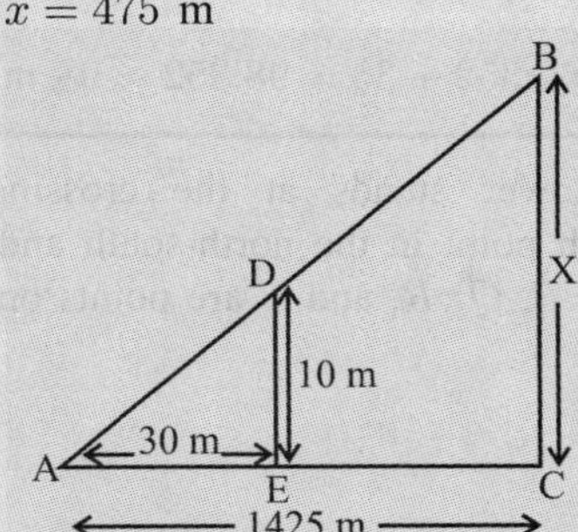

Hence, the height of the mall is 475 m.

19. From the top of a house A in a street, the angles of elevation and depression of the top and foot of another house B on the opposite side of the street are $60°$ and $45°$, respectively. If the height of house A is 36 m, then what is the height of house B? (Your answer should be nearest to an integer.)

एक गली में एक घर A के शीर्ष से, सड़क के विपरीत दिशा में एक अन्य घर B के शीर्ष और तल की ऊंचाई और अवनमन कोण क्रमशः $60°$ और $45°$ हैं। यदि घर A की ऊंचाई 36 मीटर है, तो घर B की ऊंचाई क्या है? (आपका उत्तर पूर्णांक के निकटतम होना चाहिए)

Solution:

Let the height of house $B = h$ m.

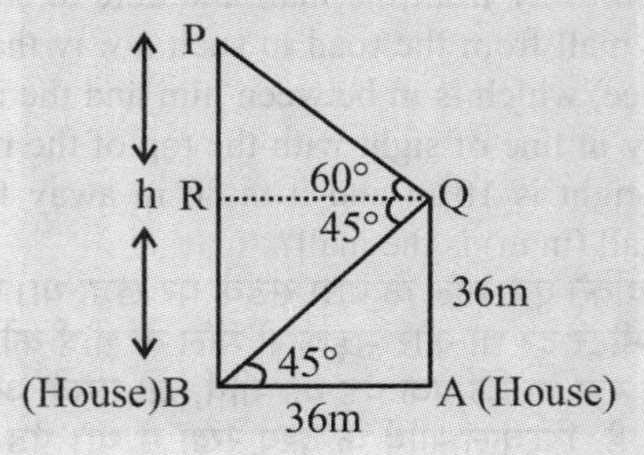

In $\triangle QAB$:

$$\tan 45° = \frac{36}{AB}$$

$$AB = 36 = RQ$$

In $\triangle PRQ$:

$$\tan 60° = \frac{PR}{QR}$$

$$\sqrt{3} = \frac{PR}{36}$$

$$PR = 36\sqrt{3} = 36 \times 1.732 = 62.352$$

$PB = PR + RB$ (where $RB = QA$)

$$PB = 62.352 + 36 = 98.352 \approx 98 \text{ m}$$

20. A clock tower stands at the crossing of two roads which point in the north-south and east-west directions. P, Q, R, and S are points on the roads due north, east, south, and west respectively, where the angles of elevation of the top of the tower are respectively α, β, γ, δ. Then $(PQ/RS)^2$ is equal to:

एक क्लॉक टॉवर दो सड़कों के चौराहे पर खड़ा है जो उत्तर-दक्षिण और पूर्व-पश्चिम दिशाओं की ओर इशारा करते हैं। P, Q, R, और S क्रमशः उत्तर, पूर्व, दक्षिण और पश्चिम की ओर जाने वाली सड़कों पर बिंदु हैं, जहां टावर के शीर्ष की ऊंचाई के कोण क्रमशः α, β, γ, δ हैं। फिर $(PQ/RS)^2$ बराबर है:

Solution:

Let the height of the clock tower $= AB$, and

$\angle PBA = \angle QBA = \angle RBA = \angle SBA = 90°$ (Since, AB is perpendicular to the ground)

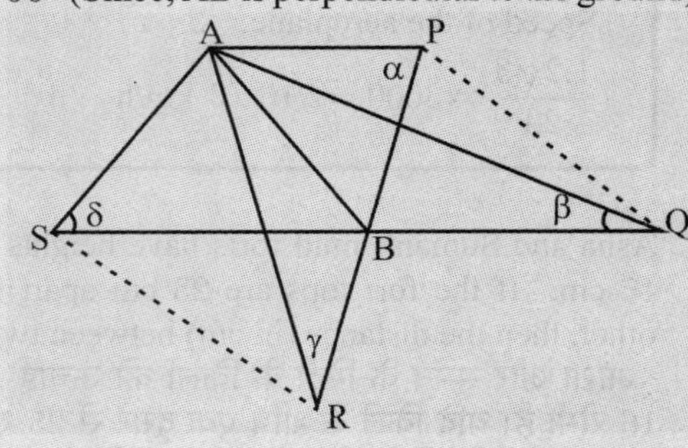

In $\triangle ABP$:

$$BP = AB \times \cot \alpha$$

In $\triangle ABQ$:

$$BQ = AB \times \cot \beta$$

In $\triangle ABR$:

$$BR = AB \times \cot \gamma$$

In $\triangle ABS$:

$$BS = AB \times \cot \delta$$

$$\frac{PQ^2}{RS^2} = \frac{BP^2 + BQ^2}{BS^2 + BR^2}$$

$$= \frac{AB^2 \times (\cot^2 \alpha + \cot^2 \beta)}{AB^2 \times (\cot^2 \delta + \cot^2 \gamma)}$$

$$= \frac{\cot^2 \alpha + \cot^2 \beta}{\cot^2 \delta + \cot^2 \gamma}$$

20 Geometry

Line and Angle/ रेखा और कोण

- **Point ():** Zero dimension figure or a circle with zero radius.
  बिंदु : शून्य आयामी आकृति या शून्य त्रिज्या वाला एक वृत्त।

Types of point/ बिन्दु के प्रकार

(i) **Collinear point/ संरेख बिन्दु**
If 3 or more than 3 points lie on a line close to or far from each other, then they are said to be collinear.
यदि 3 या 3 से अधिक बिंदु एक रेखा पर एक दूसरे के निकट या दूर स्थित हो, तो वे संरेख कहलाते है।
Ex. बिंदु P, Q, R, S संरेख बिंदु है।

(ii) **Non-collinear point/ असंरेख बिंदु**
In 3 or more points are not situated on a straight line, these all points are called non-collinear points.
यदि 3 या अधिक बिन्दु एक सीधी रेखा पर स्थित नहीं हैं, तो ये सभी बिंदु असंरेख बिंदु कहलाते हैं।

- **Line:** (One dimension figure) line is a set of points having only length with no ends.
  रेखा : (एक आयामी आकृति) रेखा बिंदुओं का एक समूह है जिसमें केवल लंबाई होती है जिसका कोई अंत नहीं होता है।

- **Line Segment:** A line with a fixed length.

रेखा खंड : एक निश्चित लंबाई वाली रेखा।

- **Ray:** A line with uni-direction length.

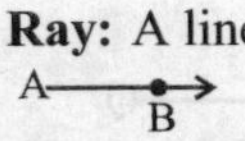

किरण: एक दिशा लंबाई वाली रेखा।

- **Parallel Lines:** Two or more lines that never intersect $L \parallel M$.
  समानांतर रेखाएँ: दो या अधिक रेखाएँ जो एक दूसरे को कभी नहीं काटती हैं। $L \parallel M$

- **Transversal Line:** A line which intersects (touches) two or more lines at distinct points is called transversal lines of the given lines.
  तिर्यक रेखा: वह रेखा जो दो या दो से अधिक रेखाओं को अलग-अलग बिंदु पर काटती (स्पर्श) है, दी गई रेखा की तिर्यक रेखा कहलाती है।

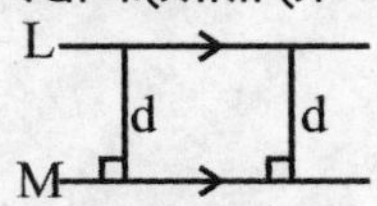

$AB \parallel CD$ and EF is a transversal line.

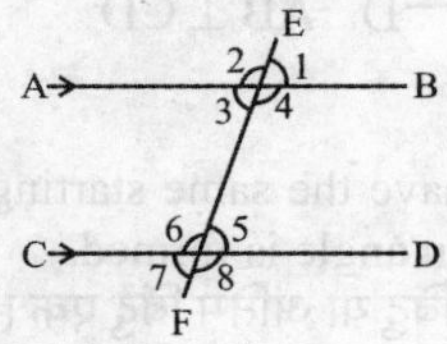

Corresponding angles/ संगत कोण :
$\angle 1 = \angle 5,\ \angle 4 = \angle 8,\ \angle 2 = \angle 6,\ \angle 3 = \angle 7$
Alternate Angles/ एकांतर कोण : $\angle 3 = \angle 5, \angle 4 = \angle 6$
$\angle 4 + \angle 5 = 180°, \angle 3 + \angle 6 = 180°$

Concurrent line/ समवर्ती रेखा

Three or more than three lines, which pass from a single point is called concurrent lines.
तीन या तीन से अधिक रेखाएं, जो एक बिंदु से होकर गुजरती हैं, समवर्ती रेखाएं कहलाती हैं।

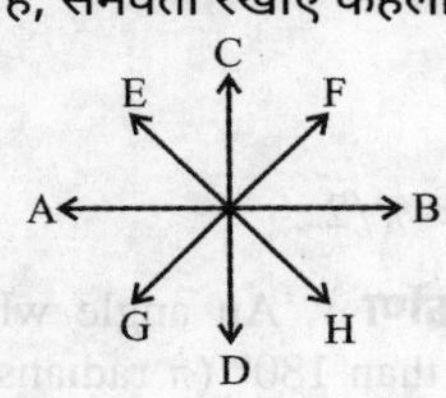

- **Intersecting lines/ प्रतिच्छेदी रेखाएँ** Two lines that intersect each other or in other words, share a common point (Called point of intersection) are called intersecting lines.
  प्रतिच्छेदी रेखाएँ: दो रेखाएँ जो एक दूसरे को प्रतिच्छेद करती हैं या दूसरे शब्दों में, एक उभयनिष्ठ बिंदु साझा करती हैं (प्रतिच्छेदी बिंदु कहा जाता है) प्रतिच्छेदी रेखाएँ कहलाती हैं।

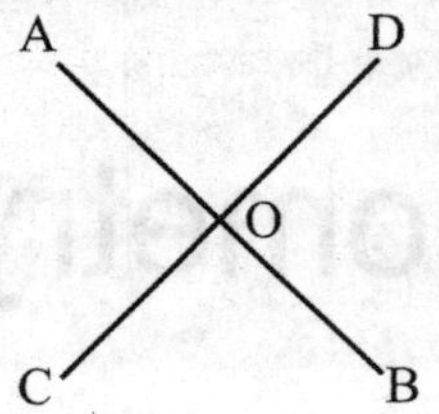

- **Perpendicular lines/ लम्ब रेखाएँ** Two lines that intersect each other at right angle (90°) are called perpendicular lines. They are denoted by the ⊥.
  लम्ब रेखाएँ: दो रेखाएँ जो एक दूसरे को समकोण (90□) पर काटती हैं, लम्ब रेखाएँ कहलाती हैं। इन्हें ” द्वारा दर्शाया जाता है।

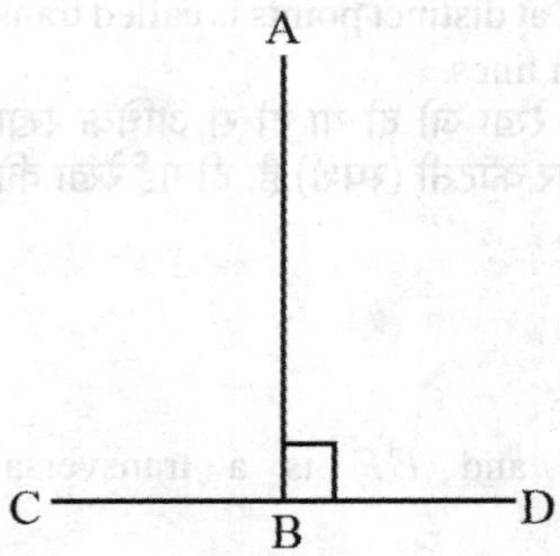

AB ⊥ CD

ANGLES/ कोण

Angle: When two rays have the same starting point or an ending point, then an angle is formed.
जब दो किरणों का प्रारंभिक बिंदु या अंतिम बिंदु एक ही हो तो एक कोण बनता है।

Types of Angles/ कोणों के प्रकार

- **Acute angle/ न्यून कोण** : An angle greater than 0° but less than 90° (i.e. $\pi/2$ radians) is an acute angle.
  न्यून कोण: 0° से बड़ा लेकिन 90° (अर्थात $\pi/2$ रेडियन) से कम कोण न्यूनकोण होता है।

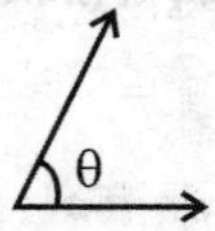

Here $0° < \theta < 90°$ or $0 < \pi/2$.

- **Obtuse angle/ अधिक कोण** : An angle which is greater than 90° and less than 180° (π radians) is an obtuse angle.
  अधिक कोण: वह कोण जो 90° से अधिक और 180° (रेडियन) से कम हो, अधिक कोण होता है।

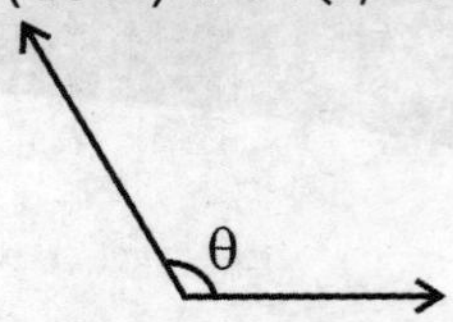

Here $90° < \theta < 180°$ or $\pi/2 < \theta < \pi$.

- **Right angle/ समकोण** : An angle equals to 90° (or $\pi/2$ radians) is a right angle.
  समकोण: 90° (या $\pi/2$ रेडियन) के बराबर का कोण समकोण होता है।

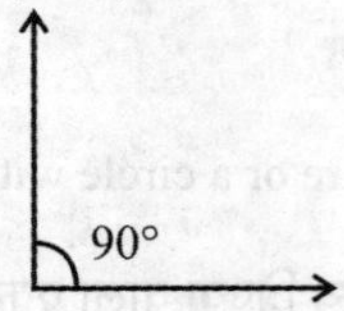

- **Straight angle/ सीधा कोण** : An angle equals to 180° (or π radians) is a straight angle.
  सीधा कोण: 180° (या π रेडियन) के बराबर का कोण सीधा कोण होता है।

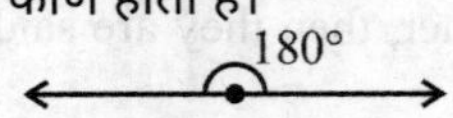

Reflex angle/ प्रतिवर्ती कोण : It is an angle greater than 180° and (π radians) but less than 360° (2π radians).
प्रतिवर्ती कोण: यह 180° और (π रेडियन) से बड़ा लेकिन 360° (2π रेडियन) से कम का कोण है।

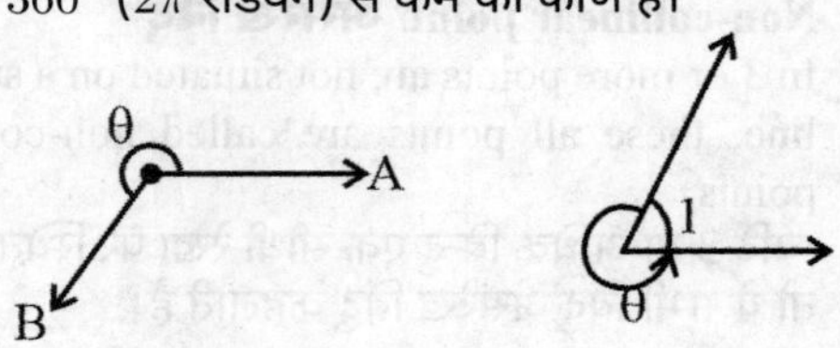

Here $180° < \theta < 360°$ or $\pi < \theta < 2\pi$.

- **Complementary angle/ पूरक कोण** : Two angles are said to be complementary if their sum is equal to 90° (or $\pi/2$ radians).
  पूरक कोण: दो कोण पूरक कहलाते हैं यदि उनका योग 90° (या $\pi/2$ रेडियन) के बराबर हो।

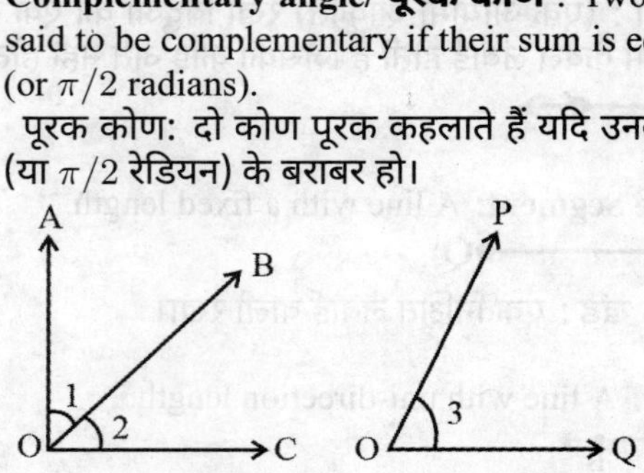

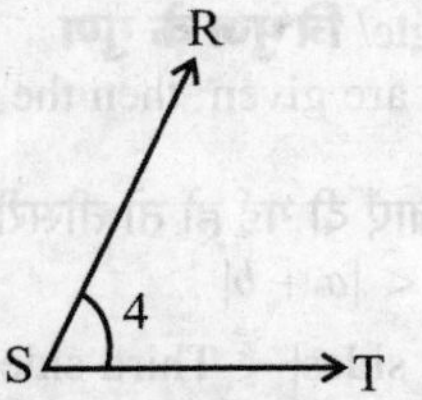

($\angle 3 + \angle 4 = 90°$)

In the above figure, $\angle 1 + \angle 2 = 90°$: Hence, they are complementary angles. Also $\angle 3 + \angle 4 = 90°$, therefore, they are also complementary angles.

उपरोक्त आकृति में, $\angle 1 + \angle 2 = 90°$: इसलिए, वे पूरक कोण हैं। साथ ही $\angle 3 + \angle 4 = 90°$, इसलिए, ये भी पूरक कोण हैं।

- **Supplementary angles/ संपूरक कोण** : If the sum of the angles is equal to 180°, then they are called supplementary angles. For example, supplementary angle of 50° is $180° - 50° = 130°$.

  संपूरक कोण: यदि कोणों का योग 180° के बराबर हो तो वे संपूरक कोण कहलाते हैं। उदाहरण के लिए, 50° का पूरक कोण $180° - 50° = 130°$ है।

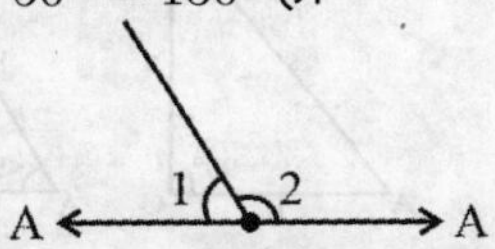

$\angle 1 + \angle 2 = 180°$

Case - I

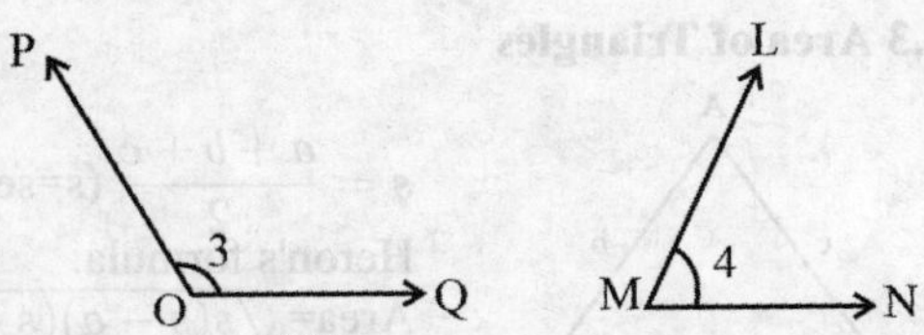

$\angle 3 + \angle 4 = 180°$ Case - II

- **Linear pair/ रैखिक युग्म** : In the case I above, $\angle 1$ and $\angle 2$ form a linear pair. Two angles form a linear pair, if they have one side common between them and the two angles are supplementary.

  रैखिक युग्म: उपरोक्त मामले □ में, $\angle 1$ और $\angle 2$ एक रैखिक युग्म बनाते हैं। दो कोण एक रैखिक युग्म बनाते हैं, यदि उनके बीच एक भुजा उभयनिष्ठ हो और दोनों कोण संपूरक हों।

- **Adjacent angles/ आसन्न कोण** : $\angle 1$ and $\angle 2$ are called adjacent angles, since they have one side common between them.

  आसन्न कोण: $\angle 1$ और $\angle 2$ को आसन्न कोण कहा जाता है, क्योंकि उनके बीच एक भुजा उभयनिष्ठ होती है।

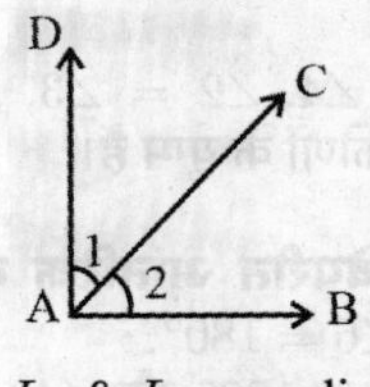

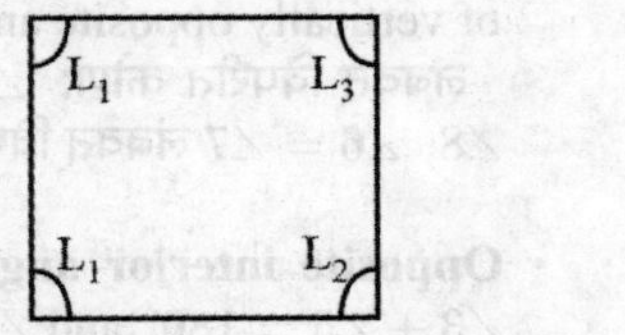

L_1 & L_2 are adjacent angles.

- **Vertically opposite angles/ लंबवत विपरीत कोण** : Let L_1 and L_2 be two intersecting lines as shown below, then $\angle 1$ and $\angle 2$ are said to be vertically opposite angles. Also, $\angle 3$ and $\angle 4$ are vertically opposite angles.

  लंबवत विपरीत कोण: मान लीजिए कि L_1 और L_2 दो प्रतिच्छेदी रेखाएं हैं जैसा कि नीचे दिखाया गया है, तो $\angle 1$ और $\angle 2$ को लंबवत विपरीत कोण कहा जाता है। साथ ही, $\angle 3$ और $\angle 4$ लंबवत विपरीत कोण हैं।

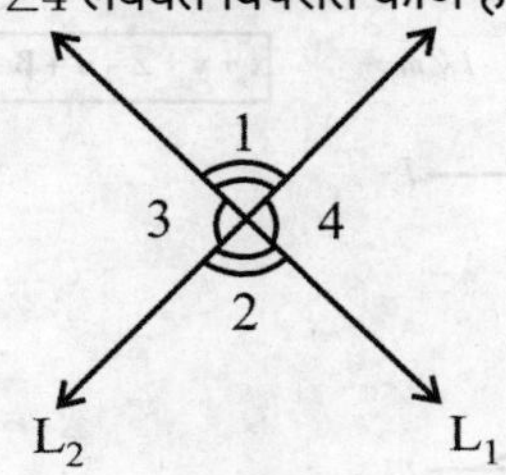

- **Angles between parallel lines/**: Let AB and CD be two parallel lines that are intercepted by a transversal T, then we have:

  समानांतर रेखाओं के बीच के कोण: मान लीजिए AB और CD दो समानांतर रेखाएँ हैं जो एक तिर्यक रेखा T द्वारा अंत:खंडित हैं, तो हमारे पास है

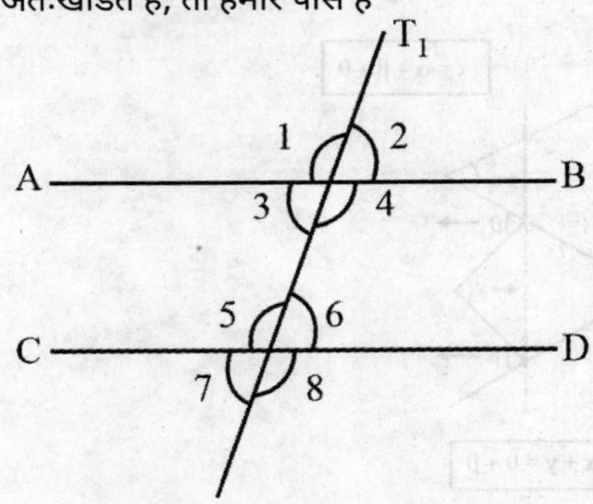

- **Corresponding angles/ संगत कोण** : $\angle 1 = \angle 5$, $\angle 2 = \angle 6$, $\angle 4 = \angle 8$, $\angle 3 = \angle 7$.

  संगत कोण: $\angle 1 = \angle 5$, $\angle 2 = \angle 6$, $\angle 4 = \angle 8$, $\angle 3 = \angle 7$ संगत कोणों के युग्म बराबर होंगे।

  Interior Alternate angles/ आंतरिक वैकल्पिक कोण : $\angle 3 = \angle 6$, $\angle 4 = \angle 5$.

- **Vertically opposite angles/ लंबवत् विपरीत कोण** : $\angle 1 = \angle 4$, $\angle 2 = \angle 3$, $\angle 5 = \angle 8$, $\angle 6 = \angle 7$ are pairs of vertically opposite angles.

लंबवत् विपरीत कोण: $\angle 1 = \angle 4$, $\angle 2 = \angle 3$, $\angle 5 = \angle 8$, $\angle 6 = \angle 7$ लंबवत् विपरीत कोणों के युग्म हैं।

- **Opposite interior angles/ विपरीत आंतरिक कोण** : $\angle 3 + \angle 5 = 180°$ and $\angle 4 + \angle 6 = 180°$.
  विपरीत आंतरिक कोण: $\angle 3 + \angle 5 = 180°$ और $\angle 4 + \angle 6 = 180°$ ।

1. ARROW CONCEPT

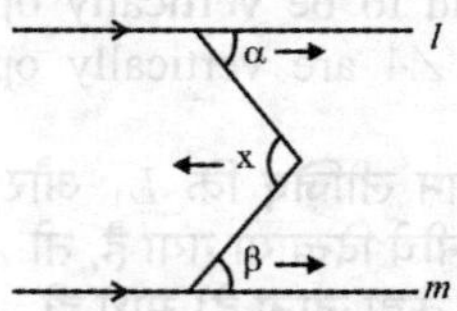

x = α + β

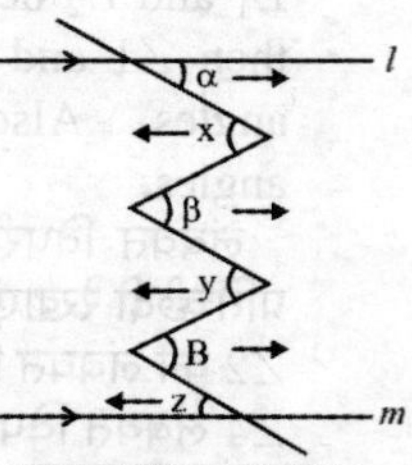

l // m

x + y + z = α + β + θ

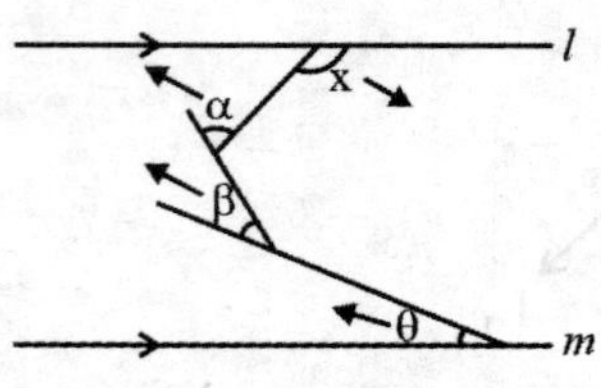

x = α + β + θ

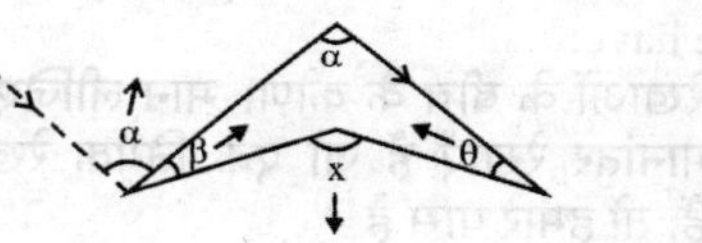

x = α + β + θ

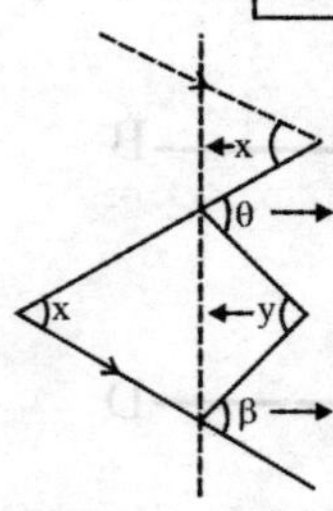

x + y = θ + β

2. TRAINGLE

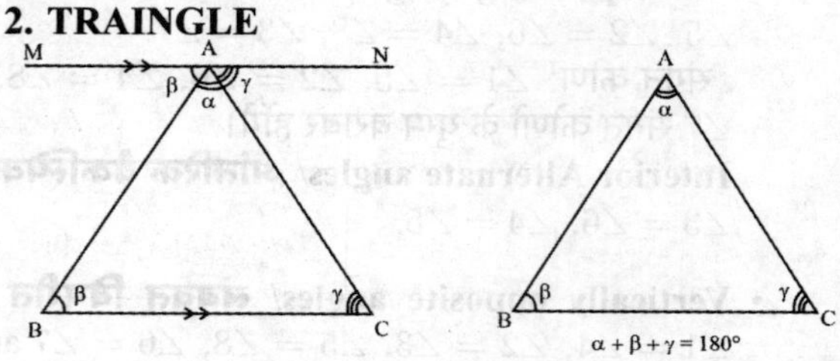

2.1 Properties of Triangle/ त्रिभुज के गुण

If two sides in a triangle are given, then the range of the third side will be:

यदि किसी त्रिभुज में दो भुजाएँ दी गई हों तो तीसरी भुजा का परिसर होगा: $|a-b| < c < |a+b|$

|Difference between two sides| < Third side < |Sum of two sides|.

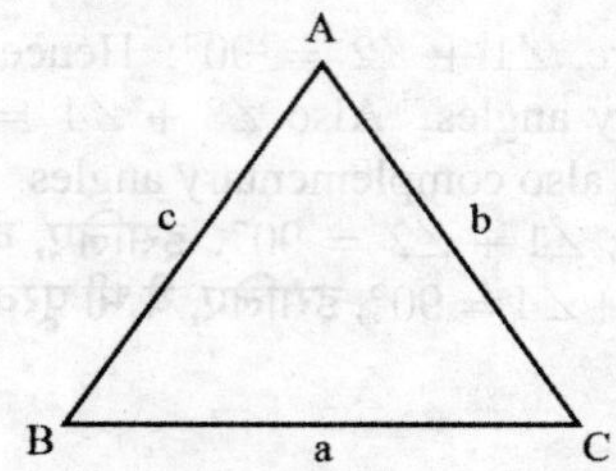

2.2 Types of Triangles/ त्रिभुजों के प्रकार

a) As per angles/ कोणों के अनुसार

OBTUSE	RIGHT ANGLED	ACUTE
One Angle > 90°	One Angle = 90°	All angles < 90°

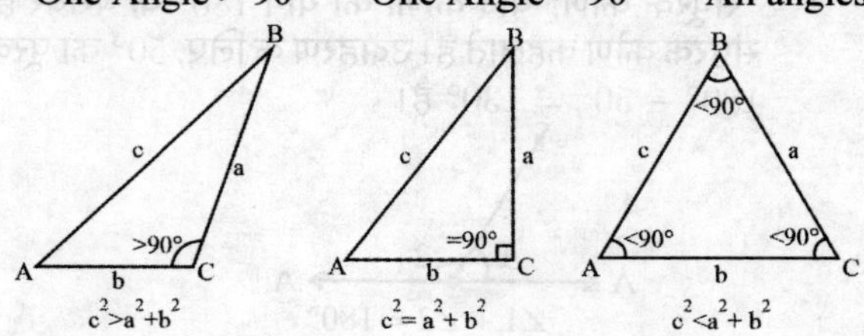

2.3 Area of Triangles

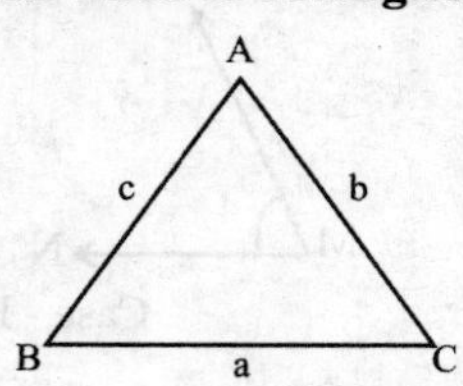

$s = \frac{a+b+c}{2}$ (s=semi-perimeter)

Heron's formula:

Area$=\sqrt{s(s-a)(s-b)(s-c)}$

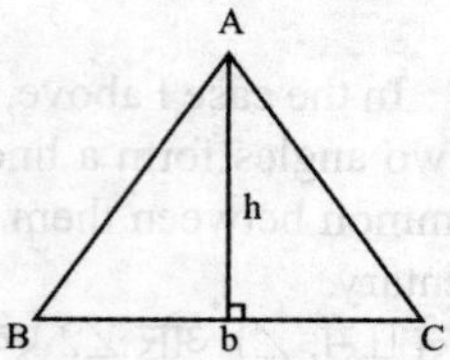

b = base, h = height

Area $= \frac{1}{2}(b \times h)$

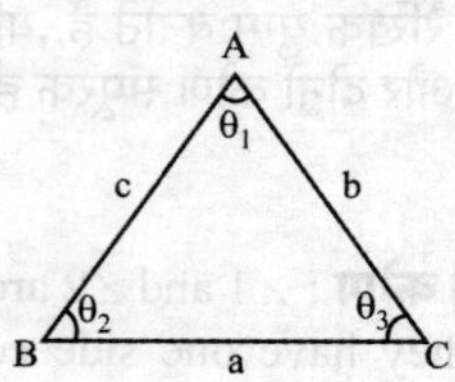

Area $= \frac{1}{2}(b \times c)sin(\theta_1)$

Area $= \frac{1}{2}(c \times a)sin(\theta_2)$

Area $= \frac{1}{2}(a \times b)sin(\theta_3)$

External angle property:

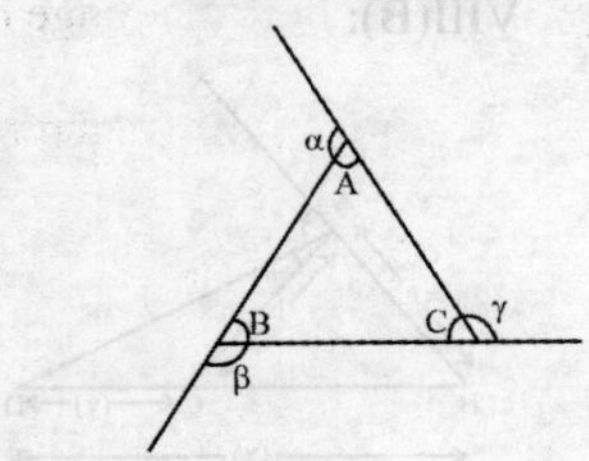

$$\alpha = B + C$$
$$\beta = A + C$$
$$\gamma = A + B$$

2.4 Median:

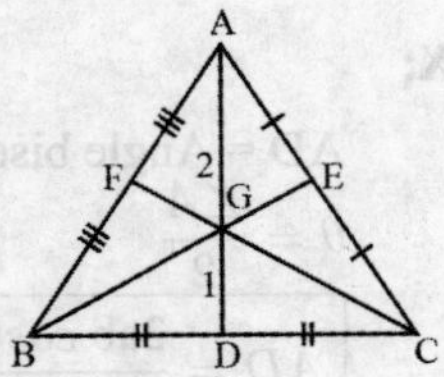

AD, BE, CF – Medians

Point of intersection (G) = Centroid

AG:GD = BG:GE = CG:GF = 2:1

$AD^2 + BE^2 + CF^2 = \frac{3}{4}[AB^2 + BC^2 + CA^2]$

Apollonius theorem:

$AB^2 + AC^2 = 2(AD^2 + BD^2)$

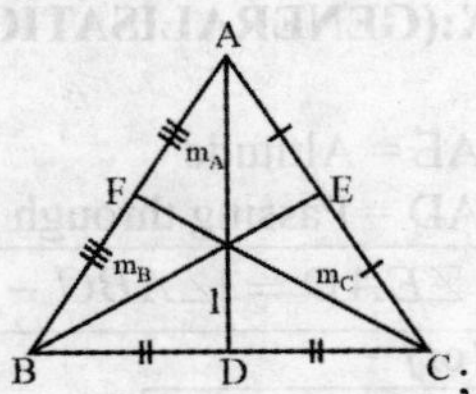

Area of $\triangle$ ABC

$= \frac{4}{3}$ [$\triangle$ medians]

$$\sigma = \frac{m_a + m_b + m_c}{2}$$

$\triangle$ medians = $\sqrt{\sigma(\sigma - m_A)(\sigma - m_B)(\sigma - m_C)}$

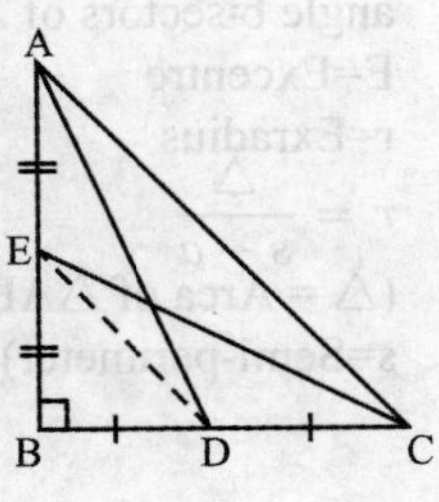

AD & CE are medians

$4(AD^2 + CE^2) = 5AC^2$

$AD^2 + CE^2 = 5ED^2$

$AD^2 + CE^2 = AC^2 + ED^2$

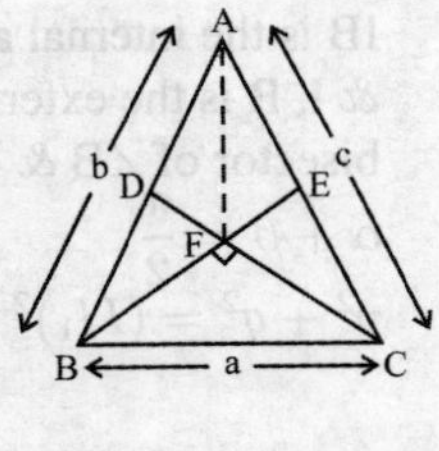

AB=b, BC=a, CA=c and BE, CD are medians

If medians are perpendicular (BE ⊥ CD):

$AB^2 + AC^2 = 5BC^2$

or $b^2 + c^2 = 5a^2$

AF = BC = a

Incentre (Angle Bisectors):

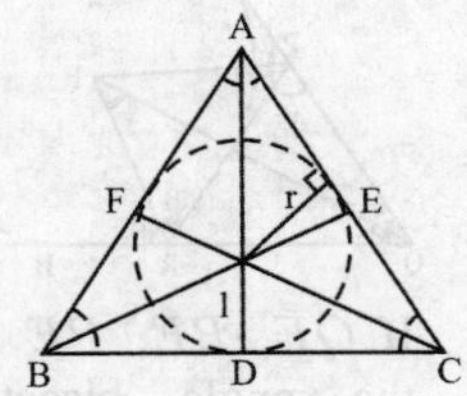

I = Incentre, r = Inradius

$r = \frac{\Delta}{s}$ [Δ = Area, s = semi-perimeter]

$\angle BIC = 90° + \frac{A}{2}$

For equilateral triangle:

$r = \frac{a}{2\sqrt{3}}$

For right angle triangle:

$r = \frac{B + P - H}{2}$

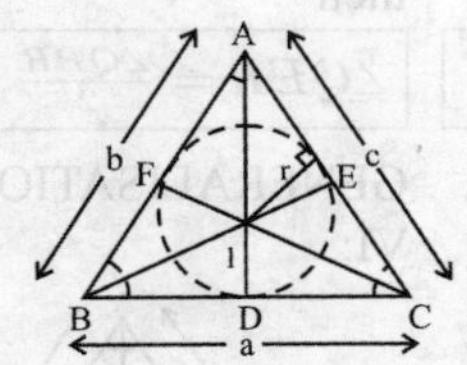

$$\frac{AI}{ID} = \frac{b+c}{a}$$
$$\frac{BI}{IE} = \frac{a+c}{b}$$
$$\frac{CI}{IF} = \frac{a+b}{c}$$

ANGLE BISECTOR THEOREM

INTERNAL

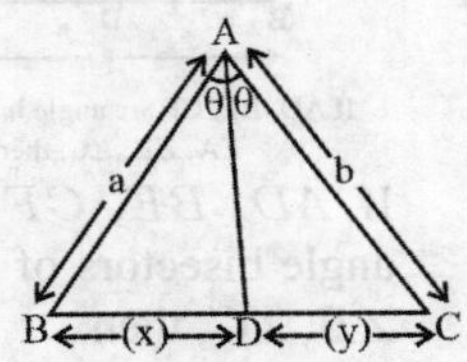

EXTERNAL:

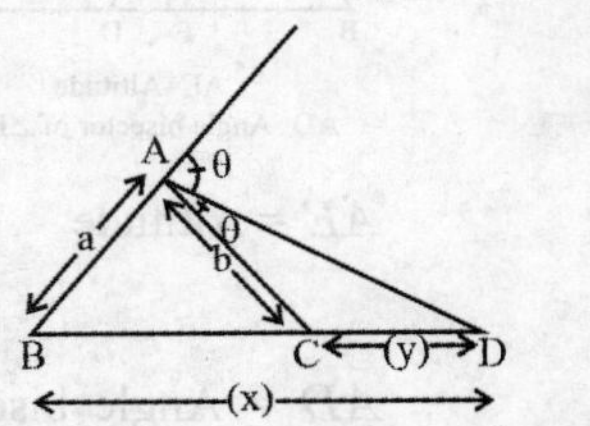

$$\frac{AB}{BD} = \frac{AC}{CD} \text{ (or) } \frac{a}{x} = \frac{b}{y} \quad \Big| \quad \frac{AB}{AC} = \frac{BD}{CD} \text{ (or) } \frac{a}{b} = \frac{x}{y}$$

SOME IMPORTANT GENERALISATION

GENERALISATION I:

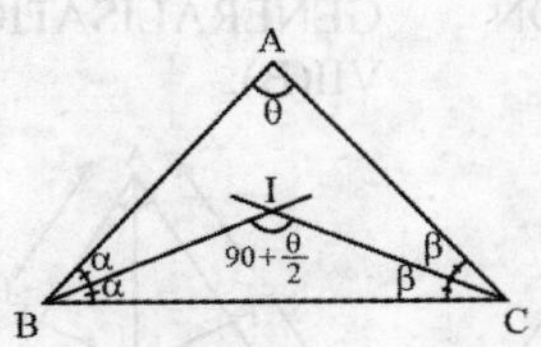

If BI & CI are angle bisectors of $\triangle ABC$ & $\triangle ACB$, I = Incentre:

$$\angle BIC = 90° + \frac{\angle BAC}{2}$$

GENERALISATION II:

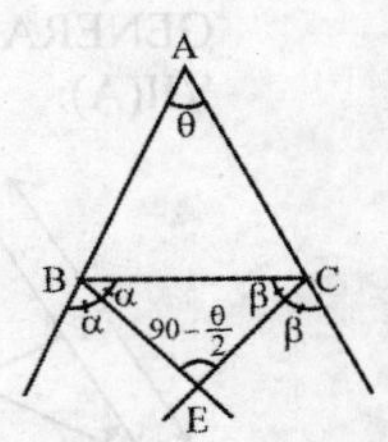

If BI & CI are external angle bisectors of $\triangle ABC$ & $\triangle ACB$, E = Excentre:

$$\angle BEC = 90° - \frac{\angle BAC}{2}$$

GENERALISATION III:

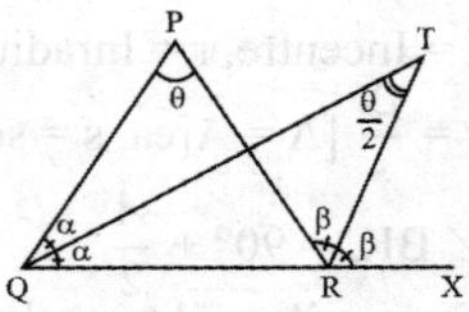

If QT is the angle bisector of $\angle PQR$ & RT is the angle bisector of $\angle PRX$, then

$$\boxed{\angle QTR = \frac{\angle QPR}{2} = \frac{\theta}{2}}$$

GENERALISATION IV:

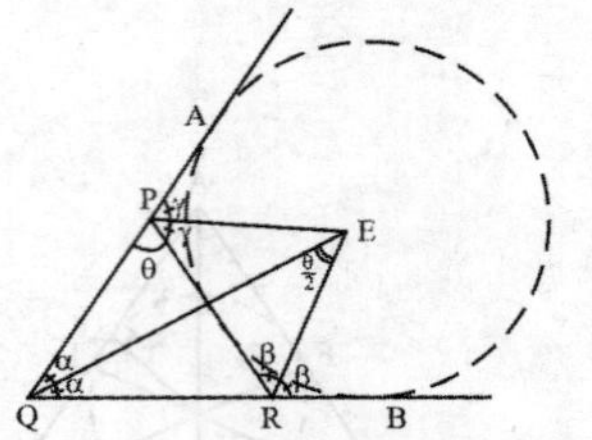

If QE, RE, PE are the angle bisectors of $\angle PQR$, $\angle PRB$, $\angle APR$, E = Excentre, then

$$\boxed{\angle QER = \frac{\angle QPR}{2} = \frac{\theta}{2}}$$

GENERALISATION V:

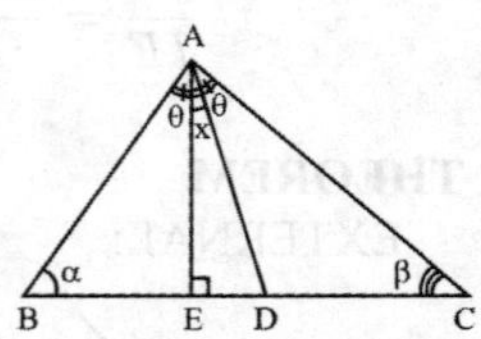

AE=Altitude
AD=Angle bisector of ∠BAC

AE = Altitude

AD = Angle bisector of $\angle BAC$

$$\boxed{\angle EAD = \frac{\angle ABC - \angle ACB}{2}}$$

(or) $\boxed{x = \frac{|\alpha-\beta|}{2}}$

GENERALISATION VI:

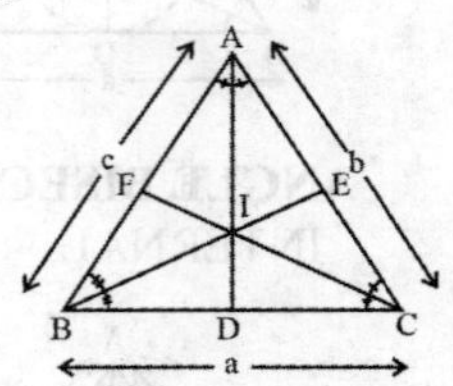

If AD, BE, CF are angle bisector of ∠A, ∠B, ∠C, then

If AD, BE, CF are angle bisectors of $\angle A$, $\angle B$, $\angle C$, then

$\boxed{\frac{AI}{ID} = \frac{b+c}{a}}$, &

$\boxed{\frac{BI}{IE} = \frac{a+c}{b}}$ &

$\boxed{\frac{CI}{IF} = \frac{a+b}{c}}$

GENERALISATION VII(A):

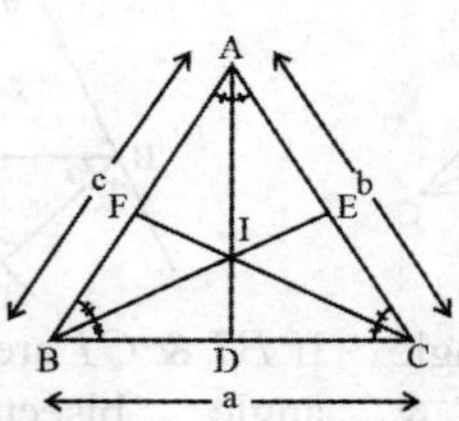

If AD, BE, CF are angle bisector of ∠A, ∠B, ∠C, then

If AD, BE, CF are angle bisectors of $\angle A$, $\angle B$, $\angle C$, then

$$\boxed{\frac{ID}{AD} + \frac{IE}{BE} + \frac{IF}{CF} = 1}$$

GENERALISATION VII(B):

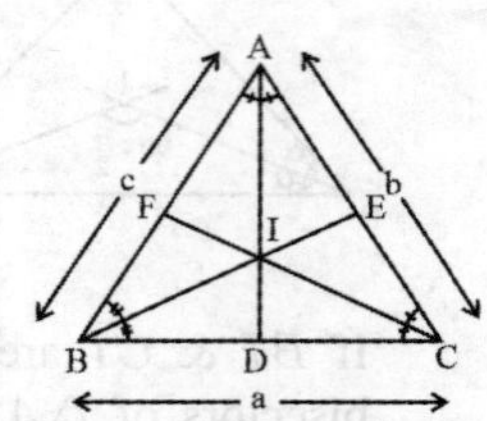

If AD, BE, CF are angle bisector of ∠A, ∠B, ∠C, then

If AD, BE, CF are angle bisectors of $\angle A$, $\angle B$, $\angle C$, then

$$\boxed{\frac{AI}{AD} + \frac{BI}{BE} + \frac{CI}{CF} = 2}$$

GENERALISATION VIII(A):

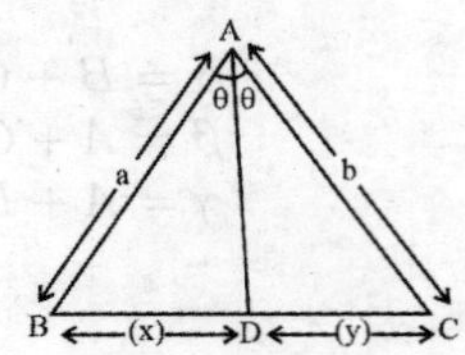

AD=Angle bisector of ∠A

AD = Angle bisector of $\angle A$

$$\boxed{AD^2 = ab - xy}$$

GENERALISATION VIII(B):

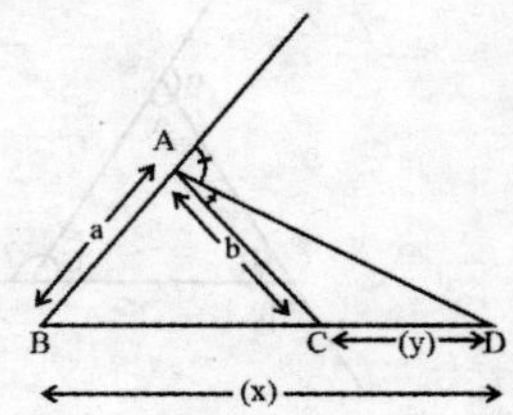

AD=Angle bisector of ∠A

AD = Angle bisector of $\angle A$

$$\boxed{AD^2 = xy - ab}$$

GENERALISATION IX:

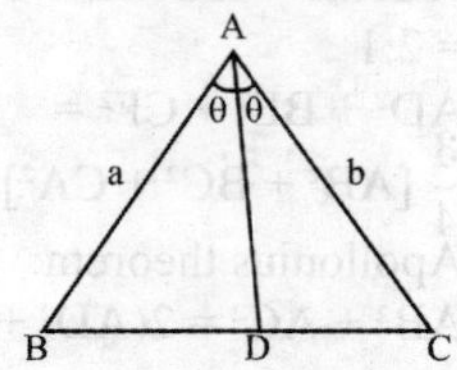

AD = Angle bisector of $\angle A$

$\theta = \frac{\angle A}{2}$

$$\boxed{AD = \frac{2ab\cos(\theta)}{a+b}}$$

GENERALISATION X:(GENERALISATION V - CASE II)

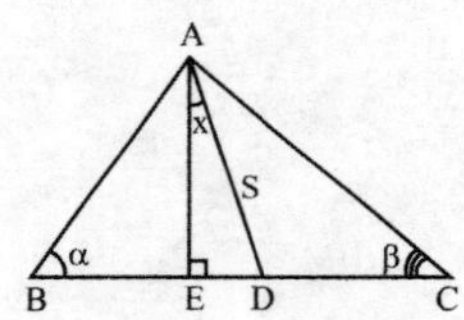

AE = Altitude

AD = Passing through S(circumcen

$\boxed{\angle EAD = |\angle ABC - \angle ACB|}$

(or)

$\boxed{x = \angle\alpha - \angle\beta}$

2.6 Excentre

1 External, 2 External Angle bisectors:

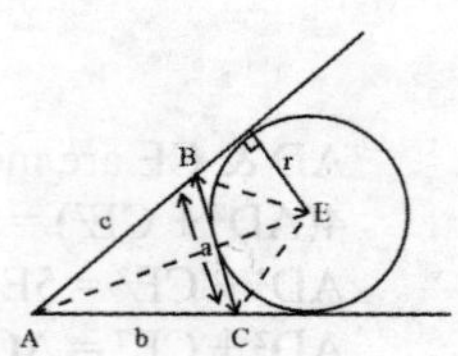

AE is the Internal angle bisector of ∠A & BE, CE are the externa angle bisectors of $\angle B$, $\angle C$

E=Excentre

r=Exradius

$r = \frac{\triangle}{s-a}$

($\triangle$ = Area of $\triangle$ABC, s=Semi-perimeter)

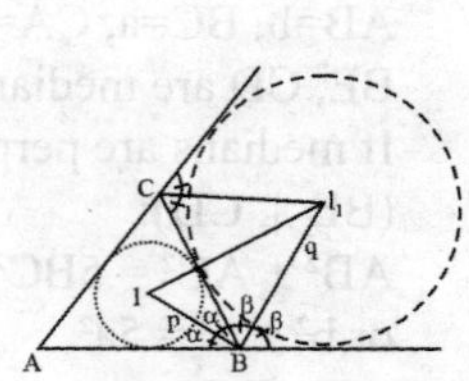

IB is the internal angle bisector & I, B is the external angle bisector of ∠B & IB = p, I_1B=q

$\alpha + \beta = \frac{\pi}{2}$

$p^2 + q^2 = (II_1)^2$

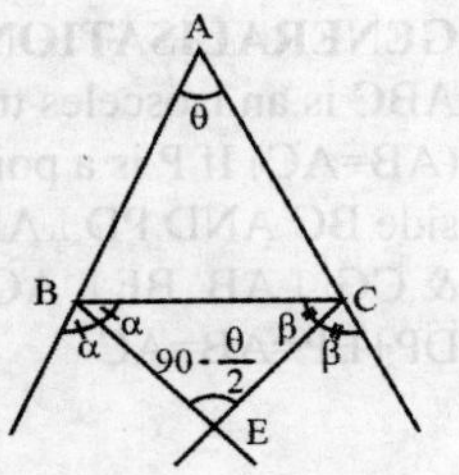

Angle at Excentre (E):

$$\angle BEC = 90° - \frac{\angle BAC}{2}$$

(or)

$$\angle BEC = 90° - \frac{\theta}{2}$$

2.7 Circumcentre (Perpendicular bisectors):

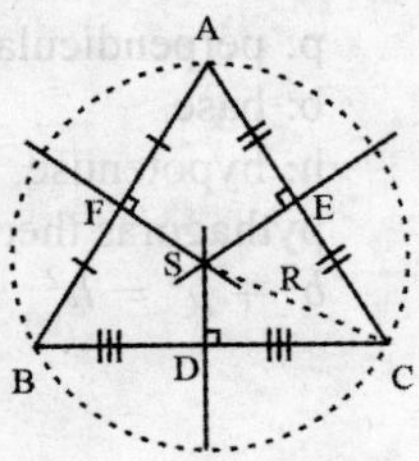

S = Circumcentre
SD, SE, SF are Perpendicular bisectors of sides BC, CA, AB.
R = Circumradius

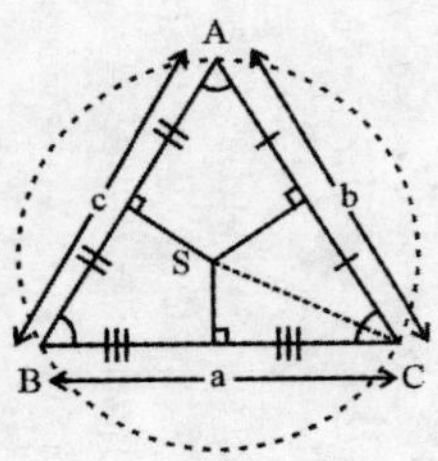

$R = \dfrac{abc}{4\Delta}$

[Δ = Area of $\triangle ABC$]

Sine law:

$$\frac{a}{\sin A} = \frac{b}{\sin B} = \frac{c}{\sin C} = 2R$$

Equilateral triangle: $R = \dfrac{a}{\sqrt{3}}$

Right angle triangle:

$R = \dfrac{\text{Hypotenuse}}{2}$

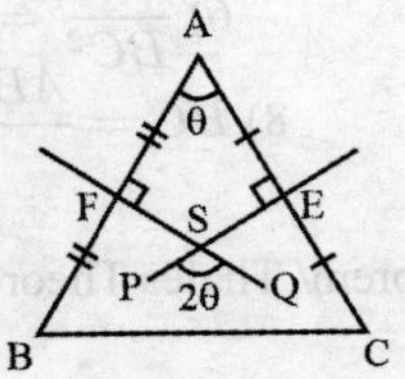

Angle at Circumcentre (S)
PE and QF are the $\perp$ bisectors of AC and AB.
$\angle PSQ = 2(\angle BAC)$
If $\angle BAC = \theta$,
then $\angle PSQ = 2\theta$.
Note: Only applicable for an equilateral triangle.

2.8 Orthocentre (Altitudes):

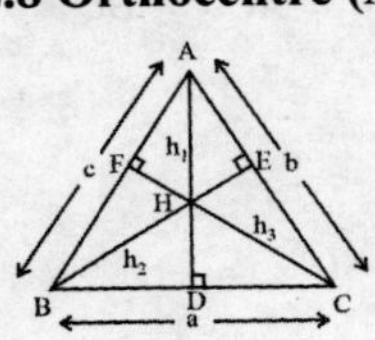

AD, BE, CF are the altitudes of ΔABC H = orthocentre.
$BH \times HE = AH \times HD = CH \times HF$
$AD = h_1,\ BE = h_2,\ CF = h_3$

$h_1 : h_2 : h_3 = \frac{1}{a} : \frac{1}{b} : \frac{1}{c}$

Height Inequality:

$$\left(\frac{1}{h_1} - \frac{1}{h_2}\right) < \frac{1}{h_3} < \left(\frac{1}{h_1} + \frac{1}{h_2}\right)$$

$$\left(\frac{h_1 h_2}{h_2 - h_1}\right) < h_3 < \left(\frac{h_1 h_2}{h_2 + h_1}\right)$$

Formula for Area of triangle when altitudes($h_1,\ h_2.h_3$) are given:

$$\frac{1}{\Delta} = \sqrt{\left(\frac{1}{h_1} + \frac{1}{h_2} + \frac{1}{h_3}\right)\left(\frac{1}{h_1} + \frac{1}{h_2} - \frac{1}{h_3}\right)\left(\frac{1}{h_1} - \frac{1}{h_2} + \frac{1}{h_3}\right)\left(-\frac{1}{h_1} + \frac{1}{h_2} + \frac{1}{h_3}\right)}$$

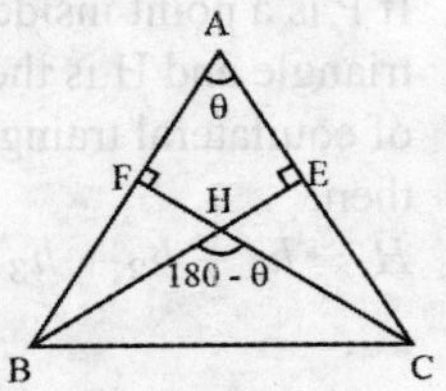

Angle at Orthocentre(H):
BE, CF are the altitudes of Δ ABC
$\angle BHC = 180 - \angle BAC$
If $\angle BAC = \theta,\ \angle BHC = 180 - \theta$

EULER'S LINE

Relation between H, G, S:

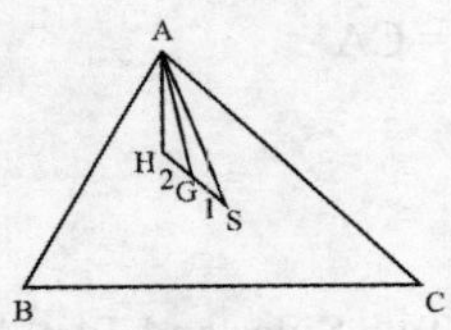

H: Orthocentre (Altitudes)
G : Centroid (Medians)
S : Circumcentre (Perpendicular bisectors)
HGS: Euler's line
HG: GS = 2 : 1

2.9 Stewart's theorem:(and it Extension):

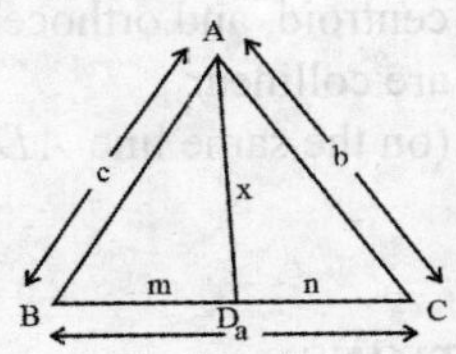

AD is any Cevian - Altitude, Medium, perpendicular Bisector, Angular Bisector
AD is dividing BC in the ratio m:n then,
$c^2 n + b^2 m = (m + n)(x^2 + mn)$

Extention 1:
Appolonius theorem:
AD is a Median [m=n] then,
$c^2 + b^2 = 2(x^2 + m^2)$

Extention 2:
Isosceles triangle:[c=b] then,
$c^2 + (x^2 + mn)$

2.10 Equilateral triangle:

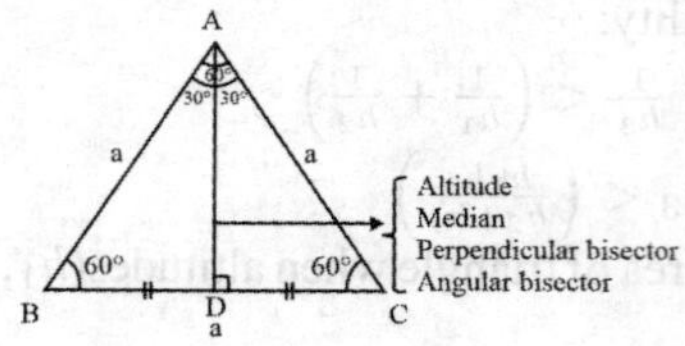

$AD = h = \frac{\sqrt{3}}{2}$

$\text{Area} = \frac{\sqrt{3}}{4}a^2$

SOME GENERALISATIONS:

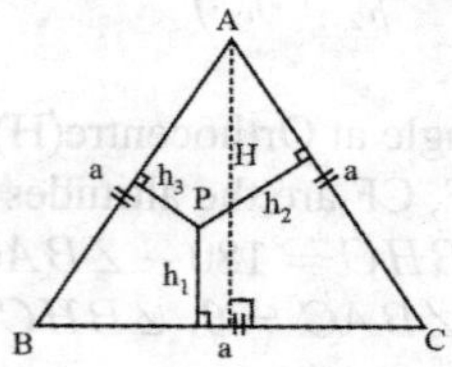

GENERALISATION I:
If P is a point inside the triangle and H is the height of equilateral traingle then,
$H = h_1 + h_2 + h_3$

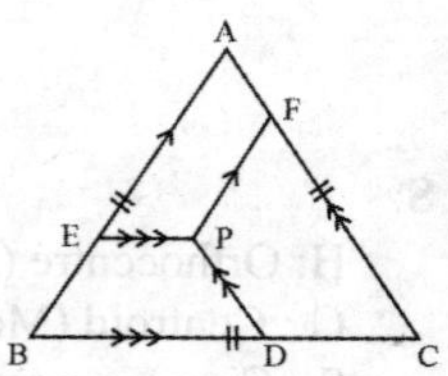

GENERALISATION II:
If P is a point inside the triangle and PF//AB, PD//AC, PE//CB then,
PD + PE + PF = AB = BC = CA

2.11 Isosceles Triangle

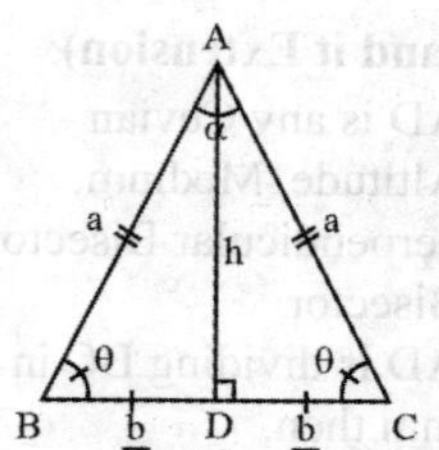

Two Sides and Two Angles are Equal

$h = \sqrt{a^2 - \frac{b^2}{4}} = \frac{\sqrt{4a^2-b^2}}{2}$

$\text{Area} = \frac{1}{2} \times b \times h = \frac{b\sqrt{4a^2-b^2}}{4}$

(or)

$\text{Area} = \frac{1}{2}a^2 \sin(\alpha)$

Note: In an isosceles triangle, incentre, circumcentre, centroid, and orthocentre are collinear (on the same line AD).

SOME GENERALISATIONS

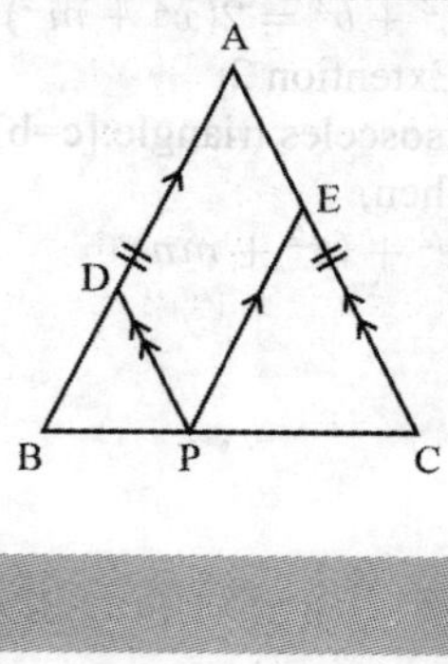

GENERALISATION I:
ABC is an isosceles triangle (AB=AC) If P is a point on the side BC AND DP//AC & EP//AB then, DP+EP=AB=AC

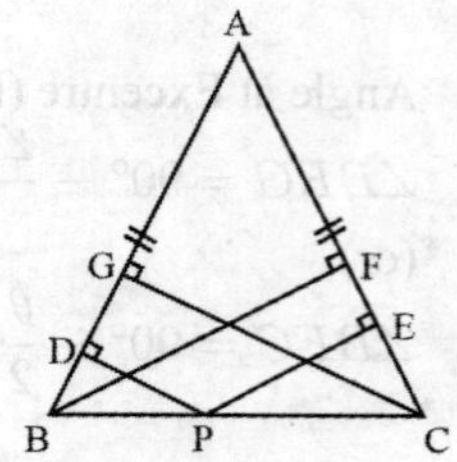

GENERALISATION II:
ABC is an isosceles triangle (AB=AC) If P is a point on the side BC AND PD⊥AB, PE⊥AC & CG ⊥AB, BF⊥ AC then, DP+EP=AB=AC

Right angle triangle

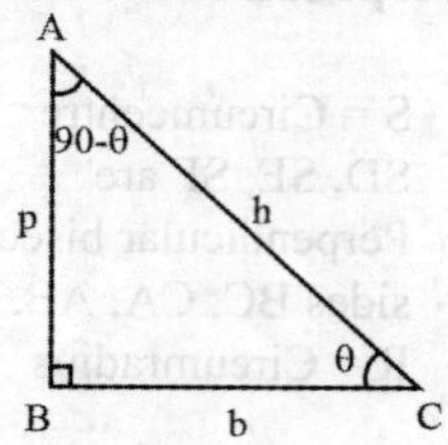

p: perpendicular
b: base
h: hypotenuse
pythagoras theorem:
$b^2 + p^2 = h^2$

VERY IMPORTANT GENERALISATION:
8 Tools: (90 in 90)

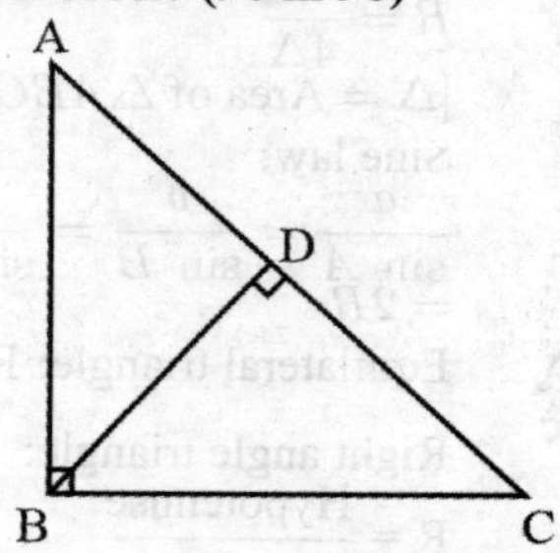

1) $BD = \frac{AB \times BC}{AC}$

2) $BD^2 = AD \times CD$

3) $\frac{1}{BD^2} = \frac{1}{AB^2} + \frac{1}{BC^2}$

4) $AB^2 = AD \times AC$

5) $BC^2 = CD \times AC$

6) $\frac{AB^2}{BC^2} = \frac{AD}{CD}$

7) $AB = \frac{BC \times BD}{CD}$

8) $BC = \frac{AB \times BD}{AD}$

SIMILARITY:
Basic Proportionally Theorem/ Thales Theorem:

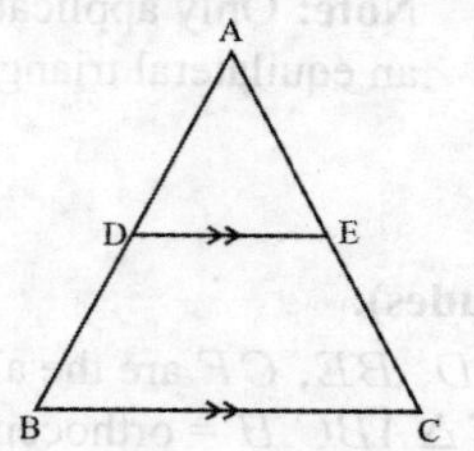

In ΔABC, if $DE \parallel BC$, then:
$\frac{AD}{DB} = \frac{AE}{EC}, \frac{DB}{AB} = \frac{EC}{AC}$
and
$\frac{AD}{AC} = \frac{AE}{AC} = \frac{DE}{BC}$

SOME IMPORTANT RESULTS:

i)

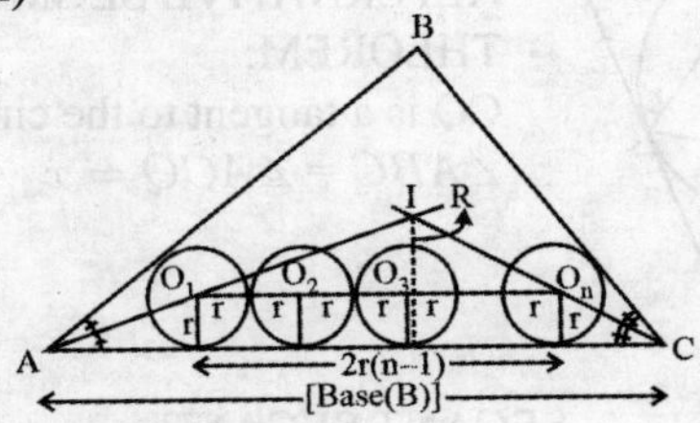

n=number of circles

R=Inradius

$$r = \frac{BR}{2R(n-1)+B}$$

(ii)

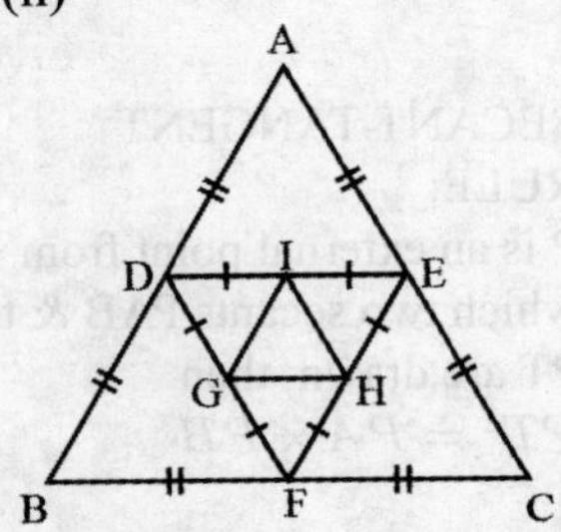

In ΔABC, D, E, F are the mid-points of AB, AC, BC and in Δ DEF, G, H, I are the mid-points of DF, FE, ED.

Sum of areas of all triangles = $\frac{4}{3}$ (Area of ΔABC)

Sum of perimeters of all triangles = 2 (Perimeter of ΔABC)

(iii)

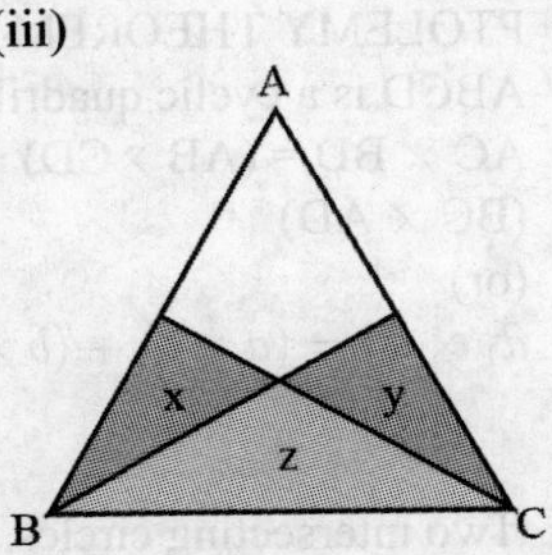

LADDER THEOREM:

$\Delta =$ Area of ΔABC

$x =$ Area of green shaded region

$y =$ Area of yellow shaded region

$z =$ Area of orange shaded region

$$\frac{1}{\Delta} = \frac{1}{x} + \frac{1}{y} = \frac{1}{x+y} + \frac{1}{y+z}$$

(iv) CIVIANS:

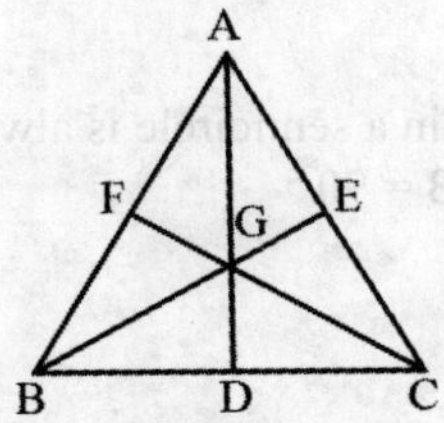

$$\frac{AF}{BF} \times \frac{BD}{DC} \times \frac{CE}{EA} = 1$$

$$\frac{GD}{AD} + \frac{GE}{BE} + \frac{GF}{CF} = 1$$

$$\frac{AG}{AD} + \frac{BG}{BE} + \frac{CG}{CF} = 2$$

$$\frac{AF}{FB} + \frac{AE}{EC} = \frac{AG}{GD}$$

Sine Rule of Cevians:

$$\frac{\sin \angle BAD \times \sin \angle ACF \times \sin \angle CBE}{\sin \angle CAD \times \sin \angle BCF \times \sin \angle ABE} = 1$$

4. Circle

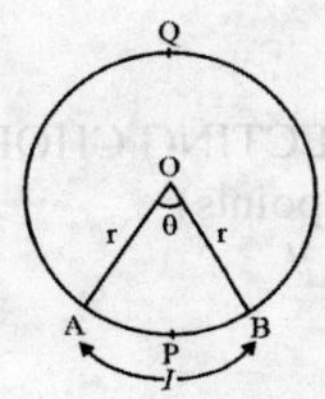

4. CIRCLE

O : Centre, r : radius, l : length of arc, θ : Angle of sector ABO

APB : Minor arc, AQB : Major arc

Circumference of circle = $2\pi r$

Area of circle = πr^2

$l = r \times \theta$

Perimeter of sector = $\frac{\theta}{360} \times 2\pi r + 2r$

Area of sector = $\frac{\theta}{360} \times \pi r^2$

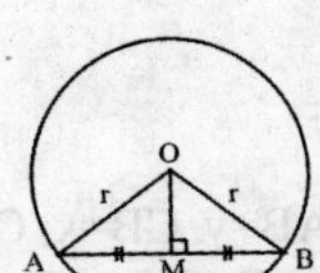

CHORD

AB is a chord of circle with centre O A perpendicular line drawn from centre O to chord AB bisects the chord.

OM ⊥ AB; AM = MB

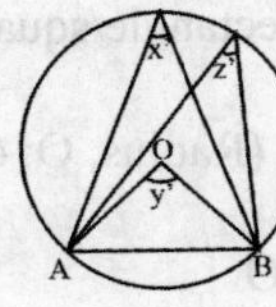

Angle subtended by a chord at the centre is double the angle substended at any point on the same side of circle.

&

Angles subtended by chord AB at any point on the same side of the circle are equal

y=2x=2z & x=z

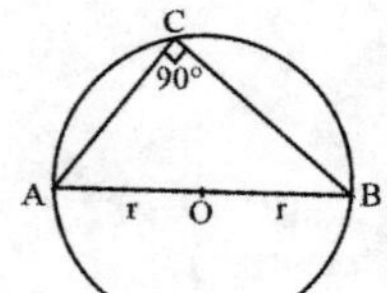

Angle in a semicircle is always 90
$\angle ACB = 90$

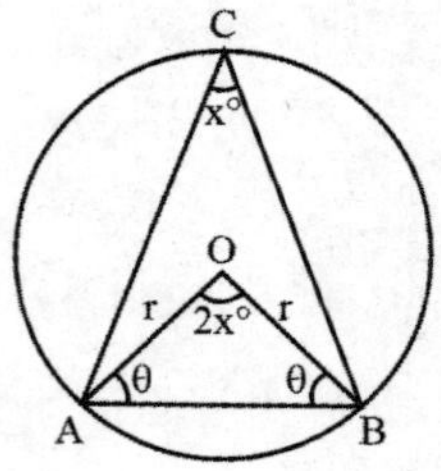

$\angle ACB + \angle OAB = \angle ACB + \angle OBA = 90$
$x + \theta = 90$

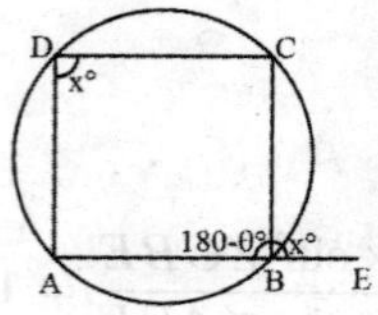

ABCD is a cyclic quadrilateral. The internal angle at a vertex and the external angle at the opposite vertex are equal
$\angle ADC = \angle CBE = x$

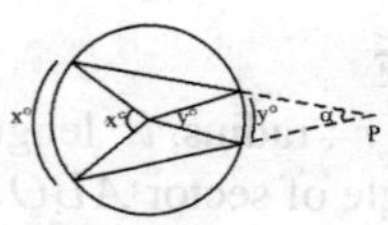

INTERSECTING CHORDS (External points):

$$\alpha = \frac{x - y}{2}$$

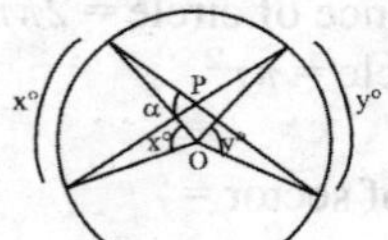

INTERSECTING CHORDS (Internal points):

$$\alpha = \frac{x + y}{2}$$

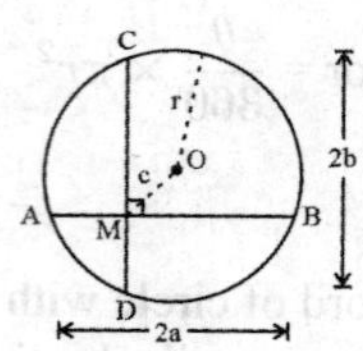

If $CD \perp AB$, AB=2b, CD=2a, OM=c & radius(r), then

$$r = \frac{\sqrt{a^2 + b^2 + c^2}}{2}$$

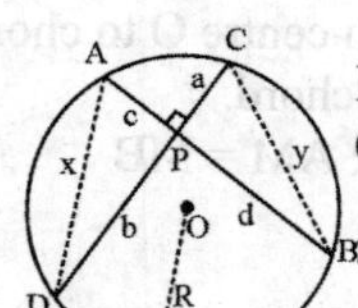

If $CD \perp AB$, AB=y, CB=x, OC=a, OD=b, OA=c, OB=d & radius(r), then

$$r = \frac{\sqrt{a^2 + b^2 + c^2}}{2}$$

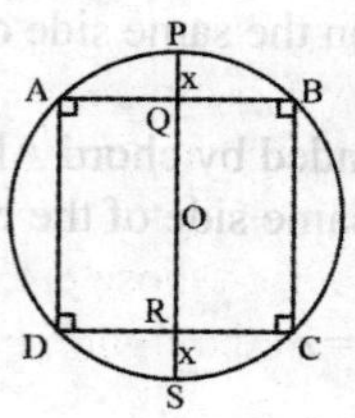

If ABCD is a rectangle/square then
PQ = RS = x
Proof: PO=OS (Radius, O: centre) & OQ = RO
PO-QO=SO-RO
PQ=RS=X

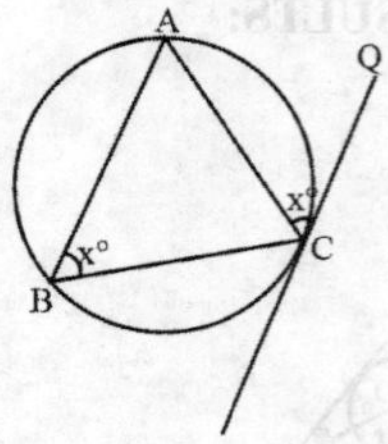

ALTERNATIVE SEGMENT THEOREM:
CQ is a tangent to the circle.
$\angle ABC = \angle ACQ = x$

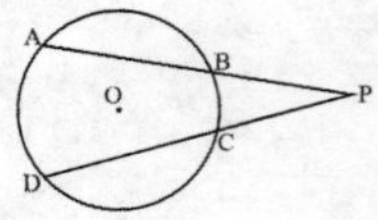

SECANT-SECANT RULE:
P is an external point from which two secants PBA & PCD are drawn, then
$\angle ABC = \angle ACQ = x$

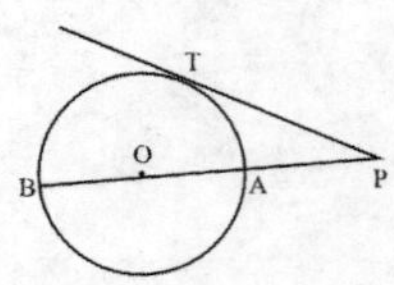

SECANT-TANGENT RULE:
P is an external point from which two secants PAB & tangent PT are drawn, then
$PT^2 = PA \times PB$

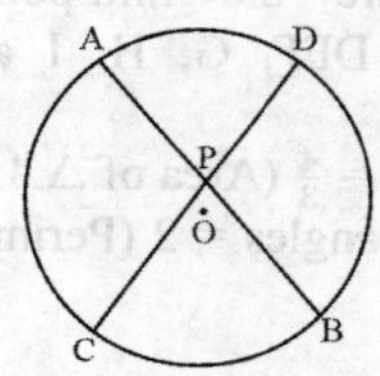

INTERSECTING CHORDS:
Two chords AB and CD intersect at a point P, then
$PA \times PB = PC \times PD$

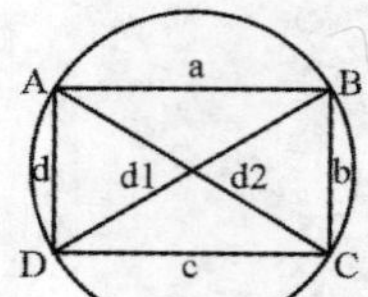

PTOLEMY THEOREM
ABCD is a cyclic quadrilateral
$AC \times BD = (AB \times CD) + (BC \times AD)$
(or)
$d_1 \times d_2 = (a \times c) + (b \times d)$

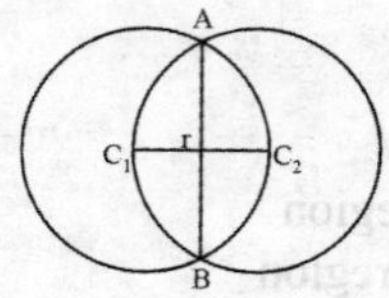

Two intersecting circles, each passing through the centre of the other
→ Radii are same (r)
$AB = r(\sqrt{3})$

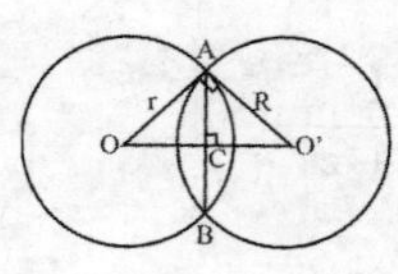

Two intersecting circles
$AC \perp OO'$ & AC=CB
→ Radii are different (r & R)

$$AB = \frac{2Rr}{OO'}$$

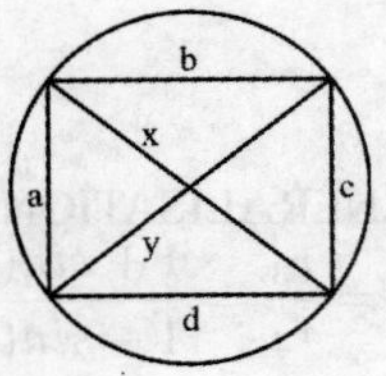

A cyclic quadrilateral with sides a, b, c, d and diagonal x & y

$$\frac{x}{y} = \frac{ad+bc}{ab+dc}$$

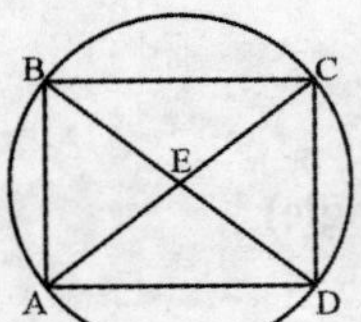

ABCD is a cyclic quadrilateral, Diagonals AC & BD intersect at E

$\dfrac{AE}{EC} = \dfrac{AB \times AD}{BC \times CD}$ If AE=EC, AB × AD = BC × CD

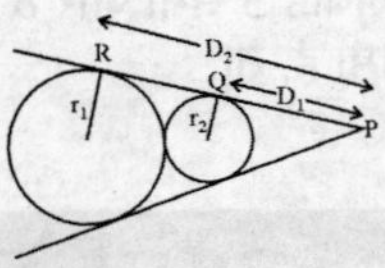

$$D_1 = \frac{2r_2\sqrt{r_1 r_2}}{r_1 - r_2}$$

$$D_1 = \frac{2r_1\sqrt{r_1 r_2}}{r_1 - r_2}$$

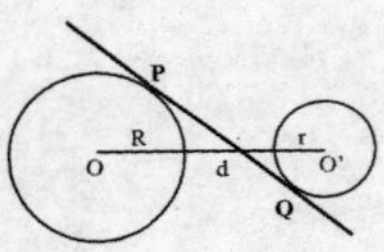

PQ-TRANSVERSE COMMON TANGENT

$$TCT = \sqrt{d^2 - (R+r)^2}$$

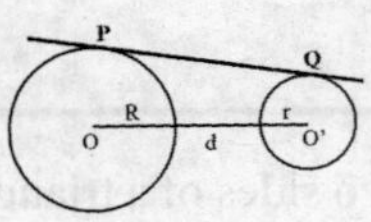

PQ-DIRECT COMMON TANGENT

$$DCT = \sqrt{d^2 - (R+r)^2}$$

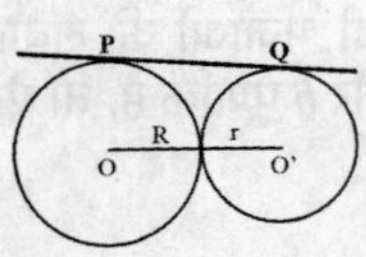

Two circles of different radii touching each other externally PQ is a Direct common tangent PQ = $2\sqrt{Rr}$

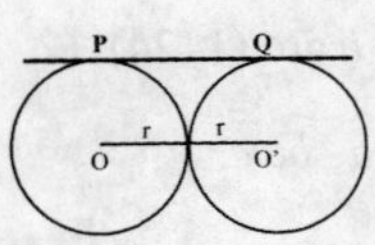

Two circles of different radii touching each other externally PQ is a Direct common tangent PQ = 2r

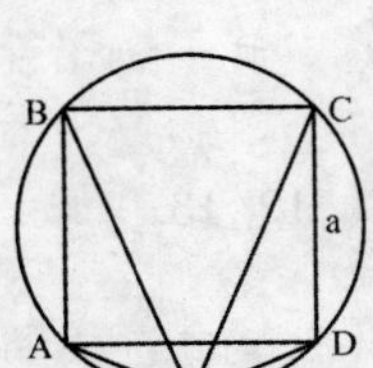

ABCD is a square (cyclic) of side a. E is a point on the circle, then

$$AE^2 + BE^2 + CE^2 + DE^2 = 4a^2$$

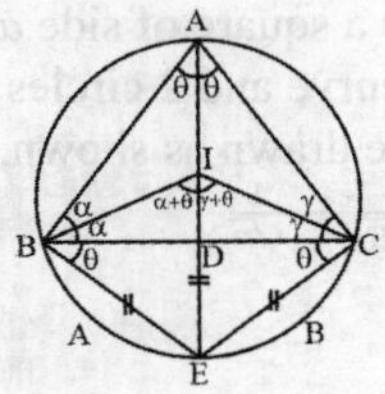

$$AB \times AC + AE \times DE = AE^2$$
and $BE = CE = IE$

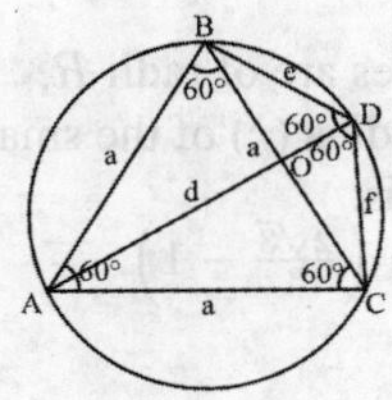

$$d = e + f$$
$$d^2 + e^2 + f^2 = 2a^2$$
$$OD = \frac{ef}{e+f}$$

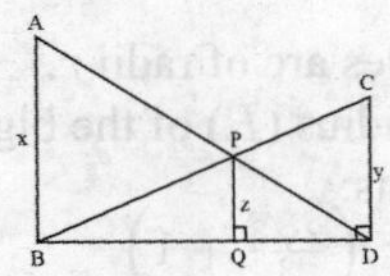

$$\frac{1}{z} = \frac{1}{x} + \frac{1}{y}$$

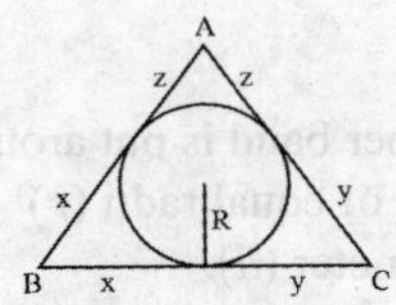

ABC is a triangle with inradius R

$$R\sqrt{\frac{xyz}{x+y+z}}$$
$$Area = \sqrt{xyz(x+y+z)}$$

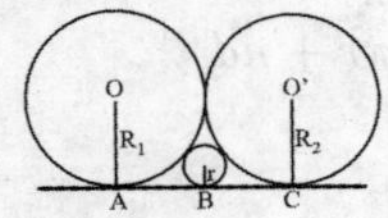

Three Tangent Circle:

$$\frac{1}{\sqrt{r}} = \frac{1}{\sqrt{R_1}} + \frac{1}{\sqrt{R_2}}$$

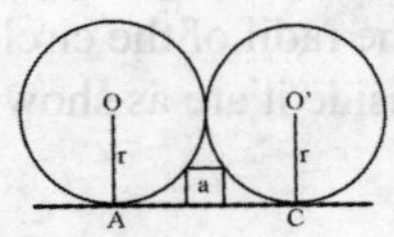

Two circles are of equal radii and touching each other externally. Side of square (a):

$$a = \frac{2}{5}r$$

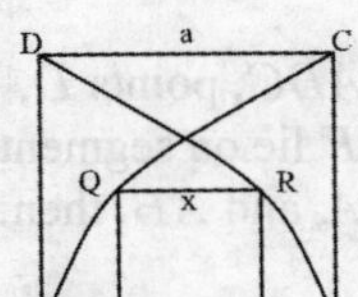

ABCD & PQRS are squares of side a & X

Curves AC, BD cut PQRS at Q & R:

$$x = \frac{3a}{5}$$

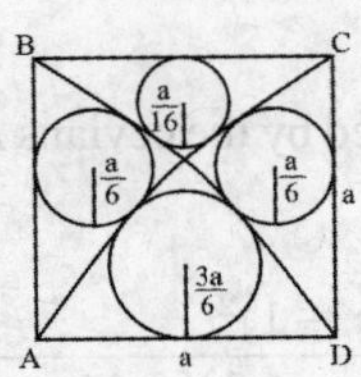

ABCD is a square of side a. AC and BD are curves, and 4 circles are drawn as shown. The radii of those circles are shown in the figure.

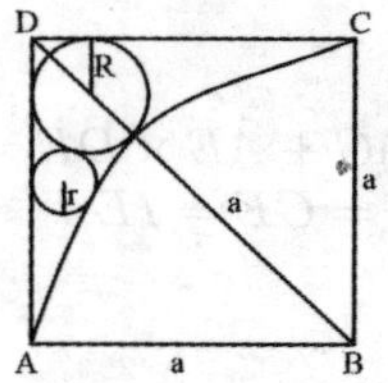

ABCD is a square of side a.
AC is a curve and 2 circles of radii r & R are drawn as shown, then:
$\frac{1}{\sqrt{r}} = \frac{1}{\sqrt{R}} + \frac{1}{\sqrt{a}}$
$r = \frac{R}{2}$

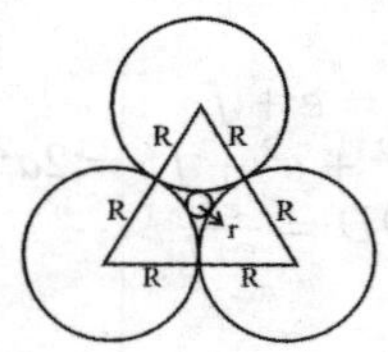

3 circles are of radii R.
The radius (r) of the smallest circle is:
$r = R\left(\frac{2\sqrt{3}}{3} - 1\right)$

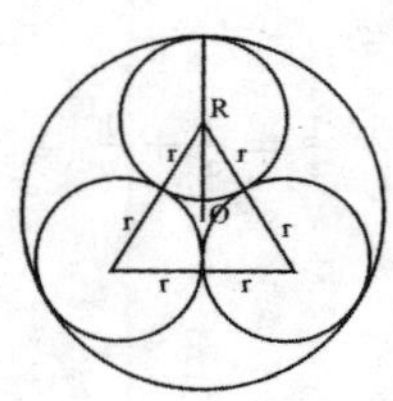

3 circles are of radii r.
The radius (R) of the biggest circle is:
$R = r\left(\frac{2\sqrt{3}}{3} + 1\right)$

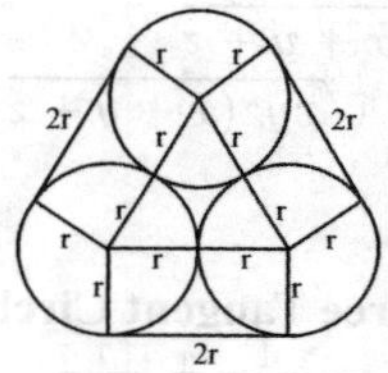

A rubber band is put around n circles of equal radii (r) or diameter (d).
Total length (L) of the rubber band is:
$L = 2\pi r + nd$

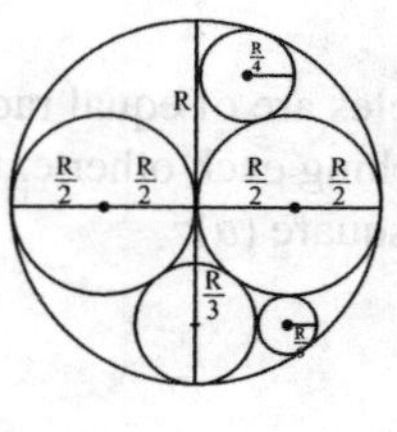

The radii of the largest circle is R, and the radii of the circles drawn inside it are as shown in the figure.

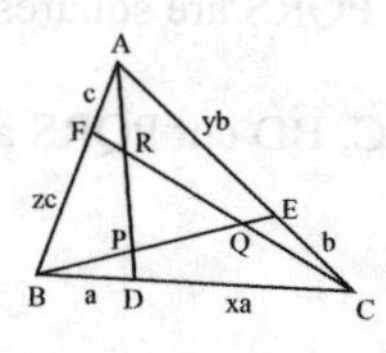

If in $\triangle ABC$, points D, E, and F lie on segments BC, CA, and AB, then:
writing
$\frac{CD}{BD} = x,\ \frac{AE}{CE} = y,\ \frac{BF}{AF} = z$
If Δ = Area of triangle ABC:

The area of the triangle formed by the cevians AD, BE & CF is:

$$= \Delta \times \left\{\frac{(xyz-1)^2}{(xy+y+1)(yz+z+1)(zx+x+1)}\right\}$$

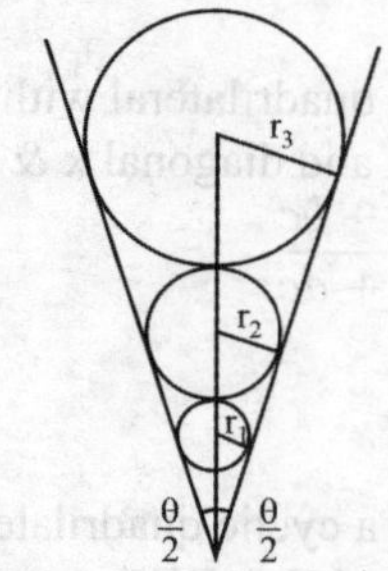

GENERALISATION:
$$\frac{r_2}{r_1} = \frac{r_3}{r_4} = \frac{1+sin(\frac{\theta}{2})}{1-sin(\frac{\theta}{2})}$$

Type 1 (Problem based on Traingle)

1. Two sides of a triangle are of length 3 cm and 8 cm. If the length of the third side is 'x' cm, then:
एक त्रिभुज की दो भुजाओं की लंबाई 3 सेमी और 8 सेमी है। यदि तीसरी भुजा की लंबाई 'x' सेमी है, तो:

Solution:

$a,\ b,\ c$ are sides of $\triangle$:
then,

$a - b < c < a + b$

$8 - 3 < c < 8 + 3$

$5 < c < 11$

2. If a and b are the length of two sides of a triangle such that the product $ab = 24$, where a and b are integers, then how many such triangles are possible?
यदि a और b एक त्रिभुज की दो भुजाओं की लंबाई हैं और गुणनफल $ab = 24$ है, जहाँ a और b पूर्णांक हैं, तो ऐसे कितने त्रिभुज संभव हैं?

Solution:

Possible values of a and b are (1, 24); (2, 12); (3, 8); (4, 6).
Sides of the triangle are $a,\ b,\ c$.
If $a = 1,\ b = 24$:
$a + b > c,\ a - b < c$
$25 > c,\ 23 < c$
Possible value of $c = 24$.
If $a = 2,\ b = 12$:
$a + b > c,\ a - b < c$
$14 > c,\ 10 < c$
Possible values of $c = 11,\ 12,\ 13$.
If $a = 3,\ b = 8$:
$a + b > c,\ a - b < c$
$11 > c,\ 5 < c$

Possible values of $c = 6, 7, 8, 9, 10$.

If $a = 4$, $b = 6$:

$a + b > c$, $a - b < c$

$10 > c$, $2 < c$

Possible values of $c = 3, 4, 5, 6, 7, 8, 9$.

So, a total of 16 triangles can be made using the above sides.

3. How many isosceles triangles with integer sides are possible such that the sum of two of the sides is 16 cm?

पूर्णांक भुजाओं वाले ऐसे कितने समद्विबाहु त्रिभुज संभव हैं जिनकी दो भुजाओं का योग 16 सेमी हो?

Solution:

There are two possibilities to solve this question:

First possibility: If the sum of equal sides is 16, then any value of the third side is possible from 1 to 15. So a total of 15 triangles are possible.

Second possibility: If the sum of two unequal sides is 16:

So, total 9 triangles are possible.

Possible values are: (1, 15, 15), (2, 14, 14), (3, 13, 13), (4, 12, 12), (5, 11, 11), (6, 10, 10), (6, 10, 6), (7, 9, 9), and (7, 9, 7).

Total number of possible triangles: $15 + 9 = 24$.

4. The perimeter of a triangle with sides of integer values is equal to 13. How many such triangles are possible?

पूर्णांक मानों वाली भुजाओं वाले त्रिभुज का परिमाप 13 के बराबर है। ऐसे कितने त्रिभुज संभव हैं?

Solution:

Let sides of the triangle be a, b, c such that

$a + b > c > |a - b|$.

Perimeter of the triangle = 13.

Possible triangles are: (1, 6, 6), (2, 5, 6), (3, 5, 5), (3, 4, 6), and (4, 4, 5).

5. In a triangle ABC, $AB = 6$ cm, $BC = 11$ cm. How many integral values of AC can be possible?

एक त्रिभुज ABC में, $AB = 6$ सेमी, $BC = 11$ सेमी। AC के कितने अभिन्न मान संभव हो सकते हैं?

Solution:

Rule of triangle: The sum of two sides of any triangle is greater than the third side, and the difference of two sides is always less than the third side.

So, $AB - BC < AC < AB + BC$:

$5 < AC < 17$.

Possible values of AC: 6, 7, 8, 9, 10, 11, 12, 13, 14, 15, 16.

Total values = 11.

6. In $\triangle ABC$, the length of BC is less than twice the length of AB by 2 cm. The length of AC exceeds the length of AB by 6 cm. The perimeter is 60 cm. The length (in cm) of the largest side of the triangle ABC is:

$\triangle ABC$ में, BC की लंबाई AB की लंबाई के दोगुने से 2 सेमी कम है। AC की लंबाई AB की लंबाई से 6 सेमी अधिक है। परिधि 60 सेमी है। त्रिभुज ABC की सबसे बड़ी भुजा की लंबाई (सेमी में) है:

Solution:

Let the length of AB be 'x'.

So, $BC = 2x - 2$, $AC = x + 6$.

According to the question:

$$x + (2x - 2) + (x + 6) = 60$$

7. In the given figure, the measure of $\angle A$ is:
दिए गए चित्र में, $\angle A$ का माप है:

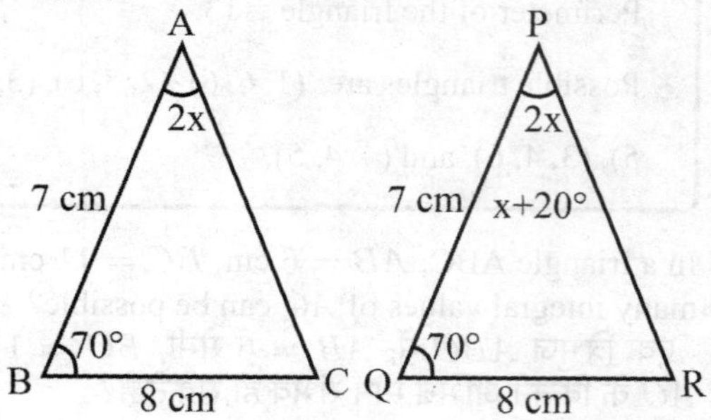

Solution:

From the Side-Angle-Side Congruency Rule,

$\triangle ABC \cong \triangle APQR$:

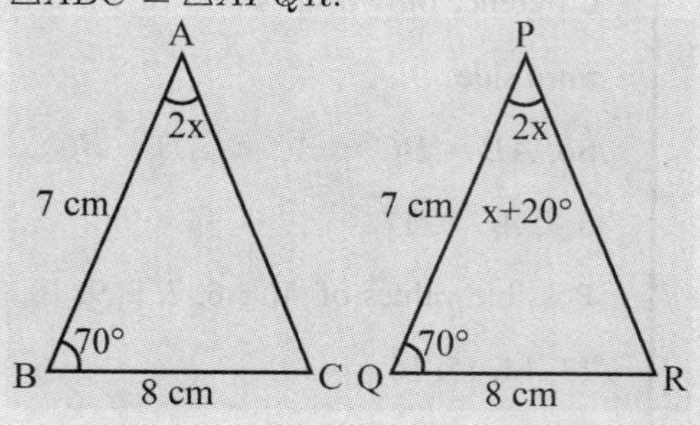

$\angle A = \angle P$

$2x = x + 20°$

$x = 20°$

$\angle A = 2x = 2 \times 20° = 40°$

8. In a triangle ABC, points P and Q are on AB and AC, respectively, such that $AP = 4$ cm, $PB = 6$ cm, $AQ = 5$ cm, and $QC = 7.5$ cm. If $PQ = 6$ cm, then find BC (in cm).
त्रिभुज ABC में, बिंदु P और Q क्रमश: AB और AC पर इस प्रकार हैं कि $AP = 4$ सेमी, $PB = 6$ सेमी, $AQ = 5$ सेमी, और $QC = 7.5$ सेमी है। यदि $PQ = 6$ सेमी है, तो BC (सेमी में) ज्ञात कीजिए।

Solution:

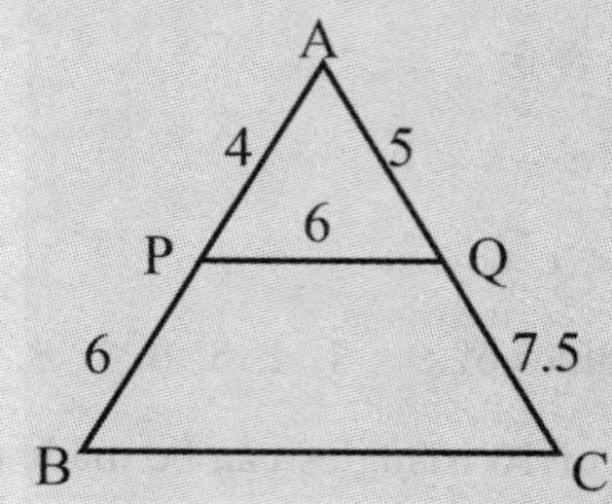

$$\frac{AP}{AB} = \frac{PQ}{BC}$$

$$\frac{4}{10} = \frac{6}{BC}$$

$BC = 15$ cm.

9. Let $\triangle ABC \sim \triangle PQR$ and $\frac{\text{ar}(\triangle ABC)}{\text{ar}(\triangle QPR)} = \frac{64}{169}$. If $AB = 10$ cm, $BC = 7$ cm, and $AC = 16$ cm, then PR (in cm) is equal to:
मान लें $\triangle ABC \sim \triangle PQR$ और $\frac{ar(\triangle ABC)}{ar(\triangle QPR)} = \frac{64}{169}$। यदि $AB = 10$ सेमी, $BC = 7$ सेमी, $AC = 16$ सेमी, तो PR (सेमी में) के बराबर है:

Solution:

$$\frac{\text{ar}(\triangle ABC)}{\text{ar}(\triangle QPR)} = \frac{64}{169}.$$

$$\frac{\text{Side}(\triangle ABC)}{\text{Side}(\triangle PQR)} = \frac{8}{13}.$$

$$\frac{AC}{PR} = \frac{\text{Side}(\triangle ABC)}{\text{Side}(\triangle PQR)}$$

$$\frac{8}{13} = \frac{16}{PR}$$

$PR = 26$ cm.

10. In $\triangle ABC$, D and E are points on BC such that $AD = AE$ and $\angle BAD = \angle CAE$. If $AB = (2p + 3)$, $BD = 2p$, $AC = (3q - 1)$, and $CE = q$, then find the value of $p + q$:
त्रिभुज ABC में, D और E BC पर इस प्रकार बिंदु हैं कि $AD = AE$ और $\angle BAD = \angle CAE$ है। यदि $AB = (2p + 3)$, $BD = 2p$, $AC = (3q - 1)$ और $CE = q$ है तो $p + q$ का मान ज्ञात कीजिए।

Solution:

$\angle ABC = \angle ACB$

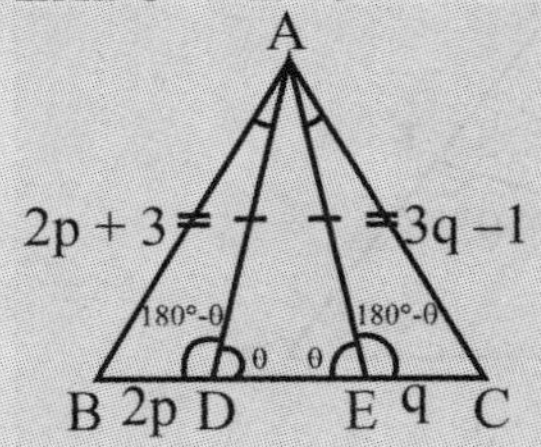

Then, $AB = AC$:

$2p + 3 = 3q - 1$

$\angle BAD = \angle EAC$ [Given]

$BD = CE$:

$2p = q$

$2p + 3 = 3(2p) - 1$

$-4p = -4$

$p = 1$

$q = 2p = 2$

$p + q = 1 + 2 = 3.$

11. In $\triangle ABC$, $DE \parallel AB$, where D and E are points on sides AC and BC, respectively. F is a point between C and D such that $EF \parallel BD$. If $AD = 15$ cm and $DC = 10$ cm, then the length of CF is:
$\triangle ABC$ में, $DE \parallel AB$, जहां D और E क्रमशः AC और BC पर बिंदु हैं। F, C और D के बीच एक बिंदु है जैसे कि $EF \parallel BD$। यदि $AD = 15$ सेमी, $DC = 10$ सेमी, तो CF की लंबाई है:

Solution:

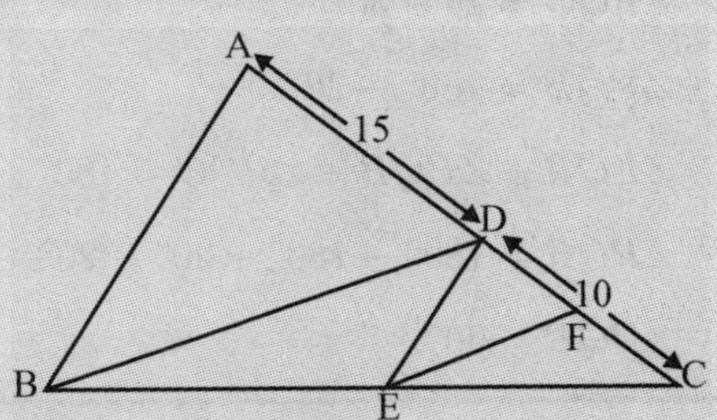

In $\triangle ABC$:

Since $DE \parallel AB$, by the Basic Proportionality Theorem:

$$\frac{CD}{AC} = \frac{CE}{CB}.$$

Again, in $\triangle CDB$:

$EF \parallel BD$:

$$\frac{CF}{CD} = \frac{CE}{CB}$$

$$\frac{CF}{10} = \frac{10}{25}$$

$CF = 4$ cm.

12. In $\triangle ABC$, D and E are the points on sides AC and BC, respectively, such that $DE \parallel AB$. F is a point on CE such that $DF \parallel AE$. If $CE = 6$ cm and $CF = 2.5$ cm, then BC is equal to:
$\triangle ABC$ में, D और E क्रमशः AC और BC भुजाओं पर इस प्रकार बिंदु हैं कि $DE \parallel AB$। F, CE पर एक बिंदु है जैसे कि $DF \parallel AE$। यदि $CE = 6$ सेमी, और $CF = 2.5$ सेमी, तो BC बराबर है:

Solution:

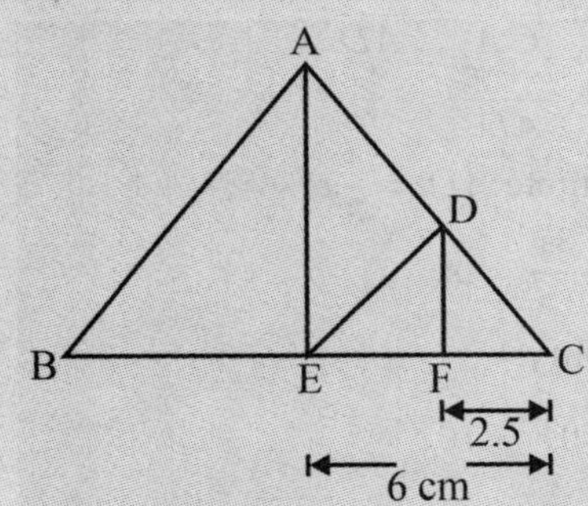

In $\triangle AEC$,

Since $DF \parallel AE$:

$$\frac{CD}{CA} = \frac{CF}{CE}$$

$$\frac{CD}{CA} = \frac{2.5}{6} = \frac{5}{12}$$

In $\triangle ABC$, $DE \parallel AB$:

$$\frac{CD}{CA} = \frac{CE}{CB}$$

$$\frac{5}{12} = \frac{6}{CB}$$

$$BC = \frac{72}{5} = 14.4 \text{ cm.}$$

13. In $\triangle ABC$, $AB = 20$ cm, $BC = 7$ cm, and $CA = 15$ cm. Side BC is produced to D such that $\triangle DAB \sim \triangle DCA$. DC is equal to:

$\triangle ABC$ में, $AB = 20$ सेमी, $BC = 7$ सेमी और $CA = 15$ सेमी। भुजा BC को D तक इस प्रकार बढ़ाया गया है कि $\triangle DAB \sim \triangle DCA$ हो। DC बराबर है:

Solution:

$\triangle DAB \sim \triangle DCA$:

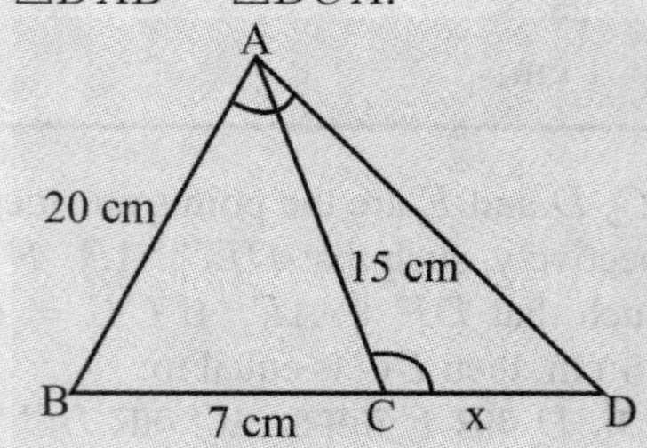

$$\frac{DA}{DC} = \frac{AB}{CA}$$

$$\frac{DA}{x} = \frac{20}{15}$$

$$DA = \frac{20}{15} \times x = \frac{4}{3}x.$$

Given: $\dfrac{AB}{CA} = \dfrac{BD}{AD}$:

$$\frac{20}{15} = \frac{7+x}{AD}.$$

Substitute $AD = \dfrac{4}{3}x$:

$$\frac{21+3x}{4} = x.$$

Simplify:

$$63 + 9x = 16x$$

$$7x = 63$$

$$x = 9 \text{ cm}.$$

Thus, $DC = 9$ cm.

14. In $\triangle PQR$, $\angle Q = 85°$ and $\angle R = 65°$. Points S and T are on the sides PQ and PR, respectively, such that $\angle STR = 95°$, and the ratio of QR and ST is 9 : 5. If $PQ = 21.6$ cm, then the length of PT is:

$\triangle PQR$ में, $\angle Q = 85°$ और $\angle R = 65°$ है। बिंदु S और T क्रमशः PQ और PR भुजाओं पर इस प्रकार हैं कि $\angle STR = 95°$ है, और QR और ST का अनुपात 9 : 5 है। यदि $PQ = 21.6$ सेमी है, तो PT की लंबाई है:

Solution:

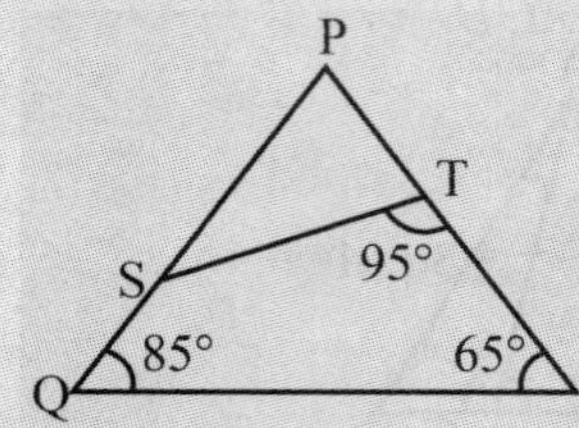

$$\angle PTS = 180° - 95° = 85°.$$

$\triangle PST \sim \triangle PRQ$:

$$\frac{ST}{QR} = \frac{PT}{PQ}.$$

$$\frac{5}{9} = \frac{PT}{21.6}.$$

$$PT = 12 \text{ cm}.$$

15. In $\triangle ABC$, points P, Q, and R are taken on AB, BC, and CA, respectively, such that $BQ = PQ$ and $QC = QR$. If $\angle BAC = 75°$, what is the measure of $\angle PQR$ (in degrees)?

$\triangle ABC$ में, बिंदु P, Q, और R को क्रमशः AB, BC, और CA पर इस प्रकार लिया जाता है, कि $BQ = PQ$ और $QC = QR$ है। यदि $\angle BAC = 75°$ है, तो $\angle PQR$ का माप (डिग्री में) क्या है?

Solution:

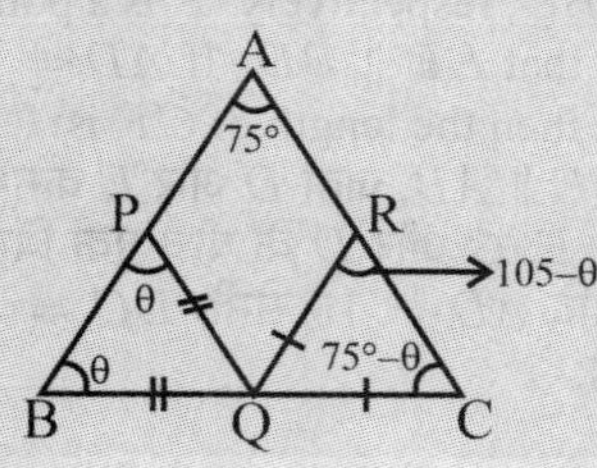

$$\angle RQC = 180° - 105° + \theta - 105° + \theta$$

$$\angle RQC = 2\theta - 30°$$

$$\angle PQB = 180° - 2\theta$$

$$\angle PQR + \angle PQB + \angle RQC = 180°$$

$$\angle PQR = 180° - 180° + 20° - 20° + 30°$$

$$\angle PQR = 30°.$$

16. An isosceles $\triangle MNP$ is inscribed in a circle. If $MN = MP = 16\sqrt{5}$ cm, and $NP = 32$ cm, what is the radius (in cm) of the circle?

एक समद्विबाहु $\triangle MNP$ एक वृत्त में अंकित है। यदि $MN = MP = 16\sqrt{5}$ सेमी, और $NP = 32$ सेमी, तो वृत्त की त्रिज्या (सेमी में) क्या है?

Solution:

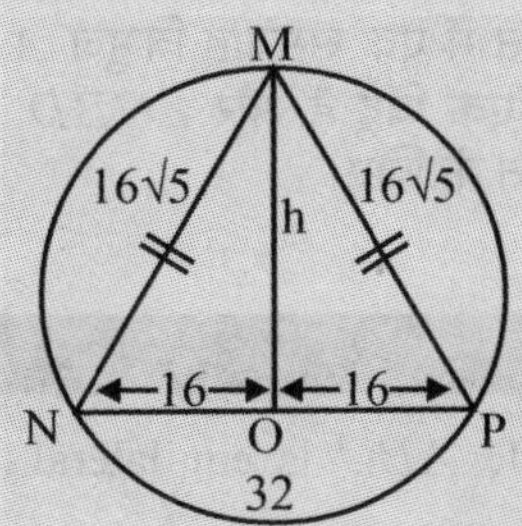

$$R = \frac{abc}{4\Delta}$$

$$\Delta = \sqrt{(16\sqrt{5})^2 - (16)^2} = \sqrt{256(5-1)} =$$

32 cm.

$$R = \frac{16\sqrt{5} \times 16\sqrt{5} \times 32}{4 \times \frac{1}{2} \times 32 \times 32}$$

$R = 20$ cm.

17. A circle is inscribed in a triangle ABC. It touches sides AB, BC, and AC at points R, P, and Q, respectively. If $AQ = 3.5$ cm, $PC = 4.5$ cm, and $BR = 7$ cm, then the perimeter (in cm) of the triangle $\triangle ABC$ is:

त्रिभुज ABC में एक वृत्त अंकित है। यह भुजाओं AB, BC, और AC को क्रमशः बिंदु R, P, और Q पर स्पर्श करता है। यदि $AQ = 3.5$ सेमी, $PC = 4.5$ सेमी और $BR = 7$ सेमी है, तो त्रिभुज $\triangle ABC$ का परिमाप (सेमी में) है:

Solution:

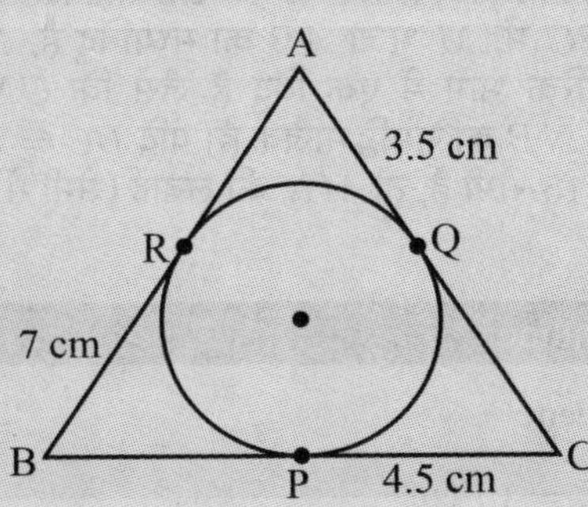

Tangents drawn from external points to a circle are equal:

$AR = AQ = 3.5$ cm

$BP = BR = 7$ cm

$CQ = CP = 4.5$ cm

Perimeter of $\triangle ABC = AB + BC + CA$:

$= (AR + BR) + (BP + CP) + (AQ + CQ)$

$= (3.5 + 7) + (7 + 4.5) + (3.5 + 4.5)$

$= 10.5 + 11.5 + 8 = 30$ cm.

18. The sides PQ and PR of $\triangle PQR$ are produced to points S and T, respectively. The bisectors of $\angle SQR$ and $\angle TRQ$ meet at point U. If $\angle QUR = 69°$, then the measure of $\angle P$ is:

$\triangle PQR$ की भुजाएँ PQ और PR को क्रमशः बिंदु S और T तक बढ़ाया जाता है। $\angle SQR$ और $\angle TRQ$ के समद्विभाजक बिंदु U पर मिलते हैं। यदि $\angle QUR = 69°$ है, तो $\angle P$ का माप है:

Solution:

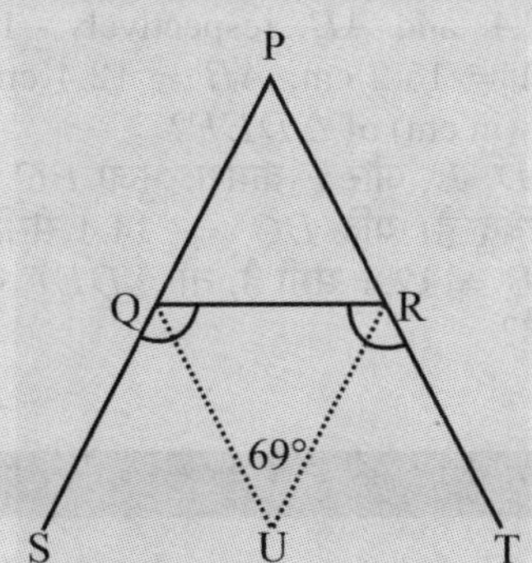

$$\angle QUR = 90° - \frac{1}{2}\angle P$$

$$69° = 90° - \frac{1}{2}\angle P$$

$$\frac{1}{2}\angle P = 21°$$

$$\angle P = 42°$$

19. In $\triangle ABC$, $AB = 7$ cm, $BC = 10$ cm, $AC = 8$ cm. If AD is the angle bisector of $\angle BAC$, where D is a point on BC, then DC (in cm) = ?

$\triangle ABC$ में, $AB = 7$ सेमी, $BC = 10$ सेमी, $AC = 8$ सेमी। यदि AD, $\angle BAC$ का कोण समद्विभाजक है, जहाँ

D, BC का एक बिंदु है, तो DC (सेमी में) = ?

Solution:

DC = x (say)

BD = (10-x)

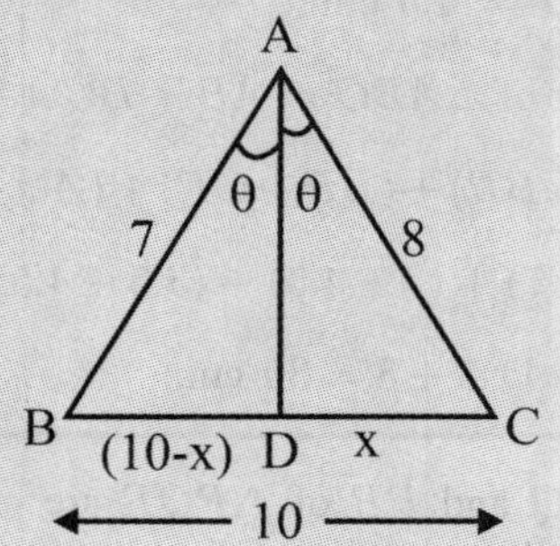

$$\frac{AB}{AC} = \frac{BD}{DC}$$

$$\frac{7}{8} = \frac{10-x}{x} \quad \text{[from angle bisector theorem]}$$

$$15x = 80$$

$$x = \frac{16}{3} \text{ cm}$$

20. In $\triangle ABC$, D, E, and F are the mid-points of sides BC, CA, and AB, respectively. If $BC = 14.4$ cm, $CA = 15.2$ cm, $AB = 12.4$ cm, what is the perimeter (in cm) of $\triangle DEF$?
$\triangle ABC$ में, D, E, और F क्रमश: भुजा BC, CA, और AB के मध्य-बिंदु हैं। यदि $BC = 14.4$ सेमी, $CA = 15.2$ सेमी, $AB = 12.4$ सेमी है, तो $\triangle DEF$ का परिमाप (सेमी में) क्या है?

Solution:

Since D, E, and F are the midpoints of sides BC, CA, and AB respectively:

AB/DE = BC/EF = AC/DF = 2

$\triangle ABC \sim \triangle DEF$.

Now:

Perimeter of $\triangle DEF = DE + EF + DF$

$$= \frac{AB}{2} + \frac{BC}{2} + \frac{AC}{2}$$

$$= \frac{12.4}{2} + \frac{14.4}{2} + \frac{15.2}{2}$$

$$= 6.2 + 7.2 + 7.6$$

$$= 21 \text{ cm}.$$

21. An equilateral triangle ABC is inscribed in a circle with center O. D is a point on the minor arc BC and $\angle CBD = 40°$. Find the measure of $\angle BCD$.
केंद्र O वाले एक वृत्त में एक समबाहु त्रिभुज ABC अंकित है। D लघु BC पर एक बिंदु है और $\angle CBD = 40°$ है। $\angle BCD$ का माप ज्ञात कीजिए।

Solution:

Given, Triangle ABC is equilateral, so all angles will be 60°.

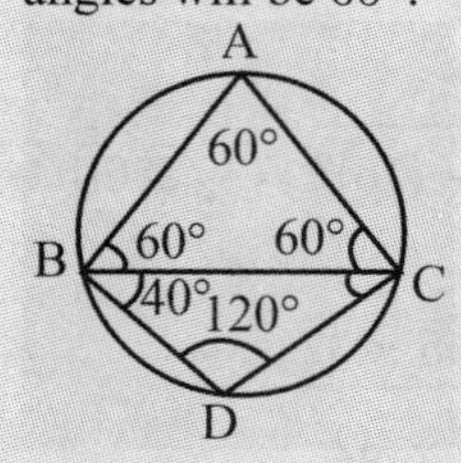

$\because ABCD$ is a cyclic quadrilateral: $\angle BAC + \angle BDC = 180°$

$\angle BDC = 120°$

In $\triangle BDC$:

$\angle BCD + \angle CBD + \angle BDC = 180°$

$\angle BCD = 180° - 40° - 120°$

$\angle BCD = 20°$

22. In $\triangle ABC$, M is the midpoint of the side AB, N is a point in the interior of $\triangle ABC$ such that CN is the bisector of $\angle C$ and $CN \perp NB$. What is the length (in cm) of MN, if $BC = 10$ cm and $AC = 15$ cm?
$\triangle ABC$ में, M भुजा AB का मध्यबिंदु है, N, $\triangle ABC$ के आंतरिक भाग में एक बिंदु है जैसे कि CN, $\angle C$ और $CN \perp NB$ का समद्विभाजक है। यदि $BC = 10$ सेमी और $AC = 15$ सेमी है, तो MN की लंबाई (सेमी में) क्या है?

Solution:

Given

$BC = 10$ cm

$AC = 15$ cm

$CN \perp NB$

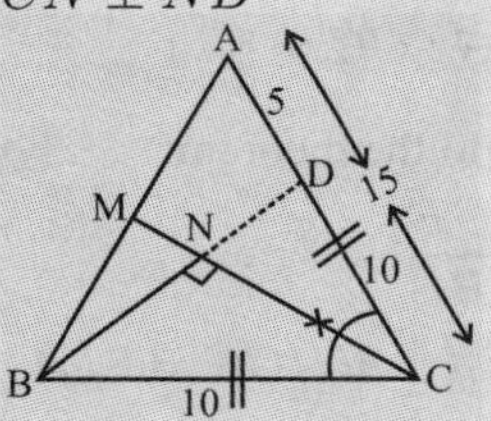

Construction: Produce BN to meet AC at point D.

Since $CN \perp BD$ and $\angle NCB = \angle NCD$:

$\angle BNC = \angle DNC = 90°$.

$BC = DC$ (corresponding sides)

$BN = ND$ (corresponding sides)

$\triangle BNC \cong \triangle DNC$:

N is the midpoint of BD.

Thus, $BC = DC = 10$ cm.

$AD = AC - DC = 5$ cm.

Now, in $\triangle ABD$:

Since M and N are the midpoints of AB and BD, respectively:

$$MN = \frac{AD}{2}$$

$$MN = \frac{5}{2} = 2.5 \text{ cm.}$$

23. In $\triangle ABC$, $\angle A = 50°$. If the bisectors of $\angle C$ and $\angle B$ meet at a point O, then $\angle BOC$ is equal to:

$\triangle ABC$ में, $\angle A = 50°$ है। यदि $\angle C$ और $\angle B$ के समद्विभाजक एक बिंदु O पर मिलते हैं, तो $\angle BOC$ बराबर है:

Solution:

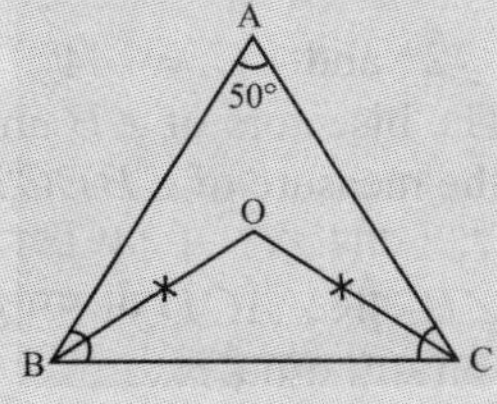

By theorem:

$$\angle BOC = 90° + \frac{\angle A}{2}$$

$$= 90° + \frac{50°}{2} = 115°.$$

24. In a triangle ABC, $AB : AC = 5 : 2$, $BC = 9$ cm. BA is produced to D, and the bisector of the angle $\angle CAD$ meets BC produced at E. What is the length (in cm) of CE?

एक त्रिभुज ABC में, $AB : AC = 5 : 2$, $BC = 9$ सेमी। BA को D तक बढ़ाया गया है, और $\angle CAD$ का समद्विभाजक BC को E पर बढ़ाया गया है। CE की लंबाई (सेमी में) क्या है?

Solution:

Given, $AB : AC = 5 : 2$, $BC = 9$ cm

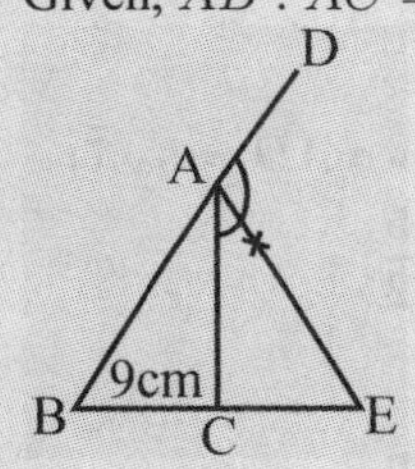

According to the given question:

$$\frac{AB}{AC} = \frac{BE}{CE} \quad \text{(bisector theorem)}$$

$$\frac{5}{2} = \frac{9 + CE}{CE}$$

$$5CE = 18 + 2CE$$

$$3CE = 18 \implies CE = 6 \text{ cm.}$$

25. $\triangle ABC$ is an equilateral triangle. D is a point on side BC such that $BD : BC = 1 : 3$. If $AD = 5\sqrt{7}$ cm, then the side of the triangle is:

$\triangle ABC$ एक समबाहु त्रिभुज है। D भुजा BC पर एक बिंदु इस प्रकार है कि $BD : BC = 1 : 3$ है। यदि $AD = 5\sqrt{7}$ सेमी, तो त्रिभुज की भुजा है:

Solution:

Given:

$BD : BC = (1 : 3) \times 2 : 6$, $AD = 5\sqrt{7}$ cm

$\triangle ABC$ is an equilateral triangle.

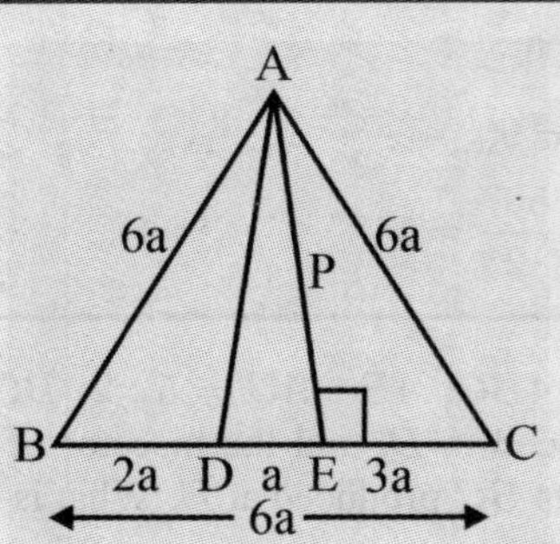

$BE = EC$

In $\triangle AEC$:

$P^2 = 36a^2 - 9a^2$

$P^2 = 27a^2$

In $\triangle ADE$:

$a^2 + P^2 = AD^2$

$25 \times 7 = 28a^2$ (since $AD = 5\sqrt{7}$)

$a^2 = \frac{25}{4} \implies a = \frac{5}{2}$

Side of the triangle:

$6a = 6 \times \frac{5}{2} = 15$ cm.

26. In $\triangle ABC$, D and E are points on sides AB and BC, respectively, such that $BD : DA = 1 : 2$ and $CE : EB = 1 : 4$. If DC and AE intersect at F, then $FD : FC$ is equal to:

$\triangle ABC$ में, D और E क्रमशः भुजाओं AB और BC पर इस प्रकार बिंदु हैं कि $BD : DA = 1 : 2$ और $CE : EB = 1 : 4$। यदि DC और AE F पर प्रतिच्छेद करते हैं, तो $FD : FC$ बराबर है:

Solution:

By Ceva theorem,

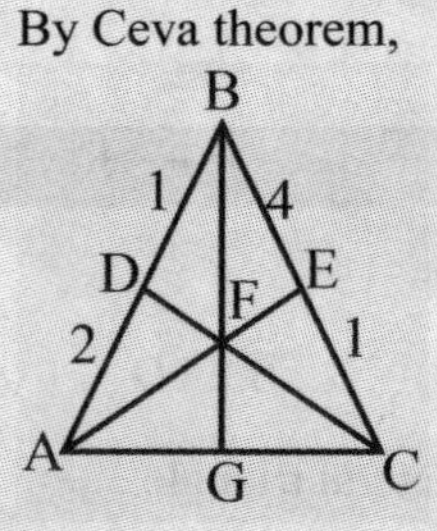

$$\frac{BD}{DA} \times \frac{AG}{GC} \times \frac{CE}{BE} = 1$$

$$\frac{AG}{GC} = \frac{8}{1}, \quad AG : GC = 8 : 1$$

Now, according to the theorem:

$$\frac{CF}{FD} = \frac{GC}{AG} + \frac{CE}{BE}$$

$$\frac{CF}{FD} = \frac{1}{8} + \frac{1}{4} \quad (\because AG/GC = 8/1)$$

$$\frac{CF}{FD} = \frac{3}{8} \quad \therefore FD : CF = 8 : 3$$

27. O is a point in the interior of $\triangle ABC$ such that $OA = 12$ cm, $OC = 9$ cm, $\angle AOB = \angle BOC = \angle COA$ and $\angle ABC = 60°$. What is the length (in cm) of OB?

O एक बिंदु है जो $\triangle ABC$ के अंदर इस प्रकार है कि $OA = 12\ cm$, $OC = 9\ cm$, $\angle AOB = \angle BOC = \angle COA$ और $\angle ABC = 60°$ है। OB की लंबाई (cm में) कितनी है?

Solution:

$\triangle BOA \sim \triangle COB$

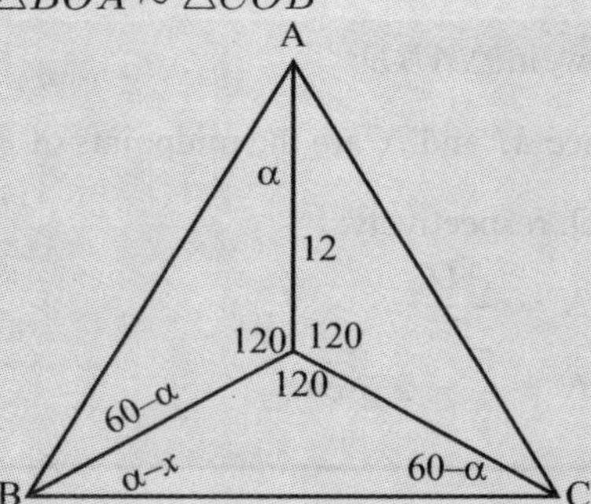

$$\frac{BO}{CO} = \frac{OA}{OB}$$

$x^2 = 12 \times 9$

$x = 6\sqrt{3}$

28. In $\triangle ABC$, $\angle B = 72°$ and $\angle C = 44°$. Side BC is produced to D. The bisectors of $\angle B$ and $\angle ACD$ meet at E. What is the measure of $\angle BEC$?

$\triangle ABC$ में, $\angle B = 72°$ और $\angle C = 44°$। भुजा BC को D तक बढ़ाया गया है। $\angle B$ और $\angle ACD$ के समद्विभाजक E पर मिलते हैं। $\angle BEC$ का माप क्या है?

Solution:

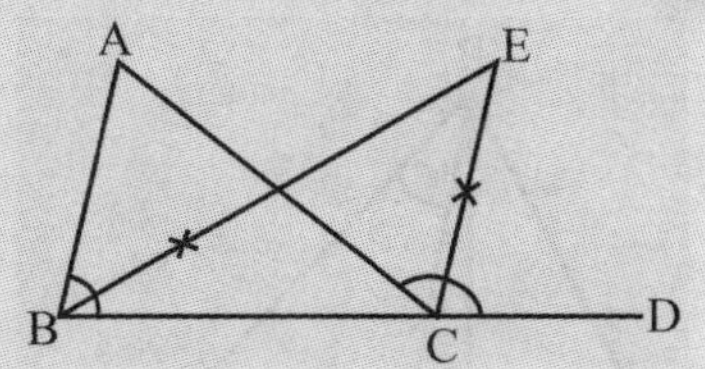

$\angle A = 180° - (72° + 44°)$

$= 64°$

$\angle BEC = \frac{\angle BAC}{2}$ (Theorem)

$= \frac{64°}{2} = 32°$

29. In the given figure, in triangle STU, ST = 8 cm, TU = 9 cm, SU = 12 cm, QU = 24 cm, SR = 32 cm and PT = 27 cm. What is the ratio of the area of triangle PQU and area of triangle PTR?

दिए गए चित्र में, त्रिभुज STU में, $ST = 8$ सेमी, $TU = 9$ सेमी, $SU = 12$ सेमी, $QU = 24$ सेमी, $SR = 32$ सेमी और $PT = 27$ सेमी। त्रिभुज PQU के क्षेत्रफल और त्रिभुज PTR के क्षेत्रफल का अनुपात क्या है?

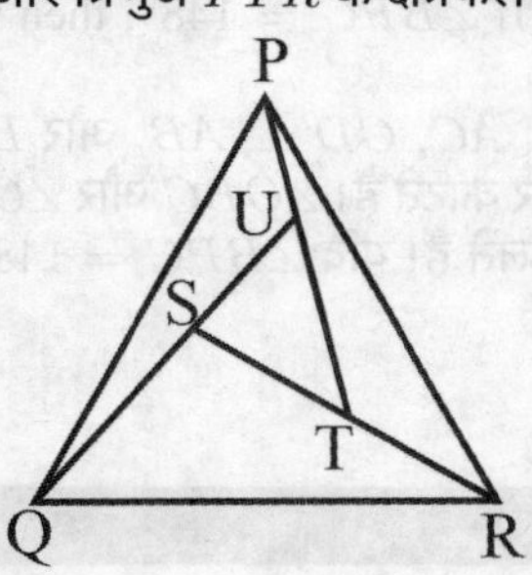

Solution:

$QU = 24$ cm

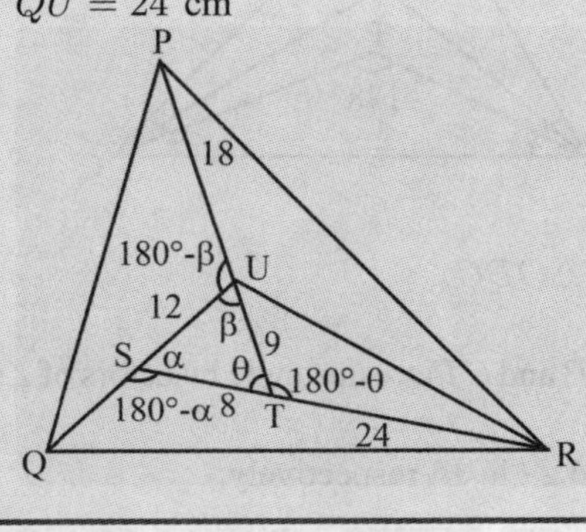

Area of triangle $= \frac{1}{2}ab \sin \theta$

$$\frac{ar(\triangle PUQ)}{ar(\triangle STU)} = \frac{\frac{1}{2} \times 18 \times 24 \times \sin(180° - \beta)}{\frac{1}{2} \times 12 \times 9 \times \sin \beta} \quad \text{(i)}$$

$$\frac{ar(\triangle PTR)}{ar(\triangle STU)} = \frac{\frac{1}{2} \times 27 \times 24 \times \sin(180° - \theta)}{\frac{1}{2} \times 8 \times 9 \times \sin \theta} \quad \text{(ii)}$$

On dividing (i) by (ii):

$$\frac{ar(\triangle PQU)}{ar(\triangle PTR)} = \frac{4}{9}$$

30. In triangle XYZ, G is the centroid. If XY = 11 cm, $YZ = 14$ cm and $XZ = 7$ cm, then what is the value (in cm) of GM?

त्रिभुज XYZ में, G केन्द्रक है। यदि XY = 11 सेमी, $YZ = 14$ सेमी और $XZ = 7$ सेमी है, तो GM का मान (सेमी में) क्या है?

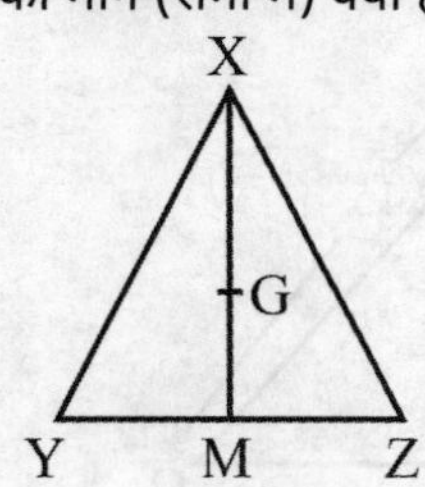

Solution:

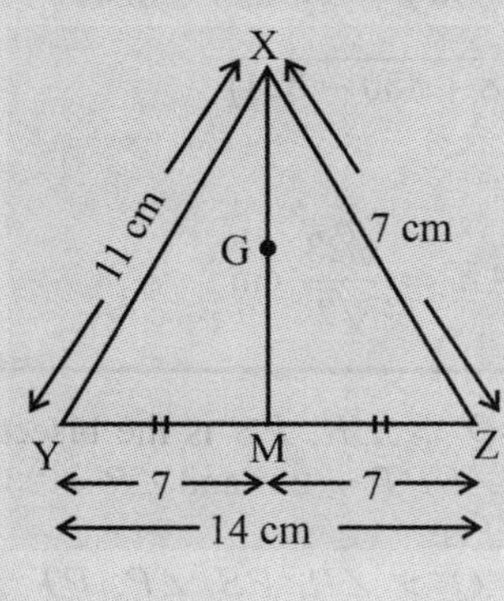

By Apollonius Theorem:

$$XY^2 + XZ^2 = 2(XM^2 + YM^2)$$

$$121 + 49 = 2(XM^2 + 49)$$

$$85 = XM^2 + 49$$

$$XM = 6 \text{ cm}$$

Centroid divides median in 2 : 1 ratio:

$$GM = 6 \times \frac{1}{3} = 2 \text{ cm}$$

31. In a triangle ABC, $AB = 12$, $BC = 18$ and $AC = 15$. The medians AX and BY intersect sides BC and AC at X and Y, respectively. If AX and BY intersect each other at O, then what is the value of OX?

एक त्रिभुज ABC में, $AB = 12$, $BC = 18$ और $AC = 15$। माध्यिकाएँ AX और BY भुजाएँ BC को काटती हैं और AC क्रमशः X और Y पर। यदि AX और BY एक दूसरे को O पर काटते हैं, तो OX का मान क्या है?

Solution:

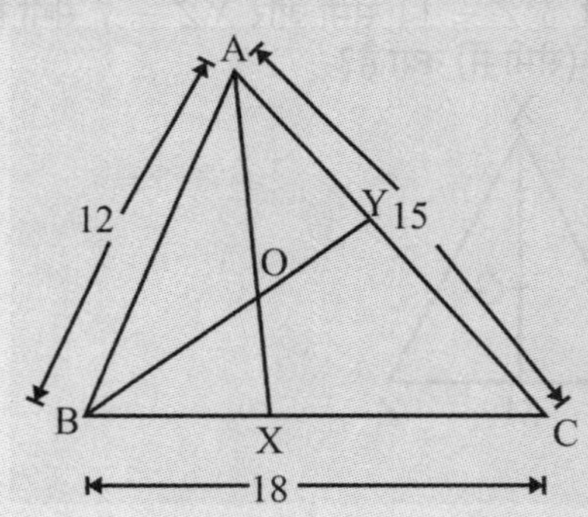

Median $AX = \frac{1}{2}\sqrt{2AB^2 + 2AC^2 - BC^2}$

$= \frac{1}{2}\sqrt{(2 \times 144 + 2 \times 225 - 324)}$

$= \frac{1}{2}\sqrt{(288 + 450 - 324)}$

$= \frac{1}{2}\sqrt{414}$

$OX = \frac{1}{3}AX = \frac{\sqrt{23}}{\sqrt{2}}$ cm

32. In $\triangle PQR$, $\angle Q > \angle R$, PS is the bisector of $\angle P$, $PT \perp QR$. If $\angle SPT = 28°$ and $\angle R = 23°$, then the measure of $\angle Q$ is:

$\triangle PQR$ में, $\angle Q > \angle R$, PS, $\angle P$, $PT \perp QR$ का समद्विभाजक है। यदि $\angle SPT = 28°$ और $\angle R = 23°$ है, तो $\angle Q$ का माप क्या है?

Solution:

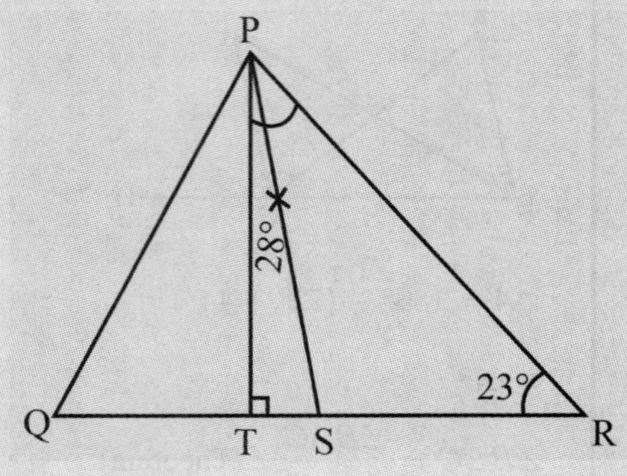

We know that:

$$\angle SPT = \frac{\angle Q - \angle R}{2}$$

$$28° = \frac{\angle Q - 23°}{2}$$

$$\angle Q = 56° + 23° = 79°$$

33. In $\triangle ABC$, $BE \perp AC$, $CD \perp AB$, and BE and CD intersect each other at O. The bisectors of $\angle OBC$ and $\angle OCB$ meet at P. If $\angle BPC = 148°$, then what is the measure of $\angle A$?

$\triangle ABC$ में, $BE \perp AC$, $CD \perp AB$, और BE और CD एक दूसरे को O पर काटते हैं। $\angle OBC$ और $\angle OCB$ के समद्विभाजक P पर मिलते हैं। यदि $\angle BPC = 148°$ है, तो $\angle A$ का माप क्या है?

Solution:

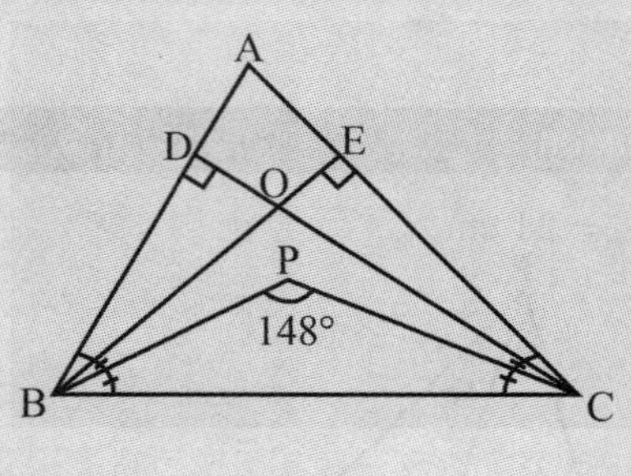

In $\triangle OBC$:

BP and CP are the angle bisectors of $\angle OBC$

and $\angle OCB$, respectively:

$\angle BPC = 90° + \frac{\angle BOC}{2}$

$148° = 90° + \frac{\angle BOC}{2}$

$\angle BOC = 116°$

$\angle DOE = 116°$ (vertically opposite angle)

In quadrilateral $ADOE$:

$\angle A = 360° - (90° + 90° + 116°)$

$= 360° - 296° = 64°$

34. In the given figure, if $AD = 3$, $DE = 4$, $AB = 12$, $BF = 2$, $FG = 6$, $BC = 10$, then the value of M/N is: (Assume M is the area of the quadrilateral $FGDE$ and N is the area of the triangle ABC).

दिए गए चित्र में, यदि $AD = 3$, $DE = 4$, $AB = 12$, $BF = 2$, $FG = 6$, $BC = 10$ है, तो M/N का मान क्या है: (मान लें: M चतुर्भुज $FGDE$ का क्षेत्रफल है और N त्रिभुज ABC का क्षेत्रफल है।)

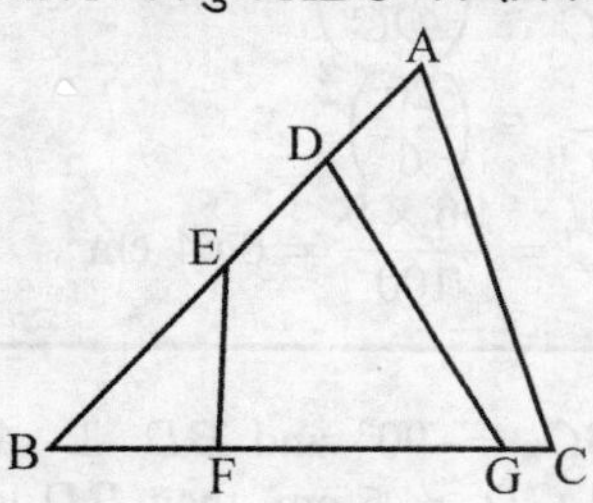

Solution:

Let $\angle ABC = \theta$:

Area of $\triangle ABC(N)$:

$= \frac{1}{2} AB \times BC \times \sin\theta$

$= \frac{1}{2} \times 12 \times 10 \times \sin\theta$

$= 60 \sin\theta$

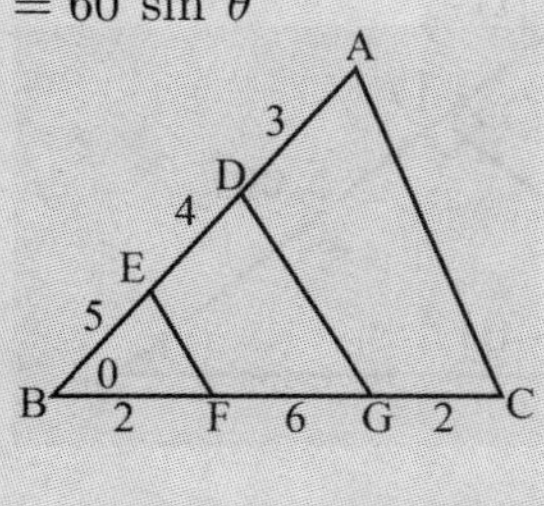

Area of quadrilateral $FGDE(M)$:

$=$ Area of $\triangle ABGD$ – Area of $\triangle BEF$

$= \frac{1}{2} \times 8 \times 9 \times \sin\theta - \frac{1}{2} \times 2 \times 5 \times \sin\theta$

$= 31 \sin\theta$

$\therefore \frac{M}{N} = \frac{31 \sin\theta}{60 \sin\theta} = \frac{31}{60}$

35. If in the following figure (not to scale), $\angle DAB + \angle CBA = 90°$, $BC = AD$, $AB = 20$ cm, $CD = 10$ cm, then the area of the quadrilateral $ABCD$ is:

यदि निम्नलिखित आकृति में (स्केल के अनुसार नहीं), $\angle DAB + \angle CBA = 90°$, $BC = AD$, $AB = 20$ सेमी, $CD = 10$ सेमी, तो चतुर्भुज $ABCD$ का क्षेत्रफल है:

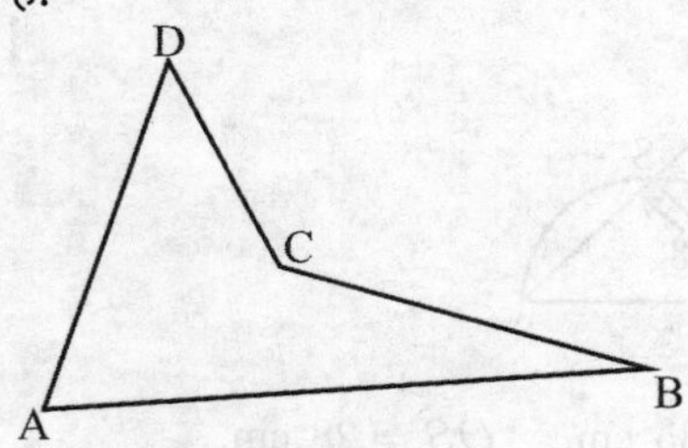

Solution:

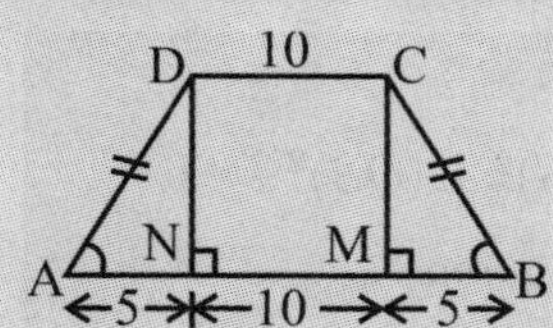

$\angle DAB + \angle CBA = 90°$

$\angle DAB = \angle CBA = 45°$

$ABCD$ is an isosceles trapezium. In $\triangle CMB$:

$\angle MBC = \angle MCB = 45°$

$MB = CM = 5$

Area of trapezium $ABCD$:

$= \frac{1}{2}(AB + CD) \times 5$

$= \frac{1}{2}(20 + 10) \times 5 = 75 \text{ cm}^2$

36. In the given figure, triangle PQR is a right-angled triangle at Q. If $PQ = 35$ cm and $QS = 28$ cm, then what is the value (in cm) of SR?

दी गई आकृति में, त्रिभुज PQR, Q पर एक समकोण त्रिभुज

है। यदि $PQ = 35$ सेमी और $QS = 28$ सेमी है, तो SR का मान (सेमी में) क्या है?

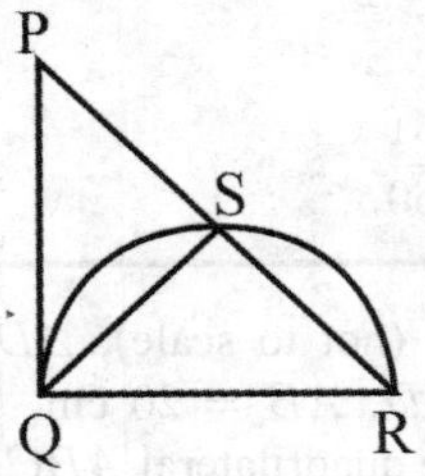

Solution:

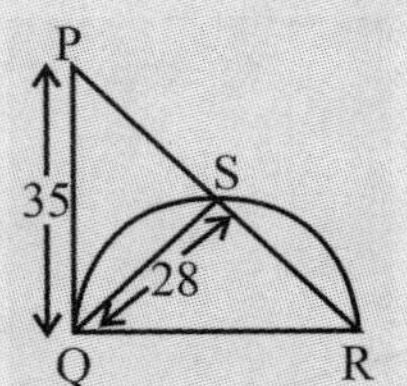

$PQ = 35$ cm, $QS = 28$ cm.

Angle drawn in a semicircle is a right angle:

$\angle QSP = 90°$.

In $\triangle QSP$:

$(PQ)^2 = (PS)^2 + (QS)^2$

$(35)^2 = PS^2 + (28)^2$

$1225 = PS^2 + 784$

$PS^2 = 441$

$PS = 21$ cm.

Using the theorem:

$(QS)^2 = PS \times SR$

$28 \times 28 = 21 \times SR$

$SR = \dfrac{28 \times 28}{21}$

$SR = 37.33$ cm.

37. In $\triangle ABC$, $\angle A = 90°$, $AB = 16$ cm, $AC = 12$ cm. D is the midpoint of AC and $DE \parallel CB$ at E. What is the area (in cm^2) of $\triangle CDE$?

$\triangle ABC$ में, $\angle A = 90°$, $AB = 16$ सेमी, $AC = 12$ सेमी। D, AC का मध्यबिंदु है और $DE \parallel CB$ E पर है। $\triangle CDE$ का क्षेत्रफल (सेमी^{2} में) क्या है?

Solution:

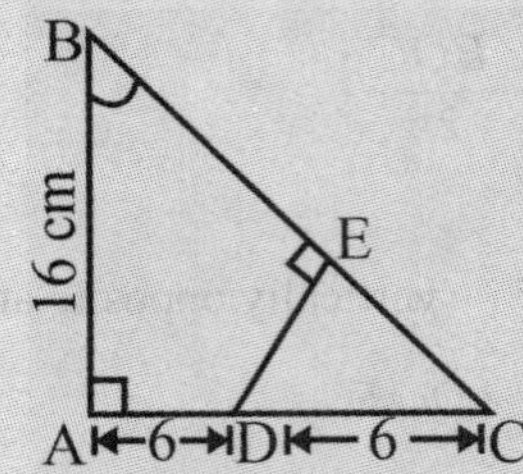

Area of $\triangle ABC = \dfrac{1}{2} \times 12 \times 16 = 96$ cm^2.

$BC^2 = AB^2 + AC^2$

$= (16)^2 + (12)^2$

$= 256 + 144$

$BC = \sqrt{400} = 20$ cm.

$\triangle ABC \sim \triangle EDC$

$\dfrac{\text{area of } \triangle ABC}{\text{area of } \triangle EDC} = \left(\dfrac{BC}{DC}\right)^2$

$\dfrac{96}{\text{area of } \triangle EDC} = \left(\dfrac{20}{6}\right)^2$

$\text{area of } \triangle EDC = \dfrac{96 \times 9}{100} = 8.64$ cm^2.

38. In $\triangle ABC$, $\angle ABC = 90°$ and $BD \perp AC$. If $AD = 4$ cm and $CD = 5$ cm, then BD is equal to:

$\triangle ABC$ में, $\angle ABC = 90°$ और $BD \perp AC$। यदि $AD = 4$ सेमी और $CD = 5$ सेमी है तो BD बराबर है:

Solution:

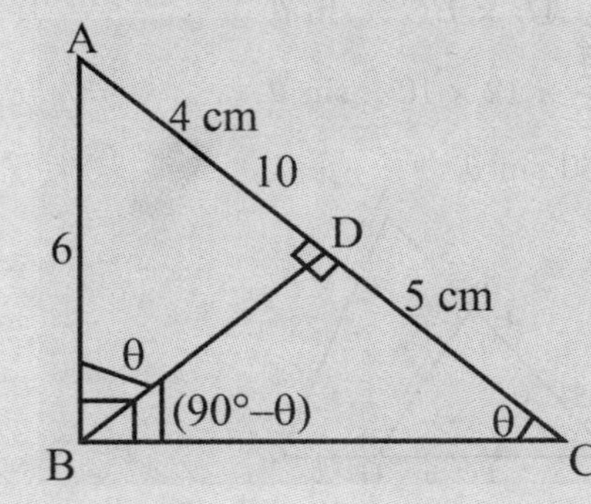

$\triangle ADB \sim \triangle BDC$:

$\dfrac{AD}{DB} = \dfrac{BD}{DC}$

$BD^2 = AD \times DC$

$BD = \sqrt{4 \times 5}$

$= 2\sqrt{5}$ cm.

39. In $\triangle ABC$, $\angle C = 90°$ and D is a point on CB such that AD is the bisector of $\angle A$. If $AC = 5$ cm and $BC = 12$ cm, then what is the length of AD?

$\triangle ABC$ में, $\angle C = 90°$ और D, CB पर एक बिंदु इस प्रकार है कि AD, $\angle A$ का समद्विभाजक है। यदि $AC = 5$ सेमी और $BC = 12$ सेमी है, तो AD की लंबाई क्या है?

Solution:

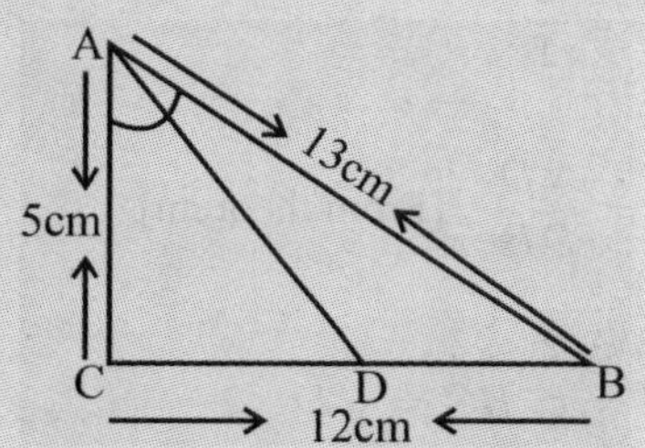

In right-angled triangle ACB:

$AB = \sqrt{5^2 + 12^2} = 13$ cm.

The bisector of an angle of a triangle divides its opposite side in the ratio of the remaining two sides:

$$\frac{DC}{DB} = \frac{AC}{AB}$$

$DC : DB = 5 : 13$

$$DC = \frac{5}{18} \times 12 = \frac{10}{3}.$$

In right-angled triangle ACD, using Pythagoras Theorem:

$$AD^2 = AC^2 + CD^2$$

$$AD^2 = 5^2 + \left(\frac{10}{3}\right)^2$$

$$AD^2 = 25 + \frac{100}{9}$$

$$AD = \sqrt{\frac{325}{9}}$$

$$AD = \frac{5\sqrt{13}}{3} \text{ cm.}$$

40. BD and CE are the medians of $\triangle ABC$, right-angled at A. If $CE = \frac{5\sqrt{13}}{2}$, $BC = 10$, the length of BD is:

BD और CE, $\triangle ABC$ की माध्यिकाएँ हैं, जो A पर समकोण हैं। यदि $CE = \frac{5\sqrt{13}}{2}$, $BC = 10$ है, तो BD की लंबाई है:

Solution:

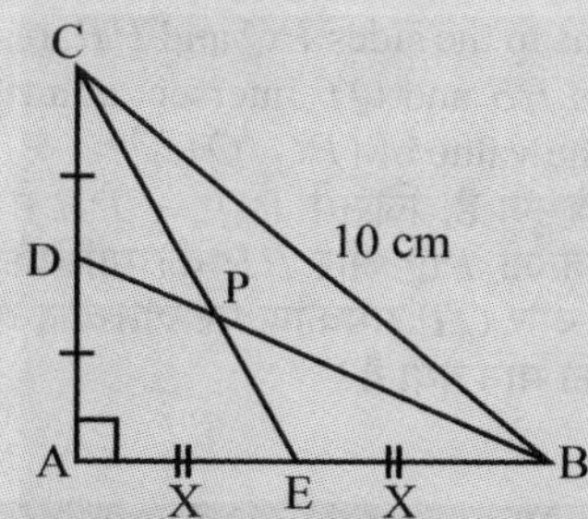

Given:

$$CE = \frac{5\sqrt{13}}{2}.$$

In $\triangle CAB$:

$$AC^2 + AB^2 = 100$$

$$AC^2 + (2x)^2 = 100$$

$$AC^2 + 4x^2 = 100 \quad \text{(i)}.$$

In $\triangle CAE$:

$$AC^2 + x^2 = \left(\frac{5\sqrt{13}}{2}\right)^2$$

$$AC^2 + x^2 = \frac{325}{4} \quad \text{(ii)}.$$

From (i) and (ii):

$$3x^2 = 100 - \frac{325}{4}$$

$$3x^2 = \frac{75}{4}$$

$$x = \frac{5}{2}.$$

$AC = \sqrt{75}$ cm (from Pythagoras theorem).

In $\triangle DAB$:

$(BD)^2 = (AB)^2 + (AD)^2$

$(BD)^2 = (5)^2 + \left(\frac{\sqrt{75}}{2}\right)^2$

$(BD)^2 = 25 + \frac{75}{4}$

$(BD)^2 = \frac{175}{4}$

$BD = \frac{5}{2}\sqrt{7}$ cm.

BD AND CE are the medium of $\triangle ABC$, right

41. PQR is a triangle such that $PQ = PR$. RS and QT are the medians to the sides PQ and PR, respectively. If the medians RS and QT intersect at a right angle, then what is the value of $(PQ/QR)^2$?

PQR एक त्रिभुज है, जिसमें $PQ = PR$ है। RS और QT क्रमश: भुजाओं PQ और PR की माध्यिकाएँ हैं। यदि माध्यिकाएँ RS और QT समकोण पर प्रतिच्छेद करती हैं, तो $(PQ/QR)^2$ का मान क्या है?

Solution:

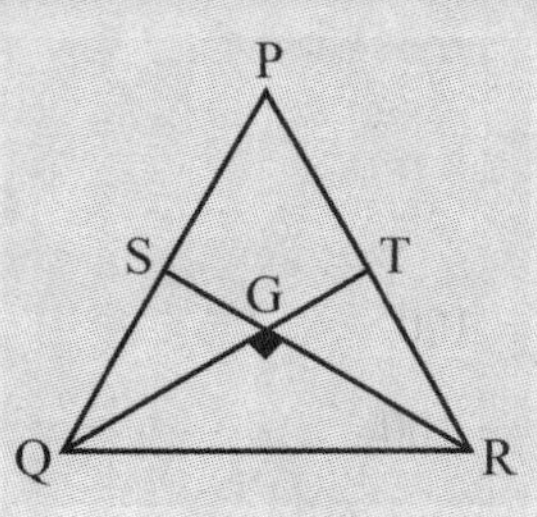

$\angle QGR = 90°$ (By theorem)

In $\triangle PQR$:

We know that:

$PQ^2 + PR^2 = 5QR^2$

$2PQ^2 = 5QR^2$ $(PQ = PR)$

$\left(\frac{PQ}{QR}\right)^2 = \frac{5}{2}$

42. In $\triangle ABC$, $\angle C = 90°$. M and N are the midpoints of sides AB and AC, respectively. CM and BN intersect each other at D and $\angle BDC = 90°$. If $BC = 8$ cm, then the length of BN is:

$\triangle ABC$ में, $\angle C = 90°$। M और N क्रमश: भुजाओं AB और AC के मध्य बिंदु हैं। CM और BN एक दूसरे को D और $\angle BDC = 90°$ पर काटते हैं। यदि $BC = 8$ सेमी है, तो BN की लंबाई है:

Solution:

D is the centroid of $\triangle ABC$:

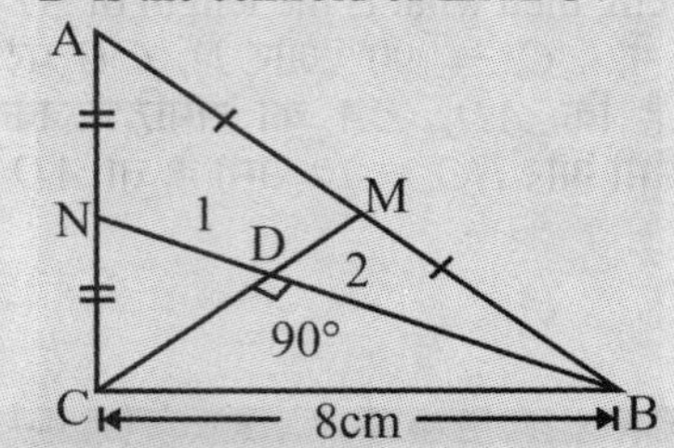

$\therefore \frac{BD}{DN} = \frac{2}{1}$

But:

$\frac{NC^2}{BC^2} = \frac{ND}{DB}$ [From theorem]

$\frac{NC}{BC} = \frac{1}{\sqrt{2}}$

If $NC = x,\ BC = \sqrt{2}x$:

$BN = \sqrt{x^2 + 2x^2} = \sqrt{3}x$

But: $\sqrt{2}x = 8$:

$x = 4\sqrt{2}$

$BN = \sqrt{3} \times 4\sqrt{2} = 4\sqrt{6}$ cm.

43. In a $\triangle ABC$, $\angle A = 90°$, if BM and CN are two medians, $\frac{BM^2 + CN^2}{BC^2}$ is equal to:

एक $\triangle ABC$ में, $\angle A = 90°$, यदि BM और CN दो माध्यिकाएँ हैं, तो $\frac{BM^2 + CN^2}{BC^2}$ का मान है:

Solution:

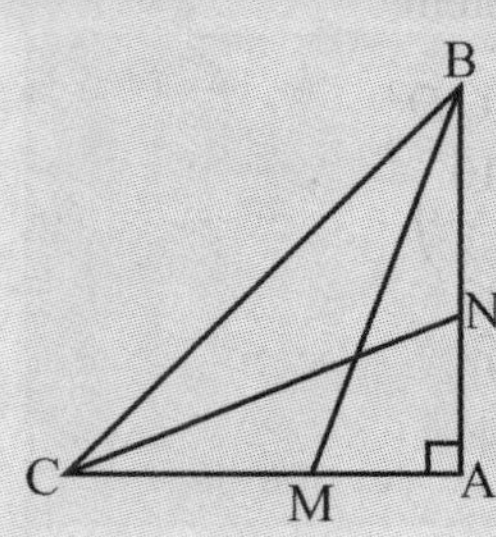

BM is the median:

$AM = CM = \frac{1}{2}AC$ (i)

CN is the median:

$AN = NB = \frac{1}{2}AB$ (ii)

In $\triangle BAC$:

$(BC)^2 = (AB)^2 + (AC)^2$

In $\triangle BAM$:

$(BM)^2 = (AB)^2 + \left(\frac{AC}{2}\right)^2$

$4BM^2 = 4AB^2 + (AC)^2$

In $\triangle NAC$:

$(CN)^2 = AN^2 + AC^2$

$(CN)^2 = \left(\frac{1}{2}AB\right)^2 + AC^2$

$4CN^2 = AB^2 + 4AC^2$

Adding equations (i) and (ii):

$\frac{BM^2 + CN^2}{BC^2} = \frac{5}{4}$

Type 2 (Problems based on Quadrilateral)

44. In a circle, $ABCD$ is a cyclic quadrilateral. AC and BD intersect each other at P. If $AB = AC$ and $\angle BAC = 48°$, then the measure of $\angle ADC$ is:

एक वृत्त में $ABCD$ एक चक्रीय चतुर्भुज है। AC और BD एक दूसरे को P पर काटते हैं। यदि $AB = AC$ और $\angle BAC = 48°$ है, तो $\angle ADC$ का माप है:

Solution:

According to the question:

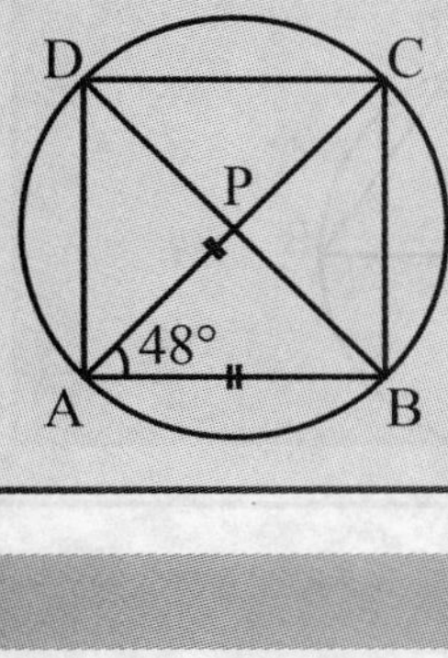

$AB = AC$ (Given)

$\triangle ACB$ is an isosceles triangle:

$\therefore \angle C = \angle B = x$

$48° + x + x = 180°$

[Sum of angles in a triangle]

$48° + 2x = 180°$

$2x = 132°$

$x = 66°$

$ABCD$ is a cyclic quadrilateral:

Sum of opposite angles $= 180°$

$\angle ADC + \angle ABC = 180°$

$\angle ADC + 66° = 180°$

$\angle ADC = 180° - 66° = 114°$

45. $ABCD$ is a cyclic quadrilateral. Sides AB and DC, when produced, meet at E, and sides AD and BC, when produced, meet at F. If $\angle ADC = 76°$ and $\angle AED = 55°$, then $\angle AFB$ is equal to:

$ABCD$ एक चक्रीय चतुर्भुज है। भुजाएँ AB और DC को बढ़ाने पर E पर मिलती हैं और भुजाएँ AD और BC को बढ़ाने पर F पर मिलती हैं। यदि $\angle ADC = 76°$ और $\angle AED = 55°$ है, तो $\angle AFB$ बराबर है:

Solution:

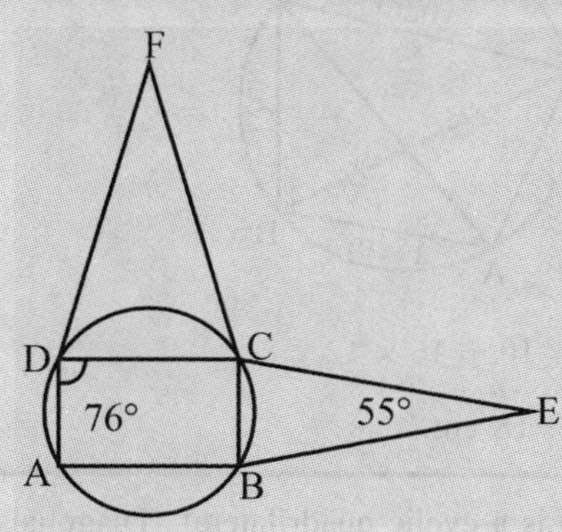

$\angle ADC + \angle ABC = 180°$

$76° + \angle ABC = 180°$

$\angle ABC = 104°$

$\angle ABC = \angle BEC + \angle ECB$

$104° = 55° + \angle ECB$

$\angle ECB = 49°$

$\angle ECB = \angle DCF$

$\angle DCF = 49°$

$\angle CDF = 180° - 76° = 104°$

In $\triangle CDF$:

$\angle DCF + \angle CFD + \angle CDF = 180°$

$49° + 104° + \angle CFD = 180°$

$\angle CFD = 27°$

$\angle CFD = \angle AFB = 27°$

46. ABCD is a cyclic quadrilateral in which AB = 15 cm, BC = 12 cm and CD = 10 cm. If AC bisects BD, then what is the measure of AD?

$ABCD$ एक चक्रीय चतुर्भुज है जिसमें AB = 15 सेमी, BC = 12 सेमी और CD = 10 सेमी है। यदि AC, BD को समद्विभाजित करता है, तो AD का माप क्या है?

Solution:

AC divides BD:

$AD \times DC = AB \times BC$ (by using theorem)

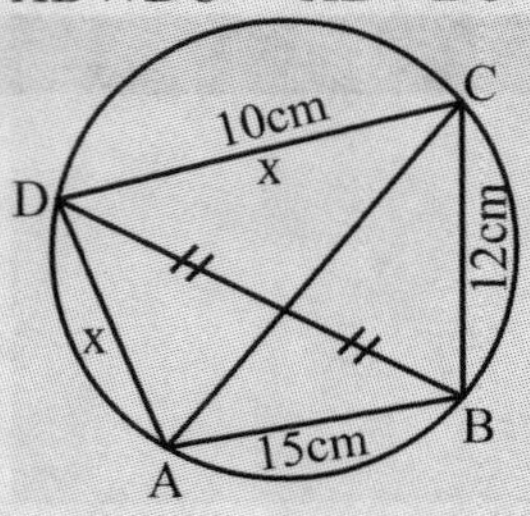

$x \times 10 = 15 \times 12$

$x = 18$ cm.

47. ABCD is a cyclic quadrilateral. Diagonals BD and AC intersect each other at E. If $\angle BEC = 138°$ and $\angle ECD = 35°$, then what is the measure of $\angle BAC$?

$ABCD$ एक चक्रीय चतुर्भुज है। विकर्ण BD और AC एक दूसरे को E पर काटते हैं। यदि $\angle BEC = 138°$ और $\angle ECD = 35°$ है, तो $\angle BAC$ का माप क्या है?

Solution:

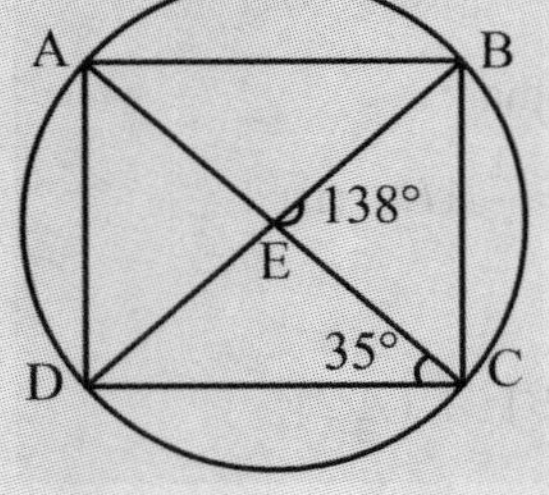

$\angle BEC = 138°$ and $\angle ECD = 35°$:

$\angle BEC$ and $\angle CED$ are on the same straight line:

$\angle CED = 180° - \angle BEC = 180° - 138° = 42°$.

In $\triangle CDE$:

$\angle CED = 42°$ and $\angle DCE = 35°$

$\angle CDE = 180° - (\angle CED + \angle DCE)$

$\angle CDE = 180° - (42° + 35°) = 103°$.

$\angle BAC$ and $\angle BDC$ are on the same arc. In a cyclic quadrilateral, angles on the same arc are always the same:

$\angle BAC = \angle CDE = 103°$.

48. ABCD is a cyclic quadrilateral such that AB is the diameter of the circle and $\angle ADC = 145°$. Then, what is the measure of $\angle BAC$?

$ABCD$ एक चक्रीय चतुर्भुज है जैसे कि AB वृत्त का व्यास है और $\angle ADC = 145°$ है, तो $\angle BAC$ का माप क्या है?

Solution:

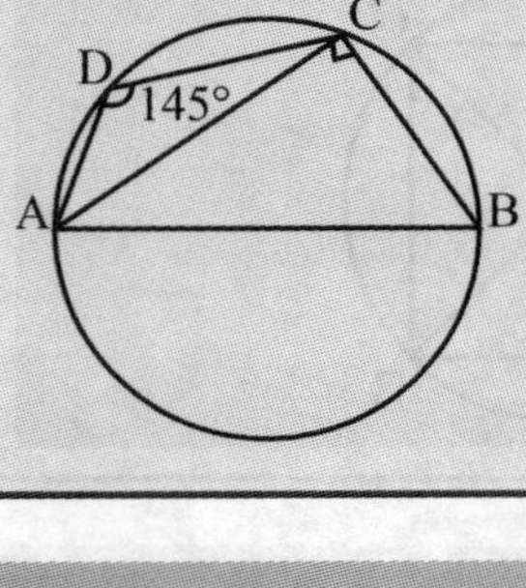

In the figure drawn:

$\angle ACB = 90°$ [Angle formed in a semicircle is always equal to (90°)].

$\angle ADC + \angle ABC = 180°$

$145° + \angle ABC = 180°$

$\angle ABC = 180° - 145° = 35°$.

In $\triangle ABC$:

$\angle BAC + \angle ABC + \angle ACB = 180°$

$\angle BAC + 35° + 90° = 180°$

$\angle BAC = 180° - 125° = 55°$.

49. In a quadrilateral $ABCD$, the bisectors of $\angle C$ and $\angle D$ meet at E. If $\angle CED = 56°$ and $\angle A = 49°$, then the measure of $\angle B$ is:
चतुर्भुज $ABCD$ में, $\angle C$ और $\angle D$ के समद्विभाजक E पर मिलते हैं। यदि $\angle CED = 56°$ और $\angle A = 49°$ है, तो $\angle B$ का माप क्या है?

Solution:

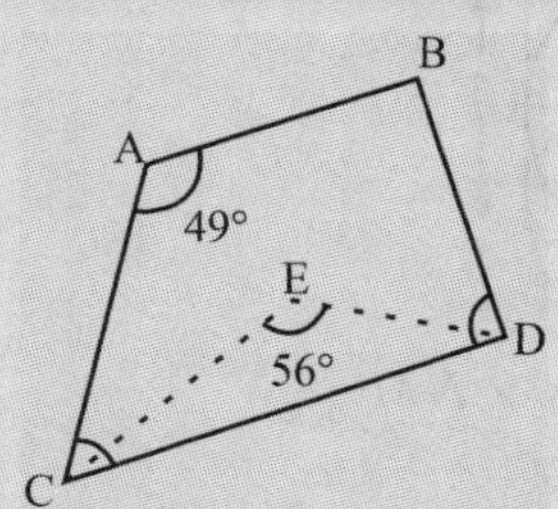

$\angle A = 49°, \quad \angle CED = 56°$:

$$\frac{\angle C}{2} + \frac{\angle D}{2} + \angle CED = 180°$$

$$\frac{\angle C}{2} + \frac{\angle D}{2} = 180° - \angle CED = 180° - 56° = 124°$$

$\angle C + \angle D = 248°$

$\angle B = 360° - (\angle A + \angle C + \angle D)$

$\angle B = 360° - (49° + 248°)$

$\angle B = 360° - 297° = 63°$.

Alternatively, by formula:

$$\angle CED = \frac{\angle A + \angle B}{2}$$

$$56° = \frac{49° + \angle B}{2}$$

$\angle B = 112° - 49° = 63°$.

50. Let $ABCD$ be a cyclic quadrilateral with sides $AB = 2$ cm, $BC = 7$ cm, $CD = 11$ cm, $DA = 14$ cm. Find the product of its diagonals.
चतुर्भुज $ABCD$ में, $AB = 2$ सेमी, $BC = 7$ सेमी, $CD = 11$ सेमी, $DA = 14$ सेमी। इसके विकर्णों का गुणनफल ज्ञात करें।

Solution:

Using Ptolemy's theorem:

$AC \times BD = (AB \times CD) + (AD \times BC)$

Substituting the values:

$AC \times BD = (2 \times 11) + (14 \times 7)$

$AC \times BD = 22 + 98 = 120$ cm.

51. ABCD is a square, and CDE is an equilateral triangle outside the square. What is the value (in degrees) of $\angle BEC$?
$ABCD$ वर्ग है और CDE वर्ग के बाहर एक समबाहु त्रिभुज है। $\angle BEC$ का मान (डिग्री में) क्या है?

Solution:

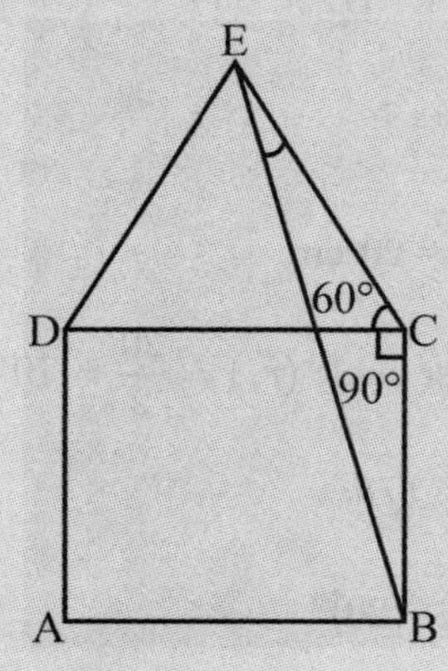

In $ABEC$:

$BC = CE$

$\angle CBE = \angle CEB$

$\therefore \angle BEC = \dfrac{180° - (90° + 60°)}{2}$

$\angle BEC = \dfrac{30°}{2}$

$\angle BEC = 15°$.

52. In the given figure, $PQRS$ is a square of side 20 cm, and SR is extended to point T. If the length of QT is 25 cm, then what is the distance (in cm) between the centers O_1 and O_2 of the two circles?

दिए गए चित्र में, $PQRS$ 20 सेमी भुजा वाला एक वर्ग है और SR को बिंदु T तक बढ़ाया गया है। यदि QT की लंबाई 25 सेमी है, तो दोनों वृत्तों के केंद्र O_1 और O_2 के बीच की दूरी (सेमी में) क्या है?

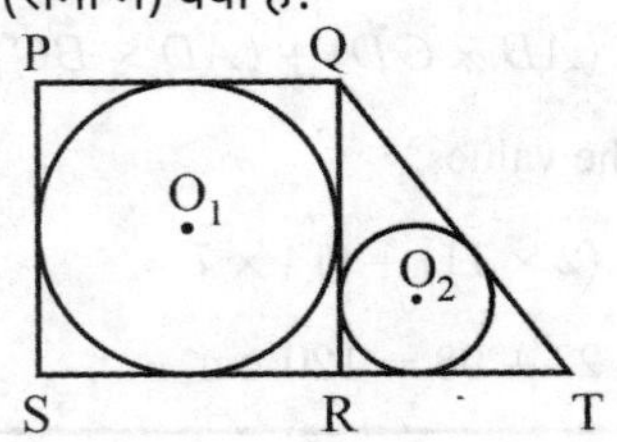

Solution:

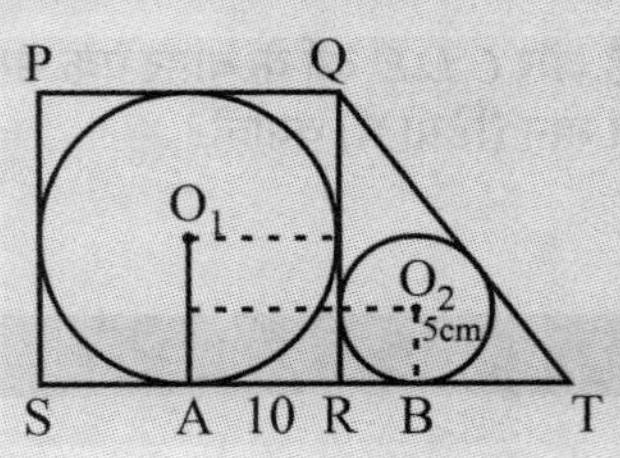

Given:

Side of square = 20 cm

Radius of larger circle $(r_1) = \dfrac{20}{2} = 10$ cm

$RT^2 = QT^2 - QR^2$

$RT^2 = (25)^2 - (20)^2$

$RT = 15$ cm

$r = \dfrac{P + B - H}{2}$

$r = \dfrac{15 + 20 - 25}{2}$

$r = 5$ cm

$(O_1, O_2)^2 = 15^2 + 5^2 = 250$

Distance between centers O_1 and O_2 = $5\sqrt{10}$ cm.

53. In the given figure, $MNOP$ is a square of side 6 cm. What is the value (in cm) of the radius of circles?

दिए गए चित्र में, $MNOP$ 6 सेमी भुजा वाला एक वर्ग है। वृत्तों की त्रिज्या का मान (सेमी में) क्या है?

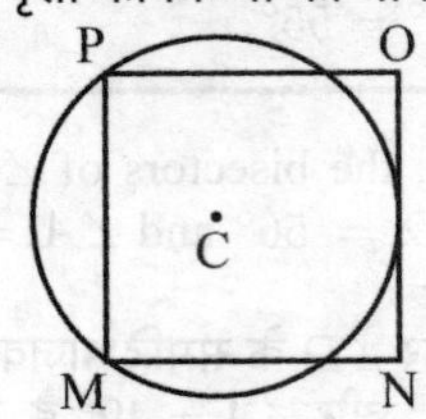

Solution:

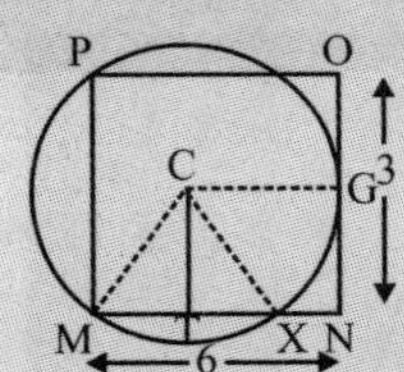

$2GN^2 = NX \times NM$

$9 = x \times 6$

$x = 1.5$ cm

$MX = 6 - 1.5 = 4.5$ cm

$TX = MT = \dfrac{4.5}{2} = 2.25$ cm

$TN = 6 - 2.25 = 3.75$ cm

Radius = 3.75 cm.

54. In the given figure, $ABCD$ is a rectangle, and P is a point on DC such that $BC = 24$ cm, $DP = 10$ cm, and $CD = 15$ cm. If AP produced intersects BC produced at Q, then find the length of AQ.

दी गई आकृति में, $ABCD$ एक आयत है और P, DC पर इस प्रकार बिंदु है कि $BC = 24$ सेमी, $DP = 10$ सेमी और $CD = 15$ सेमी है। यदि AP उत्पादित, BC को Q पर

प्रतिच्छेद करता है, तो AQ की लंबाई ज्ञात कीजिए।

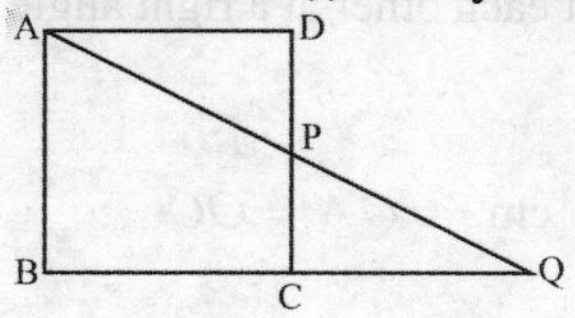

Solution:

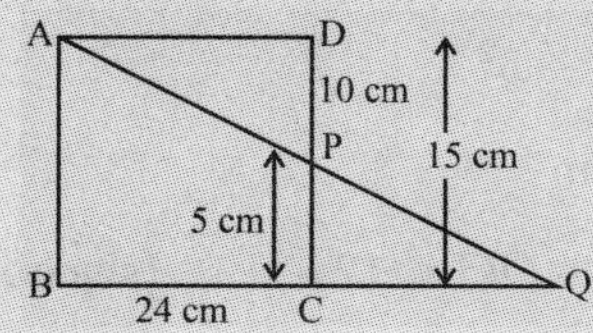

Let $CQ = x$ cm:

$$\frac{x}{x+24} = \frac{5}{15} = \frac{1}{3}$$

$3x = x + 24$

$x = 12$ cm.

$BQ = BC + CQ = 24 + 12 = 36$ cm.

$AB = 15$ cm, $AQ = ?$

$AQ^2 = AB^2 + BQ^2$

$= 15^2 + 36^2 = 1521$

$AQ = 39$ cm.

55. $PQRS$ is a rectangle. T is a point on PQ such that RTQ is an isosceles triangle and $PT = 5\ QT$. If the area of triangle RTQ is $12\sqrt{3}$ sq.cm, then the area of the rectangle $PQRS$ is:

$PQRS$ एक आयत है। PQ पर T एक बिंदु इस प्रकार है कि RTQ एक समद्विबाहु त्रिभुज है और $PT = 5\ QT$ है। यदि त्रिभुज RTQ का क्षेत्रफल $12\sqrt{3}$ वर्ग सेमी है तो आयत $PQRS$ का क्षेत्रफल है:

Solution:

Let $QT = x$ cm:

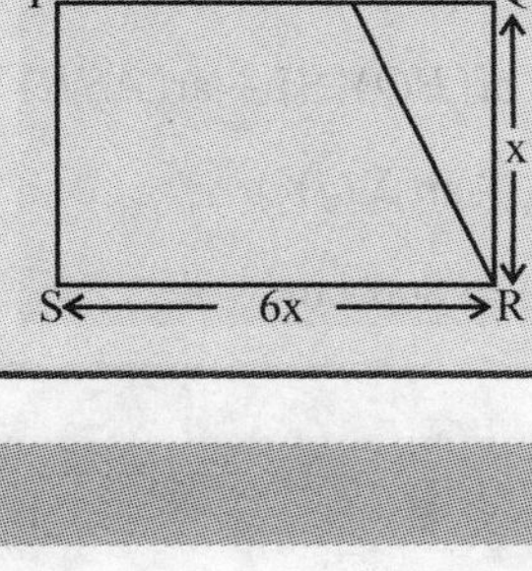

Area of $\triangle RTQ = 12\sqrt{3}$ cm²

$$\frac{1}{2} \times x \times x = 12\sqrt{3}$$

$x^2 = 24\sqrt{3}$ cm²

Area of rectangle $PQRS = QR \times SR =$

$x \times 6x = 6x^2$

$= 6 \times 24\sqrt{3}$

$= 144\sqrt{3}$ cm².

56. One diagonal of a rhombus is $8\sqrt{3}$ cm. If the other diagonal is equal to its side, then the area (in cm²) of the rhombus is:

एक समचतुर्भुज का एक विकर्ण $8\sqrt{3}$ सेमी है। यदि दूसरा विकर्ण उसकी भुजा के बराबर है, तो समचतुर्भुज का क्षेत्रफल (सेमी² में) है:

Solution:

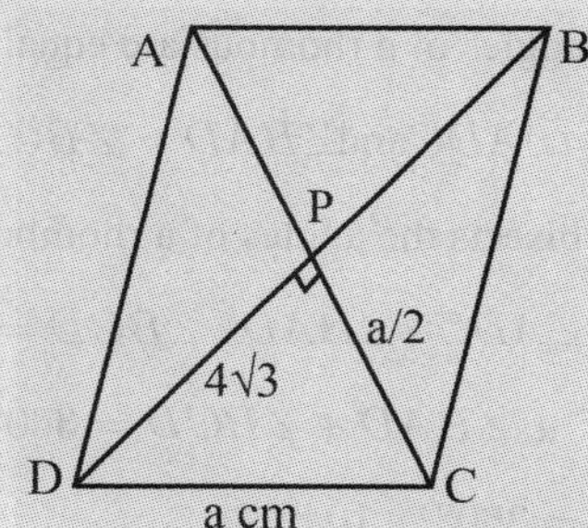

Let the length of the side of the rhombus = a cm.

According to the question:

In $\triangle DPC$:

$$a^2 - \frac{a^2}{4} = 16 \times 3$$

$$\frac{3a^2}{4} = 48$$

$a^2 = 16 \times 4 = 64$

$a = 8$ cm.

As per question:

Length of side of rhombus = Length of other diagonal of given rhombus = 8 cm.

Area of rhombus $= \frac{1}{2} \times d_1 \times d_2$

$= \frac{1}{2} \times 8\sqrt{3} \times 8 = 32\sqrt{3}$ cm^2.

57. $ABCD$ is a rhombus in which $\angle ABC = 52°$. Then find the value of $\angle ACD$.

$ABCD$ एक समचतुर्भुज है, जिसमें $\angle ABC = 52°$ है तो $\angle ACD$ का मान ज्ञात कीजिए।

Solution:

According to the question:

$ABCD$ is a rhombus.

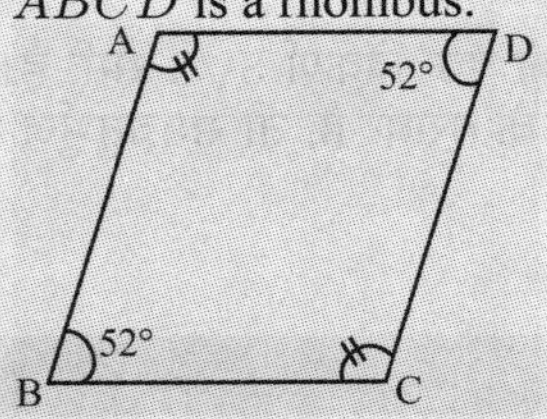

Opposite angles of a rhombus are equal:

$\angle ABC = \angle ADC$ and $\angle BAD = \angle BCD$.

Diagonal bisects the angles of a rhombus:

$\angle ABC + \angle ADC + \angle BAD + \angle BCD = 360°$

$52° + 52° + \angle BAD + \angle BCD = 360°$

$2\angle BCD = 360° - 104°$

$\angle BCD = 128°$

$\angle ACD = \frac{\angle BCD}{2} = \frac{128°}{2} = 64°$.

58. What is the area (in cm^2) of the rhombus whose side is 13 cm and its smaller diagonal is 10 cm?

उस समचतुर्भुज का क्षेत्रफल (सेमी^{2} में) क्या है जिसकी भुजा 13 सेमी है और इसका छोटा विकर्ण 10 सेमी है?

Solution:

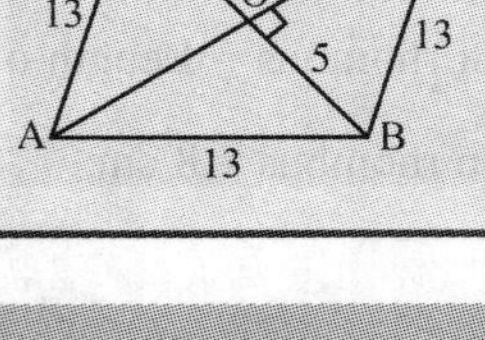

Diagonals bisect each other at a right angle of a rhombus:

$OB = OD = 5$ cm, $\quad OA = OC$

$\angle BOC = 90°$

In $\triangle AOBC$:

$OC^2 = BC^2 - OB^2$

$OC^2 = 13^2 - 5^2$

$OC = 12$ cm

$AC = AO + OC = 12 + 12$ (since $AO = OC$)

$AC = 24$ cm

Area of rhombus $= \frac{d_1 \times d_2}{2} = \frac{10 \times 24}{2} = 120$ cm^2

59. $PQRS$ is a parallelogram, and its area is 300 cm^2. Side PQ is extended to X such that $PQ = QX$. If XS intersects QR at Y, then what is the area (in cm^2) of $\triangle SYR$?

$PQRS$ एक समांतर चतुर्भुज है और इसका क्षेत्रफल 300 सेमी^{2} है। भुजा PQ को X तक इस प्रकार बढ़ाया गया है कि $PQ = QX$ है। यदि XS, QR को Y पर काटता है, तो त्रिभुज SYR का क्षेत्रफल (सेमी^{2} में) क्या है?

Solution:

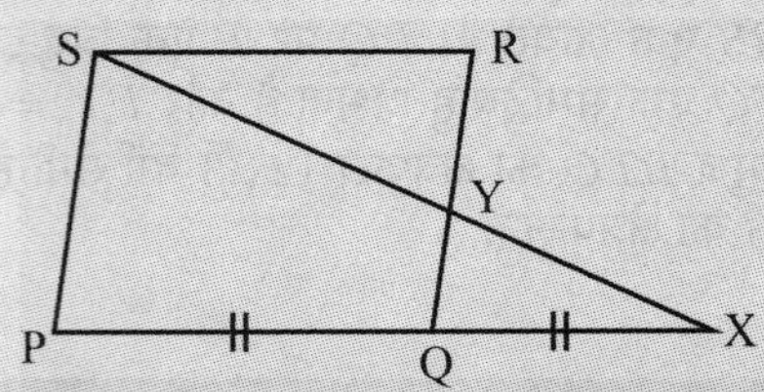

$\triangle SRY \sim \triangle XQY$:

ar$(\triangle SRY)$ = ar$(\triangle XQY)$

ar$(PQRS)$ = ar$(PQYS)$ + ar$(ASRY)$

= ar$(PQYS)$ + ar(XQY)

= ar$(\triangle XPS) = 300$ cm^2

$\triangle XQY \sim \triangle XPS$:

$$\frac{\text{ar}(\triangle XQY)}{\text{ar}(\triangle XPS)} = \left(\frac{QX}{PX}\right)^2 = \left(\frac{1}{2}\right)^2$$

$$\frac{\text{ar}(\triangle SRY)}{300} = \frac{1}{4}$$

$$\text{ar}(\triangle SYR) = 75 \text{ cm}^2$$

60. In a trapezium $PQRS$, PQ is parallel to RS, and diagonals PR and QS intersect at O. If $PQ = 4$ cm, $SR = 10$ cm, then what is the ratio of area$(\triangle POQ)$: area$(\triangle SOR)$?

एक समलंब $PQRS$ में, PQ RS के समानांतर है और विकर्ण PR और QS O पर प्रतिच्छेद करते हैं। यदि $PQ = 4$ सेमी, $SR = 10$ सेमी, तो क्षेत्रफल $(\triangle POQ)$ क्या है: क्षेत्रफल $(\triangle SOR)$?

Solution:

Given:

$PQ \parallel RS$, $PQ = 4$ cm, $SR = 10$ cm:

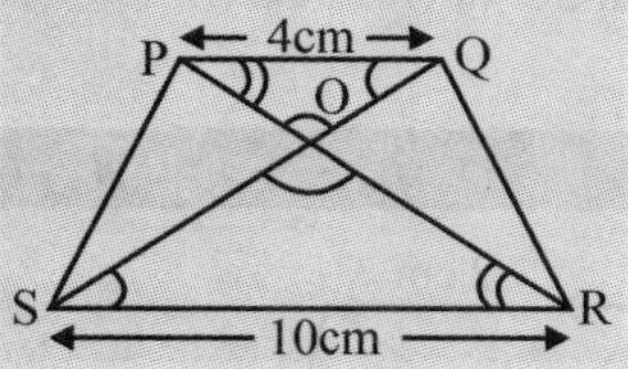

$\angle QPR = \angle PRS$, $\angle SQP = \angle QSR$, $\angle POQ = \angle SOR$

$\triangle POQ \sim \triangle ROS$:

$$\frac{\text{area of } \triangle POQ}{\text{area of } \triangle SOR} = \left(\frac{PQ}{SR}\right)^2$$

$$= \left(\frac{4}{10}\right)^2$$

$$= \left(\frac{2}{5}\right)^2$$

$$= \frac{4}{25}$$

61. A field is in the shape of a trapezium whose parallel sides are 260 m and 100 m, and non-parallel sides are each of length 100 m. What is the area (in m^2) of the field?

एक मैदान एक समलंब के आकार का है जिसकी समानांतर भुजाएँ 260 मीटर और 100 मीटर हैं, और गैर-समानांतर भुजाएँ प्रत्येक की लंबाई 100 मीटर हैं। मैदान का क्षेत्रफल (वर्ग मीटर में) क्या है?

Solution:

Let $ABCD$ is a trapezium:

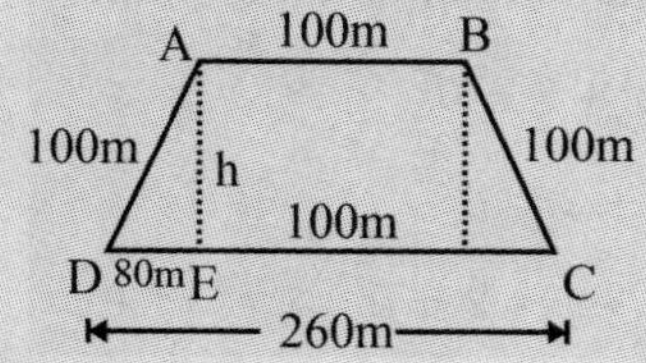

$$AE = \sqrt{(AD)^2 - (DE)^2}$$

$$AE = \sqrt{(100)^2 - (80)^2}$$

$$AE = \sqrt{10000 - 6400}$$

$$AE = \sqrt{3600}$$

$$AE = 60 \text{ m.}$$

Area of trapezium $= \frac{1}{2} \times$ Sum of parallel sides $\times$ Height

$$= \frac{1}{2} \times (100 + 260) \times 60$$

$$= \frac{1}{2} \times 360 \times 60 = 10800 \text{ m}^2.$$

62. $ABCD$ is a trapezium in which $AB \parallel DC$ and its diagonals intersect at P. If $AP = (3x - 1)$ cm, $PC = (5x - 3)$ cm, $BP = (2x + 1)$ cm and $PD = (6x - 5)$ cm, then the length of DB is:

$ABCD$ एक समलंब है जिसमें $AB \parallel DC$ और इसके विकर्ण P पर प्रतिच्छेद करते हैं। यदि $AP = (3x - 1)$ सेमी, $PC = (5x - 3)$ सेमी, $BP = (2x + 1)$ सेमी और $PD = (6x - 5)$ सेमी है, तो DB की लंबाई है:

Solution:

$\triangle PAB \sim \triangle PCD$:

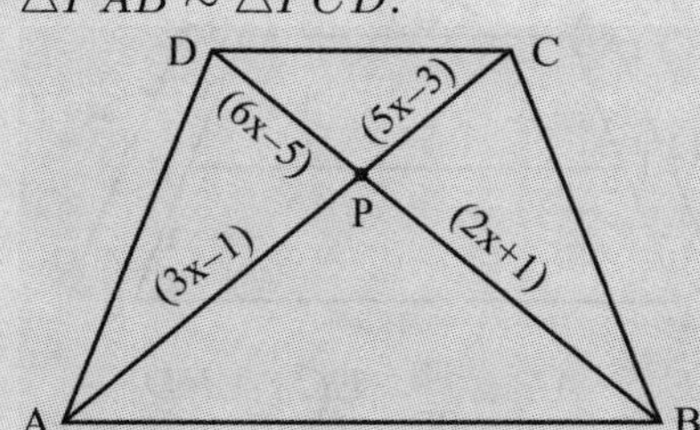

$$\frac{PA}{PC} = \frac{PB}{PD}$$

$$\frac{3x-1}{5x-3} = \frac{2x+1}{6x-5}$$

$$18x^2 - 15x - 6x + 5 = 10x^2 + 5x - 6x - 3$$

$$18x^2 - 21x + 5 = 10x^2 - x - 3$$

$$8x^2 - 20x + 8 = 0$$

$$2x^2 - 5x + 2 = 0$$

$$2x^2 - 4x - x + 2 = 0$$

$$2x(x-2) - 1(x-2) = 0$$

$$(x-2)(2x-1) = 0$$

$x = 2$ and $x = \frac{1}{2}$ (Unacceptable).

$DB = (6x-5) + (2x+1) = 8x - 4 = 16 - 4 = 12$ cm.

63. $ABCD$ is a trapezium. Sides AB and CD are parallel to each other. $AB = 6$ cm, $CD = 18$ cm, $BC = 8$ cm and $AD = 12$ cm. A line parallel to AB divides the trapezium into two parts of equal perimeter. This line cuts BC at E and AD at F. If $\frac{BE}{EC} = \frac{AF}{FD}$, then what is the value of $\frac{BE}{EC}$?

$ABCD$ ट्रैपेज़ियम है। भुजाएँ AB और CD एक दूसरे के समानांतर हैं। $AB = 6$ सेमी, $CD = 18$ सेमी, $BC = 8$ सेमी और $AD = 12$ सेमी। AB के समांतर एक रेखा समलंब को समान परिमाप वाले दो भागों में विभाजित करती है। यह रेखा BC को E पर और AD को F पर काटती है। यदि $\frac{BE}{EC} = \frac{AF}{FD}$ है, तो $\frac{BE}{EC}$ का मान क्या है?

Solution:

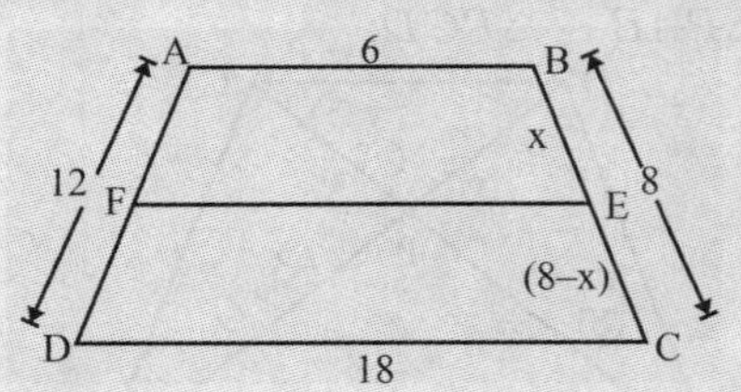

$$\frac{BE}{EC} = \frac{AF}{FD} \quad \text{or} \quad \frac{EC}{BE} = \frac{FD}{AF}$$

$$\frac{EC}{BE} + 1 = \frac{FD}{AF} + 1$$

$$\frac{8}{BE} = \frac{12}{AF} \quad \text{or} \quad \frac{8}{x} = \frac{12}{AF}$$

$$AF = \frac{3x}{2}, \quad FD = 12 - \frac{3x}{2}$$

Perimeter of $ABEF$ = Perimeter of $FDCE$:

$$AB + BE + FE + AF = CD + EC + FD$$

$$6 + x + \frac{3x}{2} = 8 - x + 18 + 12 - \frac{3x}{2}$$

$$6 + x + \frac{3x}{2} = 38 - x$$

$$5x = 32$$

$$x = \frac{32}{5}.$$

64. In a trapezium $ABCD$, $DC \parallel AB$, $AB = 12$ cm and $DC = 7.2$ cm. What is the length of the line segment joining the midpoints of its diagonals?

एक समलंब $ABCD$ में, $DC \parallel AB$, $AB = 12$ सेमी और $DC = 7.2$ सेमी। इसके विकर्णों के मध्य बिंदुओं को मिलाने वाले रेखाखंड की लंबाई क्या है?

Solution:

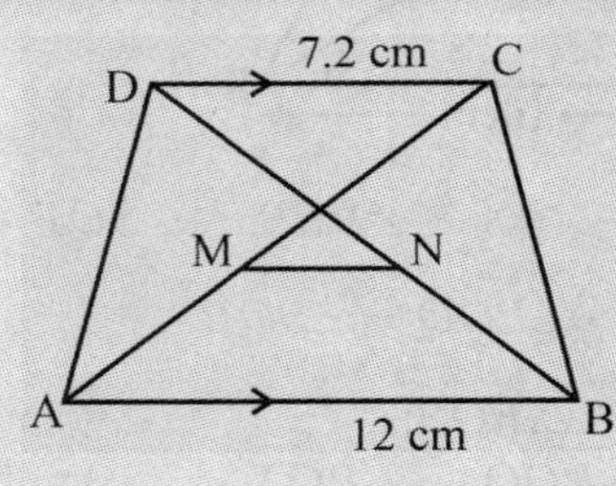

Let M and N be the midpoints of diagonals AC and BD, respectively:

$$MN = \frac{AB - DC}{2}$$

$$MN = \frac{12 - 7.2}{2} = 2.4 \text{ cm}.$$

65. A field is in the shape of a trapezium whose parallel sides are 200 m and 400 m, whereas each of the other two sides is 260 m. What is the area (in m^2) of the field?

एक मैदान एक समलंब के आकार का है जिसकी समानांतर भुजाएँ 200 मीटर और 400 मीटर लंबी हैं, जबकि अन्य दो भुजाएँ 260 मीटर लंबी हैं। मैदान का क्षेत्रफल (वर्ग मीटर में) क्या है?

Solution:

In $\triangle BMC$:

$BM \parallel AD$ and $BN \perp MC$

$a = AD = BM = 260$ m

$b = MC = 200$ m

$c = BC = 260$ m.

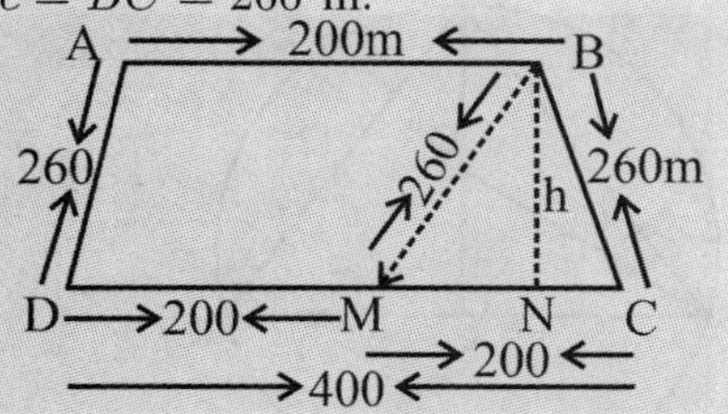

Area of $\triangle BMC$ =

$\sqrt{s(s-a)(s-b)(s-c)}$

Where $s = \dfrac{260+200+260}{2}$:

Area of $\triangle BMC$ =

$\sqrt{360 \times 100 \times 160 \times 100}$

Area of $\triangle BMC = \dfrac{1}{2} \times MC \times BN$

$24000 = \dfrac{1}{2} \times 200 \times h$

$h = 240$ m.

Area of trapezium = $\dfrac{1}{2}$ × sum of parallel sides × Height

$= \dfrac{1}{2} \times 600 \times 240$

$= 72000$ m^2.

Type 3 (Problem Based on Circle)

66. In a circle with centre O, PQ and QR are two chords such that $\angle PQR = 118°$. What is the measure of $\angle OPR$?
केंद्र O वाले एक वृत्त में, PQ और QR दो जीवाएँ इस प्रकार हैं कि $\angle PQR = 118°$ है। $\angle OPR$ का माप क्या है?

Solution:

Given,

$\angle PQR = 118°$:

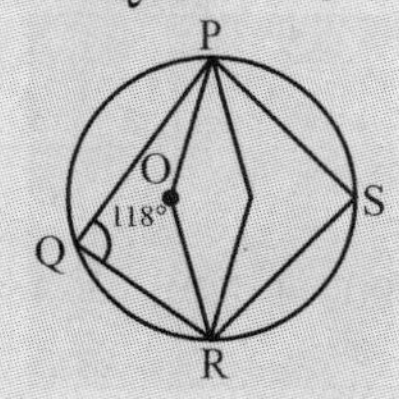

$ABCD$ is a cyclic quadrilateral, then sum of opposite angles is 180°.

$\angle PQR + \angle PSR = 180°$

$\angle PSR = 180° - 118° = 62°$

$\angle POR = \dfrac{\angle PQR - \angle PSR}{2}$

$= \dfrac{118° - 62°}{2} = \dfrac{56°}{2} = 28°$

67. AB is the diameter of a circle with centre O. C and D are two points on the circle on either side of AB such that $\angle CAB = 52°$ and $\angle ABD = 47°$. What is the difference (in degrees) between the measures of $\angle CAD$ and $\angle CBD$?
AB केंद्र O वाले वृत्त का व्यास है। AB के दोनों ओर C और D वृत्त पर दो बिंदु हैं, जैसे कि $\angle CAB = 52°$ और $\angle ABD = 47°$ है। $\angle CAD$ और $\angle CBD$ के मापों के बीच अंतर (डिग्री में) क्या है?

Solution:

Given, $\angle CAB = 52°$, $\angle ABD = 47°$:

$\angle CAD - \angle CBD = ?$

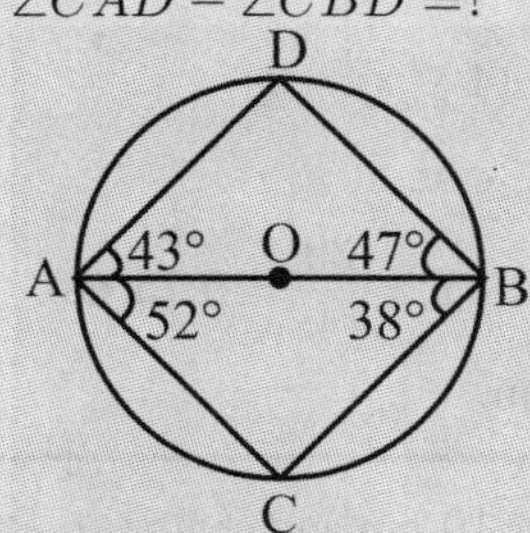

The angle subtended by a diameter on any point of the circle is 90°.

In $\triangle DAB$:

$\angle DAB = 180° - (90° + 47°) = 180° - 137° = 43°$

In $\triangle ACB$:

$\angle ABC = 180° - (90° + 52°) = 180° - 142° = 38°$

Difference between $(\angle CAD - \angle CBD) = (95° - 85°) = 10°$

68. Two chords AB and CD of a circle when produced meet at a point P outside the circle. If $AB = 6$ cm, $PB = 5$ cm, $PD = 4$ cm, then CD is equal to:

एक वृत्त की दो जीवाएँ AB और CD को जब बढ़ाया जाता है, तो वे वृत्त के बाहर एक बिंदु P पर मिलती हैं। यदि $AB = 6$ सेमी, $PB = 5$ सेमी, $PD = 4$ सेमी, तो CD बराबर है:

Solution:

Given, $AB = 6$ cm, $PB = 5$ cm, $PD = 4$ cm, let $CD = x$ cm:

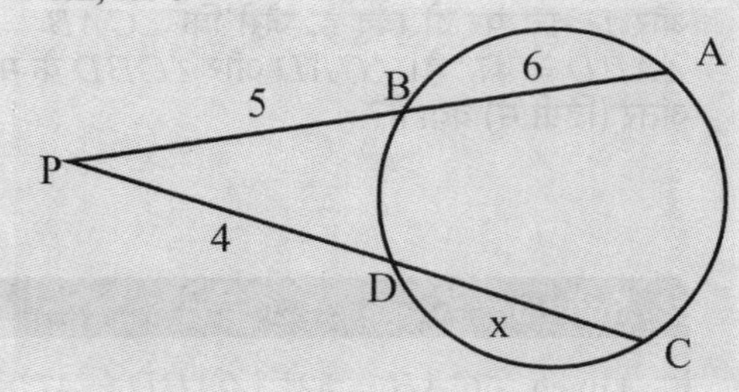

$PA \times PB = PC \times PD$

$(6 + 5) \times 5 = (4 + x) \times 4$

$55 = 16 + 4x$

$4x = 39 \quad \Rightarrow \quad x = \frac{39}{4} = 9.75$ cm

$CD = 9.75$ cm.

69. In a circle of radius 10 cm and centre O, PQ and PR are two equal chords, each of length 12 cm. What is the length (in cm) of chord QR?

10 सेमी त्रिज्या और केंद्र O वाले एक वृत्त में, PQ और PR दो समान जीवाएँ हैं, जिनमें से प्रत्येक की लंबाई 12 सेमी है। जीवा QR की लंबाई (सेमी में) क्या है?

Solution:

O is the circumcentre.

Circumradius (R):

$$10 = \frac{QR \times 12 \times 12}{4 \times \frac{1}{2} \times QR \times PS}$$

$PS = 7.2$ cm

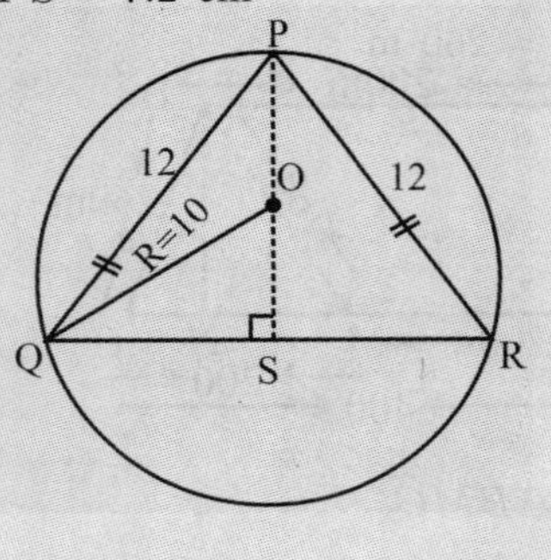

In $\triangle PSQ$:

$QS^2 = QP^2 - PS^2$

$144 - 51.84 = 92.16$

$QS = 9.6$ cm

$QR = 2 \times QS = 2 \times 9.6 = 19.2$ cm.

70. In a circle with centre O and radius 13 cm, two parallel chords are drawn on different sides of the centre. If the length of one chord is 10 cm and the distance between the two chords is 17 cm, then find the difference in lengths of the two chords (in cm).

केंद्र O और त्रिज्या 13 सेमी वाले एक वृत्त में, केंद्र के विभिन्न पक्षों पर दो समानांतर जीवाएँ खींची जाती हैं। यदि एक जीवा की लंबाई 10 सेमी है और दोनों जीवाओं के बीच की दूरी 17 सेमी है, तो दोनों जीवाओं की लंबाई में अंतर (सेमी में) ज्ञात कीजिए।

Solution:

Given, $AB = 10$ cm, $PQ = 17$ cm:

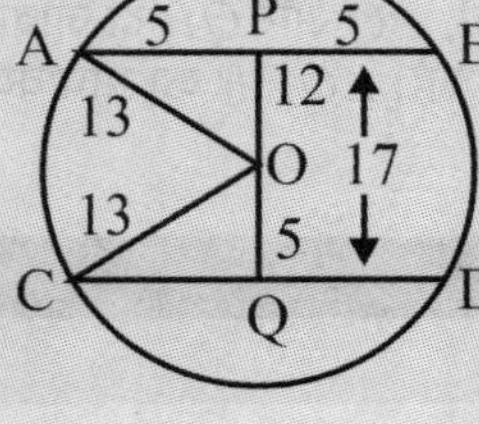

Triplet: 5, 12, 13

$CQ = 12$ cm, $CD = 24$ cm

Difference in length of the two chords:

$CD - AB = 24 - 10 = 14$ cm.

71. PQ and RS are two parallel chords of a circle of length 14 cm and 48 cm, respectively, and lie on the same side of the centre O. If the distance between the chords is 17 cm, what is the radius (in cm) of the circle?
PQ और RS क्रमश: 14 सेमी और 48 सेमी लंबाई वाले एक वृत्त की दो समानांतर जीवाएं हैं, और केंद्र O के एक ही तरफ स्थित हैं। यदि जीवाओं के बीच की दूरी 17 सेमी है, तो त्रिज्या (सेमी में) क्या है वृत्त?

Solution:

Given, $PQ = 14$ cm, $NQ = 7$ cm, $RS = 48$ cm, $MS = 24$ cm:

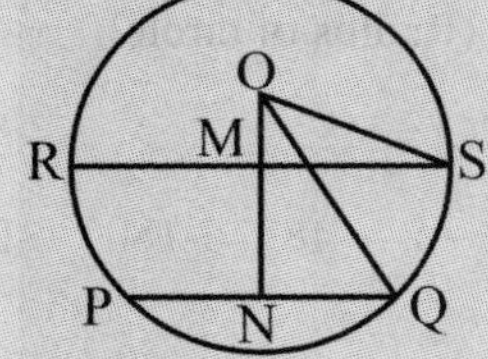

$MN = 17$ cm, $OM = x$ cm, $OS = r$ cm.

In $\triangle OMS$:

$OS^2 = OM^2 + MS^2$

$r^2 = x^2 + (24)^2$ (i).

In $\triangle ONQ$:

$OQ^2 = ON^2 + NQ^2$

$r^2 = (17 + x)^2 + (7)^2$ (ii).

From equations (i) and (ii):

$x^2 + (24)^2 = (17 + x)^2 + (7)^2$

$x^2 + 576 = 289 + x^2 + 34x + 49$

$576 - 338 = 34x$

$34x = 238$

$x = 7.$

Substitute $x = 7$ in equation (i):

$r^2 = (7)^2 + (24)^2$

$r^2 = 49 + 576$

$r = \sqrt{625} = 25$ cm.

Hence, the radius of the circle is 25 cm.

72. In a circle of diameter 20 cm, chords AB and CD are parallel to each other. BC is a diameter. If AB is 6 cm from the centre of the circle, what is the length (in cm) of the chord CD?
20 सेमी व्यास वाले एक वृत्त में, जीवा AB और CD एक दूसरे के समानांतर हैं। BC व्यास है। यदि AB वृत्त के केंद्र से 6 सेमी दूर है, तो जीवा CD की लंबाई (सेमी में) क्या है?

Solution:

Since both chords are parallel to each other, they are equidistant from the centre:

$AB = CD$.

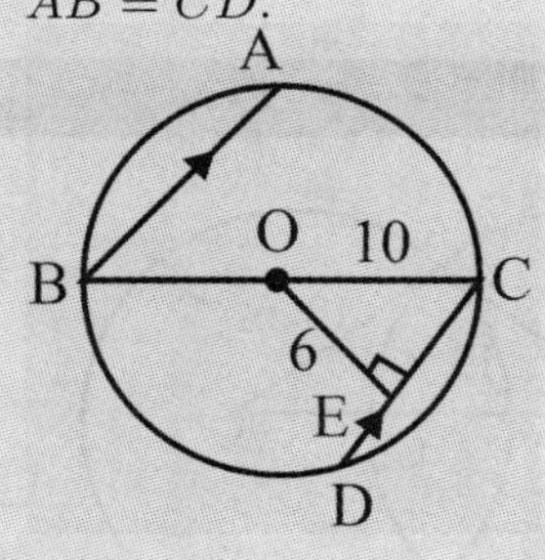

In $\triangle ODC$:

$OC^2 = OD^2 + CE^2$

$10^2 = 6^2 + CE^2$

$CE^2 = 64$

$CE = 8.$

$CD = 2 \times CE = 2 \times 8 = 16$ cm.

73. In a circle with centre O, chords PR and QS meet at the point T when produced, and PQ is a diameter. If $\angle ROS = 42°$, then what is the measure of $\angle PTQ$?
केंद्र O वाले एक वृत्त में, जीवा PR और QS बढ़ाए

जाने पर बिंदु T पर मिलते हैं, और PQ एक व्यास है। यदि $\angle ROS = 42°$ है, तो $\angle PTQ$ का माप है:

Solution:

Given, $\angle ROS = 42°$:

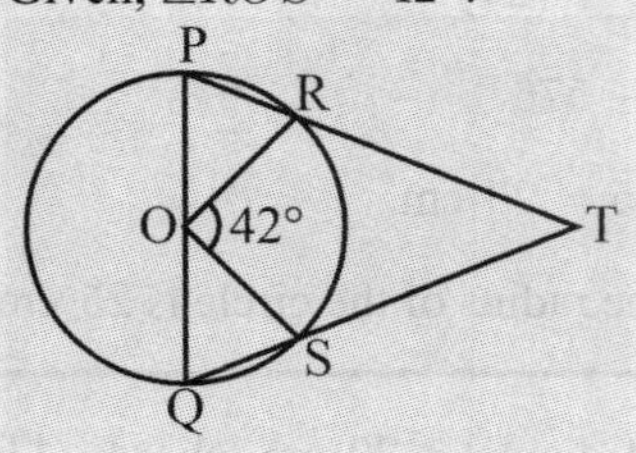

$$\angle PTQ = 90° - \frac{\angle ROS}{2}$$

$$= 90° - \frac{42°}{2}$$

$$= 90° - 21° = 69°.$$

74. Two circles of radius 13 cm and 15 cm intersect each other at points A and B. If the length of the common chord is 24 cm, then what is the distance between their centres?

13 सेमी और 15 सेमी त्रिज्या वाले दो वृत्त एक दूसरे को बिंदु A और B पर प्रतिच्छेद करते हैं। यदि उभयनिष्ठ जीवा की लंबाई 24 सेमी है, तो उनके केंद्रों के बीच की दूरी क्या है?

Solution:

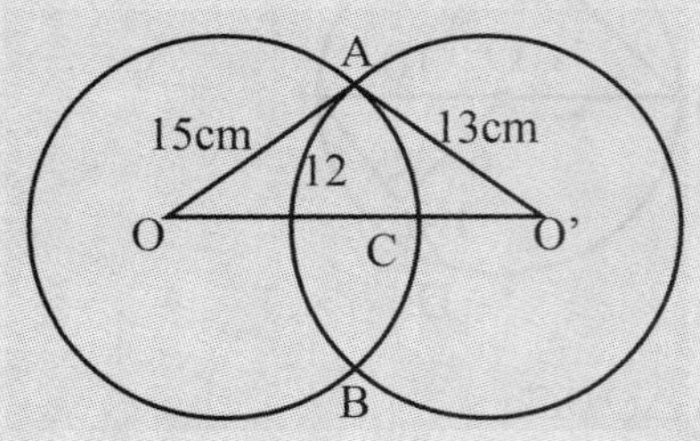

$OC = \sqrt{15^2 - 12^2}$, $O'C = \sqrt{13^2 - 12^2}$

$OC = \sqrt{81} = 9$ cm, $O'C = \sqrt{25} = 5$ cm.

Distance between their centres:

$OO' = OC + O'C = 9 + 5 = 14$ cm.

75. AB is a diameter of a circle with centre O. If C is any point on the circle such that $\angle BAC = 42°$, then find the measure of $\angle BOC$.

AB केंद्र O वाले वृत्त का व्यास है। यदि C वृत्त पर कोई बिंदु है जैसे कि $\angle BAC = 42°$ है, तो $\angle BOC$ का माप ज्ञात करें।

Solution:

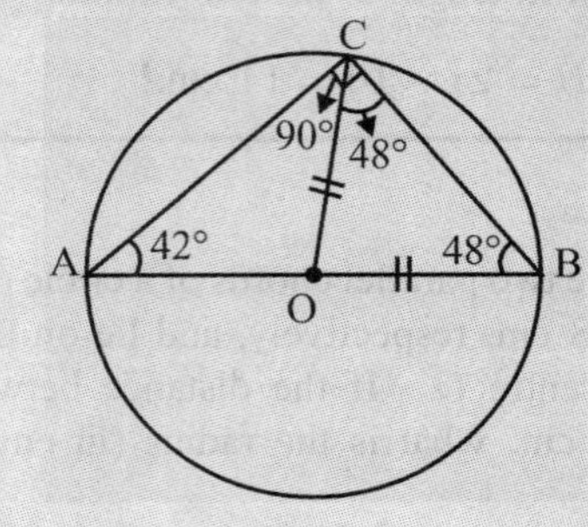

According to the question:

AB = Diameter, $\angle BAC = 42°$, $\angle ACB = 90°$.

(Angle subtended by a diameter on any point of circle is 90°):

In $\triangle ABC$:

$\angle CBA = 180° - (90° + 42°) = 48°$.

Line $OC = OB$ (Radius of circle):

$\angle OBC = \angle BCO = 48°$

$\angle BOC = 180° - 2 \times 48° = 180° - 96° = 84°$.

76. The radii of two concentric circles are 12 cm and 13 cm. AB is a diameter of the bigger circle. BD is a tangent to a smaller circle touching it at D. Find the length (in cm) of AD? (Correct to one decimal place):

दो संकेंद्रित वृत्तों की त्रिज्याएँ 12 सेमी और 13 सेमी हैं। AB बड़े वृत्त का व्यास है। BD एक छोटे वृत्त की स्पर्श रेखा है जो इसे D पर छूती है। AD की लंबाई (सेमी में) ज्ञात कीजिए? (दशमलव के एक स्थान तक शुद्ध):

Solution:

Given that-

Radius of bigger circle = 13 cm

Radius of smaller circle = 12 cm:

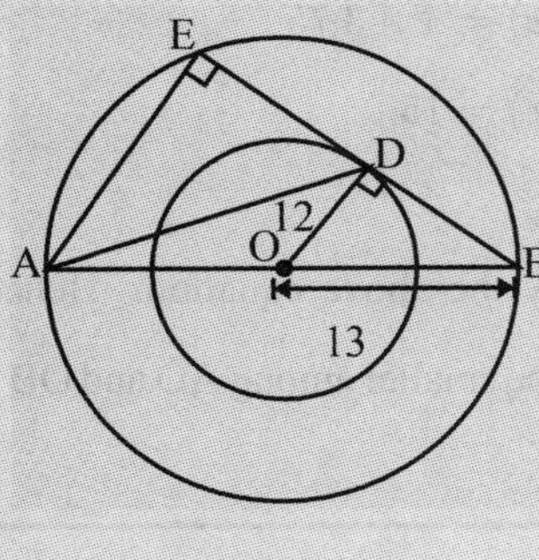

By Pythagoras theorem,

In $\triangle ODB$:

$OB^2 = OD^2 + DB^2$

$= 13^2 + 12^2 + DB^2$.

$DB = 5$ cm.

$DB = DE, \quad DB = \frac{1}{2} \times BE$

(Perpendicular drawn from the centre on a chord bisects it in two equal parts).

$DE = 5$ cm, $\quad BE = 10$ cm.

$\angle AEB = 90°$ [Angle made in semicircle].

$AB^2 = AE^2 + BE^2$

$26^2 = AE^2 + 10^2$

$AE^2 = 676 - 100$

$AE = 24$ cm.

In $\triangle ADE$:

$AD^2 = ED^2 + AE^2$

$AD^2 = 5^2 + 24^2$

$AD = \sqrt{601}$.

$AD = 24.51$ cm.

77. The vertices of a $\triangle ABC$ lie on a circle with centre O. AO is produced to meet the circle at the point P. D is a point on BC such that $AD \perp BC$. If $\angle B = 68°$ and $\angle C = 52°$, then the measure of $\angle DAP$ is:
$\triangle ABC$ के शीर्ष केंद्र O वाले एक वृत्त पर स्थित हैं। AO को बिंदु P पर वृत्त से मिलने के लिए बढ़ाया गया है। D, BC पर एक बिंदु है जैसे कि $AD \perp BC$। यदि $\angle B = 68°$ और $\angle C = 52°$ है, तो $\angle DAP$ का माप है:

Solution:

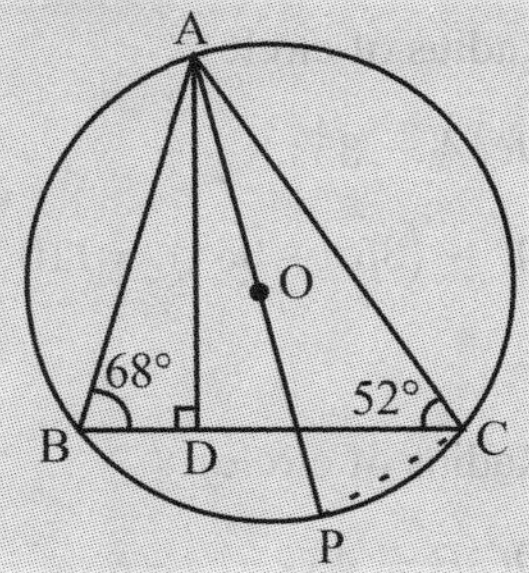

In $\triangle ABD$:

$\angle BAD = 180° - (90° + 68°)$

$= 180° - 158° = 22°$.

Angle subtended by a diameter on any point of the circle is 90°:

$\angle ACP = 90°$.

$\angle BCP = 90° - 52° = 38°$.

Angles subtended by the same arc are equal:

$\angle BAP = \angle BCP = 38°$.

$\angle DAP = 38° - 22° = 16°$.

78. Points A and B are on a circle with centre O. Point C is on the major arc AB. If $\angle OAC = 35°$ and $\angle OBC = 45°$, then what is the measure (in degrees) of the angle subtended by the minor arc AB at the centre?
बिंदु A और B केंद्र O वाले एक वृत्त पर हैं। बिंदु C प्रमुख चाप AB पर है। यदि $\angle OAC = 35°$ और $\angle OBC = 45°$ है, तो लघु चाप AB द्वारा केंद्र पर बनाए गए कोण का माप (डिग्री में) क्या है?

Solution:

Given:

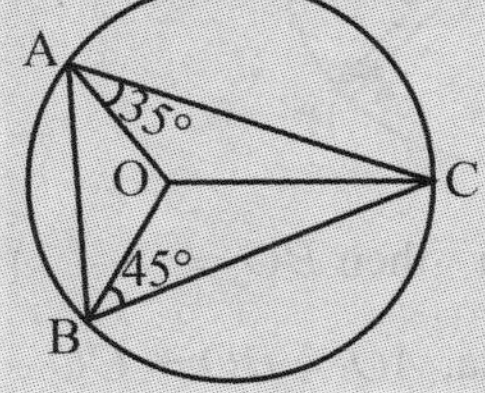

$\angle OAC = 35°$, $\angle OBC = 45°$.

In $\triangle OAC$:

$OA = OC$ (Radius of circle).

$\angle OAC = \angle OCA = 35°$.

$\angle AOC = 180° - (35° + 35°) = 110°$.

In $\triangle OBC$:

$OB = OC$ (Radius of circle).

$\angle OBC = \angle OCB = 45°$.

$\angle BOC = 180° - (45° + 45°) = 90°$.

Now:

$\angle AOB + \angle AOC + \angle BOC = 360°$

$\Rightarrow \angle AOB + 110° + 90° = 360°$

$\angle AOB = 360° - 200° = 160°$.

The angle subtended by the minor arc AB at the centre is 160°.

79. Two chords AB and CD of a circle intersect at O. If CO = 4 cm, OD = 3.75 cm, and AB = 8 cm, then what is the length (in cm) of the smaller among AO and OB?
एक वृत्त की दो जीवाएँ AB और CD O पर प्रतिच्छेद करती हैं। यदि $CO = 4$ सेमी, $OD = 3.75$ सेमी और $AB = 8$ सेमी है, तो AO और OB में से छोटी जीवा की लंबाई (सेमी में) क्या है?

Solution:

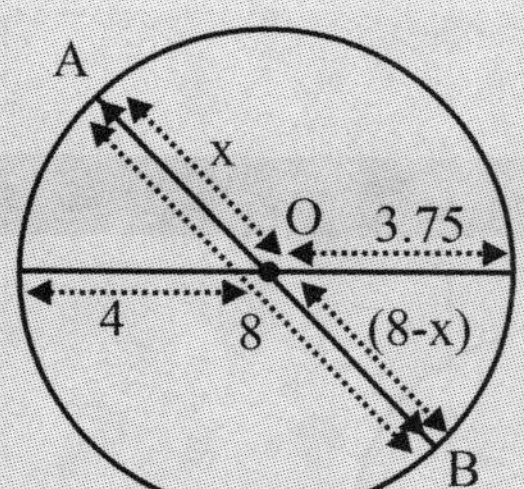

Let $AO = x$ cm, then $BO = (8 - x)$ cm.

From the formula, $AO \times BO = CO \times DO$.

$x \times (8 - x) = 4 \times 3.75$.

$x \times (8 - x) = 15$.

$x = 3$

will satisfy the above equation. Hence, the length of the smaller among AO and OB is AO = 3 cm.

80. If the length of a chord of a circle, that makes an angle of 60° with the tangent drawn at one end point of the chord, is $8\sqrt{3}$ cm, then the radius of the circle will be:
यदि किसी वृत्त की जीवा की लंबाई, जो जीवा के एक छोर पर खींची गई स्पर्शरिखा से 60° का कोण बनाती है, $8\sqrt{3}$ सेमी है, तो वृत्त की त्रिज्या होगी:

Solution:

Given

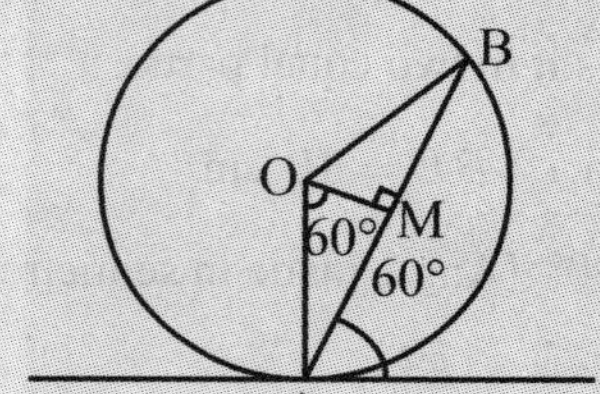

The length of the chord (AB) $= 8\sqrt{3}$ cm.

The length of $AM = \frac{AB}{2} = \frac{8\sqrt{3}}{2} = 4\sqrt{3}$.

In ΔAOM,

$\sin 60° = \frac{AM}{OA}$.

$\frac{\sqrt{3}}{2} = \frac{4\sqrt{3}}{OA}$.

$OA = 2 \times 4 = 8$ cm.

Hence, the radius of the circle is OA = 8 cm.

81. A circle touches all four sides of a quadrilateral $PQRS$. If $PQ = 11$ cm, $QR = 12$ cm and $PS = 8$ cm, then what is the length of RS?
एक वृत्त चतुर्भुज $PQRS$ की चारों भुजाओं को स्पर्श करता है। यदि $PQ = 11$ सेमी, $QR = 12$ सेमी और $PS = 8$ सेमी है, तो RS की लंबाई क्या है?

Solution:

Given:

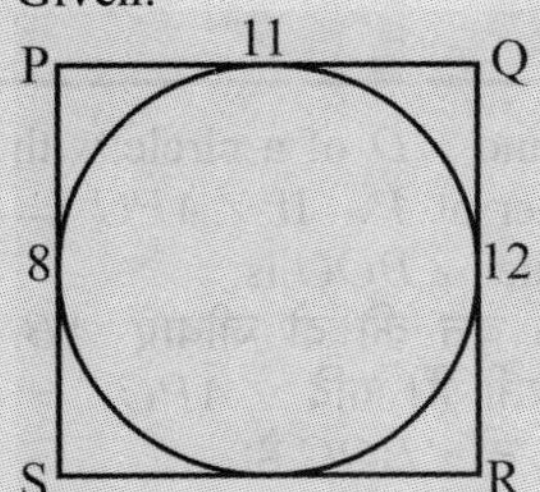

If a circle touches all four sides of a quadrilateral $PQRS$, then:

$PQ + RS = SP + QR$

$11 + RS = 8 + 12$

$RS = 20 - 11$

$RS = 9$ cm.

82. Two circles of radii 4 cm and 3 cm, respectively, touch each other externally. What is the distance (in cm) between their centres?

क्रमशः 4 सेमी और 3 सेमी त्रिज्या वाले दो वृत्त एक दूसरे को बाह्य रूप से स्पर्श करते हैं। उनके केन्द्रों के बीच की दूरी (सेमी में) क्या है?

Solution:

Given:

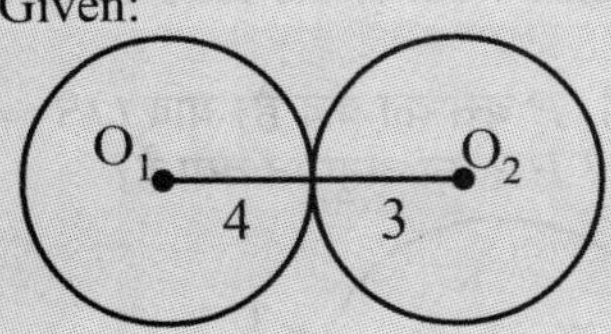

Radius of first circle = 4 cm.

Radius of second circle = 3 cm.

The distance between their centers $(O_1O_2) =$

$4 + 3 = 7$ cm.

83. A triangle $\triangle ABC$ is inscribed in a circle with centre O. AO is produced to meet the circle K and $AD \perp BC$. If $\angle B = 80°$ and $\angle C = 64°$, then the measure of $\angle DAK$ is:

एक त्रिभुज $\triangle ABC$ को केंद्र O वाले एक वृत्त में अंकित किया गया है। AO को वृत्त K और $AD \perp BC$ से मिलने के लिए बढ़ाया गया है। यदि $\angle B = 80°$ और $\angle C = 64°$ है, तो $\angle DAK$ का माप है:

Solution:

Given:

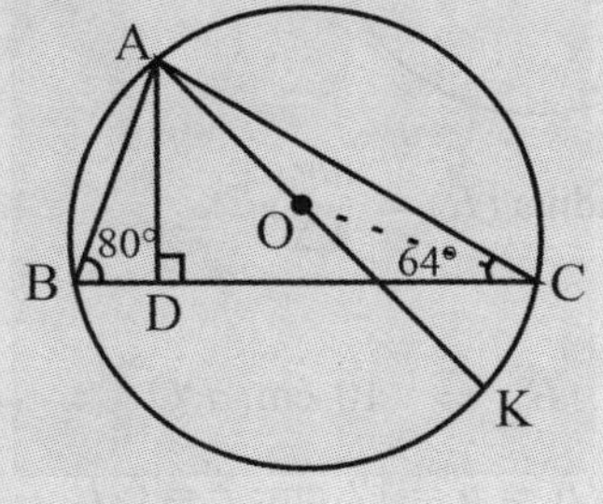

$AD \perp BC$, $\angle B = 80°$, $\angle C = 64°$.

In $\triangle ADC$:

$\angle DAC + \angle ACD + \angle CDA = 180°$

$\angle DAC + 64° + 90° = 180°$

$\angle DAC = 180° - 154° = 26°$.

Again, in $\triangle AOC$:

$\angle COA = 80° \times 2 = 160°$ [from circle property].

$\angle COA + \angle OCA + \angle CAO = 180°$

$\angle OCA + \angle CAO = 180° - 160° = 20°$.

$\angle OCA = \angle CAO = \frac{20°}{2} = 10°$ [OC = OA].

Hence, $\angle DAK = \angle DAC - \angle CAO$:

$\angle DAK = 26° - 10° = 16°$.

84. In a circle of radius 10 cm, with centre O, PO and PR are two chords each of length 12 cm. PO intersects chord QR at the points S. The length of OS is:

10 सेमी त्रिज्या वाले एक वृत्त में, जिसका केंद्र O है, PO और PR दो जीवाएँ हैं जिनमें से प्रत्येक की लंबाई 12 सेमी है। PO जीवा QR को बिंदु S पर प्रतिच्छेद करता है। OS की लंबाई है:

Solution:

We know that:

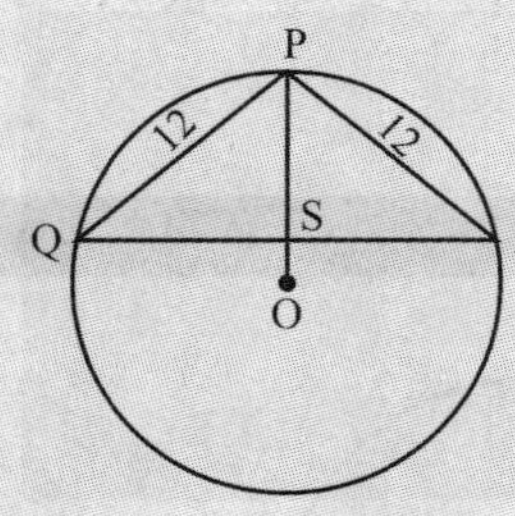

Circumradius $(R) = \dfrac{abc}{4\Delta}$.

Given:

$OP = R = 10$ cm, $PQ = a = 12$ cm, $PR = b = 12$ cm, $c = QR$.

$$10 = \frac{12 \times 12 \times QR}{4 \times \frac{1}{2} \times PS \times QR}$$

$$PS = \frac{12 \times 12}{2 \times 10} \quad PS = 7.2 \text{ cm.}$$

$$OS = OP - PS = 10 - 7.2 = 2.8 \text{ cm.}$$

85. In $\triangle ABC$, $AB = AC$. A circle drawn through B touches AC at D and intersects AB at P. If D is the midpoint of AC and $AP = 2.5$ cm, then AB is equal to:

$\triangle ABC$ में, $AB = AC$। B से होकर खींचा गया एक वृत्त AC को D पर स्पर्श करता है और AB को P पर प्रतिच्छेद करता है। यदि D, AC का मध्य बिंदु है और $AP = 2.5$ सेमी है, तो AB बराबर है:

Solution:

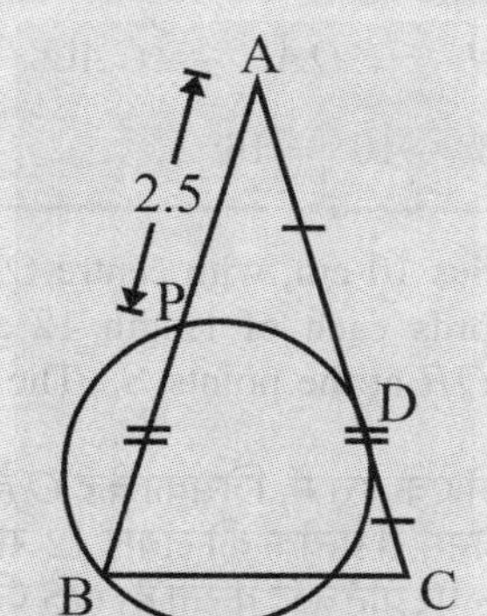

$AD^2 = AP \times AB$ (By using formula).

$\left(\dfrac{AC}{2}\right)^2 = AP \times AB$ (As $AB = AC$).

$AB^2/4 = 2.5 \times AB$

$AB = 10$ cm.

86. Two chords AB and CD of a circle with centre O intersect each other at P. If $\angle APC = 95°$ and $\angle AOD = 110°$, then $\angle BOC$ is:

केंद्र O वाले एक वृत्त की दो जीवाएँ AB और CD P पर प्रतिच्छेद करती हैं। यदि $\angle APC = 95°$ और $\angle AOD = 110°$ है, तो $\angle BOC$ है:

Solution:

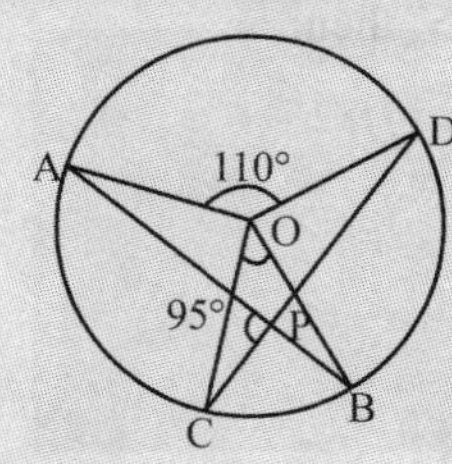

$\angle CPB = 180° - 95° = 85°$ (linear pair).

$\angle CPB = \dfrac{\angle AOD + \angle BOC}{2}$ (by using theorem).

$85° = \dfrac{110° + \angle BOC}{2}$.

$\angle BOC = 170° - 110° = 60°$.

87. In the given figure, P is the centre of the circle. If $QS = PR$, then what is the ratio of $\angle RSP$ to the $\angle TPR$?

दिए गए चित्र में, P वृत्त का केंद्र है। यदि $QS = PR$ है, तो $\angle RSP$ और $\angle TPR$ का अनुपात क्या है?

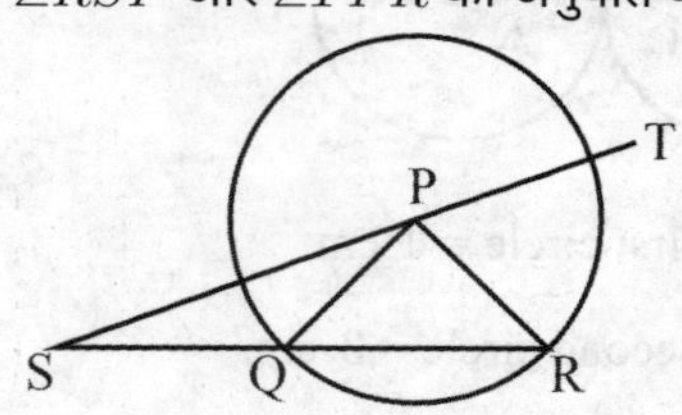

Solution:

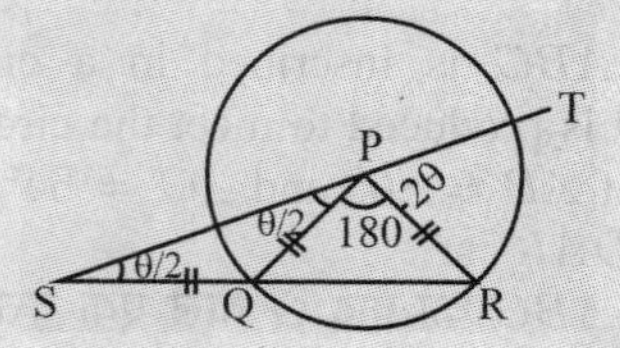

Let, $\angle PSQ = \theta/2$.

$PQ = PR = QS$.

$\angle SPQ = \theta/2$.

By exterior angle property:

$\angle PQR = \theta$.

$\angle QPR = 180° - 2\theta$.

$\angle TPR = 3\theta/2$.

$\frac{\angle RSP}{\angle TPR} = \frac{\theta/2}{3\theta/2} = 1:3$.

88. In the given figure, CD and AB are diameters of circles and AB and CD are perpendicular to each other. LQ and SR are perpendiculars to AB and CD respectively. Radius of circle is 5 cm, $PB : PA = 2 : 3$ and $CN : ND = 2 : 3$. What is the length (in cm) of SM?

दिए गए चित्र में, CD और AB वृत्तों के व्यास हैं और AB और CD एक दूसरे के लंबवत हैं। LQ और SR क्रमशः AB और CD पर लंबवत हैं। वृत्त की त्रिज्या 5 सेमी है, $PB : PA = 2 : 3$ और $CN : ND = 2 : 3$। SM की लंबाई (सेमी में) क्या है?

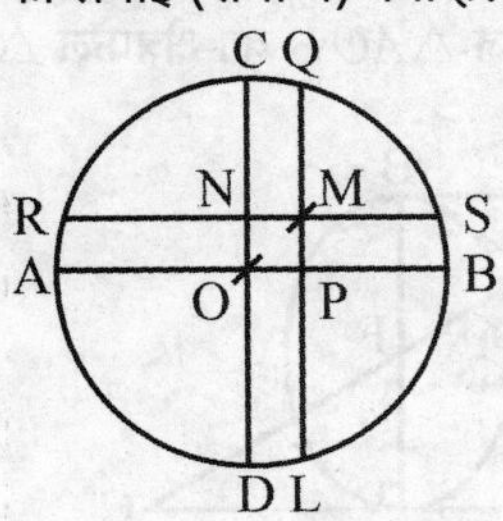

Solution:

Let, $PB = 2x$ and $PA = 3x$.

$PA + PB = AB = 10$ cm.

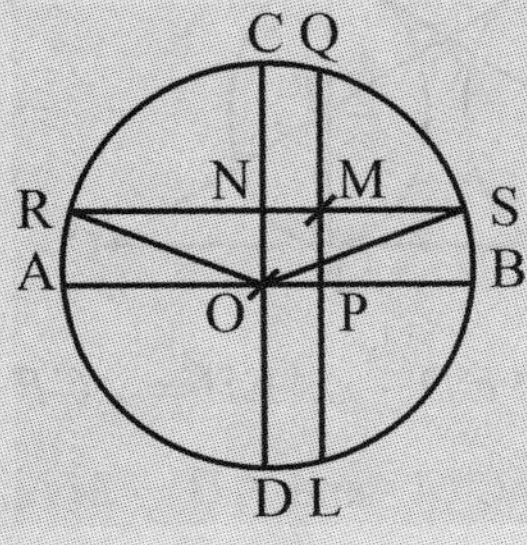

$5x = 10$ cm.

$x = 2$ cm.

$PB = 2 \times 2 = 4$ cm.

$PA = 3 \times 2 = 6$ cm.

Similarly:

$CN = 4$ cm, $ND = 6$ cm.

$ON = OC - NC = 5 - 4 = 1$ cm.

Similarly:

$PO = 6 - 5 = 1$ cm.

$NOPM$ is a square.

In $\triangle ONS$:

$SN^2 = 5^2 - 1^2$.

$SN = 2\sqrt{6}$.

$SM = SN - MN = 2\sqrt{6} - 1$ cm.

89. There are 8 equidistant points A, B, C, D, E, F, G, H (in same order) on a circle. What is the value of $\angle FDH$ (in degree)?

एक वृत्त पर 8 समान दूरी वाले बिंदु A, B, C, D, E, F, G और H (समान क्रम में) हैं। $\angle FDH$ का मान (डिग्री में) क्या है?

Solution:

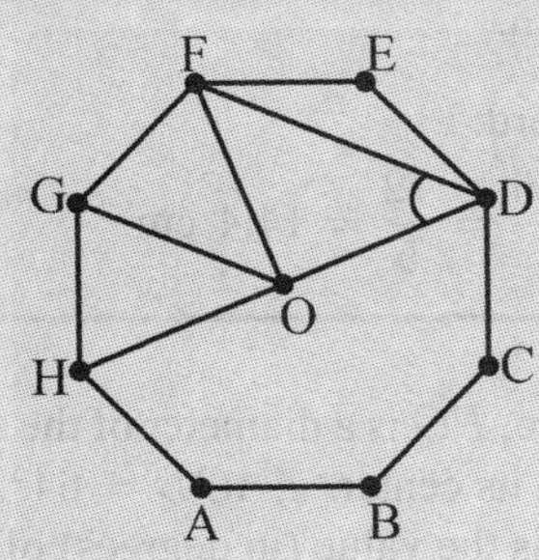

$\angle GOH = \frac{360°}{8} = 45°$.

$\angle FOH = 45° + 45° = 90°$.

Angle subtended by chord on centre is double the angle subtended by same chord on circumference:

$\angle FOH = 2 \times \angle FDH.$

$90^\circ = 2 \times \angle FDH.$

$\angle FDH = 45^\circ.$

90. Two circles are having radii 9 cm and 12 cm. The distance between their centers is 15 cm. What is the length (in cm) of their common chord?
दो वृत्तों की त्रिज्याएँ 9 सेमी और 12 सेमी हैं। उनके केन्द्रों के बीच की दूरी 15 सेमी है। उनकी उभयनिष्ठ राग की लंबाई (सेमी में) क्या है?

Solution:

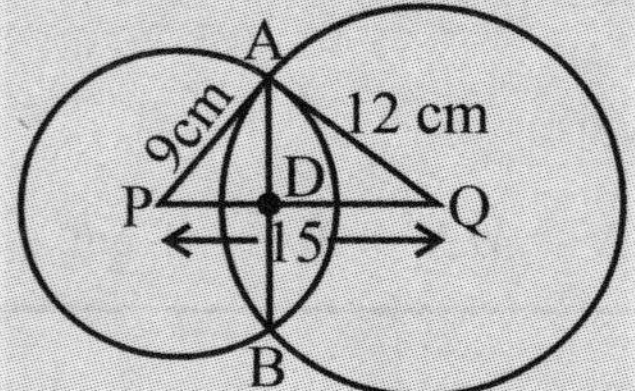

9, 12, and 15 cm are the sides of a right-angled triangle.

Area of $\triangle APQ = \frac{1}{2} \times 9 \times 12.$

$\frac{1}{2} \times PQ \times AD = \frac{1}{2} \times 9 \times 12.$

$\frac{1}{2} \times 15 \times AD = \frac{1}{2} \times 9 \times 12.$

$AD = \frac{36}{5}$ cm.

Common chord:

$AB = \frac{2 \times 36}{5} = \frac{72}{5} = 14.4$ cm.

91. In the given figure, PQ is a diameter of the semicircle $PABQ$ and O is its center. $\angle AOB = 64^\circ$. BP cuts AQ at X. What is the value (in degrees) of $\angle AXP$?
दिए गए चित्र में, PQ अर्धवृत्त $PABQ$ का व्यास है और O इसका केंद्र है। $\angle AOB = 64^\circ$। BP AQ को X पर काटता है। $\angle AXP$ का मान (डिग्री में) क्या है?

Solution:

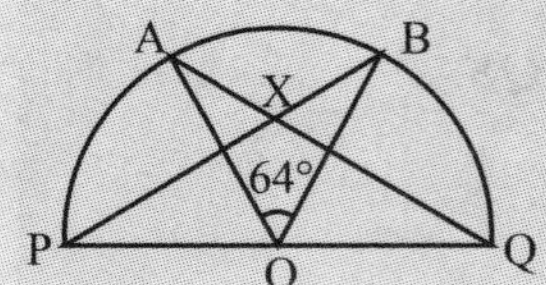

PQ is diameter.

If line AP is joined: $\angle APB = \frac{1}{2}\angle AOB = 32^\circ.$

Angle formed on the center is double the angle formed on the circumference by the same chord in the same segment. $\angle PAQ = 90^\circ$ (Angle formed in semicircle is right angle).

$\angle AXP = 180^\circ - (\angle APB + \angle PAQ).$

$\angle AXP = 180^\circ - (90^\circ + 32^\circ) = 58^\circ.$

92. In the given figure, E and F are the centers of two identical circles. What is the ratio of the area of $\triangle AOB$ to the area of $\triangle DOC$?
दिए गए चित्र में, E और F दो समान वृत्तों के केंद्र हैं। क्षेत्रफल का अनुपात क्या है? त्रिभुज $\triangle AOB$ का क्षेत्रफल $\triangle DOC$ के क्षेत्रफल से?

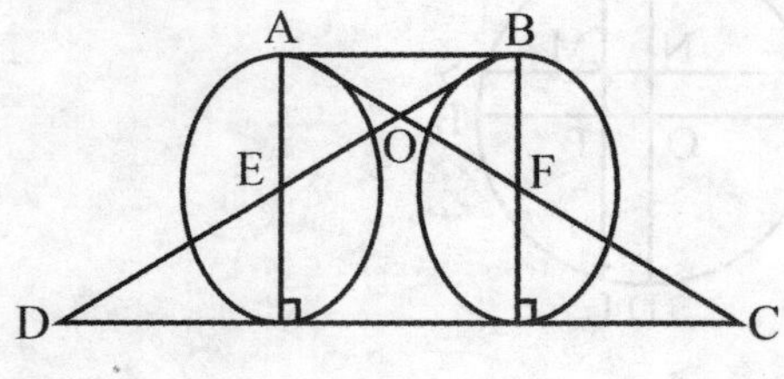

Solution:

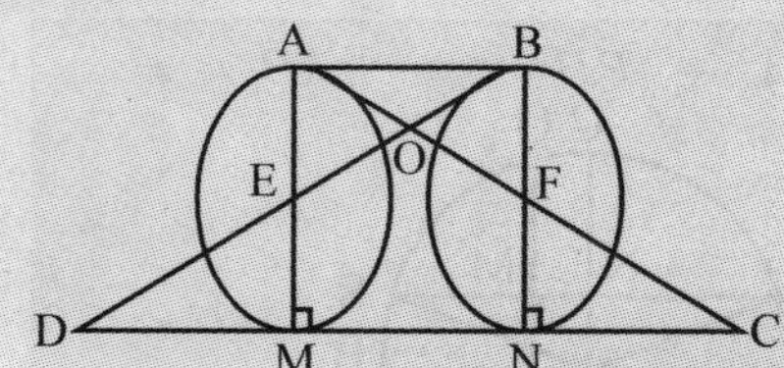

$\angle AEB = \angle MED, \quad \angle AFB = \angle CFN.$

$AB = DM = CN,$ and $AB = MN.$

$$\frac{AB}{DM+MN+NC}=\frac{AB}{DC}=\frac{1}{3}.$$

$$\frac{\text{area }\triangle ADB}{\text{area }\triangle DOC}=\frac{AB^2}{DC^2}=\left(\frac{1}{3}\right)^2=\frac{1}{9}\text{ or }1:9.$$

93. An equilateral triangle ABC is inscribed in a circle as shown in the figure. A square of the largest possible area is made inside this triangle as shown. Another circle inscribing the square is made. What is the ratio of the larger circle to the smaller circle?

जैसा कि चित्र में दिखाया गया है, एक समबाहु त्रिभुज ABC एक वृत्त में अंकित है। दिखाए गए अनुसार इस त्रिभुज के अंदर सबसे बड़े संभावित क्षेत्रफल का एक वर्ग बनाया गया है। वर्ग अंकित करते हुए एक और वृत्त बनाया गया। बड़े वृत्त और छोटे वृत्त का अनुपात क्या है?

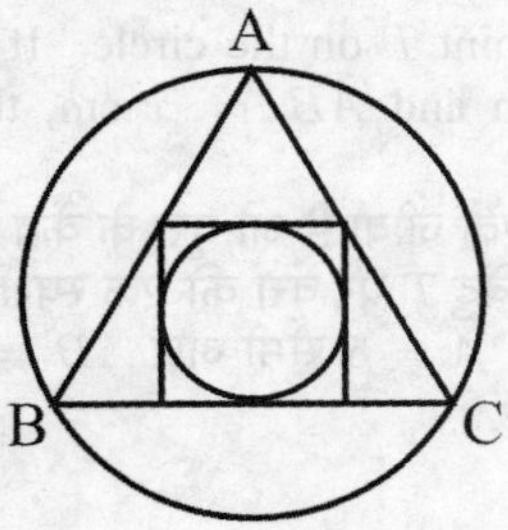

Solution:

If the side of the equilateral triangle is a, then the side of the square of maximum area drawn inside the triangle is: $x=a\sqrt{3}(2-\sqrt{3})$.

Radius of the smaller circle: $r_s=\dfrac{a\sqrt{3}(2-\sqrt{3})}{2}$.

Radius of the larger circle: $r_l=\dfrac{a}{\sqrt{3}}$.

Ratio of areas: $$\frac{r_l^2}{r_s^2}=\frac{\left(\frac{a}{\sqrt{3}}\right)^2}{\left(\frac{a\sqrt{3}(2-\sqrt{3})}{2}\right)^2}.$$

$$=\frac{a^2/3}{(3a^2(7-4\sqrt{3}))/4}.$$

$$=\frac{4}{63-36\sqrt{3}}.$$

94. In the given figure, $AB=30$ cm and $CD=24$ cm. What is the value (in cm) of MN?

दी गई आकृति में, $AB=30$ सेमी और $CD=24$ सेमी है। MN का मान (सेमी में) क्या है?

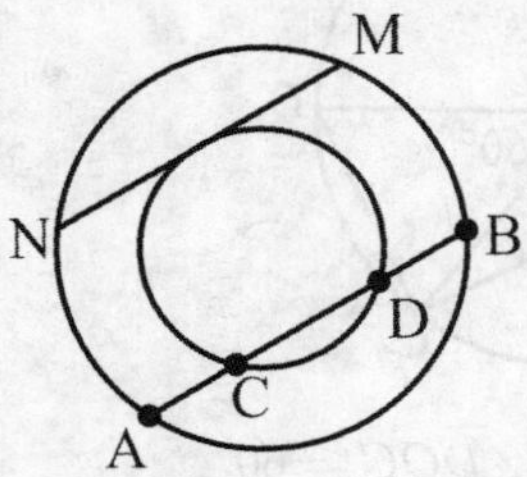

Solution:

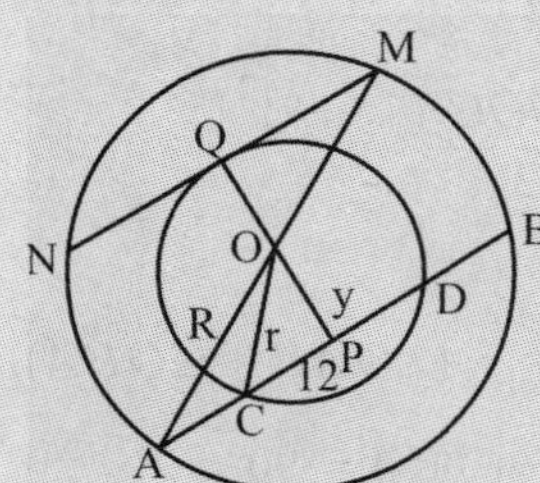

$R^2=15^2+y^2$ -------(i).

$r^2=12^2+y^2$ -------(ii).

On subtracting equations (i) and (ii): $R^2-r^2=15^2+y^2-12^2-y^2$.

$R^2-r^2=225-144$.

$R^2-r^2=81$.

$QM=\sqrt{R^2-r^2}=9$ cm.

$MN=2\times 9=18$ cm.

95. In a circle with centre O, AD is a diameter and AC is a chord. Point B is on AC such that $OB=7$ cm and $\angle OBA=60°$. If $\angle DOC=60°$, then what is the length of BC?

केंद्र O वाले वृत्त में, AD एक व्यास है और AC एक जीवा है। बिंदु B, AC पर इस प्रकार है कि $OB=7$ सेमी और $\angle OBA=60°$ है। यदि $\angle DOC=60°$ है, तो BC की लंबाई क्या है?

Solution:

Given, OB = 7 cm.

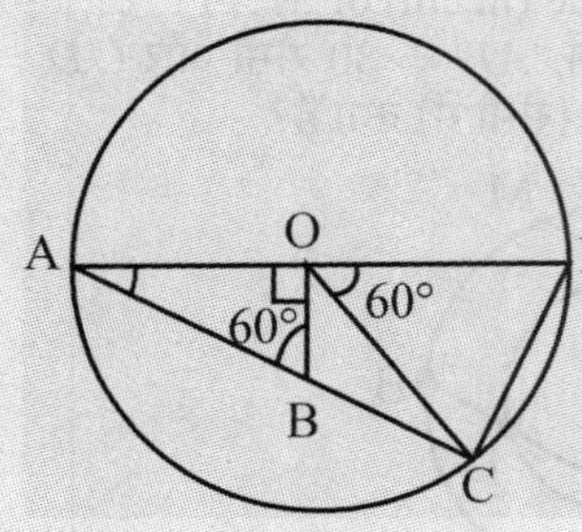

$\angle OBA = 60°$, $\angle DOC = 60°$.

On joining DC:

$\angle DOC = 60°$ (Angle formed on center).

In $\triangle DAC$:

$\angle DAC = \dfrac{\angle DOC}{2}$

$= \dfrac{60°}{2} = 30°$.

$\angle AOB = 180° - (\angle OAB + \angle OBA)$.

$= 180° - (30° + 60°) = 90°$.

In $\triangle OBC$:

$\angle OBC = 180° - \angle OBA$.

$= 180° - 60° = 120°$.

$\angle BOC = 180° - (\angle AOB + \angle DOC)$.

$= 180° - (90° + 60°) = 30°$.

$\angle BCO = 180° - (\angle OBC + \angle BOC)$.

$= 180° - (120° + 30°) = 30°$.

Again, in $\triangle OBC$:

$\angle BOC = \angle OCB$ (Corresponding angles are equal to corresponding sides).

$OB = BC$.

$BC = 7$ cm.

96. The length of the tangent drawn to a circle of radius 12 cm from a point 18 cm away from the centre of the circle is:
वृत्त के केंद्र से 18 सेमी दूर एक बिंदु से 12 सेमी त्रिज्या वाले वृत्त पर खींची गई स्पर्श रेखा की लंबाई है:

Solution:

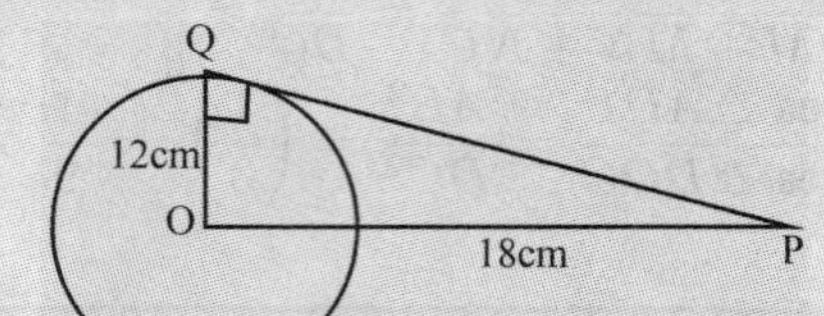

$PQ^2 = 18^2 - 12^2 = (18 - 12)(18 + 12) = 6 \times 30$.

$PQ = 6\sqrt{5}$ cm.

Hence, the length of tangent $= 6\sqrt{5}$ cm.

97. In the given figure, TB is a chord which passes through the center of the circle. PT is a tangent to the circle at the point T on the circle. If $PT = 10$ cm, $PA = 5$ cm and $AB = x$ cm, then the radius of the circle is:
दिए गए चित्र में, TB एक जीवा है जो वृत्त के केंद्र से होकर गुजरती है। PT वृत्त के बिंदु T पर वृत्त की एक स्पर्श रेखा है। यदि $PT = 10$ सेमी, $PA = 5$ सेमी और $AB = x$ सेमी है, तो वृत्त की त्रिज्या है:

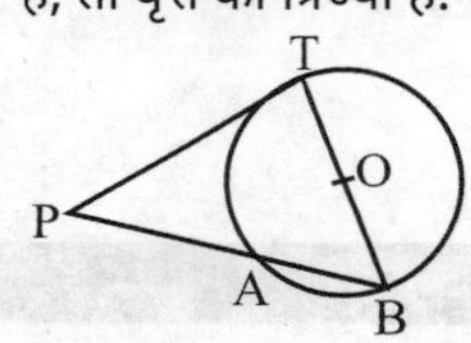

Solution:

By using the theorem, $PT^2 = PA \times PB$.

$10^2 = 5 \times (x + 5)$.

$100 = 5x + 25$.

$x = \dfrac{75}{5} = 15$ cm.

Now, in the right-angled triangle P: $PT = 10$, $PB = 5 + 15 = 20$ cm.

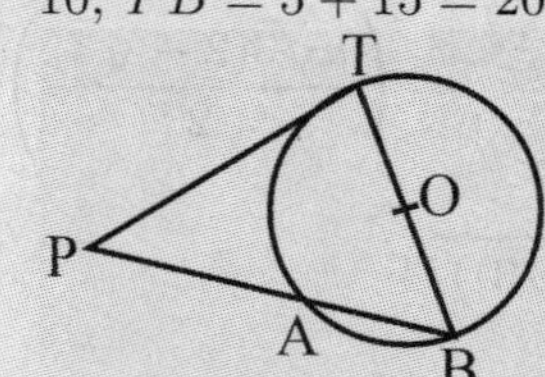

$TB = \sqrt{PB^2 - PT^2}$.

$TB = \sqrt{20^2 - 10^2}.$

$TB = \sqrt{400 - 100}.$

$TB = \sqrt{300} = 10\sqrt{3}.$

$\text{Radius} = \frac{TB}{2} = \frac{10\sqrt{3}}{2} = 5\sqrt{3}$ cm.

98. Two circles touch each other externally at T. RS is a direct common tangent to the two circles touching the circles at P and Q. $\angle TPQ = 42°$. $\angle PQT$ (in degrees) is:
दो वृत्त एक-दूसरे को बाह्य रूप से T पर स्पर्श करते हैं। RS, P और Q पर वृत्तों को स्पर्श करने वाले दो वृत्तों की सीधी उभयनिष्ठ स्पर्श रेखा है। $\angle TPQ = 42°$। $\angle PQT$ (डिग्री में) है:

Solution:

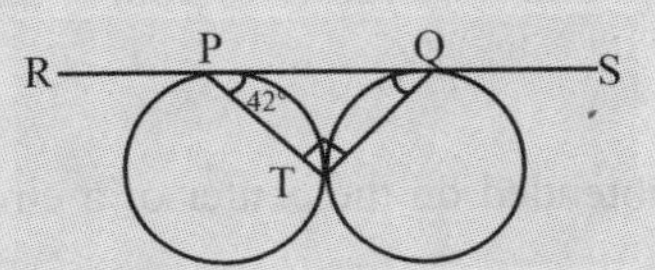

In $\triangle PTQ$: $\angle PTQ = 90°$.

$\angle TPQ = 42°$.

$\angle PTQ + \angle TPQ + \angle TQP = 180°$.

$90° + 42° + \angle TQP = 180°$.

$\angle TQP = 180° - 132° = 48°$.

99. AB is 12 cm long chord of a circle with center O and radius 10 cm. The tangents at A and B intersect at P. What is the length of OP?
केंद्र O और त्रिज्या 10 सेमी वाले एक वृत्त की AB = 12 सेमी लंबी जीवा है। A और B पर स्पर्श रेखाएँ P पर प्रतिच्छेद करती हैं। OP की लंबाई क्या है?

Solution:

AB is the chord of the circle with center O:

$OP \perp AB$ so OP bisects AB.

$AL = BL = \frac{12}{2} = 6$ cm.

$OL = \sqrt{OA^2 - AL^2}.$

$OL = \sqrt{10^2 - 6^2}.$

$OL = \sqrt{100 - 36}.$

$OL = 8$ cm.

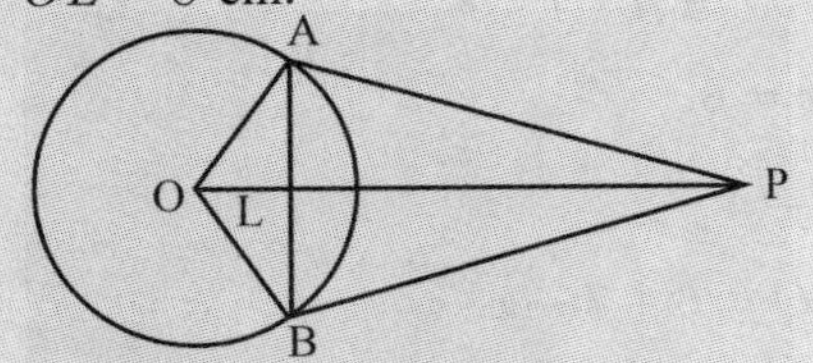

In $\triangle OAP$ and $\triangle OLA$: $\angle OAP = \angle OLA$.

$OP = OA$.

$OA = OA$ (common side).

$\triangle OAP \sim \triangle OLA$.

$\frac{OP}{OA} = \frac{OA}{OL}.$

$\frac{OP}{10} = \frac{10}{8}.$

$OP = \frac{100}{8} = 12.5$ cm.

100. $\triangle ABC$ has sides 5 cm, 6 cm, and 7 cm. AB extended touches a circle at P and AC extended touches the same circle at Q. Find the length (in cm) of AQ:
$\triangle ABC$ की भुजाएँ 5 सेमी, 6 सेमी, और 7 सेमी हैं। AB बढ़ाया गया वृत्त को P पर स्पर्श करता है और AC बढ़ाया गया उसी वृत्त को Q पर स्पर्श करता है। AQ की लंबाई (सेमी में) ज्ञात कीजिए।

Solution:

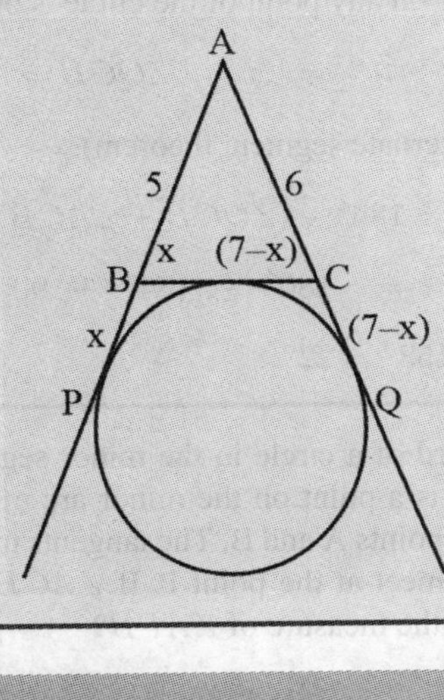

$AP = AQ.$

$5 + x = 6 + 7 - x.$

$2x = 8.$

$x = 4.$

$AQ = 6 + 7 - x.$

$AQ = 6 + 7 - 4.$

$AQ = 9$ cm.

101. AB is a diameter of the circle with centre O. The tangent at the point C on the circle meets AB produced at Q. If $\angle BAC = 34°$, then the measure of $\angle CQA$ (in degrees) will be:

AB केंद्र O वाले वृत्त का एक व्यास है। वृत्त पर बिंदु C पर स्पर्शरेखा, Q पर निर्मित AB से मिलती है। यदि $\angle BAC = 34°$ है, तो $\angle CQA$ का माप (डिग्री में) होगा:

Solution:

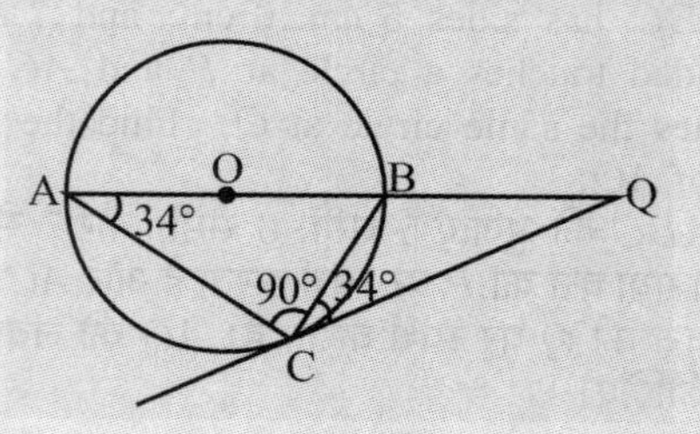

Given, $\angle BAC = 34°$.

$\angle ACB = 90°$ Angle subtended by a diameter on any point of the circle is 90°.

$\angle BAC = \angle QCB = 34°$ (Alternate segment theorem).

$\angle CQA = 180° - (\angle QCB + \angle ACB).$

$\angle CQA = 180° - (34° + 90°) = 180° - 158° = 22°.$

102. AB is a chord of a circle in the minor segment with center O. C is a point on the minor arc of the circle between the points A and B. The tangents to the circle at A and B meet at the point P. If $\angle ACB = 102°$, then what is the measure of $\angle APB$?

केंद्र O वाले लघु खंड में AB एक वृत्त की जीवा है। C, बिंदु A और B के बीच वृत्त के लघु चाप पर एक बिंदु है। A और B पर वृत्त की स्पर्शरिखाएँ बिंदु P पर मिलती हैं। यदि $\angle ACB = 102°$, तो $\angle APB$ का माप क्या है?

Solution:

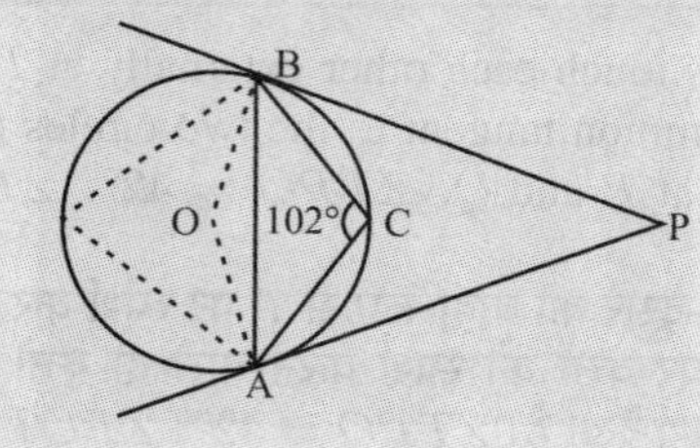

Sum of opposite angles of a cyclic quadrilateral is 180°.

$\angle ADB = 180° - \angle ACB = 180° - 102° = 78°.$

Angle subtended on the center of a circle by an arc is double the angle subtended on the circumference.

$\angle AOB = 2 \times \angle ADB = 2 \times 78° = 156°.$

$\angle AOB + \angle APB = 180°$ (By theorem).

$\angle APB = 180° - 156° = 24°.$

103. Chord AB of a circle of radius 10 cm is at a distance 8 cm from the centre O. If tangents drawn at A and B intersect at P, then the length of the tangent AP (in cm) is:

10 सेमी त्रिज्या वाले एक वृत्त की जीवा AB केंद्र O से 8 सेमी की दूरी पर है। यदि A और B पर खींची गई स्पर्शरिखाएँ P पर प्रतिच्छेद करती हैं, तो स्पर्शरिखा AP की लंबाई (सेमी में) है:

Solution:

$AQ = \sqrt{10^2 - 8^2} = \sqrt{100-64} = \sqrt{36} =$

6 cm.

$\angle APQ = \angle OAQ$ (By theorem).

$\triangle OAQ \sim \triangle APQ.$

$\frac{QO}{AQ} = \frac{AO}{AP}.$

$\frac{8}{6} = \frac{10}{AP}.$

$AP = \frac{6 \times 10}{8} = 7.5$ cm.

104. In the given figure, ABC is a right-angled triangle. $\angle ABC = 90°$ and $\angle ACB = 60°$. If the radius of the smaller circle is 2 cm, then what is the radius (in cm) of the larger circle?
दी गई आकृति में, ABC एक समकोण त्रिभुज है। $ABC = 90°$ और $ACB = 60°$। यदि छोटे वृत्त की त्रिज्या 2 सेमी है, तो बड़े वृत्त की त्रिज्या (सेमी में) क्या है?

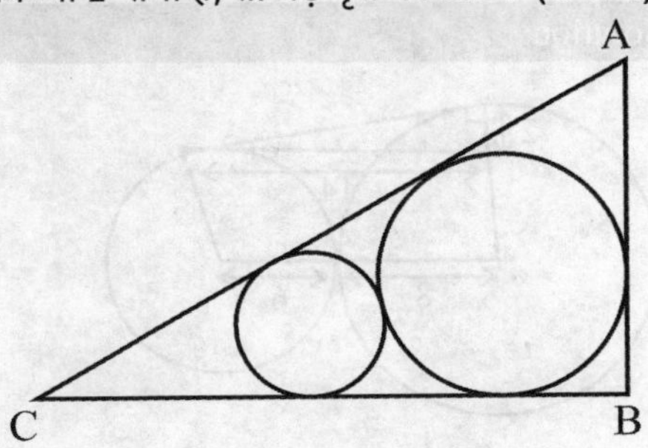

Solution:

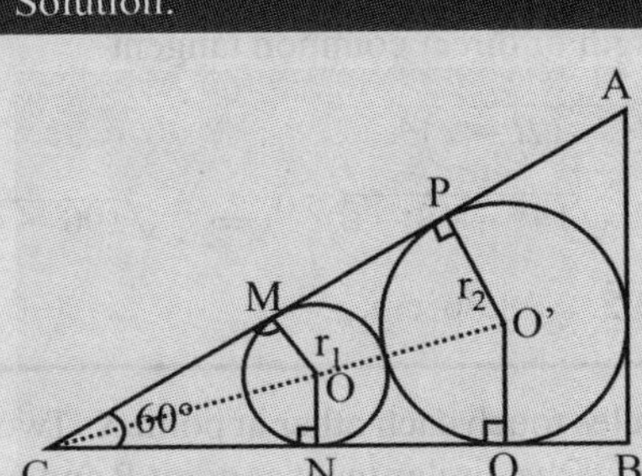

$\angle MCO = \angle PCO = \frac{\angle ACB}{2} = 30°.$

$\sin 30° = \frac{MO}{OC}.$

$\frac{1}{2} = \frac{2}{OC}.$

$OC = 4$ cm.

$\sin 30° = \frac{PO}{OC}.$

$\frac{1}{2} = \frac{r_2}{6+r_2}.$

$6 + r_2 = 2r_2.$

$r_2 = 6$ cm.

105. A circle touches the side BC of $\triangle ABC$ at D and AB and AC are produced to E and F, respectively. If $AB = 10$ cm, $AC = 8.6$ cm, and $BC = 6.4$ cm, then $BE = ?$:
एक वृत्त $\triangle ABC$ की भुजा BC को D पर स्पर्श करता है और AB और AC को क्रमश: E और F तक बढ़ाया जाता है। यदि $AB = 10$ सेमी, $AC = 8.6$ सेमी, $BC = 6.4$ सेमी, तो $BE = ?$:

Solution:

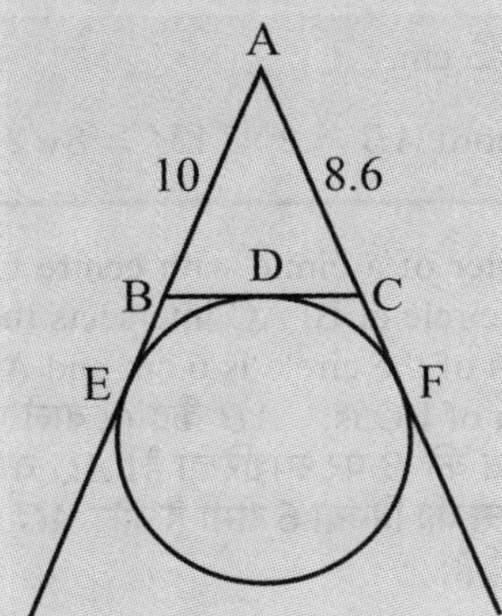

$BD = BE$ and $CD = CF$ and $AE = AF$.

$AB + BE = AC + CF.$

$AB + 2BE = AC + BC.$

$BE = \frac{AC + BC - AB}{2}.$

$BE = \frac{8.6 + 6.4 - 10}{2} = \frac{5}{2} = 2.5$ cm.

106. Two circles of radius 4 cm and 6 cm touch each other internally. What is the length (in cm) of the longest chord of the outer circle, which is also a tangent to the inner circle? 4 सेमी और 6 सेमी त्रिज्या वाले दो वृत्त एक दूसरे को आंतरिक रूप से स्पर्श करते हैं। बाहरी वृत्त की सबसे लंबी जीवा की लंबाई (सेमी में) क्या है, जो आंतरिक वृत्त की स्पर्शरेखा भी है?

Solution:

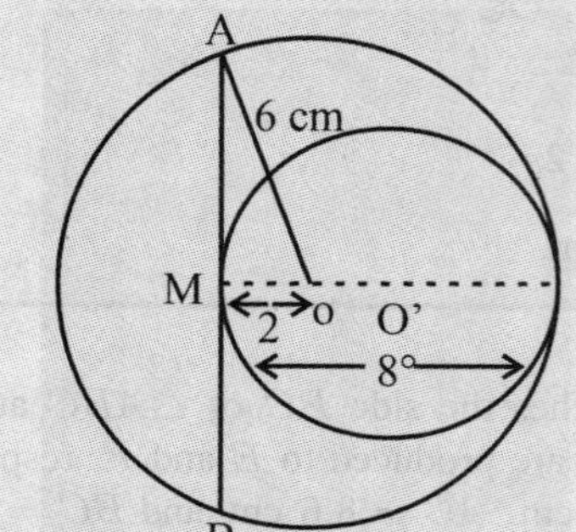

Let the centre of the larger circle be O and the smaller circle be O'. Radius of the circle is perpendicular to the tangent.

In ΔAMO, $AM^2 = OA^2 - OM^2 = 36 - 4 = 32$.

$AM = 4\sqrt{2}$ cm.

Longest chord $AB = 2 \times AM = 8\sqrt{2}$ cm.

107. AB is a diameter of a circle with centre O. CB is a tangent to the circle at B. AC intersects the circle at G. If the radius of the circle is 6 cm and AG = 8 cm, then the length of BC is: AB केंद्र O वाले एक वृत्त का व्यास है। CB वृत्त की B पर स्पर्शरेखा है। AC वृत्त को G पर काटता है। यदि वृत्त की त्रिज्या 6 सेमी है और AG = 8 सेमी है, तो BC की लंबाई है:

Solution:

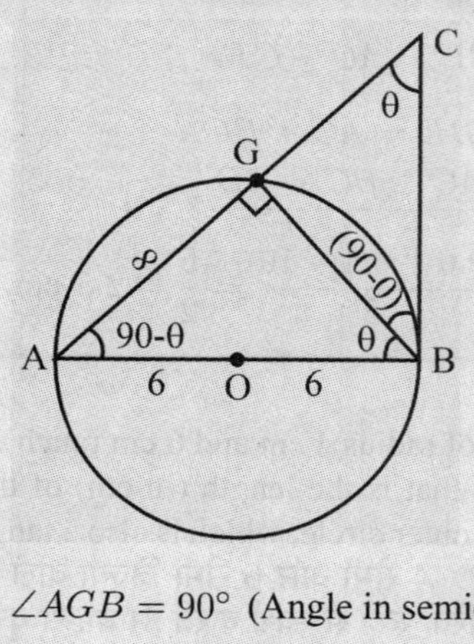

$\angle AGB = 90°$ (Angle in semicircle).

Δ AGB is a right-angled triangle.

By Pythagoras theorem,

$GB^2 = AB^2 - AG^2 = 12^2 - 8^2 = 144 - 64 = 80$.

$GB = \sqrt{80} = 4\sqrt{5}$.

$\Delta BGA \sim \Delta CGB$.

Hence, $\frac{AB}{BC} = \frac{GA}{GB}$.

$\frac{12}{BC} = \frac{8}{4\sqrt{5}}$.

$BC = \frac{12 \times 4\sqrt{5}}{8} = 6\sqrt{5}$ cm.

108. Two circles of radii 8 cm and 6 cm touch each other externally. The length of the direct common tangent is: 8 सेमी और 6 सेमी त्रिज्या वाले दो वृत्त एक दूसरे को बाह्य रूप से स्पर्श करते हैं। प्रत्यक्ष उभयनिष्ठ स्पर्शरेखा की लंबाई है:

Solution:

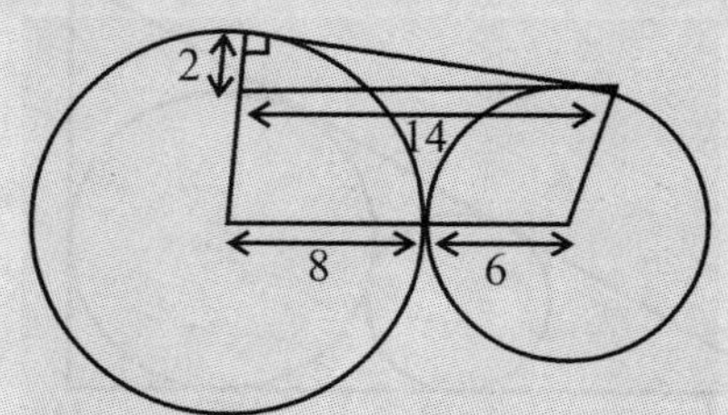

Given, $d = 8 + 6 = 14$ cm, $R = 8$ cm, $r = 6$ cm.

Length of direct common tangent $= \sqrt{d^2 - (R - r)^2}$.

$= \sqrt{14^2 - (8 - 6)^2} = \sqrt{196 - 4} = \sqrt{192} = 13.86$ cm.

109. Two circles touch each other at point X. Two common tangents of the circles meet at point P and none of the tangents passes through X. These tangents touch the larger circle at points B and C. If the radius of the larger circle is 15 cm and CP = 20 cm, then what is the radius (in cm) of the smaller circle? दो वृत्त एक दूसरे को बिंदु X पर स्पर्श करते हैं। वृत्त की दो उभयनिष्ठ स्पर्श रेखाएँ बिंदु P पर मिलती हैं और कोई भी स्पर्श रेखा इससे होकर नहीं गुजरती है। ये स्पर्शरेखाएँ बड़े वृत्त को बिंदु B और C पर स्पर्श करती हैं। यदि बड़े वृत्त की त्रिज्या 15 सेमी और CP =

20 सेमी है, तो छोटे वृत्त की त्रिज्या (सेमी में) क्या है?

Solution:

Radius of a circle is perpendicular to its tangent.

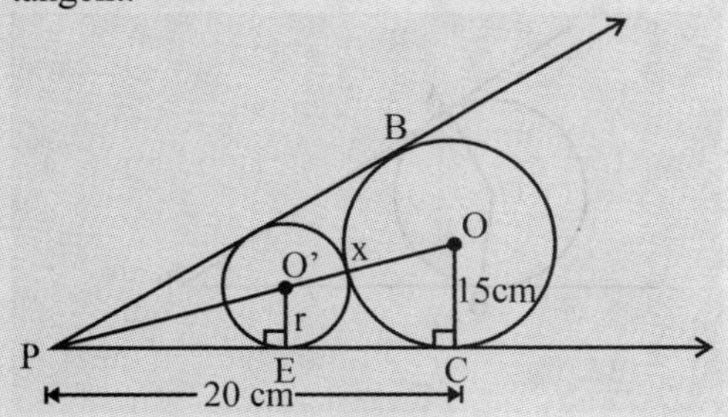

$\angle OCP = \angle O'$

$EP = 90°$.

In ΔOCP,

$OP^2 = 15^2 + 20^2 = 225 + 400 = 625$.

$OP = 25$ cm.

$O'P = 10 - r$.

By similarity theorem,

$$\frac{O'P}{OP} = \frac{r}{15}.$$

$$\frac{10 - r}{25} = \frac{r}{15}.$$

$5r = 30 - 3r$.

$8r = 30$

$r = 3.75$ cm.

110. PA and PB are two tangents drawn to two circles of radius 3 cm and 5 cm respectively. PA touches the smaller and larger circles at points X and Y respectively. PB touches the smaller and larger circles at points U and V respectively. The centres of the smaller and larger circles are O and N respectively. If ON = 12 cm, then what is the value (in cm) of PY?
PA और PB क्रमश: 3 सेमी और 5 सेमी त्रिज्या वाले दो वृत्तों पर खींची गई दो स्पर्शरेखाएँ हैं। PA छोटे और बड़े वृत्तों को क्रमश: बिंदु X और Y पर स्पर्श करता है। PB छोटे और बड़े वृत्त को क्रमश: बिंदु U और V पर स्पर्श करता है। छोटे और बड़े वृत्तों के केंद्र क्रमश: O और N हैं। यदि ON = 12 सेमी, तो PY का मान (सेमी में) क्या है?

Solution:

Let $PO = x$ cm.

$\Delta POX \sim \Delta PNY$.

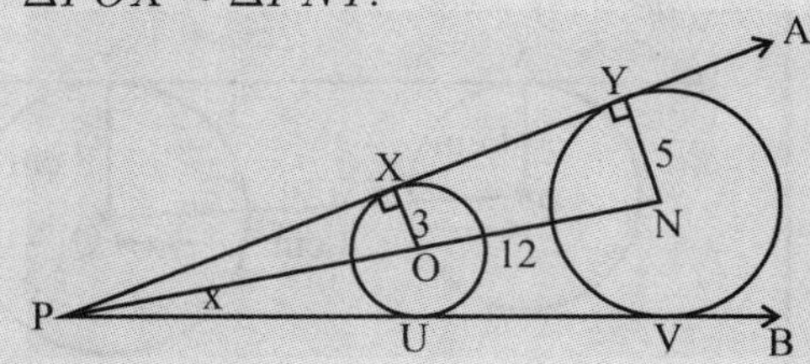

$$\frac{PO}{PN} = \frac{OX}{NY}.$$

$$\frac{x}{x+12} = \frac{3}{5}.$$

$5x = 3x + 36 \Rightarrow 2x = 36 \Rightarrow x = 18$ cm.

$PN = x + 12 = 18 + 12 = 30$ cm.

In ΔPNY, $PY^2 = PN^2 - NY^2 = 30^2 - 5^2 = 900 - 25 = 875$.

$PY = \sqrt{875} = 5\sqrt{35}$ cm.

111. The distance between the centres of two circles is 24 cm. If the radius of the two circles are 4 cm and 8 cm, then what is the sum of the lengths (in cm) of the direct common tangent and the transverse common tangent?
दो वृत्तों के केन्द्रों के बीच की दूरी 24 सेमी है। यदि दो वृत्तों की त्रिज्या 4 सेमी और 8 सेमी है, तो सीधी उभयनिष्ठ स्पशरिखा और अनुप्रस्थ उभयनिष्ठ स्पशरिखा की लंबाई (सेमी में) का योग क्या है?

Solution:

Length of direct common tangent $= \sqrt{d^2-(R-r)^2}$.

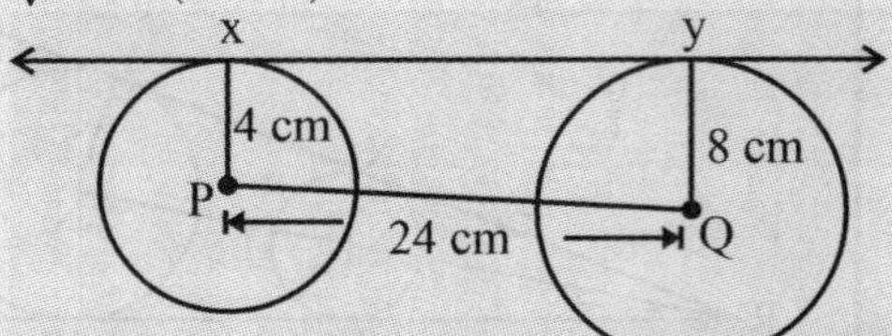

$= \sqrt{24^2-(8-4)^2} = \sqrt{576-16} = \sqrt{560} = 4\sqrt{35}$ cm.

Length of transverse common tangent $= \sqrt{d^2-(R+r)^2}$.

$= \sqrt{24^2-(8+4)^2} = \sqrt{576-144} = \sqrt{432} = 12\sqrt{3}$ cm.

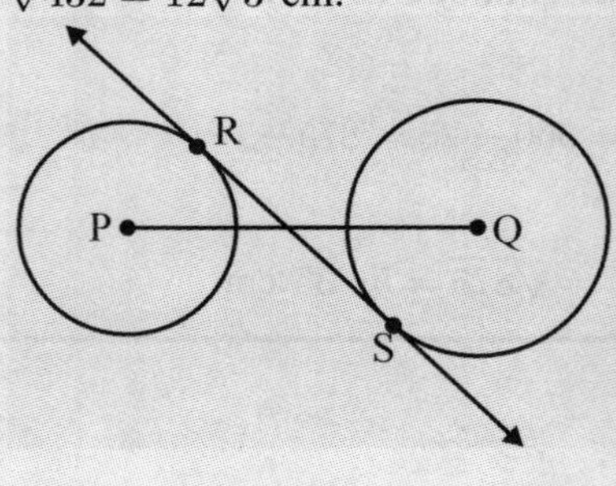

Required sum $= 4\sqrt{35} + 12\sqrt{3} = 4(\sqrt{35}+3\sqrt{3})$ cm.

112. PA and PB are tangents to a circle and O is the centre of the circle. The radius is 5 cm and PO is 13 cm. If the area of the triangle PAB is M, then the value of $\sqrt{M/15}$ is:

PA और PB वृत्त की स्पर्श रेखाएँ हैं और O वृत्त का केंद्र है। त्रिज्या 5 सेमी है और PO 13 सेमी है। यदि त्रिभुज PAB का क्षेत्रफल M है, तो $\sqrt{M/15}$ का मान है:

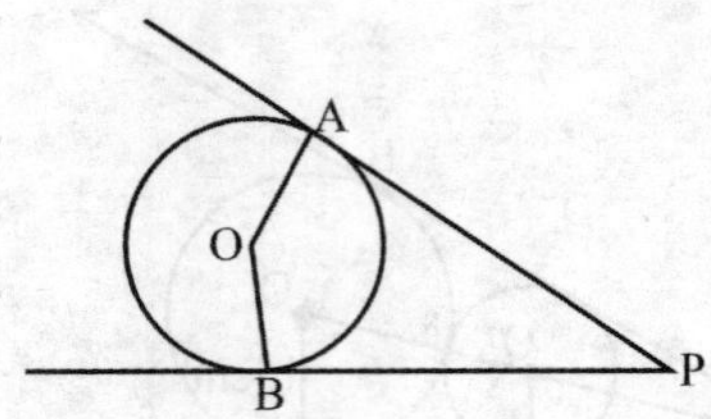

Solution:

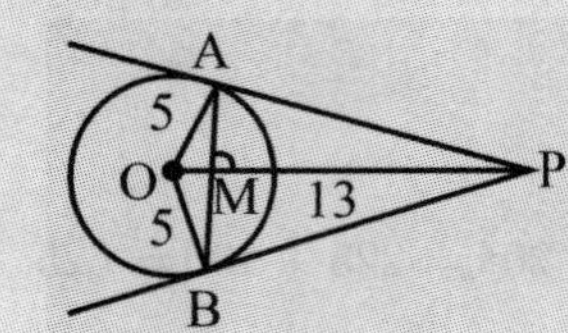

$AP = \sqrt{PO^2-OQ^2} = \sqrt{13^2-5^2} = \sqrt{169-25} = \sqrt{144} = 12$ cm.

Length of AM $= \dfrac{\text{Perpendicular} \times \text{Base}}{\text{Hypotenuse}}$

$\dfrac{5\times12}{13} = \dfrac{60}{13}$.

$AB = 2\times \text{AM} = \dfrac{120}{13}$.

In ΔOAM, $\quad AO^2 = OM^2 + AM^2$.

$25 = OM^2 + \left(\dfrac{60}{13}\right)^2 = OM^2 + \dfrac{3600}{169}$.

$OM^2 = 25 - \dfrac{3600}{169} = \dfrac{4225-3600}{169} = \dfrac{625}{169}$.

$OM = \dfrac{25}{13}$.

$MP = PO - OM = 13 - \dfrac{25}{13} = \dfrac{144}{13}$.

Area of $\Delta PAB = \dfrac{1}{2}\times\dfrac{120}{13}\times\dfrac{144}{13} = \dfrac{60\times144}{169}$.

$\sqrt{\dfrac{M}{15}} = \sqrt{\dfrac{60\times144}{169\times15}} = \sqrt{\dfrac{576}{169}} = \dfrac{24}{13}$.

$\sqrt{\dfrac{M}{15}} = \dfrac{24}{13}$.

21

Mensuration 2D & 3D

- Scalene triangle / विषमबाहु त्रिभुज

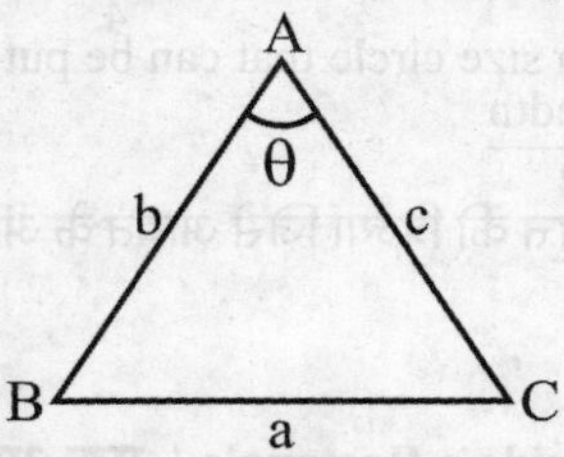

Perimeter / परिमाप $= a + b + c$

Semi-Perimeter / अर्द्धपरिमाप (s) $= \frac{a+b+c}{2}$

Area / क्षेत्रफल $(\Delta) = \sqrt{s(s-a)(s-b)(s-c)}$

Area / क्षेत्रफल $(\Delta) = \frac{1}{2}bc \sin A = \frac{1}{2}ab \sin C = \frac{1}{2}ac \sin B$

- Right angle triangle / समकोण त्रिभुज

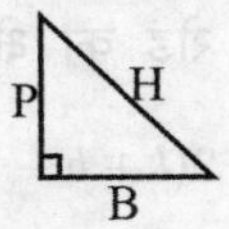

Area $(\Delta) = \frac{1}{2}PB$

Perimeter / परिमाप $= P + B + H$

- Isosceles right angle triangle / समद्विबाहु समकोण त्रिभुज

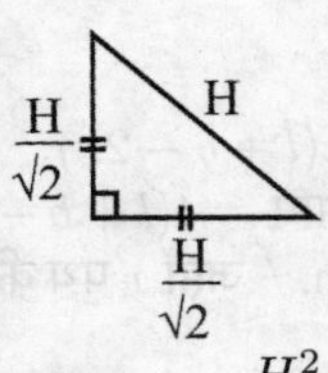

Area $(\Delta) = \frac{H^2}{4}$

Perimeter $= H + \sqrt{2}H = H(\sqrt{2}+1)$

- Equilateral triangle / समबाहु त्रिभुज

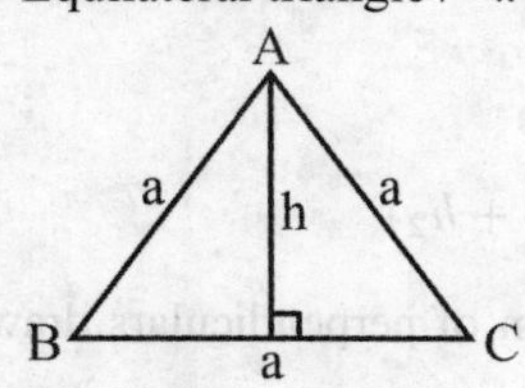

Side / भुजा $= a$

- Isosceles triangle / समद्विबाहु त्रिभुज

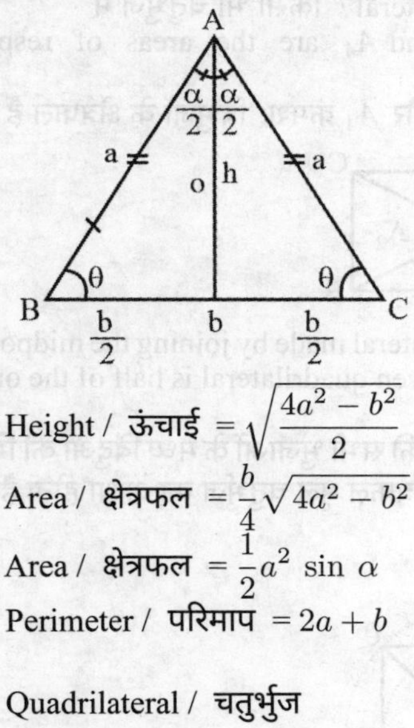

Height / ऊंचाई $= \sqrt{\frac{4a^2 - b^2}{2}}$

Area / क्षेत्रफल $= \frac{b}{4}\sqrt{4a^2 - b^2}$

Area / क्षेत्रफल $= \frac{1}{2}a^2 \sin \alpha$

Perimeter / परिमाप $= 2a + b$

Quadrilateral / चतुर्भुज

(i) Square

(ii) Rectangle

(iii) Parallelogram

(iv) Rhombus

(v) Trapezium

Quadrilateral: A closed shape and a type of polygon that has 4 sides, 4 vertices, and 4 angles.

चतुर्भुज : एक बंद आकृति और एक प्रकार का बहुभुज जिसमें 4 भुजाएँ, 4 शीर्ष और 4 कोण होते हैं।

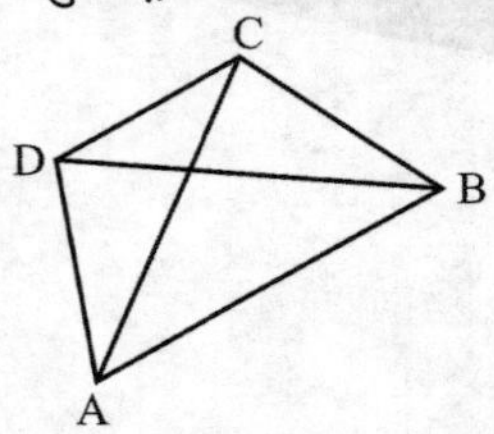

$\angle A + \angle B + \angle C + \angle D = 360°$

Area of Quadrilateral / चतुर्भुज का क्षेत्रफल

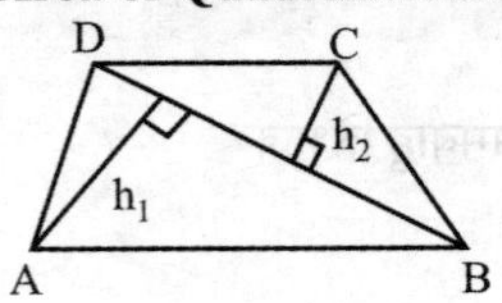

Area $= \frac{1}{2} \times BD \times (h_1 + h_2)$

$= \frac{1}{2} \times$ diagonal $\times$ (Sum of perpendiculars drawn on this diagonal)

$= \frac{1}{2} \times$ विकर्ण $\times$ (इस विकर्ण पर खींचे गए लंबों का योग)

In any Quadrilateral / किसी भी चतुर्भुज में

A_1, A_2, A_3, and A_4 are the areas of respective triangles.

A_1, A_2, A_3 और A_4 क्रमशः त्रिभुजों के क्षेत्रफल हैं।

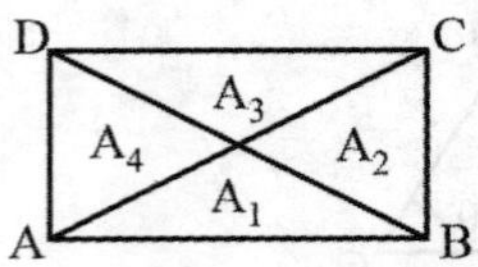

Area of quadrilateral made by joining the midpoints of all sides of a given quadrilateral is half of the original quadrilateral.

दिए गए चतुर्भुज की सभी भुजाओं के मध्य बिंदुओं को मिलाकर बने चतुर्भुज का क्षेत्रफल मूल चतुर्भुज का आधा होता है।

Square / वर्ग

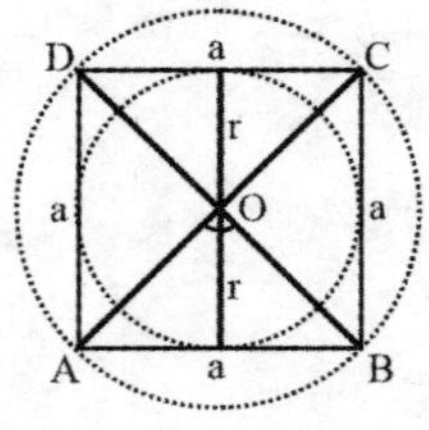

Area of AAOB = Area of ABOC = Area of ACOD = Area of AAOD $= \frac{a^2}{4}$

Perimeter / परिधि $= 4a$

Area / क्षेत्रफल $= a^2 = \frac{d^2}{2}$

Diameter / व्यास $= a\sqrt{2}$

Inradius / अंत:ज्या $(r) = \frac{a}{2}$

Rectangle / आयत

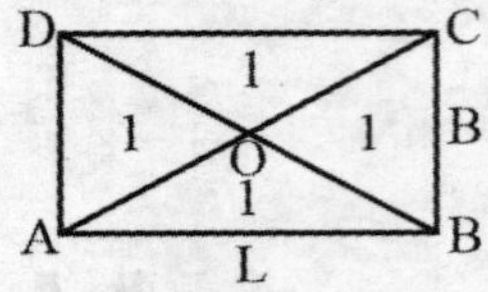

Perimeter / परिमाप $(P) = 2(L + B)$

Area / क्षेत्रफल $(A) = L \times B$

Diagonal / विकर्ण $= AC = BD = \sqrt{L^2 + B^2}$

Area of ΔAOB = Area of ΔBOC = Area of ΔCOD = Area of $\Delta AOD = \frac{LB}{4}$

Radius of maximum size circle that can be put inside the rectangle $= \frac{\text{Breadth}}{2}$

अधिकतम आकार के वृत्त की त्रिज्या जिसे आयत के अंदर रखा जा सकता है।

Path Around or Inside a Rectangle / एक आयत के चारों ओर या अंदर पथ

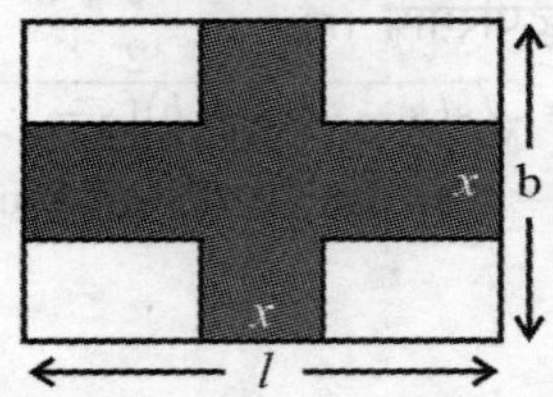

1. Crossing road inside rectangle / आयत के अंदर क्रॉसिंग सड़क

Area of crossing road / क्रॉसिंग रोड का क्षेत्र $= lx + bx - x^2 = x(l + b - x)$

Perimeter of path / पथ की परिमाप $= 2(l + b - 2x)$

2. Path inside a rectangle / एक आयत के अंदर पथ

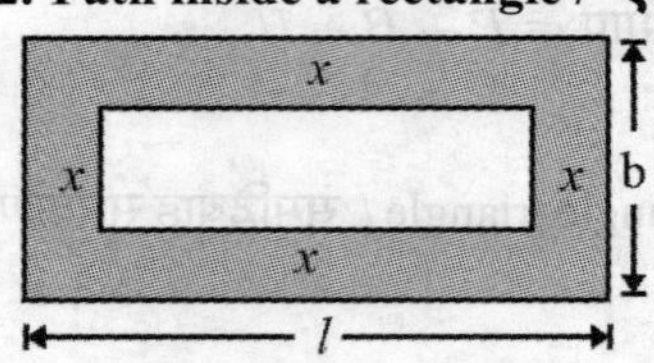

Area of path / पथ का क्षेत्र $= 2x(l + b - 2x)$

Perimeter of path / पथ की परिमाप $= 4(l + b - 2x)$

Where x is the width of the path. / जहाँ x पथ की चौड़ाई है

3. Path outside a rectangular field / एक आयताकार मैदान के बाहर पथ

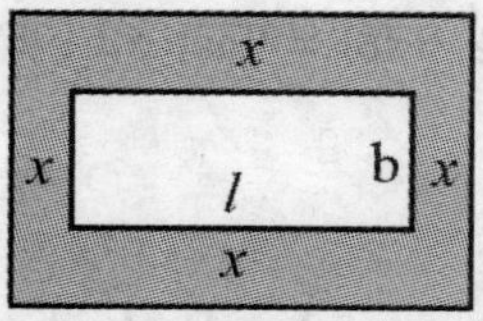

Area of path / पथ का क्षेत्रफल $= 2x(l + b + 2x)$
Perimeter of path / पथ की परिधि $= 4(l + b + 2x)$
Where x is the width of the path / जहाँ x पथ की चौड़ाई है

Parallelogram / समानान्तर चतुर्भुज

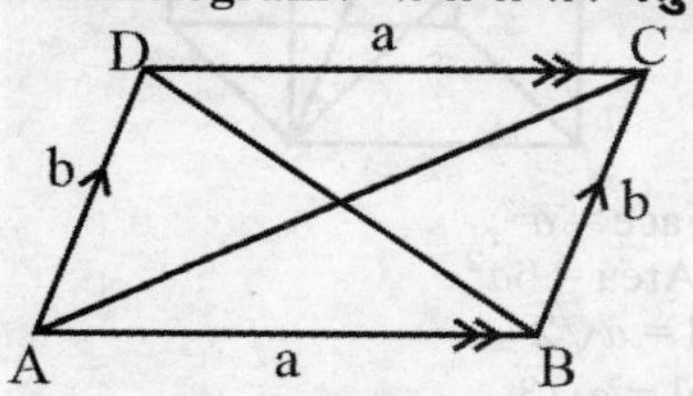

Perimeter / परिमाप $= 2(a + b)$
$AC^2 + BD^2 = 2(a^2 + b^2)$
$d_1^2 + d_2^2 = 2(a^2 + b^2)$
Area / क्षेत्रफल = Base / आधार × Height / ऊँचाई
If the length of one diagonal is d / यदि एक विकर्ण की लंबाई d है
Then Area / क्षेत्रफल $= 2\sqrt{s(s-a)(s-b)(s-d)}$
Where $s = \dfrac{(a+b+d)}{2}$ / जहाँ $s = \dfrac{(a+b+d)}{2}$

Rhombus / समचतुर्भुज

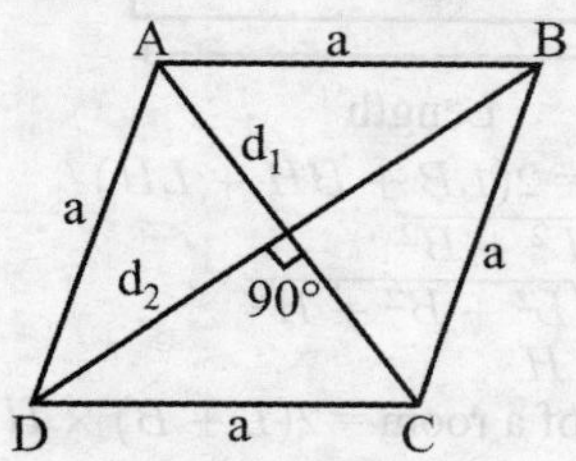

Perimeter / परिमाप $= 4a$
$4a^2 = d_1^2 + d_2^2$ (Property)
Area / क्षेत्रफल $= \dfrac{1}{2} \times d_1 \times d_2$
Area / क्षेत्रफल = Base / आधार × Height / ऊँचाई
Area of $\Delta AOB = \Delta BOC = \Delta COD = \Delta AOD$

Trapezium / समलंब चतुर्भुज

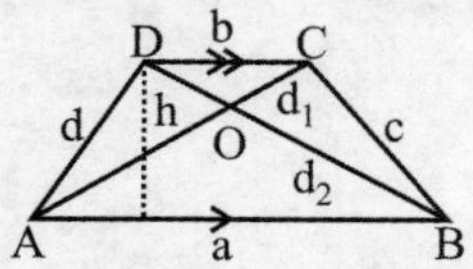

Perimeter / परिमाप $= a + b + c + d$
Area / क्षेत्रफल $= \dfrac{1}{2} \times$ (Sum of parallel sides) / समानांतर भुजाओं का योग × (Distance between them) / उनके बीच की दूरी

Area / क्षेत्रफल $= \dfrac{1}{2} \times (a + b) \times h$
$d_1^2 + d_2^2 = c^2 + d^2 + 2ab$ (Property)

Internal angle + External angle = 180°
आंतरिक कोण + बाह्य कोण = 180°

Each exterior angle of a regular polygon $= \dfrac{360°}{n}$

एक नियमित बहुभुज का प्रत्येक बाहरी कोण $= \dfrac{360°}{n}$

No. of sides / भुजाओं की संख्या $= \dfrac{360°}{n}$

No. of diagonals in a polygon $= \dfrac{n(n-3)}{2}$

एक बहुभुज में विकर्णों की संख्या $= \dfrac{n(n-3)}{2}$

Hexagon
Formed by 6 regular equilateral triangles.

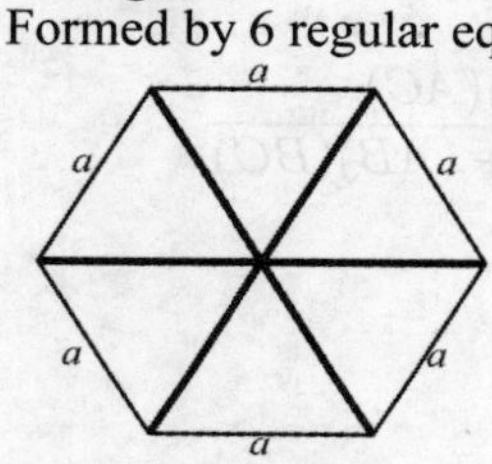

Area $= 6 \times \dfrac{\sqrt{3}}{4}a^2 = \dfrac{3\sqrt{3}}{2}a^2$
Perimeter $= 6a$
Incircle Radius $= \dfrac{\sqrt{3}}{2}a$
Circum Radius $= a$
Distance Between Two Parallel Lines $= \sqrt{3}a$

Octagon / अष्टभुज
Regular figure with 8 sides / 8 भुजाओं वाली नियमित आकृति
Let side of octagon $= a$ / मान लीजिए अष्टभुज की भुजा $= a$

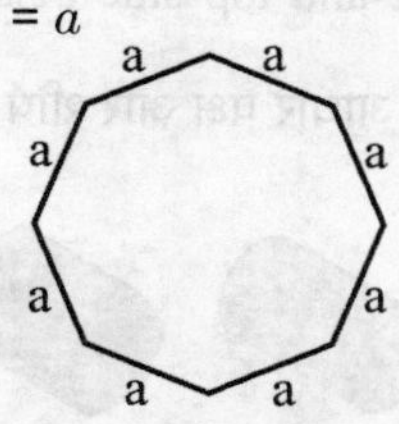

Area / क्षेत्र $= 2(\sqrt{2} + 1)a^2$
Perimeter / परिमाप $= 8a$
Inradius (r) / त्रिज्या (r) $= \dfrac{a}{2\sqrt{2} - 2}$
Circumradius (R) / परिधि (R) $= \dfrac{a}{\sqrt{2 - \sqrt{2}}} = \dfrac{\sqrt{2 + \sqrt{2}}}{2}a$
Each interior angle = 135° / प्रत्येक आंतरिक कोण =

135°
Each exterior angle = 45° / प्रत्येक बाहरी कोण = 45°
Number of diagonals = 20 / विकर्ण की संख्या = 20

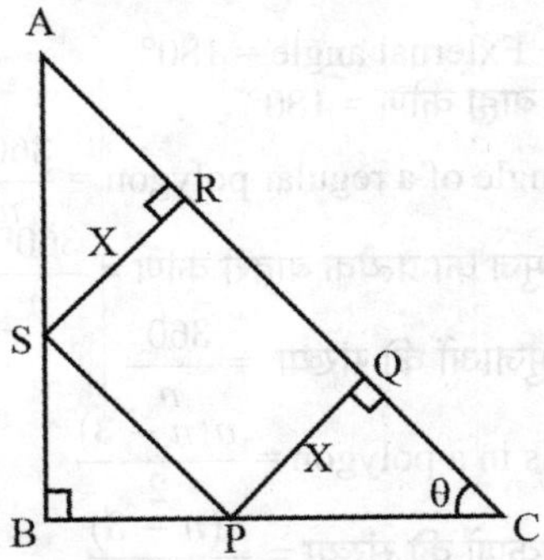

$x = \sqrt{(AR)(QC)}$

$$x = \frac{(AS)(PC)}{\sqrt{(AS)^2 + (PC)^2}}$$

$$x = \frac{(AB)(BC)(AC)}{(AB)^2 + (BC)^2 + (AB)(BC)}$$

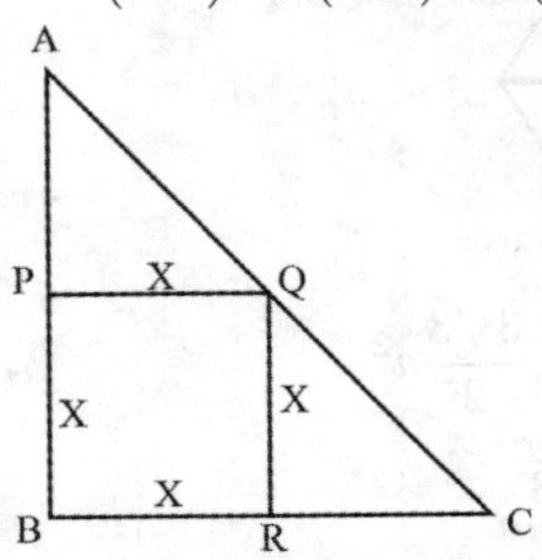

$x = \sqrt{(AP)(RC)}$

$$x = \frac{(AQ)(QC)}{\sqrt{(AQ)^2 + (QC)^2}}$$

$$x = \frac{(AB)(BC)}{AB + BC}$$

1. PRISM

A 3D shape with Base side and top side of similar shape and dimensions.
समान आकार और आयामों के आधार पक्ष और शीर्ष पक्ष के साथ एक 3डी आकार।

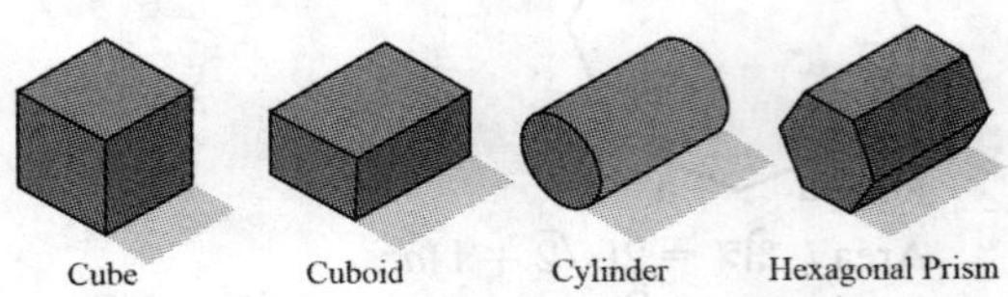

Generalised formula:
Volume = Area of Base × Height
Lateral Surface Area = Base Perimeter × Height
Total Surface Area = Lateral Surface Area +2 (Area of Base)
सामान्यीकृत सूत्र:
आयतन = आधार का क्षेत्रफल × ऊँचाई
पार्श्व सतही क्षेत्रफल = आधार परिधि × ऊँचाई
कुल सतही क्षेत्रफल = पार्श्व सतही क्षेत्रफल +2 (आधार का क्षेत्रफल)

2. Cube

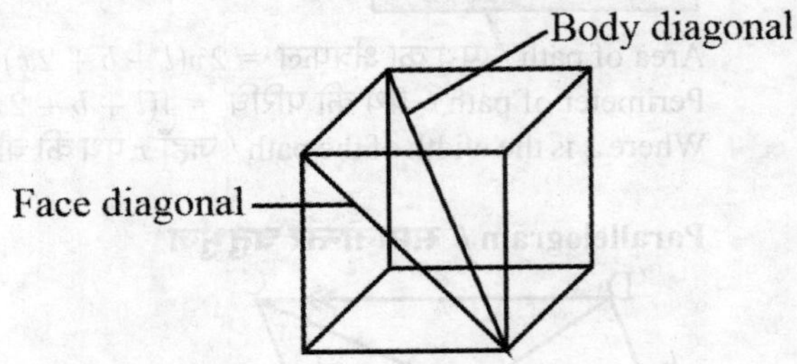

Area of each Face = a^2
Total Surface Area = $6a^2$
Face Diagonal = $a\sqrt{2}$
Body Diagonal = $a\sqrt{3}$
Volume = a^3

3. Cuboid

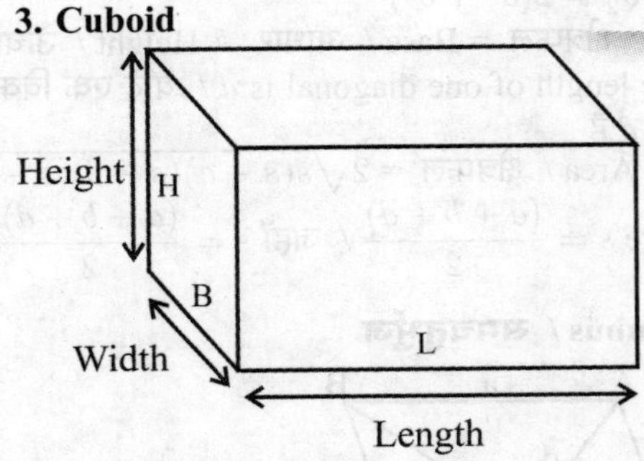

Total Surface Area = $2(LB + BH + LH)$
Face Diagonal = $\sqrt{L^2 + B^2}$
Body Diagonal = $\sqrt{L^2 + B^2 + H^2}$
Volume = $L \times B \times H$
Area of Four walls of a room = $2(L + B) \times H$

4. Cylinder

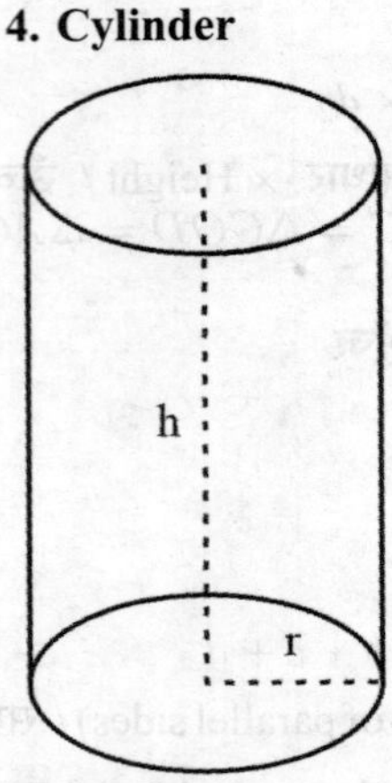

Total Surface Area = $2\pi r(r + h)$
Curved Surface Area = $2\pi rh$

Volume = $\pi r^2 h$

5. Hollow Cylinder

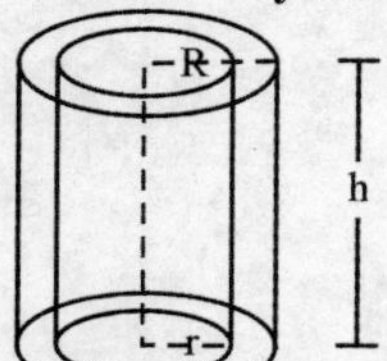

Total Surface Area = $2\pi(R+r)(h+R-r)$
Curved Surface Area = $2\pi h(R+r)$
Volume = $\pi(R^2-r^2)h$

Pyramid

A 3D shape with Base of polygon and apex or vertex at top.

बहुभुज के आधार और शीर्ष पर शीर्ष के साथ एक 3डी आकृति।

Generalised formula: / सामान्यीकृत सूत्र:

Volume = $\frac{1}{3}$ × Area of Base × Height

Lateral Surface Area = $\frac{1}{2}$ × Base Perimeter × Height

Total Surface Area = Lateral Surface Area + Area of Base

6. Cone

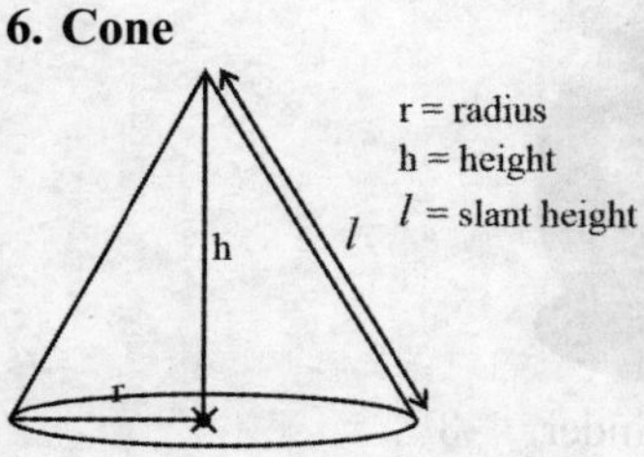

Length (l) = $\sqrt{r^2+h^2}$
Curved Surface Area = $\pi r l$
Total Surface Area = $\pi r(r+l)$
Volume = $\frac{1}{3}\pi r^2 h$

7. Sphere

Total Surface Area = $4\pi r^2$
Volume = $\frac{4}{3}\pi r^3$

8. Cutting of Sphere symmetrically from Centre: / केंद्र से गोले को सममित रूप से काटना:

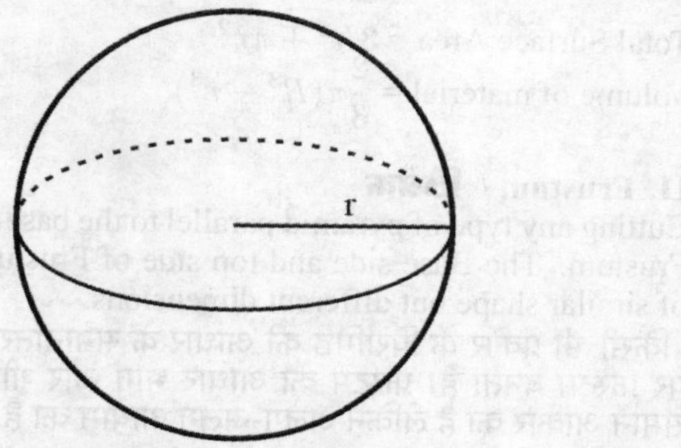

1 cut increases area by $2\pi r^2$
N cuts increase area by $2\pi r^2 N$
Total Surface Area = $4\pi r^2 + 2\pi r^2 N = 6\pi r^2$

9. Hollow Sphere (Also called as shell) / खोखला गोला (खोल भी कहा जाता है)

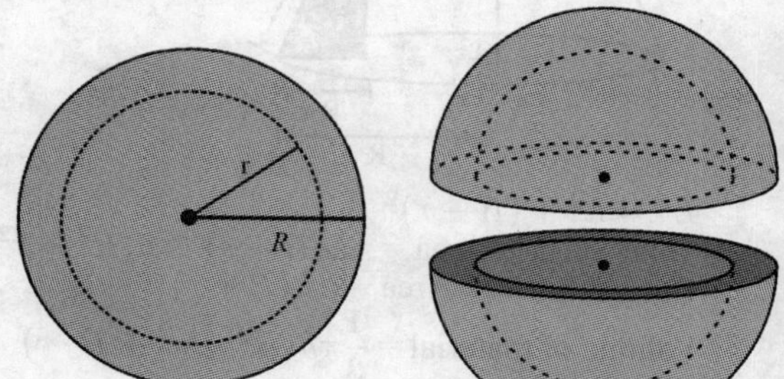

Outer Surface Area = $4\pi r^2$
Volume of material = $\frac{4}{3}\pi(R^3-r^3)$

10. Hemisphere

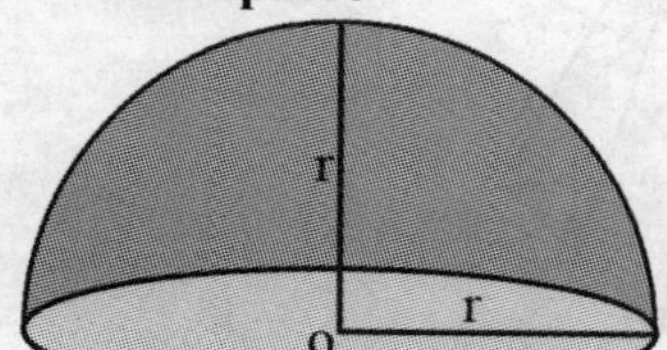

Total Surface Area = $3\pi r^2$
Curved Surface Area = $2\pi r^2$
Volume of material = $\frac{2}{3}\pi r^3$

10.1. Hollow Hemisphere (Also called as Hemisphere Shell) / खोखला गोलार्ध (इसे गोलार्ध शैल भी कहा जाता है)

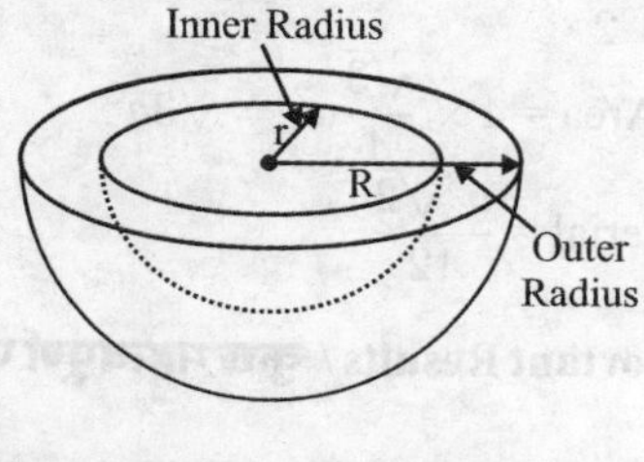

Total Surface Area = $3R^2 + \pi r^2$

Volume of material = $\frac{2}{3}\pi(R^3 - r^3)$

11. Frustum / छिन्नक

Cutting any type of pyramid parallel to the base forms Frustum. The Base side and top side of Frustum are of similar shape but different dimensions.

किसी भी प्रकार के पिरामिड को आधार के समानांतर काटने पर फ्रस्टम बनता है। फ्रस्टम का आधार भाग और शीर्ष भाग समान आकार का है लेकिन अलग-अलग आयाम का है।

11.1 Frustum of Cone / शंकु का छिन्नक

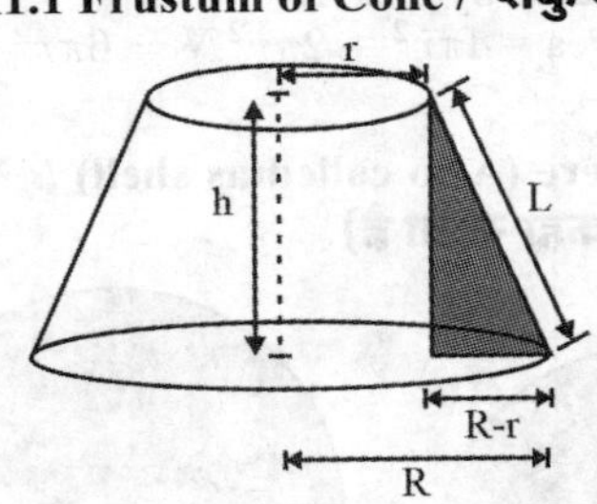

$L^2 = h^2 + (R - r)^2$

Total Surface Area = $\pi L(R + r) + n(R^2 + r^2)$

Curved Surface Area = $\pi L(R + r)$

Volume of material = $\frac{1}{3}\pi h(R^2 + r^2 + R \cdot r)$

12. Tetrahedron / चतुर्पाश्वीय

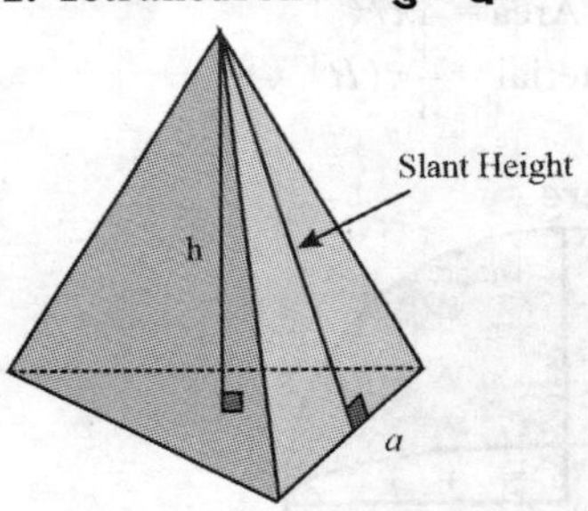

A regular pyramid with a base equilateral triangle and apex or vertex at the top. Object formed by 4 Equilateral triangles.

आधार समबाहु त्रिभुज और शीर्ष पर शीर्ष या शीर्ष वाला एक नियमित पिरामिड। 4 समबाहु त्रिभुजों से बनी वस्तु।

$h = \frac{\sqrt{2}}{\sqrt{3}}a$

Slant height = $\frac{\sqrt{3}}{2}a$

Total Surface Area = $4 \times \frac{\sqrt{3}}{4}a^2 = \sqrt{3}a^2$

Volume of material = $\frac{a^3\sqrt{2}}{12}$

13. Some Important Results / कुछ महत्वपूर्ण परिणाम

13.1 Maximum Cone inside Cylinder / सिलेंडर के अंदर अधिकतम शंकु

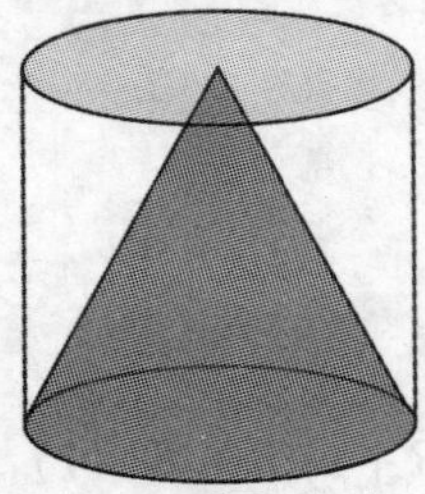

$$\frac{\text{Volume of Cylinder}}{\text{Volume of Cone}} = \frac{3}{1}$$

13.2 Maximum Cylinder inside Cube / क्यूब के अंदर अधिकतम सिलेंडर

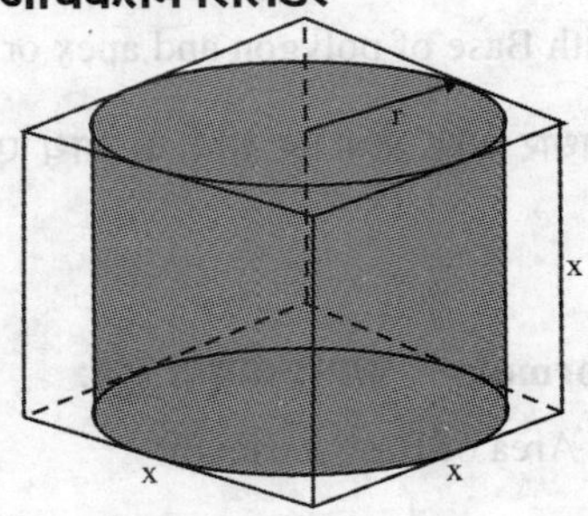

$$\frac{\text{Volume of Cube}}{\text{Volume of Cylinder}} = \frac{14}{11}$$

13.3 Maximum Sphere inside Cylinder / सिलेंडर के अंदर अधिकतम क्षेत्र

$$\frac{\text{Volume of Cylinder}}{\text{Volume of Sphere}} = \frac{3}{2}$$

13.4 Maximum Sphere inside Cube / घन के अंदर अधिकतम क्षेत्र

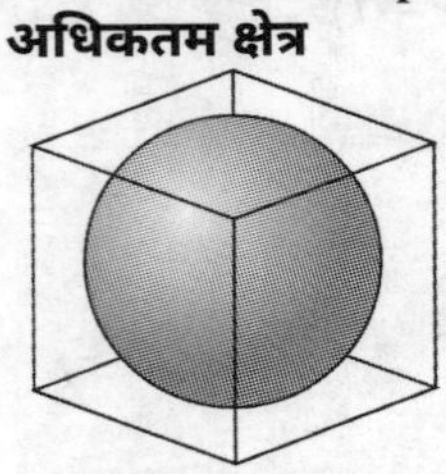

$$\frac{\text{Volume of Cube}}{\text{Volume of Sphere}} = \frac{21}{11}$$

13.5 Maximum Cube inside Cone / शंकु के अंदर

अधिकतम घन

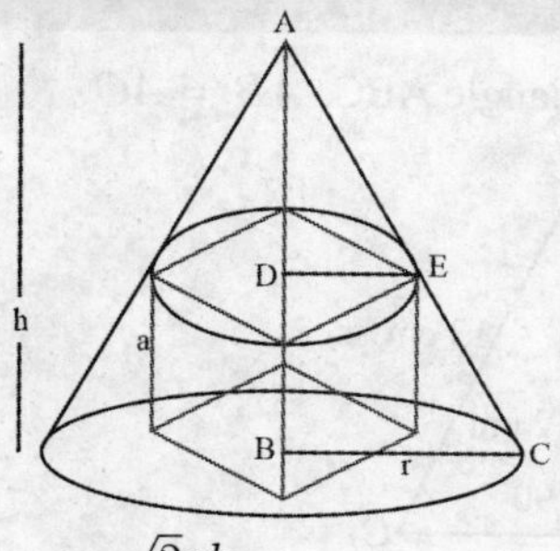

$$a = \frac{\sqrt{2}rh}{\sqrt{2}r + h}$$

13.6 Maximum Cube inside Hemisphere / गोलार्ध के अंदर अधिकतम घन

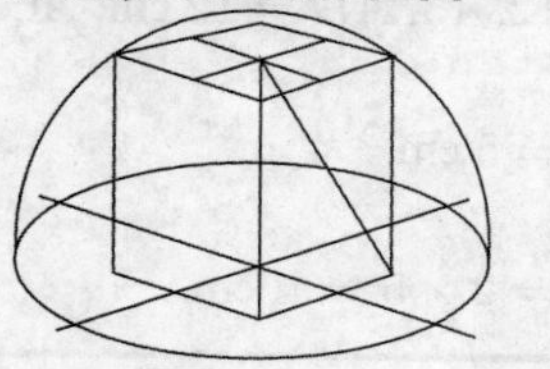

$$a = \sqrt{\frac{2}{3}}r$$

13.7 Maximum Cube inside Sphere / गोले के अंदर अधिकतम घन

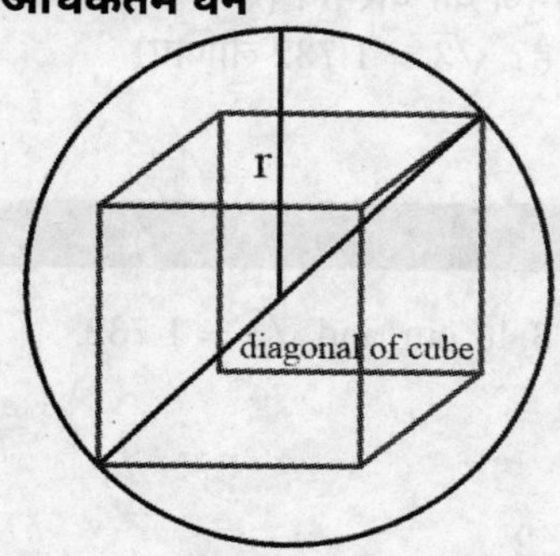

$$\frac{\text{Volume of Sphere}}{\text{Volume of Cube}} = \frac{11\sqrt{3}}{7}$$

13.8 Maximum Cylinder inside Sphere / गोले के अंदर अधिकतम सिलेंडर

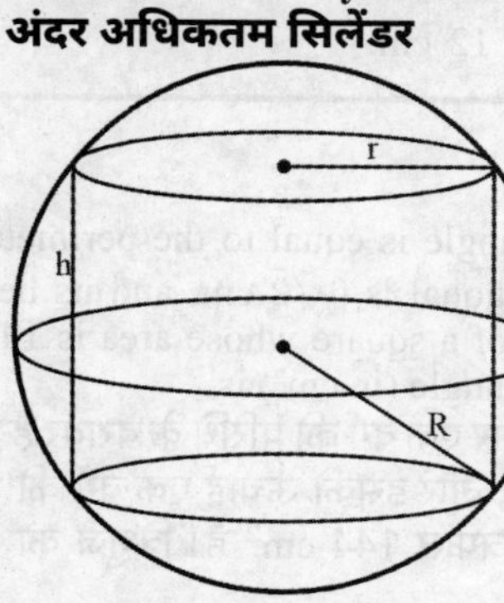

$$\frac{\text{Volume of Cylinder}}{\text{Volume of Sphere}} = \frac{1}{\sqrt{3}}$$

13.9 Maximum Cylinder inside Cone / शंकु के अंदर अधिकतम सिलेंडर

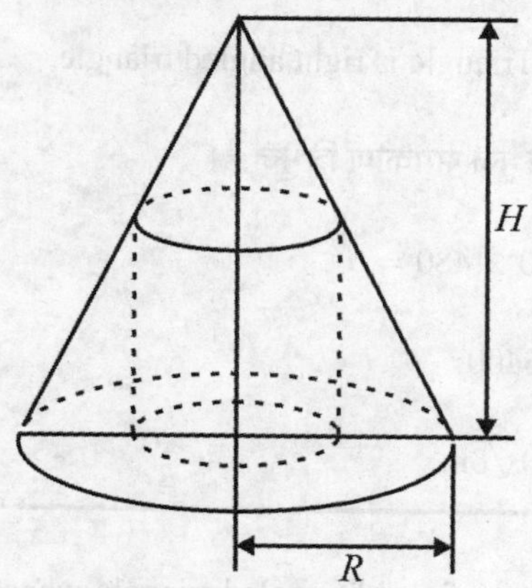

$$\frac{\text{Volume of Cylinder}}{\text{Volume of Cone}} = \frac{4}{9}$$

13.10 Maximum Cone in Sphere / गोले में अधिकतम शंकु

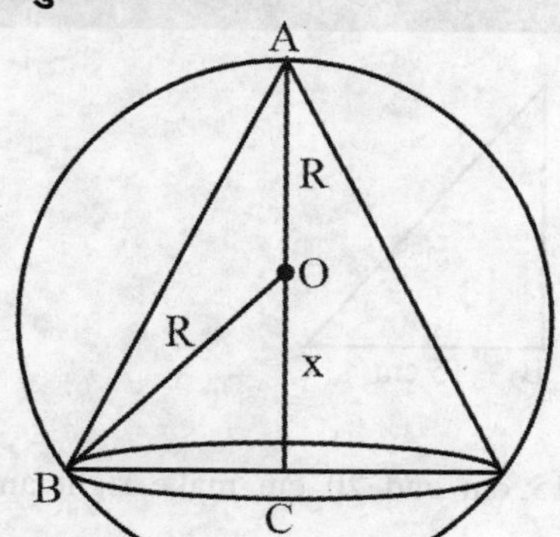

$$\frac{\text{Volume of Cone}}{\text{Volume of Sphere}} = \frac{8}{27}$$

Type 1 (Problems based on Traingle)

1. The sides of a triangular field are 360m, 480m and 600m. Its area is equal to the area of a square field. What is the side (in m) of the square field?

एक त्रिभुजाकार मैदान की भुजाएँ 360 मीटर, 480 मीटर और 600 मीटर हैं। इसका क्षेत्रफल एक वर्गाकार मैदान के क्षेत्रफल के बराबर है। वर्गाकार मैदान की भुजा (मीटर में) क्या है?

Solution:

Area of Triangle = Area of Square

Triplet (3, 4, 5) × 120

Hence, Triangle is right angled triangle.

अतः, त्रिभुज समकोण त्रिभुज है।

$\frac{1}{2} \times 360 \times 480 = a^2$

$a^2 = 86400$

$a = 120\sqrt{6}$m.

2. The perimeter of a right-angled triangle whose sides that make right angles are 15 cm and 20 cm is:
एक समकोण त्रिभुज की परिधि, जिसकी भुजाएँ समकोण बनाती हैं, 15 सेमी और 20 सेमी है।

Solution:

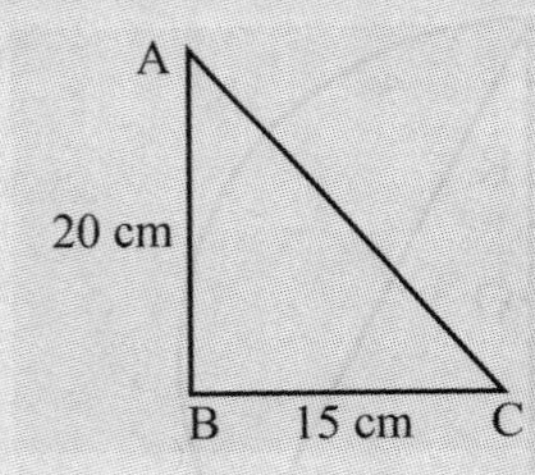

Sides 15 cm and 20 cm make right angle triangle.

By Pythagoras theorem:

Perimeter $= AB + BC + AC = 20 + 15 + 25 = 60$cm.

3. In an isosceles triangle ABC, AB = AC and AD is perpendicular to BC. If AD = 12 cm and the perimeter of $\triangle ABC$ is 36 cm, then the length of BC (in cm) as:
एक समद्विबाहु त्रिभुज ABC में, AB = AC और AD, BC पर लंबवत है। यदि AD = 12 सेमी और $\triangle ABC$ का परिमाप 36 सेमी है, तो BC की लंबाई (सेमी में) इस प्रकार है:

Solution:

In isosceles triangle ABC, $AB = AC$.

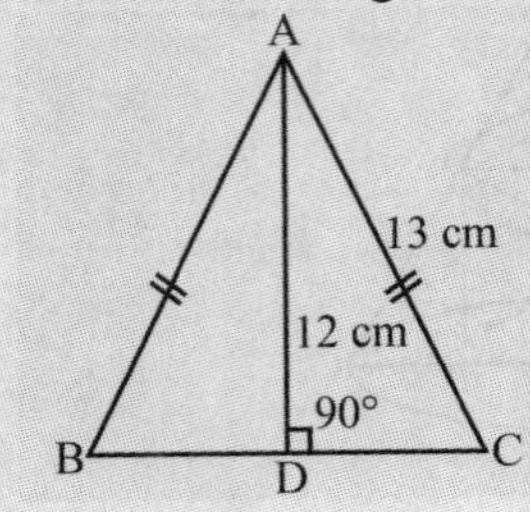

Perimeter of $\triangle ABC = 36$ cm.

Using triplet 5, 12, 13, $AD = 12$ cm, $AC = 13$ cm, and $DC = 5$ cm.

$BC = 2 \times DC = 2 \times 5 = 10$cm.

4. What is the perimeter (in cm) of an equilateral triangle whose height is 3.46 cm? Take $\sqrt{3} = 1.732$.
एक समबाहु त्रिभुज का परिमाप (सेमी में) क्या है जिसकी ऊंचाई 3.46 सेमी है? $\sqrt{3} = 1.732$ लीजिए।

Solution:

Given $h = 3.46$ cm, and $\sqrt{3} = 1.732$.

$h = \frac{\sqrt{3}a}{2}$

$a = \frac{3.46 \times 2}{1.732} = 4$ cm.

Perimeter of an equilateral triangle:

$3a = 3 \times 4 = 12$ cm.

5. The base of a triangle is equal to the perimeter of a square whose diagonal is $6\sqrt{2}$ cm, and its height is equal to the side of a square whose area is 144 cm^2. The area of the triangle (in cm^2) is:
एक त्रिभुज का आधार एक वर्ग की परिधि के बराबर है जिसका विकर्ण $6\sqrt{2}$ सेमी है, और इसकी ऊंचाई एक वर्ग की भुजा के बराबर है जिसका क्षेत्रफल 144 cm^2 है। त्रिभुज का क्षेत्रफल cm^2 में) है:

Solution:

Diagonal of the square = $6\sqrt{2}$ cm.

$a\sqrt{2} = 6\sqrt{2}$

$a = 6$ cm.

Base of the triangle = $4 \times 6 = 24$ cm.

Height of the triangle = $\sqrt{144} = 12$ cm.

Area of the triangle: $\frac{1}{2} \times 24 \times 12 = 144$ cm^2.

6. ABC is a right-angled triangle, right-angled at B, D is a point on AC such that $AD = 12$ cm and $CD = 16$ cm. If BD bisects $\angle ABC$, then the perimeter of $\triangle ABC$ will be:

ABC एक समकोण त्रिभुज है, B पर समकोण है, AC पर D इस प्रकार एक बिंदु है कि AD = 12 सेमी और CD = 16 सेमी है। यदि BD, $\angle ABC$ को समद्विभाजित करता है, तो $\triangle ABC$ का परिमाप होगा:

Solution:

According to the question:

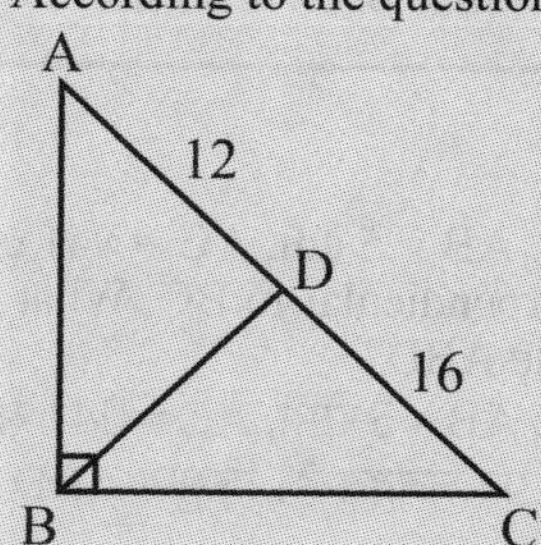

BD is the bisector of $\angle ABC$.

$\frac{BC}{AB} = \frac{16}{12} = \frac{4}{3}$.

Let the lengths of AB and BC be $3x$ and $4x$, respectively.

Using the Pythagorean theorem:

$AB^2 + BC^2 = AC^2$

$9x^2 + 16x^2 = 282$

$25x^2 = 282$

$x^2 = \frac{282}{25}$.

Then,

Perimeter $= AB + BC + AC = 3x + 4x + 5x = 12x = 67.2$ cm.

7. The in-radius and circumradius of a right-angled triangle is 3 cm and 12.5 respectively. The area of the triangle is: cm

एक समकोण त्रिभुज की अंत:त्रिज्या और परित्रिज्या क्रमश: 3 सेमी और 12.5 है। त्रिभुज का क्षेत्रफल है: सेमी

Solution:

Circumradius of a right-angled triangle $= \frac{H}{2} = 12.5$ -------- [where H = Hypotenuse]

$H = 12.5 \times 2 = 25$ cm.

And, in radius: $\frac{P + B - H}{2} = 3$

$P + B - H = 6$

$P + B = 31.$

On putting the value of H: Now, from the triplet rule, the sides of the given right-angled triangle are 24 cm, 7 cm, and 25 cm. Area of triangle $= \frac{1}{2} \times 24 \times 7 = 84$ cm^2.

8. If the perimeter of an isosceles right triangle is $15(\sqrt{2} + 1)$ cm, then the area of the triangle will be:

यदि एक समद्विबाहु समकोण त्रिभुज का परिमाप $15(\sqrt{2}+1)$ सेमी है, तो त्रिभुज का क्षेत्रफल होगा:

Solution:

Let the base and perpendicular of the isosceles right-angled triangle $= x$ cm. Then, the hypotenuse: $\sqrt{x^2+x^2} = x\sqrt{2}$.

Perimeter: $x + x + x\sqrt{2} = 15(\sqrt{2}+1)$.

$x(2+\sqrt{2}) = 15(\sqrt{2}+1)$.

$x = \dfrac{15}{\sqrt{2}\sqrt{2}}$.

Now, the area of the triangle: $\dfrac{1}{2} \times x \times x = \dfrac{x^2}{2}$.

Area $= \dfrac{1}{2} \times \left(\dfrac{15}{\sqrt{2}}\right)^2 = \dfrac{1}{2} \times \dfrac{225}{2} =$ $56.25\ \text{cm}^2$.

9. In a triangle ABC, AB = AC and the perimeter of $\triangle ABC$ is $8(2+\sqrt{2})$ cm. If the length of BC is $\sqrt{2}$ times the length of AB, then find the area of $\triangle ABC$:

त्रिभुज ABC में, AB = AC और $\triangle ABC$ का परिमाप $8(2+\sqrt{2})$ सेमी है। यदि BC की लंबाई AB की लंबाई का $\sqrt{2}$ गुना है, तो $\triangle ABC$ का क्षेत्रफल ज्ञात कीजिए।

Solution:

In $\triangle ABC$: $AB = AC$. Perimeter of triangle

$P = a + b + c$: $8(2+\sqrt{2}) = a + a + \sqrt{2}a$.

$8(2+\sqrt{2}) = a(2+\sqrt{2})$

$a = 8$.

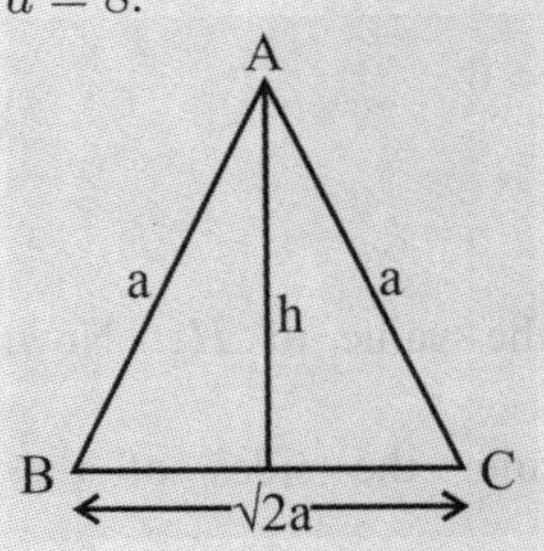

Height: $h = \sqrt{a^2 - \dfrac{a^2}{2}} = \sqrt{\dfrac{a^2}{2}}$.

Area of $\triangle ABC$: $\dfrac{1}{2} \times \text{Base} \times \text{Height} = \dfrac{1}{2} \times \sqrt{2}a \times \sqrt{\dfrac{a^2}{2}} = \dfrac{a^2}{2} = \dfrac{64}{2} = 32\ \text{cm}^2$.

10. The area of a field in the shape of a triangle with each side x metres is equal to the area of another triangular field having sides 50 m, 70 m, and 80 m. The value of x is closer to:

एक त्रिभुज के आकार के एक मैदान का क्षेत्रफल जिसकी प्रत्येक भुजा x मीटर है, दूसरे त्रिभुजाकार मैदान के क्षेत्रफल के बराबर है जिसकी भुजाएँ 50 मीटर, 70 मीटर और 80 मीटर हैं। x का मान इसके निकट है:

Solution:

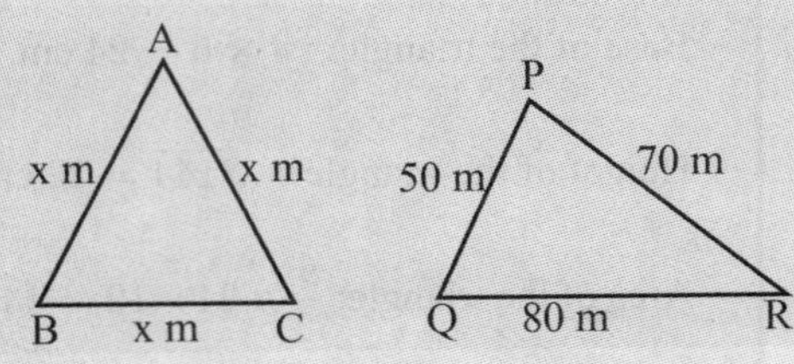

According to the question: $s = \dfrac{a+b+c}{2} = \dfrac{80+70+50}{2} = 100$.

Area of equilateral $\triangle ABC$ = Area of $\triangle PQR$: $\dfrac{\sqrt{3}}{4}x^2 = \sqrt{s(s-a)(s-b)(s-c)}$.

$\dfrac{\sqrt{3}}{4}x^2 = \sqrt{100(100-50)(100-70)(100-80)}$.

$\dfrac{\sqrt{3}}{4}x^2 = \sqrt{100 \times 50 \times 30 \times 20}$.

$x^2 = 4000$

$x = 20\sqrt{10} = 20 \times 3.16 = 63.20$ m.

11. ABC is a triangle, AB = 5 cm, AC = √41 cm and BC = 8 cm, AD is perpendicular to BC. What is the area (in cm²) of triangle ABD?

ABC एक त्रिभुज है, AB = 5 सेमी, AC = □41 सेमी और BC = 8 सेमी, AD BC पर लंबवत है। त्रिभुज ABD का क्षेत्रफल (सेमी² में) क्या है?

Solution:

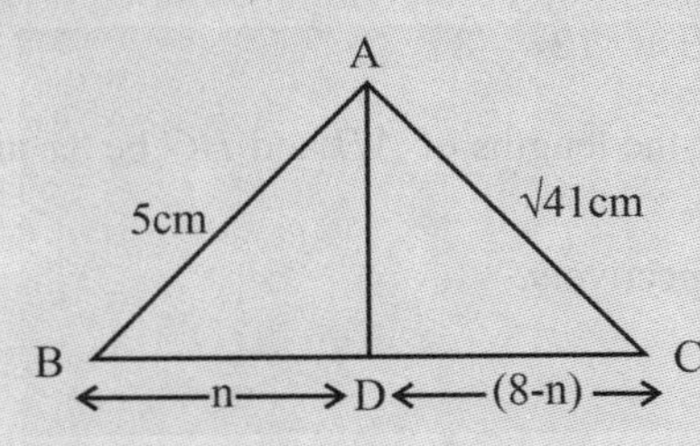

In $\triangle ABD$: $AD = \sqrt{(5^2 - n^2)}$ ----(i)

In $\triangle ADC$: $AD = \sqrt{(41 - (8 - n)^2)}$ ----(ii).

From equations (i) and (ii): $5^2 - n^2 = 41 - (8 - n)^2$.

$(8 - n)^2 - n^2 = 41 - 25$.

$64 + n^2 - 16n - n^2 = 16$

$16n = 48$

$n = 3$.

In $\triangle ABD$: $AD = \sqrt{(5^2 - 3^2)} = \sqrt{16} = 4$.

Area of $\triangle ABD = \frac{1}{2} \times BD \times AD = \frac{1}{2} \times 3 \times 4 = 6$ cm^2.

12. In the given figure, in a right-angle triangle ABC, AB = 12 cm and AC = 15 cm. A square is inscribed in the triangle. One of the vertices of the square coincides with the vertex of the triangle. What is the maximum possible area (in cm²) of the square?

दी गई आकृति में, समकोण त्रिभुज ABC में, AB = 12 सेमी और AC = 15 सेमी है। त्रिभुज में एक वर्ग अंकित है। वर्ग का एक शीर्ष त्रिभुज के शीर्ष से मेल खाता है। वर्ग का अधिकतम संभव क्षेत्रफल (सेमी² में) क्या है?

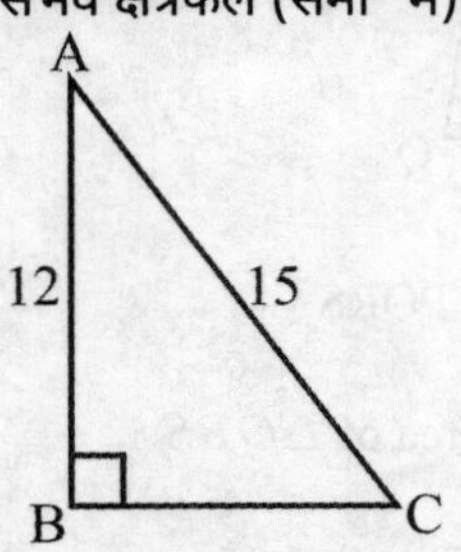

Solution:

Given $AB = 12$, $AC = 15$, and $BC = 9$:

Triplet $\rightarrow$ 9, 12, 15.

The maximum possible area of the square in the right-angle triangle $\triangle ABC$:

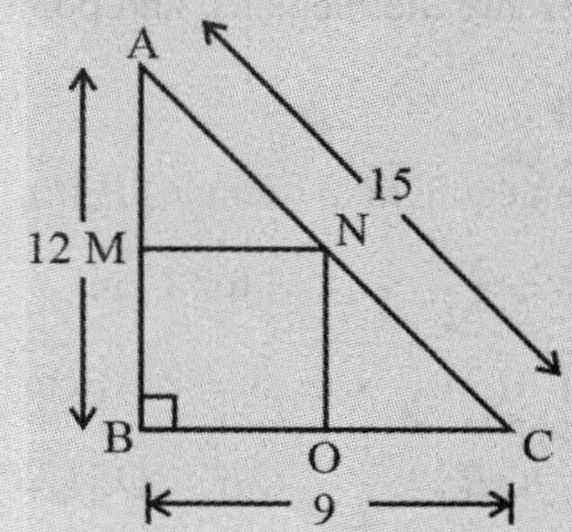

Side of square MNOB $= \frac{(12 \times 9)}{(12 + 9)} = \frac{36}{7}$.

Area of the square MNOB $= \left(\frac{36}{7}\right)^2 = \frac{1296}{49}$ cm^2.

13. In the given figure, PQR is an equilateral triangle with side as 12 cm. S and T are the mid-points of the sides PQ and PR respectively. What is the area (in cm²) of the shaded region?

दी गई आकृति में, PQR एक समबाहु त्रिभुज है जिसकी भुजा 12 सेमी है। S और T क्रमशः भुजाओं PQ और PR के मध्य बिंदु हैं। छायांकित क्षेत्र का क्षेत्रफल (सेमी² में) क्या है?

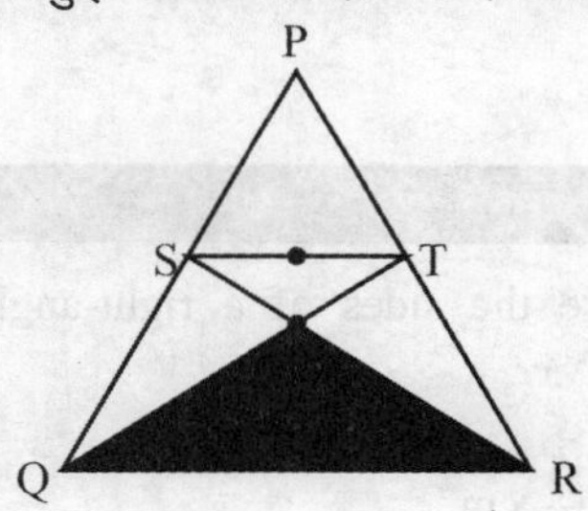

Solution:

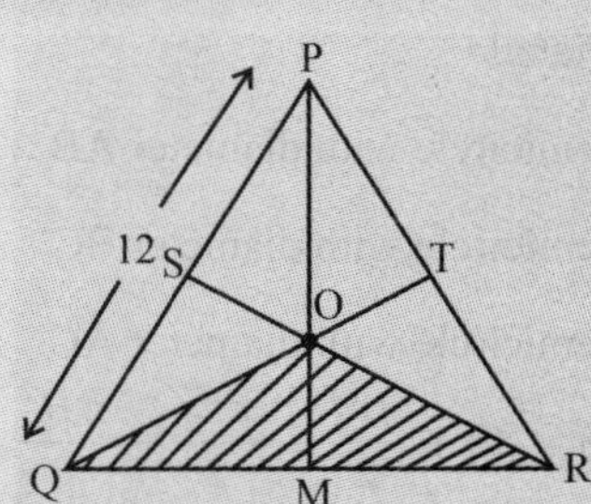

S and T are the mid-points of PQ and PR, respectively.

RS and QT are the medians, and point O is the centroid.

Height of an equilateral triangle (PM): $\frac{\sqrt{3}}{2} \times \text{side} = \frac{\sqrt{3}}{2} \times 12 = 6\sqrt{3}$ cm.

$OM = \frac{1}{3}PM = \frac{1}{3} \times 6\sqrt{3} = 2\sqrt{3}$ cm.

Area of $\triangle QOR = \frac{1}{2} \times QR \times OM = \frac{1}{2} \times 12 \times 2\sqrt{3} = 12\sqrt{3}$ cm^2.

14. In the given figure, 3 semicircles are drawn on three sides of triangle ABC. $AB = 21$ cm, $BC = 28$ cm, and $AC = 35$ cm. What is the area (in cm^2) of the shaded part?
दिए गए चित्र में, त्रिभुज ABC की तीन भुजाओं पर 3 अर्धवृत्त खींचे गए हैं। AB = 21 सेमी. BC = 28 सेमी और AC = 35 सेमी. छायांकित भाग का क्षेत्रफल (सेमी^{2} में) क्या है?

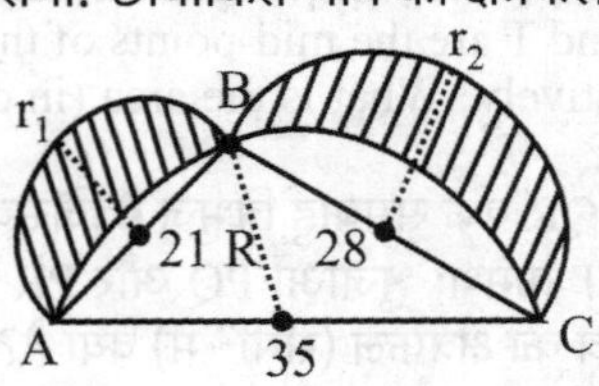

Solution:

21, 28, 35 are the sides of a right-angled triangle.

$R = \frac{AC}{2} = \frac{35}{2}$ cm.

$r_1 = \frac{AB}{2} = \frac{21}{2}$ cm, $\quad r_2 = \frac{BC}{2} = 14$ cm.

Area of the shaded part:

Area of triangle +

Area of semicircle with diameter AB +

Area of semicircle with diameter BC −

Area of semicircle with diameter AC.

$= \frac{1}{2} \times 28 \times 21 + \frac{\pi r_1^2}{2} + \frac{\pi r_2^2}{2} - \frac{\pi R^2}{2}.$

$= 294 + \frac{22}{7 \times 2}\left[\frac{441}{4} + 196 - \frac{1225}{4}\right].$

$294 + \frac{11}{7}[196 - 196] = 294$ cm^2.

Note: In this type of figure, the area of the shaded part is equal to the area of the right-angled triangle: Area of right-angled triangle $= \frac{1}{2} \times \text{base} \times \text{height} = \frac{1}{2} \times 21 \times 28 = 294$ cm^2.

Type - 2 Problem based on Quadrilaterals (Rectangle, Square, Rhombus etc)

15. In quadrilateral PQRS, RM ⊥ QS, PN ⊥ QS and QS = 6cm. If RM = 3cm and PN = 2cm, then the area of PQRS is:
चतुर्भुज PQRS, RM ⊥ QS, PN ⊥ QS और QS = 6 सेमी में। यदि RM = 3 सेमी और PN = 2 सेमी है, तो PQRS का क्षेत्रफल है:

Solution:

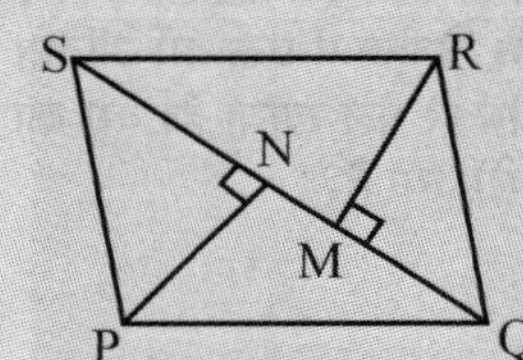

Area of quadrilateral PQRS = Area of $\triangle PQS$ + Area of $\triangle QRS$

$= \frac{1}{2} \times 6 \times 2 + \frac{1}{2} \times 6 \times 3 = 15$ cm^2.

16. In the given figure, PQR is a triangle and quadrilateral ABCD is inscribed in it. QD = 2 cm, QC = 5 cm, CR = 3 cm, BR = 4 cm, PB = 6 cm, PA = 5 cm and AD = 3 cm. What is the area (in cm^3) of the quadrilateral ABCD?
दी गई आकृति में, PQR एक त्रिभुज है और चतुर्भुज ABCD इसमें अंकित है। QD = 2 सेमी, QC = 5 सेमी, CR = 3 सेमी, BR = 4 सेमी, PB = 6 सेमी, PA = 5 सेमी और AD = 3 सेमी। चतुर्भुज ABCD का क्षेत्रफल (cm^3 में) क्या है?

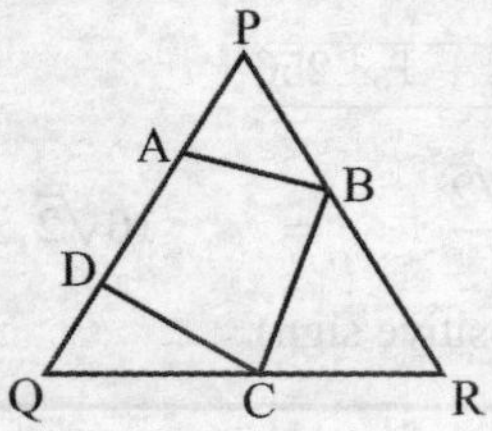

Solution:

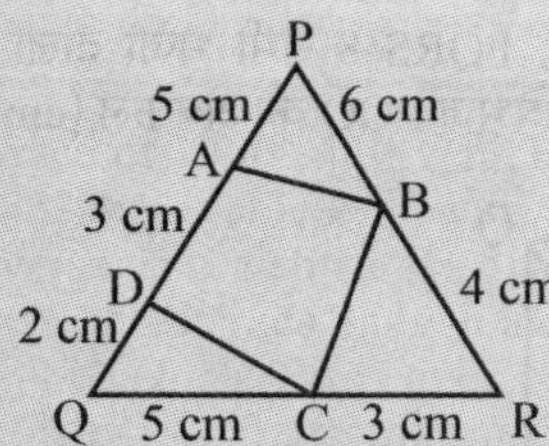

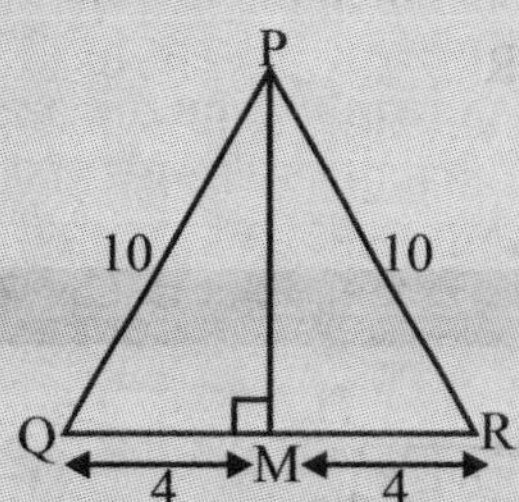

Area of triangle $= \frac{1}{2}ab \sin \theta$.

$\frac{\text{ar}(\triangle PAB)}{\text{ar}(\triangle PQR)} = \frac{\frac{1}{2} \times 5 \times 6 \sin P}{\frac{1}{2} \times 10 \times 10 \sin P} = \frac{3}{10}$.

$\frac{\text{ar}(\triangle QDC)}{\text{ar}(\triangle PQR)} = \frac{10}{80} = \frac{1}{8}$, $\frac{\text{ar}(\triangle BRC)}{\text{ar}(\triangle PQR)} = \frac{12}{80} = \frac{3}{20}$.

Let $\text{ar}(\triangle PQR) = 40$: $\text{ar}(\triangle PAB) : \text{ar}(\triangle QDC) : \text{ar}(\triangle BRC) = 12 : 5 : 6$.

$\text{ar}(\triangle PQR) = \frac{1}{2} \times 8 \times \sqrt{100 - 16} = 8\sqrt{21}$ cm^2.

Since 40 unit $= 8\sqrt{21}$, ar(ABCD) $= 40-(12+5+6) = 17$ unit $= \frac{(8\sqrt{21})}{40} \times 17 = \frac{17\sqrt{21}}{5}$ cm^2.

17. The lengths of two diagonals of a rhombus are 15 cm and 20 cm. What is the area (in cm²) of the rhombus?

एक समचतुर्भुज के दो विकर्णों की लंबाई 15 सेमी और 20 सेमी है। समचतुर्भुज का क्षेत्रफल (cm² में) क्या है?

Solution:

Given:

Diagonal $d_1 = 15$ cm,

$d_2 = 20$ cm:

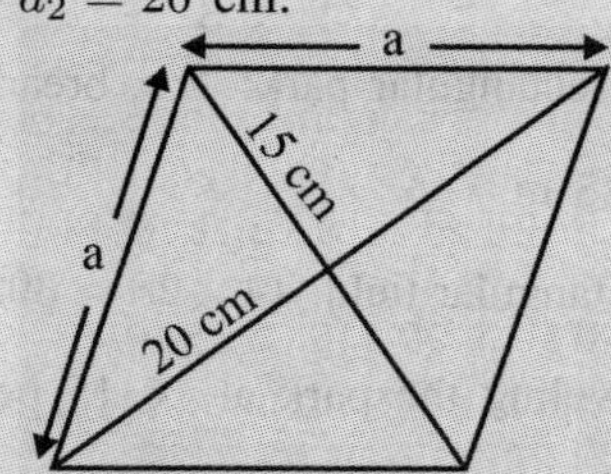

Area of the rhombus $= \frac{\text{Product of diagonals}}{2}$.

$A = \frac{d_1 \times d_2}{2} = \frac{15 \times 20}{2} = 150$ cm^2.

18. A wire is in the form of a square of side 33 cm. If the wire is molded to form a circle, then what is the radius of the circle? Use $\pi = \frac{22}{7}$.

एक तार 33 सेमी भुजा वाले एक वर्ग के आकार का है। यदि तार को ढालकर एक वृत्त बनाया जाए, तो वृत्त की त्रिज्या क्या है? $\pi = \frac{22}{7}$ लीजिए।

Solution:

Circumference of circle $=$ Perimeter of square.

$2\pi r = 4 \times 33$

$2 \times \frac{22}{7} \times r = 4 \times 33$

$r = 21$ cm.

19. The area of a square-shaped field is 1764 m^2. The breadth of a rectangular park is $\frac{1}{3}$rd the side of the square field, and its length is two times its breadth. What is the cost (in ₹) of leveling the park at ₹ 15 per m^2?

एक वर्गाकार मैदान का क्षेत्रफल 1764 वर्ग मीटर है। एक आयताकार पार्क की चौड़ाई $\frac{1}{3}$ है वर्गाकार मैदान की भुजा और इसकी लंबाई इसकी चौड़ाई से दोगुनी है। 15 प्रति वर्ग मीटर की दर से पार्क को समतल करने की लागत ₹ में) क्या है?

Solution:

Given: Area of square-shaped field $= 1764\ m^2$.

Side of square-shaped field: $\sqrt{1764} = 42$ m.

Breadth of rectangular park: $\frac{1}{3} \times 42 = 14$ m.

Length of rectangular park: $2 \times$ breadth $= 2 \times 14 = 28$ m.

Area of rectangular field: $14 \times 28 = 392\ m^2$.

Cost of leveling the park at ₹ 15 per m^2: $392 \times 15 = 5880$.

20. PQRS is a square whose side is 16 cm. What is the value of the side (in cm) of the largest octagon that can be cut from the given square?

PQRS एक वर्ग है जिसकी भुजा 16 सेमी है। दिए गए वर्ग से काटे जा सकने वाले सबसे बड़े अष्टभुज की भुजा का मान (सेमी में) क्या है?

Solution:

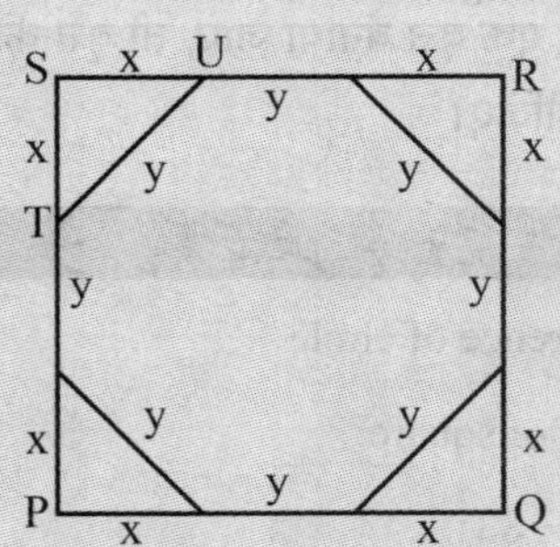

Let the side of the octagon $= y$ cm: $SR = 16$ cm.

$2x + y = 16$

$x = \frac{16 - y}{2}$.

In $\triangle STU$: $x^2 + x^2 = y^2$ ----(i).

$2\left(\frac{16 - y}{2}\right)^2 = y^2$.

$256 + y^2 - 32y = 2y^2$

$y^2 + 32y - 256 = 0$.

$y = \frac{-32 \pm \sqrt{1024 + 4 \times 256}}{2}$

$= \frac{-32 \pm 32\sqrt{2}}{2} = 16\sqrt{2} - 16$ (Taking the positive sign).

21. In the given figure, PQRS is a square of side 8 cm. $\angle PQO = 60°$. What is the area (in cm²) of the triangle POQ?

दी गई आकृति में, PQRS 8 सेमी भुजा वाला एक वर्ग है। $\angle PQO = 60°$. त्रिभुज POQ का क्षेत्रफल (cm² में) क्या है?

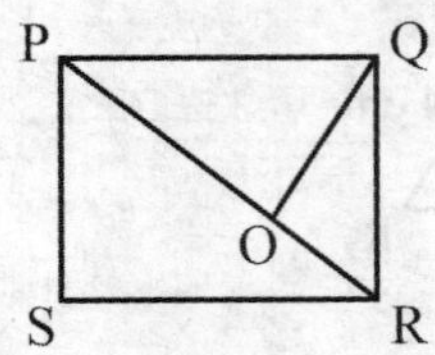

Solution:

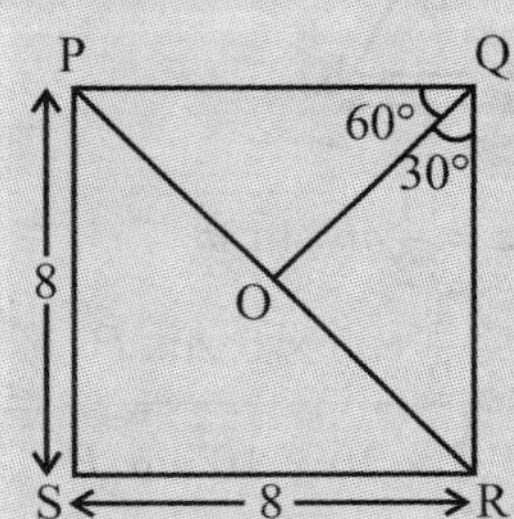

$\angle PQO = 60°$, $\angle OQR = 30°$.

$\triangle PQR = \frac{8 \times 8}{2} = 32$.

Area of $\triangle PQR$ = Area of $\triangle POQ$ + Area of $\triangle QOR$.

Area of triangle $= \frac{1}{2}ab \sin\theta$.

$32 = \frac{1}{2} \times PQ \times OQ \times \sin 60° + \frac{1}{2} \times QR \times OQ \times \sin 30°$.

$= \frac{1}{2} \times 8 \times OQ \times \frac{\sqrt{3}}{2} + \frac{1}{2} \times 8 \times OQ \times \frac{1}{2}$.

$32 = OQ(2\sqrt{3} + 2)$

$OQ = \frac{32}{2 + 2\sqrt{3}} = \frac{16}{\sqrt{3}+1} \times \frac{\sqrt{3}-1}{\sqrt{3}-1} = 8(\sqrt{3} - 1)$.

Area of $\triangle POQ = \frac{1}{2} \times 8 \times 8(\sqrt{3} - 1) \times \sin 60° = 16(\sqrt{3} - 1) \times \sqrt{3} = 16(3 - \sqrt{3})$ cm^2.

22. In the given figure, two squares of sides 8 cm and 20 cm are given. What is the area (in cm²) of the shaded part?

दी गई आकृति में, 8 सेमी और 20 सेमी भुजाओं वाले दो वर्ग दिए गए हैं। छायांकित भाग का क्षेत्रफल (cm² में) क्या है?

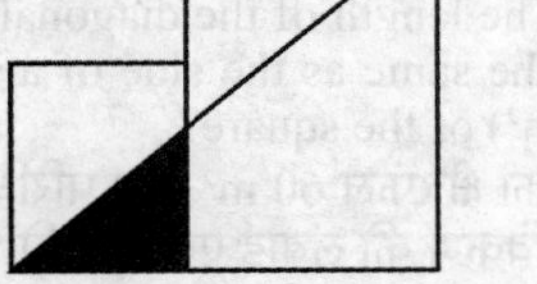

Solution:

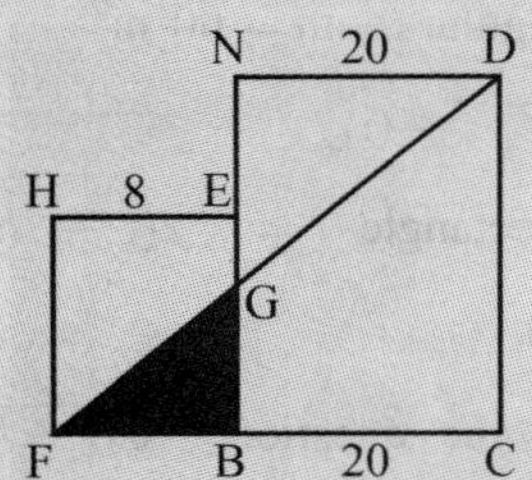

From the figure: $\triangle AFGB \sim \triangle AFDC$.

$$\frac{GB}{DC} = \frac{FB}{FC} = \frac{GB}{20} = \frac{8}{28}$$

$$GB = \frac{40}{7}.$$

The area of the shaded part $= \frac{1}{2} \times 8 \times \frac{40}{7} = \frac{160}{7}$ cm^2.

23. An equilateral triangle of side 12 cm is drawn. What is the area (in cm²) of the largest square which can be drawn inside it?

12 सेमी भुजा वाला एक समबाहु त्रिभुज बनाया गया है। इसके अंदर खींचे जा सकने वाले सबसे बड़े वर्ग का क्षेत्रफल (cm² में) क्या है?

Solution:

Let a be the side of the largest square formed inside the equilateral triangle.

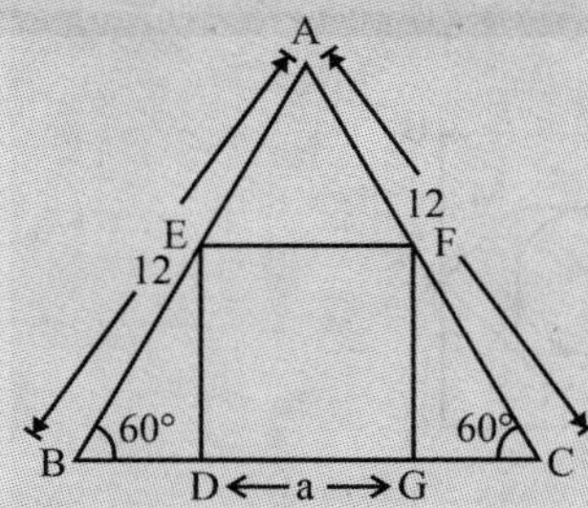

In $\triangle BED$:

$$\tan 60° = \frac{DE}{BD}$$

$$\sqrt{3} = \frac{a}{BD}$$

$$BD = \frac{a}{\sqrt{3}}.$$

Similarly:

$$GC = \frac{a}{\sqrt{3}}.$$

$$\frac{a}{\sqrt{3}} + \frac{a}{\sqrt{3}} + a = 12$$

$$a + a + a\sqrt{3} = 12\sqrt{3}.$$

$$a(2 + \sqrt{3}) = 12\sqrt{3}$$

$$a = \frac{12\sqrt{3}}{2 + \sqrt{3}} \times \frac{2 - \sqrt{3}}{2 - \sqrt{3}} = 12\sqrt{3}(2 - \sqrt{3}).$$

Area of the square $= [12\sqrt{3}(2 - \sqrt{3})]^2 = 432(4 + 3 - 4\sqrt{3}) = (3024 - 1728\sqrt{3})$ cm^2.

24. In the given figure, ABCD is a square. EFGH is a square formed by joining mid-points of sides of ABCD. LMNO is a square formed by joining mid-points of sides of EFGH. A circle is inscribed inside LMNO. If the area of the circle is 38.5 cm³, then what is the area (in cm³) of square ABCD?

दी गई आकृति में, ABCD एक वर्ग है। EFGH, ABCD की भुजाओं के मध्य बिंदुओं को मिलाने से बना एक वर्ग है। LMNO, EFGH की भुजाओं के मध्य बिंदुओं को मिलाने से बना एक वर्ग है। LMNO के अंदर एक वृत्त अंकित है। यदि वृत्त का क्षेत्रफल 38.5 सेमी^{3} है, तो वर्ग ABCD का क्षेत्रफल सेमी^{3} में) क्या है?

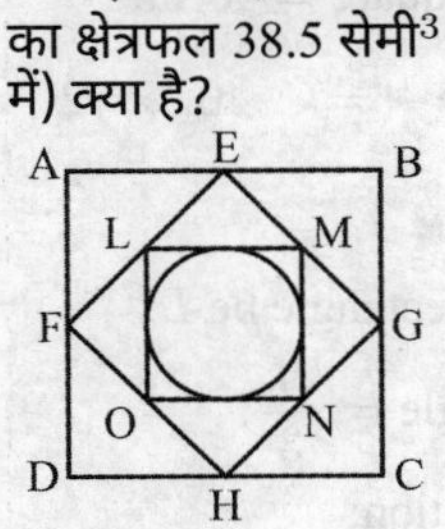

Solution:

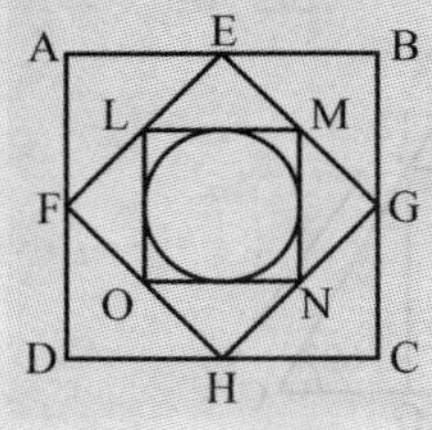

Area of the circle $= \frac{22}{7}\pi r^2$.

$38.5 = \frac{22}{7} \times r^2$

$r = \frac{7}{2}$.

Side of square LMNO $= 2r = 7$ cm.

The area of a square formed by joining the mid-points of the sides of a square is half of the original square: किसी वर्ग की भुजाओं के मध्य बिंदुओं को मिलाने से बने वर्ग का क्षेत्रफल मूल वर्ग का आधा होता है। Area of EFGH $= 49 \times 2 = 98$ cm^2.

Area of square ABCD $= 98 \times 2 = 196$ cm^2.

25. The perimeter of a square is half the perimeter of a rectangle. The perimeter of the square is 40 m. If its breadth is two-thirds of its length, then what is the area (in m³) of the rectangle?

एक वर्ग का परिमाप एक आयत के परिमाप का आधा होता है। वर्ग की परिधि 40 मीटर है। यदि इसकी चौड़ाई इसकी लंबाई की दो-तिहाई है, तो आयत का क्षेत्रफल (वर्ग मीटर में) क्या है?

Solution:

Given: Perimeter of square $= 40$ m.

Perimeter of rectangle $= 40 \times 2 = 80$ m ----(i).

Let the length of the rectangle be L:

Breadth of the rectangle $= \frac{2L}{3}$.

According to the question:

Perimeter of rectangle $= 2\left(L + \frac{2L}{3}\right)$.

$80 = 2\left(\frac{5L}{3}\right)$

$L = 24$ m.

Breadth $= \frac{2L}{3} = \frac{2 \times 24}{3} = 16$ m.

$\therefore$ Area of the rectangle $= L \times B = 24 \times 16 = 384$ m^2.

26. A rectangular room has an area of 60 m² and a perimeter of 34 m. The length of the diagonal of the rectangular room is the same as the side of a square. What is the area (in m²) of the square?

एक आयताकार कमरे का क्षेत्रफल 60 m² और परिधि 34 m है। आयताकार कमरे के विकर्ण की लंबाई एक वर्ग की भुजा के समान है। वर्ग का क्षेत्रफल (वर्ग मीटर में) क्या है?

Solution:

Area of rectangular room $= 60$ m^2

$l \times b = 60$ m^2 ----(i).

Perimeter of rectangle $= 2(l + b) = 34$ m ----(ii).

According to the question: $\sqrt{l^2 + b^2} =$ side of square.

$a^2 = l^2 + b^2$ ----(iii).

From equation (ii): $l + b = 17$ and $2lb = 289$.

$l^2 + b^2 = 289 - 120 = 169$.

Substituting $l^2 + b^2$ in equation (iii): $a^2 = 169$ m^2.

$\therefore$ Area of square $= 169$ m^2.

27. The length of a rectangle is 10 cm more than the side of a square and is 8 cm less than twice the side of the square. If the areas of both the rectangle and square are equal, then what will be the perimeter (in cm) of the rectangle?

एक आयत की लंबाई एक वर्ग की भुजा से 10 सेमी अधिक है और वर्ग की भुजा से 8 सेमी कम है। यदि आयत और वर्ग दोनों का क्षेत्रफल बराबर है, तो आयत का परिमाप सेमी में क्या होगा?

Solution:

Let each side of the square be a cm:

According to the question: $(a+10)(a-8) = a^2$.

$a^2 - 8a + 10a - 80 = a^2$

$2a = 80$

$a = 40$ cm.

Length of the rectangle $= 40 + 10 = 50$ cm.

Breadth of the rectangle $= 40 - 8 = 32$ cm.

$\therefore$ Perimeter of the rectangle $= 2(50 + 32) = 164$ cm.

28. How much iron sheet (in m²) will be needed to construct a rectangular tank measuring 10 m × 8 m × 6 m, if a circular opening of radius one metre is to be left at the top of the tank? (correct to one decimal place)
10 मीटर × 8 मीटर × 6 मीटर मापने वाले एक आयताकार टैंक के निर्माण के लिए कितनी लोहे की चादर (वर्ग मीटर में) की आवश्यकता होगी, यदि टैंक के शीर्ष पर एक मीटर त्रिज्या का एक गोलाकार उद्घाटन छोड़ा जाना है? (दशमलव के एक स्थान तक सही)

Solution:

Iron sheet needed to construct a rectangular tank $= 2(lb + bh + hl) - \pi r^2$.

$= 2(10 \times 8 + 8 \times 6 + 6 \times 10) - \pi(1)^2$.

$= 2 \times 188 - 3.14 = 376 - 3.14 = 372.86 \approx 372.9$ m^2.

29. ABCD is a quadrilateral. The diagonals of ABCD intersect at the point P. The areas of the triangles $\triangle APB$ and $\triangle CPD$ are 12 and 27, respectively. If the area of the triangle $\triangle ARD$ is 13.5 cm^2, then find the area of $\triangle DBPC$:
ABCD एक चर्तुभुज है। चर्तुभुज के विकर्ण एक दूसरे को बिंदु पर काटते है। त्रिभुज $\triangle APB$ और त्रिभुज $\triangle CPD$ के क्षेत्रफल क्रमश: 12 और 27 है। यदि त्रिभुज $\triangle ARD$ का क्षेत्रफल 13.5 m^2 है, तव त्रिभुज $\triangle BPC$ का क्षेत्रफल ज्ञात करो?

Solution:

Area of $\triangle BPC = \dfrac{27 \times 12}{13.5} = 24$ cm^2.

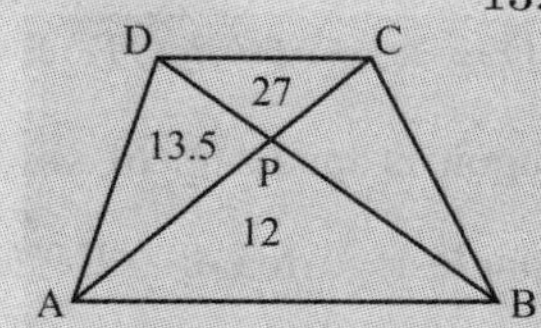

30. The length of the parallel sides of a trapezium are 51 cm and 21 cm, and the length of each of the other two sides is 39 cm. What is the area of the trapezium?
एक समलम्ब चर्तुभुज की भुजाएं जो समान्तर हैं, की लंबाई 51 सेमी और 21 सेमी है तथा बाकी दो भुजाओं में प्रत्येक की लंबाई 39 सेमी है। तो समलम्ब चर्तुभुज का क्षेत्रफल क्या है?

Solution:

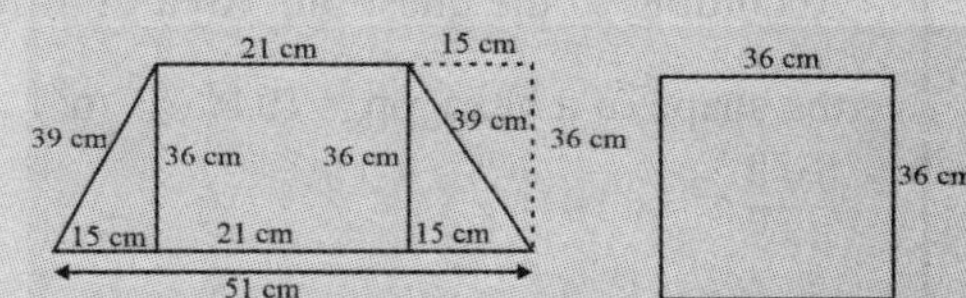

By triplets (5, 12, 13) and (15, 36, 39):

$AD = BC = 39$ cm.

Area $= 36 \times 36 = 1296$ cm^2.

Type 3 Problem based on Polygon

31. ABCDEF is a regular hexagon of side 12 cm. What is the area (in cm²) of the triangle $\triangle ECD$?
ABCDEF 12 सेमी भुजा वाला एक नियमित षट्भुज है। त्रिभुज $\triangle ECD$ का क्षेत्रफल cm^2 में) क्या है?

Solution:

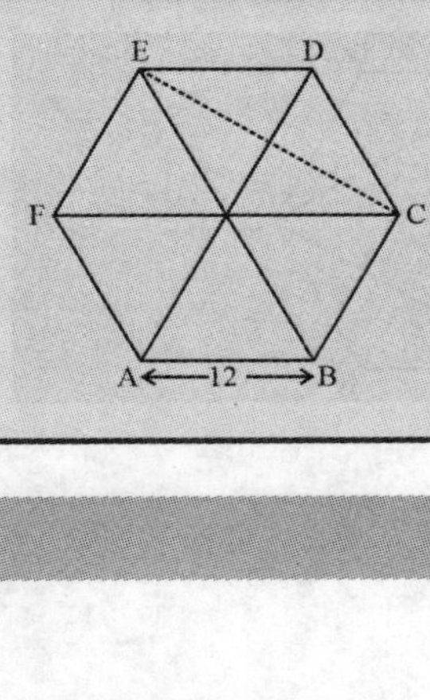

Triangles on the same base and between the same parallel lines have equal areas: Area of $\triangle ECD$ = Area of $\triangle EOD$.

A regular hexagon is made up of six equal equilateral triangles: Area of $\triangle ECD$ = $\frac{\sqrt{3}}{4} \times 12 \times 12 = 36\sqrt{3}$ cm^2.

32. The area of a field in the shape of a hexagon is $1944\sqrt{3}$ m^2. What will be the cost (in ₹) of fencing it at the rate of ₹ 11.50 per metre?

षट्कोण के आकार के एक मैदान का क्षेत्रफल 1944√3 वर्ग मीटर है। ₹ 11.50 प्रति मीटर की दर से इसकी बाड़ लगाने की लागत ₹ में) क्या होगी?

Solution:

According to the question: Area of a field in the shape of a hexagon = $1944\sqrt{3}$ m^2.

$\frac{6\sqrt{3}}{4}a^2 = 1944\sqrt{3}$

$a^2 = 324 \times 4$

$a = 36.$

Perimeter = $6a = 6 \times 36 = 216$ m.

On fencing around the hexagonal field at the rate of ₹ 11.50 per metre, the total cost: = $216 \times 11.50 = 2484$.

33. ABCDEF is a regular hexagon. What is the ratio of the area of $\triangle ACE$ and $\triangle AEF$?

ABCDEF एक नियमित षट्भुज है। त्रिभुज $\triangle ACE$ के क्षेत्रफल और त्रिभुज $\triangle AEF$ के क्षेत्रफल का अनुपात क्या है?

Solution:

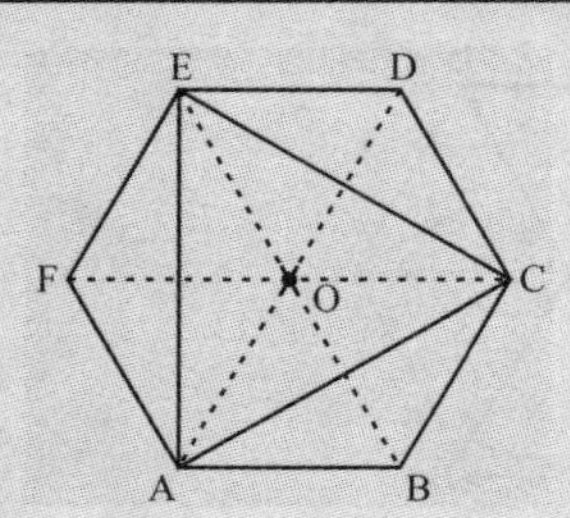

A regular hexagon is made up of six equilateral triangles: $\triangle OAB$, $\triangle OBC$, $\triangle OCD$, $\triangle ODE$, $\triangle OEF$, and $\triangle OAF$ are equilateral triangles.

Area of $\triangle ACE = \frac{1}{2} \times$ Area of six equilateral triangles.

Area of $\triangle AEF = \frac{1}{6} \times$ Area of six equilateral triangles.

ar($\triangle ACE$) : ar($\triangle AEF$) = 3 : 1.

Type 4 Problem based on Circle

34. A triangle is drawn with its vertices on the circle C such that one of its sides is a diameter of C and the other two sides have their lengths in the ratio $a : b$. If the radius of the circle is r, then the area of the triangle is:

वृत्त C पर उसके शीर्षों के साथ एक त्रिभुज इस प्रकार खींचा गया है कि उसकी एक भुजा का व्यास C है और अन्य दो भुजाओं की लंबाई का अनुपात $a : b$ है। यदि वृत्त की त्रिज्या r है, तो त्रिभुज का क्षेत्रफल है।

Solution:

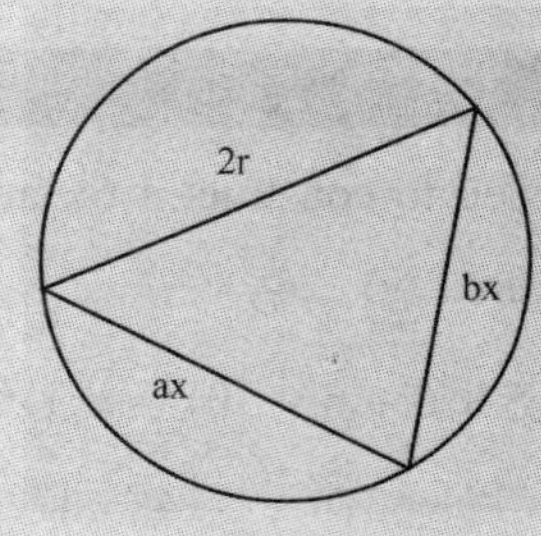

$a^2x^2 + b^2x^2 = 4r^2.$

$x^2 = \frac{4r^2}{a^2 + b^2}.$

Area of triangle $= \frac{1}{2}abx^2 = \frac{1}{2}ab \times \frac{4r^2}{a^2+b^2} = \frac{2abr^2}{a^2+b^2}.$

35. The area of a circular park is 12474 m^2. There is a 3.5 m wide path around the park. What is the area (in m²) of the path? (Take $\pi = \frac{22}{7}$)

एक गोलाकार पार्क का क्षेत्रफल 12474 वर्ग मीटर है। पार्क के

चारों ओर 3.5 मी चौड़ा रास्ता है। पथ का क्षेत्रफल (वर्ग मीटर में) क्या है? ($\pi = \frac{22}{7}$ लीजिए)

Solution:

Area of circular park = $12474\ m^2$ (given).

$\pi r^2 = 12474$

$r^2 = \frac{12474 \times 7}{22}$.

$r^2 = 81 \times 49$

$r = 9 \times 7 = 63$ m.

New radius with the width of the path (R) = $63 + 3.5 = 66.5$ m.

Area of the path = Area of the park with path − Area of the park.

$= \pi R^2 - \pi r^2 = \frac{22}{7} \times 66.5 \times 66.5 - 12474.$

$= 13898.5 - 12474 = 1424.5\ m^2.$

36. If the perimeter of a circle is 88 cm, then what is the area of the circle? ($\pi = \frac{22}{7}$)

यदि किसी वृत्त का परिमाप 88 सेमी है, तो वृत्त का क्षेत्रफल क्या है? ($\pi = \frac{22}{7}$)

Solution:

Perimeter of circle$(2\pi r) = 88$.

$2 \times \frac{22}{7} \times r = 88$

$r = 14$.

Area of circle $= \pi r^2 = \frac{22}{7} \times 14 \times 14 = 616\ m^2$.

37. A sector is cut out from a circle of diameter 42 cm. If the angle of the sector is 150°, then its area (in cm²) is: (Take $\pi = \frac{22}{7}$)

42 सेमी व्यास वाले एक वृत्त से एक त्रिज्यखंड काटा जाता है। यदि त्रिज्यखंड का कोण 150° है, तो इसका क्षेत्रफल (cm^2) में) है: (लिया गया $\pi = \frac{22}{7}$)।

Solution:

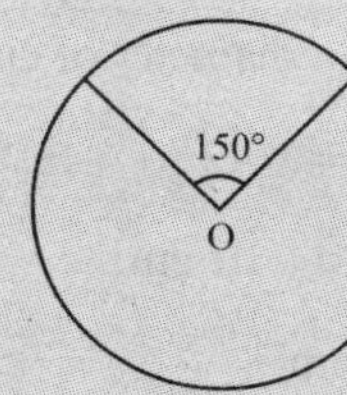

Area of sector $= \frac{\theta}{360} \times \pi r^2$.

$= \frac{150}{360} \times \frac{22}{7} \times 21 \times 21.$

$= \frac{5}{12} \times 22 \times 63 = \frac{6930}{12} = 577.5\ cm^2.$

38. The inner and outer radius of a circular track are, respectively, 29 m and 23 m. The cost of leveling the track at ₹ 7/m² is:

एक गोलाकार ट्रैक की आंतरिक और बाहरी त्रिज्या क्रमशः 29 मीटर और 23 मीटर है, ट्रैक को समतल करने की लागत 7/वर्ग मीटर है:

Solution:

Area of track $= \pi[R^2 - r^2]$.

$= \frac{22}{7}[(29)^2 - (23)^2]$.

$= \frac{22}{7} \times (29 + 23)(29 - 23)$.

$= \frac{22}{7} \times 52 \times 6$.

Cost of leveling the track $= \frac{22}{7} \times 52 \times 6 \times 7 = 6864$.

39. A rectangular sheet of length 42 cm and breadth 14 cm is cut from a circular sheet. What is the minimum area (in cm²) of the circular sheet?

42 सेमी लंबाई और 14 सेमी चौड़ाई की एक आयताकार शीट एक गोलाकार शीट से काटी जाती है। गोलाकार शीट का न्यूनतम क्षेत्रफल सेमी² में) कितना है?

Solution:

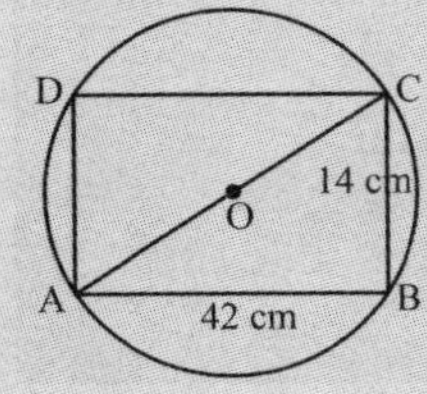

According to the question:

Diagonal of rectangle = diameter of circle.

Diagonal of rectangle $= \sqrt{42^2 + 14^2}$.

$= \sqrt{14^2(3^2 + 1^2)} = 14\sqrt{10}$ cm.

$r = \frac{14\sqrt{10}}{2} = 7\sqrt{10}$ cm.

The minimum area of a circular sheet $= \pi r^2$.

$= \frac{22}{7} \times 7\sqrt{10} \times 7\sqrt{10} = 1540\ \text{cm}^2$.

40. A wire, in the form of a circle, encloses an area 3118.5 cm^2. It is now bent to form a rectangle whose length and breadth are very nearly in the ratio 7 : 4. The length of the rectangle, in cm, is:
वृत्त के आकार का एक तार 3118.5 सेमी^{2} क्षेत्रफल को घेरता है। अब इसे एक आयत बनाने के लिए मोड़ा गया है जिसकी लंबाई और चौड़ाई लगभग 7 : 4 के अनुपात में है। आयत की लंबाई, सेमी में, है: (Take $\pi = \frac{22}{7}$)

Solution:

When the wire is bent in the form of a circle:

Area of the circle = 3118.5.

$\pi r^2 = 3118.5$.

$\frac{22}{7} \times r^2 = 3118.5$

$r^2 = \frac{3118.5 \times 7}{22}$.

$r^2 = 992.25$

$r = 31.5$.

Let the length of the rectangle = $7x$,

and the breadth of the rectangle = $4x$

Circumference of circle = Perimeter of rectangle.

$2\pi r = 2(7x + 4x)$.

$2 \times \frac{22}{7} \times 31.5 = 22x$.

$22x = 9 \times 22$

$x = 9$.

Length of the rectangle $= 7x = 7 \times 9 = 63$ cm.

41. The area of the largest triangle that can be inscribed in a semicircle of radius 6 cm is:
6 सेमी त्रिज्या वाले अर्धवृत्त में अंकित किए जा सकने वाले सबसे बड़े त्रिभुज का क्षेत्रफल है:

Solution:

According to the question:

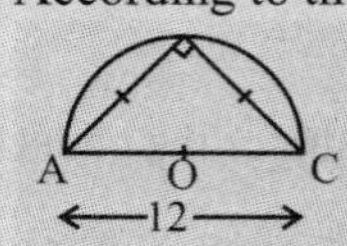

On inscribed largest triangle ABC in semicircle of radius 6cm.

6 सेमी त्रिज्या वाले अर्धवृत्त में अंकित सबसे बड़े त्रिभुज ABC पर।

$AB = BC$

$AC = 2 \times 6 = 12$ cm and

$\angle B$ = Right angle (By theorem)

Now $AB = BC = \frac{12}{\sqrt{2}} = 6\sqrt{2}$

Area of $\triangle ABC = \frac{1}{2} \times 6\sqrt{2} \times 6\sqrt{2} = 36\ \text{cm}^2$

42. The area of the largest triangle that can be inscribed in a semi-circle of radius 4 cm in square centimetres is:
वर्ग सेंटीमीटर में 4 सेमी त्रिज्या वाले अर्धवृत्त में अंकित किए जा सकने वाले सबसे बड़े त्रिभुज का क्षेत्रफल है:

Solution:

Radius of semi-circle = 4 cm

Diameter of semi-circle (AB) = 8 cm

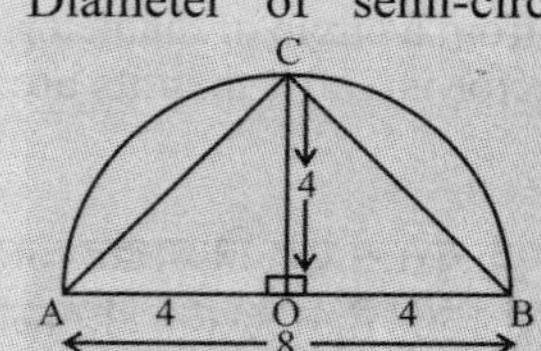

त्रिभुज की ऊंचाई अर्धवृत्त की त्रिज्या के बराबर होगी।

Area of the largest triangle $= \frac{1}{2} \times 8 \times 4 =$ 16 cm^2

43. In the given figure, PQR is a quadrant whose radius is 7 cm. A circle is inscribed in the quadrant as shown in the figure. What is the area (in cm^2) of the circle?

दिए गए चित्र में, PQR एक चतुर्थांश है जिसकी त्रिज्या 7 सेमी है। जैसा कि चित्र में दिखाया गया है, चतुर्भुज में एक वृत्त अंकित है। वृत्त का क्षेत्रफल (सेमी^{2} में) क्या है?

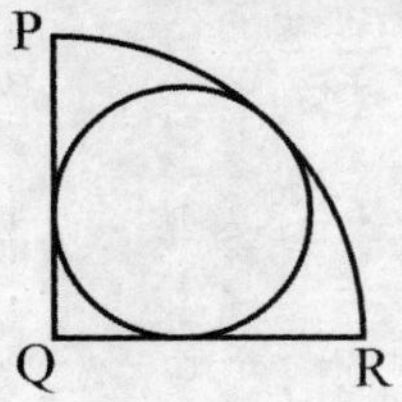

Solution:

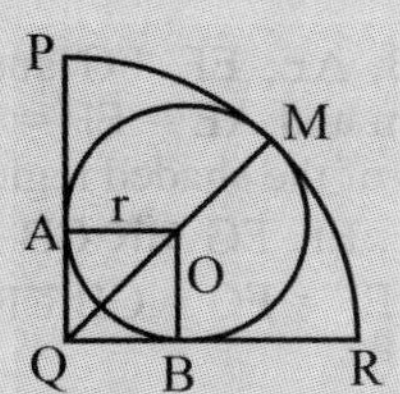

$\angle AOB = 90°$

$OA = OB = r$ (Suppose)

QM = 7 सेमी

$OQ = (7 - r)$ cm

$AOBQ$ वर्ग

$\sqrt{2}r = 7 - r$

$r(\sqrt{2} + 1) = 7$

$r = \frac{7}{\sqrt{2} - 1}$

Area of circle:

$= \frac{22}{7} \times 7 \times 7(\sqrt{2} - 1)(\sqrt{2} - 1)$

$= 22 \times 7(3 - 2\sqrt{2})$

$= (462 - 308\sqrt{2})$ cm^2

44. In the given figure, PQRS is a square whose side is 8 cm. PQS and QPR are two quadrants. A circle is placed touching both the quadrants and the square as shown in the figure. What is the area (in cm^2) of the circle?

दी गई आकृति में, PQRS एक वर्ग है जिसकी भुजा 8 सेमी है। PQS और QPR दो चतुर्थांश हैं। जैसा कि चित्र में दिखाया गया है, चतुर्भुज और वर्ग दोनों को छूते हुए एक वृत्त रखा गया है। वृत्त का क्षेत्रफल (सेमी ^{2} में) क्या है?

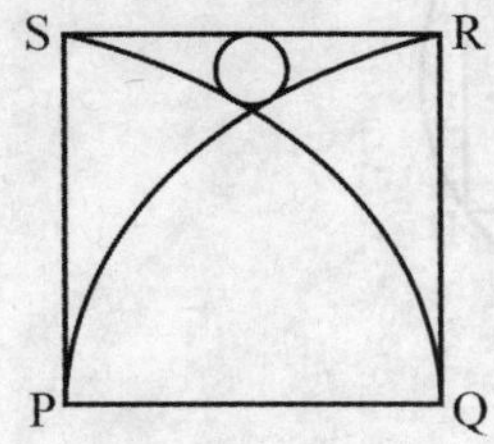

Solution:

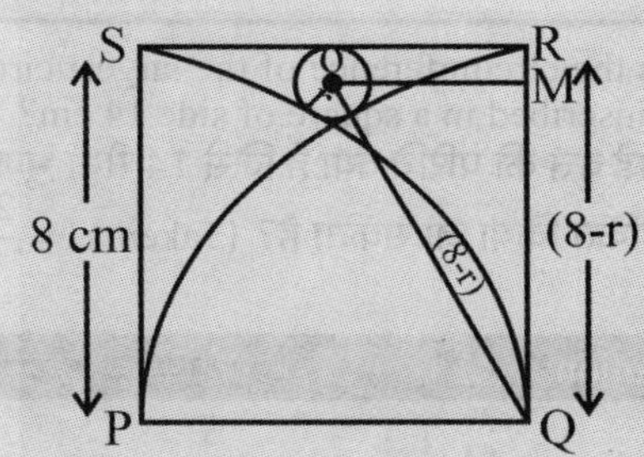

Let radius of the circle = r cm.

QM = (8 - r), OQ = (8 + r), OM = SR/2

OM $= \frac{8}{2} = 4$

In $\triangle OMQ$:

$(OM)^2 = (OQ)^2 - (MQ)^2$

$4^2 = (8 + r)^2 - (8 - r)^2$

$\Rightarrow 16 = 16 + 2r \times 16$

$\Rightarrow r = \frac{1}{2}$

Area of circle $= \pi r^2$

$= \frac{22}{7} \times \left(\frac{1}{2}\right)^2 = \frac{11}{14}$ cm^2

45. What is the area of the largest square which can be inscribed in a circle of radius 28 cm?

उस सबसे बड़े वर्ग का क्षेत्रफल कितना है जिसे 28 सेमी त्रिज्या वाले एक वृत्त में अंकित किया जा सकता है?

Solution:

Diagonal of square = Diameter of circle

$\sqrt{2}a = 2R = 2 \times 28 = 56$, Where R = radius of circle, a = side of square

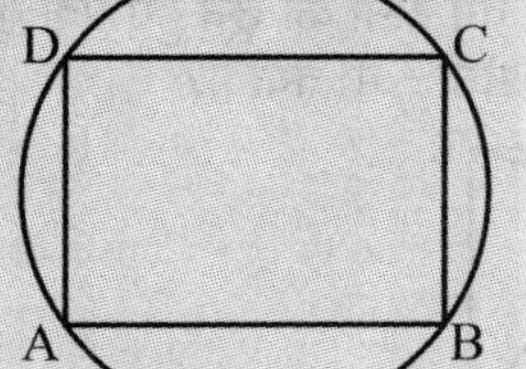

$a = \frac{56}{\sqrt{2}}$ cm

Area of square = $a^2 = \frac{56 \times 56}{2} = 28 \times 56 =$ 1568 cm^2

46. What is the circumference of the largest circle which can be inscribed in a square of side 14 cm?

सबसे बड़े वृत्त की परिधि क्या है जिसे 14 सेमी भुजा वाले एक वर्ग में अंकित किया जा सकता है? (Taken $\pi = \frac{22}{7}$)

Solution:

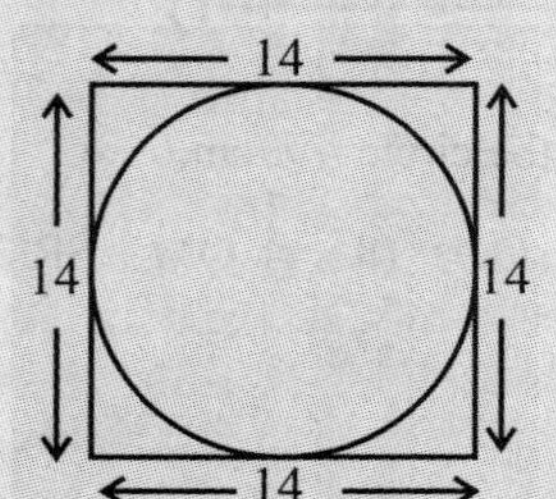

Side of square = 14 cm

Diameter of circle = 14 cm

Radius of circle = $\frac{14}{2} = 7$ cm

Then, Perimeter of circle = $2\pi r$

$= 2 \times \frac{22}{7} \times 7 = 44$ cm

47. A rectangle with the largest possible area is drawn inside a semicircle of radius 2 cm. Then, the ratio of the lengths of the largest to the smallest side of this rectangle is:

2 सेमी त्रिज्या वाले अर्धवृत्त के अंदर सबसे बड़े संभावित क्षेत्रफल वाला एक आयत खींचा गया है। फिर, इस आयत की सबसे बड़ी और सबसे छोटी भुजा की लंबाई का अनुपात है।

Solution:

Area = $x \cdot 2\sqrt{4 - x^2}$

Area = $2x\sqrt{4 - x^2}$

$A'(x) = 0$

$2\sqrt{4 - x^2} - \frac{2x \cdot 2x}{2\sqrt{4 - x^2}} = 0$

$2\sqrt{4 - x^2} = \frac{2x^2}{\sqrt{4 - x^2}}$

$8 - 2x^2 = 2x^2$

$4x^2 = 8$

$x^2 = 2$

$\sqrt{4 - x^2} = \sqrt{2}$

Ratio: $2\sqrt{2} : \sqrt{2} = 2 : 1$

48. In the given figure, AB, AE, EF, FG and GB are semicircles. AB = 56 cm and AE = EF = FG = GB. What is the area (in cm²) of the shaded region?

दिए गए चित्र में, AB, AE, EF, FG और GB अर्धवृत्त हैं। AB = 56 सेमी और AE = EF = FG = GB छायांकित क्षेत्र का क्षेत्रफल (सेमी ^{2} में) क्या है?

Solution:

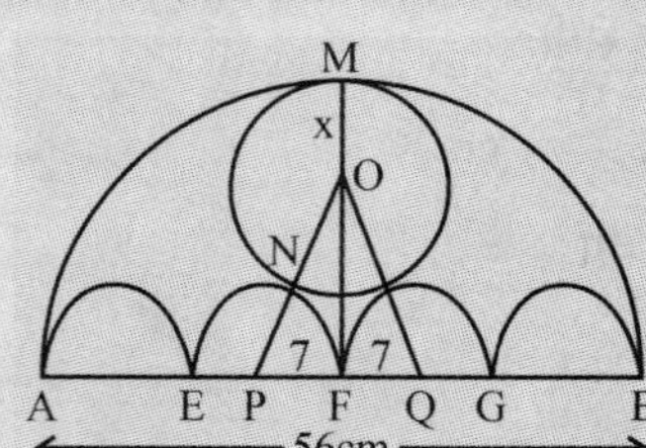

$AE = EF = FG = GB = \frac{56}{4} = 14$

Let the radius of the shaded region (MO) = x cm.

In $\triangle POF$:

$PO^2 = OF^2 + PF^2$

$$(7+x)^2 = (28-x)^2 + 7^2$$

$$49 + x^2 + 14x = 784 + x^2 - 56x + 49$$

$$70x = 784$$

$$x = \frac{784}{70} = 11.2$$

Area of the shaded region:

$$\pi x^2 = \frac{22}{7} \times 11.2 \times 11.2 = 394.24 \text{ cm}^2$$

49. In the given figure, PQRS is a rectangle, and a semicircle with SR as diameter is drawn. A circle is drawn as shown in the figure. If $QR = 7$ cm, then what is the radius (in cm) of the small circle?

दिये गये चित्र में PQRS एक आयत है और एक अर्द्धवृत्त जिसका व्यास SR है। एक अन्य वृत्त बनाया गया है जो आयत और अर्द्धवृत्त को स्पर्श करता है यदि QR = 7 सेमी तब छोटे वृत्त की त्रिज्या क्या होगी?

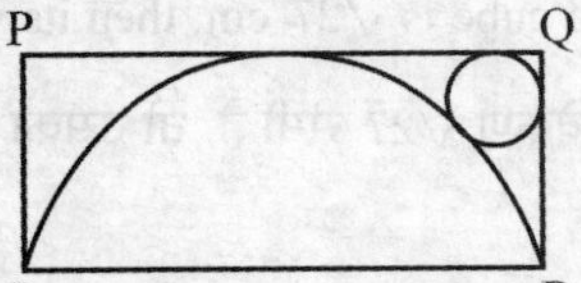

Solution:

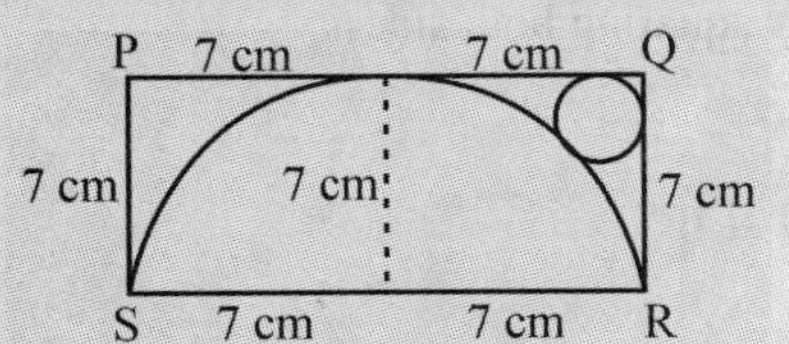

$r = a(3 - 2\sqrt{2})$

$r = 7(3 - 2\sqrt{2}) = 21 - 14\sqrt{2}$

50. In an isosceles triangle ABC, $AB = AC = 17.5$, $BC = 21$, infinite circles are made inside this triangle as shown in the figure. Find the sum of the perimeter of all the circles.

समद्विबाहु त्रिभुज ABC में, AB = AC = 17.5, BC = 21 है त्रिभुज के अंदर असंख्य वृत्त बनाये गये है तब इन सभी वृत्तो के परिमाप का योग होगा?

Solution:

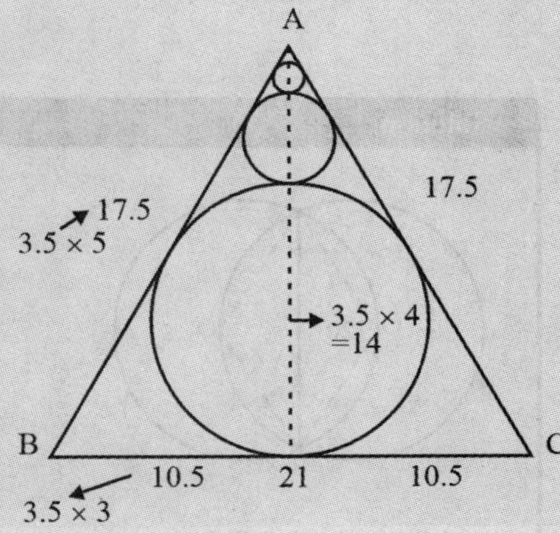

$14 = 2(r_1 + r_2 + r_3 + r_4 \ . \ . \ . \ )$

$14\pi = 2\pi(r_1 + r_2 + r_3 + r_4 \ . \ . \ . \ )$

$14 \times \frac{22}{7} = 44$ cm

51. These three circles of equal radii touch each other as shown in the figure. The radius of each circle is 1 cm. What is the area of the shaded region?

एक समान त्रिज्या वाले तीन वृत्त एक दूसरे को स्पर्श करते है तब छायांकित भाग का क्षेत्रफल क्या होगा यदि प्रत्येक वृत्त की त्रिज्या 1 सेमी है?

Solution:

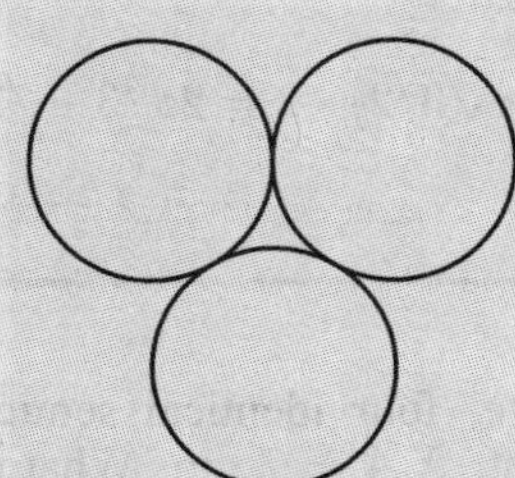

Area $= (\sqrt{3} - \pi/2)r^2$

$((1\sqrt{3} - \pi)/2)$

$23 - 7/2$ $352 - 7/2$ $25 - 2$/None

Area $= (53-)$ 82

52. Two circles of unit radii are so drawn that the center of each lies on the circumference of the other. The area of the region common to both circles is:

यूनिट त्रिज्या वाले दो वृत्त एक दूसरे को इस प्रकार काटते हैं

कि उनके केंद्र दूसरे की परिधि पर स्थित हैं। तब दोनों वृत्तों के उभयनिष्ठ क्षेत्र का क्षेत्रफल क्या होगा?

Solution:

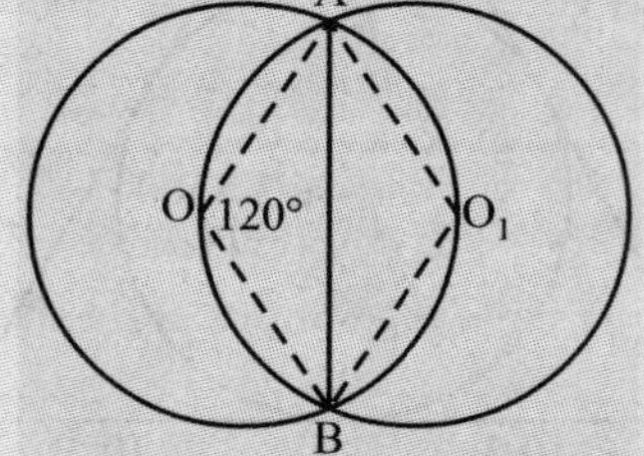

$\triangle AOO_1$ and $\triangle BOO_1$ are formed.

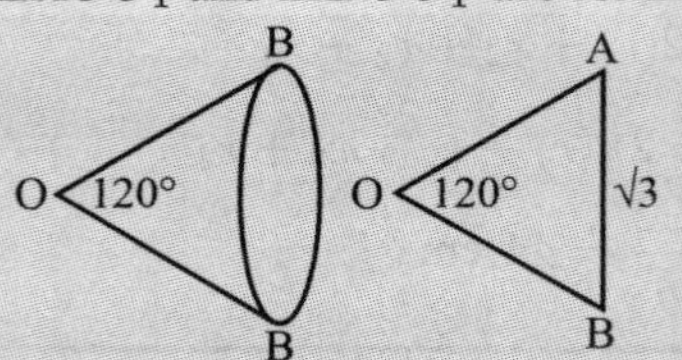

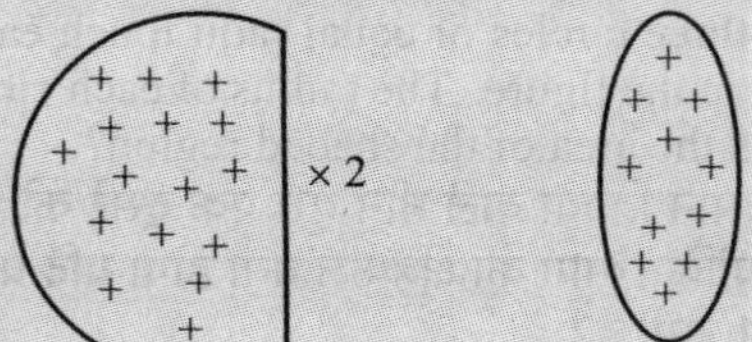

$\therefore \angle AOB = 120°$

$(\pi/3 - 1/2 \times 1 \times \sqrt{3}/2) \times 2 = 2\pi/3 - \sqrt{3}/2 = (4\pi - 3\sqrt{3})/6$

53. In the given figure, four identical semicircles are drawn in a quadrant. $XA = 7$ cm. What is the area of the shaded region?
दी गई आकृति में, चार समरूप अर्द्धवृत्तों को एक वृत्त के चतुर्थ भाग में बनाया गया है। यदि $XA = 7$ सेमी तो छायांकित भाग का क्षेत्रफल क्या है?

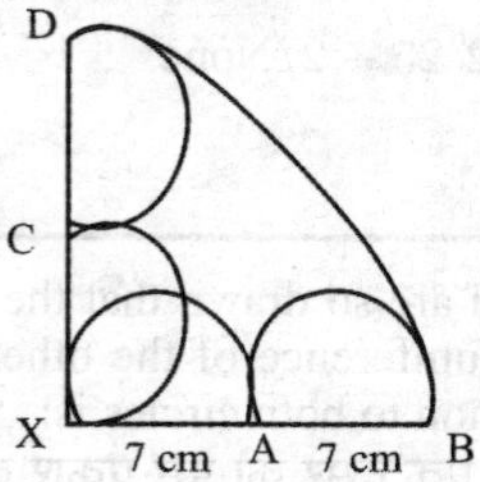

Solution:

Area of 4 semicircles $= n\left(\frac{7}{2}\right)^2 \times 4 = 154$

Area of common part $= \frac{4}{7}r^2$

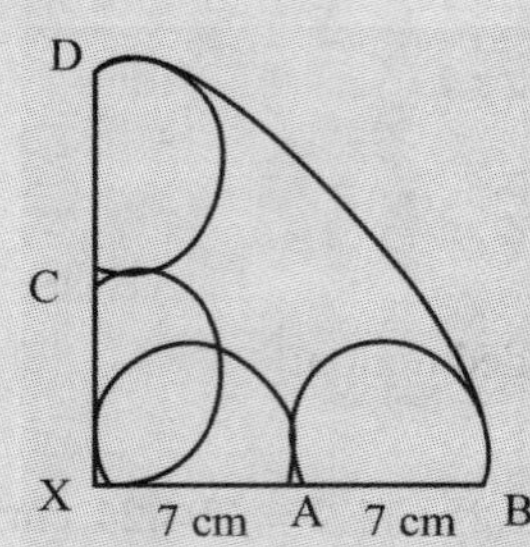

$154 \times 4 \times \frac{4}{7} \times \frac{49}{4} = 77 + \frac{4}{7} \times \frac{49}{4} = 77 + 7 = 84 \text{ cm}^2$

Type 5: Problem based on Cube & Cuboid

54. If the diagonal of a cube is $\sqrt{27}$ cm, then its volume is:
यदि किसी घन का विकर्ण $\sqrt{27}$ सेमी है, तो उसका आयतन है:

Solution:

Diagonal of cube $= \sqrt{3}a = \sqrt{27}$

By squaring both sides:

$3a^2 = 27$

$a^2 = 9$

$a = 3$

Volume of cube $= a^3 = 3^3 = 27 \text{ cm}^3$

55. If each edge of a cube is increased by 10%, then the percentage increase in its surface area is:
यदि किसी घन के प्रत्येक किनारे को 10% बढ़ा दिया जाए, तो उसके सतह क्षेत्र में प्रतिशत वृद्धि है:

Solution:

Given: Surface area of cube $= 6a^2$

$$\text{Increased percentage} = \frac{x+y+xy}{100}\%$$
$$= 10 + 10 + \frac{100}{100} = 21\%$$

56. If the volume of a cube is 64 cm^3, then find its total surface area.
यदि किसी घन का आयतन 64 सेमी^{3} है, तो उसका कुल पृष्ठीय क्षेत्रफल ज्ञात कीजिए।

Solution:

Volume of cube $= a^3$

$a^3 = 64 \Rightarrow a = 4$ cm

Total surface area of cube $= 6a^2 = 6 \times 4^2 =$ 96 cm^2

57. From the body of a solid cube of edge 7 cm, a solid sphere is removed. The volume of the remaining solid was found to be 163 cm^3. What is the diameter (in cm) of the sphere?
7 सेमी किनारे वाले एक ठोस घन के शरीर से एक ठोस गोला निकाला जाता है। शेष ठोस का आयतन 163 सेमी^{3} पाया गया। गोले का व्यास (सेमी में) क्या है? ($\pi = \frac{22}{7}$)

Solution:

According to the question:

Volume of solid left = Volume of cube − Volume of sphere

$$163 = \frac{1}{3}\left[(7)^3 - \frac{4}{3}\pi r^3\right]$$
$$343 - 163 = \frac{4}{3} \times \frac{22}{7} r^3$$
$$\frac{539}{33} = \frac{4}{3} \times \frac{22}{7} r^3$$
$$r^3 = \frac{7^3}{2^3}$$
$$r = \frac{7}{2} = 3.5 \text{ cm}$$
$$2r = 2 \times 3.5 = 7 \text{ cm}$$

Hence, the diameter of the sphere is 7 cm.

58. A metal cube of edge 18 cm is melted to form three smaller cubes, which are unequal in dimensions. If the edges of two smaller cubes are 9 cm and 15 cm, what is the surface area (in cm^2) of the third smaller cube?
18 सेमी किनारे वाले एक धातु के घन को पिघलाकर तीन छोटे घन बनाए जाते हैं, जो आयामों में असमान हैं। यदि दो छोटे घनों के किनारे 9 सेमी और 15 सेमी हैं, तो तीसरे छोटे घन का सतह क्षेत्रफल (सेमी^{2}) क्या है?

Solution:

Volume of bigger cube = 3 × [Volume of smaller cubes]

$18^3 = 9^3 + 15^3 + x^3$

$5832 = 729 + 3375 + x^3$

$x^3 = 5832 - 4104 = 1728$ cm^3

$x = 12$ cm

Surface area of third smaller cube $= 6x^2 = 6 \times 12^2 = 864$ cm^2

59. A solid cube of volume 13824 cm^3 is cut into 8 cubes of equal volumes. The ratio of the surface area of the original cube to the sum of the surface areas of three of the smaller cubes is:
13824 सेमी^{3} आयतन वाले एक ठोस घन को समान आयतन वाले 8 घनों में काटा जाता है। मूल घन के पृष्ठीय क्षेत्रफल का तीन छोटे घनों के पृष्ठीय क्षेत्रफलों के योग से अनुपात क्या है?

Solution:

Volume of first cube (a_1^3) $= 13824$ cm^3

Length of side of first cube (a_1) $= 24$ cm

Volume of second cube:

$8(a_2)^3 = 13824$

$a_2^3 = 1728$

$a_2 = 12$ cm

According to the question:

Surface area of original cube $= 6a_1^2 = 6 \times 24^2$

Surface area of three small cubes $= 3 \times 6 \times 12^2$

$$\frac{\text{Surface area of original cube}}{\text{Surface area of three small cubes}} = \frac{4}{3}$$

60. A solid cube has side 8 cm. It is cut along diagonals of the top face to get 4 equal parts. What is the total surface area (in cm^2) of each part?

एक ठोस घन की भुजा 8 सेमी है। इसे 4 बराबर भाग प्राप्त करने के लिए शीर्ष सतह के विकर्णों के साथ काटा जाता है। प्रत्येक भाग का कुल सतह क्षेत्रफल (सेमी ^{2} में) क्या है?

Solution:

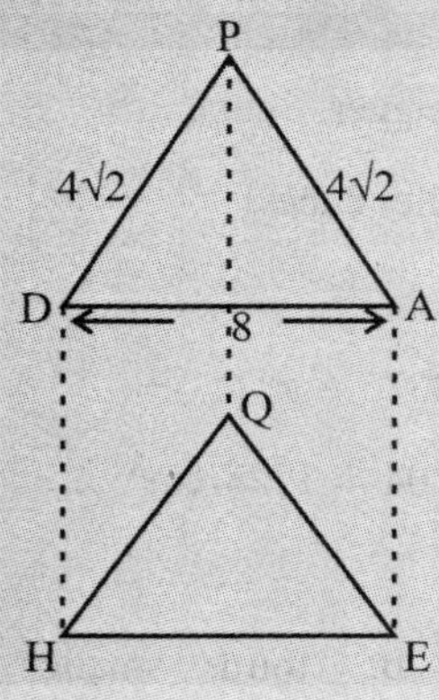

Total surface area of each part:

$=$ Area of $\triangle PDA$ + Area of $\triangle QHE$ + Area of $\triangle DHE$ + Area of $PQEA$ + Area of $PDQH$

$= \left(\frac{1}{2} \times 4\sqrt{2} \times 4\sqrt{2} + \frac{1}{2} \times 4\sqrt{2} \times 4\sqrt{2}\right) + 8 \times 8 + 4\sqrt{2} \times 8 + 8 \times 4\sqrt{2}$

$= 32 + 64 + 64\sqrt{2} = 96 + 64\sqrt{2}$ cm^2

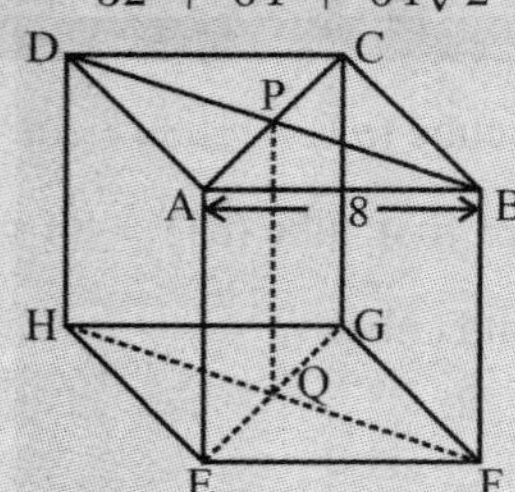

61. A cube is placed inside a cone of radius 20 cm and height 10 cm, one of its faces being on the base of the cone and vertices of the opposite face touching the cone. What is the length (in cm) of the side of the cube?
एक घन को 20 सेमी त्रिज्या और 10 सेमी ऊंचाई वाले शंकु के अंदर रखा गया है, इसका एक चेहरा शंकु के आधार पर है और विपरीत चेहरे का शीर्ष शंकु को छू रहा है। घन की भुजा की लंबाई (सेमी में) क्या है?

Solution:

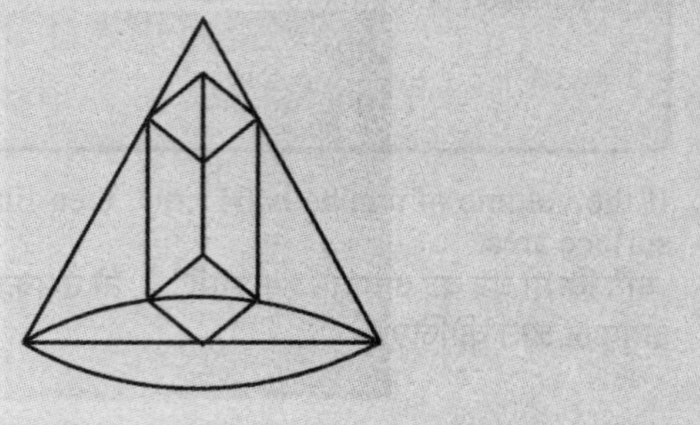

Let the side of the cube be a cm.

Length of the side of the cube inside the cone:

$\frac{\sqrt{2}rh}{h+\sqrt{2}r}$ (Formula)

$= \frac{\sqrt{2} \times 20 \times 10}{10 + \sqrt{2} \times 20}$

$= \frac{200\sqrt{2}}{10 + 20\sqrt{2}}$

$= \frac{20\sqrt{2}}{1 + 2\sqrt{2}}$

$= \frac{20 \times 1.414}{1 + 2 \times 1.414}$

$= \frac{28.28}{3.828} = 7.38 \approx 8$ cm

62. A solid cube is cut into three cuboids of the same volume. What is the ratio of the surface area of the cube to the sum of the surface areas of any two of the cuboids so formed?
एक ठोस घन को समान आयतन के तीन घनाभों में काटा जाता है। इस प्रकार बने घनाभों में से किन्हीं दो के पृष्ठीय क्षेत्रफलों के योग से घन के पृष्ठीय क्षेत्रफल का अनुपात क्या है?

Solution:

Let the side of the cube be $3a$.

By cutting a solid cube into three cuboids:

Length of the cuboid $= a$, Breadth $= 3a$,

Height $= 3a$

Ratio $= \frac{\text{Surface area of cube}}{\text{Sum of surface areas of two cuboids}}$

$= \frac{6(\text{Side})^2}{2\ [2(lb+bh+hl)]} = \frac{6 \times 9a^2}{4 \times 15a^2} = \frac{9}{10} =$

$9:10$

63. A right circular cylinder of maximum volume is cut out from a solid wooden cube. The material left is what percent of the volume (nearest to an integer) of the original cube?
अधिकतम आयतन का एक लंब वृत्तीय बेलन एक ठोस लकड़ी के घन से काटा जाता है। बचा हुआ पदार्थ मूल घन के आयतन (एक पूर्णांक के निकटतम) का कितना प्रतिशत है?

Solution:

Let the side of the cube be r cm.

Volume of the cube $= r^3$ cm^3

Maximum volume of the cylinder cut from the cube:

$= \pi\left(\frac{r}{2}\right)^2 \times r = \frac{\pi}{4}r^3$

Remaining volume $= r^3 - \frac{\pi}{4}r^3$

$= \frac{28r^3 - 22r^3}{28} = \frac{6}{28}r^3$

Required percentage $= \frac{\frac{6}{28}r^3 \times 100}{r^3} \approx 21\%$

64. A string is wound on the exterior four walls of a cube of side a cm, starting at point C and ending at point D. Find the length of the string P.
एक घन के बाहरी चार दीवारों पर एक धागा लपेटा जाता है जहाँ घन की भुजा a सेमी है। यदि धागे का प्रारंभिक सिरा बिंदु C पर और अंतिम सिरा D पर है, तब धागे की लंबाई क्या होगी?

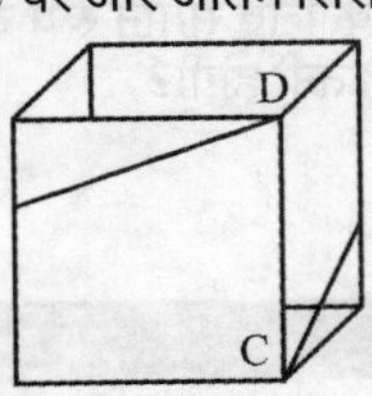

Solution:

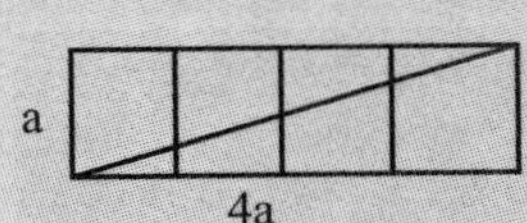

Length of string:

$= \sqrt{(4a)^2 + a^2} = \sqrt{16a^2 + a^2} = \sqrt{17a^2} = \sqrt{17}a$

65. If the length of a cuboid is 20 m, breadth is 5 m, and height is 4 m, then find the length of the diagonal of the cuboid.
यदि एक घनाभ की लंबाई 20 मीटर, चौड़ाई 5 मीटर और ऊंचाई 4 मीटर है, तो घनाभ के विकर्ण की लंबाई ज्ञात करें।

Solution:

Length of the diagonal of the cuboid:

$\sqrt{l^2 + b^2 + h^2} = \sqrt{20^2 + 5^2 + 4^2} = \sqrt{400 + 25 + 16} = \sqrt{441} = \sqrt{21 \times 21} = 21$ m

66. A rectangular water tank is 1.5 m high, 3 m long and 3.5 m wide. How much water (in litres) can it hold?
एक आयताकार पानी की टंकी 1.5 मीटर ऊंची, 3 मीटर लंबी और 3.5 मीटर चौड़ी है। इसमें कितना पानी (लीटर में) समा सकता है?

Solution:

Given, Dimensions of the rectangular tank are 1.5 m high, 3 m long, and 3.5 m wide.

Water holds in tank = Volume of rectangular tank

दिया गया है, आयताकार टैंक का आयाम 1.5 मीटर ऊंचा, 3 मीटर लंबा और 3.5 चौड़ा है। टैंक में भरा पानी = आयताकार टैंक का आयतन

$= 3 \times 3.5 \times 1.5 = 15.75$ m^3 $=$ 15750 litres [1m^3 = 1000 litres]

67. A solid cube whose each face diagonal is $128\sqrt{2}$ cm is moulded to form a cuboid. The length and breadth of the cuboid are 512 cm and 160 cm, respectively. What is the height of the cuboid?
एक ठोस घन, जिसके प्रत्येक फलक का विकर्ण $128\sqrt{2}$ सेमी है, को घनाभ बनाने के लिए ढाला गया है। एक घनाभ की लंबाई और चौड़ाई क्रमश: 512 सेमी और 160 सेमी है। घनाभ की ऊंचाई कितनी है?

Solution:

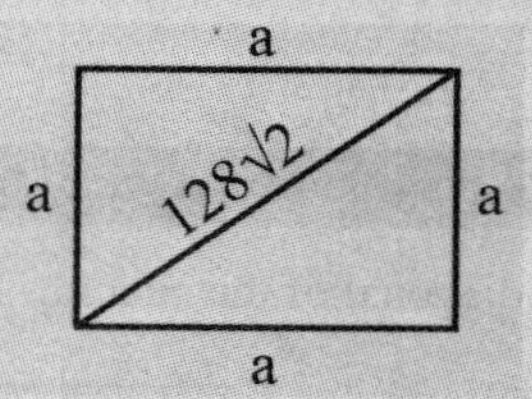

$a^2 + a^2 = (128\sqrt{2})^2$

$2a^2 = 128 \times 128 \times 2$

$a = 128$

$\text{Height of cuboid} = \dfrac{\text{Volume of cube}}{l \times b}$

$= \dfrac{128 \times 128 \times 128}{512 \times 160} = 25.6 \text{ cm}$

68. Four cubes, each of edge 5 cm, are joined end to end. What is the total surface area of the resulting cuboid?
चार घन, प्रत्येक का किनारा 5 सेमी है, एक सिरे से दूसरे सिरे तक जुड़े हुए हैं। परिणामी घनाभ का कुल सतह क्षेत्रफल क्या है?

Solution:

Length of cuboid made by joining 4 cubes of 5 cm: $5 \times 4 = 20$ cm

Breadth = 5 cm, Height = 5 cm

Total surface area of cuboid $= 2(l \times b + b \times h + h \times l)$

$= 2(20 \times 5 + 5 \times 5 + 5 \times 20) = 2 \times (100 + 25 + 100) = 2 \times 225 = 450 \text{ cm}^2$

69. The areas of three adjacent faces of a cuboidal solid block of wax are 216 cm^2, 96 cm^2, and 144 cm^2. It is melted, and 8 cubes of the same size are formed from it. What is the lateral surface area (in cm^2) of 3 such cubes?
मोम के एक घनाकार ठोस खंड के तीन आसन्न सतहों का क्षेत्रफल 216 सेमी^{2}, 96 सेमी^{2} और 144 सेमी^{2} है। इसे पिघलाया जाता है और इससे एक ही आकार के 8 घन बनते हैं। ऐसे 3 घनों का पार्श्व पृष्ठीय क्षेत्रफल (सेमी ^{2} में) क्या है?

Solution:

Given:

$l \times b = 216 \text{ cm}^2$ ---(i)

$b \times h = 96 \text{ cm}^2$ ---(ii)

$h \times l = 144 \text{ cm}^2$ ---(iii)

On multiplying equations (i), (ii), and (iii):

$l^2b^2h^2 = 216 \times 96 \times 144$

$lbh = \sqrt{2^3 \times 3^3 \times 2^5 \times 3 \times 2^4 \times 3^2} = 2^6 \times 3^3 = 1728$

Let the side of the cube $= a$ cm.

Volume of cuboid = Volume of cube:

$1728 = 8 \times a^3$

$a^3 = 216 \quad a = 6$

Lateral surface area of 3 cubes:

$= 3 \times 4a^2 = 3 \times 4(6)^2 = 432 \text{ cm}^2$

70. A 15 m deep well with radius 2.8 m is dug, and the earth taken out from it is spread evenly to form a platform of breadth 8 m and height 1.5 m. What will be the length of the platform?
2.8 मीटर त्रिज्या वाला 15 मीटर गहरा कुआँ खोदा जाता है और उसमें से निकाली गई मिट्टी को 8 मीटर चौड़ाई और 1.5 मीटर ऊंचाई का एक मंच बनाने के लिए समान रूप से फैलाया जाता है। प्लेटफार्म की लम्बाई कितनी होगी?
(Take $\pi = \dfrac{22}{7}$)

Solution:

Volume of soil taken out from well = Volume of platform.

$\dfrac{22}{7} \times 2.8 \times 2.8 \times 15 = l \times 8 \times 1.5$

$22 \times 6 \times 2.8 = l \times 12$

$l = 11 \times 2.8 = 30.8 \text{ m}$

71. The internal length of a room is two times its breadth and three times its height. The total cost of painting its four walls at the rate of ₹ 25/m² is ₹ 3,600. What is the cost of laying a carpet on its floor at the rate of ₹ 90.50/m²?

एक कमरे की आंतरिक लंबाई उसकी चौड़ाई से दो गुना और ऊंचाई से तीन गुना है। इसकी चार दीवारों को रंगने की कुल लागत ₹ 25/वर्ग मीटर की दर से ₹ 3,600 है। ₹ 90.50/m^2 की दर से इसके फर्श पर कालीन बिछाने की लागत क्या है?

Solution:

Let length of floor $= 6K$, Breadth $= 3K$,

Height $= 2K$

Area of four walls $= \dfrac{3600}{25} = 144 \text{ m}^2$

Area of four walls = Height × Perimeter:

$= 2K \times 2(6K + 3K)$

$144 = 36K^2$

$K = 2$

Area of floor $= 18K^2 = 72 \text{ m}^2$

Total cost of laying carpet $= 72 \times 90.50 = 6516$

72. A cuboid of size 100 cm × 80 cm × 60 cm is cut into eight identical parts by three cuts. What is the total surface area (in square cm) of all the eight parts?

100 सेमी × 80 सेमी × 60 सेमी आकार के एक घनाभ को तीन कट द्वारा आठ समान भागों में काटा गया। सभी आठ भागों का कुल सतह क्षेत्रफल (वर्ग सेमी में) क्या है?

Solution:

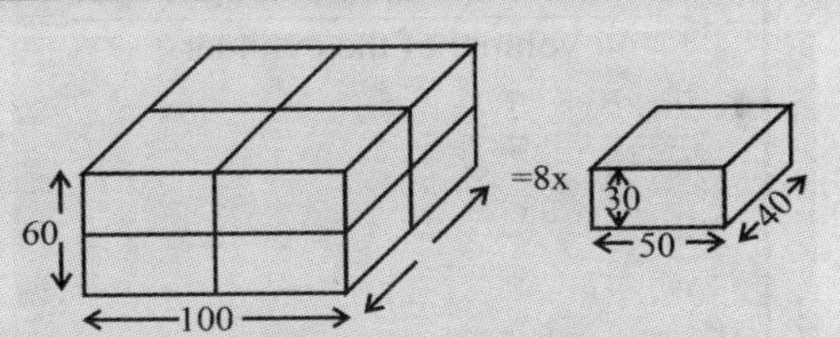

$= 8 \times [2(lb + bh + hl)]$

$= 8 \times [2(30 \times 50 + 50 \times 40 + 40 \times 30)]$

$= 8 \times [2(1500 + 2000 + 1200)]$

$= 8 \times [2(4700)] = 8 \times [9400] = 75200 \text{ cm}^2$

73. Find the longest pole that can be placed in a room 16 m long, 8 m broad, and 11 m high.

एक कमरा जो 16 मी लम्बा, 8 मी चौड़ा तथा 11 मी ऊँचा है उसमें रखी जाने वाली सबसे लंबी छड़ की लंबाई क्या होगी?

Solution:

$D = \sqrt{16^2 + 8^2 + 11^2}$

$= \sqrt{256 + 64 + 121}$

$D = \sqrt{441} = 21 \text{ m}$

74. The length and breadth of a cuboid are in 2:1, and its height is 3.5 m. If the area of its walls (including doors) is 210 m^2, then what is its volume?

यदि किसी घनाभ की लंबाई और चौड़ाई का अनुपात 2:1 है और इसकी ऊँचाई 3.5 मी है। यदि इसकी दीवारों का क्षेत्रफल (दरवाजा सहित) 210 m^2 है तो इसका आयतन क्या है?

Solution:

Let $l = 2x, \quad b = x.$

$\text{LSA} = 2(l + b)h = 210$

$2 \times 3x \times 3.5 = 210$

$x = 10$

Volume $= 20 \times 10 \times 3.5 = 700 \text{ m}^3$

75. The curved surface area and volume of a cylindrical pole are 132 m^2 and 528 m^3, respectively. What is the height (in m) of the pole? ($\pi = \dfrac{22}{7}$)

बेलनाकार खंभे का घुमावदार सतह क्षेत्र और आयतन क्रमशः 132 m^2 और 528 m^3 है। खंभे की ऊंचाई (मीटर में) क्या है? ($\pi = \dfrac{22}{7}$)

Solution:

According to the question: $2\pi rh = 132 \text{ m}^2$

$rh = \dfrac{132}{2\pi}$ ---(i)

$\pi r^2 h = 528 \text{ m}^3$

$\pi r \cdot rh = 528$ (from equation i)

$\pi r \cdot \dfrac{132}{2\pi} = 528$

$r = 8 \text{ m}$

$2\pi rh = 132 \text{ m}^2$ (given)

$h = \dfrac{132 \times 7}{2 \times 22 \times 8} = \dfrac{21}{8}$

$h = 2\dfrac{5}{8} \text{ m}$

76. A solid right-circular cylinder, whose radius of the base is 15 cm and height is 12 cm, is melted and moulded into a solid right-circular cone, whose radius of the base is 24 cm. What will be the height of this cone?

एक ठोस लंबवृत्तीय बेलन, जिसके आधार की त्रिज्या 15 सेमी और ऊंचाई 12 सेमी है, को पिघलाकर ठोस लंबवृत्ताकार शंकु में ढाला जाता है, जिसके आधार की त्रिज्या 24 सेमी है। इस शंकु की ऊंचाई क्या होगी?

Solution:

Given:

Radius of cylinder (r) = 15cm,

Height (h) = 12 cm

Radius of cone (R) = 24cm

Height = H (Let)

According to the question:

Volume of cylinder = Volume of cone (after melted and moulded):

$\pi(15)^2 \times 12 = \frac{1}{3}\pi \times 24^2 \times H$

$H = \frac{8100}{576} = 14.0625$ cm

77. The curved surface area of a right circular cylinder is 616 cm^2 and the area of its base is 38.5 cm^2. What is the volume (in cm^3) of the cylinder? (Take $\pi = \frac{22}{7}$)

एक लम्ब वृत्तीय बेलन का वक्र पृष्ठ क्षेत्रफल 616 सेमी^{2} है और इसके आधार का क्षेत्रफल 38.5 सेमी^{2} है। सिलेंडर का आयतन (सेमी^{3} में) क्या है?

Solution:

Given:

CSA = 616 cm^2, and Area of base = 38.5 cm^2

$\pi r^2 = 38.5$

$r^2 = \frac{38.5 \times 7}{22} = 12.25$

$r = 3.5$ cm

$2\pi rh = 616$ cm^2 (Given)

$2 \times \frac{22}{7} \times 3.5 \times h = 616$

$h = 28$ cm

Volume of cylinder = $\pi r^2 h$:

$\frac{22}{7} \times 3.5 \times 3.5 \times 28 = 88 \times 12.25 = 1078$ cm^3

78. Two similar jugs have heights of 8 cm and 12 cm, respectively. If the capacity of the smaller jug is 80 cm^3, what is the capacity of the bigger jug (in cm^3)?

दो समान जगों की ऊंचाई क्रमश: 8 सेमी और 12 सेमी है। यदि छोटे जग की क्षमता 80 सेमी^{3} है, तो बड़े जग की क्षमता (सेमी^{3} में) क्या है?

Solution:

According to the question:

Ratio of heights = $\frac{8}{12} = \frac{2}{3}$

Ratio of volumes = (Ratio of heights)^3

$\frac{\text{Volume of smaller jug}}{\text{Volume of bigger jug}} = \left(\frac{2}{3}\right)^3$

$\frac{80}{\text{Volume of bigger jug}} = \frac{8}{27}$

Volume of bigger jug = $\frac{80 \times 27}{8} = 270$ cm^3

79. If the ratio of the curved surface area and volume of a right circular cylinder is 5 : 7, then its radius is:

यदि एक लम्ब वृत्तीय बेलन के वक्र पृष्ठीय क्षेत्रफल और आयतन का अनुपात 5 : 7 है, तो इसकी त्रिज्या है:

Solution:

Given:

$\frac{\text{Curved surface area of the cylinder}}{\text{Volume of the cylinder}} = \frac{5}{7}$

$\frac{2\pi rh}{\pi r^2 h} = \frac{5}{7}$

$\frac{2}{r} = \frac{5}{7}$

$2 \times 7 = 5r$

$r = \frac{14}{5} = 2.8$ units

80. A solid metallic right circular cylinder has diameter 32 cm and height 9 cm. It is melted and recast into a solid sphere. What is the radius (in cm) of the sphere?

एक ठोस धात्विक लंब वृत्तीय बेलन का व्यास 32 सेमी और ऊंचाई 9 सेमी है। इसे पिघलाकर पुन: एक ठोस गोले का रूप दे दिया जाता है। गोले की त्रिज्या (सेमी में) क्या है?

Solution:

Solid metallic circular cylinder is melted and recast into a solid sphere.

Volume of sphere = Volume of right circular cylinder:

$\frac{4}{3}\pi R^3 = \pi r^2 h$

$\frac{4}{3}R^3 = 16 \times 16 \times 9$

$R^3 = 64 \times 27$

$R = 12$ cm

81. The length of a metallic pipe is 7.56 m. Its external and internal radii are 2.5 cm and 1.5 cm, respectively. If 1 cm^3 of the metal weighs 7.5 g, then the weight of the pipe is: (Take $\pi = \frac{22}{7}$)

एक धातु पाइप की लंबाई 7.56 मीटर है। इसकी बाहरी और आंतरिक त्रिज्याएँ क्रमश: 2.5 सेमी और 1.5 सेमी हैं। यदि 1 सेमी^{3} धातु का वजन 7.5 ग्राम है, तो पाइप का वजन है: ($\pi = \frac{22}{7}$)

Solution:

$R = 2.5$ cm, $r = 1.5$ cm, $l = 7.56$ m $= 7.56 \times 100 = 756$ cm

Volume of metallic pipe $= \pi(R^2 - r^2)l$:

$= \frac{22}{7} \times [(2.5)^2 - (1.5)^2] \times 756$

$= \frac{22}{7} \times (2.5 + 1.5)(2.5 - 1.5) \times 756$

$= 22 \times 4 \times 1 \times 108 = 9504$ cm^3

Weight of pipe $= \frac{9504 \times 7.5}{1000}$ kg $= 71.28$ kg

82. A right circular cylinder is formed where A = sum of total surface area and the area of the two bases, B = the curved surface area of this cylinder. If $A : B = 3 : 2$ and the volume of the cylinder is 4312 cm^3, then what is the sum of the areas (in cm^2) of the two bases of this cylinder?

एक लंब वृत्ताकार बेलन बनता है A = कुल सतह क्षेत्रफल और दो आधारों के क्षेत्रफल का योग। B = इस सिलेंडर का घुमावदार सतह क्षेत्र। यदि $A : B = 3 : 2$ और सिलेंडर का आयतन 4312 सेमी^{3} है, तो इस सिलेंडर के दोनों आधारों के क्षेत्रफल का योग (सेमी^{2}) क्या है?

Solution:

Given:

$A = 2\pi r(h + r) + 2\pi r^2$, $B = 2\pi rh$, $\frac{A}{B} = \frac{3}{2}$

$\frac{2\pi r(h+r) + 2\pi r^2}{2\pi rh} = \frac{3}{2}$

$3h = 4r + 2h$

$\frac{h}{r} = \frac{4}{1}$

$\pi r^2 h = 4312$

$\frac{22}{7} \times r^2 \times 4r = 4312$

$r^3 = \frac{4312 \times 7}{22 \times 4} = 49$

$r = 7$ cm

Sum of the areas of two bases $= 2\pi r^2$:

$= 2 \times \frac{22}{7} \times 49 = 308$ cm^2

83. The ratio of the curved surface area and total surface area of a right circular cylinder is 2 : 5. If the total surface area is 3080 cm^2, then what is the volume (in cm^3) of the cylinder?

एक लम्ब वृत्तीय बेलन के वक्र पृष्ठ क्षेत्रफल और कुल पृष्ठ क्षेत्रफल का अनुपात 2 : 5 है। यदि कुल सतह क्षेत्रफल 3080 सेमी^{2} है, तो सिलेंडर का आयतन (सेमी^{3}) कितना है?

Solution:

$\frac{2\pi rh}{2\pi r(r+h)} = \frac{2}{5}$

$\frac{h}{r+h} = \frac{2}{5}$

$5h = 2r + 2h \implies 3h = 2r$

$h = \frac{2}{3}r$ ---(i)

Total surface area $= 3080$ cm^2:

$2\pi r(r + h) = 3080$

$r(r + \frac{2}{3}r) = \frac{3080 \times 7}{22 \times 2}$

$$r \times \frac{5r}{3} = 490 \implies r^2 = 98 \times 3 \implies r = 7\sqrt{6} \text{ cm}$$

$$h = \frac{14}{3}\sqrt{6} \text{ cm}$$

Volume of cylinder:

$$\frac{22}{7} \times 7\sqrt{6} \times 7\sqrt{6} \times \frac{14}{3} = 4312\sqrt{6} \text{ cm}^3$$

84. A cylinder of radius 4.5 cm and height 12 cm just fits in another cylinder completely with their axes perpendicular. What is the radius (in cm) of the second cylinder?

त्रिज्या 4.5 सेमी और ऊंचाई 12 सेमी का एक सिलेंडर दूसरे सिलेंडर में पूरी तरह से अपनी धुरी के लंबवत फिट बैठता है। दूसरे सिलेंडर की त्रिज्या (सेमी) में क्या है?

Solution:

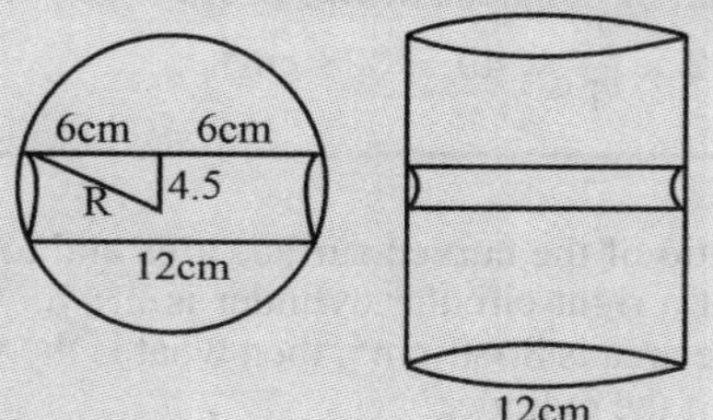

Radius of the second cylinder (R):

$$R^2 = (12)^2 + (4.5)^2$$

$$R^2 = 144 + 20.25 = 164.25$$

$$R = 7.5 \text{ cm}$$

85. A right circular cylinder has height 28 cm and radius of base 14 cm. Two hemispheres of radius 7 cm each are cut from each of the two bases of the cylinder. What is the total surface area (in cm²) of the remaining part?

एक लम्ब वृत्तीय बेलन की ऊंचाई 28 सेमी और आधार की त्रिज्या 14 सेमी है। बेलन के दोनों आधारों में से प्रत्येक से 7 सेमी त्रिज्या वाले दो अर्धगोले काटे जाते हैं। शेष भाग का कुल पृष्ठीय क्षेत्रफल (सेमी ^{2} में) क्या है?

Solution:

Radius of cylinder (R) = 14 cm, h = 28 cm

Radius of hemisphere (r) = 7 cm

Area of upper and lower surface after cutting the hemispheres:

$$= 2\pi(R^2 - 2r^2) = 2\pi(14 \times 14 - 2 \times 7 \times 7)$$

$$= 2 \times \frac{22}{7} \times 98 = 616 \text{ cm}^2$$

Area of curved surface of cylinder $= 2\pi Rh$:

$$= 2 \times \frac{22}{7} \times 14 \times 28 = 2464 \text{ cm}^2$$

Area of curved surface of the hemispheres $= 8\pi r^2$:

$$= 8 \times \frac{22}{7} \times 7 \times 7 = 1232 \text{ cm}^2$$

Surface area of the remaining part:

$$= 616 + 1232 + 2464 = 4312 \text{ cm}^2$$

86. N solid metallic spherical balls are melted and recast into a cylindrical rod whose radius is 3 times that of a spherical ball and height is 4 times the radius of a spherical ball. The value of N is:

N ठोस धातु की गोलाकार गेंदों को पिघलाया जाता है और एक बेलनाकार छड़ में बदल दिया जाता है, जिसकी त्रिज्या गोलाकार गेंद की त्रिज्या से 3 गुना और ऊंचाई गोलाकार गेंद की त्रिज्या से 4 गुना होती है। N का मान:

Solution:

Let the radius of the solid metallic sphere be r:

$N\times$ Volume of one sphere = Volume of cylindrical rod

$$N \times \frac{4}{3}\pi r^3 = \pi(3r)^2 \cdot 4r$$

$$N = 9 \times 3 = 27$$

87. The radius of the base of a right circular cylinder is increased by 20%. By what percent should its height be reduced so that its volume remains the same as before?

एक लम्ब वृत्तीय बेलन के आधार की त्रिज्या 20% बढ़ जाती है। इसकी ऊंचाई कितने प्रतिशत कम की जानी चाहिए ताकि इसका आयतन पहले जैसा ही रहे?

Solution:

Let the radius of the base of the cylinder be r:

Radius after increment $= r \cdot \dfrac{120}{100} = \dfrac{6r}{5}$

Initial volume of the cylinder $= \pi r^2 h$

New volume of the cylinder:

$$= \pi \left(\frac{6r}{5}\right)^2 \cdot h_1 = \frac{36}{25}\pi r^2 h_1$$

Since volume is constant:

$$\pi r^2 h = \frac{36}{25}\pi r^2 h_1$$

$$\frac{h_1}{h} = \frac{25}{36}$$

Required decrement percentage:

$$= \frac{36-25}{36} \times 100 = \frac{11}{36} \times 100 = 30\frac{5}{9}\%$$

88. A solid cylinder has radius of base 14 cm and height 15 cm. 4 identical cylinders are cut from each base as shown in the given figure. Height of small cylinder is 5 cm. What is the total surface area (in cm²) of the remaining part?

एक ठोस बेलन के आधार की त्रिज्या 14 सेमी और ऊँचाई 15 सेमी है। दिए गए चित्र में दिखाए अनुसार प्रत्येक आधार से 4 समान सिलेंडर काटे जाते हैं। छोटे बेलन की ऊँचाई 5 सेमी है। शेष भाग का कुल पृष्ठीय क्षेत्रफल (सेमी^{2}) क्या है?

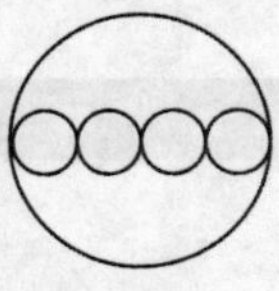

Solution:

Radius of base of solid cylinder $(r_1) = 14$ cm

Height of solid cylinder $(h_1) = 15$ cm

Radius of small cylinder $(r_2) = \dfrac{14}{4} = \dfrac{7}{2}$ cm

Height of small cylinder $(h_2) = 5$ cm

Total surface area:

$$= 2\pi r_1(h_1 + r_1) + 8 \times 2\pi r_2 h_2$$

$$= 2 \times \frac{22}{7} \times 14(15+14) + 8 \times 2 \times \frac{22}{7} \times \frac{7}{2} \times 5$$

$$= 44(58) + 44(20) = 44 \times 78 = 3432 \text{ cm}^2$$

89. The radius of the base of a solid cylinder is 7 cm, and its height is 21 cm. It is melted and converted into small bullets. Each bullet is of the same size, consisting of two parts: a cylinder and a hemisphere on one of its bases. The total height of the bullet is 3.5 cm, and the radius of the base is 2.1 cm. Approximately how many complete bullets can be obtained?

एक ठोस बेलन के आधार की त्रिज्या 7 सेमी है और इसकी ऊंचाई 21 सेमी है। इसे पिघलाकर छोटी-छोटी गोलियों में बदल दिया जाता है। प्रत्येक गोली एक ही आकार की है। प्रत्येक गोली में दो भाग होते हैं अर्थात। इसके एक आधार पर एक बेलन और एक अर्धगोला। गोली की कुल ऊंचाई 3.5 सेमी और आधार की त्रिज्या 2.1 सेमी है। लगभग कितनी पूर्ण गोलियाँ प्राप्त की जा सकती हैं?

Solution:

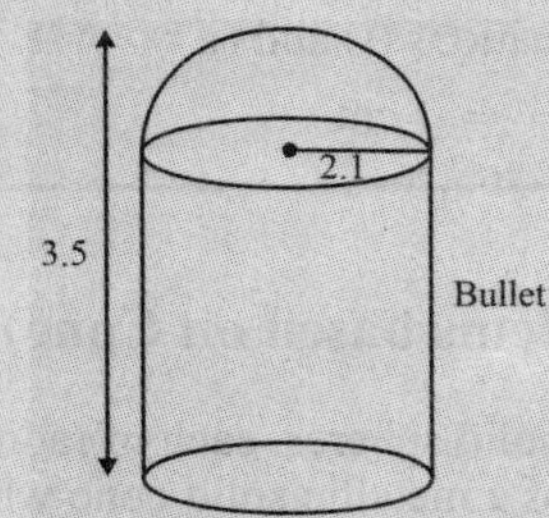

Let the total number of bullets $= n$.

Height of the cylindrical part of the bullet (h):

$h = 3.5 - 2.1 = 1.4$ cm

Total volume of the solid cylinder: Volume of solid cylinder = n × Volume of one bullet

$$\pi R^2 H = n \times \left[\pi \times 2.1^2 \times 1.4 + \frac{2}{3}\pi \times 2.1^3\right]$$

$$\pi \times 7^2 \times 21 = n \times \pi \times 2.1^2[1.4 + 1.4]$$

$$n = \frac{7 \times 7 \times 21}{2.1 \times 2.1 \times 2.8} = 83.33 \approx 83$$

90. A hollow cylinder is made of metal. The difference between the outer and inner curved surface area of this cylinder is 352 cm^2. The height of the cylinder is 28 cm. If the total surface area of this hollow cylinder is 2640 cm^2, then what are the inner and outer radii (in cm)?

एक खोखला बेलन धातु का बना होता है। इस सिलेंडर के बाहरी और आंतरिक घुमावदार सतह क्षेत्र के बीच का अंतर 352 सेमी^{2} है। बेलन की ऊँचाई 28 सेमी है। यदि इस खोखले सिलेंडर का कुल सतह क्षेत्रफल 2640 सेमी^{2} है, तो आंतरिक और बाहरी त्रिज्या (सेमी में) क्या हैं?

Solution:

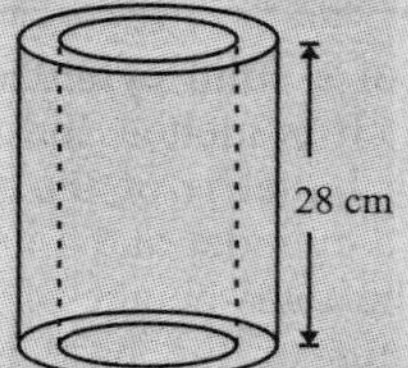

Let the outer and inner radii be R and r, respectively.

$2\pi(R-r)h = 352\ \text{cm}^2$

$2 \times \frac{22}{7} \times (R-r) \times 28 = 352$

$R - r = 2$ ---(i)

Total surface area:

$2\pi(R+r)(R-r+h) = 2640$

$2 \times \frac{22}{7} \times (R+r)(2+28) = 2640$

$R + r = 14$ ---(ii)

By solving equations (i) and (ii):

$R = 8$ cm, $\quad r = 6$ cm

Type 6 (Problems based on Cone)

91. How many small solid spheres, each of radius 5 mm, can be made out of a metallic solid cone whose base has a radius of 21 cm and height 30 cm?

एक धात्विक ठोस शंकु, जिसके आधार की त्रिज्या 21 सेमी और ऊंचाई 30 सेमी है, से 5 मिमी त्रिज्या वाले कितने छोटे ठोस गोले बनाए जा सकते हैं?

Solution:

Given:

Radius of sphere $(r) = 5$ mm

Radius of cone $(R) = 21$ cm $= 210$ mm

Height of cone $(H) = 30$ cm $= 300$ mm

According to the question:

Volume of cone $= N \times$ Volume of sphere

$\frac{1}{3}\pi R^2 H = N \times \frac{4}{3}\pi r^3$

$\pi \times 210^2 \times 300 = N \times 4 \times 5^3$

$210 \times 210 \times 300 = N \times 500$

$441 \times 60 = N$

$N = 26460$

92. A container in the shape of a right circular cone, whose radius and depth are equal, gets completely filled by 128000 spherical droplets, each of diameter 2 mm. What is the radius (in cm) of the container?

एक लंब वृत्ताकार शंकु के आकार का एक कंटेनर, जिसकी त्रिज्या और गहराई बराबर है, 128000 गोलाकार बूंदों से पूरी तरह भर जाता है, प्रत्येक का व्यास 2 मिमी है। कंटेनर की त्रिज्या (सेमी में) क्या है?

Solution:

Let the radius of the cone be r, then the height will also be r. According to the question:

Volume of cone $=$ Volume of one spherical drop $\times$ 128000

$\frac{1}{3}\pi r^2 r = 128000 \times \frac{4}{3}\pi \left(\frac{2}{20}\right)^3$ (as 1 cm = 10 mm)

$r^3 = 4 \times \frac{1}{1000} \times 128000$

$r^3 = 512$

$r = 8$ cm

93. A heap of wheat is in the form of a cone whose base diameter is 8.4 m and height is 1.75 m. The heap is to be covered by canvas. What is the area (in m^2) of the canvas required?

गेहूं का एक ढेर एक शंकु के आकार का है जिसका आधार व्यास 8.4 मीटर और ऊंचाई 1.75 मीटर है। ढेर को कैनवास से ढक देना है। कैनवास का आवश्यक क्षेत्रफल (वर्ग मीटर) कितना है?

Solution:

Given:

Diameter of cone $= 8.4$ m, Radius of cone $= \frac{8.4}{2} = 4.2$ m, Height of cone $(h) = 1.75$ m

Curved surface area of cone $= \pi r l$, where $l =$ slant height:

$l = \sqrt{r^2 + h^2} = \sqrt{(4.2)^2 + (1.75)^2}$

$l = \sqrt{17.64 + 3.0625} = \sqrt{20.7025} = 4.55$ m

Curved surface area of cone:

$= \pi r l = \frac{22}{7} \times 4.2 \times 4.55$

$= 22 \times 6 \times 4.55 = 60.06\ m^2$

94. If the diameter of the base of a cone is 18 cm and its curved surface area is $424\ \frac{2}{7}\ cm^2$, then its height will be:

यदि एक शंकु के आधार का व्यास 18 सेमी है और इसका घुमावदार सतह क्षेत्र $424\ \frac{2}{7}$ सेमी^{2} है, तो इसकी ऊंचाई होगी:

Solution:

Given:

Curved surface area of cone $= 424\ \frac{2}{7}\ cm^2$,

Diameter of cone = 18 cm

Radius of cone $= \frac{\text{diameter}}{2} = \frac{18}{2} = 9$ cm

Curved surface area of cone $= \pi r l = \frac{2970}{7}\ cm^2$

$\frac{22}{7} \times 9 \times l = \frac{2970}{7}$

$l = \frac{2970}{22 \times 9} = 15$ cm

Height of cone:

$h = \sqrt{l^2 - r^2}$

$= \sqrt{15^2 - 9^2} = \sqrt{225 - 81} = \sqrt{144} =$ 12 cm

95. The ratio of the height and the diameter of a right circular cone is 6 : 5, and its volume is 2200 cm^3. What is its slant height?

एक लंब वृत्तीय शंकु की ऊंचाई और व्यास का अनुपात 6 : 5 है और इसका आयतन 2200 सेमी^{3} है। इसकी तिरछी ऊंचाई क्या है?

Solution:

According to the question:

$\frac{h}{2r} = \frac{6}{5}$

$\frac{h}{r} = \frac{12}{5}$

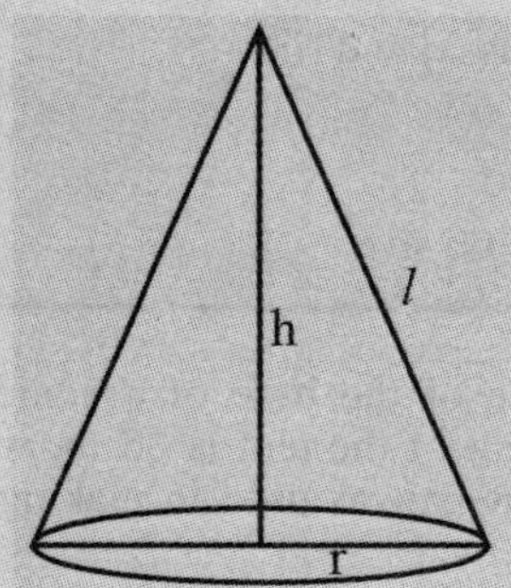

Let $h = 12x$, $r = 5x$:

$l = \sqrt{r^2 + h^2} = \sqrt{(5x)^2 + (12x)^2} = \sqrt{25x^2 + 144x^2} = 13x$

Volume of cone:

$\frac{1}{3}\pi r^2 h = \frac{2200}{7}$

$\frac{1}{3} \times \frac{22}{7} \times (5x)^2 \times 12x = \frac{2200}{7}$

$x^3 = \frac{2200 \times 3 \times 7}{22 \times 25 \times 12} = 1$

$x = 1$

$l = 13$ cm

96. A conical tent has to accommodate 25 persons. Each person must have 4 m^2 of space on the ground and 80 m^3 of air to breathe. Find the height of the tent.

एक शंक्वाकार तंबू में 25 व्यक्ति रह सकते हैं। प्रत्येक व्यक्ति के पास सांस लेने के लिए जमीन पर 4 वर्ग मीटर जगह और 80 घन मीटर हवा होनी चाहिए। तम्बू की ऊँचाई ज्ञात कीजिए।

Solution:

According to the question:

Space for 1 person = 4 m^2

Space for 25 persons = $25 \times 4 = 100\ m^2$

Air for breathing of 1 person = 80 m^3

Air for breathing of 25 persons = $80 \times 25 = 2000\ m^3$

Area of base $= \pi r^2 = 100\ m^2$ ---(i)

Volume of tent $= \frac{\pi r^2 h}{3} = 2000\ m^3$ ---(ii)

From equations (i) and (ii):

$$\frac{\pi r^2 h}{3 \cdot \pi r^2} = \frac{2000}{100}$$

$$h = 20 \times 3 = 60 \text{ m}$$

97. The circumference of the base of a conical tent is 66 m. If the height of the tent is 36 m, what is the area (in m^2) of the canvas used in making the tent? (Take $\pi = \frac{22}{7}$)

एक शंक्वाकार तंबू के आधार की परिधि 66 मीटर है। यदि तंबू की ऊंचाई 36 मीटर है, तो तंबू बनाने में प्रयुक्त कैनवास का क्षेत्रफल (वर्ग मीटर) क्या है? लिया गया $\pi = \frac{22}{7}$

Solution:

Perimeter of base = 66 m:

$$2 \times \frac{22}{7} \times r = 66$$

$$r = \frac{21}{2} \text{ m}$$

Slant height (l):

$$l = \sqrt{\left(\frac{441}{4} + 1296\right)} = 37.5 \text{ m}$$

Area of canvas:

$$= \frac{22}{7} \times \frac{21}{2} \times 37.5 = 1237.5 \text{ m}^2$$

98. Four identical cones each of radius 10.5 cm and height 14 cm are cut from a cuboid of dimensions 30 cm × 32 cm × 40 cm (base of each cone lies on the surface of cuboid). What is the total surface area (in cm^2) of the remaining solid?

30 सेमी × 32 सेमी × 40 सेमी आयाम वाले एक घनाभ से 10.5 सेमी त्रिज्या और 14 सेमी ऊंचाई वाले चार समान शंकु काटे जाते हैं (प्रत्येक शंकु का आधार घनाभ की सतह पर स्थित होता है)। शेष ठोस का कुल सतह क्षेत्रफल (सेमी^{2}) क्या है?

Solution:

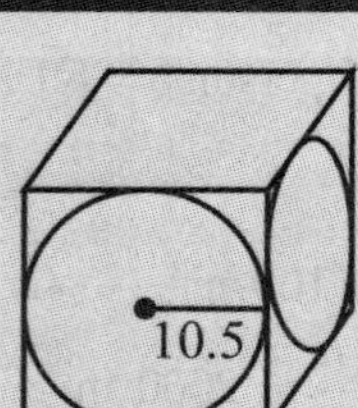

Four cones are made on the lateral surface.

Slant height of cone (l):

$$l = \sqrt{\left(\frac{21}{2}\right)^2 + 14^2} = \frac{35}{2} \text{ cm}$$

Total surface area of remaining solid:

$$= 2(lb + bh + hl) + 4\pi rl - 4\pi r^2$$

$$= 2[960 + 1280 + 1200] + 4 \times \frac{22}{7} \times 10.5\ (17.5 - 10.5)$$

$$= 2 \times 3440 + 4 \times \frac{22}{7} \times 10.5 \times 7$$

$$= 6880 + 924 = 7804 \text{ cm}^2$$

99. If the radius of the base of a cone is doubled, and the volume of the new cone is three times the volume of the original cone, then what will be the ratio of the height of the original cone to that of the new cone?

यदि किसी शंकु के आधार की त्रिज्या दोगुनी कर दी जाए और नए शंकु का आयतन मूल शंकु के आयतन का तीन गुना हो जाए, तो मूल शंकु की ऊंचाई और नए शंकु की ऊंचाई का अनुपात क्या होगा?

Solution:

Radius of original cone (r_1) = r cm

Height of original cone (h_1):

Radius of new cone (r_2) = $2r$ cm

Height of new cone (h_2):

According to the question:

$$\frac{1}{3}\pi r_2^2 h_2 = 3 \times \frac{1}{3}\pi r_1^2 h_1$$

$$4r^2 h_2 = 3r^2 h_1$$

$$\frac{h_1}{h_2} = \frac{4}{3}$$

100. A conical vessel whose internal base radius is 18 cm and height 60 cm is full of liquid. The entire liquid of the vessel is emptied into a cylindrical vessel with an internal radius of 15 cm. The height (in cm) to which the liquid rises in the cylindrical vessel is?

एक शंक्वाकार बर्तन जिसका आंतरिक आधार त्रिज्या 18 सेमी और ऊंचाई 60 सेमी है, एक तरल से भरा है। बर्तन का पूरा तरल पदार्थ 15 सेमी आंतरिक त्रिज्या वाले एक बेलनाकार बर्तन में खाली कर दिया जाता है। बेलनाकार पात्र में द्रव जिस ऊँचाई (सेमी) में तक चढ़ता है वह कितनी है?

Solution:

Given:

Inner radius of conical pot $(r) = 18$ cm,

Height of conical pot $(h) = 60$ cm

The liquid fills the cylindrical vessel from the conical vessel. The volume of the liquid remains the same:

$$\frac{1}{3}\pi(18)^2(60) = \pi(15)^2h$$
$$h = \frac{18 \times 18 \times 60}{3 \times 15 \times 15} = 28.8 \text{ cm}$$

101. The volume of a circular cone is equal to the volume of a sphere whose radius is half of the radius of the base of the cone. What is the ratio of the radius of the base of the cone to its height?

एक वृत्ताकार शंकु का आयतन उस गोले के बराबर होता है जिसकी त्रिज्या शंकु के आधार की त्रिज्या की आधी होती है। शंकु के आधार की त्रिज्या और उसकी ऊंचाई का अनुपात क्या है?

Solution:

Let the radius of the base of the perpendicular cone $= 2x$ and the radius of the sphere $= x$.

According to the question:

Volume of perpendicular cone $=$ Volume of sphere

$$\frac{1}{3}\pi r_1^2 h = \frac{4}{3}\pi r_2^3$$
$$\frac{1}{3}\pi(2x)^2 h = \frac{4}{3}\pi x^3$$
$$4x^2h = 4x^3$$
$$x/h = 1/1$$

Hence, the ratio of the radius of the base of the cone to its height:

$$\frac{2x}{h} = 2/1 \text{ or } 2:1$$

102. The volume of a perpendicular circular cone is 1232 cm^3 and its height is 24 cm. What is the area of its curved surface?

लंब वृत्ताकार शंकु का आयतन 1232 सेमी^{3} है और इसकी ऊँचाई 24 सेमी है। इसकी घुमावदार सतह का क्षेत्रफल कितना है? $(\pi = \frac{22}{7})$

Solution:

Let the radius and height be r and h:

According to the question:

$$\frac{1}{3}\pi r^2 h = 1232 \quad (\text{given } h = 24)$$
$$\frac{1}{3} \times \frac{22}{7} \times r^2 \times 24 = 1232$$
$$r^2 = 49$$
$$r = 7$$

Slant height of cone (l):

$$l = \sqrt{r^2 + h^2} = \sqrt{49 + 576} = 25 \text{ cm}$$

Area of curved surface of cone:

$$\pi rl = \frac{22}{7} \times 7 \times 25 = 550 \text{ cm}^2$$

103. The radii of the two circular faces of the frustum of a cone of height 14 cm are 5 cm and 2 cm. What is its volume in cm^3?

14 सेमी ऊँचाई वाले शंकु के छिन्नक के दो गोलाकार फलकों की त्रिज्याएँ 5 सेमी और 2 सेमी हैं। इसका आयतन (सेमी^{3}) में क्या है? $(\pi = \frac{22}{7})$

Solution:

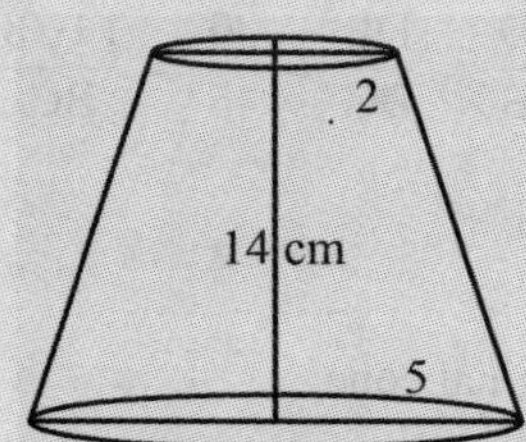

Volume of frustum of cone:

$$= \frac{1}{3}\pi\ (R^2 + r^2 + Rr)\ h$$
$$= \frac{1}{3} \times \frac{22}{7} \times 14\ [5^2 + 2^2 + 5 \times 2]$$
$$= \frac{1}{3} \times \frac{22}{7} \times 14 \times 39$$
$$= 572 \text{ cm}^3$$

104. The height of a cone is 45 cm. It is cut at a height of 15 cm from its base by a plane parallel to its base.

If the volume of the smaller cone is 18480 cm^3, then what is the volume (in cm^3) of the original cone?

एक शंकु की ऊँचाई 45 सेमी है। इसे इसके आधार से 15 सेमी की ऊंचाई पर इसके आधार के समानांतर एक समतल द्वारा काटा जाता है। यदि छोटे शंकु का आयतन 18480 सेमी^{3} है, तो मूल शंकु का आयतन (सेमी^{3}) में क्या है?

Solution:

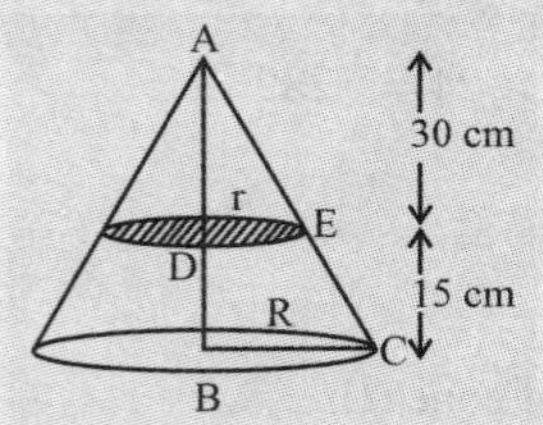

Volume of small cone=18480

$\frac{1}{3}\pi r^2 h = 18480$

$\frac{1}{3} \times \frac{22}{7} \times r^2 \times 30 = 18480$

$r^2 = 12 \times 7 \times 7$

$r = 14\sqrt{3}$

Using similarity:

$\frac{\Delta ADE}{\Delta ABC} = \frac{\text{AD}}{\text{AB}} = \frac{\text{DE}}{\text{BC}} = \frac{30}{45} = \frac{14\sqrt{3}}{\text{BC}}$

$R = 21\sqrt{3}$ cm

Volume of original cone:

$\frac{1}{3}\pi R^2 H = \frac{1}{3} \times \frac{22}{7} \times (21\sqrt{3})^2 \times 45 =$

62370 cm^3

105. A cone of radius 90 cm and height 120 cm stands on its base. It is cut into 3 parts by 2 cuts parallel to its base such that the height of the three parts (from top to bottom) are in the ratio 1 : 2 : 3. What is the total surface area (in cm^2) of the middle part?

90 सेमी त्रिज्या और 120 सेमी ऊंचाई का एक शंकु इसके आधार पर खड़ा है। इसे इसके आधार के समानांतर 2 कट द्वारा 3 भागों में इस प्रकार काटा जाता है कि तीनों भागों की ऊंचाई (ऊपर से नीचे तक) 1 : 2 : 3 के अनुपात में हो। मध्य भाग का कुल सतह क्षेत्रफल (सेमी^{2}) क्या है?

Solution:

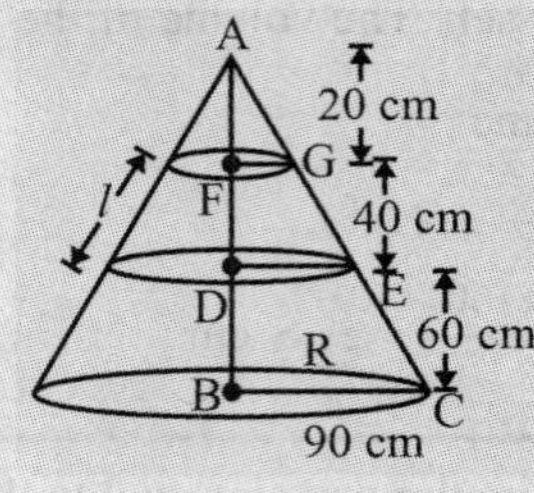

Using similarity for $\Delta AFG \sim \Delta ABC$:

$\frac{20}{120} = \frac{\text{FG}}{90}$

$\text{FG} = 15 \text{ cm} = r$

Using similarity for $\Delta ADE \sim \Delta ABC$:

$\frac{60}{120} = \frac{\text{DE}}{90}$

$\text{DE} = 45 \text{ cm} = R$

Slant height (l):

$l = \sqrt{h^2 + (R-r)^2} = \sqrt{40^2 + (45-15)^2} = \sqrt{1600+900} =$ 50 cm

Surface area of the middle part:

$= \pi(r+R)l + \pi r^2 + \pi R^2$

$= \pi[60 \times 50 + 225 + 2025]$

$= \frac{22}{7} \times 5250 = 16500 \text{ cm}^2$

106. The radius of the base of a hollow cone is 8 cm and its height is 15 cm. A sphere of the largest radius is put inside the cone. What is the ratio of the radius of the base of the cone to the radius of the sphere?

एक खोखले शंकु के आधार की त्रिज्या 8 सेमी है और इसकी ऊंचाई 15 सेमी है। शंकु के अंदर सबसे बड़ी त्रिज्या का एक गोला रखा गया है। शंकु के आधार की त्रिज्या और गोले की त्रिज्या का अनुपात क्या है?

Solution:

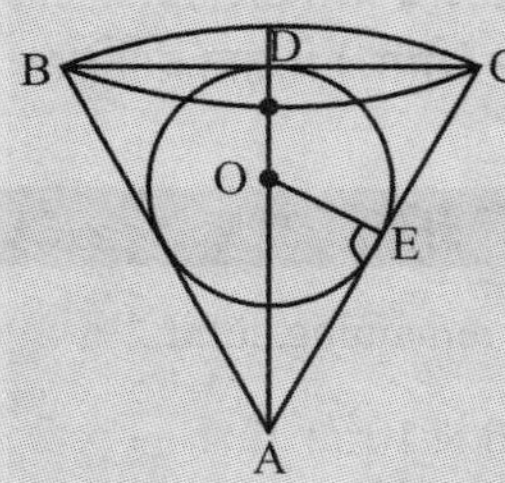

Let $OD = OE = r$ cm. Radius

$$AC^2 = AD^2 + DC^2$$

$$AC = \sqrt{15^2 + 8^2} = 17 \text{ cm}$$

Using similarity $\triangle AEO \sim \triangle ADC$:

$$\frac{EO}{DC} = \frac{AO}{AC}$$

$$\frac{r}{8} = \frac{15 - r}{17}$$

$$17r = 120 - 8r$$

$$r = \frac{120}{25} = \frac{24}{5}$$

Required ratio:

$$\frac{8}{\frac{24}{5}} = 5 : 3$$

107. A sector of radius 10.5 cm with the central angle 120° is folded to form a cone by joining the two bounding radii of the sector. What is the volume (in cm^3) of the cone so formed?

10.5 सेमी त्रिज्या वाले एक त्रिज्यखंड को केंद्रीय कोण 120° के साथ मोड़कर त्रिज्यखंड की दो सीमाबद्ध त्रिज्याओं को जोड़कर एक शंकु बनाया जाता है। इस प्रकार बने शंकु का आयतन (सेमी^{3}) में क्या है?

Solution:

The length of the arc of the radial section will be the circumference of the base of the cone, and the radius of the sector will be the slant height of the cone.

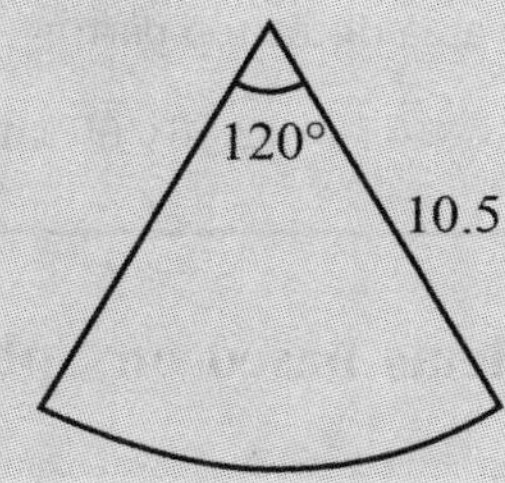

$$l = 2\pi r \quad \text{and} \quad \frac{2\pi R\theta}{360°} = 2\pi r$$

$$\frac{10.5 \times 120°}{360°} = r$$

$$r = 3.5 = \frac{7}{2} \text{ cm}$$

$$h^2 = \sqrt{l^2 - r^2} = \sqrt{\left(\frac{21}{2}\right)^2 - \left(\frac{7}{2}\right)^2} = \sqrt{\frac{392}{4}} = \sqrt{98}$$

$$h = 7\sqrt{2} \text{ cm}$$

Volume of cone formed:

$$\frac{1}{3}\pi r^2 h = \frac{1}{3}\pi \times \frac{49}{4} \times 7\sqrt{2} = \frac{343\sqrt{2}}{12}\pi \text{ cm}^3$$

108. The diameter of the base of a solid cone is 24 cm and its height is 21 cm. It is cut into three parts by two cuts which are parallel to its base. The cuts are at the height of 7 cm and 14 cm from the base, respectively. Find the curved surface area (in cm^2) of the top part. (Use $\pi = 3.14$)

एक ठोस शंकु के आधार का व्यास 24 सेमी है और इसकी ऊँचाई 21 सेमी है। इसे दो कटों द्वारा तीन भागों में काटा जाता है जो इसके आधार के समानांतर हैं। कट क्रमशः आधार से 7 सेमी और 14 सेमी की ऊँचाई पर हैं। शीर्ष भाग का वक्र पृष्ठीय क्षेत्रफल (सेमी^{2}) ज्ञात कीजिए। ($\pi = 3.14$ का उपयोग करें)

Solution:

Given:

$D = 24$ cm, $r = 12$ cm, $h = 21$ cm. For the top part:

$$h = 7 \text{ cm}, \quad r = \frac{12}{3} = 4 \text{ cm}$$

$$l = \sqrt{7^2 + 4^2} = \sqrt{65}$$

Curved surface area of the top part:

$\pi r l = \frac{22}{7} \times 4 \times \sqrt{65} = 12.56\sqrt{65}\ \text{cm}^2$

Type 7 (Problems based on Sphere & Hemisphere)

109. Find the total surface area of a solid hemisphere whose radius is 4.2 cm. (Take $\pi = \frac{22}{7}$)

एक ठोस अर्धगोले का कुल पृष्ठीय क्षेत्रफल ज्ञात कीजिए जिसकी त्रिज्या 4.2 सेमी है। ($\pi = \frac{22}{7}$)

Solution:

Radius (R) of the solid hemisphere $= 4.2$ cm:

Total surface area of a solid hemisphere $=$ $3\pi R^2$

$= 3 \times \frac{22}{7} \times 4.2 \times 4.2$

$= 66 \times 0.6 \times 4.2 = 166.32\ \text{cm}^2$

110. If the surface area of a sphere is $1386\ \text{cm}^2$, then find the radius of the sphere.

यदि किसी गोले का पृष्ठीय क्षेत्रफल 1386 सेमी^{2} है, तो गोले की त्रिज्या क्या है?

Solution:

Given:

TSA of sphere $= 1386\ \text{cm}^2$. Radius of sphere $=?$:

TSA of sphere $= 4\pi r^2$

$\Rightarrow 4\pi r^2 = 1386$

$\Rightarrow 4 \times \frac{22}{7} \times r^2 = 1386$

$r^2 = \frac{1386 \times 7}{22 \times 4} = 21^2$

$r = \frac{21}{2} = 10.5$ cm

111. A hemispherical depression of diameter 4 cm is cut out from each face of a cubical block of sides 10 cm. Find the surface area of the remaining solid (in cm^2). Use $\pi = \frac{22}{7}$.

10 सेमी भुजाओं वाले एक घनाकार ब्लॉक के प्रत्येक फलक से 4 सेमी व्यास का एक अर्धगोलाकार गड्ढा काटा जाता है। शेष ठोस का पृष्ठीय क्षेत्रफल (सेमी^{2}) में ज्ञात कीजिए।

Solution:

Surface area of remaining solid:

$6a^2 + 6\pi r^2 = 6 \times 10^2 + 6 \times \frac{22}{7} \times 2^2$

$= 600 + \frac{528}{7} = \frac{4200 + 528}{7} = \frac{4728}{7} =$

$675\ \frac{3}{7}\ \text{cm}^2$

112. The radius of a sphere is 9 cm. It is melted and drawn into a wire of radius 0.3 cm. The length of the wire is:

एक गोले की त्रिज्या 9 सेमी है। इसे पिघलाकर 0.3 सेमी त्रिज्या का एक तार बनाया जाता है। तार की लंबाई है:

Solution:

Volume of sphere $=$ Volume of wire:

$\frac{4}{3}\pi R^3 = \pi r^2 h$

$\frac{4}{3} \times 9 \times 9 \times 9 = 0.3 \times 0.3 \times h$

$h = 10800$ cm $= 108$ m

113. What is the volume (in cm^3) of a spherical shell whose inner and outer radii are respectively 2 cm and 3 cm?

एक गोलाकार खोल का आयतन (सेमी^{3}) में क्या है जिसकी आंतरिक और बाहरी त्रिज्याएँ क्रमशः 2 सेमी और 3 सेमी हैं?

Solution:

Volume of a spherical shell:

$\frac{4}{3}\pi\ (R^3 - r^3)$

Where $R = 3$ cm, $r = 2$ cm:

$\frac{4}{3}\pi\ (3^3 - 2^3) = \frac{4}{3}\pi\ (27 - 8) = \frac{76\pi}{3}\ \text{cm}^3$

114. A solid metallic hemisphere of radius 6.3 cm is melted and recast into a right circular cylinder of radius 9 cm. What is the height (in cm, correct to one decimal place) of the cylinder?

6.3 सेमी त्रिज्या वाला ठोस धातु अर्धगोला पिघलाया जाता है और इसे 9 सेमी त्रिज्या के एक ठोस बेलन में ढाला जाता है। बेलन की ऊंचाई (सेमी) में कितनी होगी (एक दशमलव स्थान तक)?

Solution:

Volume of the hemisphere = Volume of the cylinder:

$\frac{2}{3}\pi R^3 = \pi r^2 h$

$\frac{2}{3}\pi \times 6.3 \times 6.3 \times 6.3 = \pi \times 9 \times 9 \times h$

$h = \frac{2 \times 6.3 \times 6.3 \times 6.3}{9 \times 9 \times 3} = \frac{500.094}{243}$

$h = 2.1$ cm

115. The ratio of the total surface area and volume of a sphere is 1 : 7. This sphere is melted to form small spheres of equal size. The radius of each small sphere is $\frac{1}{6}$th the radius of the large sphere. What is the sum (in cm^2) of the curved surface areas of all small spheres?

एक गोले के कुल सतह क्षेत्रफल और आयतन का अनुपात 1 : 7 है। इस गोले को पिघलाकर समान आकार के छोटे गोले बनाए जाते हैं। प्रत्येक छोटे गोले की त्रिज्या बड़े गोले की त्रिज्या का $\frac{1}{6}$ है। सभी छोटे गोले के घुमावदार सतह क्षेत्रों का योग (सेमी^{2}) में क्या है?

Solution:

$\frac{\text{Total surface area of sphere}}{\text{Volume of sphere}} = \frac{1}{7}$

$\frac{4\pi R^2}{\frac{4}{3}\pi R^3} = \frac{1}{7}$

$\frac{3}{R} = \frac{1}{7}$

$R = 21$ cm

Radius of small sphere $r = \frac{21}{6}$:

$n \times$ Volume of smaller sphere = Volume of large sphere

$n \times \frac{4}{3}\pi \left(\frac{21}{6}\right)^3 = \frac{4}{3}\pi \times 21^3$

$n = 216$ small spheres

Area of curved surface of all small spheres = $216 \times 4\pi r^2$

$= 216 \times 4 \times \frac{22}{7} \times \frac{21}{6} \times \frac{21}{6} = 33264$ cm^2

116. A sphere of radius 21 cm is cut into 8 identical parts by cuts (1 cut along each axis). What will be the total surface area (in cm^2) of each part?

21 सेमी त्रिज्या वाले एक गोले को कट द्वारा 8 समान भागों में काटा जाता है (प्रत्येक अक्ष के अनुदिश 1 कट)। प्रत्येक भाग का कुल पृष्ठीय क्षेत्रफल (सेमी^{2}) में क्या होगा?

Solution:

By each cut, the sphere will be divided into 8 identical parts, and the surface area will increase by $2\pi r^2$ by each cut. Hence,

Total surface area $= 4\pi r^2 + 3 \times 2\pi r^2 = 10\pi r^2$

Total surface area of each part $= \frac{10\pi r^2}{8} = \frac{5}{4} \times \frac{22}{7} \times 21 \times 21 = \frac{55 \times 63}{2} = 1732.5$ cm^2

117. A hemispherical dome is open from its base and is made of iron. Thickness of the dome is 3.5 m. Total cost of painting the dome's outer curved surface is Rs. 2464. If the rate of painting is Rs. 8 per m^2, then what is the volume (in m^3) of iron used in making the dome?

एक अर्धगोलाकार गुंबद अपने आधार से खुला है और लोहे का बना होता है। गुंबद की मोटाई 3.5 मीटर है। गुंबद की बाहरी घुमावदार सतह को पेंट करने की कुल लागत रु। 2464 है। यदि पेंटिंग की दर रु। 8 प्रति मीटर^2 है, तो गुंबद बनाने में प्रयुक्त लोहे का आयतन (मीटर^3) में क्या है?

Solution:

Area of outer curved surface of dome:

$\frac{2464}{8} = 308$ m^2

$2\pi R^2 = 308$

$R = 7$ m,

$r = R -$ thickness $= 7 - 3.5 = 3.5$ m

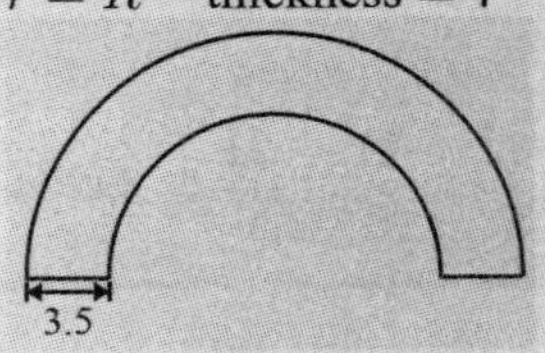

Volume of iron used in making the dome:

$$\frac{2}{3}\pi\ (R^3 - r^3) = \frac{2}{3}\pi\ \left(7^3 - \left(\frac{7}{2}\right)^3\right)$$

$$= \frac{44}{21} \times 7^3 \times \frac{7}{8} = 628.83 \text{ m}^3$$

118. A metallic hemispherical bowl is made up of steel. The total steel used in making the bowl is 342 cm^3. The bowl can hold 144 cm^3 water. What is the thickness (in cm) of the bowl and the curved surface area (in cm^2) of the outer side?
एक धात्विक अर्धगोलाकार कटोरा स्टील का बना होता है। कटोरा बनाने में प्रयुक्त कुल स्टील 342 सेमी^{3} है। कटोरे में 144 सेमी^{3} पानी आ सकता है। कटोरे की मोटाई (सेमी) और बाहरी सतह का घुमावदार सतह क्षेत्रफल (सेमी^{2}) में क्या है?

Solution:

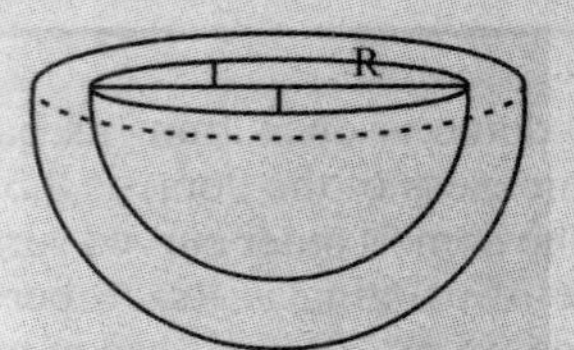

Volume of bowl (inner hemisphere):

$$\frac{2}{3}\pi r^3 = 144\pi$$

$$r^3 = 216$$

$$r = 6 \text{ cm}$$

Volume of steel used:

$$\frac{2}{3}\pi\ (R^3 - r^3) = 342$$

$$R^3 - 6^3 = 513$$

$$R^3 = 729$$

$$R = 9 \text{ cm}$$

Thickness of bowl:

$$R - r = 9 - 6 = 3 \text{ cm}$$

Curved surface area of outer side:

$$2\pi R^2 = 2\pi(9)^2 = 162\pi = 162 \text{ cm}^2$$

119. Three solid metallic spheres whose radii are 1 cm, x cm, and 8 cm, are melted and recast into a single solid sphere of diameter 18 cm. The surface area (in cm^2) of the sphere with radius x cm is:
तीन ठोस धातु के गोले जिनकी त्रिज्याएँ 1 सेमी, x सेमी और 8 सेमी हैं, को पिघलाकर 18 सेमी व्यास वाले एक ठोस गोले में ढाला जाता है। x सेमी त्रिज्या वाले गोले का सतह क्षेत्रफल (सेमी^{2}) क्या है?

Solution:

Sum of volume of small metallic spheres =

Volume of new sphere:

$$\frac{4}{3}\pi\ (1^3 + x^3 + 8^3) = \frac{4}{3}\pi\ \left(\frac{18}{2}\right)^3$$

$$x^3 + 513 = 729$$

$$x^3 = 216$$

$$x = 6$$

Surface area of sphere with radius x cm:

$$4\pi x^2 = 4\pi(6)^2 = 144\pi \text{ cm}^2 = 144 \text{ cm}^2$$

120. A right circular cone of largest volume is cut out from a solid wooden hemisphere. The remaining material is what percentage of the volume of the original hemisphere?
सबसे बड़े आयतन का एक लंब वृत्तीय शंकु एक ठोस लकड़ी के गोलार्ध से काटा जाता है। शेष सामग्री मूल गोलार्ध के आयतन का कितना प्रतिशत है?

Solution:

A right circular cone of largest volume is cut out from a solid wooden hemisphere.

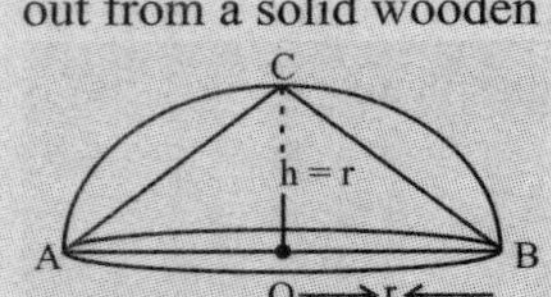

Hence, the radius of the cut-out cone will equal the radius of the hemisphere, and the height of the cone will also equal the radius of the hemisphere.

The percentage of the remaining material of the volume of the original hemisphere:

$= \frac{\frac{2}{3}\pi r^3 - \frac{1}{3}\pi r^3}{\frac{2}{3}\pi r^3} \times 100$

$= \frac{1}{3} \times \frac{3}{2} \times 100 = 50\%$

121. The base of a right prism is a triangle with sides 16 cm, 30 cm, and 34 cm. Its height is 32 cm. The lateral surface area (in cm^2) and the volume (in cm^3) are, respectively:
एक समकोण प्रिज्म का आधार एक त्रिभुज है जिसकी भुजाएँ 16 सेमी, 30 सेमी और 34 सेमी हैं। इसकी ऊंचाई 32 सेमी है। पार्श्व सतह क्षेत्र (सेमी^{2} में) और आयतन (सेमी^{3} में) क्रमशः हैं:

Solution:

Given, sides of the triangle are 16 cm, 30 cm, and 34 cm, and height = 32 cm.

We know that:

Lateral surface area of prism (LSA) = Perimeter of base × height, and

Volume of prism = Area of base × height.

Now,

LSA = (Perimeter of base) × height

$= (16+30+34) \times 32 = 80 \times 32 = 2560 \text{ cm}^2$

Again,

Volume of prism $= \frac{1}{2} \times$ Base × Height × Height of prism

$= \frac{1}{2} \times 16 \times 30 \times 32 = 7680 \text{ cm}^3$

122. A prism has a regular hexagonal base with side 6 cm. If the total surface area of the prism is $216\sqrt{3}$ cm^2, then what is the height (in cm) of the prism?
एक प्रिज्म का आधार नियमित षट्कोणीय है जिसकी भुजा 6 सेमी है। यदि प्रिज्म का कुल सतह क्षेत्रफल $216\sqrt{3}$ सेमी^{2} है, तो प्रिज्म की ऊंचाई (सेमी) में क्या है?

Solution:

Side of regular hexagon = 6 cm. Let height = h.

Area of lateral surface of prism = Perimeter of base ×height:

$6 \times 6 \times h = 36h \text{ cm}^2$

Area of base of prism $= 6 \times \frac{\sqrt{3}}{4} \times 6 \times 6 = 54\sqrt{3} \text{ cm}^2$.

Total surface area of prism = Area of lateral surface $+2 \times$ Area of base:

$216\sqrt{3} = 36h + 2 \times 54\sqrt{3}$

$216\sqrt{3} = 36h + 108\sqrt{3}$

$108\sqrt{3} = 36h$

$h = 3\sqrt{3}$ cm

123. A regular hexagonal base prism has height 8 cm and side of base 4 cm. What is the total surface area (in cm^2) of the prism?
एक नियमित षट्कोणीय आधार प्रिज्म की ऊंचाई 8 सेमी है और आधार की भुजा 4 सेमी है। प्रिज्म का कुल सतह क्षेत्रफल (सेमी^{2}) में क्या है?

Solution:

Total surface area of prism = Perimeter of base × height + 2 × Area of base

$$= (6 \times 4) \times 8 + 2 \times \frac{3\sqrt{3}}{2} \times (4)^2$$
$$= 192 + 48\sqrt{3} = 48(4 + \sqrt{3})\ \text{cm}^2$$

124. The base of a pyramid is an equilateral triangle of side 10 m. If the height of the pyramid is $40\sqrt{3}$ m, then the volume of the pyramid is:
एक पिरामिड का आधार 10 मीटर भुजा वाला एक समबाहु त्रिभुज है। यदि पिरामिड की ऊंचाई $40\sqrt{3}$ मीटर है, तो पिरामिड का आयतन है:

Solution:

Volume of pyramid $= \frac{1}{3} \times$ Area of base $\times$ Height

$$= \frac{1}{3} \times \frac{\sqrt{3}}{4} \times 10^2 \times 40\sqrt{3}$$
$$= 1000\ \text{m}^3$$

125. A regular pyramid has a square base. The height of the pyramid is 22 cm, and the side of its base is 14 cm. The volume of the pyramid is equal to the volume of a sphere. What is the radius (in cm) of the sphere?
एक नियमित पिरामिड का आधार वर्गाकार होता है। पिरामिड की ऊंचाई 22 सेमी है और इसके आधार की भुजा 14 सेमी है। पिरामिड का आयतन एक गोले के आयतन के बराबर है। गोले की त्रिज्या (सेमी में) क्या है?

Solution:

Let the radius of the sphere = R.

According to the question, Volume of the pyramid = Volume of the sphere.

$\frac{1}{3} \times$ Area of base $\times$ Height $= \frac{4}{3} \times \frac{22}{7} \times R^3$

$\frac{1}{3} \times 14 \times 14 \times 22 = \frac{4}{3} \times \frac{22}{7} \times R^3$

$R^3 = 7 \times 7 \times 7$

$R = 7$ cm

126. A regular triangular pyramid is cut by 2 planes which are parallel to its base. The planes trisect the altitude of the pyramid. Volume of top, middle, and bottom part is V_1, V_2, and V_3, respectively. What is the value of $V_1 : V_2 : V_3$?
एक नियमित त्रिकोणीय पिरामिड को 2 समतल द्वारा काटा गया है जो इसके आधार के समानांतर हैं। समतल पिरामिड की ऊँचाई को त्रिविभाजित करते हैं। शीर्ष, मध्य और निचले भाग का आयतन क्रमशः V_1, V_2 और V_3 है। $V_1 : V_2 : V_3$ का मान क्या है?

Solution:

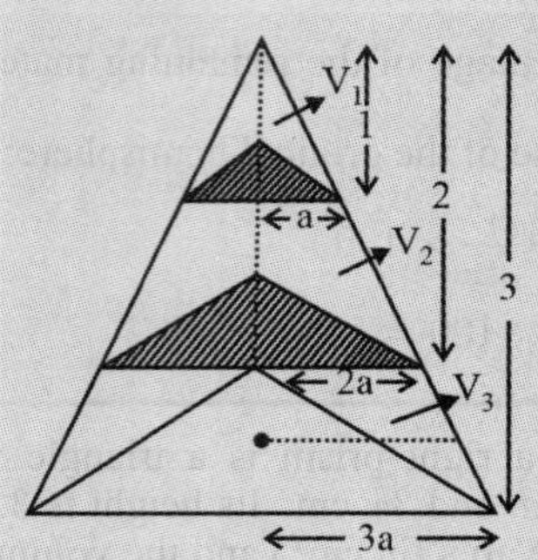

Let the side of base of the smallest pyramid = a.

Ratio of volume of smallest, medium, and largest prism:

$$\frac{1}{3} \times \frac{\sqrt{3}}{4} \times a^2 \times 1 : \frac{1}{3} \times \frac{\sqrt{3}}{4} \times (2a)^2 \times 2 : \frac{1}{3} \times \frac{\sqrt{3}}{4} \times (3a)^2 \times 3$$
$$= 1 : 8 : 27$$
$$\therefore V_1 : V_2 : V_3 = 1 : 7 : 19$$

127. A regular square pyramid has side of its base 20 cm and height 45 cm. It is melted and recast into regular triangular pyramids of equilateral base of side 10 cm and height $10\sqrt{3}$ cm. What are the total numbers of regular triangular pyramids?
एक नियमित वर्गाकार पिरामिड जिसकी आधार भुजा 20 सेमी और ऊँचाई 45 सेमी है, को पिघलाकर समबाहु आधार भुजा 10 सेमी और ऊँचाई $10\sqrt{3}$ सेमी वाले नियमित त्रिकोणीय पिरामिड में ढाला गया है। नियमित त्रिभुजाकार पिरामिडों की कुल संख्या कितनी है?

Solution:

Volume of pyramid built on the base of square $= N \times$ Volume of pyramid built on the base of the triangle

$\frac{1}{3} \times 20 \times 20 \times 45 = N \times \frac{1}{3} \times \frac{\sqrt{3}}{4} \times 10 \times 10 \times 10\sqrt{3}$

$N = \frac{2 \times 2 \times 2 \times 3}{1} = 24$

128. A pyramid has a square base, whose side is 8 cm. If the height of the pyramid is 16 cm, then what is the total surface area (in cm^2) of the pyramid?

एक पिरामिड का आधार वर्गाकार है, जिसकी भुजा 8 सेमी है। यदि पिरामिड की ऊंचाई 16 सेमी है, तो पिरामिड का कुल सतह क्षेत्रफल (सेमी^{2}) में क्या है?

Solution:

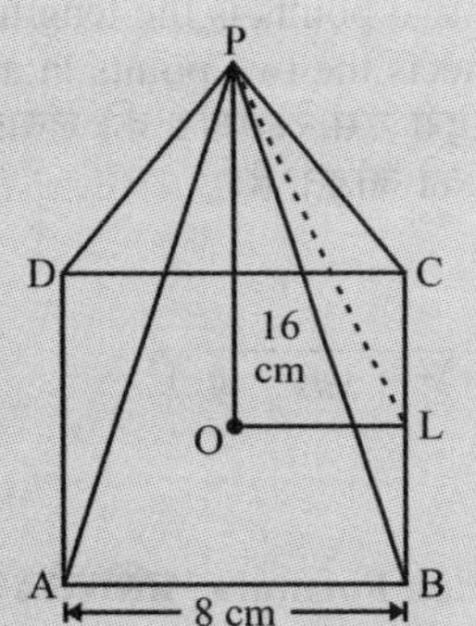

Slant height $l = \sqrt{16^2 + 4^2} = \sqrt{272} = 4\sqrt{17}$ cm

Total surface area of pyramid = Curved surface area + Area of base

$= \frac{1}{2} \times$ Perimeter of base $\times$ Slant height + Area of base

$= \frac{1}{2} \times (4 \times 8) \times 4\sqrt{17} + (8)^2$

$= 64\sqrt{17} + 64 = 64(\sqrt{17} + 1)$ cm^2

129. The base of a right pyramid is an equilateral triangle with area $16\sqrt{3}$ cm^2. If the area of one of its lateral faces is 30 cm^2, then its height (in cm) is:

एक समकोण पिरामिड का आधार एक समबाहु त्रिभुज है जिसका क्षेत्रफल $16\sqrt{3}$ सेमी^{2} है। यदि इसके एक पार्श्व फलक का क्षेत्रफल 30 सेमी^{2} है, तो इसकी ऊंचाई (सेमी) में है:

Solution:

Area of equilateral triangle = $16\sqrt{3}$ cm^2

$\frac{\sqrt{3}}{4}a^2 = 16\sqrt{3}$

$a^2 = 64$

$a = 8$ cm

Curved surface area of right pyramid = $\frac{1}{2} \times$ Perimeter of base $\times$ Slant height:

$3 \times 30 = \frac{1}{2} \times (3 \times 8) \times l$

$l = \frac{15}{2}$ cm

Radius of incircle $r = \frac{a}{2\sqrt{3}} = \frac{8}{2\sqrt{3}} = \frac{4}{\sqrt{3}}$ cm.

$h^2 = l^2 - r^2 = \frac{225}{4} - \frac{16}{3} = \frac{675 - 64}{12} = \frac{611}{12}$

$h = \sqrt{\frac{611}{12}}$ cm

22 Coordinate Geometry

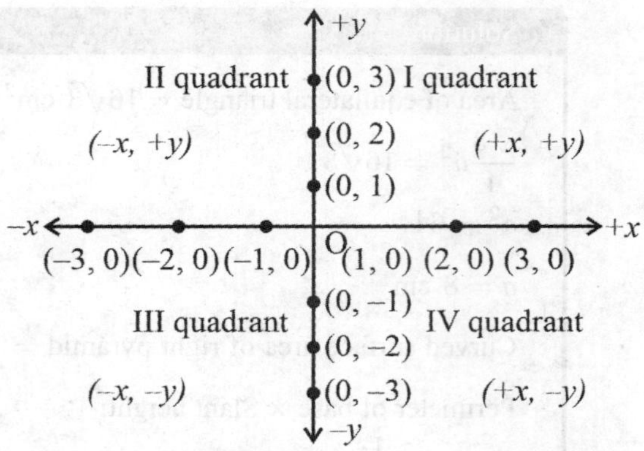

- O (origin) (मूल) = (0, 0)
  P = (x, y) → Ordinate
  ↓
  abscissa
  Equation of x-axis ⇒ y = 0
  x-अक्ष का समीकरण ⇒ y = 0
  Equation of y-axis ⇒ x = 0
  y-अक्ष का समीकरण ⇒ x = 0

Mid-Point Formula (मध्य बिंदु सूत्र)

A mid point is the middle point of a line segment which is equidistant from both the end points. एक मध्य बिंदु एक रेखा खंड का मध्य बिंदु होता है जो दोनों अंत बिंदुओं से समान दूरी पर होता है।

Mid point/ (मध्य बिंदु)

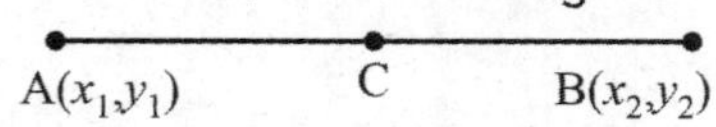

Mid point: $C = \left(\dfrac{x_1 + x_2}{2}, \dfrac{y_1 + y_2}{2}\right)$

Distance Formula (दूरी सूत्र)

Distance between two points is the length of the line segment that connects the two points in a plane. दो बिंदुओं के बीच की दूरी उस रेखाखंड की लंबाई है जो एक समतल में दो बिंदुओं को जोड़ती है।

$A = (x_1, y_1)$,
$B = (x_2, y_2)$,
$AB = \sqrt{(x_2 - x_1)^2 + (y_2 - y_1)^2}$

Point of Intersection Formula / प्रतिच्छेद बिंदु का सूत्र

(A) Internal Division / आंतरिक विभाजन →

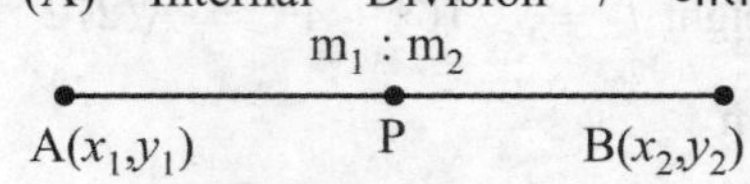

$$P = \left(\frac{m_1x_2 + m_2x_1}{m_1 + m_2}, \frac{m_1y_2 + m_2y_1}{m_1 + m_2}\right)$$

(B) External Division / बाह्य विभाजन → AP:PB=m_1:m_2

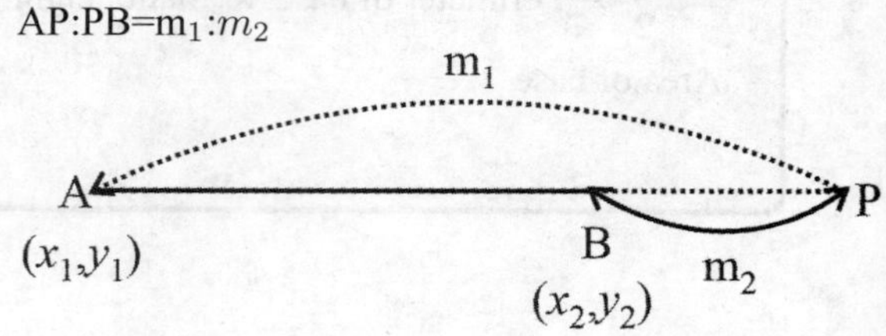

$AP : PB = m_1 : m_2$,

$$P = \left(\frac{m_1x_2 - m_2x_1}{m_1 - m_2}, \frac{m_1y_2 - m_2y_1}{m_1 - m_2}\right)$$

Slope of a Line (रेखा की ढलान)

- The slope of a line is a measure of its steepness. It is the change in y-coordinate with respect to the change in x-coordinate. किसी रेखा का ढाल उसकी ढाल का माप होता है। यह y निर्देशांक में परिवर्तन के संबंध में x निर्देशांक में परिवर्तन है।

Slope of a line / रेखा की ढलान (m) = $\tan\theta$,

$$\tan\theta = \frac{y_2 - y_1}{x_2 - x_1}$$

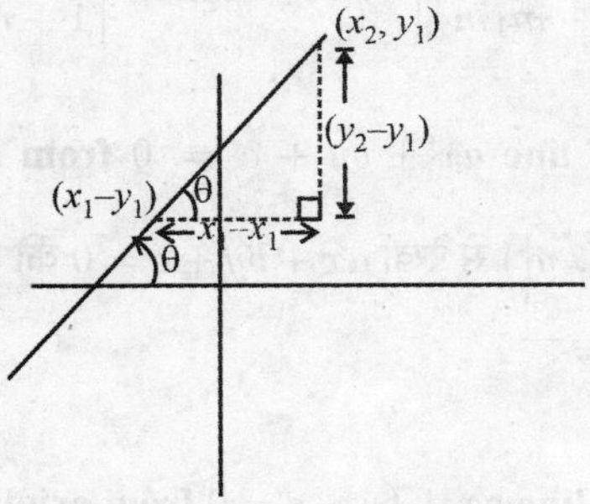

• Standard Equation of a Line / एक रेखा का मानक समीकरण

$ax + by + c = 0$,
slope $= -\dfrac{a}{b}$

Point from of a Line / एक रेखा से बिंदु

Slope of line AP = slope of line AB. रेखा AP का

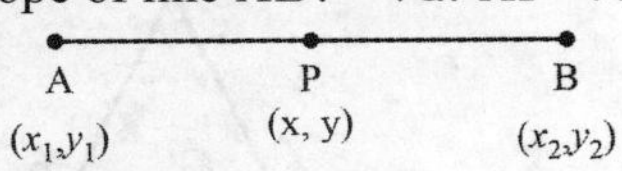

ढाल = रेखा AB का ढाल
$y - y_1 = m(x - x_1)$

Slope / ढलान = m,
Intercept on y-axis / y-अक्ष पर अवरोधन = c,
$y - c = m(x - 0)$,
$y = mx + c$

• If two lines are parallel to each other, their slope will be equal $m_1 = m_2$. यदि दो रेखाएँ एक दूसरे के समांतर हों तो उनकी प्रवणता बराबर होगी

$$\frac{-a_1}{b_1} / = \frac{-a_2}{b_2}$$
$$\frac{a_1}{a_2} = \frac{b_1}{b_2}$$

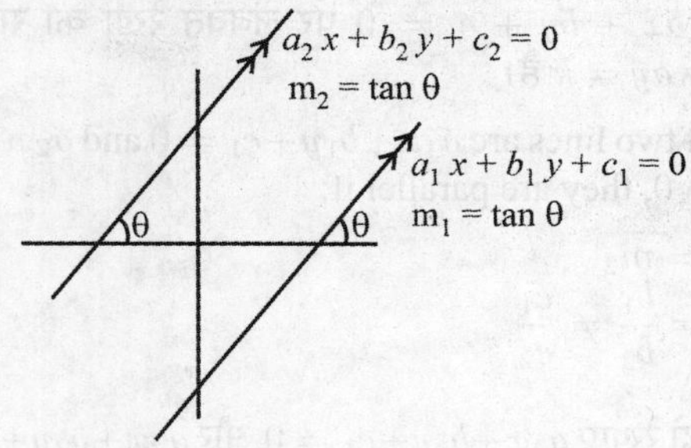

• If two lines are perpendicular to each other: यदि दो रेखाएँ एक दूसरे के लंबवत हों:

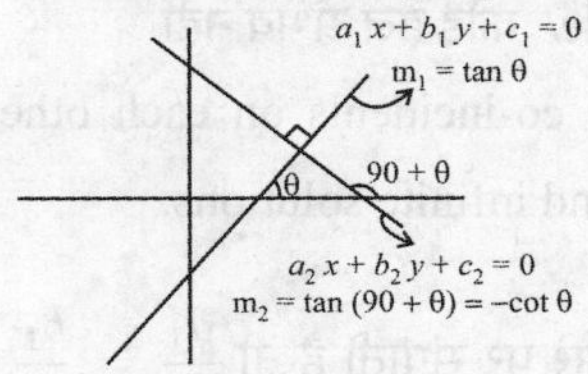

$\therefore m_1m_2 = \tan\theta(-\cot\theta)$
$\therefore m_1m_2 = -1$

Intercept Form of a Line (एक रेखा का अवरोधन रूप)

• Intercept at x-axis = a x-अक्ष पर अंत: खंड a

Intercept at y-axis = b y-अक्ष पर अवरोधन (अंत: खंड) b

$$\frac{x}{a} + \frac{y}{b} = 1$$

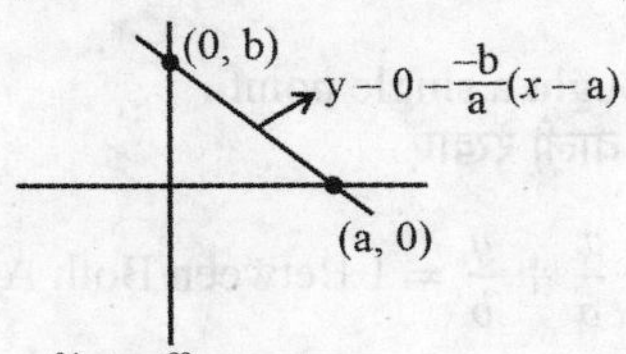

$$-\frac{y}{b} = \frac{x}{a} - 1,$$
$$1 = \frac{x}{a} + \frac{y}{b}$$

• Equation of a Line Parallel to Line $ax + by + c = 0$

Equation: $ax + by = k$

रेखा $ax + by + c = 0$ के समांतर रेखा का समीकरण $ax + by = k$ है।

Equation of a Line Perpendicular to Line $ax+by+c = 0$

Equation: $bx - ay = k$

रेखा $ax + by + c = 0$ पर लंबवत रेखा का समीकरण $bx - ay = k$ है।

(a) If two lines are $a_1x + b_1y + c_1 = 0$ and $a_2x + b_2y + c_2 = 0$, they are parallel if:

$m_1 = m_2$,
$$\frac{a_1}{a_2} = \frac{b_1}{b_2} \neq \frac{c_1}{c_2}$$

यदि दो रेखाएँ $a_1x + b_1y + c_1 = 0$ और $a_2x + b_2y + c_2 = 0$ एक दूसरे के समानांतर हैं तो $m_1 = m_2$, $\frac{a_1}{a_2} = \frac{b_1}{b_2} \neq \frac{c_1}{c_2}$

No intersecting point/ कोई प्रतिच्छेदन बिंदु नहीं
No solution possible/ कोई हल संभव नहीं

(b) If two line are co-incidents on each other then $\frac{a_1}{a_2} = \frac{b_1}{b_2} = \frac{c_1}{c_2}$ and infinite solutions.

यदि दो रेखाएँ एक दूसरे पर संपाती है तो $\frac{a_1}{a_2} = \frac{b_1}{b_2} = \frac{c_1}{c_2}$ अनंत हल संभव है।

(c) If two lines intersect at single point then $m_1 \neq m_2$ $\frac{b_1}{b_2}$ unique solution possible.

यदि दो रेखाएँ एक बिंदु पर प्रतिच्छेद करती हैं तो $m_1 \neq m_2$ $\frac{b_1}{b_2}$ अद्वितीय हल संभव है।

Concurrent Lines (समवर्ती रेखाएँ)

- Line passing throught a single point
  एक बिंदु से गुजरने वाली रेखाएँ
- Intercept by Line $\frac{x}{a} + \frac{y}{b} = 1$ Between Both Axes $= \sqrt{a^2 + b^2}$

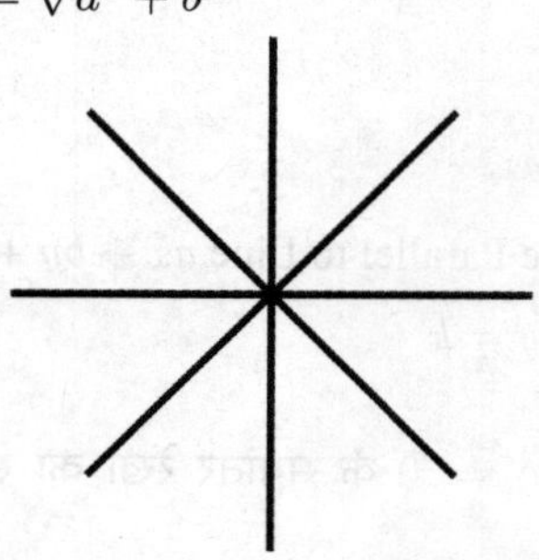

Line $ax + by = 0$ always passes through origin:
रेखा $ax + by = 0$ हमेशा मूल बिंदु से होकर गुजरती है।

Angle Between Two Lines:
दो रेखाओं के बीच का कोण

$\theta \rightarrow$ acute angle
न्यून कोण
मान लीजिए कि दो रेखाओं के बीच का कोण θ और उनकी प्रवणताएँ m_1 और m_2 हैं, तब

$$\tan\theta = \left|\frac{m_1 - m_2}{1 + m_1m_2}\right|, \quad \theta = \tan^{-1}\left|\frac{m_1 - m_2}{1 + m_1m_2}\right|$$

Distance of line $ax + by + c = 0$ from a point (x_1, y_1):
एक बिंदु (x_1, y_1) से रेखा $ax + by + c = 0$ की दूरी
$$\frac{|ax_1 + by_1 + c|}{\sqrt{a^2 + b^2}}$$

Distance of line $ax + by + c = 0$ from origin:
मूल बिंदु से रेखा $ax + by + c = 0$ की दूरी
$$\frac{|c|}{\sqrt{a^2 + b^2}}$$

Distance between two parallel lines:
दो समानांतर रेखाओं के बीच की दूरी:
$ax + by + c_1 = 0, \quad ax + by + c_2 = 0$
Distance: $\frac{|c_1 - c_2|}{\sqrt{a^2 + b^2}}$

Centroid and Incentre/ केन्द्रक और अन्तःकेंद्र

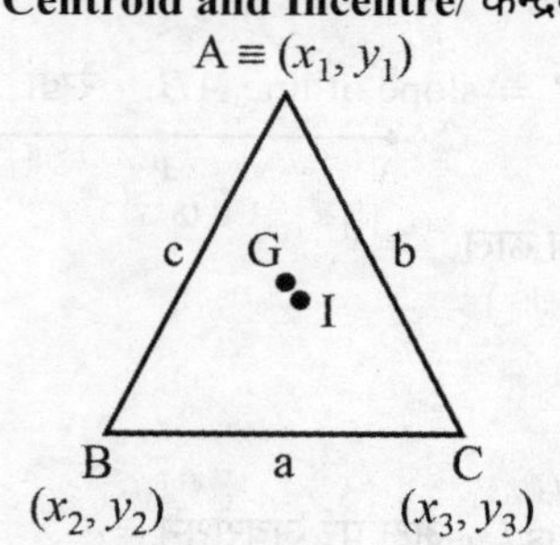

$G \rightarrow$ Centroid /, $\quad I \rightarrow$ Incentre /
$$G = \left(\frac{x_1 + x_2 + x_3}{3}, \frac{y_1 + y_2 + y_3}{3}\right)$$
$$I = \left(\frac{ax_1 + bx_2 + cx_3}{a + b + c}, \frac{ay_1 + by_2 + cy_3}{a + b + c}\right)$$

Equation of Circle/ वृत्त का समीकरण
Where (a, b) are co-ordinates of center of circle and r is the radius. जहाँ (a, b) वृत्त के केंद्र के निर्देशांक हैं और r त्रिज्या है।
Equation of circle / वृत्त का समीकरण
$(x - a)^2 + (y - b)^2 = r^2$

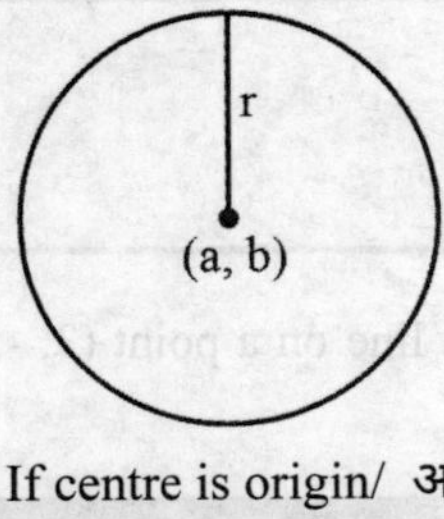

If centre is origin/ अगर केंद्र मूल है:

$x^2 + y^2 = r^2$

Area of a Triangle/ त्रिभुज का क्षेत्रफल

The area of a triangle the Co-ordinate of whose vertices are $(x_1 y)$, (x_2, y_2) and (x_3, y_3) is

एक त्रिभुज का क्षेत्रफल जिसके शीर्षों के निर्देशांक हैं

$\frac{1}{2}|x_1(y_2 - y_3) + x_2(y_3 - y_1) + x_3(y_1 - y_2)|$

Type 1 (Problems based on Traingle)

1. The graph of the equations $5x - 2y + 1 = 0$ and $4y - 3x + 5 = 0$, intersect at the point $P(\alpha, \beta)$. What is the value of $(2\alpha - 3\beta)$?

Solution:

On solving the given equation, $5x - 2y = -1$ --------(1)

$-3x + 4y = -5$ --------(2)

$10x - 4y = -2$(Multiply by 2 in equation (1))

$10x - 4y - (-3x + 4y) = -2 - (-5)$

$7x = -7$

$x = -1$

Putting the value of x in equation (1):

$-2y = -1 + 5$

$y = -2$

The given equations intersect at point $P(\alpha, \beta)$.

Hence,

$\alpha = -1$, $\beta = -2$

$(2\alpha - 3\beta) = -2 + 6 = 4$

2. What is the reflection of the point $(5, -3)$ in the line $y = 3$?

Solution:

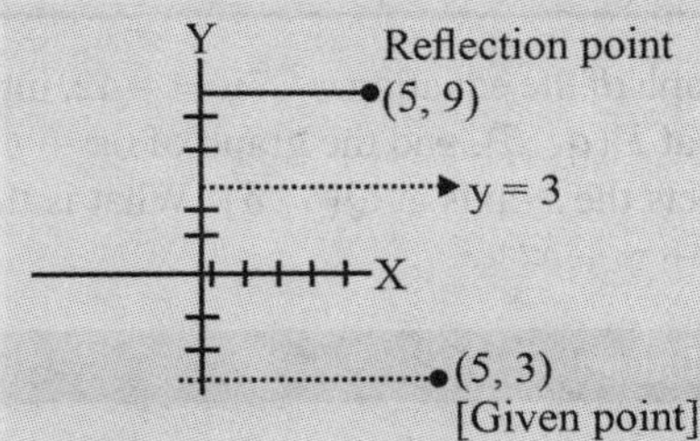

Hence the reflection of the point $(5, -3)$ will be $(5, 9)$ with respect to $y = 3$.

3. What is the area (in square units) of the triangular region enclosed by the graphs of the equations $x + y = 3$, $2x + 5y = 12$, and the x-axis?

Solution:

$x + y = 3$ (1)

$2x + 5y = 12$ (2)

Intersects the x-axis, then

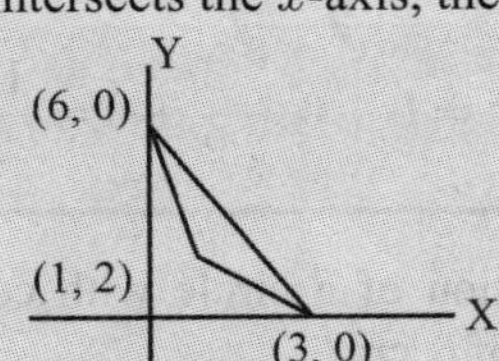

$y = 0$ --------(3)

On solving equation (1) and (2):

$x = 1, y = 2$ (1, 2)

On solving equation (2) and (3):

$x = 6, y = 0$ (6, 0)

On solving equation (3) and (1):

$x = 3,\ y = 0$ (3, 0)

Area of the triangular region:

$\text{Area} = \frac{1}{2}|x_1(y_2-y_3)+x_2(y_3-y_1)+x_3(y_1-y_2)|$

$= \frac{1}{2}|1(0-0)+6(0-2)+3(2-0)|$

$= \frac{1}{2}|-12+6| = 3$ square units

4. The graph of the equation $x-7y = -42$, intersects the y-axis at $P(\alpha,\ \beta)$, and the graph of $6x + y - 15 = 0$, intersects the x-axis at $Q(\gamma,\ \delta)$. What is the value of $\alpha+\beta+\gamma+\delta$?

Solution:

$x-7y = -42$ (i) Intersects the y-axis, then

$x = 0$:

$0-7y = -42$

$y = 6$

Hence, $P(\alpha,\ \beta) = (0,\ 6)$.

For $6x + y - 15 = 0$ (iii), intersects the x-axis, then $y = 0$:

$6x + 0 - 15 = 0$

$x = \frac{5}{2}$

Hence, $Q(\gamma,\ \delta) = \left(\frac{5}{2},\ 0\right)$.

$\alpha+\beta+\gamma+\delta = 0+6+\frac{5}{2}+0 = \frac{17}{2}$

5. What is the equation of a circle with centre at the origin and radius of 6 cm?

Solution:

Centre of the circle = (0, 0)

Radius of the circle = 6 cm

The required equation of the circle is:

$(x-0)^2 + (y-0)^2 = 6^2$

$x^2 + y^2 = 36$

$x^2 + y^2 - 36 = 0$

6. The equation of a straight line on a point (3, -5) and slope 2 is:

Solution:

The equation of the line passing through the point $(x,\ y)$ and the slope m:

$(y - y_1) = m(x - x_1)$

Point $(3,\ -5)$, slope (m) 2 (given):

$(y+5) = 2\cdot(x-3)$

$y + 5 = 2x - 6$

$2x - y = 11$

$2x - y - 11 = 0$

7. Find the equation of a circle whose diameter has end points (4, 3) and (−2, 1).

Solution:

If the ends of the diameter are $(x_1,\ y_1)$ and $(x_2,\ y_2)$, the equation of the circle is:

$(x-x_1)(x-x_2)+(y-y_1)(y-y_2) = 0$

Ends of the diameter of the circle are (4, 3) and (−2, 1):

$(x-4)(x+2)+(y-3)(y-1) = 0$

$x^2 - 4x + 2x - 8 + y^2 - y - 3y + 3 = 0$

$x^2 + y^2 - 2x - 4y - 5 = 0$

$x^2 + y^2 - 2x - 4y = 5$

8. The equation of a circle with centre (1, −2) and radius 4 cm is:

Solution:

Centre of the circle (1, −2),

Radius $(r) = 4$ cm:

The equation of the circle is:

$(x - h)^2 + (y - k)^2 = r^2$

$(x - 1)^2 + (y + 2)^2 = 16$

$x^2 - 2x + 1 + y^2 + 4y + 4 = 16$

$x^2 + y^2 - 2x + 4y + 5 = 16$

$x^2 + y^2 - 2x + 4y = 11$

9. Find the value of k such that the equations $2x + 3y + 11 = 0$ and $6x + ky + 33 = 0$ represent coincident lines.

Solution:

For coincident lines:

$\frac{a_1}{a_2} = \frac{b_1}{b_2} = \frac{c_1}{c_2}$

$\frac{2}{6} = \frac{3}{k}$

$k = 9$

10. Find the relation between x and y such that the point (x, y) is equidistant from $(6, 2)$ and $(4, 6)$.

Solution:

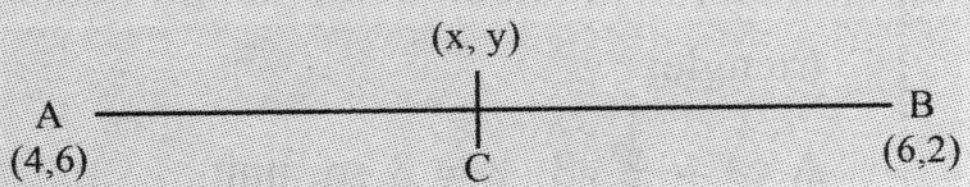

Point C is equidistant from points A and B:

$AC = BC$

$\sqrt{(x - 4)^2 + (y - 6)^2} = \sqrt{(6 - x)^2 + (2 - y)^2}$

Squaring both sides:

$(x - 4)^2 + (y - 6)^2 = (6 - x)^2 + (2 - y)^2$

$x^2 + 16 - 8x + y^2 + 36 - 12y = 36 + x^2 - 12x + 4 + y^2 - 4y$

$-8x + 12x - 12y + 4y = 40 - 52$

$4x - 8y = -12$

$x - 2y = -3$

11. The system of equations $2x + ky = 11$ and $5x - 7y = 5$ has no solution when $k =$ ___.

Solution:

The system of equations:

$2x + ky = 11, 5x - 7y = 5$

For no solution:

$\frac{a_1}{a_2} = \frac{b_1}{b_2}$

$\frac{2}{5} = \frac{k}{-7}$

$5k = -14$

$k = -\frac{14}{5}$

12. Find the points of intersection of the line $4x - 3y = 10$ and the circle $x^2 + y^2 - 2x + 4y - 20 = 0$.

Solution:

To find the points of intersection of the line $4x - 3y = 10$ and the circle $x^2 + y^2 - 2x + 4y - 20 = 0$:

Substitute $x = \frac{3y + 10}{4}$ into the equation of the circle:

$25y^2 + 100y - 300 = 0$

$(y - 2)(y + 6) = 0$

The values of $y = -6, 2$, and the corresponding values of $x = -2, 4$.

So, the points of intersection are: $(4, 2)$ and $(-2, -6)$.

13. Find the equation of the tangents to the circle $x^2 + y^2 = 9$ at $x = 2$.

Solution:

$x^2 + y^2 = 9$ at $x = 2$:

$y = \pm\sqrt{5}$

So now, putting the value in all the equations, we find that the equations are:

$2x + \sqrt{5}y = 9$

$2x - \sqrt{5}y = 9$

14. Find the coordinates of the point which will divide the line joining the points (2, 4) and (7, 9) internally in the ratio 1 : 2.

Solution:

Given points are (2, 4) and (7, 9), $m_1 : m_2 = 1 : 2$:

$$X = \frac{m_2x_1 + m_1x_2}{m_1 + m_2} = \frac{2 \cdot 2 + 1 \cdot 7}{3} = \frac{11}{3}$$

$$Y = \frac{m_2y_1 + m_1y_2}{m_1 + m_2} = \frac{2 \cdot 4 + 1 \cdot 9}{3} = \frac{17}{3}$$

Point is $\left(\frac{11}{3}, \frac{17}{3}\right)$.

15. The angle between two circles $x^2 + y^2 - 12x - 6y + 41 = 0$ and $x^2 + y^2 + kx + 6y - 59 = 0$ is 45°. Find the value of k.

Solution:

Suppose θ is the angle between the circles:

$x^2 + y^2 - 12x - 6y + 41 = 0$

and

$x^2 + y^2 - kx - 6y - 59 = 0$

$g_1 = -6,\ f_1 = -3,\ c_1 = 41$

$g_2 = \frac{k}{2},\ f_2 = -3,\ c_2 = -59$

$$\cos 45° = \frac{c_1 + c_2 - 2g_1g_2 - f_1f_2}{2r_1r_2}$$

$$\cos 45° = \frac{41 - 59 - 2(-6)\frac{k}{2} - 2(-3)(-3)}{\sqrt{36 + 9 - 41} \cdot \sqrt{\frac{k^2}{4} + 9 + 59}}$$

$$\frac{1}{\sqrt{2}} = \frac{-18 + 6k + 18}{2 \cdot 2 \cdot \sqrt{\frac{k^2}{4} + 68}}$$

Squaring and cross-multiplying:

$$4\left(\frac{k^2}{4} + 68\right) = 18k^2$$

$k^2 + 272 = 18k^2$

$17k^2 = 272$

$k^2 = \frac{272}{17} = 16$

$k = \pm 4$

16. What is the area of the triangle formed by the coordinate axes and the line $2x + 3y = 6$?

Solution:

Equation of line: $2x + 3y = 6$

Divide both sides by 6:

$\frac{x}{3} + \frac{y}{2} = 1$

x-intercept = 3, y-intercept = 2.

The x-intercept will be the base of the triangle and the y-intercept will be the height of the triangle:

Area $= \frac{1}{2} \cdot 3 \cdot 2 = 3$ sq. units

17. A line passes through two points (3, 4) and (4, 5). What is the angle of the slope of this line?

Solution:

Given points are (3, 4) and (4, 5):

Slope of line $= \tan\theta = \frac{y_2 - y_1}{x_2 - x_1} = \frac{5 - 4}{4 - 3} = 1$

$\tan\theta = 1 = \tan 45°$

$\theta = 45°$

18. If the length of the tangent from (2, 5) to $x^2 + y^2 -$

$5x + 4y + k = 0$ is $\sqrt{37}$ units, then the value of k is:

Solution:

Given point is (2, 5),

Length of tangent = $\sqrt{37}$:

$\sqrt{2^2 + 5^2 - 5 \cdot 2 + 4 \cdot 5 + k} = \sqrt{37}$

$\sqrt{39 + k} = \sqrt{37}$

$k = -2$

19. If the distance between two points $(x, 7)$ and $(1, 15)$ is 10 units, then the possible values of x are:

Solution:

Distance formula:

$\sqrt{(x_2 - x_1)^2 + (y_2 - y_1)^2}$

$10 = \sqrt{(1 - x)^2 + (15 - 7)^2}$

$10 = \sqrt{(x^2 + 1 - 2x) + 64}$

$100 = x^2 + 1 - 2x + 64$

$x^2 - 2x - 35 = 0$

$x = 7, -5$

20. The equation of a line which passes through a point $(1, 5)$ and is perpendicular to the line $x - y = 5$ is:

Solution:

Given line: $x - y = 5$

Slope = $-\dfrac{a}{b} = 1$

Slope of the perpendicular line = −1.

Equation of line passing through (1, 5):

$y - 5 = -1(x - 1)$

$y - 5 = -x + 1$

$x + y = 6$

23 Permutation & Combination

Permutation: A permutation is an arrangement in a definite order of a number of objects taken some or all at a time. Mathematically, the number of ways of arranging n distinct objects in a row taking r $(0 < r < n)$ at a time is denoted by $P(n, r)$ or ${}^nP_r$.
क्रमचय:- क्रमचय, एक निश्चित क्रम में कई वस्तुओं की व्यवस्था है, जिसमें एक समय में कुछ या सभी वस्तुएं ली जाती हैं। गणितीय रूप से, एक समय में $r(0 < r < n)$ लेते हुए n अलग-अलग वस्तुओं को एक पंक्ति में व्यवस्थित करने के तरीकों की संख्या को $P(n, r)$ या ${}^nP_r$ द्वारा दर्शाया जाता है।

$${}^nP_r = \frac{n!}{(n-r)!}$$

Combination: Each of the different groups or selections which can be made by some or all of a number of given things without reference to the order of the things in each group is called a combination. Mathematically, the number of combinations of n different things taken r at a time is nCr.
संयोजन:- प्रत्येक समूह या चयन जो प्रत्येक समूह में चीजों के क्रम के संदर्भ के बिना दी गई चीजों में से कुछ या सभी द्वारा बनाया जा सकता है, संयोजन कहलाता है। गणितीय रूप से एक समय में r लिए गए n अलग-अलग चीजों के संयोजनों की संख्या ${}^nC_r$ है।

$$nCr = \frac{n!}{r!(n-r)!}$$

When Alphabets are repeated, we use this formula:
जब अक्षर दोहराए जाते हैं तो हम इस सूत्र का उपयोग करते हैं

The number of ways (तरीकों की संख्या) $= \dfrac{n!}{p! \times q! \times r!}$

n = Total number of alphabets (अक्षरों की कुल संख्या)
p = First type repeated alphabets (पहले दोहराए गए अक्षरों का प्रकार)
q = Second type repeated alphabets (दूसरे प्रकार के दोहराए गए अक्षर)
r = Third type repeated alphabets (तीसरे प्रकार के दोहराए गए अक्षर)

1. Find the number of 4 letter words, with or without meaning, which can be formed out of the letters of the word ROSE, where the repetition of the letters is not allowed?
ROSE शब्द के अक्षरों से अर्थ सहित या अर्थहीन 4 अक्षर वाले शब्दों की संख्या ज्ञात कीजिए, जहाँ अक्षरों की पुनरावृत्ति की अनुमति नहीं है?

Solution:

Arrangements of 4 letters:
For the first letter, we have 4 options (since all letters are available).
For the second letter, we have 3 options (since one letter has already been used).
Thus, the total arrangements for 4-letters word can be calculated as $4 \times 3 \times 2 \times 1 = 24$.

2. Given 4 flags of different colours, how many different signals can be generated, if a signal requires the use of 2 flags one below the other?
विभिन्न रंगों के 4 झंडे दिए गए हैं, कितने अलग-अलग सिग्नल उत्पन्न किए जा सकते हैं, यदि एक सिग्नल के लिए एक के नीचे एक 2 झंडों का उपयोग करना आवश्यक हो?

Solution:

For 1st flag = 4 options
For 2nd flag = 3 options (since one flag is used, Repetition is not allowed)
No. of ways = $4 \times 3 = 12$

3. How many 2 digit even numbers can be formed from the digits $1, 2, 3, 4, 5$ if the digits can be repeated?

अंक 1, 2, 3, 4, 5 से कितनी 2 अंकीय सम संख्याएँ बनाई जा सकती हैं, यदि अंकों की पुनरावृत्ति हो सकती है?

Solution:

For Even no, Unit digit must be divisible by 2

Possible Unit digits = 2 and 4 (2 numbers)

For 10's place, we have 5 options (since Repetition is allowed)

No. of possible 2 digit even numbers = $5 \times 2 = 10$

4. How many 3-digit numbers can be formed from the digits 1, 2, 3, 4 and 5 assuming that Repetition of the digits is allowed?
   अंक 1, 2, 3, 4 और 5 से कितनी 3 -अंकीय संख्याएँ बनाई जा सकती हैं, यह मानते हुए कि अंकों की पुनरावृत्ति की अनुमति है?

Solution:

No. of Possible 3-digit numbers (Repetition is Allowed) = $5 \times 5 \times 5$

No. of Possible 3-digit numbers = 125

5. How many 3-digit numbers are possible where all the 3 digits are even?
   ऐसी कितनी 3 अंकीय संख्याएँ संभव हैं जिनमें सभी 3 अंक सम हों?

Solution:

Even Digits are = 0, 2, 4, 6, 8

Repetition is allowed

At unit place = 0, 2, 4, 6, 8 can come

At 10's place = 0, 2, 4, 6, 8 can come

At 100's place = 2, 4, 6, 8 can come

No. of possible numbers = $4 \times 5 \times 5 = 100$

6. Find the number of different signals that can be generated by arranging at least 2 flags in order (one below the other) on a vertical staff, if five different flags are available?
   यदि पांच अलग-अलग झंडे उपलब्ध हों, तो एक ऊर्ध्वाधर डंडे पर कम से कम 2 झंडों को क्रम में (एक के नीचे एक) व्यवस्थित करके उत्पन्न किए जा सकने वाले विभिन्न संकेतों की संख्या ज्ञात कीजिए?

Solution:

Arrangements for 2 flags:

For the first flag, we have 5 options (since all flags are available).

For the second flag, we have 4 options (since one flag has already been used).

Thus, the total arrangements for 2 flags can be calculated as $5 \times 4 = 20$

so on for others,

Arrangements for 3 flags = $5 \times 4 \times 3 = 60$

Arrangements for 4 flags = $5 \times 4 \times 3 \times 2 = 120$

Arrangements for 5 flags = $5 \times 4 \times 3 \times 2 \times 1 = 120$

Total = $120 + 120 + 60 + 20 = 320$

7. A coin is tossed 3 times and the outcomes are recorded. How many possible outcomes are there?
   एक सिक्का 3 बार उछाला जाता है और परिणाम रिकॉर्ड किए जाते हैं। कितने संभावित परिणाम हैं?

Solution:

Outcome of Coin = Head or Tail

Coin is tossed 3 times, Every time we have 2 options either it will be Head or Tail.

No. of Possible Outcomes = $2 \times 2 \times 2 = 8$

8. A bag contains 20 balls, 8 balls are green, 7 are white, and 5 are red. What is the minimum number of balls that must be picked up from the bag blind-folded (without replacing any of it) to be assured of picking at least one ball of each colour?

एक बैग में 20 गेंदें हैं, 8 गेंदें हरी हैं, 7 सफ़ेद हैं और 5 लाल हैं। प्रत्येक रंग की कम से कम एक गेंद चुनने के लिए, आंखों पर पट्टी बांधकर (बिना किसी गेंद को बदले) बैग से कम से कम कितनी गेंदें निकाली जानी चाहिए?

Solution:

Since,

8 Green balls + 7 White balls = 15 balls

7 White balls + 5 Red balls = 12 balls

8 Green balls + 5 Red balls = 13 balls

Now, if we pick 15 balls, they may be white, green, or red but if we pick 16 balls, then it's certain that there will be at least one ball of each colour.

9. How many numbers can be made with the digits 3, 4, 5, 6, 7, 8 lying between 3000 and 4000 which are divisible by 5 while repetition of any digit is not allowed in any number?

3000 और 4000 के बीच के अंक 3, 4, 5, 6, 7, 8 से कितनी संख्याएँ बनाई जा सकती हैं जो 5 से विभाज्य हों, जबकि किसी भी संख्या में किसी भी अंक की पुनरावृत्ति की अनुमति नहीं है?

Solution:

Since, No. is lying between 3000 and 4000

At 1000's place = 3 will be there

And divisible by 5

then At unit's place, 5 will be there

Now for the 100's place, only 4 numbers are there

& for the 10's place, 3 numbers will be there

(since Repetition is not allowed)

Total Possible numbers $= 4 \times 3 = 12$

10. All the six letters of the name SACHIN are arranged to form different words without repeating any letter in any one word. The words so formed are then arranged as in a dictionary. What will be the position of the word SACHIN in that sequence?

SACHIN नाम के सभी छह अक्षरों को किसी भी एक शब्द में किसी भी अक्षर को दोहराए बिना अलगअलग शब्द बनाने के लिए व्यवस्थित किया गया है। फिर इस तरह से बने शब्दों को शब्दकोश के अनुसार व्यवस्थित किया गया है। उस क्रम में SACHIN शब्द का स्थान क्या होगा?

Solution:

Out of the given letters in the word SACHIN, S is the last letter in the alphabetical order to start a word.

If the word starts with A, then A can be kept fixed and the remaining letters can be arranged in 5! ways.

Similarly,

Number of words starting with C = 5!

Number of words starting with H = 5!

Number of words starting with I = 5!

Number of words starting with N = 5!

Now, when the word starts with S, then SACHIN is the first word in the alphabetical order to follow up.

So, Position of the word SACHIN = 5(5!) + 1 = 601

11. In how many different ways can four books A, B, C and D be arranged one above another in a vertical order such that the books A and B are never in continuous position?

चार पुस्तकों A, B, C और D को एक के ऊपर एक ऊर्ध्वाधर क्रम में कितने विभिन्न तरीकों से व्यवस्थित किया जा सकता है, ताकि पुस्तकें A और B कभी भी निरंतर स्थिति में न हों?

Solution:

Let us take books A and B as one i.e., they are always continuous.

Now, number of books $= 4 - 2 + 1 = 3$

These three books can be arranged in 3! ways and also A and B can be arranged in 2 ways among themselves.
So, number of ways when books A and B are always continuous = $2 \times 3!$
Total number of ways of arrangement of A, B, C and $D = 4!$
Hence, number of ways when A and B are never continuous = Total number of ways - number of ways when A and B are always continuous = $4! - 2 \times 3! = 12$

12. When ten persons shake hands with one another, in how many ways is it possible?
जब दस व्यक्ति एक दूसरे से हाथ मिलाते हैं तो यह कितने तरीकों से संभव है?

Solution:

First person can shake hands with the other 9, i.e., in 9 ways.
Second person can shake hands with the remaining 8 persons and so on.
Total no. of hands shaken = (9+8+. . .+1) = $9 \times \frac{10}{2} = 45$

13. In how many different ways can six players be arranged in a line such that two of them, Ajit and Mukherjee, are never together?
छह खिलाड़ियों को एक पंक्ति में कितने अलग-अलग तरीकों से व्यवस्थित किया जा सकता है कि उनमें से दो, अजीत और मुखर्जी, कभी एक साथ न हों?

Solution:

Total number of ways of arrangement for six players = 6!
Let us take Ajit and Mukherjee as one entity.
So now there are $(6 - 2 + 1 = 5)$ players.

These 5 players can be arranged in 5! ways and Ajit and Mukherjee can be arranged among themselves in 2! ways.
Thus, number of ways when Ajit and Mukherjee are always together = $5! \times 2!$
Hence, number of ways when they are never together = Total number of ways - number of ways when they are always together
$6! - (5! \times 2!) = 6 \times 5! - (5! \times 2!) = 5!(6-2) = 480$

14. Find the number of permutations of the letters of the word ALLAHABAD?
शब्द ALLAHABAD के अक्षरों के क्रमचयों की संख्या ज्ञात कीजिए?

Solution:

In the word ALLAHABAD,
$n = 9$ (Total letters in ALLAHABAD)
$p = 4$ (number of A's)
$q = 2$ (number of L's)
The number of arrangements = $\frac{9!}{(4! \times 2!)}$
The number of arrangements = $\frac{(9 \times 8 \times 7 \times 6 \times 5 \times 4!)}{(2 \times 1) \times 4!}$
The number of arrangements = $\frac{(9 \times 8 \times 7 \times 6 \times 5)}{2}$
The number of arrangements = 756
∴ Word ALLAHABAD can arrange in 7560 ways

15. In how many ways can 4 red, 3 yellow, and 2 green discs be arranged in a row if the discs of the same colour are indistinguishable?
4 लाल, 3 पीली और 2 हरी डिस्कों को एक पंक्ति में कितने तरीकों से व्यवस्थित किया जा सकता है, यदि एक ही रंग की डिस्कों में अंतर नहीं किया जा सकता है?

Solution:

Number of discs = 4 red + 3 yellow + 2 green

Number of Arrangements = $\frac{9!}{(4! \times 3! \times 2!)}$

$= \frac{9 \times 8 \times 7 \times 6 \times 5 \times 4!}{4! \times (3 \times 2 \times 1) \times (2 \times 1)}$

$= \frac{9 \times 8 \times 7 \times 6 \times 5}{12}$

$= 1260$

Hence, Total Number of Arrangements = 1260

16. Find the number of different 8-letter arrangements that can be made from the letters of the word DAUGHTER so that
(i) All vowels occur together.
(ii) All vowels do not occur together.
DAUGHTER शब्द के अक्षरों से 8 अक्षरों की कितनी विभिन्न व्यवस्थाएँ बनाई जा सकती हैं, ताकि
(i) सभी स्वर एक साथ आएँ।
(ii) सभी स्वर एक साथ न आएँ।

Solution:

(i) Total number of letters in DAUGHTER = 8

Vowels in DAUGHTER = A, U, E

Since all vowels occur together, assume (AUE) as a single object.

So, our letters become (AUE), D, G, H, T, R

Arranging 3 vowels

Since vowels are coming together, they can be AUE, UAE, EAU, and so on.

Number of Permutation of 3 vowels = 3! = 6 ways

Arranging 6 letters

Number of Permutation of 6 letters = 6! = 720

Thus, Total number of Arrangements = 720 × 6 = 4320

(ii) Total number of words in DAUGHTER = 8

Total number of Permutation of 8 letters = 8!

Number of Permutation where all vowels come together = 4320

Thus, number of Permutation in which all vowels are never together = Total number of Permutation - Number of Permutation where all vowels come together

$8! - 4320 = 40320 - 4320 = 36000$

17. Find the number of arrangements of the letters of the word INDEPENDENCE. In how many of arrangements do the words begin with I and end in P?
INDEPENDENCE शब्द के अक्षरों की व्यवस्था की संख्या ज्ञात कीजिए। कितनी व्यवस्थाओं में शब्द I से शुरू होते हैं और P पर खत्म होते हैं?

Solution:

Let's fix I and P at the extreme ends.

I _ _ _ _ _ _ P

Between I and P, 10 letters are there (2 D, 4 E, and 3 N).

Since letters are repeating, we use the formula: $\left(\frac{n!}{p! \times q! \times r!}\right)$

Where $n = 10, p = 2, q = 4$, and $r = 3$.

Required number of arrangements = $\frac{10!}{(2! \times 4! \times 3!)} = 12600$

18. In how many of the distinct permutations of the letters in MISSISSIPPI do the four I's not come together?
मिसिसिपी में अक्षरों के कितने अलग-अलग क्रमपरिवर्तनों में चार 'आई' एक साथ नहीं आते हैं?

Solution:

Total number of permutations of 4 I's not coming together = Total permutations - Total permutations of I's coming together

In MISSISSIPPI ($n = 11$):

There are 4 I, 4 S, 2 P, and 1 M.

Total permutations = $\frac{11!}{4! \times 4! \times 2!}$

Total permutations = 34650

Total permutations of I coming together:

Now, taking 4 I's as one

MISSISSIPPI $\rightarrow$ MSSSSPP(IIII).

$n = 8$: 4 S, 2 P, 1 M, and 1 (IIII).

Number of permutations with 4 I's together = $\frac{8!}{4! \times 2!} = 840$

Now,

Total number of permutations of 4 I's not coming together = Total permutations - Total permutations of I's coming together

$= 34650 - 840 = 33810$

19. How many 3-digit natural numbers (without repetition of digits) are there such that each digit is odd and the number is divisible by 5?
ऐसी कितनी 3-अंकीय प्राकृतिक संख्याएँ (अंकों की पुनरावृत्ति के बिना) हैं जिनमें प्रत्येक अंक विषम है और संख्या 5 से विभाज्य है?

Solution:

We have odd numbers as 1, 3, 5, 7, 9.

Unit digit of 3-digit number divisible by 5 will be 5.

For the remaining two positions, we have 4 numbers: 1, 3, 7, 9. So, for arranging these numbers, we use permutation as:

$$4P2 = \frac{4!}{(4-2)!} = \frac{(4 \times 3 \times 2 \times 1)}{(2 \times 1)} = 12$$

20. Find the number of 4-digit numbers that can be formed using the digits 1, 2, 3, 4, 5 if no digit is repeated. How many of these will be even?
अंक 1, 2, 3, 4, 5 का उपयोग करके बनाई जा सकने वाली 4-अंकीय संख्याओं की संख्या ज्ञात कीजिए, यदि कोई अंक दोहराया नहीं गया है। इनमें से कितनी सम संख्याएँ होंगी?

Solution:

We have numbers 1, 2, 3, 4, 5.

For an even number

Unit digit will be either 2 or 4.

Since repetition is not allowed,

No. of possible 4-digit even numbers (2 at unit place) $= 4 \times 3 \times 2 = 24$

No. of possible 4-digit even numbers (4 at unit place) $= 4 \times 3 \times 2 = 24$

Total possible numbers $= 24 + 24 = 48$

21. How many words, with or without meaning, can be made from the letters of the word MONDAY, assuming that no letter is repeated, if all letters are used but the first letter is a vowel?
शब्द MONDAY के अक्षरों से अर्थ सहित या अर्थहीन कितने शब्द बनाए जा सकते हैं, यह मानते हुए कि कोई भी अक्षर दोहराया नहीं गया है, यदि सभी अक्षरों का उपयोग किया गया है लेकिन पहला अक्षर स्वर है?

Solution:

First letter should be a vowel (a, e, i, o, u).

Vowels in MONDAY are O and A.

Assuming the first letter is "O"

If first letter is O, word will be of the form O_.

Number of letters left = 5.

Number of different words starting with O = 5! = 120.

Similarly,

If the first letter is A

Number of different words = 120.

Thus,

Required number of words = 120 + 120 = 240.

22. In how many ways can the letters of the word PERMUTATIONS be arranged if there are always 4 letters between P and S?

शब्द PERMUTATIONS के अक्षरों को कितने तरीकों से व्यवस्थित किया जा सकता है यदि P और S के बीच हमेशा 4 अक्षर हों?

Solution:

Total letters = 12.

Position of P and S:

P _ _ _ _ S _ _ _ _ _ _

_ P _ _ _ _ S _ _ _ _ _

_ _ P _ _ _ _ S _ _ _ _

_ _ _ P _ _ _ _ S _ _ _

So, there are total 7 cases but that is when P is before S.

There can also be a case where S is before P.

Thus, total cases = $7 \times 2 = 14$ cases.

Now, let's find the permutation.

$n = 10$ (2 T's are there).

Number of arrangements of the other letters = $\frac{10!}{2!}$.

Thus, total number of arrangements = $14 \times \frac{10!}{2!} = 14 \times 10! = 2540160$.

23. A committee of 3 persons is to be constituted from a group of 2 men and 3 women.
(i) In how many ways can this be done?
(ii) How many of these committees would consist of 1 man and 2 women?
2 पुरुषों और 3 महिलाओं के समूह से 3 व्यक्तियों की एक समिति गठित की जानी है।
(i) यह कितने तरीकों से किया जा सकता है?
(ii) इनमें से कितनी समितियों में 1 पुरुष और 2 महिलाएँ होंगी?

Solution:

(i) We have to select a committee of 5 different persons taken 3 at a time.

Here, $n = 5$, $r = 3$

The required number of ways = ${}^nC_r = {}^5C_3$

$= \frac{5!}{3! \cdot 2!} = 5 \times 2 = 10$ ways

(ii) Total: 2 men and 3 women

Committee would consist of 1 man and 2 women

Number of ways = ${}^2C_1 \times {}^3C_2$

$= \left(\frac{2!}{1! \cdot 1!}\right) \times \left(\frac{3!}{2! \cdot 1!}\right) = 2 \times 3 = 6$ ways

24. What is the number of ways of choosing 4 cards from a pack of 52 playing cards? In how many of these:
(i) Four cards are of the same suit.
(ii) Four cards belong to four different suits.
(iii) Are face cards.
(iv) Two are Red cards and two are Black cards.
(v) Cards are of the same colour.
52 ताश के पत्तों के एक पैकेट में से 4 पत्ते चुनने के कितने तरीके हैं? इनमें से कितने:
(i) चार पत्ते एक ही सूट के हैं
(ii) चार पत्ते चार अलग-अलग सूट के हैं
(iii) फेस कार्ड हैं
(iv) दो लाल कार्ड हैं और दो काले कार्ड हैं
(v) पत्ते एक ही रंग के हैं

Solution:

There will be as many ways of choosing 4 cards from 52 cards as there are combinations of 52 different things, taken 4 at a time.

(i) There are four suits - Diamond, Club, Spade, and Heart, and there are 13 cards of each suit.

Therefore, there are ${}^{13}C_4$ ways of choosing 4 Diamonds, 4 Clubs, 4 Spades, and 4 Hearts.

Thus, the required number of ways = ${}^{13}C_4 + {}^{13}C_4 + {}^{13}C_4 + {}^{13}C_4 = 4 \times \frac{13!}{4! \times 9!} = 2860$.

(ii) There are 13 cards in each suit.

Therefore, there are $^{13}C_1$ ways of choosing 1 card from 13 cards of Diamonds, Clubs, Spades, and Hearts.

The required number of ways =

$^{13}C_1 \times ^{13}C_1 \times ^{13}C_1 \times ^{13}C_1 = 13^4$.

(iii) There are 12 face cards and 4 are to be selected out of these 12 cards.

This can be done in $^{12}C_4$ ways.

The required number of ways $= \dfrac{12!}{4! \times 8!} =$ 495.

(iv) There are 26 red cards and 26 black cards.

Therefore required number of ways =

$^{26}C_2 \times ^{26}C_2$

$\left(\dfrac{26!}{2! \times 24!}\right) \times \left(\dfrac{26!}{2! \times 24!}\right)$

$(26 \times \frac{25}{2}) \times (26 \times \frac{25}{2})$

105625

(v) 4 red cards can be selected out of 26 red cards in $^{26}C_4$ ways.

4 black cards can be selected out of 26 black cards in $^{26}C_4$ ways.

The required number of ways =

$^{26}C_4 + ^{26}C_4$

$\left(\dfrac{26!}{4! \times 22!}\right) \times \left(\dfrac{26!}{4! \times 22!}\right)$

$(26 \times 25 \times 23) + (26 \times 25 \times 23)$

$= 14950 + 14950 = 29900$

25. How many chords can be drawn through 21 points on a circle?
एक वृत्त पर 21 बिंदुओं से होकर कितनी जीवाएँ खींची जा सकती हैं?

Solution:

Number of points = 21.

Hence, $n = 21$.

A chord connects the circle at 2 points. Hence, r = 2

Number of chords from 21 points = $^{21}C_2$.

$= \dfrac{21!}{2! \times 19!}$

$= \dfrac{21 \times 20 \times 19!}{2! \times 1 \times 19!}$

$= 21 \times \dfrac{20}{2}$

$= 210$ chords

26. There are 10 points in a plane, no three of which are collinear. Find the number of triangles formed by joining them?
एक समतल में 10 बिंदु हैं, जिनमें से कोई भी तीन संरेख नहीं हैं। उन्हें मिलाकर बनने वाले त्रिभुजों की संख्या ज्ञात कीजिए?

Solution:

A triangle is defined by selecting 3 points from the available points.

The number of ways to choose 3 points from 10 can be calculated using the combination formula: $^{10}C_3$

$\dfrac{10!}{3! \times 7!}$

$\dfrac{10 \times 9 \times 8 \times 7!}{3 \times 2 \times 7!} = 120$

Number of triangles formed by joining them = 120.

27. In how many ways can 6 persons be selected from 4 officers and 8 constables, if at least one officer is to be included?
4 अधिकारियों और 8 कांस्टेबलों में से 6 व्यक्तियों का चयन कितने तरीकों से किया जा सकता है, यदि कम से कम एक अधिकारी को शामिल किया जाना है?

Solution:

6 persons have to be selected from 4 officers and 8 constables.

At least 1 officer is to be selected.

Possible cases:

Case 1: 1 officer and 5 constables = ${}^4C_1 \times {}^8C_5$

Case 2: 2 officers and 4 constables = ${}^4C_2 \times {}^8C_4$

Case 3: 3 officers and 3 constables = ${}^4C_3 \times {}^8C_2$

Case 4: 4 officers and 2 constables = ${}^4C_4 \times {}^8C_2$

Total possible ways = Total Possible ways:

${}^4C_1 \times {}^8C_5 + {}^4C_2 \times {}^8C_4 + {}^4C_3 \times {}^8C_3 + {}^4C_4 \times {}^8C_2$

$= \left(\frac{4!}{1!\cdot 3!}\right) \times \left(\frac{8!}{5!\cdot 3!}\right) + \left(\frac{4!}{2!\cdot 2!}\right) \times \left(\frac{8!}{4!\cdot 4!}\right) + \left(\frac{4!}{1!\cdot 3!}\right) \times \left(\frac{8!}{3!\cdot 5!}\right) + \left(\frac{4!}{4!}\right) \times \left(\frac{8!}{2!\cdot 6!}\right)$

$= 4 \times 8 \times 7 + 6 \times 7 \times 2 \times 5 + 4 \times 8 \times 7 + 4 \times 7$

$= 224 + 420 + 224 + 28$

$= 896$

28. How many different sums can be formed with the denominations 50, 100, 200, 500 and 2,000 taking at least three denominations at a time?

50, 100, 200, 500 और 2,000 मूल्यवर्गों से एक समय में कम से कम तीन मूल्यवर्गों को लेकर कितनी अलग-अलग राशियाँ बनाई जा सकती हैं?

Solution:

Let each denomination be considered separately.

Total denominations = 5.

According to the question, taking at least three denominations at a time.

Possible cases:

Case 1: 3 denominations out of 5 in ${}^5C_3$ ways = 10 ways.

Case 2: 4 denominations out of 5 in ${}^5C_4$ ways = 5 ways.

Case 3: 5 denominations out of 5 in ${}^5C_5$ ways = 1 way.

Total ways = $10 + 5 + 1 = 16$.

29. If 2 boys and 2 girls are to be arranged in a row so that the girls are not next to each other, how many possible arrangements are there?

यदि 2 लड़के और 2 लड़कियाँ को एक पंक्ति में इस प्रकार व्यवस्थित किया जाए कि लड़कियाँ एक दूसरे के बगल में न हों, तो कितनी व्यवस्थाएं संभव हैं?

Solution:

Total number of people = 4 (2 boys and 2 girls).

Total number of possible ways for arrangements = $4! = 24$.

Number of possible ways in which both girls are next to each other:

G1G2B1B2, G1G2B2B1, G2G1B1B2, G2G1B2B1 (4 ways when the first two places have girls arranged).

Similarly, on _ GG _, _ _ GG (8 more ways).

Number of possible ways where both girls are next to each other = 12.

Number of possible ways where girls are not next to each other = $24 - 12 = 12$ ways.

30. A selection is to be made for one post of Principal and two posts of Vice-Principal. Amongst the six candidates called for the interview, only two are eligible for the post of Principal while they all are

eligible for the post of Vice-Principal. The number of possible combinations of selectees is?

प्रिंसिपल के एक पद और वाइस प्रिंसिपल के दो पदों के लिए चयन किया जाना है। साक्षात्कार के लिए बुलाए गए छह उम्मीदवारों में से केवल दो प्रिंसिपल के पद के लिए पात्र हैं जबकि वे सभी वाइस प्रिंसिपल के पद के लिए पात्र हैं। चयनित उम्मीदवारों के संभावित संयोजनों की संख्या है?

Solution:

Number of ways to select Principal = ${}^2C_1$

Number of ways to select Vice-Principal = ${}^5C_2$.

Total number of ways = ${}^2C_1 \times {}^5C_2. = 2 \times 10 = 20$

Number of possible combinations = $2 \times 10 = 20$

31. Amit has five friends: 3 girls and 2 boys. Amit's wife also has 5 friends: 3 boys and 2 girls. In how many maximum number of different ways can they invite 2 boys and 2 girls such that two of them are Amit's friends and two are his wife's?

अमित के पाँच दोस्त हैं: 3 लड़कियाँ और 2 लड़के। अमित की पत्नी के भी 5 दोस्त हैं: 3 लड़के और 2 लड़कियाँ। वे 2 लड़कों और 2 लड़कियों को अधिकतम कितने तरीकों से आमंत्रित कर सकते हैं, ताकि उनमें से दो अमित के दोस्त हों और दो उसकी पत्नी के?

Solution:

There can be three cases:

Case 1: Amit (1 boy & 1 girl) and Wife (1 boy & 1 girl)

Number of ways = ${}^2C_1 \times {}^3C_1 \times {}^3C_1 \times {}^2C_1 = 36$.

Case 2: Amit (2 boys) and Wife (2 girls)

Number of ways = ${}^2C_2 \times {}^2C_2 = 1$.

Case 3: Amit (2 girls) and Wife (2 boys)

Number of ways = ${}^3C_2 \times {}^3C_2 = 9$.

Total number of ways = $36 + 1 + 9 = 46$.

32. Three dice (each having six faces with each face having one number from 1 to 6) are rolled. What is the number of possible outcomes such that at least one dice shows the number 2?

तीन पासे (प्रत्येक के छह फलक हैं, जिनमें से प्रत्येक पर 1 से 6 तक एक अंक अंकित है) फेंके जाते हैं। संभावित परिणामों की संख्या क्या है, जिससे कम से कम एक पासा संख्या 2 दर्शाता है?

Solution:

There can be 3 cases possible:

Case 1: When one die shows 2.

The die which shows 2 can be selected out of the 3 dice in ${}^3C_1$ ways.

Remaining 2 dice can have any 5 numbers except 2.

So, number of ways for them = ${}^5C_1$ each.

So, the number of ways when one die shows $2 = {}^3C_1 \times {}^5C_1 \times {}^5C_1 = 75$.

Case 2: When two dice show 2.

Two dice showing 2 can be selected out of the 3 dice in ${}^3C_2$ ways and the rest one can have any 5 numbers except 2. so number of ways for the remaining 1 dice = 5.

So, number of ways, when two dices show 2 $= {}^3C_2 \times 5 = 15$

Case 3: When all three dice show 2

When three dices show 2 then these can be selected in ${}^3C_3 = 1$ way.

Hence, total number of ways = $75 + 15 + 1 = 91$.

33. A school teacher has to select the maximum possible number of different groups of 3 students out of a total of 6 students. In how many groups any particular student will be included?

एक स्कूल शिक्षक को कुल 6 छात्रों में से 3 छात्रों के विभिन्न समूहों की अधिकतम संभव संख्या का चयन करना है। किसी विशेष छात्र को कितने समूहों में शामिल किया जाएगा?

Solution:

Suppose any particular student is always selected.

Now, remaining 2 students are to be selected out of the remaining 5 students.

It can be done in ${}^5C_2$ ways $= \dfrac{5!}{2! \times 3!} = 10$ ways.

34. How many three-digit numbers can be generated from 1, 2, 3, 4, 5, 6, 7, 8, 9 such that the digits are in ascending order?
1, 2, 3, 4, 5, 6, 7, 8, 9 से कितनी तीन-अंकीय संख्याएँ उत्पन्न की जा सकती हैं, जिनमें अंक बढ़ते क्रम में हों?

Solution:

Any 3 digits out of 9 can be selected in ${}^9C_3$ ways.

Now, these three digits can be arranged among themselves in ascending order in only 1 way.

Hence, total number of ways $= {}^9C_3 \times 1 = 84$.

35. Groups each containing 3 boys are to be formed out of 5 boys A, B, C, D, E such that no one group contains both C and D together. What is the maximum number of such different groups?
5 लड़कों - A, B, C, D और E में से 3-3 लड़कों वाले समूह इस प्रकार बनाए जाने हैं कि किसी भी समूह में C और D दोनों एक साथ न हों। ऐसे विभिन्न समूहों की अधिकतम संख्या क्या है?

Solution:

Total number of arrangements, when any 3 boys are selected out of 5 $= {}^5C_3 = \dfrac{5!}{3! \times 2!} = 10$.

Now, when groups contain both C and D, their selection is fixed, and the remaining 1 boy can be selected out of the remaining 3 boys.

It can be done in ${}^3C_1$ ways $= \dfrac{3!}{1! \times 2!} = 3$.

So, the number of groups, when none contains both C and D = Total number of arrangements - Number of arrangements when group contains both C and $D = 10 - 3 =$ 7 groups.

36. How many words, with or without meaning, each of 3 vowels and 2 consonants can be formed from the letters of the word INVOLUTE?
शब्द INVOLUTE के अक्षरों से 3 स्वरों और 2 व्यंजनों वाले कितने शब्द, अर्थ सहित या अर्थहीन, बनाए जा सकते हैं?

Solution:

Word: INVOLUTE

Vowels = I, O, U, E

Consonants = N, V, L, T

Number of ways of selecting 3 vowels & 2 consonants $= {}^4C_3 \times {}^4C_2$

$= \left(\dfrac{4!}{3! \times 1!}\right) \times \left(\dfrac{4!}{2! \times 2!}\right) = 4 \times 6 = 24$

Number of arrangements of 5 letters $= 5! = 120$.

Thus, total number of words = Number of ways of selecting × Number of arrangements $= 24 \times 120 = 2880$.

37. A group consists of 4 girls and 7 boys. In how many ways can a team of 5 members be selected if the team has:
(i) No Girl
(ii) At least 3 Girls
एक समूह में 4 लड़कियाँ और 7 लड़के हैं। 5 सदस्यों वाली एक टीम का चयन कितने तरीकों से किया जा सकता है, यदि टीम में:
(i) कोई लड़की नहीं है
(ii) कम से कम 3 लड़कियाँ हैं

Solution:

(i) Team of 5 members to be selected which has no girl

So, we have to take 5 boys from 7 boys

Number of ways $= {}^7C_5 = \frac{7!}{5! \cdot 2!}$

$= \frac{7 \times 6 \times 5!}{2 \times 5!} = 21$ ways

(ii) Team of 5 members to be selected which has at least 3 girls

Case (1): 2 boys & 3 girls

Number of ways $= {}^7C_2 \times {}^4C_3 = 84$ ways

Case (2): 1 boy & 4 girls

Number of ways $= {}^7C_1 \times {}^4C_4 = 7$ ways

Total Number of ways $= 7 + 84 = 91$ ways

38. Find the number of words with or without meaning which can be made using all the letters of the word AGAIN. If these words are written as in a dictionary, what will be the 50th word?

AGAIN शब्द के सभी अक्षरों का उपयोग करके अर्थ सहित या अर्थहीन शब्दों की संख्या ज्ञात कीजिए। यदि इन शब्दों को शब्दकोश के अनुसार लिखा जाए, तो 50 वाँ शब्द क्या होगा?

Solution:

"AGAIN" - (2 A, 1 G, 1 I & 1 N)

In Dictionary, letters appear alphabetically.

Case 1.

Word starts with letter A (A _ _ _ _)

Number of ways for arranging other 4 letters after A = 4! = 24.

Case 2.

Word starts with letter G (Repetition of A)

Number of ways for arranging other 4 letters after G $= \frac{4!}{2!} = 12$.

Case 3.

Word starts with letter I (Repetition of A)

Number of ways for arranging other 4 letters after I $= \frac{4!}{2!} = 12$. Now, total number of words $= 24 + 12 + 12 = 48$.

Now word will start with letter N.

49th word = NAAGI,

50th word = NAAIG.

Hence, NAAIG will be the 50th word.

39. How many numbers greater than 1000000 can be formed by using the digits 1, 2, 0, 2, 4, 2, 4?

अंक 1, 2, 0, 2, 4, 2, 4 का उपयोग करके 1000000 से बड़ी कितनी संख्याएँ बनाई जा सकती हैं?

Solution:

1, 2, 0, 2, 4, 2, 4 (One 0, three 2's, two 4's).

Case 1: When the number starts with 1. We have to arrange 0, three 2's, and two 4's.

Number of ways $= \frac{6!}{3! \times 2!} = 60$.

Case 2: When the number starts with 2. We have to arrange One 0, One 1, two 2's, and two 4's.

Number of ways $= \frac{6!}{2! \times 2!} = 180$.

Case 3: When the number starts with 4. We have to arrange One 0, One 1, three 2's, and one 4's.

Number of ways $= \frac{6!}{3!} = 120$.

Total number of ways $= 60 + 180 + 120$.

Total number of ways = 360.

40. How many words, with or without meaning, can be formed using all the letters of the word EQUATION at a time so that the vowels and consonants occur together?

EQUATION शब्द के सभी अक्षरों का प्रयोग करके एक साथ कितने शब्द, अर्थ सहित या अर्थहीन, बनाए जा सकते हैं, ताकि स्वर और व्यंजन एक साथ हों?

Solution:

EQUATION

Vowels = A, E, I, O, U =

So, Number of ways vowels can be arranged = 5! = 120.

Consonants = N, Q, T

So, Number of ways consonants can be arranged = 3! = 6.

Two Possibilities Either start with vowel or start with consonants

So, Total number of ways in which vowels & consonants occurr together = 2 × (number of ways vowels arrange) × (number of ways consonants arrange)

$2 \times 120 \times 6 = 1440$ ways.

41. If the different permutations of all the letters of the word EXAMINATION are listed as in a dictionary, how many words are there in this list before the first word starting with E?
 यदि EXAMINATION शब्द के सभी अक्षरों के विभिन्न क्रमचयों को शब्दकोश के अनुसार सूचीबद्ध किया जाए, तो इस सूची में E से शुरू होने वाले पहले शब्द से पहले कितने शब्द हैं?

Solution:

EXAMINATION

The letters before E in the word are only A.

Words in the list before the word starting with E = words starting with letter A.

EXAMINATION

(Repeated letters are 2 I's & 2 N's)

Total words before E are $\dfrac{10!}{2! \times 2!} = 907200$.

42. Determine the number of 5-card combinations out of a deck of 52 cards if each selection of 5 cards has exactly one king?
 52 कार्डों के एक डेक में से 5-कार्ड संयोजनों की संख्या निर्धारित करें यदि 5 कार्डों के प्रत्येक चयन में ठीक एक राजा है?

Solution:

There are 4 king cards out of 52.

We have to select 1 king from 4 king cards.

Number of ways of selecting King card = ${}^4C_1 = 4$ ways.

Now, select other 4 cards from remaining 48 cards.

Number of ways = ${}^{48}C_4 = 194580$.

Total number of ways = $4 \times 194580 = 778320$ ways.

43. It is required to seat 5 men and 4 women in a row so that the women occupy the even places. How many such arrangements are possible?
 5 पुरुषों और 4 महिलाओं को एक पंक्ति में इस प्रकार बैठाना आवश्यक है कि महिलाएं सम स्थानों पर बैठें। ऐसी कितनी व्यवस्थाएँ संभव हैं?

Solution:

There are total 9 people, and women will occupy even places.

Places 2, 4, 6, and 8 will be occupied by women, and the rest will be occupied by men.

The seating arrangement will be as follows: 4 women can sit in 4 positions.

Total number of ways = ${}^4P_4 = 24$.

5 men can sit in 5 positions.

Total number of ways = ${}^5P_5 = 120$.

Thus, total number of arrangements = $24 \times 120 = 2880$ ways.

44. The letters of the word OUGHT are written in all possible orders and these words are written out as in a dictionary. Find the rank of the word TOUGH in this dictionary.
 OUGHT शब्द के अक्षरों को सभी संभावित क्रम में लिखा गया है और इन शब्दों को शब्दकोश की तरह लिखा गया है। इस शब्दकोश में TOUGH शब्द का स्थान ज्ञात कीजिए।

Solution:

Let's arrange the letters of OUGHT in alphabetical order: "G, H, O, T, U".

Number of words starting with $G = 4!$.

Number of words starting with $H = 4!$.

Number of words starting with $O = 4!$.

Number of words starting with TG = 3!.

Number of words starting with TH = 3!.

Number of words starting with TOG = 2!.

Number of words starting with TOH = 2!.

Total number of words till now $= 3 \times 4! + 2 \times 3! + 2 \times 2!$

$= 72 + 12 + 4 = 88$.

Then, the 89th word will be TOUGH.

45. On a railway route between two places A and B, there are 20 stations on the way. If 4 new stations are to be added, how many types of new tickets will be required if each ticket is issued for a one-way journey?
दो स्थानों A और B के बीच एक रेलवे मार्ग पर, रास्ते में 20 स्टेशन हैं। यदि 4 नए स्टेशन जोड़े जाने हैं, तो प्रत्येक टिकट एकतरफा यात्रा के लिए जारी किए जाने पर कितने प्रकार के नए टिकटों की आवश्यकता होगी?

Solution:

For (10 + A + B) = 12 stations, no of Tickets required, when 4 New Stations are added for one way Journey $= 12 \times 4 = 48$.
Also, each 4 new stations requires $(16 - 1) = 15$ New tickets for one-way journeys.
Therefore, No. of tickets for 4 new stations $= 15 \times 4 = 60$.
Hence, Total new tickets $= 60 + 48 = 108$.

46. How many numbers lie between 300 and 500 in which 4 comes only one time?
300 और 500 के बीच कितनी संख्याएँ हैं जिनमें 4 केवल एक बार आता है?

Solution:

Case 1: When 4 is at the hundredth place. Remaining two places can be filled through any of the numbers 0 to 9 except 4.

So, the number of ways $= 1 \times 9 \times 9 = 81$.

Case 2: When 4 is at the units and 3 is at the hundredth place. Then tens place can be filled through any of the numbers 0 to 9 except 4.

So, the number of ways = 9.

Case 3: When 4 is at the tens place and 3 is at the hundredth place. The units place can be filled through any of the numbers 0 to 9 except 4.

So, the number of ways = 9.

Here, either case I , case II or case III is possible.

Hence, total number of ways $= 81 + 9 + 9 = 99$.

47. In how many ways can a batsman score exactly 25 runs by scoring single runs, fours, and sixes only, irrespective of the sequence of scoring shots?
एक बल्लेबाज केवल एक रन, चौका और छक्का लगाकर, स्कोरिंग शॉट्स के क्रम पर ध्यान दिए बिना, कितने तरीकों से ठीक 25 रन बना सकता है?

Solution:

Case 1: Four sixes: (1, 0, 4)
(four sixes and one single) = 1 way.
Case 2: 3 sixes:
(7,0,3) (7 Single & 3 sixes) & (3,1,3) (3 singles , 1 four & 3 sixes) = 2 ways
Case 3: 2 sixes:
(13, 0, 2), (9, 1, 2), (5, 2, 2), (1, 3, 2) = 4 ways.

Case 4: 1 six:

(19, 0, 1), (15, 1, 1), (11, 2, 1), (7, 3, 1), (3, 4, 1) = 5 ways.

Case 5: 0 six:

(25, 0, 0), (21, 1, 0), (17, 2, 0), (13, 3, 0), (9, 4, 0), (5, 5, 0), (1, 6, 0) = 7 ways.

Each bracket represents (singles, fours, sixes)

Total number of ways: $1+2+4+5+7 = 19$

Therefore, there are 19 different ways for a batsman to score exactly 25 runs using combinations of singles, fours, and sixes.

48. How many distinct 8-digit numbers can be formed by rearranging the digits of the number 11223344 such that odd digits occupy odd positions and even digits occupy even positions?
संख्या 11223344 के अंकों को इस प्रकार पुनर्व्यवस्थित करके कितनी अलग-अलग 8-अंकीय संख्याएँ बनाई जा सकती हैं कि विषम अंक विषम स्थान पर हों और सम अंक सम स्थान पर हों?

Solution:

The number 11223344 has 4 odd digits (1, 1, 3, 3) and 4 even digits (2, 2, 4, 4).

In an 8-digit number, there are 4 odd positions (1st, 3rd, 5th, 7th) and 4 even positions (2nd, 4th, 6th, 8th).

The 4 odd digits can be arranged in the 4 odd positions in $\frac{4!}{2! \times 2!} = 6$ ways.

Similarly, the 4 even digits can be arranged in the 4 even positions in $\frac{4!}{2! \times 2!} = 6$ ways.

Therefore, the total number of distinct 8-digit numbers is $6 \times 6 = 36$.

49. 2 men and 1 woman board a bus in which 5 seats are vacant. One of these five seats is reserved for ladies. A woman may or may not sit on the seat reserved for ladies but a man cannot sit on the seat reserved for ladies. In how many different ways can the five seats occupied by these passengers?
2 पुरुष और 1 महिला एक बस में चढ़ते हैं जिसमें 5 सीटें खाली हैं। इन पाँच सीटों में से एक सीट महिलाओं के लिए आरक्षित है। महिलाएँ महिलाओं के लिए आरक्षित सीट पर बैठ सकती हैं या नहीं भी बैठ सकती हैं, लेकिन पुरुष महिलाओं के लिए आरक्षित सीट पर नहीं बैठ सकते। इन पाँचों सीटों पर इन यात्रियों द्वारा कितने अलग-अलग तरीकों से कब्जा किया जा सकता है?

Solution:

There can be two cases:

Case 1: Lady occupies the reserved seat.

Fixing one seat for the lady, the first man can occupy any of the remaining 4 seats in 4 ways, and the second man can occupy any of the remaining 3 seats in 3 ways.

Hence, the number of ways $= 1 \times 4 \times 3 = 12$.

Case 2: Lady does not occupy the reserved seat.

Leaving the reserved seat, the first man can occupy any of the 4 seats in 4 ways. The second man can occupy any of the remaining 3 seats in 3 ways. The lady can occupy any of the remaining 2 seats in 2 ways.

Hence, the number of ways $= 4 \times 3 \times 2 = 24$.

Thus, total number of ways $= 12 + 24 = 36$.

50. Each of the 3 persons is to be given some identical items such that the product of the numbers of items received by each of the three persons is equal to 30. In how many maximum different ways can this distribution be done?

तीनों व्यक्तियों को कुछ समान वस्तुएं इस प्रकार दी जानी हैं कि तीनों व्यक्तियों द्वारा प्राप्त वस्तुओं की संख्या का गुणनफल 30 के बराबर हो। यह वितरण अधिकतम कितने विभिन्न तरीकों से किया जा सकता है?

Solution:

Suppose three people have been given a, b, and c number of items. Then, $a \times b \times c = 30$.

Now, There can be 5 cases.

Case 1: When one of them is given 30 items and the rest two 1 item each. So, number of ways for $(30 \times 1 \times 1) = \frac{3!}{2!} = 3$ ways.

Case 2: Similarly, number of ways for $(10 \times 3 \times 1) = 3! = 6$ ways.

Case 3: Number of ways for $(15 \times 2 \times 1) = 3! = 6$ ways.

Case 4: Number of ways for $(6 \times 5 \times 1) = 3! = 6$ ways.

Case 5: Number of ways for $(5 \times 3 \times 2) = 3! = 6$ ways.

Here, either of these 5 cases are possible.

Hence, total number of ways $= 3 + 6 + 6 + 6 + 6 = 27$.

24 Probability

Trial - Performing an experiment is called a trial. The number of times an experiment is repeated is called the number of trials.
Sample Space - The set of all possible outcomes of a random experiment is called the sample space of the experiment and it is denoted by S.
Event - A subset of the sample space associated with a random experiment is called event or case.
Let n be the total number of trials. The empirical Probability (P(E) of an event E is given by
माना परीक्षणों की कुल संख्या n है। एक घटना E की अनुभवजन्य संभावना (P(E) द्वारा दी गई है

$$P(E) = \frac{\text{No. of trials in which the event happened}}{\text{The total no. of trials}}$$

Note- The probability of each event lies between 0 and 1.
प्रत्येक घटना की संभावना 0 और 1 के बीच होती है।
Sum of all the probabilities of an event is 1.
किसी घटना की सभी संभावनाओं का योग 1 है।

Concept of Coin

1. When one coin is tossed then S = H, T $\rightarrow 2^1$ outcomes = 2 outcomes
2. When two coins are tossed then S = HH, TT, HT, TH $\rightarrow$ 2^2 outcomes = 4 outcomes
3. When three coins are tossed then S = HHH, TTT, HTT, THT, TTH, THH, HTH, HHT $\rightarrow 2^3$ outcomes = 8 outcomes
4. When four coins are tossed then S = HHHH, TTTT, HTTT, THTT, TTHT, TTTH, HHHH, HHHT, HHTH, THHH, HTHH, THTH, TTHH, HHTT, THHT, HTTH $\rightarrow$ 2^4 outcomes = 16 outcomes

(i) When single coin is tossed

1 Coin
Head Tail

(ii) When two coins are tossed: $\rightarrow 2^2 = 4$ outcomes

2 coins
H, H I, II — H, T I, II — T, H I, II — T, T I, II

2^2 = 4 total outcomes

(iii) When three coins are tossed: 2^3 outcomes = 8 outcomes

3 coins
HHH, HHT, THH, THH, HTT, THT, TTH, TTT

Concept of Dice

1. When a die is thrown once then S = 1, 2, 3, 4, 5, 6 n(S) = 6
2. When two dice are thrown together or A die is thrown twice then
S = (1, 1), (1, 2), (1, 3), (1, 4), (1, 5), (1, 6)
(2, 1), (2, 2), (2, 3), (2, 4), (2, 5), (2, 6)
(3, 1), (3, 2), (3, 3), (3, 4), (3, 5), (3, 6)
(4, 1), (4, 2), (4, 3), (4, 4), (4, 5), (4, 6)
(5, 1), (5, 2), (5, 3), (5, 4), (5, 5), (5, 6)
(6, 1), (6, 2), (6, 3), (6, 4), (6, 5), (6, 6) n(S) = 6 x 6 =36
Dice: Number are written on each face 1, 2, 3, 4, 5, 6

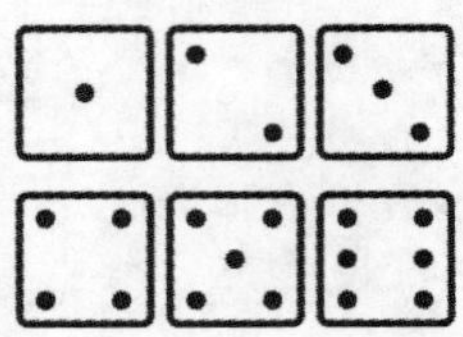

1 Dice = 6^1 outcomes ↓ 1, 2, 3, 4, 5, 6

Total number of outcomes: 6

Concept of Cards

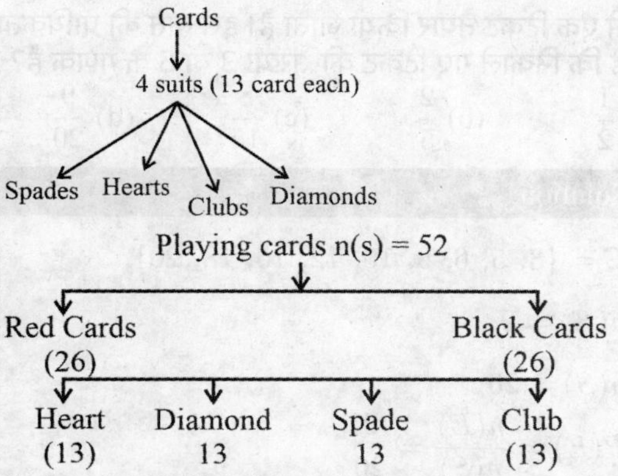

Each suit contains 1 ace, 1 king, 1 queen, 1 jack and nine number cards 2, 3, 4, 5, 6, 7, 8, 9, 10

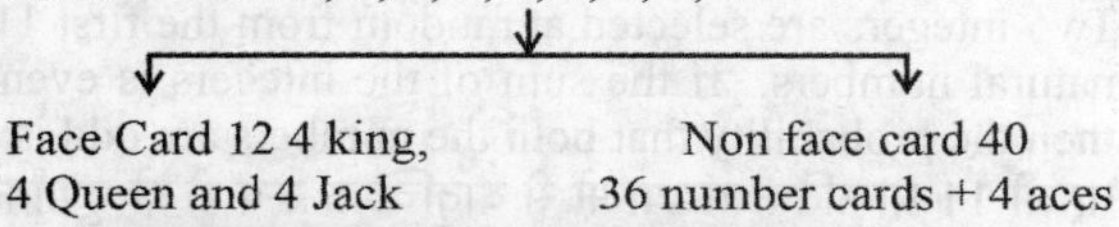

Spades Hearts Diamonds Clubs

Card Type	Hindi Name
Spade (Black color card)	हुकुम
Heart (Red color card)	पान
Club (Black color card)	चिड़ी
Diamond (Red color card)	ईंट
Jack	गुलाम
Queen	बेगम
King	बादशाह
Ace	इक्का

Important Figures in Cards/ कार्ड में कुछ महत्त्वपूर्ण आंकड़े

- Total cards: 52
- Red cards = Black cards: 26 each
- Heart cards (पान) = Diamond cards (ईंट) = Spade cards (हुकुम) = Club cards (चिड़ी)= 13 each
- Total King = Total Queen = Total Jack = Total Ace: 4 each
- Red King = Red Queen = Red Jack = Red Ace: 2 each
- Black King = Black Queen = Black Jack = Black Ace: 2 each
- Some pattern for all cards (2 to 10) (Number cards):
    - Total cards = 36 , Red cards = 18 , Black cards = 18
- King of Spade/Heart/Diamond/Club = 1 each
- Queen of Spade/Heart/Diamond/Club = 1 each
- Jack of Spade/Heart/Diamond/Club = 1 each
- Ace of Spade/Heart/Diamond/Club = 1 each
- Number cards (2 to 10) of Spade/Heart/Diamond/Club = 1 each
- Total Face cards = Jack(4) + King(4) + Queen(4) = 12 cards
- Red Face cards = 2 (Jack) + 2 (King) + 2 (Queen) = 6
- Black Face cards = 2 (Jack) + 2 (King) + 2 (Queen) = 6
- Face cards of Spade = 1 (Jack) + 1 (King) + 1 (Queen) = 3
- Face cards of Heart = 1 (Jack) + 1 (King) + 1 (Queen) = 3
- Face cards of Diamond = 1 (Jack) + 1 (King) + 1 (Queen) = 3

Concept of Odds/Favour

Odds are used to describe the chance of an Event occurring. The Odds are the ratios that compare the number of ways the Event can occur with the number of ways the Event cannot occur.

The Odds in Favour: The ratio of the number of ways that an outcome can occur compared to how many ways it cannot occur.

Odds in Favour = Number of Successes : Number of Failures

The Odds Against: The ratio of the number of ways that an outcome cannot occur compared to how many ways it can occur.

Odds Against = Number of Failures : Number of Successes

Concept of Set Theory

- $P(A \cup B) = P(A) + P(B) - P(A \cap B)$
- $P(A \cup B \cup C) = P(A) + P(B) + P(C) - P(A \cap B) - P(B \cap C) - P(A \cap C) + P(A \cap B \cap C)$
- $P(A') = 1 - P(A)$
- $P(B') = 1 - P(B)$
- $P(A' \cap B') = 1 - P(A \cup B)$
- $P(A' \cup B') = 1 - P(A \cap B)$

Concept of Conditional Probability

The Conditional Probability of an Event E, Given the occurrence of the Event F is given by

$$P(E \mid F) = \frac{P(E \cap F)}{P(F)}, \quad P(F) \neq 0$$

$$P(F \mid E) = \frac{P(F \cap E)}{P(E)}, \quad P(E) \neq 0$$

- If E and F are Mutually Exclusive Events $P(A \cap B) = 0$
- If E and F are Independent Events $P(E \cap F) = P(E) \times P(F)$

Type 1 (Basic)

1. A letter of the English alphabet is chosen at random. Probability of getting a vowel is:
अंग्रेजी वर्णमाला का एक अक्षर यादृच्छिक रूप से चुना जाता है। स्वर प्राप्त करने की प्रायिकता क्या है?

(a) $\frac{5}{26}$ (b) $\frac{6}{25}$ (c) $\frac{1}{4}$ (d) $\frac{5}{21}$

Solution:

$$P(E) = \frac{n(E)}{n(S)} = \frac{^{5}C_1}{^{26}C_1} = \frac{5}{26}$$

2. In a lottery, there are 10 prizes and 25 blanks. A lottery is drawn at random. What is the probability of getting a prize?
एक लॉटरी में, 10 पुरस्कार और 25 रिक्त स्थान होते हैं। एक लॉटरी यादृच्छिक रूप से तैयार की जाती है। पुरस्कार प्राप्त करने की संभावना क्या है?

(a) $\frac{1}{10}$ (b) $\frac{2}{5}$ (c) $\frac{2}{7}$ (d) $\frac{5}{7}$

Solution:

$n(E) = 10,\ \ n(S) = 10 + 25 = 35,$

$$P(E) = \frac{n(E)}{n(S)} = \frac{^{10}C_1}{^{35}C_1} = \frac{10}{35} = \frac{2}{7}$$

3. Tickets numbered 1 to 20 are mixed up and then a ticket is drawn at random. What is the probability that the ticket drawn bears a number which is a multiple of 3?
1 से 20 नंबर के टिकट को मिलाया जाता है और फिर यादृच्छिक रूप से एक टिकट तैयार किया जाता है। इस बात की प्रायिकता क्या है कि निकाले गए टिकट पर एक संख्या है जो 3 का गुणज है?

(a) $\frac{3}{10}$ (b) $\frac{3}{20}$ (c) $\frac{2}{5}$ (d) $\frac{1}{2}$

Solution:

$E = \{3, 6, 9, 12, 15, 18\},$

$n(E) = 6,$

$n(S) = 20$

$$P(E) = \frac{n(E)}{n(S)} = \frac{^{6}C_1}{^{20}C_1} = \frac{6}{20} = \frac{3}{10}$$

4. Tickets numbered 1 to 20 are mixed up and then a ticket is drawn at random. What is the probability that the ticket drawn has a number which is a multiple of 3 or 5?
1 से 20 नंबर के टिकट को मिलाया जाता है और फिर यादृच्छिक रूप से एक टिकट तैयार किया जाता है। इस बात की प्रायिकता क्या है कि निकाले गए टिकट की संख्या 3 या 5 के गुणक है?

(a) $\frac{1}{2}$ (b) $\frac{2}{5}$ (c) $\frac{8}{15}$ (d) $\frac{9}{20}$

Solution:

$E = \{3, 5, 6, 9, 10, 12, 15, 18, 20\},$

$n(E) = 9,$

$n(S) = 20$

$$P(E) = \frac{n(E)}{n(S)} = \frac{9}{20}$$

5. Two integers are selected at random from the first 11 natural numbers. If the sum of the integers is even, then the probability that both the numbers are odd is:
पहली 11 प्राकृतिक संख्याओं से यादृच्छिक रूप से दो पूर्णांक चुने जाते हैं। यदि पूर्णांकों का योग सम है, तो दोनों संख्याओं के विषम होने की प्रायिकता क्या है?

(a) $\frac{13}{212}$ (b) $\frac{3}{5}$ (c) $\frac{4}{9}$ (d) $\frac{5}{11}$

Solution:

Even + Even = Even

Odd + Odd = Even

11

5 (even) 6 (odd)

$$P(E) = \frac{n(E)}{n(S)} = \frac{^{6}C_2}{^{5}C_2 + {}^{6}C_2} = \frac{30}{50} = \frac{3}{5}$$

6. Two letters selected from the word 'TENDULKAR'. The probability that both are vowels is:
'TENDULKAR' शब्द से चुने गए दो अक्षर। दोनों के स्वर होने की प्रायिकता क्या है?

(a) $\frac{2}{9}$ (b) $\frac{1}{12}$ (c) $\frac{5}{9}$ (d) $\frac{5}{12}$

Solution:

$S = \{T, E, N, D, U, L, K, A, R\},$

$E = \{E, U, A\}$

$$P(E) = \frac{n(E)}{n(S)} = \frac{^{3}C_2}{^{9}C_2} = \frac{1}{12}$$

7. In a class, there are 15 boys and 10 girls. Three students are selected at random. The probability that

1 girl and 2 boys are selected is:
एक कक्षा में, 15 लड़के और 10 लड़कियाँ हैं। तीन छात्रों को यादृच्छिक रूप से चुना जाता है। 1 लड़की और 2 लड़कों के चुने जाने की प्रायिकता क्या है?

(a) $\frac{21}{46}$ (b) $\frac{25}{17}$ (c) $\frac{1}{50}$ (d) $\frac{3}{25}$

Solution:

Total students $= 15 + 10 = 25$,

$n(S) = {}^{25}C_3$

$n(E) = {}^{10}C_1 \times {}^{15}C_2$

$P(E) = \frac{n(E)}{n(S)} = \frac{{}^{10}C_1 \times {}^{15}C_2}{{}^{25}C_3} = \frac{21}{46}$

8. Four persons are chosen at random from a group of 3 men, 2 women, and 4 children. The chance that exactly 2 of them are children is:
चार व्यक्तियों को 3 पुरुषों, 2 महिलाओं और 4 बच्चों के एक समूह से यादृच्छिक रूप से चुना जाता है। संभावना है कि उनमें से 2 बच्चे हैं:

(a) $\frac{1}{9}$ (b) $\frac{1}{5}$ (c) $\frac{1}{12}$ (d) $\frac{10}{21}$

Solution:

Total persons $= 3 + 2 + 4 = 9$

No of ways selecting 2 children out $= {}^{4}C_2$

No of ways selecting 2 persons out of $= {}^{5}C_2$

$n(E) = {}^{4}C_2 \times {}^{5}C_2$,

$n(S) = {}^{9}C_4$

$P(E) = \frac{n(E)}{n(S)} = \frac{{}^{4}C_2 \times {}^{5}C_2}{{}^{9}C_4} = \frac{10}{21}$

9. Five salesmen A, B, C, D, and E of a company are considered for a three-member trade delegation to represent the company at an international trade conference. What is the probability that A gets selected?
एक कंपनी के पांच सेल्समैन A, B, C, D और E को एक अंतरराष्ट्रीय व्यापार सम्मेलन में कंपनी का प्रतिनिधित्व करने के लिए तीन सदस्यीय व्यापार प्रतिनिधिमंडल के लिए माना जाता है। A के चुने जाने की प्रायिकता क्या है?

(a) $\frac{3}{5}$ (b) $\frac{2}{5}$ (c) $\frac{4}{5}$ (d) $\frac{1}{5}$

Solution:

No of ways selecting A $= {}^{1}C_1$

No of ways selecting two other members out of 4 $= {}^{4}C_2$

$n(E) = {}^{1}C_1 \times {}^{4}C_2$,

$n(S) = {}^{5}C_3$

$P(E) = \frac{n(E)}{n(S)} = \frac{{}^{1}C_1 \times {}^{4}C_2}{{}^{5}C_3} = \frac{3}{5}$

10. If 9 students are standing on a circular path, then the probability that 2 of them are always standing together is:
यदि 9 छात्र एक वृत्ताकार पथ पर खड़े हैं, तो उनमें से 2 के हमेशा एक साथ खड़े रहने की प्रायिकता क्या है?

(a) $\frac{1}{4}$ (b) $\frac{2}{7}$ (c) $\frac{7}{8}$ (d) $\frac{1}{3}$

Solution:

Total =

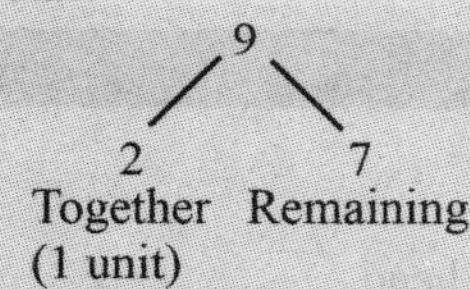

Favorable position $= n(E) = 2! \times 7!$

Total position $= n(S) = 8!$

$P(E) = \frac{n(E)}{n(S)} = \frac{2! \times 7!}{8!} = \frac{1}{4}$

11. Two unbiased coins are tossed simultaneously. Find the probability of getting one head?
दो निष्पक्ष सिक्के एक साथ उछाले जाते हैं। एक चित प्राप्त करने की प्रायिकता ज्ञात कीजिये?

(a) $\frac{1}{2}$ (b) $\frac{1}{3}$ (c) $\frac{2}{3}$ (d) $\frac{1}{4}$

Solution:

S = {(H, H), (H, T), (T, H), (T, T)},

n(S) = 4

E = {(H, T), (T, H)},

n(E) = 2

$P(E) = \frac{n(E)}{n(S)} = \frac{2}{4} = \frac{1}{2}$

12. In a simultaneous throw of two coins, the probability of getting at least one head is:
दो सिक्कों को एक साथ फेंकने पर, कम से कम एक चित प्राप्त करने की प्रायिकता क्या है?

(a) $\frac{1}{2}$ (b) $\frac{1}{3}$ (c) $\frac{2}{3}$ (d) $\frac{3}{4}$

Solution:

S = {(H, H), (H, T), (T, H), (T, T)},

n(S) = 4 E = {(H, H), (H, T), (T, H)},

n(E) = 3 $P(E) = \frac{n(E)}{n(S)} = \frac{3}{4}$

13. Two unbiased coins are tossed simultaneously. Find the probability of getting at most one head?
दो निष्पक्ष सिक्के एक साथ उछाले जाते हैं। अधिकतम एक चित प्राप्त करने की प्रायिकता ज्ञात कीजिये?

(a) $\frac{3}{8}$ (b) $\frac{1}{4}$ (c) $\frac{1}{2}$ (d) $\frac{3}{4}$

Solution:

$n(S) = 4$

E = almost one head

$E = \{(H,\ T),\ (T,\ H),\ (T,\ T)\},$

$n(E) = 3$

$P(E) = \frac{n(E)}{n(S)} = \frac{3}{4}$

14. Two unbiased coins are tossed simultaneously. Find the probability of getting no head?
दो निष्पक्ष सिक्के एक साथ उछाले जाते हैं। चित न होने की प्रायिकता ज्ञात कीजिए?

(a) $\frac{3}{4}$ (b) $\frac{1}{4}$ (c) $\frac{3}{8}$ (d) $\frac{1}{2}$

Solution:

$n(S) = 4$

$E = \{(T,\ T)\},$

$n(E) = 1$

$P(E) = \frac{n(E)}{n(S)} = \frac{1}{4}$

Note: When 3 coins are tossed simultaneously or 1 coin is tossed three times, the sample space will be: S = {(H, H, H), (H, H, T), (H, T, H), (H, T, T), (T, H, H), (T, H, T), (T, T, H), (T, T, T)},

n(S) = 8

15. A coin is tossed 3 times. What is the probability that they will fall two heads and one tail?
एक सिक्का 3 बार उछाला जाता है। इस बात की प्रायिकता क्या है कि वे दो चित और एक पट गिराएंगे?

(a) $\frac{1}{3}$ (b) $\frac{1}{2}$ (c) $\frac{1}{4}$ (d) $\frac{3}{8}$

Solution:

$n(S) = 8$

$E = \{(H,\ H,\ T),\ (H,\ T,\ H),\ (T,\ H,\ H)\},$

$n(E) = 3$

$P(E) = \frac{n(E)}{n(S)} = \frac{3}{8}$

16. A coin is tossed 3 times. The probability of getting a head and a tail alternately is:
एक सिक्का 3 बार उछाला जाता है। बारी-बारी से चित और पट प्राप्त करने की संभावना क्या है?

(a) $\frac{1}{4}$ (b) $\frac{1}{8}$ (c) $\frac{1}{2}$ (d) $\frac{3}{8}$

Solution:

$n(S) = 8$

$E = \{(H,\ T,\ H),\ (T,\ H,\ T)\},$

$n(E) = 2$

$P(E) = \frac{n(E)}{n(S)} = \frac{2}{8} = \frac{1}{4}$

17. Three unbiased coins are tossed together. Find the probability of getting two heads:
तीन निष्पक्ष सिक्के एक साथ उछाले जाते हैं। दो चित प्राप्त करने की प्रायिकता ज्ञात कीजिये?

(a) $\frac{1}{4}$ (b) $\frac{3}{4}$ (c) $\frac{3}{8}$ (d) $\frac{2}{3}$

Solution:

$n(S) = 8$

$E = \{(H, H, T), (H, T, H), (T, H, H)\}$,

$n(E) = 3$

$P(E) = \frac{n(E)}{n(S)} = \frac{3}{8}$

18. Three unbiased coins are tossed. What is the probability of getting at least 2 heads?
तीन निष्पक्ष सिक्के उछाले जाते हैं। कम से कम 2 चित प्राप्त करने की प्रायिकता क्या है?

(a) $\frac{1}{4}$ (b) $\frac{1}{2}$ (c) $\frac{1}{3}$ (d) $\frac{1}{8}$

Solution:

$n(S) = 8$

E = {(H, H, H), (H, H, T), (H, T, H), (T, H, H)},

n(E) = 4

$P(E) = \frac{n(E)}{n(S)} = \frac{4}{8} = \frac{1}{2}$

19. Three unbiased coins are tossed. What is the probability of getting at most two heads?
तीन निष्पक्ष सिक्के उछाले जाते हैं। अधिकतम दो चित के होने की प्रायिकता क्या है?

(a) $\frac{3}{4}$ (b) $\frac{1}{4}$ (c) $\frac{3}{8}$ (d) $\frac{7}{8}$

Solution:

$n(S) = 8$

E = getting almost two heads

E = {(H, H, T), (H, T, H), (H, T, T), (T, H, H), (T, H, T), (T, T, H), (T, T, T)},

n(E) = 7

$P(E) = \frac{n(E)}{n(S)} = \frac{7}{8}$

Or:

$P(E) = 1 - P(E')$,

$P(E') = \frac{1}{8}$

$P(E) = 1 - \frac{1}{8} = \frac{7}{8}$

20. A coin is tossed thrice and outcomes are recorded. What will be the expectation of exactly 2 heads?
एक सिक्का तीन बार उछाला जाता है और परिणाम दर्ज किए जाते हैं। ठीक 2 चित की अपेक्षा क्या होगी?

(a) $\frac{3}{8}$ (b) $\frac{3}{4}$ (c) $\frac{1}{8}$ (d) None

Solution:

$n(S) = 8$
E = expectation of exactly 2 heads
$E = \{(H, H, T), (H, T, H), (T, H, H)\}$,
$n(E) = 3$
$P(E) = \frac{n(E)}{n(S)} = \frac{3}{8}$

21. An unbiased coin is tossed 3 times, if the third toss gets head what is the probability of getting at least one more head?
एक निष्पक्ष सिक्के को 3 बार उछाला जाता है, यदि तीसरी उछाल में चित आता है तो कम से कम एक और चित आने की प्रायिकता क्या है?

(a) $\frac{3}{4}$ (b) $\frac{1}{4}$ (c) $\frac{1}{2}$ (d) $\frac{1}{3}$

Solution:

S = Third toss gets head = {(H, H, H), (H, T, H), (T, H, H), (T, T, H)},
n(S) = 4
E = Getting at least one more head in the first two tosses = {(H, H, H), (H, T, H), (T, H, H)},
n(E) = 3
$P(E) = \frac{n(E)}{n(S)} = \frac{3}{4}$

22. Five coins are tossed once. What is the probability of getting at most four tails?
पांच सिक्के एक बार उछाले जाते हैं। अधिकतम चार पट प्राप्त करने की प्रायिकता क्या है?

(a) $\frac{31}{32}$ (b) $\frac{15}{16}$ (c) $\frac{29}{32}$ (d) $\frac{7}{8}$

Solution:

S = five coins are tossed
$n(S) = 2^5 = 32$
E' = Getting all tails = {(T, T, T, T, T)},
n(E') = 1
$P(E') = \frac{n(E')}{n(S)} = \frac{1}{32}$
$P(E) = 1 - P(E') = 1 - \frac{1}{32} = \frac{31}{32}$

23. The number of possible outcomes, when a coin is tossed 6 times is:

जब एक सिक्के को 6 बार उछाला जाता है, तो संभावित परिणामों की संख्या क्या है?
(a) 36 (b) 64 (c) 12 (d) None

Solution:

S = a coin is tossed 6 times
$n(S) = 2^6 = 64$

24. The probability of drawing a red card from a deck of playing cards is:
ताश खेलने के डेक से लाल कार्ड खींचने की संभावना क्या है?
(a) $\frac{2}{18}$ (b) $\frac{1}{13}$ (c) $\frac{1}{4}$ (d) $\frac{1}{2}$

Solution:

$n(S) = 52$
$n(E) = 26$(Red cards)
$P(E) = \frac{n(E)}{n(S)} = \frac{26}{52} = \frac{1}{2}$

25. One card is drawn from a pack of 52 cards, each of the 52 cards being equally likely to be drawn. Find the probability that the card drawn is an ace:
52 पत्तों की गड्डी में से एक पत्ता निकाला जाता है। 52 पत्तों में से प्रत्येक के निकाले जाने की संभावना समान होती है। प्रायिकता ज्ञात कीजिए कि निकाला गया कार्ड इक्का है?
(a) $\frac{1}{13}$ (b) $\frac{1}{12}$ (c) $\frac{2}{13}$ (d) $\frac{3}{28}$

Solution:

$n(S) = {}^{52}C_1$
$n(E) = {}^{4}C_1$(4 aces) $P(E) = \frac{n(E)}{n(S)} = \frac{{}^4C_1}{{}^{52}C_1} = \frac{4}{52} = \frac{1}{13}$

26. One card is drawn at random from a pack of 52 cards. What is the probability that the card drawn is a face card? 52 पत्तों की एक गड्डी में से यादृच्छिक रूप से एक पत्ता निकाला जाता है। इसकी क्या प्रायिकता है कि निकाला गया कार्ड एक अंकित कार्ड है?
(a) $\frac{1}{22}$ (b) $\frac{2}{39}$ (c) $\frac{1}{26}$ (d) $\frac{2}{51}$

Solution:

n(S) = ${}^{52}C_1$,
E = {drawn card is a face card},
n(E) = ${}^{12}C_1$,
$P(E) = \frac{n(E)}{n(S)} = \frac{{}^{12}C_1}{{}^{52}C_1} = \frac{12}{52} = \frac{3}{13}$

27. One card is drawn from a pack of 52 cards, each of the 52 cards being equally likely to be drawn. Find the probability that the card drawn is red and a king:
52 पत्तों की गड्डी में से एक पत्ता निकाला जाता है, 52 पत्तों में से प्रत्येक के निकाले जाने की संभावना समान होती है। प्रायिकता ज्ञात कीजिए कि निकाला गया कार्ड लाल और बादशाह है?
(a) $\frac{1}{13}$ (b) $\frac{2}{13}$ (c) $\frac{1}{26}$ (d) $\frac{1}{52}$

Solution:

$n(S) = {}^{52}C_1$,
$n(E) = {}^{2}C_1$
(2 red kings) $P(E) = \frac{n(E)}{n(S)} = \frac{{}^2C_1}{{}^{52}C_1} = \frac{2}{52} = \frac{1}{26}$

28. A card is drawn at random from a pack of 52 cards, the probability of getting a queen of club or a king of heart is:
52 पत्तों की एक गड्डी में से यादृच्छिक रूप से एक पत्ता निकाला जाता है, चिड़ी की बेगम या पान का बादशाह मिलने की प्रायिकता है:
(a) $\frac{4}{13}$ (b) $\frac{1}{4}$ (c) $\frac{1}{13}$ (d) $\frac{1}{26}$

Solution:

$n(S) = {}^{52}C_1$,
$n(E) = {}^1C_1 + {}^1C_1$
(1 queen of club + 1 king of heart) $P(E) = \frac{n(E)}{n(S)} = \frac{{}^1C_1 + {}^1C_1}{{}^{52}C_1} = \frac{2}{52} = \frac{1}{26}$

29. From a pack of 52 cards, one card is drawn at random. What is the probability that the card drawn is a ten or a spade?
52 पत्तों की एक गड्डी में से एक पत्ता यादृच्छिक रूप से निकाला जाता है। इसकी क्या प्रायिकता है कि निकाला गया कार्ड दस या हुकुम है?

(a) $\frac{4}{13}$ (b) $\frac{1}{4}$ (c) $\frac{1}{13}$ (d) $\frac{1}{26}$

Solution:

$n(S) = {}^{52}C_1$,

$n(E) = {}^{3}C_1 + {}^{13}C_1$

(3 tens + 13 spades)

$P(E) = \frac{n(E)}{n(S)} = \frac{{}^{3}C_1 + {}^{13}C_1}{{}^{52}C_1} = \frac{3+13}{52} = \frac{16}{52} = \frac{4}{13}$

30. The probability that a card drawn from a pack of 52 cards will be a diamond or a king is:
52 पत्तों की गड्डी से निकाला गया पत्ता ईंट या बादशाह होगा इसकी प्रायिकता है:

(a) $\frac{2}{13}$ (b) $\frac{4}{13}$ (c) $\frac{1}{13}$ (d) $\frac{1}{52}$

Solution:

$n(S) = {}^{52}C_1$

E = drawn card is a Diamond or a king

$n(E) = {}^{13}C_1 + {}^{3}C_1$

$P(E) = \frac{{}^{13}C_1 + {}^{3}C_1}{{}^{52}C_1} = \frac{13+3}{52} = \frac{16}{52} = \frac{4}{13}$

31. One card is drawn at random from a pack of 52 cards. What is the probability that the card drawn is either a red card or a king?
52 पत्तों की एक गड्डी में से यादृच्छिक रूप से एक पत्ता निकाला जाता है। इसकी क्या प्रायिकता है कि निकाला गया कार्ड या तो लाल कार्ड है या बादशाह?

(a) $\frac{1}{2}$ (b) $\frac{6}{13}$ (c) $\frac{7}{13}$ (d) $\frac{27}{52}$

Solution:

$n(S) = {}^{52}C_1$

E = drawn card is either a red or a king

$n(E) = {}^{26}C_1 + {}^{2}C_1$

$P(E) = \frac{{}^{26}C_1 + {}^{2}C_1}{{}^{52}C_1} = \frac{26+2}{52} = \frac{28}{52} = \frac{7}{13}$

32. From a pack of 52 cards, two cards are drawn together at random. What is the probability of both the cards being kings?
52 पत्तों की एक गड्डी में से यादृच्छिक रूप से दो पत्ते एक साथ निकाले जाते हैं। दोनों कार्डों के बादशाह होने की प्रायिकता क्या है?

(a) $\frac{1}{15}$ (b) $\frac{25}{57}$ (c) $\frac{35}{256}$ (d) $\frac{1}{221}$

Solution:

$n(S) = {}^{52}C_2$

E = both cards being kings

$n(E) = {}^{4}C_2$

$P(E) = \frac{{}^{4}C_2}{{}^{52}C_2} = \frac{6}{1326} = \frac{1}{221}$

33. Two cards are drawn together from a pack of 52 cards. The probability that one is a spade and one is a heart is:
52 पत्तों की एक गड्डी से दो पत्ते एक साथ निकाले जाते हैं। एक के हुक्म और एक के पान होने की प्रायिकता है:

(a) $\frac{3}{20}$ (b) $\frac{29}{34}$ (c) $\frac{47}{100}$ (d) $\frac{13}{102}$

Solution:

$n(S) = {}^{52}C_2$

E = one card is a spade and one is a heart

$n(E) = {}^{13}C_1 \times {}^{13}C_1$

$P(E) = \frac{{}^{13}C_1 \times {}^{13}C_1}{{}^{52}C_2} = \frac{13 \times 13}{(52 \times 51)/2} = \frac{13}{102}$

34. Two cards are drawn from a pack of well-shuffled cards. Find the probability that one is a club and the other is a king.
अच्छी तरह से फेटे गए पत्तों की एक गड्डी से दो पत्ते निकाले जाते हैं। प्रायिकता ज्ञात कीजिए कि एक चिड़ी है और दूसरा बादशाह।

(a) $\frac{1}{52}$ (b) $\frac{1}{26}$ (c) $\frac{1}{13}$ (d) $\frac{1}{2}$

Solution:

$n(S) = {}^{52}C_2$

E = one is a club and the other is a king

Case I: When the king is taken from the club

$n(E_1) = {}^{12}C_1 \times {}^{1}C_1$

Case II: When the king is not taken from the club

$n(E_2) = {}^{13}C_1 \times {}^{3}C_1$

$n(E) = n(E_1) + n(E_2)$

$n(E) = {}^{12}C_1 \times {}^{1}C_1 + {}^{13}C_1 \times {}^{3}C_1 = 12 + 39 = 51$

$P(E) = \dfrac{n(E)}{n(S)} = \dfrac{51}{{}^{52}C_2} = \dfrac{51}{(52 \times 51)/2} = \dfrac{1}{26}$

35. Two cards are drawn from a pack of 52 cards. The probability that out of 2 cards, one card is red and one card is black is:

52 पत्तों की एक गड्डी से दो पत्ते निकाले जाते हैं। 2 कार्डों में से एक कार्ड लाल और एक कार्ड काला होने की प्रायिकता है:

(a) $\dfrac{1}{2}$ (b) $\dfrac{13}{25}$ (c) $\dfrac{25}{51}$ (d) $\dfrac{26}{51}$

Solution:

$n(S) = {}^{52}C_2$

E = one card is red and one is black

$n(E) = {}^{26}C_1 \times {}^{26}C_1$

$P(E) = \dfrac{{}^{26}C_1 \times {}^{26}C_1}{{}^{52}C_2} = \dfrac{26}{51}$

36. Two cards are drawn together from a pack of 52 cards. The probability that either both are red or both are kings is:

52 पत्तों की एक गड्डी से दो पत्ते एक साथ निकाले जाते हैं। संभावना है कि या तो दोनों लाल हैं या दोनों बादशाह हैं:

(a) $\dfrac{7}{13}$ (b) $\dfrac{3}{26}$ (c) $\dfrac{63}{221}$ (d) $\dfrac{55}{221}$

Solution:

$n(S) = {}^{52}C_2$

E = either both are red or both are kings

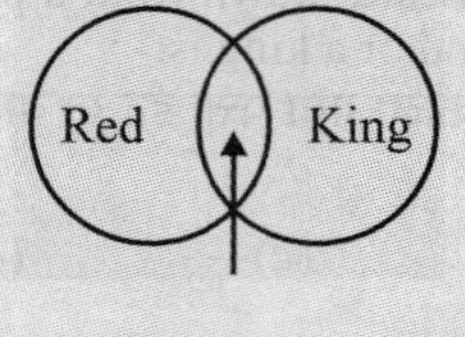

$n(E) = {}^{26}C_2 + {}^{4}C_2 - {}^{2}C_2$

$P(E) = \dfrac{{}^{26}C_2 + {}^{4}C_2 - {}^{2}C_2}{{}^{52}C_2} = \dfrac{55}{221}$

37. Two cards are drawn at random from a pack of 52 cards. What is the probability that either both are black or both are queens?

52 पत्तों की एक गड्डी में से यादृच्छिक रूप से दो पत्ते निकाले जाते है। इसकी क्या प्रायिकता है कि या तो दोनों काली हैं या दोनों बेगम हैं?

(a) $\dfrac{52}{221}$ (b) $\dfrac{55}{190}$ (c) $\dfrac{55}{221}$ (d) $\dfrac{19}{221}$

Solution:

$n(S) = {}^{52}C_2$

E = either both are black or both are queens

$n(E) = {}^{26}C_2 + {}^{4}C_2 - {}^{2}C_2$

$P(E) = \dfrac{{}^{26}C_2 + {}^{4}C_2 - {}^{2}C_2}{{}^{52}C_2} = \dfrac{55}{221}$

38. From a pack of 52 cards, 3 cards are drawn. What is the probability that one is ace, one is queen and one is jack?

52 पत्तों की एक गड्डी में से 3 पत्ते निकाले जाते है। इसकी क्या प्रायिकता है कि एक इक्का है, एक बेगम है और एक गुलाम है?

(a) $\dfrac{19}{5525}$ (b) $\dfrac{16}{5525}$ (c) $\dfrac{17}{5525}$ (d) $\dfrac{7}{5525}$

Solution:

$n(S) = {}^{52}C_3$

E = one is ace, one is queen, and one is jack

$n(E) = {}^4C_1 \times {}^4C_1 \times {}^4C_1$

$P(E) = \frac{{}^4C_1 \times {}^4C_1 \times {}^4C_1}{{}^{52}C_3} = \frac{16}{5525}$

39. On average, every one out of 10 telephones is found busy. Six telephone numbers are selected at random. Find the probability that four of them will be busy.

औसतन 10 में से हर एक टेलीफोन व्यस्त पाया जाता है। छह टेलीफोन नंबर यादृच्छिक रूप से चुने गए हैं। उनमें से चार के व्यस्त होने की प्रायिकता ज्ञात कीजिए।

(a) $\frac{21 \times 81}{10^6}$ (b) $\frac{30 \times 81}{10^6}$
(c) $\frac{15 \times 81}{10^6}$ (d) $\frac{45 \times 81}{10^6}$

Solution:

A = telephone is dialed, and it is busy

A' = telephone is dialed, and it is not busy

$P(A) = \frac{1}{10}$,

$P(A') = \frac{9}{10}$

Using the binomial distribution,

Required probability $= {}^6C_4 \times \left(\frac{1}{10}\right)^4 \times \left(\frac{9}{10}\right)^2 = \frac{15 \times 81}{10^6}$

40. Find the probability of getting 5 exactly twice in 7 throws of a die.

एक पासे को 7 बार फेंकने पर ठीक दो बार 5 आने की प्रायिकता ज्ञात कीजिए।

(a) $\frac{7}{12} \times \left(\frac{5}{6}\right)$ (b) $\frac{7}{12} \times \left(\frac{6}{5}\right)$
(c) $\frac{12}{7} \times \left(\frac{5}{6}\right)$ (d) $\frac{12}{7} \times \left(\frac{6}{5}\right)$

Solution:

A = getting 5

A' = not getting 5

$P(A) = \frac{1}{6}$,

$P(A') = \frac{5}{6}$

Using the binomial distribution,

Required probability $= {}^7C_2 \times \left(\frac{1}{6}\right)^2 \times \left(\frac{5}{6}\right)^5 = \frac{7}{12} \times \left(\frac{5}{6}\right)^5$

41. The probability that a student is not a swimmer is $\frac{1}{5}$. Then the probability that out of five students, four are swimmers is:

एक छात्र के तैराक न होने की प्रायिकता $\frac{1}{5}$ है। तो पाँच विद्यार्थियों में से चार के तैराक होने की प्रायिकता है:

(a) $\left(\frac{4}{5}\right)^4$ (b) $\frac{256}{3125}$ (c) $\left(\frac{4}{5}\right)^3$ (d) none

Solution:

A = the student is not a swimmer

A' = the student is a swimmer

$P(A') = \frac{4}{5}$

Using binomial distribution

42. In a box containing 100 bulbs, 10 are defective. The probability that out of a sample of 5 bulbs, none is defective is:

100 बल्बों वाले एक बॉक्स में 10 खराब हैं। 5 बल्बों के नमूने में से किसी के भी ख़राब न होने की प्रायिकता है:

(a) $\left(\frac{9}{10}\right)^5$ (b) $\frac{1}{10}$ (c) $\frac{9}{10}$ (d) $\left(\frac{1}{2}\right)^6$

Solution:

A = bulb is defective

A' = bulb is not defective

$P(A') = \frac{9}{10}$

Using binomial distribution

43. The probability of a man hitting the target is $\frac{1}{4}$. If he fires 7 times, the probability of hitting the target exactly six times is:

एक व्यक्ति द्वारा लक्ष्य को भेदने की प्रायिकता $\frac{1}{4}$ है। यदि वह 7 बार फायर करता है, तो लक्ष्य पर ठीक छह बार वार करने की प्रायिकता:

(a) $216 \times \left(\frac{1}{4}\right)^6 \times \left(\frac{3}{4}\right)$ (b) $7^6 \times \frac{3}{4} \times \left(\frac{1}{4}\right)$

(c) $7^6 \times \left(\frac{1}{2}\right)^6 \times \left(\frac{1}{2}\right)$ (d) None

Solution:

A = the man hits the target

A' = the man does not hit the target

$P(A') = \frac{3}{4}$

Using binomial distribution

44. A person can hit a target 5 times out of 8 shots. If he fires 10 shots, what is the probability that he will hit the target twice?

एक व्यक्ति 8 में से 5 वार किसी लक्ष्य पर निशाना साध सकता है। यदि वह 10 गोलियाँ चलाता है, तो इसकी क्या प्रायिकता है कि वह लक्ष्य पर दो बार वार करेगा?

(a) $\frac{1135 \times 3^8}{8^{10}}$ (b) $\frac{1165 \times 3^8}{8^{10}}$

(c) $\frac{1175 \times 3^8}{8^{10}}$ (d) $\frac{1125 \times 3^8}{8^{10}}$

Solution:

A = person can hit a target

A' = person cannot hit a target

$P(A') = \frac{3}{8}$

Using binomial distribution

45. Ten eggs are drawn successively with replacement from a lot containing 10% defective eggs. Find the probability that there is at least one defective egg.

10% दोषपूर्ण अंडों वाले लॉट से प्रतिस्थापन के साथ क्रमिक रूप से दस अंडे निकाले जाते हैं। कम से कम एक अंडा ख़राब होने की प्रायिकता ज्ञात कीजिए।

(a) $1 - \left(\frac{9}{10}\right)^9$ (b) $1 - \left(\frac{1}{10}\right)^9$

(c) $1 - \left(\frac{9}{10}\right)^{10}$ (d) $1 - \left(\frac{1}{10}\right)^{10}$

Solution:

A = egg is defective

A' = egg is not defective

$P(A) = \frac{1}{10}$,

$P(A') = \frac{9}{10}$

Required probability = 1 − P(no egg defective)

46. A bag consists of 10 balls each marked with one of the digits 0 to 9. If four balls are drawn successively with replacement from the bag, what is the probability that none is marked with the digit 0?

एक बैग में 10 गेंदें हैं जिनमें से प्रत्येक पर 0 से 9 अंकों में से एक अंकित है। यदि बैग से प्रतिस्थापन के साथ लगातार चार गेंदें निकाली जाती हैं, तो क्या संभावना है कि किसी पर भी अंक 0 अंकित नहीं है?

(a) $\left(\frac{9}{10}\right)^4$ (b) $\left(\frac{9}{10}\right)^3$

(c) $\left(\frac{1}{10}\right)^4$ (d) $\left(\frac{1}{10}\right)^3$

Solution:

A = ball is marked with digit zero

A' = ball is not marked with digit zero

$P(A) = \frac{1}{10}$,

$P(A') = \frac{9}{10}$

Using binomial distribution

Required probability $= (9/10)^4$

47. There are 5% defective items in a large bulk of items. What is the probability that a sample of 10 items will include not more than one defective item?

बड़ी मात्रा में वस्तुओं में 5% दोषपूर्ण वस्तुएँ हैं। इसकी क्या प्रायिकता है कि 10 वस्तुओं के नमूने में एक से अधिक दोषपूर्ण वस्तुएँ शामिल नहीं होंगी?

(a) $\frac{29}{29} \times \left(\frac{19}{20}\right)^9$ (b) $\frac{20}{29} \times \left(\frac{19}{20}\right)^9$

(c) $\frac{19}{20} \times \left(\frac{29}{20}\right)^9$ (d) $\frac{19}{20} \times \left(\frac{20}{19}\right)^9$

Solution:

A = item is defective

A' = item is not defective

$P(A) = \frac{1}{20}$,

$P(A') = \frac{19}{20}$

Case I: when zero items are defective

$P(E_1) = (19/20)^{10}$

Case II: when one item is defective

$P(E_2) = {}^{10}C_1 \times (1/20)^1 \times (19/20)^9$

Required probability $= P(E) = (19/20)^{10} + (10 \times 19^9)/20^{10} = \frac{29}{20} \times (19/20)^9$

48. Find the probability of throwing at most 2 sixes in 6 throws of a single die.

एक ही पासे के 6 थ्रो में अधिकतम 2 छक्के लगाने की प्रायिकता ज्ञात कीजिए।

(a) $\frac{7}{3} \times \left(\frac{5}{6}\right)^5$ (b) $\frac{3}{7} \times \left(\frac{5}{6}\right)^5$

(c) $\frac{3}{7} \times \left(\frac{6}{5}\right)^5$ (d) none

Solution:

A = getting 6 in a throw of dice

A' = not getting 6 in a throw of dice

$P(A) = \frac{1}{6}$

$P(A') = \frac{5}{6}$

Probability of getting at most two sixes =

$P(\text{zero six}) + P(\text{one six}) + P(\text{two sixes})$

Using binomial distribution:

$P = {}^6C_0 \times (5/6)^6 \times (1/6)^0 + {}^6C_1 \times (5/6)^5 \times (1/6)^1 + {}^6C_2 \times (5/6)^4 \times (1/6)^2$

$= (5/6)^4 \left[\frac{25}{36} + \frac{30}{36} + \frac{15}{36}\right]$

$= \frac{70}{36} \times (5/6)^4 = \frac{7}{3} \times (5/6)^5$

49. In a single throw of a die, what is the probability of getting a number greater than 4?

एक पासे को एक बार फेंकने पर 4 से अधिक संख्या प्राप्त होने की प्रायिकता क्या है?

(a) $\frac{1}{2}$ (b) $\frac{1}{3}$ (c) $\frac{2}{3}$ (d) $\frac{1}{4}$

Solution:

$S = \{1, 2, 3, 4, 5, 6\}$,

$n(S) = 6$

E = getting a number greater than 4 = {5, 6}

$n(E) = 2$

$P(E) = \frac{n(E)}{n(S)} = \frac{2}{6} = \frac{1}{3}$

50. An unbiased die is thrown. What is the probability of getting an even number?

एक निष्पक्ष पासा फेंका जाता है। सम संख्या प्राप्त होने की प्रायिकता क्या है?

(a) 1/2 (b) 2/5 (c) 1/3 (d) 2/3

Solution:

$S = \{1, 2, 3, 4, 5, 6\}$,

$n(S) = 6$

E = getting an even number = {2, 4, 6}

$n(E) = 3$

$P(E) = \frac{n(E)}{n(S)} = \frac{3}{6} = \frac{1}{2}$

51. An unbiased die is thrown. What is the probability of getting an even number or a multiple of 3?

एक निष्पक्ष पासा फेंका जाता है। एक सम संख्या या 3 का गुणज प्राप्त होने की प्रायिकता क्या है?

(a) $\frac{1}{3}$ (b) $\frac{2}{3}$ (c) $\frac{1}{2}$ (d) $\frac{1}{6}$

Solution:

$S = \{1, 2, 3, 4, 5, 6\}$,

$n(S) = 6$

E = getting an even number or a multiple of 3

$E = \{2, 3, 4, 6\}$,

$n(E) = 4$

$P(E) = \frac{n(E)}{n(S)} = \frac{4}{6} = \frac{2}{3}$

52. In a simultaneous throw of two dice, what is the probability of getting a total of 7?
दो पासों को एक साथ फेंकने पर कुल 7 आने की प्रायिकता क्या है?
(a) $\frac{1}{6}$ (b) $\frac{1}{4}$ (c) $\frac{2}{3}$ (d) $\frac{3}{4}$

Solution:

$n(S) = 36$

E = getting a total of 7

E = {(1,6), (2,5), (3,4), (4,3), (5,2), (6,1)},

n(E) = 6

$P(E) = \frac{n(E)}{n(S)} = \frac{6}{36} = \frac{1}{6}$

53. In a simultaneous throw of a pair of dice, find the probability of getting the total more than 7.
पासों के एक जोड़े को एक साथ फेंकने पर कुल योग 7 से अधिक आने की प्रायिकता ज्ञात कीजिए।
(a) $\frac{1}{2}$ (b) $\frac{5}{12}$ (c) $\frac{7}{15}$ (d) $\frac{3}{12}$

Solution:

$n(S) = 36$

E = getting the total more than 7

E = {(2,6), (3,5), (3,6), (4,4), (4,5), (4,6), (5,3), (5,4), (5,5), (5,6), (6,2), (6,3), (6,4), (6,5), (6,6)},

n(E) = 15

$P(E) = \frac{n(E)}{n(S)} = \frac{15}{36} = \frac{5}{12}$

54. In a simultaneous throw of two dice, what is the probability of getting a doublet?
दो पासों को एक साथ फेंकने पर दोहरा पासा आने की प्रायिकता क्या है?
(a) $\frac{1}{6}$ (b) $\frac{1}{4}$ (c) $\frac{2}{3}$ (d) $\frac{3}{7}$

Solution:

$n(S) = 36$

E = getting a doublet

E = {(1,1), (2,2), (3,3), (4,4), (5,5), (6,6)},

n(E) = 6

$P(E) = \frac{n(E)}{n(S)} = \frac{6}{36} = \frac{1}{6}$

55. In a single throw of two dice, find the probability that neither a doublet nor a total of 8 will appear.
दो पासों को एक बार फेंकने पर इसकी प्रायिकता ज्ञात कीजिए कि न तो एक भी दोहरा और न ही कुल 8 आएगा।
(a) $\frac{7}{15}$ (b) $\frac{5}{18}$ (c) $\frac{13}{18}$ (d) $\frac{3}{16}$

Solution:

$n(S) = 36$

E' = getting either a doublet or a total of 8

E' = {(1,1), (2,2), (3,3), (4,4), (5,5), (6,6), (2,6), (3,5), (5,3), (6,2)},

n(E') = 10

$n(E) = n(S) - n(E') = 36 - 10 = 26$

$P(E) = \frac{n(E)}{n(S)} = \frac{26}{36} = \frac{13}{18}$

56. In a simultaneous throw of two dice, what is the probability of getting a total of 10 or 11?
दो पासों को एक साथ फेंकने पर कुल 10 या 11 आने की प्रायिकता क्या है?
(a) $\frac{1}{4}$ (b) $\frac{1}{6}$ (c) $\frac{7}{12}$ (d) $\frac{5}{36}$

Solution:

$n(S) = 36$

E = getting a total of 10 or 11

E = {(4,6), (5,5), (5,6), (6,4), (6,5)},

n(E) = 5

$P(E) = \frac{n(E)}{n(S)} = \frac{5}{36}$

57. Two dice are thrown simultaneously. What is the probability of getting two numbers whose product is even?

दो पासे एक साथ फेंके जाते हैं। ऐसी दो संख्याएँ प्राप्त होने की प्रायिकता क्या है जिनका गुणनफल सम है?

(a) $\frac{1}{2}$ (b) $\frac{3}{4}$ (c) $\frac{3}{8}$ (d) $\frac{5}{16}$

Solution:

$n(S) = 36$

E = getting two numbers whose product is even

E = {(1,2), (1,4), (1,6), (2,1), (2,2), (2,3), (2,4), (2,5), (2,6), (3,2), (3,4), (3,5), (4,1), (4,2), (4,3), (4,4), (4,5), (4,6), (5,2), (5,4), (5,6), (6,1), (6,2), (6,3), (6,3), (6,4), (6,5), (6,6)}

$n(E) = 27$

$P(E) = \frac{n(E)}{n(S)} = \frac{27}{36} = \frac{3}{4}$

58. Two dice are tossed. The probability that the total score is a prime number is:

दो पासे उछाले गए। संभावना है कि कुल स्कोर एक अभाज्य संख्या है:

(a) $\frac{1}{6}$ (b) $\frac{5}{12}$ (c) $\frac{1}{2}$ (d) $\frac{7}{9}$

Solution:

$n(S) = 36$

E = getting total score a prime number

E = {(1,1), (1,2), (1,4), (1,6), (2,1), (2,3), (2,5), (3,2), (3,4), (4,1), (4,3), (5,2), (5,6), (6,1), (6,5)}

$n(E) = 15$

$P(E) = \frac{n(E)}{n(S)} = \frac{15}{36} = \frac{5}{12}$

59. Two dice are thrown together. What is the probability that the sum of the numbers on the two faces is divisible by 4 or 6?

दो पासे एक साथ फेंके जाते हैं। इसकी क्या प्रायिकता है कि दोनों सतह पर आई संख्या का योग 4 या 6 से विभाजित हो?

(a) $\frac{7}{18}$ (b) $\frac{14}{35}$ (c) $\frac{8}{18}$ (d) $\frac{7}{35}$

Solution:

$n(S) = 36$

E = getting the sum of the numbers divisible by 4 or 6

E = {(1,3), (1,5), (2,2), (2,4), (2,6), (3,1), (3,3), (3,5), (4,2), (4,4), (5,1), (5,3), (6,2), (6,6)}

$n(E) = 14$

$P(E) = \frac{n(E)}{n(S)} = \frac{14}{36} = \frac{7}{18}$

60. What is the probability of getting at least one six in a single throw of three unbiased dice?

तीन निष्पक्ष पासों को एक बार फेंकने पर कम से कम एक छक्का लगने की प्रायिकता क्या है?

(a) $\frac{1}{36}$ (b) $\frac{91}{256}$ (c) $\frac{13}{256}$ (d) $\frac{43}{256}$

Solution:

$n(S) = 6^3 = 216$

E = getting at least one six

$n(E') = 5 \times 5 \times 5 = 125$

$n(E) = 216 - 125 = 91$

$P(E) = \frac{n(E)}{n(S)} = \frac{91}{216}$

61. Four dice are thrown simultaneously. Find the probability that all of them show the same face.

चार पासे एक साथ फेंके जाते हैं। प्रायिकता ज्ञात कीजिए कि उन सभी का सतह एक जैसा है।

(a) $\frac{1}{216}$ (b) $\frac{1}{36}$ (c) $\frac{2}{216}$ (d) $\frac{4}{216}$

Solution:

$n(S) = 6^4 = 1296$

E = getting the same number on all faces

E = {(1,1,1,1), (2,2,2,2), (3,3,3,3), (4,4,4,4), (5,5,5,5), (6,6,6,6)}

$n(E) = 6$

$P(E) = \frac{n(E)}{n(S)} = \frac{6}{1296} = \frac{1}{216}$

62. A dice is thrown twice and the sum of the appearing numbers is 10. Then the probability that the number 5 has appeared at least once is:
एक पासे को दो बार फेंका जाता है और आने वाली संख्याओं का योग 10 है। तब संख्या 5 कम से कम एक बार आने की प्रायिकता है:
(a) $\frac{1}{4}$ (b) $\frac{1}{3}$ (c) $\frac{2}{3}$ (d) $\frac{1}{2}$

Solution:

S = sum of appearing numbers is 10

$S = \{(4, 6), (5, 5), (6, 4)\}$,

$n(S) = 3$

E = number 5 has appeared at least once

$E = \{(5, 5)\}$,

$n(E) = 1$

$P(E) = \frac{n(E)}{n(S)} = \frac{1}{3}$

63. A bag contains 6 black and 8 white balls. One ball is drawn at random. What is the probability that the ball drawn is white?
एक बैग में 6 काली और 8 सफेद गेंदें हैं। एक गेंद यादृच्छिक रूप से निकाली जाती है। इसकी क्या प्रायिकता है कि निकाली गई गेंद सफेद है?
(a) $\frac{3}{4}$ (b) $\frac{4}{7}$ (c) $\frac{1}{8}$ (d) $\frac{3}{7}$

Solution:

Black balls = 6

White balls = 8

Total balls = 14

$n(S) = {}^{14}C_1$,

$n(E) = {}^{8}C_1$

$P(E) = \frac{n(E)}{n(S)} = \frac{{}^{8}C_1}{{}^{14}C_1} = \frac{8}{14} = \frac{4}{7}$

64. A box contains 6 white, 2 black, and 3 red balls. If a ball is drawn at random, what is the probability that it will not be white?
एक डिब्बे में 6 सफेद, 2 काली और 3 लाल गेंदें हैं। यदि एक गेंद यादृच्छिक रूप से निकाली जाती है, तो इसकी क्या प्रायिकता है कि वह सफेद नहीं होगी?
(a) $\frac{5}{6}$ (b) $\frac{6}{5}$ (c) $\frac{5}{11}$ (d) $\frac{6}{11}$

Solution:

White balls $= 6$,

Black balls $= 2$,

Red balls $= 3$,

Total balls $= 11$

$n(S) = {}^{11}C_1$,

$n(E) = {}^{5}C_1$

$P(E) = \frac{n(E)}{n(S)} = \frac{{}^{5}C_1}{{}^{11}C_1} = \frac{5}{11}$

65. A box contains 6 white, 8 black, 4 red, and 3 blue marbles. If one marble is drawn at random, what is the probability that it is black or blue?
एक बक्से में 6 सफेद, 8 काले, 4 लाल और 3 नीले पत्थर हैं। यदि एक कंचा यादृच्छिक रूप से निकाला जाता है, तो इसकी क्या प्रायिकता है कि वह काला या नीला है?
(a) $\frac{5}{21}$ (b) $\frac{1}{21}$ (c) $\frac{8}{21}$ (d) $\frac{11}{21}$

Solution:

White marbles $= 6$,

Black marbles $= 8$,

Red marbles $= 4$,

Blue marbles $= 3$,

Total $= 21$

$n(S) = {}^{21}C_1$,

$n(E) = {}^{8}C_1 + {}^{3}C_1$

$P(E) = \frac{n(E)}{n(S)} = \frac{{}^{8}C_1 + {}^{3}C_1}{{}^{21}C_1} = \frac{8+3}{21} = \frac{11}{21}$

66. A box contains 4 red balls, 5 green balls, and 6 white balls. A ball is drawn at random from the box. What is the probability that the ball drawn is either red or green?
एक डिब्बे में 4 लाल गेंदें, 5 हरी गेंदें और 6 सफेद गेंदें हैं। बॉक्स से यादृच्छिक रूप से एक गेंद निकाली जाती है। इसकी क्या प्रायिकता है कि निकाली गई गेंद या तो लाल या हरी है?
(a) $\frac{2}{5}$ (b) $\frac{3}{5}$ (c) $\frac{1}{5}$ (d) $\frac{7}{15}$

Solution:

Red balls = 4,

Green balls = 5,

White balls = 6,

Total balls = 15

$n(S) = {}^{15}C_1,$

$n(E) = {}^{4}C_1 + {}^{5}C_1$

$P(E) = \frac{n(E)}{n(S)} = \frac{{}^{4}C_1 + {}^{5}C_1}{{}^{15}C_1} = \frac{4+5}{15} = \frac{9}{15} = \frac{3}{5}$

67. In a box, there are 8 red, 7 blue, and 6 green balls. One ball is picked up randomly. What is the probability that it is neither red nor green?

एक डिब्बे में 8 लाल, 7 नीली और 6 हरी गेंदें हैं। एक गेंद यादृच्छिक रूप से उठाई जाती है। इसकी क्या प्रायिकता है कि यह न तो लाल है और न ही हरा?

(a) $\frac{2}{3}$ (b) $\frac{3}{4}$ (c) $\frac{7}{19}$ (d) $\frac{8}{21}$

Solution:

Red balls = 8,

Blue balls = 7,

Green balls = 6,

Total balls = 21

$n(S) = {}^{21}C_1,$

$n(E) = {}^{7}C_1$

$P(E) = \frac{n(E)}{n(S)} = \frac{{}^{7}C_1}{{}^{21}C_1} = \frac{7}{21} = \frac{1}{3}$

68. A bag contains 2 red, 3 green, and 2 blue balls. Two balls are drawn at random. What is the probability that none of the balls drawn is blue?

एक बैग में 2 लाल, 3 हरी और 2 नीली गेंदें हैं। दो गेंदें यादृच्छिक रूप से निकाली जाती हैं। इसकी क्या प्रायिकता है कि निकाली गई कोई भी गेंद नीली नहीं है?

(a) $\frac{10}{21}$ (b) $\frac{11}{21}$ (c) $\frac{2}{7}$ (d) $\frac{5}{7}$

Solution:

Red balls = 2,

Green balls = 3,

Blue balls = 2,

Total balls = 7

E = none of the two balls drawn is blue

$n(E) = {}^{5}C_2,$

$n(S) = {}^{7}C_2$

$P(E) = \frac{n(E)}{n(S)} = \frac{{}^{5}C_2}{{}^{7}C_2} = \frac{10}{21}$

69. A bag contains 6 white and 4 red balls. Three balls are drawn at random. What is the probability that one ball is red and the other two are white?

एक बैग में 6 सफेद और 4 लाल गेंदे हैं। तीन गेंदें यादृच्छिक रूप से निकाली जाती हैं। इसकी क्या प्रायिकता है कि एक गेंद लाल है और बाकी दो सफेद हैं?

(a) $\frac{1}{2}$ (b) $\frac{1}{12}$ (c) $\frac{3}{10}$ (d) $\frac{7}{12}$

Solution:

White balls = 6,

Red balls = 4,

Total balls = 10

E = one ball is red and the other two are white

$n(E) = {}^{4}C_1 \times {}^{6}C_2,$

$n(S) = {}^{10}C_3$

$P(E) = \frac{n(E)}{n(S)} = \frac{{}^{4}C_1 \times {}^{6}C_2}{{}^{10}C_3} = \frac{1}{2}$

70. A bag contains 4 white, 5 red, and 6 blue balls. Three balls are drawn at random from the bag. The probability that all of them are red is:

एक बैग में 4 सफेद, 5 लाल और 6 नीली गेंदें हैं। बैग से यादृच्छिक रूप से तीन गेंदें निकाली जाती हैं। उनमें से सभी के लाल होने की प्रायिकता है:

(a) $\frac{1}{22}$ (b) $\frac{3}{22}$ (c) $\frac{2}{91}$ (d) $\frac{2}{77}$

Solution:

White balls = 4,

Red balls = 5,

Blue balls = 6,

Total balls = 15

E = all three balls are red

$n(E) = {}^5C_3$,

$n(S) = {}^{15}C_3$

$$P(E) = \frac{n(E)}{n(S)} = \frac{{}^5C_3}{{}^{15}C_3} = \frac{2}{91}$$

71. A box contains 10 black and 10 white balls. The probability of drawing two balls of the same color is:
एक डिब्बे में 10 काली और 10 सफेद गेंदें हैं। एक ही रंग की दो गेंदें निकालने की प्रायिकता है:

(a) $\frac{9}{19}$ (b) $\frac{9}{38}$ (c) $\frac{10}{19}$ (d) $\frac{5}{19}$

Solution:

Black balls = 10,

White balls = 10,

Total balls = 20

E = drawing two balls of the same color = either both are black or both are white

$n(E) = {}^{10}C_2 + {}^{10}C_2$,

$n(S) = {}^{20}C_2$

$$P(E) = \frac{n(E)}{n(S)} = \frac{{}^{10}C_2 + {}^{10}C_2}{{}^{20}C_2} = \frac{9}{19}$$

72. A box contains 6 white balls and 7 black balls. Two balls are drawn at random. What is the probability that both of them are of different colors?
एक डिब्बे में 6 सफेद गेंदें और 7 काली गेंदें हैं। दो गेंदें यादृच्छिक रूप से निकाली जाती हैं। इसकी क्या प्रायिकता है कि वे दोनों अलग-अलग रंग की हों?

(a) $\frac{4}{13}$ (b) $\frac{2}{13}$ (c) $\frac{6}{13}$ (d) $\frac{7}{13}$

Solution:

White balls = 6,

Black balls = 7,

Total balls = 13

E' = both balls are of the same color

$n(E') = {}^6C_2 + {}^7C_2$,

$n(S) = {}^{13}C_2$

$$P(E') = \frac{n(E')}{n(S)} = \frac{{}^6C_2 + {}^7C_2}{{}^{13}C_2} = \frac{6}{13}$$

$$P(E) = 1 - P(E') = 1 - \frac{6}{13} = \frac{7}{13}$$

73. A box contains 5 green, 4 yellow, and 3 white balls. Three balls are drawn at random. What is the probability that they are not of the same color?
एक डिब्बे में 5 हरी, 4 पीली और 3 सफेद गेंदें हैं। तीन गेंदें यादृच्छिक रूप से निकाली जाती हैं। इसकी क्या प्रायिकता है कि वे एक ही रंग के नहीं हैं?

(a) $\frac{3}{44}$ (b) $\frac{3}{55}$ (c) $\frac{52}{55}$ (d) $\frac{41}{44}$

Solution:

Green balls = 5,

Yellow balls = 4,

White balls = 3,

Total balls = 12

E' = all three balls are of the same color

$n(E') = {}^5C_3 + {}^4C_3 + {}^3C_3$,

$n(S) = {}^{12}C_3$

$$P(E') = \frac{n(E')}{n(S)} = \frac{{}^5C_3 + {}^4C_3 + {}^3C_3}{{}^{12}C_3} = \frac{3}{44}$$

$$P(E) = 1 - P(E') = 1 - \frac{3}{44} = \frac{41}{44}$$

74. A box contains 20 electric bulbs, out of which 4 are defective. Two bulbs are chosen at random from this box. The probability that at least one of these is defective is:
एक बॉक्स में 20 बिजली के बल्ब हैं, जिनमें से 4 ख़राब हैं। इस बॉक्स से यादृच्छिक रूप से दो बल्ब चुने जाते हैं। इनमें से कम से कम एक के ख़राब होने की प्रायिकता क्या है?

(a) $\frac{4}{19}$ (b) $\frac{7}{19}$ (c) $\frac{12}{19}$ (d) $\frac{21}{95}$

Solution:

Defective bulbs = 4,

Non-defective bulbs = 16,

Total bulbs = 20

E' = none is defective

$n(E') = {}^{16}C_2$,

$n(S) = {}^{20}C_2$

$$P(E') = \frac{n(E')}{n(S)} = \frac{{}^{16}C_2}{{}^{20}C_2} = \frac{12}{19}$$

$$P(E) = 1 - P(E') = 1 - \frac{12}{19} = \frac{7}{19}$$

75. A box contains 10 black and 5 white balls. Two balls are drawn from the box one after the other without replacement. What is the probability that both drawn balls are black?

एक डिब्बे में 10 काली और 5 सफेद गेंदें हैं। बिना प्रतिस्थापन के एक के बाद एक बॉक्स से दो गेंदें निकाली जाती हैं। इसकी क्या प्रायिकता है कि दोनों निकाली गई गेंदें काली हैं?

(a) $\frac{2}{7}$ (b) $\frac{3}{7}$ (c) $\frac{4}{7}$ (d) $\frac{5}{7}$

Solution:

Black balls = 10,

White balls = 5,

Total balls = 15

E = both drawn balls are black (without replacement case)

$$P(E) = \frac{10}{15} \times \frac{9}{14} = \frac{3}{7}$$

76. Three cards are drawn successively, without replacement, from a pack of 52 well-shuffled cards. What is the probability that the first two cards are kings and the third card drawn is an ace?

अच्छी तरह से फेंटे गए 52 पत्तों की एक गड्डी से बिना प्रतिस्थापन के तीन पत्ते क्रमिक रूप से निकाले जाते हैं। इसकी क्या प्रायिकता है कि पहले दो कार्ड किंग हों और निकाला गया तीसरा कार्ड इक्का हो?

(a) $\frac{2}{5525}$ (b) $\frac{3}{5525}$ (c) $\frac{4}{5525}$ (d) $\frac{1}{5525}$

Solution:

$n(S) = 52$

E = first two cards are kings and the third card drawn is an ace (without replacement case)

Total kings = 4,

Total aces = 4

$$P(E) = \frac{4}{52} \times \frac{3}{51} \times \frac{4}{50} = \frac{2}{5525}$$

77. Two cards are drawn at random and without replacement from a pack of 52 playing cards. Find the probability that both the cards are black.

52 ताश के पत्तों की एक गड्डी से दो पत्ते यादृच्छिक रूप से और बिना प्रतिस्थापन के निकाले जाते हैं। दोनों कार्डों के काले होने की प्रायिकता ज्ञात कीजिए।

(a) $\frac{21}{102}$ (b) $\frac{22}{102}$ (c) $\frac{25}{102}$ (d) $\frac{26}{102}$

Solution:

$n(S) = 52$

E = both drawn cards are black (without replacement case)

Total black cards = 26

$$P(E) = \frac{26}{52} \times \frac{25}{51} = \frac{25}{102}$$

78. A box of oranges is inspected by examining three randomly selected oranges drawn without replacement. If all three oranges are good, the box is approved for sale; otherwise, it is rejected. Find the probability that a box containing 15 oranges, out of which 12 are good and 3 are bad, will be approved for sale.

संतरे के एक डिब्बे का निरीक्षण बिना प्रतिस्थापन के निकाले गए तीन यादृच्छिक रूप से चयनित संतरे की जांच करके किया जाता है। यदि तीनों संतरे अच्छे हैं, तो डिब्बा बिक्री के लिए अनुमोदित कर दिया जाता है, अन्यथा इसे अस्वीकार कर दिया जाता है। प्रायिकता ज्ञात कीजिए कि एक बॉक्स जिसमें 15 संतरे हैं, जिनमें से 12 अच्छे और 3 खराब हैं, बिक्री के लिए स्वीकृत हो जाएगा।

(a) $\frac{44}{91}$ (b) $\frac{43}{91}$ (c) $\frac{42}{91}$ (d) $\frac{41}{91}$

Solution:

Good oranges = 12,

Bad oranges = 3,

Total oranges = 15

E = all three oranges are good (without replacement case)

$$P(E) = \frac{12}{15} \times \frac{11}{14} \times \frac{10}{13} = \frac{44}{91}$$

79. Two balls are drawn at random with replacement from a box containing 10 black and 8 red balls. Find the probability that
(i) both balls are red.
(ii) first ball is black and second is red.
(iii) one of them is black and the other is red.
10 काली और 8 लाल गेंदों वाले एक बॉक्स से प्रतिस्थापन के साथ यादृच्छिक रूप से दो गेंदें निकाली जाती हैं। प्रायिकता ज्ञात कीजिए कि
(i) दोनों गेंदें लाल हैं।
(ii) पहली गेंद काली है और दूसरी लाल है।
(iii) उनमें से एक काला है और दूसरा लाल है।

Solution:

Black balls = 10,

Red balls = 8,

Total balls = 18

With replacement case:

(i) Probability that both balls are red:

$$P = \frac{8}{18} \times \frac{8}{18} = \frac{16}{81}$$

(ii) Probability that the first ball is black and the second is red:

$$P = \frac{10}{18} \times \frac{8}{18} = \frac{20}{81}$$

(iii) Probability that one of them is black and the other is red:

$$P = \frac{10}{18} \times \frac{8}{18} + \frac{8}{18} \times \frac{10}{18} = \frac{40}{81}$$

80. In a class, 30% of the students offered English, 20% offered Hindi, and 10% offered both. If a student is selected at random, what is the probability that he has offered English or Hindi?
एक कक्षा में, 30% छात्र अंग्रेजी चुनते हैं, 20% हिंदी चुनते हैं और 10% दोनों चुनते हैं। यदि एक छात्र को यादृच्छिक रूप से चुना जाता है, तो क्या संभावना है कि उसने अंग्रेजी या हिंदी चुनी है?

(a) $\frac{2}{5}$ (b) $\frac{3}{4}$ (c) $\frac{3}{5}$ (d) $\frac{3}{10}$

Solution:

Student offered English = 30%,

Student offered Hindi = 20%,

Student offered both = 10%

Using Venn diagram:

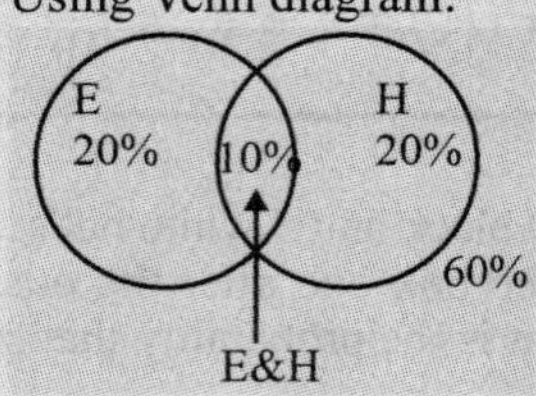

$$\text{Required probability} = \frac{40\%}{100\%} = \frac{2}{5}$$

81. In an entrance test graded on the basis of two examinations, the probability of a randomly chosen student passing the first examination is 0.8, and the probability of passing the second examination is 0.7. The probability of passing at least one of them is 0.95. What is the probability of passing both?
एक प्रवेश परीक्षा में, जिसे दो परीक्षाओं के आधार पर वर्गीकृत किया जाता है, यादृच्छिक रूप से चुने गए छात्र की पहली परीक्षा उत्तीर्ण करने की संभावना 0.8 है और दूसरी परीक्षा उत्तीर्ण करने की संभावना 0.7 है। उनमें से कम से कम एक के उत्तीर्ण होने की प्रायिकता 0.95 है। दोनों के उत्तीर्ण होने की प्रायिकता क्या है?

(a) 0.65 (b) 0.55 (c) 0.50 (d) 0.60

Solution:

Given:

Probability of passing the first exam (a) = 0.8,

Probability of passing the second exam (b) = 0.7

Probability of passing at least one of them (a + b + c) = 0.95

(i)

Using Venn diagram:

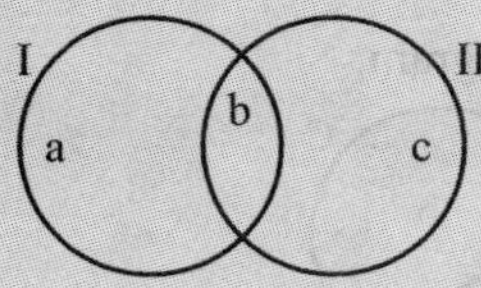

$a + c = 0.8,$

$b + c = 0.7,$

$a + b + 2c = 1.5$

(ii)

From (i) and (ii):

$a + b + c = 0.95,$

$a + b + 2c = 1.5$

$c = 0.55$

The probability of passing both $= c = 0.55$

82. The probability that a student will pass the final examination in both English and Hindi is 0.5, and the probability of passing neither is 0.1. If the probability of passing the English examination is 0.75, what is the probability of passing the Hindi examination?

एक छात्र के अंग्रेजी और हिंदी दोनों में अंतिम परीक्षा उत्तीर्ण करने की संभावना 0.5 है और दोनों में से किसी में भी उत्तीर्ण होने की संभावना 0.1 है। यदि अंग्रेजी परीक्षा उत्तीर्ण करने की संभावना 0.75 है, तो हिंदी परीक्षा उत्तीर्ण करने की संभावना क्या है?

(a) 0.60 (b) 0.65 (c) 0.70 (d) 0.55

Solution:

Given probabilities:

Pass in both English and Hindi (c) $= 0.5,$

Pass in neither exam (d) $= 0.1,$

Pass in English exam (a + b) $= 0.75$

Let pass in Hindi exam $= b + c.$

Using Venn diagram:

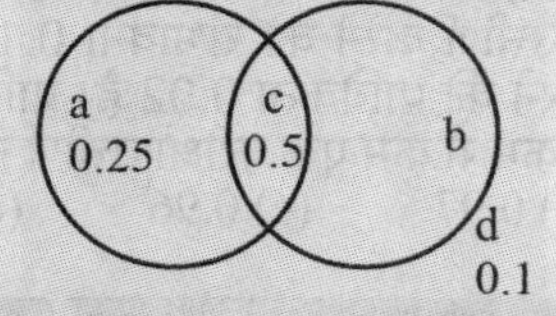

$a + c = 0.75$

$a + 0.5 = 0.75$

$a = 0.25$

Since:

$a + b + c + d = 1$

$0.25 + b + 0.5 + 0.1 = 1$

$b = 0.15$

Required probability$(b + c) = 0.15 + 0.5 = 0.65$

83. Two students Anil and Ashima appeared in an examination. The probability that Anil will qualify the examination is 0.05, and that Ashima will qualify the examination is 0.10. The probability that both will qualify the examination is 0.02. Find the probability that both Anil and Ashima will not qualify the examination.

दो छात्र अनिल और आशिमा एक परीक्षा में उपस्थित हुए। अनिल के परीक्षा उत्तीर्ण करने की संभावना 0.05 है और आशिमा के परीक्षा उत्तीर्ण करने की संभावना 0.10 है। दोनों के परीक्षा उत्तीर्ण करने की प्रायिकता 0.02 है। प्रायिकता ज्ञात कीजिए कि अनिल और आशिमा दोनों परीक्षा उत्तीर्ण नहीं करेंगे।

(a) 0.87 (b) 0.86 (c) 0.88 (d) 0.89

Solution:

Given probabilities:

Anil will qualify the exam = 0.05,

Ashima will qualify the exam = 0.10,

Both will qualify the exam = 0.02

Using Venn diagram:

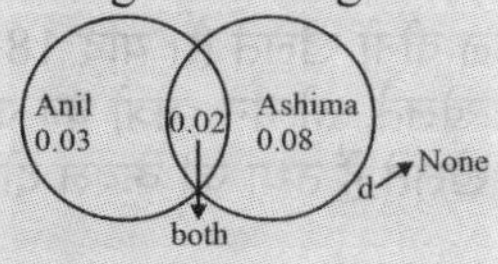

$0.03 + 0.02 + 0.08 + d = 1$

$d = 0.87$

Probability that both Anil and Ashima will not qualify the examination = d = 0.87

84. Two students Anil and Ashima appeared in an examination. The probability that Anil will qualify the examination is 0.05 and that Ashima will qualify the examination is 0.10. The probability that both will qualify the examination is 0.02. Find the probability that at least one of them will not qualify the examination.

दो छात्र अनिल और आशिमा एक परीक्षा में उपस्थित हुए। अनिल के परीक्षा उत्तीर्ण करने की संभावना 0.05 है और आशिमा के परीक्षा उत्तीर्ण करने की संभावना 0.10 है। दोनों के परीक्षा उत्तीर्ण करने की प्रायिकता 0.02 है। प्रायिकता ज्ञात कीजिए कि उनमें से कम से कम एक परीक्षा उत्तीर्ण नहीं करेगा।
(a) 0.98 (b) 0.97 (c) 0.96 (d) 0.95

Solution:

Given probabilities are:

Anil will qualify the exam = 0.05,

Ashima will qualify the exam = 0.10,

Both will qualify the exam = 0.02

Using Venn diagram:

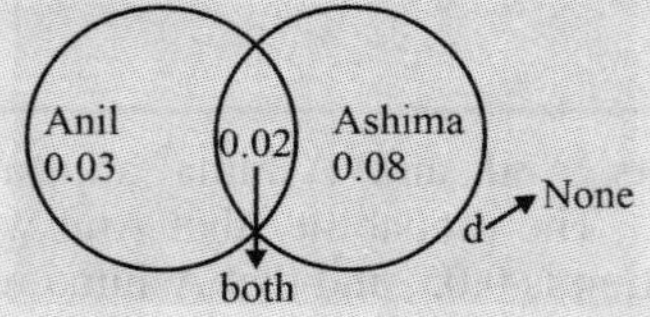

Required probability = 1 - (Probability of both qualifying the exam)

$= 1 - 0.02 = 0.98$

85. In a class of 50 students, 20 play hockey, 15 play cricket, and 11 play football. 7 play both hockey and cricket, 4 play cricket and football, and 5 play hockey and football. If 18 students don't play any of these given sports, find the ratio of students who play all of these 3 sports to those who play exactly two of these sports.

50 छात्रों की एक कक्षा में 20 छात्र हॉकी, 15 छात्र क्रिकेट और 11 छात्र फुटबॉल खेलते हैं। 7 छात्र हॉकी तथा क्रिकेट दोनों खेलते हैं। 4 छात्र क्रिकेट और फुटबॉल दोनों खेलते हैं। 5 छात्र हॉकी और फुटबॉल दोनों खेलते हैं। यदि 18 छात्र ऐसे हैं जो कोई भी खेल नहीं खेलते, तो उन छात्रों का अनुपात ज्ञात कीजिए जो सभी खेल खेलते हैं तथा जो केवल दो खेल खेलते हैं।
(a) 1:3 (b) 1:7 (c) 1:5 (d) 4:3

Solution:

Given:
Total students = 50,
Hockey players = 20,
Cricket players = 15,
Football players = 11

Hockey and Cricket players = 7,

Cricket and Football players = 4,

Hockey and Football players = 5

Do not play any game = 18

Using Venn diagram:

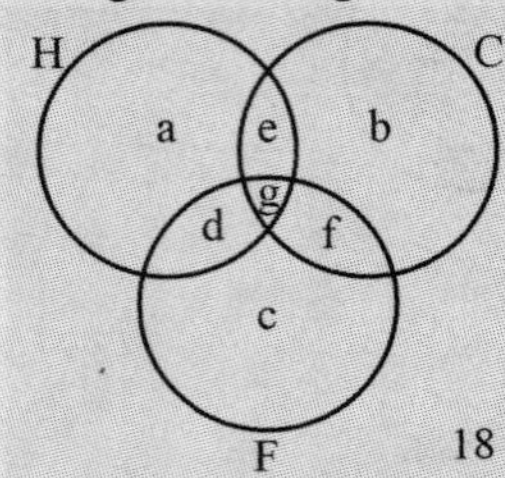

$a + b + c + d + e + f + g = 50 - 18 = 32$

(i)

From the Venn diagram:

$a+b+c+2(d+e+f)+3g = 20+15+11 = 46$

(ii)

And:

$d + e + f + 3g = 4 + 7 + 5 = 16$

(iii)

Using (i) and (ii):

$a + b + c + 2(d + e + f) + 3g = 46$

$a + b + c + d + e + f + g = 32$

$\therefore d + e + f + 2g = 14$

(iv)

Using (iii) and (iv):

$d + e + f + 3g = 16$

$d + e + f + 2g = 14$

$\therefore g = 2$

and

$d + e + f = 10$

Hence, the required ratio:

$g : (d + e + f) = 2 : 10 = 1 : 5$

86. In class 11 of a school, 40 students opted for Physics, 17 for Biology, and 20 for Chemistry. If the total number of students in class 11 was 60, and all these students chose at least one of the three subjects mentioned here, and exactly five of these students chose all the three subjects, then what is the probability that a student chosen at random from class 11th will have chosen exactly one subject out of these three subjects?

एक स्कूल की कक्षा 11 में, 40 छात्रों ने भौतिकी, 17 ने जीवविज्ञान और 20 ने रसायन विज्ञान चुना। यदि कक्षा 11 में छात्रों की कुल संख्या 60 थी, और इन सभी छात्रों ने यहां उल्लिखित तीन विषयों में से कम से कम एक को चुना, और इनमें से ठीक पांच छात्रों ने सभी तीन विषयों को चुना, तो क्या संभावना है कि कक्षा 11वीं से यादृच्छिक रूप से चुना गया छात्र इन तीन विषयों में से बिल्कुल एक विषय चुनेगा?

(a) 0.85 (b) 0.90 (c) 0.75 (d) 0.80

Solution:

Total students = 60,
Physics students = 40,
Biology students = 17,
Chemistry students = 20
Students who choose all three subjects = 5
Using Venn diagram:

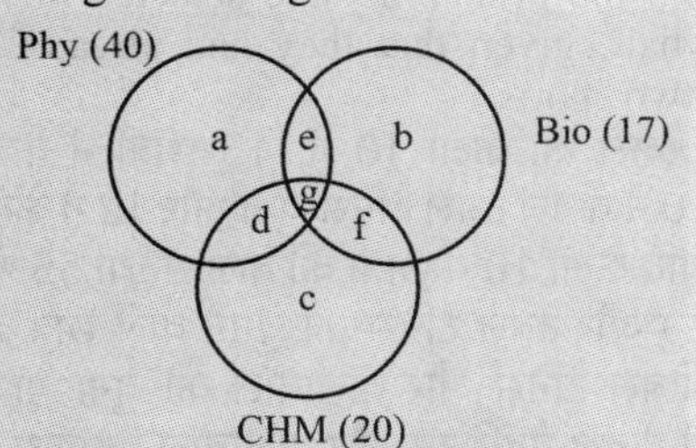

Given: $g = 5$
ATQ: $a + b + c + 2(d + e + f) + 3g = 77$
(i) $a + b + c + d + e + f + g = 60$
(ii)
From (ii):
$d + e + f + 2g = 17$
$d + e + f + 10 = 17$
$d + e + f = 7$
From (i) and (ii):
$a + b + c + 7 + 5 = 60$
$a + b + c = 48$

Required probability: $\dfrac{a+b+c}{\text{Total}} = \dfrac{48}{60} = \dfrac{4}{5} = 0.80$

87. In class 10 of a school, 13 students opted for only Badminton, 10 opted for only Table Tennis, and 12 for only Volleyball. If the total number of students in class 10 was 55, and students who opted for only Badminton and Table Tennis are x, those who opted for only Table Tennis and Volleyball are x, and those who opted for only Badminton and Volleyball are $x - 3$. Students who opted for all of these are $x + 3$. Find the probability that the selected member did not choose Volleyball.

एक स्कूल की कक्षा 10 में, 13 छात्रों ने केवल बैडमिंटन चुना, 10 ने केवल टेबल टेनिस चुना और 12 ने केवल वॉलीबॉल चुना। यदि कक्षा 10 में छात्रों की कुल संख्या 55 थी और केवल बैडमिंटन और टेबल टेनिस को चुनने वाले छात्र x हैं, जिन्होंने केवल टेबल टेनिस और वॉलीबॉल को चुना वे x हैं, और जिन्होंने केवल बैडमिंटन और वॉलीबॉल को चुना वे $x - 3$ हैं। इन सभी को चुनने वाले छात्र $x + 3$ हैं। चयनित सदस्य ने वॉलीबॉल नहीं चुना, इसकी प्रायिकता ज्ञात कीजिए।

(a) 0.85 (b) 0.75 (c) 0.80 (d) 0.90

Solution:

Students opted only Badminton = 13,

Students opted only Table Tennis = 10,

Students opted only Volleyball = 12

Students opted only Badminton and Table Tennis = x,

Students opted only Table Tennis and Volleyball = x,

Students opted only Badminton and Volleyball = x - 3

Students opted all three sports = $x + 3$,

Total students = 55

Using Venn diagram:

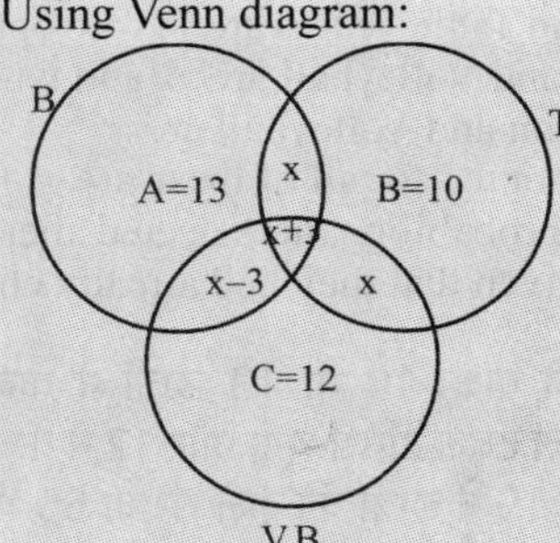

$13 + 10 + 12 + x + x + (x - 3) + (x + 3) = 55$

$35 + 4x = 55$

$4x = 20$

x = 5

Probability the selected member did not choose Volleyball:

$$P = \frac{a+b+x}{\text{Total}} = \frac{25}{55}$$

88. In class 10 of a school, 13 students opted for only Badminton, 10 opted for only Table Tennis, and 12 for only Volleyball. If the total number of students in class 10 was 55 and students who opted for only Badminton and Table Tennis are x, who opted for only Table Tennis and Volleyball are x, and who opted for only Badminton and Volleyball are $x-3$. Students who opted for all of these are $x+3$.
(a) Find the probability the selected member did not choose Volleyball.

एक स्कूल की कक्षा 10 में, 13 छात्रों ने केवल बैडमिंटन चुना, 10 ने केवल टेबल टेनिस चुना और 12 ने केवल वॉलीबॉल चुना। यदि कक्षा 10 में छात्रों की कुल संख्या 55 थी और केवल बैडमिंटन और टेबल टेनिस को चुनने वाले छात्र x हैं, जिन्होंने केवल टेबल टेनिस और वॉलीबॉल को चुना था वे x हैं और जिन्होंने केवल बैडमिंटन और वॉलीबॉल को चुना था वे $x-3$ हैं। इन सभी को चुनने वाले छात्र $x+3$ हैं। (a) प्रायिकता ज्ञात कीजिए कि चयनित सदस्य ने वॉलीबॉल नहीं चुना।

Solution:

Probability the selected member did not choose Volleyball $= \dfrac{a+b+x}{\text{Total}} = \dfrac{25}{55}$

89. In class 10 of a school, 13 students opted for only Badminton, 10 opted for only Table Tennis, and 12 for only Volleyball. If the total number of students in class 10 was 55 and students who opted for only Badminton and Table Tennis are x, who opted for only Table Tennis and Volleyball are x, and who opted for only Badminton and Volleyball are $x-3$.
Students who opted for all of these are $x+3$.
(b) Find the probability the selected member chose Badminton, given that they had already chosen Table Tennis.

एक स्कूल की कक्षा 10 में, 13 छात्रों ने केवल बैडमिंटन चुना, 10 ने केवल टेबल टेनिस चुना और 12 ने केवल वॉलीबॉल चुना। यदि कक्षा 10 में छात्रों की कुल संख्या 55 थी और केवल बैडमिंटन और टेबल टेनिस को चुनने वाले छात्र x हैं, जिन्होंने केवल टेबल टेनिस और वॉलीबॉल को चुना था वे x हैं और जिन्होंने केवल बैडमिंटन और वॉलीबॉल को चुना था वे $x-3$ हैं। इन सभी को चुनने वाले छात्र $x+3$ हैं। (b) प्रायिकता ज्ञात कीजिए कि चयनित सदस्य ने बैडमिंटन चुना है, यह देखते हुए कि उन्होंने पहले ही टेबल टेनिस चुना है।

Solution:

Probability the selected member chose Badminton, given that they had already chosen Table Tennis:

$$P(E) = \frac{n(E)}{n(S)}$$

$n(E) = x + x + 3 = 5 + 8 = 13,$

$n(S) = x+x+3+x+10 = 5+8+5+10 = 28$

$$P(E) = \frac{n(E)}{n(S)} = \frac{13}{28}$$

90. In class 10 of a school, 13 students opted for only Badminton, 10 opted for only Table Tennis, and 12 for only Volleyball. If the total number of students in class 10 was 55 and students who opted for only Badminton and Table Tennis are x, who opted for only Table Tennis and Volleyball are x, and who opted for only Badminton and Volleyball are $x-3$. Students who opted for all of these are $x+3$.
(c) Find the probability the selected member chose Volleyball, given that they had already chosen the other two sports.

एक स्कूल की कक्षा 10 में, 13 छात्रों ने केवल बैडमिंटन चुना, 10 ने केवल टेबल टेनिस चुना और 12 ने केवल वॉलीबॉल चुना। यदि कक्षा 10 में छात्रों की कुल संख्या 55 थी और केवल बैडमिंटन और टेबल टेनिस को चुनने वाले छात्र x हैं, जिन्होंने केवल टेबल टेनिस और वॉलीबॉल को चुना था वे x हैं और जिन्होंने केवल बैडमिंटन और वॉलीबॉल को चुना था वे $x-3$ हैं। इन सभी को चुनने वाले छात्र $x+3$ हैं। (c) प्रायिकता ज्ञात कीजिए कि चयनित सदस्य ने वॉलीबॉल चुना, यह देखते हुए कि उन्होंने पहले ही अन्य दो खेल चुन लिए थे।

Solution:

Probability that selected member chose Volleyball, given that they had already chosen the other two sports:

$$P(E) = \frac{n(E)}{n(S)}$$

$n(S) = x + 3 + x = 5 + 3 + 5 = 13,$

$n(E) = x + 3 = 5 + 3 = 8$

$$P(E) = \frac{n(E)}{n(S)} = \frac{8}{13}$$

91. In class 10 of a school, 13 students opted for only Badminton, 10 opted for only Table Tennis, and 12 for only Volleyball. If the total number of students in class 10 was 55 and students who opted for only badminton and table tennis are x, who opted for only table tennis and volleyball are x, and who opted for only Badminton and Volleyball are $x - 3$. Students who opted for all of these are $x + 3$. Find the probability that the selected member chose badminton, given that they only chose one sport.

एक स्कूल की कक्षा 10 में, 13 छात्रों ने केवल बैडमिंटन चुना, 10 ने केवल टेबल टेनिस चुना और 12 ने केवल वॉलीबॉल चुना। यदि कक्षा 10 में छात्रों की कुल संख्या 55 थी और, केवल बैडमिंटन और टेबल टेनिस को चुनने वाले छात्र x है, जिन्होंने केवल टेवल टेनिस और वॉलीबॉल को चुना था वे x हैं और जिन्होंने केवल बैडमिंटन और वॉलीबॉल को चुना था वे $x - 3$ हैं। इन सभी को चुनने वाले छात्र $x + 3$ हैं।

(d) प्रायिकता ज्ञात कीजिए कि चयनित सदस्य ने बैडमिंटन चुना है, जबकि उन्होंने केवल एक ही खेल चुना है।

Solution:

Probability that the selected member chose badminton, given that they only chose one sport: $P(E) = \frac{n(E)}{n(S)}$

$n(S) = 13 + 10 + 12 = 35$

$n(E) = 13$

$P(E) = \frac{n(E)}{n(S)} = \frac{13}{35}$

92. The probability of Kalam and Kaleem solving a specific problem independently is 0.40 and 0.75 respectively. If both are trying to solve the problem independently, find the probability that exactly one of them solves it.

कलाम और कलीम द्वारा किसी विशिष्ट समस्या को स्वतंत्र रूप से हल करने की प्रायिकता क्रमशः 0.40 और 0.75 है। यदि दोनों स्वतंत्र रूप से समस्या को हल करने का प्रयास कर रहे हैं, तो प्रायिकता ज्ञात कीजिए कि उनमें से कोई एक इसे हल कर लेता है।

(a) $\frac{11}{20}$ (b) $\frac{9}{20}$ (c) $\frac{3}{10}$ (d) $\frac{1}{10}$

Solution:

Probability that Kalam will solve the problem:

$P(A) = 0.40 = \frac{2}{5}$

$P(A') = 1 - \frac{2}{5} = \frac{3}{5}$

Probability that Kaleem will solve the problem: $P(B) = 0.75 = \frac{3}{4}$

$P(B') = 1 - \frac{3}{4} = \frac{1}{4}$

Exactly one of them solves the problem means either Kalam solves but not Kaleem or Kaleem solves but not Kalam.

Required probability: $P = \frac{2}{5} \times \frac{1}{4} + \frac{3}{4} \times \frac{3}{5} = \frac{11}{20}$

93. A speaks the truth 5 times out of 7 and B speaks 8 times out of 9. What is the probability that they will contradict each other in stating the same fact?

A 7 में से 5 बार सच बोलता है और B 9 में से 8 बार सच बोलता है। क्या संभावना है कि वे एक ही तथ्य बताते समय एक-दूसरे का खंडन करेंगे?

(a) $\frac{1}{4}$ (b) $\frac{1}{9}$ (c) $\frac{1}{3}$ (d) $\frac{1}{7}$

Solution:

Probability that A speaks the truth: $P(A) = \frac{5}{7}$

$P(A') = 1 - \frac{5}{7} = \frac{2}{7}$

Probability that B speaks the truth: $P(B) = \frac{8}{9}$

$P(B') = 1 - \frac{8}{9} = \frac{1}{9}$

Required probability: $\frac{5}{7} \times \frac{1}{9} + \frac{8}{9} \times \frac{2}{7} = \frac{21}{63} = \frac{1}{3}$

94. A speaks truth in 75% cases and B in 80% of the cases. In what percentage of cases are they likely to contradict each other narrating the same incident?

A 75% मामलों में सच बोलता है और B 80% मामलों में सच बोलता है। कितने प्रतिशत मामलों में उनके एक ही घटना का वर्णन करते हुए एक-दूसरे का खंडन करने की संभावना है?

(a) 5% (b) 15% (c) 35% (d) 45%

Solution:

Probability that A speaks the truth: $P(A) = \frac{75}{100} = \frac{3}{4}$

$P(A') = 1 - \frac{3}{4} = \frac{1}{4}$

Probability that B speaks the truth: $P(B) = \frac{80}{100} = \frac{4}{5}$

$P(B') = 1 - \frac{4}{5} = \frac{1}{5}$

Required probability: $\frac{3}{4} \times \frac{1}{5} + \frac{4}{5} \times \frac{1}{4} = \frac{7}{20} =$ 35%

95. A man and his wife appear in an interview for two vacancies in the same post. The probability of the husband's selection is $\frac{1}{7}$ and the probability of the wife's selection is $\frac{1}{5}$. What is the probability that only one of them is selected?

एक आदमी और उसकी पत्नी एक ही पद पर दो रिक्तियों के लिए साक्षात्कार में उपस्थित होते हैं। पति के चयन की संभावना $\frac{1}{7}$ है और पत्नी के चयन की संभावना $\frac{1}{5}$ है। इसकी क्या प्रायिकता है कि उनमें से केवल एक का चयन किया जाए?

(a) $\frac{4}{5}$ (b) $\frac{2}{7}$ (c) $\frac{8}{15}$ (d) $\frac{4}{7}$

Solution:

Probability of husband's selection: $P(A) = \frac{1}{7}$

$P(A') = 1 - \frac{1}{7} = \frac{6}{7}$

Probability of wife's selection: $P(B) = \frac{1}{5}$

$P(B') = 1 - \frac{1}{5} = \frac{4}{5}$

Only one of them is selected means either the husband gets selected but not the wife, or the wife gets selected but not the husband.

Required probability: $\frac{1}{7} \times \frac{4}{5} + \frac{1}{5} \times \frac{6}{7} = \frac{10}{35} = \frac{2}{7}$

96. The probability of solving a problem by three students A, B, and C are $\frac{3}{7}, \frac{5}{9}, \frac{1}{5}$ respectively. What is the probability that the problem is solved?

तीन छात्रों A, B और C द्वारा एक समस्या को हल करने की प्रायिकता क्रमशः $\frac{3}{7}, \frac{5}{9}, \frac{1}{5}$ है। इसकी क्या प्रायिकता है कि समस्या हल हो जायेगी?

(a) $\frac{64}{315}$ (b) $\frac{32}{315}$ (c) $\frac{251}{315}$ (d) $\frac{155}{315}$

Solution:

Probability that A will solve the problem: $P(A) = \frac{3}{7}$

$P(A') = 1 - \frac{3}{7} = \frac{4}{7}$

Probability that B will solve the problem: $P(B) = \frac{5}{9}$

$P(B') = 1 - \frac{5}{9} = \frac{4}{9}$

Probability that C will solve the problem: $P(C) = \frac{1}{5}$

$P(C') = 1 - \frac{1}{5} = \frac{4}{5}$

Let P(E') be the probability that the problem is not solved: P(E') = $\frac{4}{7} \times \frac{4}{9} \times \frac{4}{5} = \frac{64}{315}$

Hence, $P(E) = 1 - P(E') = 1 - \frac{64}{315} = \frac{251}{315}$

97. Three children took part in a racing competition in their school with their respective probabilities to reach the finishing point being $\frac{1}{3}, \frac{1}{5}, \frac{1}{4}$ respectively. What is the probability that at least one of them will finish the race?

तीन बच्चों ने अपने स्कूल में रेसिंग प्रतियोगिता में भाग लिया और अंतिम बिंदु तक पहुँचने की उनकी संभावनाएँ क्रमशः $\frac{1}{3}, \frac{1}{5}, \frac{1}{4}$ थीं। इसकी क्या प्रायिकता है कि उनमें से कम से कम एक दौड़ पूरी कर लेगा?

(a) $\frac{3}{5}$ (b) $\frac{2}{5}$ (c) $\frac{2}{3}$ (d) $\frac{1}{10}$

Solution:

Given: $P(A) = \frac{1}{3}, \ P(B) = \frac{1}{5}, \ P(C) = \frac{1}{4}$

Solution:

$P(A') = 1 - \frac{1}{3} = \frac{2}{3}$,

$P(B') = 1 - \frac{1}{5} = \frac{4}{5}$,

$P(C') = 1 - \frac{1}{4} = \frac{3}{4}$

Let P(E) be the probability that at least one of them will finish the race.

$P(E')$ = Probability that no one finishes the race.

$P(E') = \frac{2}{3} \times \frac{4}{5} \times \frac{3}{4} = \frac{2}{5}$

Hence, $P(E) = 1 - P(E') = 1 - \frac{2}{5} = \frac{3}{5}$

98. A man is known to speak the truth $\frac{3}{4}$ times. He throws a die and reports that it is a six. Find the probability that it is actually a six.

एक आदमी 4 में से 3 बार सच बोलने के लिए जाना जाता है। वह पासा फेंकता है और रिपोर्ट करता है कि यह छक्का है। इसकी प्रायिकता ज्ञात कीजिए कि यह वास्तव में छक्का है।

(a) $\frac{3}{8}$ (b) $\frac{7}{8}$ (c) $\frac{4}{8}$ (d) $\frac{1}{8}$

Solution:

Probability that the man speaks the truth: $P(A) = \frac{3}{4}$,

$P(A') = 1 - \frac{3}{4} = \frac{1}{4}$

Probability of getting a six: $P(B) = \frac{1}{6}$,

$P(B') = 1 - \frac{1}{6} = \frac{5}{6}$

Using Bayes' Theorem:

$$P(\text{Actually six}) = \frac{\frac{3}{4} \times \frac{1}{6}}{\frac{3}{4} \times \frac{1}{6} + \frac{1}{4} \times \frac{5}{6}} = \frac{3}{8}$$

99. A man is known to speak the truth $\frac{3}{5}$ times. He throws a die and reports that it is '1'. Find the probability that it is actually 1.

एक आदमी 5 में से 3 बार सच बोलने के लिए जाना जाता है। वह पासा फेंकता है और रिपोर्ट करता है कि यह 1 है। इसकी प्रायिकता ज्ञात कीजिए कि यह वास्तव में 1 है

(a) $\frac{3}{13}$ (b) $\frac{1}{8}$ (c) $\frac{4}{8}$ (d) $\frac{1}{13}$

Solution:

Probability that man speaks truth: $P(A) = \frac{3}{5}$,

$P(A') = 1 - \frac{3}{5} = \frac{2}{5}$

Probability of getting '1': $P(B) = \frac{1}{6}$,

$P(B') = 1 - \frac{1}{6} = \frac{5}{6}$

Using Bayes' theorem:

Probability of getting actually '1':

$$\frac{\frac{3}{5} \times \frac{1}{6}}{\frac{3}{5} \times \frac{1}{6} + \frac{2}{5} \times \frac{5}{6}} = \frac{3}{13}$$

100. A man speaks truth $\frac{2}{3}$ times. He picks one of the natural numbers in the set $S = \{1, 2, 3, 4, 5, 6, 7\}$ and reports that it is even. The probability that it is actually even is:

एक आदमी 3 में से 2 बार सच बोलता है। वह सेट $S = \{1, 2, 3, 4, 5, 6, 7\}$ में से एक प्राकृतिक संख्या चुनता है और रिपोर्ट करता है कि यह सम है। संभावना यह है कि यह वास्तव में सम है।

(a) $\frac{2}{5}$ (b) $\frac{1}{5}$ (c) $\frac{1}{10}$ (d) $\frac{3}{5}$

Solution:

Probability that man speaks truth: $P(A) = \frac{2}{3}$,

$P(A') = 1 - \frac{2}{3} = \frac{1}{3}$

$S = \{1, 2, 3, 4, 5, 6, 7\}$

Probability of selecting an even number:

$P(B) = \frac{3}{7}$,

$P(B') = 1 - \frac{3}{7} = \frac{4}{7}$

Using Bayes' theorem:

Probability that the picked number is actually

$$\text{an even number} = \frac{\frac{2}{3} \times \frac{3}{7}}{\frac{2}{3} \times \frac{3}{7} + \frac{1}{3} \times \frac{4}{7}} = \frac{3}{5}$$

101. All the letters of the word NORMALITY are jumbled and possible five-letter words are formed without repetition of letters. From those newly formed words, a word is drawn at random. What is the probability that the word drawn starts with the letter M and ends with the letter R?

NORMALITY शब्द के सभी अक्षर अव्यवस्थित होते हैं और अक्षरों की पुनरावृत्ति के बिना संभावित पाँच अक्षर शब्द बनते हैं। उन नवगठित शब्दों से, एक शब्द पादृच्छिक रूप से खींचा जाता है। इस बात की प्रायिकता क्या है कि खींचा गया शब्द M अक्षर से शुरू होता है और अक्षर R के साथ समाप्त होता है?

(a) $\frac{7}{9!}$ (b) $\frac{1}{72}$ (c) $\frac{7}{72}$ (d) $\frac{2!}{9!}$

Solution:

Word = $\{N, O, R, M, A, L, I, T, Y\}$

Total number of ways five-letter words can be formed = n(S) = $9 \times 8 \times 7 \times 6 \times 5$

Number of five-letter words starting with M and ending with R: n(E) = $7 \times 6 \times 5$

Required probability: $P(E) = \frac{n(E)}{n(S)}$

$= \frac{7 \times 6 \times 5}{9 \times 8 \times 7 \times 6 \times 5} = \frac{1}{72}$

102. If the odds in favour of any random event A are 5:6, then the odds against the event are ?

(a) $\frac{6}{5}$ (b) $\frac{5}{11}$ (c) $\frac{6}{11}$ (d) $\frac{11}{6}$

Solution:

Odds in Favour = Num of Success/Num of Failures

Odds in Favour = 5 : 6

success = 5

Fail = 6

Odds Against = $\frac{\text{Num of Failures}}{\text{Num of Success}} = \frac{6}{5}$

103. A jewelry box contains 5 white pearls, 2 gold rings, and 6 silver rings. What are the odds of drawing a white pearl from the jewelry box?

(a) $\frac{5}{8}$ (b) $\frac{3}{8}$ (c) $\frac{5}{3}$ (d) $\frac{3}{5}$

Solution:

Probability (Pearl) = $\frac{5}{13}$

Success = 5

Fail = 8

Odds in Favour = $\frac{5}{8}$

104. A bag contains 8 red balls and x blue balls, the odd against drawing a blue ball are 2:5. What is the value of x?

(a) 20 (b) 10 (c) 5 (d) 15

Solution:

P (Blue ball) = $\frac{x}{x+8}$

success = x

fail = 8

Odds Against = $\frac{8}{x} = \frac{2}{5}$

$40 = 2x$

$x = 20$ ans...

105. The odds against an event is 4:5 and the odds in favour of another event is 3:7. If both the events are independent, then the probability that at least one of the events will happen is?

(a) $\frac{31}{45}$ (b) $\frac{77}{90}$ (c) $\frac{1}{6}$ (d) $\frac{5}{6}$

Solution:

Odds Against $(A) = \frac{4}{5}$

$P(A)$ Success = $\frac{5}{9}$, $P(A)$ Fail = $\frac{4}{9}$

Odds Favour (B) = $\frac{3}{7}$

$P(B)$ Success = $\frac{3}{10}$, $P(B)$ Fail = $\frac{7}{10}$

$P(\text{At least one happen}) = 1 - \frac{4}{9} \times \frac{7}{10}$

$1 - \frac{14}{45} = \frac{31}{45}$ ans....

106. The odds against a certain event are 5 to 2 and the odds in favour of another event independent of the former are 6 to 5. The probability that none of the events will happen is

(a) $\frac{25}{77}$ (b) $\frac{12}{77}$ (c) $\frac{52}{77}$ (d) $\frac{5}{77}$

Solution:

Odds Against $(A) = \frac{5}{2}$

$P(A)$ Success $= \frac{2}{7}$, $P(A)$ Fail $= \frac{5}{7}$

Odds Favour (B) $= \frac{6}{5}$

$P(B)$ Success $= \frac{6}{11}$, $P(B)$ Fail $= \frac{5}{11}$

$P(\text{None happen}) = \frac{5}{7} \times \frac{5}{11} = \frac{25}{77}$

107. In solving any problem, odds against A are 4 to 3 and in favour of B in solving the same is 7 to 5. Then what is the odd in favour that the problem will be solved?

(a) $\frac{16}{5}$ (b) $\frac{5}{16}$ (c) $\frac{16}{21}$ (d) $\frac{5}{21}$

Solution:

Odds Against $(A) = \frac{4}{3}$

$P(A)$ Fail $= \frac{4}{7}$

Odds Favour (B) $= \frac{7}{5}$

$P(B)$ Fail $= \frac{5}{12}$

$P(\text{None Solved}) = \frac{4}{7} \times \frac{5}{12} = \frac{5}{21}$

$P(\text{Solved}) = 1 - \frac{5}{21} = \frac{16}{21}$

Odds in Favour $= \frac{16}{5}$

108. A bag contains 5 Green balls and 4 white balls. 3 balls are drawn. Find the odds against these being all green.

(a) $\frac{37}{5}$ (b) $\frac{5}{37}$ (c) $\frac{5}{42}$ (d) $\frac{37}{42}$

Solution:

$P(\text{Green}) = \frac{\binom{5}{3}}{\binom{9}{3}}$

$P(\text{Green}) = \frac{5!/(3!2!)}{9!/(3!6!)} = \frac{5}{42}$

Odds Against $= \frac{37}{5}$

109. A card is drawn from a pack of 52 cards and a gambler bets that it is a spade or an ace. Which one of the following are the odds against his winning this bet?

(a) $\frac{9}{4}$ (b) $\frac{4}{9}$ (c) $\frac{4}{13}$ (d) $\frac{13}{4}$

Solution:

Spades = 13, Aces = 4, Ace of Spades = 1

Total Cards = 13 + 4 − 1 = 16

$P = \frac{16}{52} = \frac{4}{13}$

Odds Against $= \frac{9}{4}$

110. If $P(A \cup B) = \frac{5}{9}$, $P\ (A' \cup B') = \frac{13}{27}$, and $P(A) = \frac{11}{18}$, then the odds against the event B are?

(a) $\frac{38}{17}$ (b) $\frac{7}{57}$ (c) $\frac{29}{25}$ (d) $\frac{47}{7}$

Solution:

$P\ (A' \cup B') = 1 - P(A \cap B)$

$P(A \cap B) = 1 - \frac{13}{27} = \frac{14}{27}$

$P(A \cup B) = P(A) + P(B) - P(A \cap B)$

$\frac{5}{9} = \frac{11}{18} + x - \frac{14}{27}$

$x = \frac{14}{27} - \frac{1}{18} = \frac{28-3}{54} = \frac{25}{54}$

Odds Against $B = \frac{29}{25}$

111. Given $P(A) = \frac{3}{5}$ and $P(B) = \frac{1}{5}$, find $P(A \cup B)$, if A and B are mutually exclusive events.

(a) $\frac{3}{5}$ (b) $\frac{4}{5}$ (c) $\frac{2}{5}$ (d) 0

Solution:

$P(A) = \frac{3}{5}$, $P(B) = \frac{1}{5}$

$P(A \cap B) = 0$ (Mutually Exclusive)

$P(A \cup B) = P(A) + P(B) - P(A \cap B)$

$P(A \cup B) = \frac{3}{5} + \frac{1}{5} - 0 = \frac{4}{5}$

112. A and B are events such that $P(A) = 0.42$, $P(B) = 0.48$. Find the value of $P\ (A') + P\ (B')$.

(a) 1.1 (b) 1.0 (c) 1.2 (d) 1.3

Solution:

$P\ (A') = 1 - P(A) = 1 - 0.42 = 0.58$

$P\ (B') = 1 - P(B) = 1 - 0.48 = 0.52$

$P\ (A') + P\ (B') = 0.58 + 0.52 = 1.10$

113. A and B are two events such that $P(A) = 0.54$, $P(B) = 0.69$, and $P(A \cap B) = 0.35$. Find $P\ (A' \cap B') + P\ (A' \cup B')$.
(a) 0.77 (b) 0.75 (c) 0.78 (d) 0.79

Solution:

$P(A \cup B) = P(A) + P(B) - P(A \cap B)$

$P(A \cup B) = 0.54 + 0.69 - 0.35 = 0.88$

$P\ (A' \cap B') = 1 - P(A \cup B) = 1 - 0.88 = 0.12$

$P\ (A' \cup B') = 1 - P(A \cap B) = 1 - 0.35 = 0.65$

$P\ (A' \cap B') + P\ (A' \cup B') = 0.12 + 0.65 = 0.77$

114. Let A and B be Independent Events with $P(A) = 0.3$ and $P(B) = 0.4$. Find $P(A \cup B) + P(A \cap B)$.
(a) 0.70 (b) 0.65 (c) 0.75 (d) 0.80

Solution:

When Two Events are independent, then

$P(A \cap B) = P(A) \times P(B)$

$P(A \cap B) = 0.3 \times 0.4 = 0.12$

$P(A \cup B) = P(A) + P(B) - P(A \cap B)$

$P(A \cup B) = 0.3 + 0.4 - 0.12 = 0.58$

$P(A \cup B) + P(A \cap B) = 0.58 + 0.12 = 0.70$

115. Given Two Independent Events A and B such that $P(A) = 0.3$, $P(B) = 0.6$. Find the value of $P\ (A' \cup B) - P\ (A \cap B')$.
(a) 0.75 (b) 0.76 (c) 0.78 (d) 0.74

Solution:

$P(A) = 0.3,\ P(B) = 0.6$

$P(A \cap B) = P(A) \times P(B) = 0.3 \times 0.6 = 0.18$

$P\ (A \cap B') = P(A) - P(A \cap B)$

$P\ (A \cap B') = 0.3 - 0.18 = 0.12$

$P\ (A' \cup B) = 1 - P\ (A \cap B') = 1 - 0.12 = 0.88$

$P\ (A' \cup B) - P\ (A \cap B') = 0.88 - 0.12 = 0.76$

116. If A and B are Events such that $P(A) = 0.6$, $P(B) = 0.3$, and $P(A \cap B) = 0.2$, Find $P(A \mid B)$ and $P(B \mid A)$.
(a) $\frac{1}{3}$ (b) $\frac{2}{3}$ (c) $\frac{1}{2}$ (d) $\frac{3}{4}$

Solution:

$P(A|B) = \dfrac{P(A \cap B)}{P(B)}$

$P(A|B) = \dfrac{0.2}{0.3} = \dfrac{2}{3}$

$P(B|A) = \dfrac{P(B \cap A)}{P(A)}$

$P(B|A) = \dfrac{0.2}{0.6} = \dfrac{1}{3}$

$P(A|B) - P(B|A) = \dfrac{2}{3} - \dfrac{1}{3} = \dfrac{1}{3}$

117. If A and B are Two Events such that $P(A) = 0.4$, $P(B) = 0.8$, and $P(B \mid A) = 0.6$, find $P(A \cup B)$.
(a) 0.86 (b) 0.96 (c) 0.69 (d) 0.68

Solution:

$P(B \mid A) = \dfrac{P(B \cap A)}{P(A)}$

$0.6 = \dfrac{P(B \cap A)}{0.4}$

$P(B \cap A) = 0.24$

$P(A \cup B) = P(A) + P(B) - P(A \cap B)$

$P(A \cup B) = 0.4 + 0.8 - 0.24 = 0.96$

118. Two Events E and F are Independent. If $P(E) = 0.3$, $P(E \cup F) = 0.5$, Then $P(E \mid F) - P(F \mid E)$ equals?

(a) $\frac{1}{70}$ (b) $\frac{2}{35}$ (c) $\frac{1}{56}$ (d) $\frac{4}{5}$

Solution:

$P(E) = 0.3,\ P(E \cup F) = 0.5$

Independent Events: $P(E \cap F) = P(E) \times P(F)$

$P(E \cap F) = 0.3 \times P(F)$

$P(E \cup F) = P(E) + P(F) - P(E \cap F)$

$0.5 = 0.3 + P(F) - 0.3 \times P(F)$

$0.7 \times P(F) = 0.2$

$P(F) = \frac{2}{7}$

$$P(E \mid F) - P(F \mid E) = \frac{0.3 \times P(F)}{P(F)} - \frac{0.3 \times P(F)}{0.3}$$

$$= \frac{3}{10} - \frac{2}{7}$$

$$= \frac{1}{70}$$

25

Statistics

Statistics - It is the discipline that involves collecting, organizing, analyzing, interpreting, and presenting data.
सांख्यिकी - यह वह अनुशासन है जिसमें डेटा एकत्र करना, व्यवस्थित करना, विश्लेषण करना, व्याख्या करना और प्रस्तुत करना शामिल है।

In this chapter, You will study about three measures, i.e., mean, median, and mode from ungrouped data to that of grouped data. We will also discuss the concept of cumulative frequency, the cumulative frequency distribution, and how to draw cumulative frequency curves, called ogives.
इस अध्याय में, आप तीन मापों के बारे में अध्ययन करेंगे, अर्थात, माध्य, माध्यिका और अवर्गीकृत डेटा से समूहीकृत डेटा का मोड। हम संचयी आवृत्ति की अवधारणा, संचयी आवृत्ति वितरण और संचयी आवृत्ति वक्र, जिन्हें तोरण कहा जाता है, कैसे बनाया जाए, इस पर भी चर्चा करेंगे।

Mean of Grouped Data / समूहीकृत डेटा का मतलब

The mean (or average) of observation, as we know, is the sum of the value of all the observations divided by the total number of observations. From Class IX, recall that if $x_1, x_2, \ldots, x_n$ are observations with respective frequencies $f_1, f_2, \ldots, f_n$, then this means observation x_1 occurs f_1 times, x_2 occurs f_2 times, and so on.
जैसा कि हम जानते हैं, अवलोकन का माध्य (या औसत), सभी अवलोकनों के मूल्य के योग को अवलोकनों की कुल संख्या से विभाजित करने पर प्राप्त होता है। कक्षा IX से, याद रखें कि यदि $x_1, x_2, \ldots, x_n$ संबंधित आवृत्तियों $f_1, f_2, \ldots, f_n$ के साथ अवलोकन हैं तो इसका मतलब है कि अवलोकन x_1 f_1 बार होता है, x_2 f_2 बार होता है, इत्यादि।

Now, the sum of the values of all the observations $= f_1x_1 + f_2x_2 + \cdots + f_nx_n$, and the number of observations $= f_1 + f_2 + \cdots + f_n$.
So, the mean $\bar{x}$ of the data is given by:

$$\bar{x} = \frac{f_1x_1 + f_2x_2 + \cdots + f_nx_n}{f_1 + f_2 + \cdots + f_n}$$

OR

$$\bar{x} = \frac{\sum_{i=1}^{n} f_ix_i}{\sum_{i=1}^{n} f_i}$$

अब, सभी अवलोकनों के मूल्यों का योग $= f_1x_1 + f_2x_2 + \cdots + f_nx_n$, और अवलोकनों की संख्या $= f_1 + f_2 + \cdots + f_n$।
इसलिए, डेटा का माध्य $\bar{x}$ निम्नलिखित है:

$$\bar{x} = \frac{f_1x_1 + f_2x_2 + \cdots + f_nx_n}{f_1 + f_2 + \cdots + f_n}$$

या

$$\bar{x} = \frac{\sum_{i=1}^{n} f_ix_i}{\sum_{i=1}^{n} f_i}$$

Median - It is a measure of central tendency which gives the value of the middle-most observation in the data. For finding the median of ungrouped data, we first arrange the data values of the observations in ascending order.
माध्यिका - यह केंद्रीय प्रवृत्ति का एक माप है जो डेटा में मध्यतम अवलोकन का मूल्य बताता है। अवर्गीकृत डेटा का माध्यिका ज्ञात करने के लिए, हम पहले प्रेक्षणों के डेटा मानों को आरोही क्रम में व्यवस्थित करते हैं।

(i) If the number of observations is odd:
यदि प्रेक्षणों की संख्या विषम है: Median $= \left(\frac{n+1}{2}\right)$ th term

(ii) If the number of observations is even:
यदि प्रेक्षणों की संख्या सम है: Median $= \frac{\left(\frac{n}{2}\right) \text{ th term} + \left(\frac{n}{2}+1\right) \text{ th term}}{2}$

(iii) Median formula for grouped data: Median of grouped data is the median of the data that is continuous and in the form of a frequency distribution.
समूहीकृत आँकड़ों के लिए माध्यिका सूत्र: Median $= l + \left[\frac{\frac{n}{2} - cf}{f}\right] h$

Where: n = total number of observations (total frequency) कुल आवृत्ति
cf = cumulative frequency of the preceding class

माध्यिका वर्ग से पहले वर्ग की संचयी आवृत्ति
f = frequency of the median class
वर्ग माध्यिका की आवृत्ति

h = size of each class
l = lower boundary of the median class
वर्ग माध्यिका की निचली सीमा

1. Find the mean for the data
1, 1, 1, 2, 4, 5, 5, 7, 8, 12.
डेटा 1, 1, 1, 2, 4, 5, 5, 7, 8, 12 का माध्य ज्ञात कीजिए।

Solution:

Sum of Data = $1+1+1+2+4+5+5+7+8+12 = 46$

$\text{Mean} = \frac{46}{10} = 4.6$

2. If the mean of the following data is 11, find the value of 'k'.
यदि निम्नलिखित डेटा का माध्य 11 है, तो 'k' का मान ज्ञात करें।
11, 19, 5, 10, k, 13, 12, 8, 15, 14

Solution:

Mean = 11

Sum of Data = $11+19+5+10+k+13+12+8+15+14$

Sum = $107 + k$

$11 = \frac{107+k}{10}$

$110 = 107 + k$

$k = 3$

3. Mean of first ten odd natural numbers is:
प्रथम दस विषम प्राकृतिक संख्याओं का माध्य है:

Solution:

Sum of First Ten Odd Natural Numbers = $1+3+5+7+9+11+13+15+17+19$

Sum = 100

$\text{Mean} = \frac{100}{10} = 10$

4. Find the value of r such that the mean of the first r odd natural numbers is $\frac{r^2}{16}$.
r का मान इस प्रकार ज्ञात करें कि पहली r विषम प्राकृतिक संख्याओं का माध्य हो $\frac{r^2}{16}$।

Solution:

Sum of First r Odd Natural Numbers = $1+3+5+7\ldots r$

Formula: Sum of First r Odd Natural Numbers = r^2

$S = r^2$

$\text{Mean} = \frac{r^2}{r} = \frac{r^2}{16}$

$r = 16$

5. The mean of three numbers is 32. The range of this data set is 28 while the difference between the two smallest numbers is 8. The greatest of the three numbers is:
तीन संख्याओं का माध्य 32 है। इस डेटा सेट की सीमा 28 है जबकि दो सबसे छोटी संख्याओं के बीच का अंतर 8 है। तीन संख्याओं में से सबसे बड़ी संख्या है।

Solution:

Let the numbers be $x,\ x+8,\ x+28$:

$\text{mean} = \frac{3x+36}{3} = 32$

$x + 12 = 32$

$x = 20$

$x + 28 = 20 + 28 = 48$

6. 10 is the mean of a set of 7 observations and 5 is the mean of another set of 3 observations. The mean of the combined set is:
7 प्रेक्षणों के एक सेट का माध्य 10 है और 3 प्रेक्षणों के दूसरे सेट का माध्य 5 है। संयुक्त समुच्चय का माध्य है:

Solution:

$$\text{Combined mean} = \frac{10 \cdot 7 + 5 \cdot 3}{10}$$

$$\text{Mean} = \frac{70 + 15}{10}$$

$$\text{Mean} = \frac{85}{10} = 8.5$$

7. The marks obtained by 30 students of Class X of a certain school in a Mathematics paper consisting of 100 marks are presented in the table below. Find the mean of the marks obtained by the students.

100 अंकों के गणित के पेपर में एक निश्चित स्कूल के दसवीं कक्षा के 30 छात्रों द्वारा प्राप्त अंक नीचे दी गई तालिका में प्रस्तुत किए गए हैं। विद्यार्थियों द्वारा प्राप्त अंकों का माध्य ज्ञात कीजिए।

Marks obtained x	10	20	36	40	50	56
No. of students f	1	1	3	4	3	2

60	70	72	80	88	92	95
4	4	1	1	2	3	1

Solution:

Recall that to find the mean marks, we require the product of each x with the corresponding frequency f_i. So, let us put them in a column.

Marks obtained (x_i)	Number of students (f_i)	$f_i x_i$
10	1	10
20	1	20
36	3	108
40	4	160
50	3	150
56	2	112
60	4	240
70	4	280
72	1	72
80	1	80
88	2	176
92	3	276
95	1	95
Total:	$\sum f_i = 30$	$\sum f_i x_i = 1779$

$$\bar{x} = \frac{\sum f_i x_i}{\sum f_i} = \frac{1779}{30} = 59.3$$

Therefore, the mean marks obtained is 59.3.

8. The table given below shows the number of students having obtained different marks.

नीचे दी गई तालिका विभिन्न अंक प्राप्त करने वाले छात्रों की संख्या दर्शाती है।

Marks	Number of Students
9-11	6
11-13	5
13-15	2
15-17	2
17-19	5

What is the mean marks per student?

प्रति छात्र का औसत अंक क्या है?

Solution:

Marks	No. of Stu.	x_i	$f_i x_i$
9-11	6	$(9+11)/2 = 10$	60
11-13	5	$(11+13)/2 = 12$	60
13-15	2	$(13+15)/2 = 14$	28
15-17	2	$(15+17)/2 = 16$	32
17-19	5	$(17+19)/2 = 18$	90
Total:	$\sum f_i = 20$		$\sum f_i x_i = 270$

$$\text{The mean marks per student} = \frac{\sum f_i x_i}{\sum f_i} = \frac{270}{20} = 13.5$$

9. Find the median value of the following ungrouped data: 26, 12, 15, 42, 36, 16.

निम्नलिखित अवर्गीकृत डेटा का माध्य मान ज्ञात कीजिए: 26, 12, 15, 42, 36, 16

Solution:

To find the median of ungrouped data, we first arrange the data in ascending order:

12, 15, 16, 26, 36, 42

There are an even number of data, so:

$$\text{Median} = \frac{16 + 26}{2} = 21$$

10. The median of observations $k - \frac{3}{2}$, $k + 2$, $k - 1$, $k + 4$, $k + \frac{1}{2}$, $k - 3$, $k + 4$.

Solution:

Let $k = 1$:
$1-\frac{3}{2}, 1+2, 1-1, 1+4, 1+\frac{1}{2}, 1-3, 1+4$
$-\frac{1}{2}, 3, 0, 5, \frac{3}{2}, -2, \frac{11}{2}$ Arranging in ascending order: $-2, -\frac{1}{2}, 0, \frac{3}{2}, 3, 5, 5.5$
Hence, Median $= \frac{3}{2} = k + \frac{1}{2}$

11. Given data is arranged in an ascending order. If the median is 10, then find the value of p.
दिए गए डेटा को आरोही क्रम में व्यवस्थित किया गया है। यदि माध्यिका 10 है, तो p का मान ज्ञात कीजिए।
$3, 5, 6, 2p + 3, 3p + 2, 15, 25, 51$

Solution:

$3, 5, 6, 2p + 3, 3p + 2, 15, 25, 51$
Median of the given data = 10

$$\frac{(2p+3)+(3p+2)}{2} = 10$$

$$\frac{5p+5}{2} = 10$$

$$5(p+1) = 20$$

$$p+1 = 4$$

$$p = 3$$

12. In the data given below, the number of sixes and the batsmen who hit them has been shown.
नीचे दिए गए आंकड़ों में छक्कों की संख्या और उन्हें मारने वाले बल्लेबाजों की संख्या दर्शाई गई है।

Number of Sixes	Number of Batsman
1	2
2	3
3	1
4	3
5	2

What is the median of the number of sixes?

Solution:

No. of Sixes (x_i)	No. of Batsman (f_i)	Cumulative Frequency (cf)
1	2	2
2	3	$2+3=5$
3	1	$5+1=6$
4	3	$6+3=9$
5	2	$9+2=11$
	Cumulative mean	$11/2 = 5.5$

Since 5.5 lies in the 6th cumulative frequency zone, the Median of sixes will be 3.

13. A survey regarding the heights (in cm) of 51 girls of Class X of a school was conducted and the following data was obtained:
एक स्कूल की दसवीं कक्षा की 51 लड़कियों की ऊंचाई (सेमी में) के संबंध में एक सर्वेक्षण किया गया और निम्नलिखित डेटा प्राप्त किया गया:

Height (in cm)	Number of Girls
Less than 140	4
Less than 145	11
Less than 150	29
Less than 155	40
Less than 160	46
Less than 165	51

find the maximum height.

Solution:

To calculate the median height, we need to find the class intervals and their corresponding frequencies.

Class Intervals	Frequency	Cumulative Frequency
Below 140	4	4
140-145	7	11
145-150	18	29
150-155	11	40
155-160	6	46
160-165	5	51

Now, $n = 51$. So, $\frac{n}{2} = \frac{51}{2} = 25.5$.
This observation lies in the class $145 - 150$.
Then:
l (the lower limit) = 145,
cf (the cumulative frequency of the class preceding $145 - 150$) = 11,
f (the frequency of the median class $145 - 150$) = 18,

h (the class size) = 5.

Using the formula:

$$\text{Median} = l + \left(\frac{\frac{n}{2} - cf}{f}\right) \cdot h$$

$$\text{Median} = 145 + \left(\frac{25.5 - 11}{18}\right) \cdot 5$$

$$\text{Median} = 145 + \frac{72.5}{18} = 149.03$$

So, the median height of the girls is 149.03 cm.

14. Find the Median.

Class Interval	Frequency
40-45	2
45-50	3
50-55	8
55-60	6
60-65	6
65-70	3
70-75	2

Solution:

C.I	f	c.f
40-45	2	2
45-50	3	5
50-55	8	13 → C.F.
55-60	6	19 (Median class)
60-65	6	25
65-70	3	28
70-75	2	30

$$M = l + \left(\frac{\frac{n}{2} - cf}{f}\right) \cdot h = 55 + \left(\frac{30/2 - 13}{6}\right) \cdot 5$$

$$= 55 + \frac{6}{3}$$

$$= 56.66$$

Mode: A mode is that value among the observations which occurs most often, that is, the value of the observation having the maximum frequency.

In a grouped frequency distribution, it is not possible to determine the mode by looking at the frequencies. Here, we can only locate a class with the maximum frequency, called the modal class. The mode is a value inside the modal class, and is given by the formula:

$$\text{mode} = l \left(\frac{f_1 - f_0}{f_1 - f_0 - f_2}\right) \times h$$

where l = lower limit of the modal class,

h = size of the class interval (assuming all class sizes to be equal),

f_1 = frequency of the modal class,

f_0 = frequency of the class preceding the modal class,

f_2 = frequency of the class succeeding the modal class.

15. Mode of the given data is:
दिए गए डेटा का मोड है।
2, 5, 5, 7, 2, 6, 8, 6, 9, 6

Solution:

The mode is the number that is repeated most often. Here 6 is repeated the maximum number of times (3 times), so the mode will be 6.

16. If the mode of the following data is 12, then find the value of k: 11, 15, 8, 9, k, 11, 12, 12, 15, 14.

Solution:

For mode = 12, the repetition of 12 should be at least 3 times because 11 is repeated 2 times.

So, $k = 12$.

17. A survey conducted on 20 households in a locality by a group of students resulted in the following frequency table for the number of family members in a household:
छात्रों के एक समूह द्वारा एक इलाके में 20 घरों पर किए गए सर्वेक्षण के परिणामस्वरूप एक घर में परिवार के सदस्यों की संख्या के लिए निम्नलिखित आवृत्ति तालिका प्राप्त हुई:

Family Size	Number of Families
1-3	7
3-5	8
5-7	2
7-9	2
9-11	1

Find the mode of this data.

Solution:

Here, the maximum class frequency is 8, and the class corresponding to this frequency is 3 – 5. So, the modal class is 3 – 5.

Now

modal class = 3-5, lower limit (l) of modal class = 3, class size (h) = 2

frequency (f_1) of the modal class = 8,

frequency (f_0) of class preceding the modal class = 7,

frequency (f_2) of class succeeding the modal class = 2.

Now, let us substitute these values in the formula:

$$\text{mode} = l\left(\frac{f_1 - f_0}{f_1 - f_0 - f_2}\right) \cdot h$$

$$= 3 + \left(\frac{8-7}{2 \cdot 8 - 7 - 2}\right) \cdot 2 = 3 + \frac{2}{7} = 3.286$$

Therefore, the mode of the data above is 3.286.

18. The mode of the following data is 36. What is the value of x?
निम्नलिखित डेटा का मोड 36 है। x का मान क्या है?

Class	Frequency
0-10	13
10-20	10
20-30	10
30-40	16
40-50	x
50-60	8

$$\text{mode} = l\left(\frac{f_1 - f_0}{f_1 - f_0 - f_2}\right) \cdot h$$

Where: l = Lower limit of the modal class
f_1 = Frequency of modal class
f_0 = Frequency of the previous class
f_2 = Frequency of the next class
h = Class size

Solution:

$$36 = 30 + \left(\frac{16 - 10}{2(16) - 10 - x}\right) \cdot 10$$

$$36 - 30 = \frac{6}{22 - x} \cdot 10$$

$$6 = \frac{60}{22 - x}$$

$$22 - x = 10$$

$$x = 12$$

Variance and Standard Deviation

Variance (विचरण, प्रसरण):

Variance is the expected value of the squared variation of a random variable from its mean value. OR Variance is the measure of how data points differ from the mean.

$$\text{Variance}(X) = \frac{\sum_{i=1}^{n}(x_i - \bar{x})^2}{n}$$

Where: $\bar{x}$ = Mean of the observations, $\bar{x} = \frac{x_1 + x_2 + x_3 + \ldots + x_n}{n}$, n = Number of observations. **Standard Deviation (मानक विचलन):**

Standard deviation is the positive square root of the variance. It is abbreviated as SD and denoted by σ. It tells how much the values deviate from the mean value.

$$\text{Standard deviation} = \sqrt{\text{Variance}}$$

Formula:

$$\sigma = \sqrt{\frac{\sum_{i=1}^{n}(x_i - \bar{x})^2}{n}}$$

Where x_i = Value in the data distribution,
$\bar{x}$ = Mean of the observations,
n = Number of observations.

19. Find the variance of the following data points: 6, 7, 5, 9, 12, 15.

निम्नलिखित डेटा बिंदुओं का विचरण ज्ञात कीजिए: 6, 7, 5, 9, 12, 15।

Solution:

Given data: 6, 7, 5, 9, 12, 15

Mean $(\bar{x}) = \frac{6+7+5+9+12+15}{6} = \frac{54}{6} = 9$

Number of terms $(n) = 6$

Variance: Variance $= \frac{\sum(x - \bar{x})^2}{n}$

$= \frac{(6-9)^2 + (7-9)^2 + (5-9)^2 + (9-9)^2 +}{6}$
$\frac{(12-9)^2 + (15-9)^2}{6}$

$= \frac{9+4+16+0+9+36}{6} = \frac{74}{6} = \frac{37}{3}$

20. Calculate the standard deviation for the following data: 3, 4, 5, 6, 7.

निम्नलिखित डेटा के लिए मानक विचलन की गणना करें: 3, 4, 5, 6, 7

Solution:

Given data: 3, 4, 5, 6, 7

Mean $(\bar{x}) = \frac{3+4+5+6+7}{5} = 5$

Total number of terms $(n) = 5$

$\text{SD} = \sqrt{\frac{\sum(x - \bar{x})^2}{n}}$

$= \sqrt{\frac{(3-5)^2 + (4-5)^2 + (5-5)^2 + (6-5)^2}{5}}$
$\sqrt{\frac{+(7-5)^2}{5}}$

$= \sqrt{\frac{2^2 + 1^2 + 0 + 1^2 + 2^2}{5}} = \sqrt{\frac{10}{5}} = \sqrt{2}$

21. Find the variance and standard deviation for the following data:
निम्नलिखित डेटा के लिए विचरण और मानक विचलन ज्ञात करें:

x_i	4	8	11	17	20	24	32
f_i	3	5	9	5	4	3	1

Solution:

Presenting the data in tabular form, we get:

x_i	f_i	f_ix_i	$x_i - \bar{x}$	$(x_i - \bar{x})^2$	$f_i(x_i - \bar{x})^2$
4	3	12	-10	100	300
8	5	40	-6	36	180
11	9	99	-3	9	81
17	5	85	3	9	45
20	4	80	6	36	144
24	3	72	10	100	300
32	1	32	18	324	324
Total:	30	420	N=30	$\sum_{r=1}^{7} f_ix_i = 420$	$\sum_{r=1}^{7} f_i(x_i - \bar{x})^2 = 1374$

Therefore $\bar{x} = \frac{\sum_{r=1}^{7} f_ix_i}{N} = \frac{420}{30} = 14$

Variance: $\sigma^2\ \sigma^2 = \frac{1}{N}\sum_{r=1}^{7} f_i(x_i - \bar{x})^2 = \frac{1}{30} \cdot 1374 = 45.8$

Standard Deviation: $\sigma\ \sigma = \sqrt{\sigma^2} = \sqrt{45.8} = 6.77$

Miscellaneous

22. If the median and mean are 36 and 35 respectively, find the mode.
यदि माध्यिका और माध्य क्रमशः 36 और 35 हैं, तो बहुलक ज्ञात कीजिए।

Solution:

$3 \cdot \text{Median} = \text{Mode} + 2 \cdot \text{Mean}$

$\Rightarrow 3 \cdot 36 = \text{Mode} + 2 \cdot 35$

$\Rightarrow 108 = \text{Mode} + 70$

$\text{Mode} = 108 - 70 = 38$

23. If mean is 40 and standard deviation is 5, then C.V (Coefficient of variation) is: / यदि माध्य 40 है और मानक विचलन 5 है तो C.V (भिन्नता का गुणांक) है।

Solution:

We know the formula for Coefficient of Variation: $\text{C.V} = \frac{\sigma}{\mu} \cdot 100$

Where σ = standard deviation of the given data set and μ = mean of the given data set Now mean is 40 and standard deviation is 5 then C.V (Coefficient of variation) is

$\text{C.V} = \frac{5}{40} \cdot 100\%$

$= \frac{1}{8} \cdot 100\% = 12.5\%$

24. In a frequency distribution, the mid value of a class is 12 and its width is 6. The lower limit of the class is:
बारंबारता वितरण में, एक वर्ग का मध्य मान 12 है और इसकी चौड़ाई 6 है। वर्ग की निचली सीमा है:

Solution:

In a frequency distribution, the mid value of a class is 12 and its width is 6.

As we know:

$\text{Lower limit} = \text{Mid value} - \frac{\text{Width}}{2}$

So, the lower limit of the class is $= 12 - \frac{6}{2} = 12 - 3 = 9$

25. The following are the weights (in kg) of 25 students:
25 छात्रों के वजन (किलो में) निम्नलिखित हैं।
58, 55, 53, 50, 53, 51, 52, 54, 53, 52, 54, 53, 58, 53, 59, 55, 53, 52, 51, 54, 53, 59, 55, 53, 52

What is the range of the given data?
दिए गए डेटा की सीमा क्या है?

Solution:

The weights of 25 students are:

58, 55, 53, 50, 53,

51, 52, 54, 53, 52, 54, 53, 58, 53, 59, 55, 53,

52, 51, 54, 53, 59, 55, 53, 52.

So, the range of the given data is (Max weight - Min weight) = (59 - 50) = 9

26

Data Interpretation

1. The given table show the production on units of different product in different months.
दी गई तालिका विभिन्न महीनों में विभिन्न उत्पादों की इकाइयों पर उत्पादन दर्शाती है।

Product	March	June	September	December
Z	15	18	12	19
Y	18	13	17	8
X	14	17	19	16
W	12	14	11	15

Y's total production is what percentage of Z's total production?
Y का कुल उत्पादन Z के कुल उत्पादन का कितना प्रतिशत है?
(a) 88.5% (b) 89.5%
(c) 87.5% (d) 114.28%

Solution:

Y's total production $= 18+13+17+8 = 56$

Z's total production $= 15+18+12+19 = 64$

$\therefore \frac{56}{64} \times 100 = 87.5\%$

2. The given table shows the number of new employees who joined in different categories of employees in an organization and also the number of employees from these categories who left the organization every year since the foundation of the organization in 2015. Study the table and answer the questions that follow:
दी गई तालिका एक संगठन में विभिन्न श्रेणियों के कर्मचारियों में शामिल होने वाले नए कर्मचारियों की संख्या और 2015 में संगठन की स्थापना के बाद से हर साल संगठन छोड़ने वाले इन श्रेणियों के कर्मचारियों की संख्या को दर्शाती है। तालिका का अध्ययन करें और निम्नलिखित प्रश्नों के उत्तर दें

Year	Business analysts		Accountants		Sale representatives		Peons	
	Joined	Left	Joined	Left	Joined	Left	Joined	Left
2015	80	0	35	0	50	0	73	0
2016	63	21	26	16	46	20	34	23
2017	78	28	15	5	38	15	48	29
2018	42	39	20	38	34	0	14	12
2019	67	44	14	24	22	30	21	30

What was the total number of accountants working in the organization in the year 2019?
वर्ष 2019 में संगठन में कार्यरत लेखाकारों की कुल संख्या कितनी थी?
(a) 110 (b) 83
(c) 27 (d) 193

Solution:

Required answer $= (35 + 26 + 15 + 20 + 14) - (0 + 16 + 5 + 38 + 24) = 27.$

3. The following tables shows the production of different types of two wheelers from 1993 to 1998.
निम्नलिखित तालिका 1993 से 1998 तक विभिन्न प्रकार के दो पहिया वाहनों के उत्पादन को दर्शाती है।

Number of two wheelers (in 1000s)						
Type	1993	1994	1995	1996	1997	1998
A	36	34	40	35	37.5	40
B	20	22	25	23	19.5	18
C	14	22	16	25	29	35
D	60	62	67.5	75	76	80
E	40	45	48	50	80	105
F	45	52	55	60	57.5	56
Total	215	237	251.5	268	299.5	334

What is the percentage in the total production of all types of two wheelers in 1998 in comparison to 1996 (rounded off to the nearest integer)?
1996 की तुलना में 1998 में सभी प्रकार के दो पहिया वाहनों के कुल उत्पादन का प्रतिशत क्या है (निकटतम पूर्णांक तक पूर्णांकित)?

(a) 27% (b) 25%
(c) 24% (d) 26%

Solution:

Total production in 1996 = 268

Total production in 1998 = 334

Percentage increase = $\frac{66}{268} \times 100 = 24.62 \approx 25\%$

4. The following table shows the production of different types of two wheelers from 1993 to 1998.
निम्नलिखित तालिका 1993 से 1998 तक विभिन्न प्रकार के दो पहिया वाहनों के उत्पादन को दर्शाती है।

Number of two wheelers (in 1000s)						
Year Type	1993	1994	1995	1996	1997	1998
A	36	34	40	35	37.5	40
B	20	20	22	23	19.5	18
C	14	22	16	25	29	35
D	60	62	67.5	75	76	80
E	40	45	48	50	80	105
F	45	52	55	60	57.5	56
Total	215	237	251.5	268	299.5	334

In which year is the percentage increase in the total production minimum in comparison to the previous year? किस वर्ष कुल उत्पादन में प्रतिशत वृद्धि पिछले वर्ष की तुलना में न्यूनतम है?

(a) 1994 (b)1995
(c) 1996 (d) 1997

Solution:

$\% \uparrow / \downarrow$ in 1994 = $\frac{22}{215} \times 100 = 10.23\%$

$\% \uparrow / \downarrow$ in 1995 = $\frac{14.5}{237} \times 100 = 6.11\%$

$$\% \uparrow / \downarrow \text{ in } 1996 = \frac{16.5}{251.5} \times 100 = 6.56\%$$

$$\% \uparrow / \downarrow \text{ in } 1997 = \frac{31.5}{268} \times 100 = 11.75\%$$

Hence, 1995 is the correct option.

5. The pie chart given below shows the number of television sold by 8 different companies. The total number of television sold by all these 8 companies is 21000. The number of television sold by a particular company is shown as a percent of total number of television sold by all these 8 companies.

नीचे दिया गया पाई चार्ट 8 अलग-अलग कंपनियों द्वारा बेचे गए टेलीविज़न की संख्या दर्शाता है। इन सभी 8 कंपनियों द्वारा बेचे गए टेलीविज़न की कुल संख्या 21000 है। किसी विशेष कंपनी द्वारा बेचे गए टेलीविज़न की संख्या को इन सभी 8 कंपनियों द्वारा बेचे गए टेलीविज़न की कुल संख्या के प्रतिशत के रूप में दर्शाया गया है।

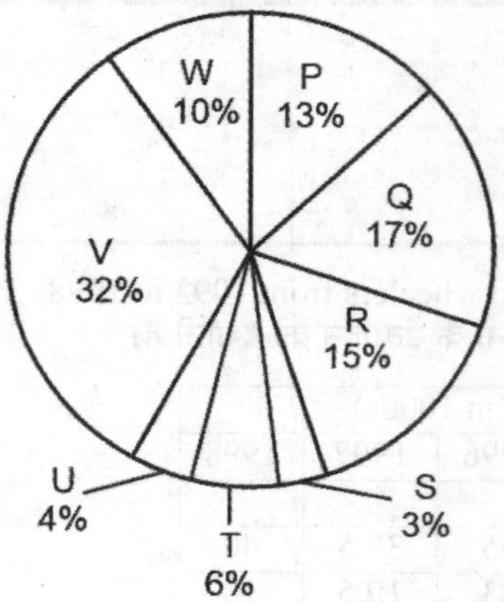

J1 = The value of average number of television sold by P and W

J2 = The value of average number of television sold by Q, R and V.

What is the value of $J2 - J1$?

(J2 - J1) का मान क्या है?

(a) 5630 (b) 2046

(c) 11025 (d) 3260

Solution:

$$J_1 = \frac{10+13}{2} = \frac{23}{2}\%$$

$$J_2 = \frac{17+15+3+32}{4} = \frac{67}{4}\%$$

$$J_2 - J_1 = \frac{67}{4} - \frac{23}{2} = \frac{21}{4}\%$$

$$\therefore 21000 \times \frac{21}{4 \times 100} = 11025$$

6. The table given below shows the expenditure of two companies in different years.

नीचे दी गई तालिका दो कंपनियों के विभिन्न वर्षों में व्यय को दर्शाती है।

years	Companies	
	L	M
P	200	500
Q	150	350
R	300	250
S	100	150
T	400	100

J1 = the total expenditure of company L in all the 5 years

J2 = the total expenditure of company M in all the 5 years

What is the value of J1:J2?

J1:J2 का मान क्या है?

(a) 27 : 23 (b) 21 : 29

(c) 23 : 27 (d) 29 : 23

Solution:

$$J1 = 200 + 150 + 300 + 100 + 400 = 1150$$

$$J2 = 500 + 350 + 250 + 150 + 100 = 1350$$

$$J1 : J2 = 1150 : 1350 = 23 : 27$$

7. The pie chart given below shows the number of cars sold by 8 different companies. The total number of cars sold by all these 8 companies are 10,000. Number of cars sold by a particular company is shown as a percent of total number of cars sold by all these 8 companies.

नीचे दिया गया पाई चार्ट 8 अलग-अलग कंपनियों द्वारा बेची गई कारों की संख्या दर्शाता है। इन सभी 8 कंपनियों द्वारा बेची गई कारों की कुल संख्या 10,000 है। किसी विशेष कंपनी द्वारा

बेची गई कारों की संख्या को इन सभी 8 कंपनियों द्वारा बेची गई कारों की कुल संख्या के प्रतिशत के रूप में दर्शाया गया है।

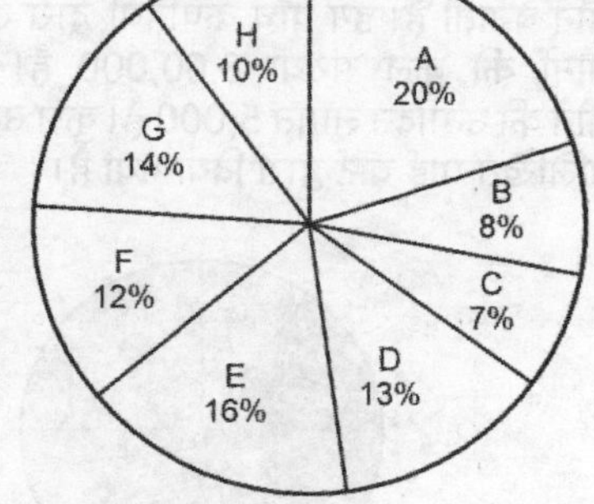

Which of the following statement is correct?
निम्नलिखित में से कौन सा कथन सही है?

I. The average number of cars sold by company B, F, and H are 100
I. कंपनी B, F और H द्वारा बेची गई कारों की औसत संख्या 100 है II. The ratio of number of cars sold by A to the number of cars sold by E are 4:5
II. A द्वारा बेची गई कारों की संख्या का E द्वारा बेची गई कारों की संख्या से अनुपात 4 : 5 है

(a) only I (b) only II
(c) neither I nor II (d) both I and II

Solution:

I. average number of Cars sold by B, F, and H

are $= \dfrac{(8+12+10)\times 10000}{100\times 3} = 1000$

II. $A : E$

20% of 10000 : 16% of 10000 = 5:4

Hence, neither I nor II is correct.

8. The given table shows the consumption of grains in terms of kilogram of a small village per day over the given set of years. Study the table and answer the question that follows
दी गई तालिका दिए गए वर्षों में प्रति दिन एक छोटे से गाँव में किलोग्राम के हिसाब से अनाज की खपत को दर्शाती है। तालिका का अध्ययन करें और नीचे दिए गए प्रश्न का उत्तर दें

Year	Rice	Wheat	Maize	Barley
1998 – 200	181	50	35	26
2003 – 2005	167	44	39	17
2008 – 2010	169	38	49	25
2013 – 2015	150	45	51	21
2018 – 2020	147	54	63	24

For wheat grains, for the given set of years, the highest consumption is approximately how much percent of that of all grains consumption per day in such a set of years?
दिए गए वर्षों में गेहूँ के दानों की अधिकतम खपत, ऐसे वर्षों में प्रतिदिन सभी अनाजों की खपत का लगभग कितना प्रतिशत है?

(a) 13.52% (b) 18.75%
(c) 15.32% (d) 17.85%

Solution:

Year	Total Consumption	Wheat	%
1998 – 200	292	50	17.12 %
2003 – 2005	267	44	16.48 %
2008 – 2010	281	38	13.52 %
2013 – 2015	267	45	16.85 %
2018 – 2020	288	54	18.75 %

Hence, highest consumption = 18.75%

9. Study the given pie-chart and answer the question that follows. The pie-chart represents the spending of a state on various sports during a particular year.
दिए गए पाई-चार्ट का अध्ययन करें और नीचे दिए गए प्रश्नों के उत्तर दें। यह पाई-चार्ट एक विशेष वर्ष के दौरान विभिन्न खेलों पर एक राज्य के खर्च को दर्शाता है।

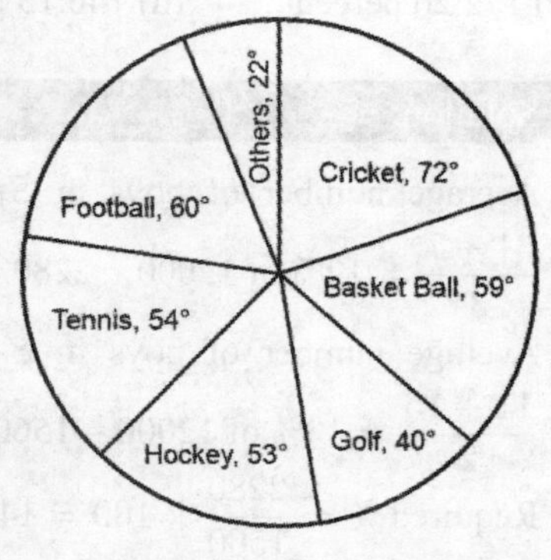

How much percent more is spent on cricket than on tennis (correct up to 2 decimal places)?
क्रिकेट पर टेनिस की तुलना में कितना प्रतिशत अधिक खर्च किया जाता है (2 दशमलव स्थानों तक सही)?

(a) 33.33% (b) 32.75%
(c) 30.35% (d) 28.25%

Solution:

$$\text{Required } \% = \frac{72 - 54}{54} \times 100 = 33.33\%$$

10. The pie chart given below shows the number of boys in 8 schools. The total number of boys in all these 8 schools is 12000. Number of boys in a particular school is shown as a percent of total number of boys in all these 8 schools.
नीचे दिया गया पाई चार्ट 8 स्कूलों में लड़कों की संख्या दर्शाता है। इन सभी 8 स्कूलों में लड़कों की कुल संख्या 12000 है। किसी विशेष स्कूल में लड़कों की संख्या इन सभी 8 स्कूलों में लड़कों की कुल संख्या के प्रतिशत के रूप में दर्शाई गई है।

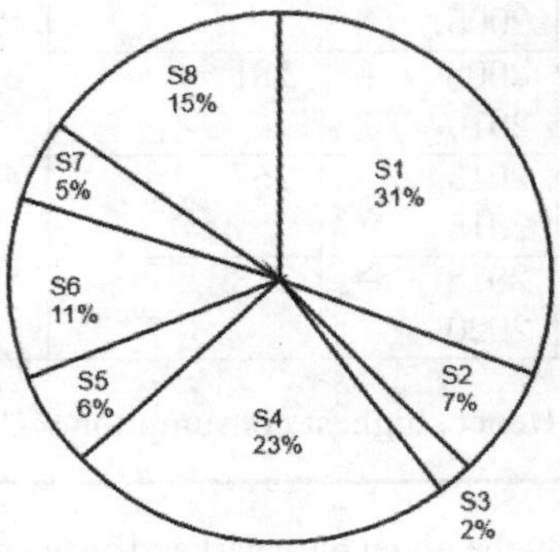

The average number of boys in S1 and S2 is what percent of the average number of boys in S6 and S8?
S 1 और S 2 में लड़कों की औसत संख्या S 6 और S 8 में लड़कों की औसत संख्या का कितना प्रतिशत है?
(a) 133.33 percent (b) 192.24 percent
(c) 172.26 percent (d) 146.15 percent

Solution:

Average number of boys in S_1 and S_2 =

$$\frac{31+7}{2} = 19\% \text{ of } 12000 = 2280$$

Average number of boys in S_6 and S_8 =

$$\frac{11+15}{2} = 13\% \text{ of } 12000 = 1560$$

$$\text{Required } \% = \frac{2280}{1560} \times 100 = 146.15\%$$

11. Five different companies, A, B, C, D, and E, make robotic toys. The total number of robotic toys produced by these five companies is 8,00,000. The cost of production of each robotic toy is 5,000. The distribution of the total production is given by the following pie chart.
पाँच अलग-अलग कंपनियाँ A, B, C, D और E रोबोटिक खिलौने बनाती हैं। इन पाँच कंपनियों द्वारा उत्पादित रोबोटिक खिलौनों की कुल संख्या 8,00,000 है। प्रत्येक रोबोटिक खिलौने की उत्पादन लागत 5,000 है। कुल उत्पादन का वितरण निम्नलिखित पाई चार्ट द्वारा दिया गया है।

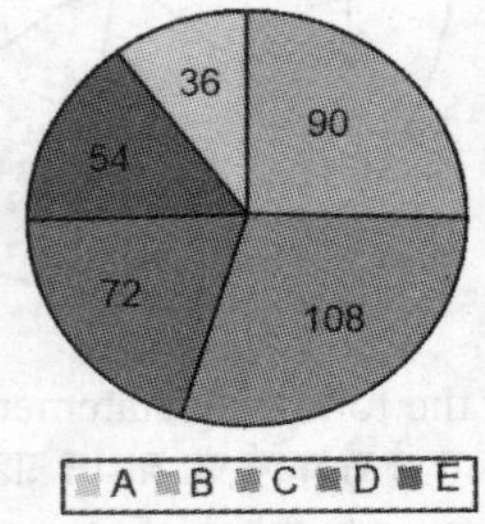

What is the total cost of production of the given item by companies A and E together?
कंपनी A और E द्वारा दी गई वस्तु के उत्पादन की कुल लागत क्या है?
(a) 190 crores (b) 85 crores
(c) 96 crores (d) 140 crores

Solution:

Total cost of production by A and E =

$$\frac{(90+36)}{360} \times 800000 \times 5000 = 140 \text{ crores}$$

12. The given bar graph shows the data of production of pulses (in lakh tonnes) by three different countries X, Y, and Z over the years. Study the bar graph and answer the question that follows.
दिया गया बार ग्राफ तीन अलग-अलग देशों X, Y और Z द्वारा पिछले कुछ वर्षों में दालों के उत्पादन (लाख टन में) के आंकड़ों को दर्शाता है। बार ग्राफ का अध्ययन करें और नीचे दिए गए प्रश्नों के उत्तर दें।

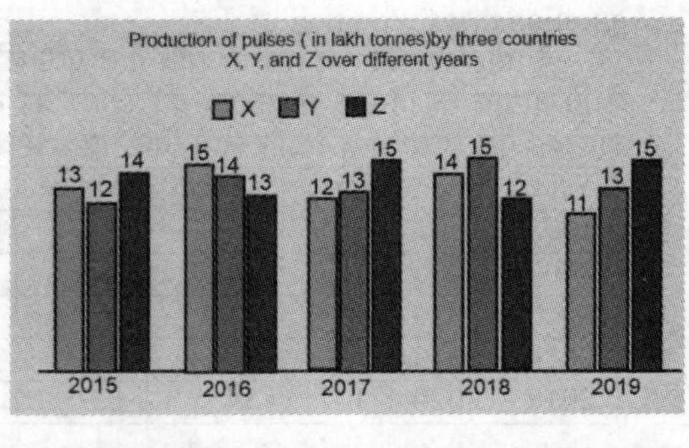

For which country is the average production (in lakh tonnes) for five years the maximum?

किस देश का पांच वर्षों का औसत उत्पादन (लाख टन में) अधिकतम है?

(a) X (b) Z
(c) Y (d) Y and Z both

Solution:

Average production of X = $\frac{13+15+12+14+11}{5} = \frac{65}{5} = 13$

Average production of Y = $\frac{12+14+13+15+13}{5} = \frac{67}{5} = 13.4$

Average production of Z = $\frac{14+13+15+12+15}{5} = \frac{69}{5} = 13.8$

Hence, Z has the maximum production.

13. The table given below shows the marks obtained by two students in 6 subjects.

नीचे दी गई तालिका दो विद्यार्थियों द्वारा 6 विषयों में प्राप्त अंकों को दर्शाती है।

subject	Students	
	A	B
P	180	90
Q	240	420
R	360	390
S	300	210
T	60	270
U	330	120

What is the difference between the total marks obtained by A and B in all the 6 subjects?

सभी 6 विषयों में A और B द्वारा प्राप्त कुल अंकों के बीच क्या अंतर है?

(a) 40 (b) 30
(c) 20 (d) 50

Solution:

Total marks obtained by A = 180 + 240 + 360 + 300 + 60 + 330 = 1470

Total marks obtained by B = 90 + 420 + 390 + 210 + 270 + 120 = 1500

Difference = 1500 − 1470 = 30

14. The pie chart given below shows the sales of 7 different companies. The total sales of all these 7 companies is 2160. The sale of a particular company is shown in terms of degree with respect to the total sales of all these 7 companies.

नीचे दिया गया पाई चार्ट 7 अलग-अलग कंपनियों की बिक्री दर्शाता है। इन सभी 7 कंपनियों की कुल बिक्री 2160 है। किसी विशेष कंपनी की बिक्री को इन सभी 7 कंपनियों की कुल बिक्री के संबंध में डिग्री के संदर्भ में दिखाया गया है।

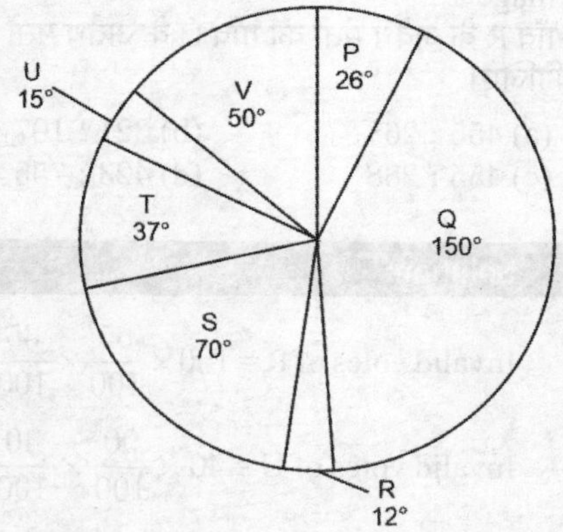

The total sales of P, R, and S are how much percent less than the total sales of Q, T, and U?

P, R और S की कुल बिक्री Q, T और U की कुल बिक्री से कितने प्रतिशत कम है?

(a) 56.42% (b) 24.42%
(c) 46.53% (d) 32.32%

Solution:

Total sales of P, R, S = 26 + 12 + 70 = 108

Total sales of Q, T, U = 150 + 37 + 15 = 202

Required % = $\frac{202 - 108}{202} \times 100 = 46.53\%$

15. The following table gives information of panchayat elections held in six villages P, Q, R, S, T and U.

निम्नलिखित तालिका छह गांवों P, Q, R, S, T और U में हुए पंचायत चुनावों की जानकारी देती है।

Village	Total number of votes (in hundreds)	Votes Polled (in %)	Valid Votes (in %)
P	50	80	80
Q	60	75	80
R	100	65	65
S	80	60	70
T	60	80	90
U	40	90	60

Hint:

(a) Percentage of votes polled = $\dfrac{\text{Total votes polled}}{\text{Total number of votes}} \times 100$

(b) Percentage of valid votes = $\dfrac{\text{Total valid votes}}{\text{Total votes polled}} \times 100$

Find the ratio of invalid votes of village R to that of village U?
गाँव R के अवैध मतों का गाँव U के अवैध मतों से अनुपात ज्ञात कीजिए।

(a) 455 : 268 (b) 321 : 107
(c) 455 : 288 (d) 423 : 755

Solution:

$$\text{Invalid votes of R} = 100 \times \frac{65}{100} \times \frac{35}{100} = \frac{2275}{100}$$

$$\text{Invalid votes of U} = 40 \times \frac{90}{100} \times \frac{40}{100} = \frac{1440}{100}$$

$$\text{Ratio R : U} = \frac{2275}{100} : \frac{1440}{100} = 455 : 288$$

16. The import and export of a country (in lakhs rupees) during a certain period of time is given in the following bar-graph. Read the bar-graph carefully and answer the following question.
किसी निश्चित अवधि के दौरान किसी देश का आयात और निर्यात (लाखों रुपए में) निम्नलिखित बार-ग्राफ में दिया गया है। बार-ग्राफ को ध्यान से पढ़ें और निम्नलिखित प्रश्न का उत्तर दें।

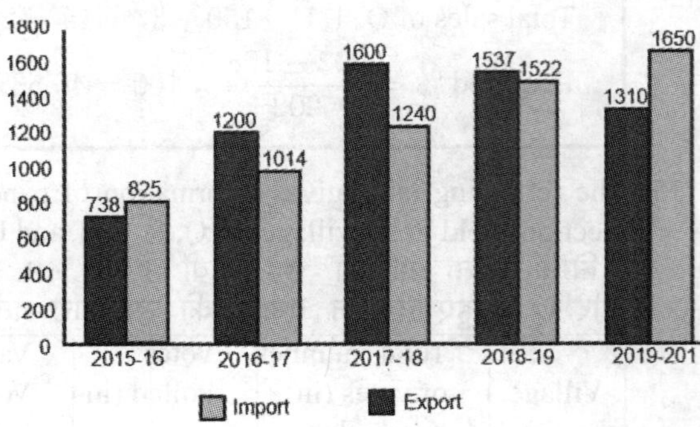

What is the average of exports during the given period of time? दी गई समयावधि के दौरान निर्यात का औसत क्या है?

(a) Rs.1,250.2 lakhs (b) Rs.1,240.0 lakhs
(c) Rs.1,248.2 lakhs (d) Rs.1,115.4 lakhs

Solution:

Required Average = $\dfrac{825 + 1014 + 1240 + 1522 + 1650}{5}$

Required Average = $\dfrac{6251}{5} = 1250.2$ lakhs

17. Who scored the least marks in aggregate of all the subjects?
सभी विषयों के कुल योग में सबसे कम अंक किसने प्राप्त किये?

Name of the student	Subject		
	English	Hindi	Maths
A	54	65	67
B	52	88	78
C	48	48	67
D	41	87	60
E	74	70	55

(a) D (b) E
(c) A (d) C

Solution:

A's score = 54 + 65 + 67 = 186

B's score = 52 + 88 + 78 = 218

C's score = 48 + 48 + 67 = 163

D's score = 41 + 87 + 60 = 188

E's score = 74 + 70 + 55 = 199

Hence, C scored the least marks.

18. Study the table given below and answer the questions that follow.
नीचे दी गई तालिका का अध्ययन करें और प्रश्नों के उत्तर दें।

Countries	Number of Tests	Positive cases	Number of patients for Ventilators
The US	14000	12%	20%
Spain	12000	8%	12.5%
Italy	10000	12%	16%
China	1000	11%	15%

According to the table given above, which country has the maximum number of positive cases?
उपरोक्त तालिका के अनुसार, किस देश में पॉजिटिव मामलों की संख्या अधिकतम है?

(a) The US (b) China
(c) Spain (d) Italy

Solution:

Positive cases in the US $= \frac{14000 \times 12}{100} =$ 1680

Positive cases in Spain $= \frac{12000 \times 8}{100} = 960$

Positive cases in Italy $= \frac{10000 \times 12}{100} = 1200$

Positive cases in China $= \frac{10000 \times 11}{100} =$ 1100

Hence, the US has the maximum positive cases.

19. Study the given table and answer the question that follows.
The table shows the number of customers for different services in different cities.
दी गई तालिका का अध्ययन करें और नीचे दिए गए प्रश्नों के उत्तर दें। तालिका विभिन्न शहरों में विभिन्न सेवाओं के लिए ग्राहकों की संख्या दर्शाती है।

City → / Services ↓	X	Y	Z
A	250	175	350
B	220	190	240
C	260	200	270
D	245	185	330

In which service is the average number of customers per city the highest?
किस सेवा में प्रति शहर ग्राहकों की औसत संख्या सबसे अधिक है?

(a) C (b) A
(c) D (d) B

Solution:

In A: $\frac{250 + 175 + 350}{3} = 541.67$

In B: $\frac{220 + 190 + 240}{3} = 490$

In C: $\frac{260 + 200 + 270}{3} = 550$

In D: $\frac{245 + 185 + 330}{3} = 540$

Hence, in service 'C' the average number of customers per city is the highest.

20. Various expenditures incurred by a publishing company for publishing a book in 2018 are given in the following pie-chart. Study the chart and answer the question.
2018 में एक पुस्तक प्रकाशित करने के लिए एक प्रकाशन कंपनी द्वारा किए गए विभिन्न व्यय निम्नलिखित पाई-चार्ट में दिए गए हैं। चार्ट का अध्ययन करें और प्रश्न का उत्तर दें।

Expenditures of the company

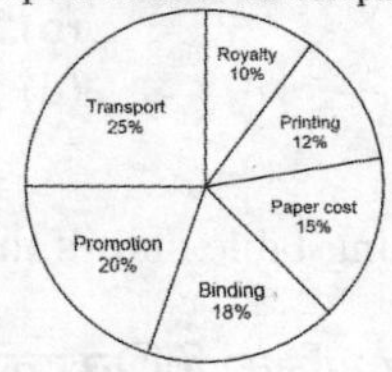

Price printed on a book is 15% above the cost price. If the price printed on a book is Rs 942, then the cost of paper for a single copy in is (rounded off to one decimal place).
किसी पुस्तक पर छपा मूल्य, क्रय मूल्य से 15% अधिक है। यदि पुस्तक पर छपा मूल्य 942 रुपये है, तो एक प्रति के लिए कागज़ की लागत (एक दशमलव स्थान तक पूर्णांकित) है।

(a) Rs 122.9 (b) Rs 188.5
(c) Rs 182.5 (d) Rs 220.6

Solution:

Cost price of book $= \frac{942}{115} \times 100 = 819.13$

Cost of paper $= 819.13 \times \frac{3}{20} = 122.9$ Rs

21. The pie chart given below shows the number of bikes sold by 8 different companies. The total number of bikes sold by all these 8 companies is 2000. Number of bikes sold by a particular company is shown as a percent of total number of bikes sold by all these 8 companies.
नीचे दिया गया पाई चार्ट 8 अलग-अलग कंपनियों द्वारा बेची गई बाइक की संख्या दर्शाता है। इन सभी 8 कंपनियों द्वारा बेची गई बाइक की कुल संख्या 2000 है। किसी विशेष कंपनी द्वारा बेची गई बाइक की संख्या को इन सभी 8 कंपनियों द्वारा बेची गई बाइक की कुल संख्या के प्रतिशत के रूप में दर्शाया गया है।

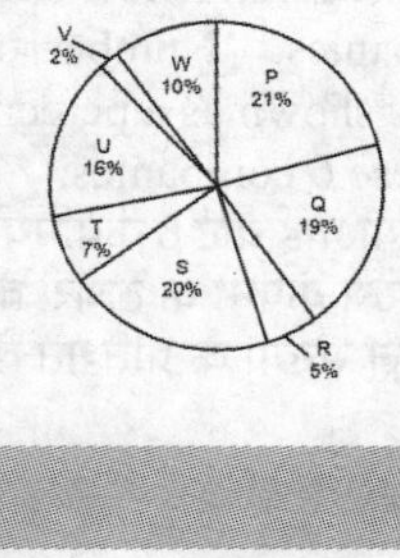

What is the difference between the average number of bikes sold by P, Q, R and S and the average number of bikes sold by T, U, V and W?

P, Q, R और S द्वारा बेची गई बाइकों की औसत संख्या तथा T, U, V और W द्वारा बेची गई बाइकों की औसत संख्या के बीच क्या अंतर है?

(a) 150 (b) 200

(c) 175 (d) 125

Solution:

$$\text{Difference} = \frac{2119+5+20}{4} - \frac{7+16+2+10}{4}$$

$$= \frac{65}{4} - \frac{35}{4} = \frac{30}{4} = 7.5$$

$$\frac{30}{4} \times \frac{2000}{100} = 150$$

22. The following table shows the number of different items in different shops and their respective selling prices per unit.

निम्नलिखित तालिका विभिन्न दुकानों में विभिन्न वस्तुओं की संख्या और प्रति इकाई उनके संबंधित विक्रय मूल्यों को दर्शाती है।

Shops	Total no. of items	AC:Cooler:Fan	Selling price per unit		
			Cooler	AC	Fan
A	5000	4:5:1	8000	25000	8500
B	1800	3:2:4	10000	20000	16000
C	3400	6:4:7	6000	42000	15000
D	3600	4:2:3	12000	32000	8000
E	4000	5:1:4	8000	26500	12200
F	1210	2:4:5	11000	28000	11100

Find the percentage of total revenue which comes from Cooler from shop E, considering all given items are being sold from shop E and from all the given shops only given three items are being sold. (Rounded off to three decimal places)

दुकान E से कूलर से प्राप्त कुल राजस्व का प्रतिशत ज्ञात कीजिए, यह मानते हुए कि सभी दी गई वस्तुएं दुकान □ से बेची जा रही हैं और सभी दी गई दुकानों से केवल दी गई तीन वस्तुएं बेची जा रही हैं। तीन दशमलव स्थानों तक पूर्णांकित)

(a) 4.226% (b) 3.516%

(c) 10.45% (d) 8.910%

Solution:

Total number of items in E = 4000

Number of AC = $\frac{5}{10} \times 4000 = 2000$

Number of cooler = $\frac{1}{10} \times 4000 = 400$

Number of fan = $\frac{4}{10} \times 4000 = 1600$

Total revenue = $2000 \times 26500 + 400 \times 8000 + 1600 \times 12200 = 75720000$

Required % = $\frac{3200000}{75720000} \times 100 = 4.226\%$

23. The pie chart given below shows the number of tyres of 6 companies. Number of tyres of a particular company is shown as a percentage of total number of tyres of these 6 companies.

नीचे दिया गया पाई चार्ट 6 कंपनियों के टायरों की संख्या दर्शाता है। किसी विशेष कंपनी के टायरों की संख्या इन 6 कंपनियों के टायरों की कुल संख्या के प्रतिशत के रूप में दर्शाई गई है

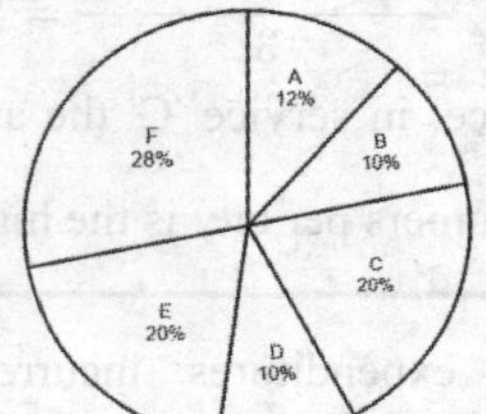

What is the sum of central angles formed by sectors representing tyres of company F and A?
कंपनी F और A के टायरों का प्रतिनिधित्व करने वाले क्षेत्रों द्वारा गठित केंद्रीय कोणों का योग क्या है?

(a) 108°
(b) 84°
(c) 144°
(d) 112°

Solution:

$F + A = 28\% + 12\% = 40\%$

If $100\% = 360°$

Then $40\% = 144°$

24. A soft drink company prepares drinks of three different flavours- A, B and C. The production of the three flavours over a period of five years from 2010 to 2014 has been expressed in the given bar graph. Study the graph and answer the question that follows: Production of three different flavours A, B and C (in lakh bottles) by a company over five years.
The difference between the average production of flavour A in 2012, 2013, and 2014 and the average production of flavour C in 2012, 2013, and 2014 is:
1.5 एक सॉफ्ट ड्रिंक कंपनी तीन अलग-अलग फ्लेवर- A, B और C के ड्रिंक्स बनाती है। 2010 से 2014 तक पाँच वर्षों की अवधि में तीनों फ्लेवर का उत्पादन नीचे दिए गए बार ग्राफ में दर्शाया गया है। ग्राफ का अध्ययन करें और निम्नलिखित प्रश्नों के उत्तर दें:
पांच वर्षों में एक कंपनी द्वारा तीन अलग-अलग फ्लेवर A, B और C का उत्पादन (लाख बोतलों में)। 2012, 2013 और 2014 में फ्लेवर A के औसत उत्पादन और 2012, 2013 और 2014 में फ्लेवर C के औसत उत्पादन के बीच का अंतर है:

(a) 10 lakh bottles
(b) 12 lakh bottles
(c) 18 lakh bottles
(d) 15 lakh bottles

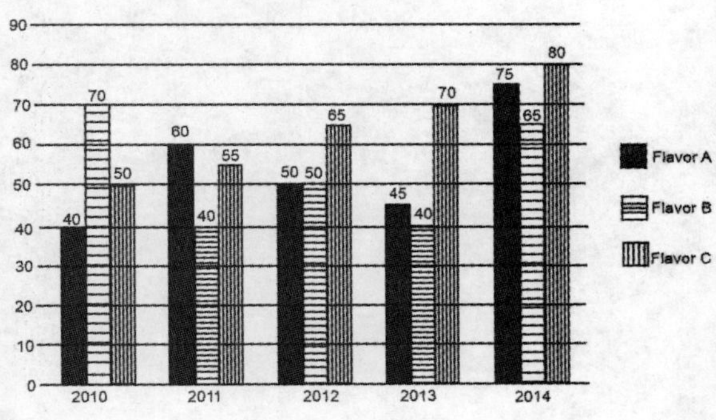

Solution:

Average production of flavour A in 2012, 2013, and 2014 $= \frac{50 + 45 + 75}{3} = \frac{170}{3}$

Average production of flavour C in 2012, 2013, and 2014 $= \frac{65 + 70 + 80}{3} = \frac{215}{3}$

Required difference $= \frac{215 - 170}{3} = 15$

Hence, the difference is 15 lakh bottles.

25. The given line graph shows the production (in tonnes) and the sales (in tonnes) of a company. How much more is the production in 2015 and 2017 taken together than the sales in 2016 and 2018 taken together (in tonnes)?
दिया गया रेखा ग्राफ एक कंपनी के उत्पादन (टन में) और बिक्री (टन में) को दर्शाता है। 2015 और 2017 में कुल उत्पादन, 2016 और 2018 में कुल बिक्री (टन में) से कितना अधिक है?

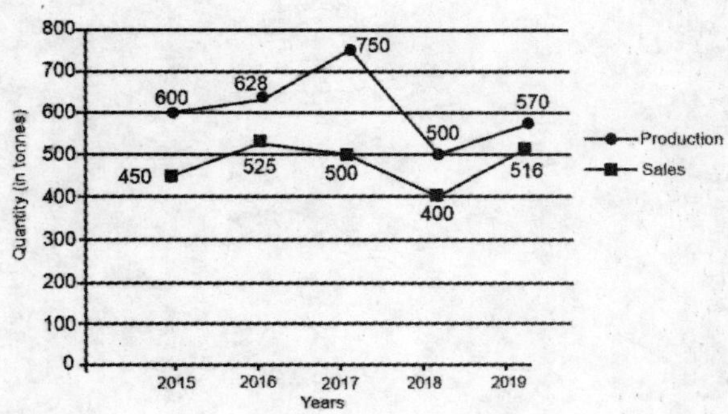

(a) 425
(b) 328
(c) 434
(d) 178

Solution:

Total production in 2015 and 2017 $= 600 + 750 = 1350$

Total sales in 2016 and 2018 $= 525 + 400 = 925$

Required increase $= 1350 - 925 = 425$